Advances in Intelligent Systems and Computing

Volume 813

The series "Advances in Intelligent Systems and Computing" contains publications on theory, applications, and design methods of Intelligent Systems and Intelligent Computing. Virtually all disciplines such as engineering, natural sciences, computer and information science, ICT, economics, business, e-commerce, environment, healthcare, life science are covered. The list of topics spans all the areas of modern intelligent systems and computing such as: computational intelligence, soft computing including neural networks, fuzzy systems, evolutionary computing and the fusion of these paradigms, social intelligence, ambient intelligence, computational neuroscience, artificial life, virtual worlds and society, cognitive science and systems, Perception and Vision, DNA and immune based systems, self-organizing and adaptive systems, e-Learning and teaching, human-centered and human-centric computing, recommender systems, intelligent control, robotics and mechatronics including human-machine teaming, knowledge-based paradigms, learning paradigms, machine ethics, intelligent data analysis, knowledge management, intelligent agents, intelligent decision making and support, intelligent network security, trust management, interactive entertainment, Web intelligence and multimedia.

The publications within "Advances in Intelligent Systems and Computing" are primarily proceedings of important conferences, symposia and congresses. They cover significant recent developments in the field, both of a foundational and applicable character. An important characteristic feature of the series is the short publication time and world-wide distribution. This permits a rapid and broad dissemination of research results.

More information about this series at http://www.springer.com/series/11156

Ajith Abraham · Paramartha Dutta
Jyotsna Kumar Mandal · Abhishek Bhattacharya
Soumi Dutta
Editors

Emerging Technologies in Data Mining and Information Security

Proceedings of IEMIS 2018, Volume 2

Editors
Ajith Abraham
Machine Intelligence Research Labs
Auburn, WA, USA

Abhishek Bhattacharya
Institute of Engineering and Management
Kolkata, West Bengal, India

Paramartha Dutta
Department of Computer and System Sciences
Visva-Bharati University
Santiniketan, West Bengal, India

Soumi Dutta
Institute of Engineering and Management
Kolkata, West Bengal, India

Jyotsna Kumar Mandal
Department of Computer Science and Engineering
University of Kalyani
Kalyani, India

ISSN 2194-5357 ISSN 2194-5365 (electronic)
Advances in Intelligent Systems and Computing
ISBN 978-981-13-1497-1 ISBN 978-981-13-1498-8 (eBook)
https://doi.org/10.1007/978-981-13-1498-8

Library of Congress Control Number: 2018947481

This Springer imprint is published by the registered company Springer Nature Singapore Pte Ltd.
The registered company address is: 152 Beach Road, #21-01/04 Gateway East, Singapore 189721, Singapore

Foreword

Welcome to the Springer International Conference on Emerging Technologies in Data Mining and Information Security (IEMIS 2018) held from February 23 to 25, 2018, in Kolkata, India. As a premier conference in the field, IEMIS 2018 provides a highly competitive forum for reporting the latest developments in the research and application of information security and data mining. We are pleased to present the proceedings of the conference as its published record. The theme of this year is Crossroad of Data Mining and Information Security, a topic that is quickly gaining traction in both academic and industrial discussions because of the relevance of privacy-preserving data mining model (PPDM).

IEMIS is a young conference for research in the areas of information and network security, data sciences, big data, and data mining. Although 2018 is the debut year for IEMIS, it has already witnessed a significant growth. As evidence of that, IEMIS received a record of 532 submissions. The authors of submitted papers come from 35 countries. The authors of accepted papers are from 11 countries.

We hope that this program will further stimulate research in information security and data mining and provide the practitioners with better techniques, algorithms and tools for deployment. We feel honored and privileged to serve the best recent developments in the field of WSDM to the readers through this exciting program.

Kolkata, India

Bimal Kumar Roy
General Chair, IEMIS 2018
Indian Statistical Institute

Preface

This volume presents the proceedings of the International Conference on Emerging Technologies in Data Mining and Information Security, IEMIS 2018, which took place at the University of Engineering and Management, Kolkata, India, from February 23 to 25, 2018. The volume appears in the series "Advances in Intelligent Systems and Computing" (AISC) published by Springer Nature, one of the largest and most prestigious scientific publishers. It is one of the fastest growing book series in their program. AISC series publication is meant to include various high-quality and timely publications, primarily proceedings of relevant conferences, congresses, and symposia but also monographs, on the theory, applications, and implementations of broadly perceived modern intelligent systems and intelligent computing, in their modern understanding, i.e., including not only tools and techniques of artificial intelligence (AI), and computational intelligence (CI)—which includes data mining, information security, neural networks, fuzzy systems, evolutionary computing, as well as hybrid approaches that synergistically combine these areas—but also topics such as multiagent systems, social intelligence, ambient intelligence, Web intelligence, computational neuroscience, artificial life, virtual worlds and societies, cognitive science and systems, perception and vision, DNA- and immune-based systems, self-organizing and adaptive systems, e-learning and teaching, human-centered and human-centric computing, autonomous robotics, knowledge-based paradigms, learning paradigms, machine ethics, intelligent data analysis, and various issues related to "big data," security, and trust management. These areas are at the forefront of science and technology and have been found useful and powerful in a wide variety of disciplines such as engineering, natural sciences, computer, computation and information sciences, ICT, economics, business, e-commerce, environment, health care, life science, and social sciences. The AISC book series is submitted for indexing to ISI Conference Proceedings Citation Index (now run by Clarivate), Ei Compendex, DBLP, SCOPUS, Google Scholar, and SpringerLink, and many other indexing services around the world. IEMIS 2018 is a debut annual conference series organized by the School of Information Technology, under the aegis of Institute of Engineering and Management, India. Its idea came from the heritage of the other two cycles of the events, IEMCON and UEMCON, which were organized by the Institute of Engineering and Management under the leadership of Prof. (Dr.) Satyajit Chakrabarti.

In this volume of "Advances in Intelligent Systems and Computing," we would like to present the results of studies on the selected problems of data mining and information security. Security implementation is the contemporary answer to the new challenges in threat evaluation of complex systems. Security approach in theory and engineering of complex systems (not only computer systems and networks) is based on multidisciplinary attitude to information theory, technology, and maintenance of the systems working in real (and very often unfriendly) environments. Such a transformation has shaped the natural evolution in the topical range of subsequent IEMIS conferences, which can be seen over the recent years. Human factors likewise infest the best digital dangers. Workforce administration and digital mindfulness are fundamental for accomplishing all-encompassing cybersecurity. This book will be of an extraordinary incentive to a huge assortment of experts, scientists, and understudies concentrating on the human part of the Internet, and for the compelling assessment of safety efforts, interfaces, client-focused outline and plan for unique populaces, especially the elderly. We trust this book is instructive yet much more than it is provocative. We trust it moves, driving peruser to examine different inquiries, applications and potential arrangements in making sheltered and secure plans for all.

The Program Committee of IEMIS 2018, its organizers, and the editors of this proceedings would like to gratefully acknowledge the participation of all the reviewers who helped to refine the contents of this volume and evaluated the conference submissions. Our thanks go to Prof. Bimal Kumar Roy, Dr. Ajith Abraham, Dr. Sheng-Lung peng, Dr. Detlef Streitferdt, Dr. Shaikh Fattah, Dr. Celia Shahnaz, Dr. Swagatam Das, Dr. Niloy Ganguly, Dr. K. K. Shukla, Dr. Nilanjan Dey, Dr. Florin PopentiuVladicescu, Dr. Dewan Md. Farid, Dr. Saptarshi Ghosh, Dr. Rita Choudhury, Dr. Asit Kumar Das, Prof. Tanupriya Choudhury, Prof. Arijit Ghosal, Prof. Rahul Saxena, Prof. Monika Jain, Dr. Aakanksha Sharaff, Prof. Dr. Sajal Dasgupta, Prof. Rajiv Ganguly, and Prof. Sukalyan Goswami.

Thanking all the authors who have chosen IEMIS 2018 as the publication platform for their research, we would like to express our hope that their papers will help in further developments in design and analysis of engineering aspects of complex systems, being a valuable source material for scientists, researchers, practitioners, and students who work in these areas.

Auburn, USA	Ajith Abraham
Santiniketan, India	Paramartha Dutta
Kalyani, India	Jyotsna Kumar Mandal
Kolkata, India	Abhishek Bhattacharya
Kolkata, India	Soumi Dutta

Organizing Committee

Patron

Prof. (Dr.) Satyajit Chakrabarti, Institute of Engineering and Management, India

Conference General Chair

Dr. Bimal Kumar Roy, Indian Statistical Institute, Kolkata, India

Convener

Dr. Subrata Saha, Institute of Engineering and Management, India
Abhishek Bhattacharya, Institute of Engineering and Management, India

Co-convener

Sukalyan Goswami, University of Engineering and Management, India
Krishnendu Rarhi, Institute of Engineering and Management, India
Soumi Dutta, Institute of Engineering and Management, India
Sujata Ghatak, Institute of Engineering and Management, India
Dr. Abir Chatterjee, University of Engineering and Management, India

Keynote Speakers

Dr. Ajith Abraham, Machine Intelligence Research Labs (MIR Labs), USA
Dr. Fredric M. Ham, IEEE Life Fellow, SPIE Fellow, and INNS Fellow, Melbourne, USA
Dr. Sheng-Lung Peng, National Dong Hwa University, Hualien, Taiwan

Dr. Shaikh Fattah, Editor, IEEE Access & CSSP (Springer), Bangladesh
Dr. Detlef Streitferdt, Technische Universität Ilmenau, Germany
Dr. Swagatam Das, Indian Statistical Institute, Kolkata, India
Dr. Niloy Ganguly, Indian Institute of Technology Kharagpur, India
Dr. K. K. Shukla, IIT (B.H.U.), Varanasi, India
Dr. Nilanjan Dey, Techno India College of Technology, Kolkata, India
Dr. Florin Popentiu Vladicescu, "UNESCO Chair in Information Technologies", University of Oradea, Romania
Dr. Celia Shahnaz, senior member of IEEE, fellow member of Institution of Engineers, Bangladesh (IEB)

Technical Program Committee Chair

Dr. J. K. Mondal, University of Kalyani, India
Dr. Paramartha Dutta, Visva-Bharati University, India
Abhishek Bhattacharya, Institute of Engineering and Management, India

Technical Program Committee Co-chair

Dr. Satyajit Chakrabarti, Institute of Engineering and Management, India
Dr. Subrata Saha, Institute of Engineering and Management, India
Dr. Kamakhya Prasad Ghatak, University of Engineering and Management, India
Dr. Asit Kumar Das, IIEST, Shibpur, India

Editorial Board

Dr. Ajith Abraham, Machine Intelligence Research Labs (MIR Labs), USA
Dr. J. K. Mondal, University of Kalyani, India
Dr. Paramartha Dutta, Visva-Bharati University, India
Abhishek Bhattacharya, Institute of Engineering and Management, India
Soumi Dutta, Institute of Engineering and Management, India

Advisory Committee

Dr. Mahmoud Shafik, University of Derby
Dr. MohdNazri Ismail, National Defence University of Malaysia
Dr. Bhaba R. Sarker, Louisiana State University
Dr. Tushar Kanti Bera, University of Arizona, USA
Dr. Shirley Devapriya Dewasurendra, University of Peradeniya, Sri Lanka
Dr. Goutam Chakraborty, Professor and Head of the Intelligent Informatics Lab, Iwate Prefectural University, Japan
Dr. Basabi Chakraborty, Iwate Prefectural University, Japan

Dr. Kalyanmoy Deb, Michigan State University, East Lansing, USA
Dr. Vincenzo Piuri, University of Milan, Italy
Dr. Biswajit Sarkar, Hanyang University, Korea
Dr. Raj Kumar Buyya, the University of Melbourne
Dr. Anurag Dasgupta, Valdosta State University, Georgia
Dr. Prasenjit Mitra, the Pennsylvania State University
Dr. Esteban Alfaro-Cortés, University of Castilla–La Mancha, Spain
Dr. Ilkyeong Moon, Seoul National University, South Korea
Dr. Izabela Nielsen, Aalborg University, Denmark
Dr. Prasanta K. Jana, IEEE Senior Member, Indian Institute of Technology (ISM), Dhanbad
Dr. Gautam Paul, Indian Statistical Institute, Kolkata
Dr. Malay Bhattacharyya, IIEST, Shibpur
Dr. Sipra Das Bit, IIEST, Shibpur
Dr. Jaya Sil, IIEST, Shibpur
Dr. Asit Kumar Das, IIEST, Shibpur
Dr. Saptarshi Ghosh, IIEST, Shibpur, IIT KGP
Dr. Prof. Hafizur Rahman, IIEST, Shibpur
Dr. C. K. Chanda, IIEST, Shibpur
Dr. Asif Ekbal, Associate Dean, IIT Patna
Dr. Sitangshu Bhattacharya, IIIT Allahabad
Dr. Ujjwal Bhattacharya, CVPR Unit, Indian Statistical Institute
Dr. Prashant R.Nair, Amrita Vishwa Vidyapeetham (University), Coimbatore
Dr. Tanushyam Chattopadhyay, TCS Innovation Labs, Kolkata
Dr. A. K. Nayak, Fellow and Honorary Secretarý CSI
Dr. B. K. Tripathy, VIT University
Dr. K. Srujan Raju, CMR Technical Campus
Dr. Dakshina Ranjan Kisku, National Institute of Technology, Durgapur
Dr. A. K. Pujari, University of Hyderabad
Dr. Partha Pratim Sahu, Tezpur University
Dr. Anuradha Banerjee, Kalyani Government Engineering College
Dr. Amiya Kumar Rath, Veer Surendra Sai University of Technology
Dr. Kandarpa Kumar Sharma, Gauhati University
Dr. Amlan Chakrabarti, University of Calcutta
Dr. Sankhayan Choudhury, University of Calcutta
Dr. Anjana Kakoti Mahanta, Gauhati University
Dr. Subhankar Bandyopadhyay, Jadavpur University
Dr. Debabrata Ghosh, Calcutta University
Dr. Rajat Kr. Pal, University of Calcutta
Dr. Ujjwal Maulik, Jadavpur University
Dr. Himadri Dutta, Kalyani Government Engineering College, Kalyani
Dr. Brojo Kishore Mishra, C. V. Raman College of Engineering (Autonomous), Bhubaneswar
Dr. S. VijayakumarBharathi, Symbiosis Centre for Information Technology (SCIT)
Dr. Govinda K., VIT University, Vellore
Dr. Ajanta Das, University of Engineering and Management, India

Technical Committee

Dr. Vincenzo Piuri, University of Milan, Italy
Dr. Mahmoud Shafik, University of Derby
Dr. Bhaba Sarker, Louisiana State University
Dr. Mohd Nazri Ismail, University Pertahanan National Malaysia
Dr. Tushar Kanti Bera, Yonsei University, Seoul
Dr. Birjodh Tiwana, LinkedIn, San Francisco, California
Dr. Saptarshi Ghosh, IIEST, Shibpur, IIT KGP
Dr. Srimanta Bhattacharya, Indian Statistical Institute, Kolkata
Dr. Loingtam Surajkumar Singh, NIT Manipur
Dr. Sitangshu Bhattacharya, IIIT Allahabad
Dr. Sudhakar Tripathi, NIT Patna
Dr. Chandan K. Chanda, IIEST, Shibpur
Dr. Dakshina Ranjan Kisku, NIT Durgapur
Dr. Asif Ekbal, IIT Patna
Dr. Prasant Bharadwaj, NIT Agartala
Dr. Somnath Mukhopadhyay, Hijli College, Kharagpur, India, and Regional Student Coordinator, Region II, Computer Society of India
Dr. G. Suseendran, Vels University, Chennai, India
Dr. Sumanta Sarkar, Department of Computer Science, University of Calgary
Dr. Manik Sharma, Assistant Professor, DAV University, Jalandhar
Dr. Rita Choudhury, Gauhati University
Dr. Kuntala Patra, Gauhati University
Dr. Helen K. Saikia, Gauhati University
Dr. Debasish Bhattacharjee, Gauhati University
Dr. Somenath Sarkar, University of Calcutta
Dr. Sankhayan Choudhury, University of Calcutta
Dr. Debasish De, Maulana Abul Kalam Azad University of Technology
Dr. Buddha Deb Pradhan, National Institute of Technology, Durgapur
Dr. Shankar Chakraborty, Jadavpur University
Dr. Durgesh Kumar Mishra, Sri Aurobindo Institute of Technology, Indore, Madhya Pradesh
Dr. Angsuman Sarkar, Secretary, IEEE EDS Kolkata Chapter, Kalyani Government Engineering College
Dr. A. M. SUDHAKARA, University of Mysore
Dr. Indrajit Saha, National Institute of Technical Teachers' Training and Research Kolkata
Dr. Bikash Santra, Indian Statistical Institute (ISI)
Dr. Ram Sarkar, Jadavpur University
Dr. Priya Ranjan Sinha Mahapatra, Kalyani University
Dr. Avishek Adhikari, University of Calcutta

Dr. Jyotsna Kumar Mandal, Kalyani University
Dr. Manas Kumar Sanyal, Kalyani University
Dr. Atanu Kundu, Chairman, IEEE EDS, Heritage Institute of Technology, Kolkata
Dr. Chintan Kumar Mandal, Jadavpur University
Dr. Kartick Chandra Mondal, Jadavpur University
Mr. Debraj Chatterjee, Manager, Capegemini
Mr. Gourav Dutta, Cognizant Technology Solutions
Dr. Soumya Sen, University of Calcutta
Dr. Soumen Kumar Pati, St. Thomas' College of Engineering and Technology
Mrs. Sunanda Das, Neotia Institute of Technology Management and Science, India
Mrs. Shampa Sengupta, MCKV Institute of Engineering, India
Dr. Brojo Kishore Mishra, C. V. Raman College of Engineering (Autonomous), Bhubaneswar
Dr. S. Vijayakumar Bharathi, Symbiosis Centre for Information Technology, Pune
Dr. Govinda K., Vellore Institute of technology
Dr. Prashant R. Nair, Amrita University
Dr. Hemanta Dey, IEEE Senior Member
Dr. Ajanta Das, University of Engineering and Management, India
Dr. Samir Malakar, MCKV Institute of Engineering, Howrah
Dr. Tanushyam Chattopadhyay, TCS Innovation Lab, Kolkata
Dr. A. K. Nayak, Indian Institute of Business Management, Patna
Dr. B. K. Tripathy, Vellore Institute of Technology, Vellore
Dr. K. Srujan Raju, CMR Technical Campus, Hyderabad
Dr. Partha Pratim Sahu, Tezpur University
Dr. Anuradha Banerjee, Kalyani Government Engineering College
Dr. Amiya Kumar Rath, Veer Surendra Sai University of Technology, Odisha
Dr. S. D. Dewasurendra
Dr. Arnab K. Laha, IIM Ahmedabad
Dr. Kandarpa Kumar Sarma, Gauhati University
Dr. Ambar Dutta, BIT Mesra and Treasurer—CSI Kolkata
Dr. Arindam Pal, TCS Innovation Labs, Kolkata
Dr. Himadri Dutta, Kalyani Government Engineering College, Kalyani
Dr. Tanupriya Choudhury, Amity University, Noida, India
Dr. Praveen Kumar, Amity University, Noida, India

Organizing Chairs

Krishnendu Rarhi, Institute of Engineering and Management, India
Sujata Ghatak, Institute of Engineering and Management, India
Dr. Apurba Sarkar, IIEST, Shibpur, India

Organizing Co-chairs

Sukalyan Goswami, University of Engineering and Management, India
Rupam Bhattacharya, Institute of Engineering and Management, India

Organizing Committee Convener

Dr. Sajal Dasgupta, Vice-Chancellor, University of Engineering and Management

Organizing Committee

Subrata Basak, Institute of Engineering and Management, India
Anshuman Ray, Institute of Engineering and Management, India
Rupam Bhattacharya, Institute of Engineering and Management, India
Abhijit Sarkar, Institute of Engineering and Management, India
Ankan Bhowmik, Institute of Engineering and Management, India
Manjima Saha, Institute of Engineering and Management, India
Biswajit Maity, Institute of Engineering and Management, India
Soumik Das, Institute of Engineering and Management, India
Sreelekha Biswas, Institute of Engineering and Management, India
Nayantara Mitra, Institute of Engineering and Management, India
Amitava Chatterjee, Institute of Engineering and Management, India
Ankita Mondal, Institute of Engineering and Management, India

Registration Chairs

Abhijit Sarkar, Institute of Engineering and Management, India
Ankan Bhowmik, Institute of Engineering and Management, India
Ankita Mondal, Institute of Engineering and Management, India
Ratna Mondol, Institute of Engineering and Management, India

Publication Chairs

Dr. Debashis De, Maulana Abul Kalam Azad University of Technology, India
Dr. Kuntala Patra, Gauhati University, India

Publicity and Sponsorship Chair

Dr. J. K. Mondal, University of Kalyani, India
Dr. Paramartha Dutta, Visva-Bharati University, India
Abhishek Bhattacharya, Institute of Engineering and Management, India
Biswajit Maity, Institute of Engineering and Management, India
Soumik Das, Institute of Engineering and Management, India

Treasurer and Conference Secretary

Dr. Subrata Saha, Institute of Engineering and Management, India
Rupam Bhattacharya, Institute of Engineering and Management, India
Krishnendu Rarhi, Institute of Engineering and Management, India

Hospitality and Transport Chair

Soumik Das, Institute of Engineering and Management, India
Nayantara Mitra, Institute of Engineering and Management, India
Manjima Saha, Institute of Engineering and Management, India
Sreelekha Biswas, Institute of Engineering and Management, India
Amitava Chatterjee, Institute of Engineering and Management, India

Web Chair

Samrat Goswami
Samrat Dey

About this Book

This book features research papers presented at the International Conference on Emerging Technologies in Data Mining and Information Security (IEMIS 2018) held at the University of Engineering and Management, Kolkata, India, from February 23 to 25, 2018.

Data mining is currently a well-known topic in mirroring the exertion of finding learning from information. It gives the strategies that enable supervisors to acquire administrative data from their heritage frameworks. Its goal is to distinguish legitimate, novel, and possibly valuable and justifiable connection and examples in information. Information mining is made conceivable by the very nearness of the expansive databases.

Information security advancement is an essential part to ensure open and private figuring structures. With the use of information development applications in all cases, affiliations are ending up more aware of the security risks to their benefits. Notwithstanding how strict the security techniques and parts are, more affiliations are getting the chance to be weak to a broad assortment of security breaks against their electronic resources. Network-intrusion area is a key protect part against security perils, which have been growing in rate generally.

This book comprises high-quality research work by academicians and industrial experts in the field of computing and communication, including full-length papers, research-in-progress papers and case studies related to all the areas of data mining, machine learning, Internet of things (IoT) and information security.

Contents

About the Editors

Dr. Ajith Abraham received his Ph.D. from Monash University, Melbourne, Australia, and his M.Sc. from Nanyang Technological University, Singapore. His research and development experience includes over 25 years in the industry and academia, spanning different continents such as Australia, America, Asia, and Europe. He works in a multidisciplinary environment involving computational intelligence, network security, sensor networks, e-commerce, Web intelligence, Web services, computational grids, and data mining, applied to various real-world problems. He has authored/co-authored over 350 refereed journal/conference papers and chapters, and some of the papers have also won the best paper awards at international conferences and also received several citations. Some of the articles are available in the ScienceDirect Top 25 hottest articles → http://top25.sciencedirect.com/index.php?cat_id=6&subject_area_id=7.

He has given more than 20 plenary lectures and conference tutorials in these areas. He serves the editorial board of several reputed international journals and has also guest-edited 26 special issues on various topics. He is actively involved in the hybrid intelligent systems (HISs), intelligent systems design and applications (ISDA), and information assurance and security (IAS) series of international conferences. He is General Co-Chair of the Tenth International Conference on Computer Modeling and Simulation (UKSIM'08), Cambridge, UK; Second Asia International Conference on Modeling and Simulation (AMS 2008), Malaysia; Eighth International Conference on Intelligent Systems Design and Applications (ISDA'08), Taiwan; Fourth International Symposium on Information Assurance and Security (IAS'07), Italy; Eighth International Conference on Hybrid Intelligent Systems (HIS'08), Spain; Fifth IEEE International Conference on Soft Computing as Transdisciplinary Science and Technology (CSTST'08), Cergy Pontoise, France; and Program Chair/Co-chair of Third International Conference on Digital Information Management (ICDIM'08), UK, and Second European Conference on Data Mining (ECDM 2008), the Netherlands.

He is Senior Member of IEEE, IEEE Computer Society, IEE (UK), ACM, etc. More information is available at: http://www.softcomputing.net.

Dr. Paramartha Dutta was born in 1966. He completed his bachelor's and master's degrees in statistics from the Indian Statistical Institute, Calcutta, in the years 1988 and 1990, respectively. He afterward completed his M.Tech. in computer science from the same institute in 1993 and Ph.D. in Engineering from the Bengal Engineering and Science University, Shibpur, in 2005. He has served in the capacity of research personnel in various projects funded by Government of India, which are done by Defence Research Development Organization, Council of Scientific and Industrial Research, Indian Statistical Institute, etc. He is now Professor in the Department of Computer and System Sciences of the Visva-Bharati University, West Bengal, India. Prior to this, he served Kalyani Government Engineering College and College of Engineering in West Bengal as Full-Time Faculty Member. He remained associated as Visiting/Guest Faculty of several universities/institutes such as West Bengal University of Technology, Kalyani University, Tripura University.

He has co-authored eight books and has also seven edited books to his credit. He has published more than two hundred technical papers in various peer-reviewed journals and conference proceedings, both international and national, several chapters in edited volumes of reputed international publishing houses like Elsevier, Springer-Verlag, CRC Press, John Wiley. He has guided six scholars who had already been awarded their Ph.D. apart from one who has submitted her thesis. Presently, he is supervising six scholars for their Ph.D. program.

He is Co-Inventor of ten Indian patents and one international patent, which are all published apart from five international patents which are filed but not yet published.

He, as investigator, could implement successfully the projects funded by All India Council for Technical Education, Department of Science and Technology of the Government of India. He has served/serves in the capacity of external member of Board of Studies of relevant departments of various universities encompassing West Bengal University of Technology, Kalyani University, Tripura University, Assam University, Silchar. He had the opportunity to serve as the expert of several interview boards organized by West Bengal Public Service Commission, Assam University, Silchar; National Institute of Technology, Arunachal Pradesh; Sambalpur University, etc.

He is Life Fellow of the Optical Society of India (FOSI), Institution of Electronics and Telecommunication Engineers (IETE), Institute of Engineering (FIE), Life Member of Computer Society of India (LMCSI), Indian Science Congress Association (LMISCA), Indian Society for Technical Education (LMISTE), Indian Unit of Pattern Recognition and Artificial Intelligence (LMIUPRAI)—the Indian affiliate of the International Association for Pattern Recognition (IAPR), and Senior Member of Associated Computing Machinery (SMACM), and Institution of Electronics and Electrical Engineers (SMIEEE), USA.

Dr. Jyotsna Kumar Mandal received his M.Tech. in computer science from University of Calcutta in 1987. He was awarded Ph.D. in computer science and engineering by Jadavpur University in 2000. Presently, he is working as Professor of computer science and engineering and Former Dean, Faculty of Engineering, Technology and Management, Kalyani University, Kalyani, Nadia, West Bengal, for two consecutive terms since 2008. He is Ex-Director, IQAC, Kalyani University, and Chairman, CIRM, Kalyani University. He was appointed as Professor in Kalyani Government Engineering College through Public Service Commission under the Government of West Bengal. He started his career as Lecturer at NERIST, under MHRD, Government of India, Arunachal Pradesh, in September 1988. He has teaching and research experience of 30 years. His areas of research are coding theory, data and network security; remote sensing and GIS-based applications, data compression, error correction, visual cryptography, and steganography. He has guided 21 Ph.D. scholars, 2 scholars have submitted their Ph.D. thesis, and 8 are pursuing. He has supervised 03 M.Phil. and more than 50 M.Tech. dissertations and more than 100 M.C.A. dissertations. He is Chief Editor of CSI Journal of Computing and Guest Editor of MST Journal (SCI indexed) of Springer Nature. He has published more than 400 research articles, out of which 154 articles are in various international journals. He has published 5 books from LAP Germany and one from IGI Global. He was awarded A. M. Bose Memorial Silver Medal and Kali Prasanna Dasgupta Memorial Silver Medal in M.Sc. by Jadavpur University. India International Friendship Society (IIFS), New Delhi, conferred "Bharat Jyoti Award" for his meritorious service, outstanding performance, and remarkable role in the field of computer science and engineering on August 29, 2012. He received "Chief Patron" Award from CSI India in 2014. International Society for Science, Technology and Management conferred "Vidyasagar Award" in the Fifth International Conference on Computing, Communication and Sensor Network on December 25, 2016. ISDA conferred Rastriya Pratibha Award in 2017.

Abhishek Bhattacharya is Assistant Professor of computer application department at the Institute of Engineering and Management, India. He did his Masters in computer science from the Biju Patnaik University of Technology and completed his M.Tech. in computer science from BIT, Mesra. He remained associated as Visiting/Guest Faculty of several universities/institutes in India. He has three books to his credit. He has published twenty technical papers in various peer-reviewed journals and conference proceedings, both international and national, and chapters in edited volumes of the reputed international publishing houses. He has teaching and research experience of 13 years. His areas of research are data mining, network security, mobile computing, and distributed computing. He is the reviewer of a couple of journals of IGI Global, Inderscience Publications, and Journal of Information Science Theory and Practice, South Korea.

He is Member of International Association of Computer Science and Information Technology (IACSIT), Universal Association of Computer and Electronics Engineers (UACEE), International Association of Engineers (IAENG),

Internet Society as a Global Member (ISOC), the Society of Digital Information and Wireless Communications (SDIWC) and International Computer Science and Engineering Society (ICSES); Technical Committee Member of CICBA 2017, 52nd Annual Convention of Computer Society of India (CSI 2017), International Conference on Futuristic Trends in Network and Communication Technologies (FTNCT-2018), ICIoTCT 2018, ICCIDS 2018, and Innovative Computing and Communication (ICICC-2018); Advisory Board Member of ISETIST 2017.

Soumi Dutta is Assistant Professor at the Institute of Engineering and Management, India. She is also pursuing her Ph.D. in the Department of Computer Science and Technology, Indian Institute of Engineering Science and Technology, Shibpur. She received her B.Tech. in information technology and her M.Tech. in computer science securing first position (gold medalist), both from Techno India Group, India. His research interests include social network analysis, data mining, and information retrieval. She is Member of several technical functional bodies such as the Michigan Association for Computer Users in Learning (MACUL), the Society of Digital Information and Wireless Communications (SDIWC), Internet Society as a Global Member (ISOC), International Computer Science and Engineering Society (ICSES). She has published several papers in reputed journals and conferences.

Part I
Data Analytics

Sentiment Analysis of Tweet Data: The Study of Sentimental State of Human from Tweet Text

S. M. Mazharul Hoque Chowdhury, Priyanka Ghosh, Sheikh Abujar, Most. Arina Afrin and Syed Akhter Hossain

Abstract Sentiment analysis is a hot topic today. The purpose of this research is finding out sentimental state of a person or a group of people using data mining. The target of this research is building a user friendly interface for general people, so that they will be able to see the analysis report very easily. This analysis process contains both supervised and unsupervised learning, which is a hybrid process. Analysis is done based on keywords, which is defined by the user. User is able to set the number of tweets he/she wants to analyze. We used web-based library for the system. The system is tested and found satisfactory result.

Keywords Emotion · Polarity · R · Sentiment analysis · Scores · Social media Twitter · Word cloud

1 Introduction

At present, social media or websites like Facebook, Twitter, Google + are top-visited website and part and parcel of most of the people around all over the world. They communicate to each other, post their opinion and reviews, share their likes and

---

S. M. M. H. Chowdhury (✉) · P. Ghosh · S. Abujar · Most. Arina Afrin · S. Akhter Hossain
Department of Computer Science and Engineering, Daffodil International University, Dhanmondi, Dhaka 1205, Bangladesh
e-mail: mazharul2213@diu.edu.bd

P. Ghosh
e-mail: priyanka2378@diu.edu.bd

S. Abujar
e-mail: sheikh.cse@diu.edu.bd

Most. Arina Afrin
e-mail: afrin15-2964@diu.edu.bd

S. Akhter Hossain
e-mail: aktarhossain@daffodilvarsity.edu.bd

A. Abraham et al. (eds.), *Emerging Technologies in Data Mining and Information Security*, Advances in Intelligent Systems and Computing 813,
https://doi.org/10.1007/978-981-13-1498-8_1

dislikes and even build up communities in those sites. So every day they actually generate tons of data or petabytes of data in cloud.

So now it has become very difficult to control and analyze such big amount of data for different purpose. So to solve this problem the term machine learning and big data analysis came. They can simplify data and make it trackable. Sentiment analysis is a part of big data analysis. It is like a branch of a large tree. So for analyzing sentiment we are going to use big data analysis process. The main focus of sentiment analysis is getting the real meaning of text in an organized approaches, methods and techniques [1]. To tackle the problems of classifying the sentiments several different methods has been applied and under research [2]. Among them we worked with machine learning with supervised learning [3] and unsupervised learning [4] approach.

Therefore, we considered Twitter's tweets as our data source and built a system that can analyze and generate outputs. There are four types of outputs. They are—emotion, polarity, word cloud, and scores. Between all of them emotion, polarity, and word cloud uses supervised learning and scores uses unsupervised learning.

2 Literature Review

Sentiment analysis is a computational process to classify human emotion in different classes. Sentiment analysis is possible for text, audio, video, or image. But each of them will require different way, approach and algorithm. As we worked on text analysis, so this paper is containing data about only text mining. But if we want to start sentiment analysis, at first we need to know about the sentiment analysis and its approaches.

Sentiment can be considered as feeling, attitude, emotion or opinion. Computationally identifying and processing some text, we can do Sentiment analysis to determine whether the writer's emotion on a particular topic or a subject or a product or a person is positive, negative, or neutral. So using natural language processing categorizing, classifying or statistically viewing some analytical report from a bunch of text is generally considered as sentiment analysis.

We can define sentiment analysis in main three levels [5]. They are:

i. Document level sentiment analysis
ii. Sentence level sentiment analysis
iii. Phrase level sentiment analysis

Here we are going describe those levels of sentiment analysis:

Document-level sentiment analysis: In the document level sentiment analysis we will consider a single document containing a single topic. So this level of sentiment analysis is not applicable for analyzing different forum or blog sites. The main challenge for document level analysis is there can be some texts that are not relevant to the topic. So whenever we apply this, we have to remove irrelevant sentences. For this level analysis we can use both supervised and unsupervised learning. Any supervised learning algorithm like naïve bays [6] or vector machine and unsupervised learning algorithm like clustering or k-means algorithm can be used for this classification.

Sentence level sentiment analysis: In the sentence level sentiment analysis we will consider every sentence in the document. So this level of sentiment analysis is applicable for every kind of document as general document or forum or blog sites. Here by determining the positive and negative words we will be able to determine if the sentence is a positive or a negative sentence. Sentence level sentiment analysis is not applicable for the complex sentences. Like document level sentiment analysis, we can apply both supervised and unsupervised learning in sentence level sentiment analysis.

Phrase level sentiment analysis: In the phrase level sentiment analysis we will consider opinion or sentiment related words. This level of analysis is a pinpointed approach for the sentiment analysis. In some cases exact opinion can be extracted on a specific topic from the text. But this level analysis faces the problem of negation and long range dependencies. Those words who appear near to each other are considered as a phrase in here.

Between all of them sentence level analysis is applied for this research. As sentence level analysis can be used for both supervised and unsupervised learning, so use of both supervised and unsupervised learning gives better result of the analysis.

Sentiment analysis is mainly important for business companies, political groups and different social organizations. By collecting customer review and satisfaction business companies can determine their next business strategy and success of their business. Political groups can manipulate people using current statistics and most discussed topics, so they will be able to know what they want from them. On the other hand, different organizations will be able to determine the support for them for a particular task or their work. Those are some examples for the importance of sentiment analysis. So we can say that by analyzing sentiment we can be beneficial in our practical life.

Modern era brought a very large field for computer scientists to work with data for determining sentiment of people. All around the world a lot of people are working on sentiment analysis right now. Continuous research brought us new features and more accurate result. Research done by Pang and Lee [7] brought so many different approaches like detection of human expression; classifying sentences as positive, negative or neutral; detection of subjective and objective sentences; classifying human emotion in different classes like anger, happy, sad, etc.; application of sentiment analysis in different sectors [8].

Hatzivassiloglou and McKeown [9] and Esuli and Sebastiani [10] worked with the polarity detection from the phrases. Yu and Hatzivassiloglou [11] and Kim and Hovy [12] worked on the character limitation and determined that the twitter message analysis is much similar to the sentence level sentiment analysis [13].

But at recent time so many people are working on different topics of sentiment analysis like twitter, facebook, newspaper, blogs, novels, etc. For example, Safa Ben Hamouda and Jalel Akaichi [14] worked on sentiment classification from Facebook for Arabic spring era. Wang et al. [15] was working on issues of data analysis of social media with new algorithms; Bright et al. [16] worked on the use of social media for research. Mäntylä et al. [17] worked on evolution of sentiment analysis. Padmaja et al. [18] worked on evolution of sentiment analysis and negation in newspaper.

All those researchers worked on those topics based on supervised and unsupervised learning. Even some cases hybrid methods are applicable for sentiment analysis. Application of different algorithms makes sentiment analysis process easier.

Sentiment analysis based on text faces different kind of difficulties. According to Professor Bing Liu from University of Chicago department of computer science, being sure about the accuracy for the analysis is difficult and it depends on the level of analysis, number of data sets, measurement and so on [19].

One of the most common problems is different meaning for the same text. There are some other problems as well like different languages, shortcut words written by the writer, typing mistakes, etc. So while working with those types of problems it is really difficult to decide how to solve them [20].

On the other hand, negation words or sentences like "it is not good" can create problems for algorithms to find out accurate result to determine whether it is a positive or a negative word by showing a positive word as negative and a negative word as positive.

3 Experiments and Observation

This application works based on the tweets posted by the twitter users like persons, channels, etc. So the term tweet data means the tweet posts are available in the twitter as data. Application will collect data for analysis from twitter through a twitter API. So it is needed to create a twitter API for the data access of the application without any complexity.

3.1 Big Data and R Platform

Big data is not only one single oversized data. It is a large set of structured and unstructured data. Big data is needed for them where traditional data analysis method cannot work. Big data faces the challenge of data search, storage, understanding, processing, visualization, transfer, etc. Big data analysis method refers advanced analysis of data. It can find out values from a large data set.

There are several popular platforms for big data analysis like Hadoop, MapReduce, Apache storm, R, HPCC, Weka, etc. In here R platform is used for the processing because of the statistical output. R platform is well known for generating good statistical output through its different well established libraries [21]. Using and maintaining those libraries are quite easy and simple. For example, R can connect to the Twitter API using the consumer and access keys. Outputs like plot, emotion or polarity statistics can be done using those libraries. So because of those features R platform is selected for this application.

3.2 Algorithms Used in Here

In this research, both supervised and unsupervised learning is used. For supervised learning Naïve Bayes algorithm and for unsupervised learning Lexicon-based analyzer is used. Use of this kind of hybrid algorithm can provide better output. So the reason of generation of this many output is providing a clear picture of the report to the user.

Use of Naïve Bayes algorithm generated emotion class, polarity class, and word-cloud which are relevant to the emotion class. On the other hand use of lexicon-based analyzer generated score bar diagram. Which come from the score generated by analyzing each tweet. So that it will be easy to find out strongly positive and strongly negative tweets amount.

3.3 Resource and Data Analysis

As previously described that twitter is used as the social networking website and R platform for the analysis. So now brief discussion of this is given below.

To collect data from twitter at first creating a twitter API using our twitter account is needed. Without the application the system will not be able to get access and collect data from twitter. The created account provides some keys and tokens named consumer key, consumer, Consumer Secret, access token, access token secret. Those are unique for every single account in the twitter available. Those keys and tokens will be required for the connection between twitter and R platform.

When creation of twitter application is done, process for sentiment analysis starts. As only R software is not that much user friendly so Rstudio must be installed in the computer. Because RStudio provides a user-friendly interface to its user which is easier also.

The plan for the application was like, it will first connect with twitter. If the connection is successful then it will collect data from twitter. After collecting data it will go for the analysis process. The analysis process is divided in four steps. Each step will generate one output. They are emotion, polarity, word cloud and scores. According to the number of tweets the application has collected, the time for the processing can vary. When all the processing is done those data will be visualized in the application page. Though the application was built in the R platform, a library named Shiny gave it a user interface. So after running the application it will open in a browser.

Here in Fig. 1 it is showing the data flow diagram of the application:

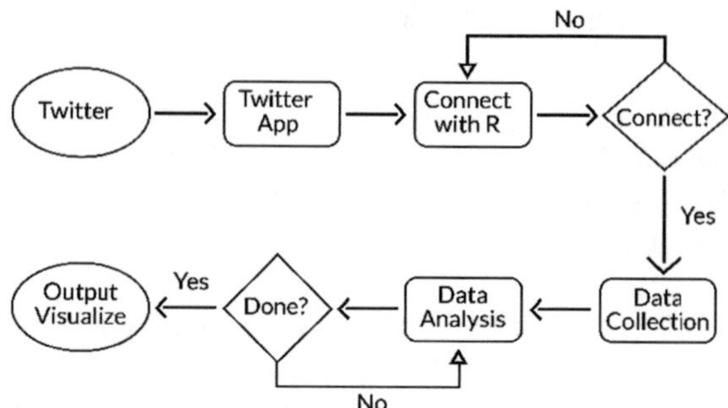

Fig. 1 Flow of data analysis process

3.4 Experimental Result

For better understanding of the scenario this application generates four types of output. Each of those results have their own way of analysis. So the user will be able to see the results from different angles. The application works with both supervised and unsupervised learning. It uses supervised learning for calculating the emotion class, polarity class and word cloud. To show the impact of unsupervised learning the application generated Score bar diagram. Polarity and score bar diagram works in the same way. But the target was showing the difference between outputs of same category. Because they show a little difference on the output while analyzing the text. Let us see the detail of those outputs.

Emotion can be defined as psychological state of human. For the basic division we can divide emotion into six basic classes. They are anger, fear, disgust, surprise, happiness, and sadness [22]. As the emotion flow throws someone's thoughts, expression, or behave. So it is possible to simply identify it using the basic text analysis. Because every emotion-related word can be defined in different classes. Naïve Bayes algorithm was used to calculate the emotion class, which is a supervised learning method. Considering the application, when it can find most of the words are related to happiness in the sentence, it will include it to the happiness class. If the words are found in the sentence are related to sadness, then it will consider the sentence as a sad sentence and sadness class will get point. On the other hand if any sentence does not contain any emotion related word then it will be added to the unknown emotion class.

The application is showing the that output according to the point's tweets are having for the each class. For simple calculation, the score for each tweet considered as 1 point. It does not matter in which class it belongs to, it will get the same point. If the tweet contains different emotional state then it will consider the probability of the emotions and it will be calculated according to that result. User will be able to see in the figure that on the right side application is showing tweet numbers. On the right each color representing the emotional state, it is also added in the footer part. On the top of the figure the keyword is visible for which user got this output. An example figure of emotion class is given below.

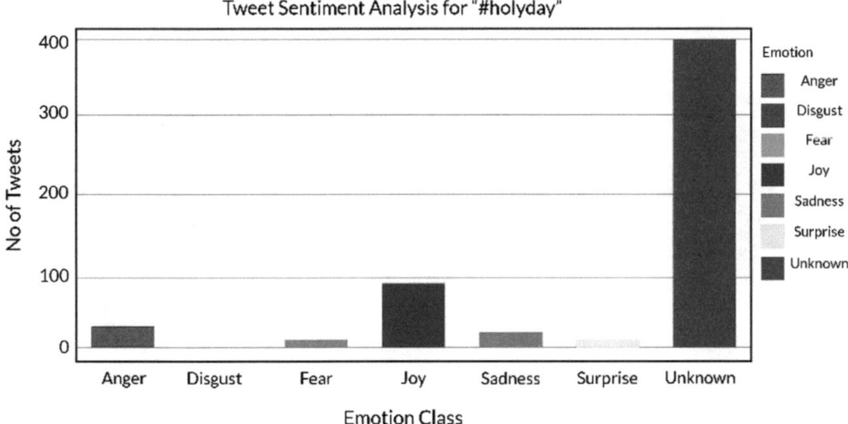

Fig. 2 Emotion class

From the following Fig. 2 we can see that for the search keyword "#holiday" emotional state "joy" having highest value. That means most of the emotion related tweets expressing joy for "holiday". Application will not consider unknown class in any state of emotion.

Polarity can be defined as a state of emotion which defines the positivity and negativity of text. Polarity can be divided into three classes, they are positive, negative, and neutral. Polarity depends on the words are used in the tweet as they are in the negative words list or in the positive words list. Considering the application it counted every single positive or negative word as 1 point. If a sentence has three positive words then it will have three points. On the other hand, if a sentence or tweet has three negative words then it will have –3 points.

For polarity class sentiment analysis Naïve Bayes algorithm like emotion class is used. So this time the application used built-in library for the positive and negative texts. When user runs the program, the program matches every single word with the positive and negative words. If match found, it add point either positive side or in negative side. It keeps track of this calculation for the final determination of polarity. This process is also known as "bag of words". After calculating the positive and negative value of a tweet we will apply following equation:

$$\text{Tweet score} = \text{positive value} - \text{negative value} \tag{1}$$

For example, we have a tweet "That was a pretty good party."

For this user will get the score,

$$\text{Positive} = \text{pretty (1)} + \text{good (1)} = 2 \tag{2}$$

$$\text{And, Negative} = 0 \tag{3}$$

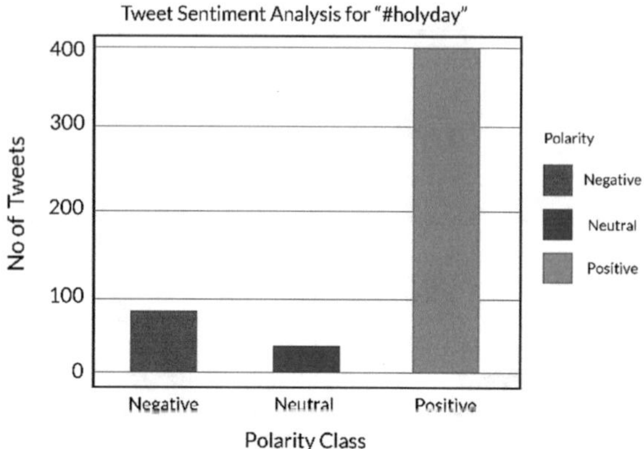

Fig. 3 Polarity class

As in Eq. 3 there are no negative words.

So, the Tweet score,

$$T = 2 - 0 = 2 \tag{4}$$

In Fig. 3 for polarity class user will be able to see that the application marked positive tweets bar diagram using green color, negative tweets bar diagram using red color and neutral bar is sky blue. On the top it is showing the search keyword.

An example figure of polarity class is given above.

From the figure we can see that for the search keyword "holiday" we got most of the tweets as positive. So that it can be said that most of the people are having a positive feelings about the holiday. But still some people are having a bad day in a holiday.

Word cloud is nothing but an image created by some words. But the interesting fact is in a word cloud every words view represents something. For proper explanation it can be said that every words size, boldness and frequency depends on the appearance in the text we are analyzing and have their own meaning. Top appeared words will stay in the center of the word cloud and they will be bigger and larger in size. In some cases they are bold.

In here Naïve Bayes algorithm is used for this word cloud also. Because the word cloud application created in here is related to the emotion class analysis. If someone look closely then he will be able to see that application is showing only those emotion class related words in the word cloud, which we got in the emotion class analysis. Though the colors are not matching because of the algorithm applied for the analysis. Word cloud emotion class will change every time depending on the words the application found. If any emotion class exists in the word cloud, then it will be plotted in the emotion class analysis. So it can be simply said that emotion class analysis depends on the word cloud.

An example of word cloud figure is given below (Fig. 4).

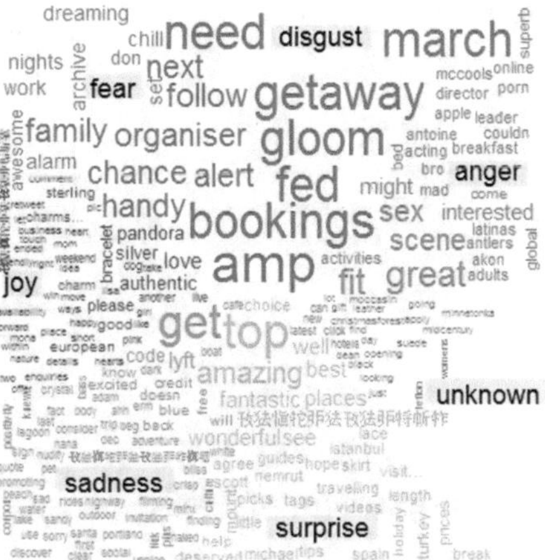

Fig. 4 Word cloud

Form the figure we can see that for the search keyword "holiday" words like booking, getaway, top, etc., are the largest words. So it can be said that those are most frequently used words in the collected tweet texts.

But the application applied manual process instead of libraries. Two separate text files are already created in the same folder of the application code file. One of them contains positive words and another is containing negative words. Like the polarity class this process will also scan each word and match them. The calculation system is also same in here. But in here the application did not divide it into positive, negative, and neutral, instead of that it divided them into scores. That is why it is named as Score Bar diagram. The task of score bar diagram is quite similar as the polarity class. Even they show same kind of output, which is positive, negative and neutral.

The application calculated the score for each tweet separately and found some numbers like –2, –1, 0, 1, 2, 3, etc. All of them are representing the value of different tweets. Using this process the application will be able to find out strongly positive and strongly negative tweets. The higher value tweets get, the more strong side (positive or negative) it is representing. Here 0 represents the neutral, greater than 0 represents positive and smaller than 0 represents negative tweets. In the figure user will be able to see the number of tweets having which value and how many tweets are strongly representing its kind.

An example of Score Bar Diagram is given below (Fig. 5).

From the figure we will be able to see that for the search keyword "holiday" strongly positive tweets are having score 4. But maximum tweets are having the score 0, which means they are neutral and this is the difference between polarity class and Score bar and the output is also changed.

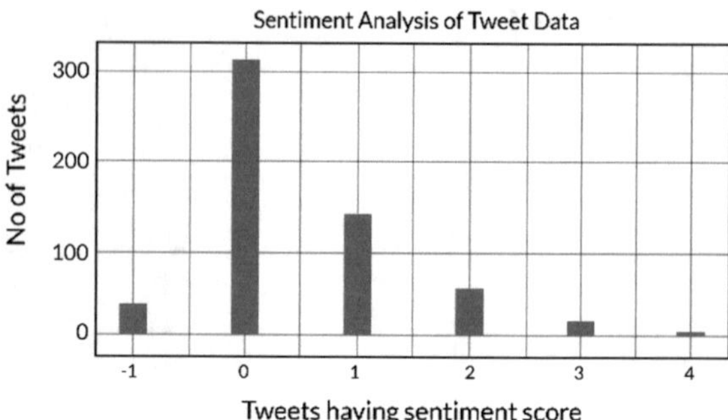

Fig. 5 Score bar diagram

4 Comparison of Methods

Like most other sentiment analysis tools in this research also used Naive Bays Algorithm. But with Naive Bays, Lexical analysis was also included due to increase the validity of this research. In most of the research, for example a research done by V. A. Kharde and S. S. Sonawane indicates that Naive bays is more accurate and useful for text analysis and sentiment analysis [23]. On the other hand, Kiruthika M.'s team, Maria Karanasou's team and so many other researchers worked on only on the polarity of the text [24]. Main focus of this research was not only on the polarity but also on the emotion of the text and the word cloud with sentiment polarity score system increased the uniqueness of this research. High accuracy and more output reports of this research can be implemented in the industry sector for business analysis, product review and other related type of operations.

5 Conclusion

Considering the current growth of technology all over the world, people are getting more and more dependent on computer and Internet. They are getting used to buy products online rather than going to a shop. So whenever they decide to purchase anything, at first they try to take a look on product review. If the review is not good enough people will not buy it. Same way, if a political party does not get enough public support on a task or their revolution, they will not be able to succeed.

So the only way to get success is determining public sentiment. Our research tried to focus on that particular spot. We focused on the streamline of current way of analysis and find out generative results by removing searching limitations. So getting output for different keywords will be possible in a single application. So right now our goal is positive development of accuracy of the algorithms applied for sentiment analysis. Business companies are providing huge fund for the further research on sentiment analysis to determine customer satisfaction. So we can hope that we will see good results very soon.

Acknowledgements This research was developed under Daffodil International University and it was our final year research based project. Our special thanks goes to DIU - NLP and Machine Learning Research LAB for their help and support. We would like to thank Subhenur Latif, Lecturer, Department of Computer Science and Engineering, Daffodil International University for her proper instructions. Any error in this research paper is our own and should not tarnish the reputations of these esteemed persons.

References

1. Krishna, D.S., Kulkarni, G.A., Mohan A.: Sentiment analysis-time variant analytics. Int. J. Adv. Res. Comput. Sci. Softw. Eng. **5**(3) (2015). ISSN: 2277 128X
2. Celikyilmaz, A., Hakkani-Tur, D., Feng, J.: Probabilistic model based sentiment analysis of twitter messages. In: Spoken Language Technology Workshop (SLT), pp. 79–84, 2010 IEEE (2010)
3. Muhammad, I., Yan, Z.: Supervised machine learning approaches: a survey. Int. J. Digit. Curation 133. **5**(3), 946–952 (2015). DOI:10.21917
4. Sathya, R., Abraham, A.: Comparison of supervised and unsupervised learning algorithms for pattern classification. Int. J. Adv. Res. Artif. Intell. **2**(2) (2013)
5. Varghese, R., Jayasree, M.: A survey on sentiment analysis and opinion mining. Int. J. Res. Eng. Technol. **2** (2013). eISSN: 2319-1163, pISSN: 2321-7308
6. Naive Bayes for Machine Learning. http://machinelearningmastery.com/naive-bayes-for-machine-learning
7. Pang, B., Lee, L.: Opinion mining and sentiment analysis. Found. Trends Inform. Retr. **2**(1–2), 1–135 (2008). https://doi.org/10.1561/1500000001
8. Svetlana, K., Zhu, X., Mohammad, S.M.: Sentiment analysis of short informal texts Svetlana. J. Artif. Intell. Res. **50**, 723–762 (2014)
9. Hatzivassiloglou, V., McKeown, K.R.: Predicting the semantic orientation of adjectives. In: Eighth Conference on European chapter of the Association for Computational Linguistics archive, pp. 174–181, Madrid, Spain (1997)
10. Esuli, A., Sebastiani, F.: SentiWordNet: a publicly available lexical resource for opinion mining. In: Proceeding of the 5th Conference on Language Resources and Evaluation, Genoa, Italy (2006)
11. Yu, H., Hatzivassiloglou, V.: Towards answering opinion questions: separating facts from opinions and identifying the polarity of opinion sentences. In: Conference on Empirical Methods in Natural Language Processing, Sapporo, Japan, 19372, 129–136 (2003). https://doi.org/10.3115/1119355.11
12. Kim, S.M., Hovy, E.: Determining the sentiment of opinions. In: 20th International Conference on Computational Linguistics, no. 1367, Geneva, Switzerland (2004)
13. Kouloumpis, E., Wilson, T., Moore, J.: Twitter sentiment analysis: the good the bad and the OMG!. In: Proceedings of the Fifth International AAAI Conference on Weblogs and Social Media, pp. 538–541, Barcelona, Spain (2011)
14. Hamouda, S.B., Akaichi, J.: Social networks' text mining for sentiment classification: the case of Facebook' statuses updates in the "Arabic Spring" Era. Int. J. Appl. Innov. Eng. Manage. **2**(5). ISSN: 2319 – 4847 (2013)
15. Wang, Z., Tong, V.J.C., Chan, D.: Issues of social data analytics with a new method for sentiment analysis of social media data. In: IEEE 6th International Conference on Cloud Computing Technology and Science, Singapore, Singapore (2014). eISBN. 978-1-4799-4093-6, pISBN: 978-1-4799-4092-9
16. Bright, J., Margetts, H., Hale, S., Yasseri, T.: The use of social media for research and analysis: a feasibility study. In: Department for Work and Pensions, London, England (2014). ISBN: 978-1-78425-407-0

17. Mäntylä, M.V., Graziotin, D., Kuutila, M.: The Evolution of Sentiment Analysis—A Review of Research Topics, Venues, and Top Cited Papers (2016)
18. Padmaja, S., Fatima, S.S., Bandu, S.: Evaluating sentiment analysis methods and identifying scope of negation in newspaper articles. Int. J. Adv. Res. Artif. Intell. **3**(11) (2014)
19. What are the applications of sentiment analysis?, https://www.quora.com/What-are-the-applications-of-sentiment-analysis-Why-is-it-in-so-much-discussion-and-demand
20. Vohra, S., Teraiya, J.: Applications and challenges for sentiment analysis: a survey.Int. J. Eng. Res. Technol. **2**(2) (2013). e-ISSN: 2278-0181
21. Why use the R Language?, http://www.burns-stat.com/documents/tutorials/why-use-the-r-language
22. What Are Emotions and the Types of Emotional Responses?, https://www.verywell.com/what-are-emotions-2795178
23. Kharde, V.A.: Sentiment analysis of Twitter data: a survey of techniques. Int. J. Comput. Appl. **139**(11) (2016)
24. Karanasou, M., Ampla, A., Doulkeridus, C., Halkidi, M.: Scalable and real-time sentiment analysis of tweet data. In: IEEE 16th International Conference on Data Mining Workshops, Barcelona, Spain (2016). ISSN: 2375–9259, https://doi.org/10.1109/ICDMW.2016.0138

Data Analytic Techniques with Hardware-Based Encryption for High-Profile Dataset

M. Sharmila Begum and A. George

Abstract Data analytics is the science of extracting patterns, trends, and actionable information from large sets of data. The growing nature of data in from different servers with in consistent data formats like structured, semi-structured and unstructured data. Traditional IT infrastructure is simply not able to meet the demands of this new "Data Analytics" landscape. For these reasons, many enterprises are turning to the Hadoop (open source projects) as a potential solution to this unmet commercial need. As the amount of data especially unstructured data collected by organizations and enterprises explodes, Hadoop is emerging rapidly as one of the primary options for storing and performing operations on that data. The secondary problem for data analytics is security, this rapid increase in usage of Internet, drastic change in acceptance of people using social media applications that allow users to create contents freely and amplify the already huge web volume. In today's businesses, there are few things to keep in mind while beginning big data and analytics innovation projects. The need of secured data analytics tool is mandatory for the business world. So, in the proposed model, major intention of work is to develop the two-pass security-enabled data analytics tool. This work concentrates on two different ends of current business worlds need namely attribute-based analytical report generation and better security model for clients. This proposed work for Key generation and data analytics is the process of generating keys is used to encrypt and decrypt whatever data need to be analyzed. The work is to develop the two-pass security-enabled data analytics tool. The software and hardware keys are programmed and embedded in the kit. When the user inserts the software and hardware key, the unique key will be generated in 1024 bit key size. This will provide high level of authentication for the data to be analyzed. The data analytics part is performed with attribute-based constraints

M. Sharmila Begum (✉)
Department of Computer Science and Engineering, Periyar Maniammai Institute of Science and Technology, Thanjavur, Tamil Nadu, India
e-mail: sharmilagaji@gmail.com

A. George
Department of Mathematics, Periyar Maniammai Institute of Science and Technology, Thanjavur, Tamil Nadu, India
e-mail: amalanathangeorge@gmail.com

A. Abraham et al. (eds.), *Emerging Technologies in Data Mining and Information Security*, Advances in Intelligent Systems and Computing 813,
https://doi.org/10.1007/978-981-13-1498-8_2

enabled data extraction. This model is given better performance than the existing data analytical tools in both security and sensitive report generation.

Keywords Data analytic framework · Data security kit · Secured embedded system

1 Introduction

The research model mainly targeted at protecting high-profile data set. This research ideology is the very first initiative in the state of art in the field of big data analytics along with the use of embedding technology. But most of the previous related algorithms encountered some problems such as password cracking, time taken for encryption and decryption. Password cracking is major issue for just password protected security system. Software-based security system is only password protected thus if some gain full access to user's password then it may able to interact with whole system. This cannot consider as secure if user's privacy is compromised. Hardware devices have the ability to tie a system or an object together with its users to the physical world. Thus by adapting hardware-based security the system becomes more efficient and secured as compared to software-based security. This research work describes the customized algorithm along with hardware key and randomization technique; its main goal is to keep the data secure from unauthorized access

1.1 Motivation

Privacy protection of high-profile data is the challenging issue as the information are always hacked by unauthorized users. So privacy and security is the major consideration of this proposed Embedded Security model along with the Integrated Data Analytics model. Following measures are taken into account while developing the framework for the benefit of user:

- Provided privacy protection for high-profile data is considered as the main objective of the proposed work.
- Ensured the data security as the analysis is carried out through data analytic integrated framework along with the hardware kit.
- Improved authentication level to protect access by unauthorized user
- Enabled security and privacy in data analytic framework

1.1.1 **Confidentiality (Data privacy)**: Foremost goal of this project is maintaining confidentiality. The data set and result analysis of data framework and the messages are concealed by encoding it. The message is encrypted using a unique key formed using password, dongle serial number and key. At any time when the information needed the same can be decrypted by providing the same unique key. Hence data confidentiality is ensured.

1.1.2 **Data Integrity**: Maintaining originality of the information is another important factor that is taken into account when developing the framework. The proposed work protects the information by encrypting the database with the secured Key. The same database can be retrieved only when the unique key is provided. Thus the originality of the information stored is maintained with utmost care.

1.1.3 **Authentication**: Protecting the information from the hackers is major concern and hence the framework is protected with the authentication using unique key. This unique key cannot be altered manually as this will be generated from the firmware number of the USB device used, Password programmed and Keys generated. This Key is used both in encryption and decryption process. Hence the information can be accessed only through the authenticated source of the framework.

- In previous research work the framework provides low security. The privacy of the data will be used by unauthorized users. So that the high profile database can be easily hacked.
- Software-based encryption is only available in existing framework. Hardware-based encryption runs independently of the operating system so it is not exposed to these types of security flaws.

2 Methodology

2.1 Proposed Infrastructure

The ultimate goal of this research is to develop a highly secured data analytics tool. This tool focuses mainly on attribute based analytical report generation and an efficient security system (Figs. 1 and 2).

2.1.1 Data Analytic Framework Design

The system is provided with rule generation algorithm for analyzing the data from whole data container (Fig. 3).

- Creating the data set with a new template.datx
- Mapping and reducing of attributes
- Mapped data are processed into the database
- Data filtering based on threshold value using new algorithm open grid rule generation algorithm
- History of processing data and log files can be retrieved in the framework.

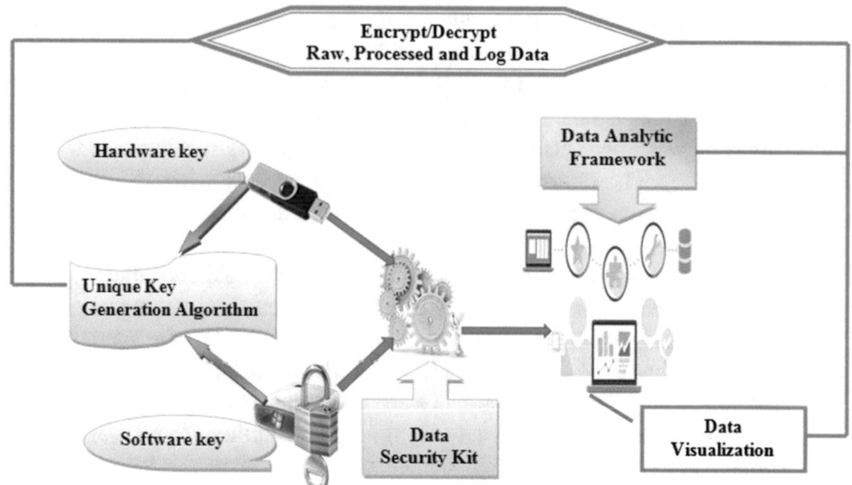

Fig. 1 System infrastructure

2.1.2 Securing High Profile Data

The intention to develop this system is to process and store high-profile data in an efficient and secured way (Fig. 4).

Experimental Setup

- Windows OS is taken as the platform
- Here C# is the framework language and the necessary components related to C# like .NET Framework 4.0, Visual Studio 2012 are installed and configured in the PC.
- Embedded framework which is PIC 16F877a Microcontroller is embedded in the PCB (Printed Circuit Board)
- The Integrated Analytical Framework is connected to the PC via the dedicated COM port (Fig. 5).

Software Key
A software key generator is a computer program that generates a product key, such as a serial number, necessary to activate the data analytic framework.

PIC 16F877a Microcontroller
The PIC microcontroller PIC16f877a embedded kit the embedded security device will recognize the USB stick and read the firmware data from it and embed with 16 bit software key in PIC microcontroller to generate a unique key using the combination of the USB firmware and software key

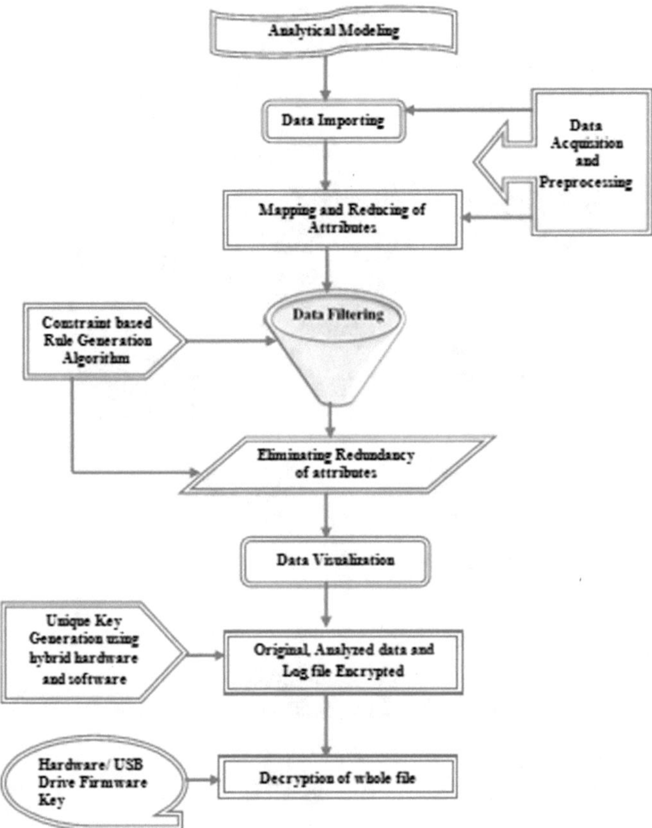

Fig. 2 Data flow diagram

Hardware Key
A USB hardware key is a physical computing device that safeguards and manages digital keys for strong authentication and provides crypto processing.
Unique Key Generation using Software Key and Hardware Key
Key generation is the process of generating keys in cryptography. A key is used to encrypt and decrypt whatever data is being encrypted/decrypted. The software and hardware key are programmed and installed in the embedded kit. When the users insert the embedded kit the unique key will be generated in 1024 bit key size. This will provide high level of authentication (Figs. 6 and 7 and Table 1)

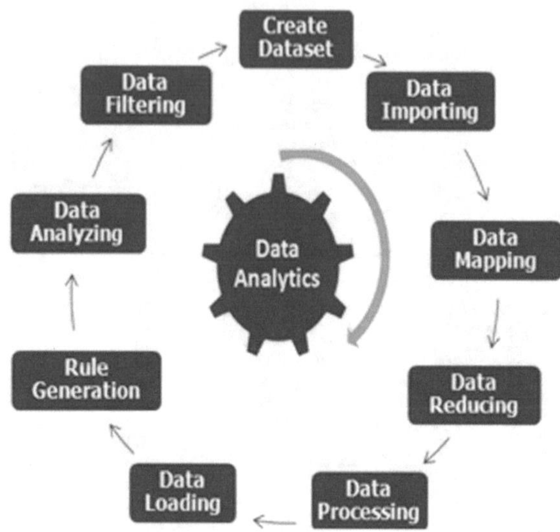

Fig. 3 Data analysis in the framework

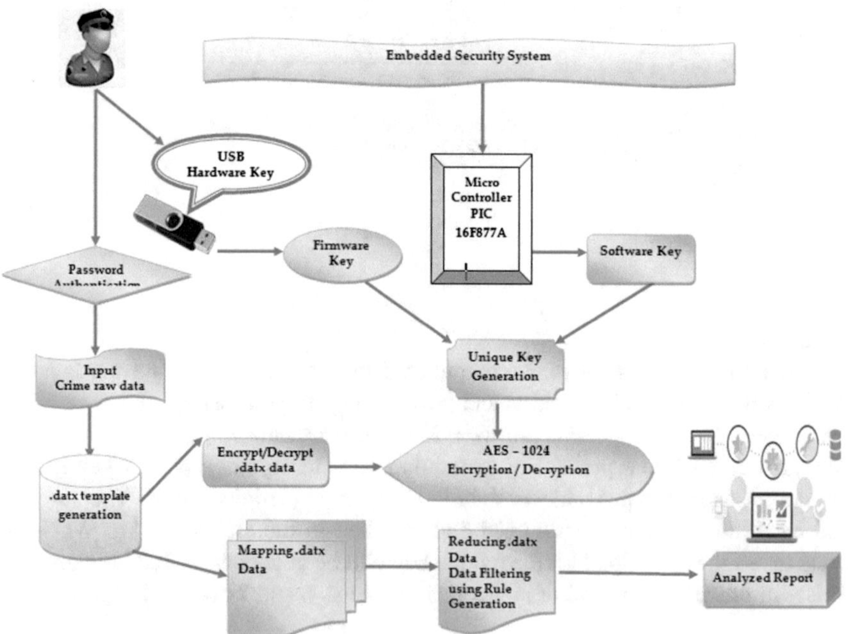

Fig. 4 Integrated analytical frameworks with embedded security

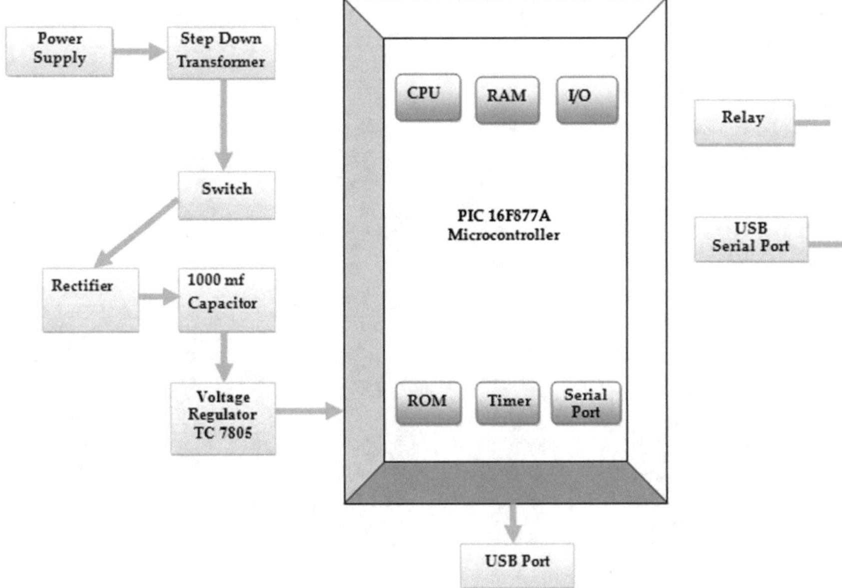

Fig. 5 Block diagram for embedded data security kit

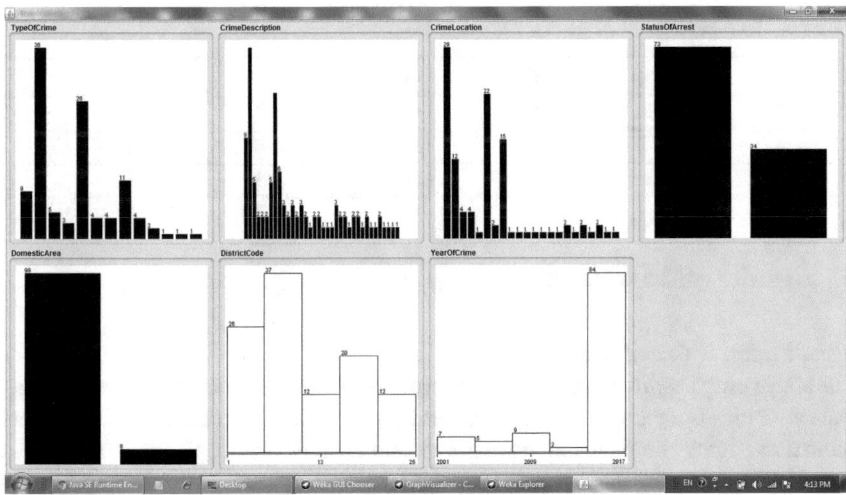

Fig. 6 Visualization of crime dataset and its attributes

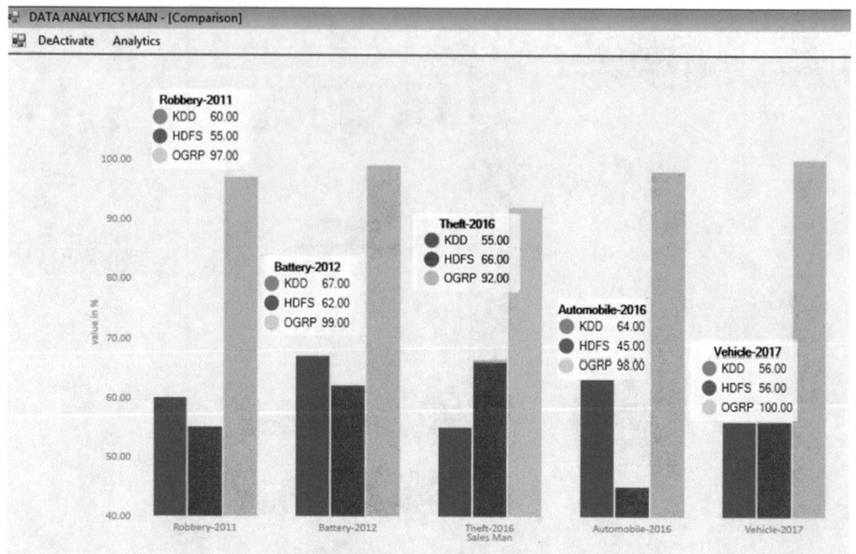

Fig. 7 Algorithm comparison

Table 1 Legends KDD process: knowledge discovery databases, HDFS process: Hadoop distributed file system, open grid rule generation

S.NO	Primary type	Year	KDD process (%)	HDFS process (%)	Open grid rule process (%)
1	Robbery	2011	60	55	97
2	Battery	2012	67	62	99
3	Theft	2016	55	66	92
4	Automobile	2016	64	45	98

3 Results and Discussions

Visualization of Crime Dataset

The above graph estimates the analyzing process of the attributes by their weight age. Data analytics is a technology-enabled strategy for gaining richer, deeper, and more accurate insights about customers, partners and business environment. This process is handled by clearing the raw data from the data base. When a data mapping process is handled it is reviewed that data has been reduced and there is no data redundancy

4 Conclusion

There are many frameworks available to analyze the dataset in a secured way but this work has focused on security and privacy problems in big data analytics framework. Data analytics is a promising technology in crime department and integrate the analytical framework using embedded system to secure the dataset. There will be increasing demand in data analytics framework including security of data. Generating the key by using the combination of both software and hardware will not be hack the dataset by anyone.

References

1. Abad, F., Saeed, A., Hamidi, H.: An architecture for security and protection of big data. Int. J. Eng. Trans. A: Basics **30**(10), 1479–1486 (2017)
2. Maluf, D.A., Bell, G.D., Knight, C., Tran, P., La, T., Lin, J., McDermott, B., Pell, B.: "NASA-XDB-IPG: ExtensibleDatabase—Information Grid" Global Grid Forum8 (2003)
3. Kaisler, S., Armour, F., Espinosa, J.A.: Introduction to big data: challenges, opportunities, and realities minitrack. In: 47th Hawaii International Conference on 2014, pp. 728–728 (2014)
4. Doan, A.H., Naughton, J.F., Baid, A., Chai, X., Chen, F., Chen, T., Chu, E., DeRose, P., Gao, B., Gokhale, C., Huang, J., Shen, W., Vuong, B.-Q.: The Case for a Structured Approach to Managing Unstructured Data"
5. Iwashita, M., Nishimatsu, K., Shimogawa, S.: Semantic analysis method for unstructured data in telecom services. In: IEEE13. IEEE 13th International Conference on Data Mining Workshops, pp. 789–795. https://doi.org/10.1109/icdmw.2008.79
6. Ohlhorst, F.: Big Data Analytics: Turning Big Data into Big Money. Wiley, USA (2012)
7. Gärtner, M., Rauber, A., Berger, H.: Bridging structured and unstructured data via hybrid semantic search and interactive ontology-enhanced query formulation. Knowl. Inf. Syst. 1–32 (2013). https://doi.org/10.1007/s10115-013-0678-y
8. Diakopoulos, N., Naaman, M., Kivran-Swaine, F.: Diamonds in the rough: social media visual analytics for journalistic inquiry. In: Proceedings of IEEE Conference on Visual Analytics Science and Technology, pp. 115–122 (2010)

Exploring Student Migration in Rural Region of Bangladesh

Arafat Bin Hossain, Sakibul Mowla, Sheikh Nabil Mohammad and Md. R. Amin

Abstract In developing countries like Bangladesh, there are places from where students tend to move to other places from their localities to acquire their higher secondary certificate education (HSC) when they finish their secondary school certificate education (SSC). The places where these movements are occurring frequently and factors that are related to it is unknown. It is important to sort out the factors because it can work as suggesting factors to start the overall educational development scheme. In this study, raw data from Sylhet Education Board containing information of 3,08,875 students from Sylhet division to produce a refined dataset containing many properties of students. Descriptive statistics were applied to define a threshold distance. The students were split into two groups on the basis of whether students travelled beyond or below the threshold distance. Comparing the HSC grades between both the groups and the migration frequency and with the help of student t test and effective data visualization, locations were categorized to sort out the locations where migration is more frequent and the migration is yielding good grade points.

Keywords Student t-test · Data visualization · Student migration · Educational development · Data mining

---

A. B. Hossain · S. Mowla · S. N. Mohammad (✉) · Md. R. Amin
Shahjalal University of Science and Technology, Sylhet, Bangladesh
e-mail: nabil-cse@sust.edu

A. B. Hossain
e-mail: arafatbhossain@gmail.com

S. Mowla
e-mail: sakibulmowla@gmail.com

Md. R. Amin
e-mail: shajib-cse@sust.edu

A. Abraham et al. (eds.), *Emerging Technologies in Data Mining and Information Security*, Advances in Intelligent Systems and Computing 813,
https://doi.org/10.1007/978-981-13-1498-8_3

1 Introduction

In remote areas of a country like Bangladesh, students often requires to migrate to another location when to acquire their Higher Secondary Certificate examination (HSC) once they complete their Secondary School Certificate Examination (SSC). Our work deals with data from education board of a district of Bangladesh named as Sylhet and explores the movements of students within the Upazila under the districts to extract interesting information about migration scenarios of those areas. In this paper, words like *movement* and *migration* depicts the similar meaning while *location* and *Upazila* will be often used in different sections interchangeably that is, they depict the similar meaning. To ensure that migration is a problem, which is extant in Sylhet division, we tried to communicate and undertake personal interviews with officials from Sylhet Education board and different institutions. Personnel A from Education Board told us that: "*Students don't want to go to their nearby colleges; they want to migrate. Main reason behind this migration is lack of quality colleges*". Teacher 1 from a Degree College mentioned that: "*Financially well to do families try to send their children into a good college, at least to a Zilla Sadar college if he/she does well in the secondary exam. If there were some good colleges in their thana, they would less likely migrate in my opinion. 2–3% of the students who finished secondary education from Fenchuganj, migrated to other thana for college education.*" A survey was also conducted on 46 HSC students from different Upazila. Some of them were migrants and some were not. They filled up a survey form answering questions about migration and its impact on them. The main aim was to know what they think of the reasons behind migration and what would be their suggestions. The tables illustrate the results of the survey (Table 1).

To sum up all these, it is quite evident that education quality is surely a factor that is driving the problem. All of the personnel from Education Board confirmed that although the Education Board has all the information about the students, there has not yet been any analysis of Upazila-based migration while choosing a place to set up a college. They also appreciated the idea of taking migration into consideration and said that that'd recommend the findings from the research while placing a new college to the ministry if it's realistic.

2 Related Study

Student's t-test is an effective approach when one requires to compare the means of two groups each of which is being treated with different treatments. It is frequently used in pain study [1]. Independent t-test is used when two groups under consideration is not dependent on each other. T-tests like one-sample location test and two-sample location tests are the ones that are used quite often. Independent sample t-tests are often called unpaired t test as the samples being compared are non-overlapping [2]. The null hypothesis it is concerned with is that- mean of two groups are equal. Such

Table 1 Reasons opined by migrants and non-migrants as their reason of migration and stay respectively

	Reasons	Number of participants who opined
For migration (As commented by migrant participants)	*Bad quality college*	11
	Scarcity of quality teachers	7
	Lack of quality education	8
For not migrating (As commented by non-migrant participants)	*Inadequate lab facility*	3
	Problems due to distance from family	6
	Expensive living cost	9
	Expensive transportation cost	4
	To Stay near home	22
	Reduce transportation cost	10
	Bad economic condition, can't move	3

tests are often called Welch's t-test as they disregard the assumption that variance of two groups are equal [3]. The t-test returns a test statistic and p-value using which the null hypothesis can be proved true false. The use of p-value as 0.05 can be used to determine the significance of difference between two means [4].

The idea of Kernel Density Estimation is an effective means of estimating probability density function of a random variable in order to make statistical inferences about a population of definite size. It is also termed as Parzen-Rosenblatt window method [5, 6] and is readily used in signal processing and econometrics. The idea of KDE is of great importance to our study as it is able to generalize the distance moved by students by representing them through a probability density function showing the picture of the migration of a region graphically. It can be hugely helpful in making inferences about population through data visualization.

3 Related Works

Marc Frenette performed a study on 2002 that was done using the Survey of Labor and Income Dynamics (SLID) and a small (publicly available) database of Canadian university postal codes, information of family income, parental education, and other factors to analyze if distance is a associated factor with University participation for students living within and 'out of commuting distance'. They used the straight-line distance from homes of high school students prior to graduation and the nearest University using postal code conversion file and showed that students living out of the commuting distance are far less likely to attend than students living within commuting distance. It showed that distance plays an important role in the relationship between university participation and its other correlates such as family income and sex [7].

Marc Frenette performed another study on June, 2003. The study basically asks and answer these three questions—First, do students who live too far to attend university "make-up" for this disadvantage by attending college (if one is nearby)? Second, how does this uptake in college participation differ by class of income? And finally, does distance to school deter students from attending college? He showed that Students living beyond commuting distance from a college are far less likely to attend college, especially if they are from a lower income family. The uptake in college participation in outlying areas mainly occurs within groups of students who are from lower and middle-income families—those most negatively affected by living far away from universities. Nevertheless, very few students live beyond commuting distance from a college [8].

4 Data

4.1 Datasets

Student Dataset: The main resource of data was collected from Education Board of Sylhet. All required information for 3,08,875 students for the year 2001 to 2015 was collected in csv format which was converted to csv for storing and processing in the database we created. There were two separate collections of data-1. It contained SSC information for students 2. It contained HSC information for students. Rows in each collection comprised of columns such as- *name, father's name, mother's name, gender, gpa, exam location, exam year, subject group, institution code*.

Geo-location Data: Resource of data containing details about institutions was collected from Education Board of Sylhet. The names of each school or college was collected using Google Map Geocoding API [9] was used to extract out the latitude and longitudes of each institutions.

These names of institutions, i.e., schools and colleges then passed through the Google Geocoding API [9] and after extracting the response, we get the latitude and longitudes of each institutions. As there were a major number of institutions, for which no response was found from Google Map Distance Matrix API [10].

Rows contain important attributes for each institution (schools and colleges) such as- *institution name, institution code, latitude, longitude.*

2011 Census Data: Data of population census for each location we used for the research were collected from Statistics Bureau of Bangladesh and Education Management Information System (Bangladesh Bureau of Statistics [11] and Education Management Information System [12] respectively). Rows contain important attributes for each location such as—*total population, total number of colleges for each location, ratio of [number of colleges/population size].*

4.2 Data Processing

The primary steps for data processing was mainly a two-step process.

(a) The goal of this step is to produce a dataset containing information of both HSC and SSC in a single row. It has been done by joining the table from SSC student database to that of HSC student database using *student's name, mother's name, father's name.* After this step, we have all necessary information about HSC and SSC for a student packed in a single dataset rather than in separate collection. The attributes in new dataset are- *name, father's name, mother's name, gender, gpa ssc exam location, hsc exam location, hsc exam year, ssc exam year, hsc subject group, ssc_subject group, institution code for both school and college.*
(b) The dataset is fed to a python program. The program checks institution code for both school and college in each row, gets the corresponding latitude and longitude from Geolocation dataset and calculates distances between school and college. The new attributes in dataset are same as that in step (a) but now it contains a more important information-"*distance between school and college*". Hence, the output of this extensive data processing yield a dataset which contains attributes as such: *name, father's name, mother's name, gender, gpa, ssc exam location, hsc exam location, hsc exam year, ssc exam year, hsc subject group, ssc_subject group, institution code for both school and college, distance between school and college.* This very dataset is used for the primary research.

5 Methodology

5.1 Defining Threshold Distance

From Fig. 1 we see that, considering the movement of students within their own Upazila, the graph shows average distance moved by students. In this way we got

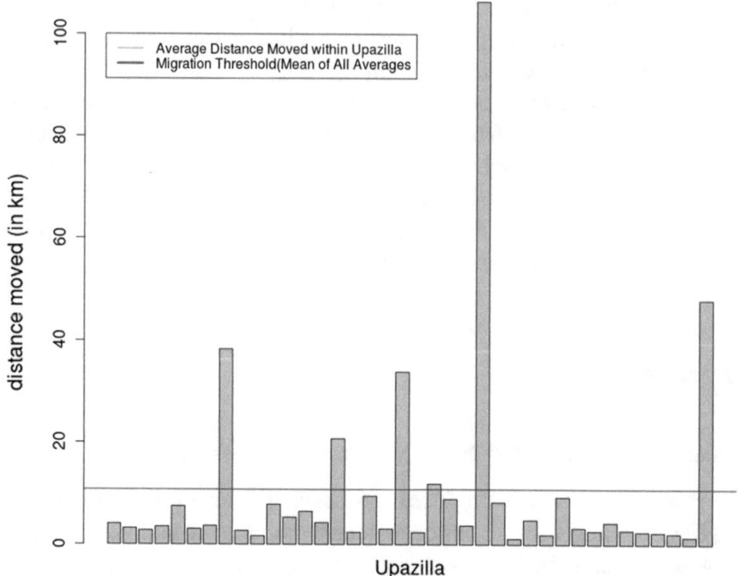

Fig. 1 Figure showing the average distance of movements for students at all 38 Upazilas in Sylhet

38 averages the average of which is 10.8 km, which we considered as migration threshold distance.

The kernel density plot in Fig. 2 compares the probability of students travelling at various distances for two locations. It implies that—high peaks at smaller distance in urban region means less number of students should be migrating from there while high peaks at lower distance means less students are migrating. Since the high peaks are to the right of our defined migration threshold for rural region, it labels most of students in this region as migrants. For the same reason, it labels less number of students as migrants and more as non-migrants in urban region. Therefore, our generic threshold is able to reflect the real scenario and can be considered as pragmatic.

5.2 *Labeling an Upazila*

Our aim was to understand if students from Upazila is significantly leaving or staying and if the result is significantly improving, worsening or remaining unaffected. In order to do so, for each Upazila, the data was summarized for 11 years. We prepared a summary like this for each Upazila. In Table 2, the pattern for the table of an Upazila is shown as an example.

Here is what each column represents:

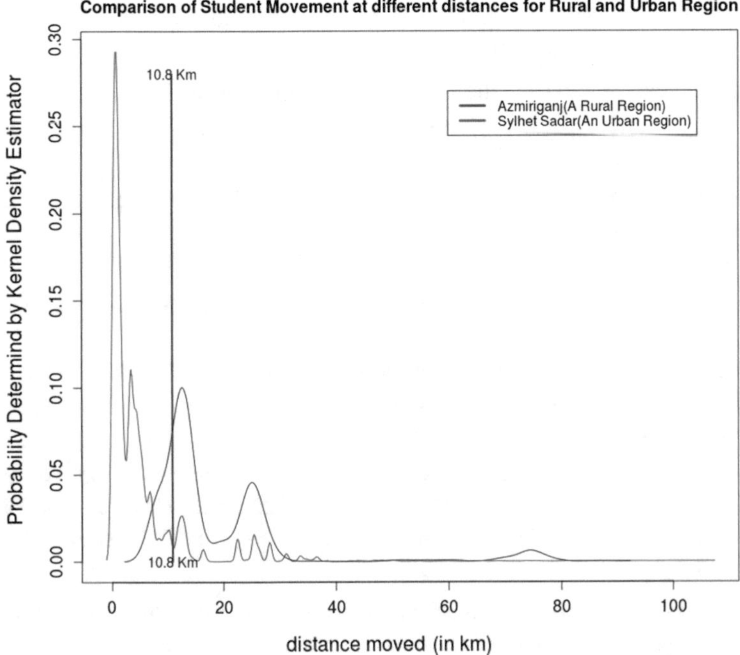

Fig. 2 Figure showing Kernel Density Plot illustrating the comparison of student's movement at different distances with respect to migration threshold for a particular rural and urban region in Sylhet

Table 2 Sample of a table for a single Upazila showing the summary of each Upazila's migration and HSC result scenario (Single row data is used for example)

Year of HSC examination	Number of students left, L	Number of students stayed, S	Mean.L	Mean.S	P (Mean.L!= Mean.S)	Students left (as %)
2001	–	–	–	–	–	–

Note There are 38 tables like these. Each table has 11 rows

Number of Student Left: Number of Students left from an Upazila to pursue Higher Secondary Education. In other words, they will be also be called as *migrants* throughout this writing.

Number of Student Stayed: Number of Students stayed in the Upazila to pursue Higher Secondary Education. In other words, they will be also be called *non-migrants* throughout this writing.

Mean.L: Average HSC GPA of migrants.

Mean.S: Average HSC GPA of non-migrants.

P(Mean.L ! = Mean.S): It is the measure of the probability that Mean.L and Mean.S is different from each other. For each year of data, student t-test is applied on HSC

GPAs for the two groups (migrants and non-migrants) to come up with a t-statistic and corresponding p-value for it.

There are certain interpretations from P (Mean.L !-Mean.S):

- If p-value<0.05 and Mean.L>Mean.S-HSC GPA for migrants are greater than non-migrants. We labelled it as "***Improved***"
- If p-value<0.05 and Mean.L<Mean.S-HSC GPA for non-migrants are greater than migrants. We called it "***Deteriorated***".
- If p-value>=0.05-There is no change of result. We termed it as "***Unchanged***".

Percentage of Students Left: Migrants as a percentage of total number of Students.

Now, a simple counting technique is used to profile an Upazila. We basically wanted to come up with four counts by observing 11 rows of data form Upazila tables like Table 1-

1. The number of times, in 11 years, there were occurrences such that **Percentage of Students Left>=50%** and **Percentage of Students Left<50**. (2st Column)
2. The number of times there were occurrences of "***Improved***", "***Deteriorated***" and "***Unchanged***".

Using 10 years of data and these four counts, we came up with a conclusion about whether students from an Upazila is indeed significantly leaving or Staying and what is overall impact of result due to the migration, if any. These four counts were used to produce a summary table like Table 3.

The values in the two columns *Overall Significance of Migration* and *Overall Impact on HSC Result* are deduced using the logic below:

1. **Overall Significance of migration** is-

   SL(Significantly Left)-if overall occurrences of migration>=50 is greater than or equal to 6, which is median year of 11 years.
   SS(Significantly Stayed)-if overall occurrences of migration>=50 is less than 6, which is median year of 11 years.

2. **Overall impact on HSC result** is-

   I(Improved)-if A>B and A>C;
   NE(No effect)-if B>A and B>C
   D(Deteriorated)-C>A and C>B

Based on these observations, calculations and labels we came up with five categories of Upazila which are discussed in the result section.

6 Results

If we look at various combination of *Overall Significance of Migration-Overall Impact on HSC Result* due to the migration, we can end up in five different com-

Table 3 Table showing migration scenario of Upazila of different categories (Results of 12 Upazila where migration is very significant is only shown in the table)

Upazila	Number of occurrence that migration >=50%	Number of times grade points of migrants improved, A	Number of times grade points of migrants deteriorated, B	Number of times grade points were unchanged for migrants, C	Overall Significance of Migration	Overall impact on HSC result
Chhatak	11	4	0	3	SL	I
Baniachang	11	3	1	7	SL	NE
S. Sunamganj	11	5	0	6	SL	NE
Azmiriganj	11	4	0	6	SL	I
Tahirpur	9	5	0	5	SL	I
Jagannathpur	8	5	0	6	SL	I
Madhabpur	8	1	0	9	SL	NE
Lakhai	7	4	0	6	SL	NE
Doarabazar	6	4	1	4	SL	I
Bishwanath	6	4	0	5	SL	NE
Churnarughat	6	9	0	1	SL	I
Sylhet Sadar	0	2	6	3	SS	D

SL-*Significantly Left*, **SS**-*Significantly Stayed*, **I**-*Improved*, **NE**-*No effect*, **D**-*Deteriorated*

binations—*Significantly Left-Improved*, *Significantly Left-No Effect*, *Significantly Stayed-Improved*, *Significantly Stayed-No Effect*, *Significantly Stayed-Deteriorated*. After we group all the Upazila by different combinations, we get interesting outcomes. It is because each and every Upazila can be related to any one of the combinations while each combination provides useful information about the education scenario of the Upazila. Upazilas from where students *left significantly* and *improved* their result give an insight about the quality of the students of those areas-which is, of course, good. As shown in the last row of Table 4, Upazilas from where students are not leaving very often and those who are staying are rather doing well gives an insight about the education quality of those areas.

However, there are greater number of Upazilas where *Significantly Stayed* but the Results of those who left *Improved.* This fact alone provides a huge important information about the average scenario of rural regions. The students who are leaving are getting good results, be the migration significant in amount or not. The movement of good quality students can be linked with migration.

Table 4 Different Categories of Upazila that we acquired while Labeling an Upazila

Categories	Upazila
Upazilas from where Students ***Significantly Left*** *and Results were* ***Improved***	Chhatak, Chunarughat, Azmiriganj, Tahirpur, Jagannathpur, Doarabazar
Upazilas from where Students ***Significantly Left*** *and the were* ***No Effect*** *on Results*	Baniachang, South Sunamganj, Lakhai, Madhabpur, Bishwanath
Upazilas where Students ***Significantly Stayed*** *but the Results of those who left* ***Improved***	Balaganj, Dharma Pasha, Rajnagar, Fenchuganj, Zakiganj, Jamalgonj, Nabiganj, Compani Ganj, Derai, Sunamganj Sadar, Kanaighat, Kulaura, South Surma, Golapganj, Baralekha, Beani Bazar
Upazilas where Students ***Significantly Stayed*** *but there were* ***No Effect*** *on results for those who left*	Shulla, Goainghat, Bishwambharpur, Sreemangal, Jury, Moulavi Bazar Sadar, Kamolganj, Bahubal, Habiganj Sadar, Jaintapur
Students ***Significantly Stayed*** *and result* ***deteriorated***	Sylhet Sadar

Table 5 Population to Institution Ratio for Upazila based on 2011 Population Census Data. Population size for Upazila is taken form Bangladesh Bureau of Statistics [11] while Number of Institution is taken from Education Management Information System [12]

Upazila	Population	Number of institution	Number of institution per 75,000 population
Lakhai	319016	1	0.50
Doarabazar	228460	3	0.98
S. Sunamganj	183881	1	0.40
Sylhet Sadar	829103	21	1.89

7 Discussion

Table 5 shows that some Upazila have capacity problems because the value of the ratio is less than 1. Among those, first three Upazila from the table match with our *Significantly Left* category. The ratio is a mathematical form of what officials from education board informed us-"*There should be at least 1 institution per 75000 population*". The link between the ratio and different categories of Upazila is discussed below.

If we observe information from Table 4, all the locations with "Sadar"(City) are seen to be labeled with *Significantly Stayed*. It might be because of the relative better educational facilities in these places than any other places. Sylhet Sadar is only one of those kinds of location from where students constantly do worse if they leave. It sounds reasonable because according to data from institution management data center, there are total of 21 schools and colleges in Sylhet(highest among the Upazilas) in 2011 [11]. If we combine it with Table 5, we see that Sylhet Sadar has the highest ratio. From Table 5 it can be observed that *Lakhai*, *South Sunamganj* and

Doarabazar are the places from where students are significantly migrating and these places also tend to have capacity problems. Therefore, capacity can be one of the factors that is driving the migration.

However, it is worth mentioning that there are more Upazila, which falls to *Significantly Stayed-Improved* category than any other category around. It gives us two useful information- 1. Migration is not a common scenario for most of the Upazila. 2. Students are getting good grades due to migration from these areas. This alone not only gives us knowledge about the trend of movement of students in different rural region but also the direction of movement of specific quality of students. If we trace our study back to the survey results in Table 1, we already know that non-migrants opined transportation cost, poor economic condition as vital reasons for not wanting to move out of their localities. Therefore, it also gives us a slight image of the socioeconomic conditions of rural areas like these. As a result, while capacity not being a problem and migration being not a common scenario for most of the upazilas, good grades can be linked with migration for at least most of the Upazilas taking lack of education quality into consideration.

8 Conclusions

Presently, there are no systems or set of system which is detailed to the extent our procedures are and can explore the migration scenario, identify the factors closely related to migration for a set of locations. Our system is able to map all the HSC and SSC data for a particular student, create a profile and label each location with their corresponding migration scenario. By intuition, we can say that *Upazilas from where Students **Significantly Left** and Results were **Improved*** are reservoirs of good students who are lagging behind due to lack of good local education quality. While some upazila have capacity problems, education quality is a common problem for most of the Upazila. Migration is not a common scenario for most of the Upazila and students are getting good result grades due to migration given.

9 Acknowledgements

We would like to thank Mr. Md. Atiqur Rahman, System Analyst of Sylhet Education Board for helping us to collect the data from Sylhet Education Board for this research. We would also like to thank Mr. Sabir Ismail for helping us in the process of merging student's HSC and SSC information. All procedures performed in studies involving human participants were in accordance with the ethical standards of the institutional and/or national research committee and with the 1964 Helsinki declaration and its later amendments or comparable ethical standards.

References

1. Yim, K.H., Nahm, F.S., Han, K.A., Park, S.Y.: Analysis of statistical methods and errors in the articles published in the Korean journal of pain. Korean J. Pain **23**, 35–41 (2010)
2. Barbara, F.: High-Yield Behavioural Science (High-Yield Series). Hagerstwon, MD: Lippincott Williams & Wilkins (2008)
3. Derrick, B., Toher, D., White, P.: Why Welch's test is Type 1 error robust. Quant. Methods Psychol. **12**(1), 30–38 (2016)
4. Fisher, Ronald A.: Statistical Methods for Research Workers, p. 43. Oliver and Boyd, Edinburgh, UK (1925)
5. Rosenblatt, M.: Remarks on some nonparametric estimates of a density function. Ann. Math. Stat. **27**(3), 832 (1956). https://doi.org/10.1214/aoms/1177728190
6. Parzen, E.: On estimation of a probability density function and mode. Ann. Math. Stat. **33**(3), 1065 (1962). https://doi.org/10.1214/aoms/1177704472.JSTOR2237880
7. Frenette, M.: Too Far to Go On?: Distance to School and University Participation. Statistics Canada, Ottawa, ON (2002)
8. 2003. Access to College and University: Does Distance Matter? Statistics Canada, Ottawa, ON (2003)
9. Google Geocoding API. https://developers.google.com/maps/documentation/geocoding/start
10. Google Map Distance Matrix API. https://developers.google.com/maps/documentation/distance-matrix/intro
11. Bangladesh Bureau of Statistics. http://www.bbs.gov.bd
12. Education Management Information System. http://www.emis.gov.bd

Analysis on Lightning News and Correlation with Lightning Imaging Sensor (LIS) Data

Sunny Chowdhury, K. M. Sajjadul Islam, Sheikh Nabil Mohammad, Md. R. Amin and Choudhury Wahid

Abstract Due to the geographic location of Bangladesh, heavy rainfall with the thunderstorm is a common phenomenon. Every year lightning causes lots of casualties. There are very few sources available for lightning-related information. Global Hydrology Resource Centre (GHRC) has a huge collection of lightning data from 1998 to April 2015. The other source of lightning is the news incidents published in Newspapers. This paper explores the effects of lightning in Bangladesh during the last several years, by analyzing lightning news from newspapers and Lightning Imaging Sensor (LIS) data from Global Hydrology Resource Center (GHRC). We analyzed LIS data and news data to find the correlation between these datasets. Finally, we report lightning-prone areas with time when lightning mostly occurs in Bangladesh based on those results.

1 Introduction

Atmospheric lightning is a natural phenomenon which is caused by the separation of positive and negative charges in a cloud. There is an ample supply of both positive and negative electric charges in the air due to the ionization from cosmic rays and natural radioactivity [1]. It is an abrupt flow of static electric charges that occurs when

S. Chowdhury · K. M. S. Islam · S. N. Mohammad (✉) · Md. R. Amin
Shahjalal University of Science and Technology, Sylhet 3114, Bangladesh
e-mail: nabil-cse@sust.edu

S. Chowdhury
e-mail: sunnycse248@gmail.com

K. M. S. Islam
e-mail: s.i.sajjad.islam@gmail.com

Md. R. Amin
e-mail: shajib-cse@sust.edu

C. Wahid
Metropolitan University, Sylhet 3100, Bangladesh
e-mail: mwahid@metrouni.edu.bd

A. Abraham et al. (eds.), *Emerging Technologies in Data Mining and Information Security*, Advances in Intelligent Systems and Computing 813,
https://doi.org/10.1007/978-981-13-1498-8_4

oppositely charged regions of the atmosphere temporarily equalize themselves. This discharge is generally known as the lightning flash. A typical lightning bolt contains about 30 kA to about 400 kA current and carries about 15–400 C electric charges [2]. The typical duration a lightning flash is about 0.2 s to few microseconds [3]. The enormous amount of electricity contained in a lightning is very lethal and a great threat to lives and property due to its statistical nature.

Bangladesh is a densely populated country in South Asia with approximately 164 million [4]. According to the reports published in The Daily Star between January 2010 and May 2015, each year at about 117 people get killed by lightning strikes. The exact number is much higher [5]. Rising number of fatal lightning strikes is responsible for the country's population decrease and disappearance of many tall trees. Frequent use of metal equipment in open fields, cell phone use during the thunderstorms, standing near metal cell phone towers or electric power towers, taking shelter under trees during electrical storms and so on are accountable for increasing lightning-related deaths in Bangladesh [6].

Lightning news contain details of the lightning incidents such as casualties, places, times, and so on whereas LIS data provide information about lightning properties namely Radiance, flash time, latitude, longitude, events. So we are trying to make a relation between newspaper data and LIS data to find the Radiance value for which most of the people die and make assumptions based on that Radiance value.

2 Methodology

First of all, we crawl news reports from five different Bangla newspapers from 2007 to 2015 based on our keywords and create a news database. We prepare another database of all Upazilas of Bangladesh (a local government entity smaller than a district). Then we parse place names from news using upazila database as well as dates and make a processed news database. Moreover, we prepare the LIS database using GHRC dataset and by Reverse Geocoding we extract place names from each LIS lat-long. Finally, we correlate LIS data with news data to find the matched LIS. Here, Fig. 1 shows the full process.

2.1 Study Area

Bangladesh is located between 24° 00′ N latitude and 90° 00′ E longitude. Its topography is characterized by marshlands and never rises over 10 m above sea level [7]. Pre-monsoon (March–May), Monsoon (June–September), Post-monsoon (October–November) and winter (December–February) are four distinct seasons in Bangladesh. However, summer and winter monsoons have dominance on its climate [8] with difference in rainfall and temperature. 71% of the total rainfall occurs dur-

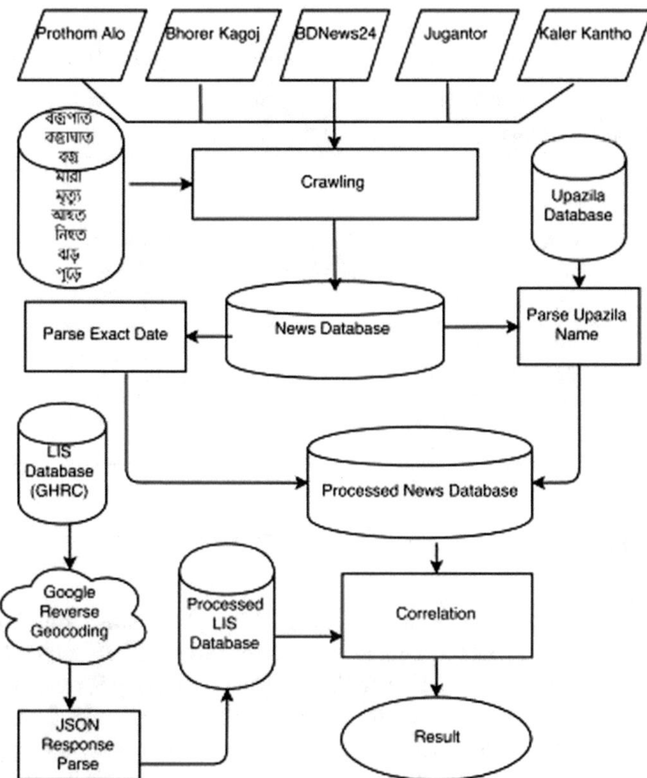

Fig. 1 Process flow

ing monsoon season, with the highest occurrence in the northeastern part due to the influence of Meghalaya Plateau [9].

2.2 *LIS Data*

LIS was launched in 1997 with the Tropical Rainfall Measuring Mission (TRMM) to record lightning activity between 35°N and 35°S latitude [10]. It is an improvement over the Optical Transient Detector (OTD) sensor [11], in terms of detection sensitivity [12]. It can efficiently detect weak and daytime lightning flashes [12]. LIS detection efficiency range is between 69% near local noon to 88% nightly [13]. LIS efficiently identifies the spatial location of lightning, the time of events and radiant energy of lightning [14]. Global Hydrology Resource Centre (GHRC) [15] has an enormous collection LIS data from 1998 to April 2015 [16]. We collected a total of 223935 LIS data by a Space-Time Domain search in GHRC.

Table 1 Upazila database

English name	Bengali name	District name English	District name Bengali
Bagerhat Sadar	বাগেরহাট সদর	Bagerhat	বাগেরহাট
Chitalmari Upazila	চিতলমারী	Bagerhat	বাগেরহাট

Table 2 Newspaper database

Newspaper name	Number of report	Keywords (death) নিহত, মারা, মৃত্যু, প্রাণহানি	Keywords (injured) আহত, হতাহত, দগ্ধ,ঝলসে, পুড়ে, অসুস্থ
Prothom Alo	393	268	202
BDNews24	80	54	57
Bhorer Kagoj	35	27	16
Jugantar	18	13	11
Kaler Kantha	15	11	11

2.3 Reverse Geocoding and Upazila Database

For matching each news with LIS data, we find the place name from latitude and longitude provided by LIS data using Reverse Geocoding API of Google Maps [17]. We successfully parsed 37644 LIS data for Bangladesh. During Reverse Geocoding, we ignored the place names which do not represent any recognized area. We also create a database with all 491 upazilas of Bangladesh. Table 1 shows sample of the Upazila Database. The names of the upazilas are taken from two different sources [18, 19]. To avoid confusions in names, for example, 'মধুখালি' and 'মধুখালী' (Madhukhali), all possible combinations are taken.

2.4 Newspaper Data Collection

We have crawled lightning news from five popular Bangladeshi newspapers [20–24] using Web Crawlers [25]. During our crawling, we have used several keywords (indicates in Table 2) for searching in news text and title to find only lightning news.

2.4.1 News Normalization

As we have taken lightning news from Bangla newspapers, we faced some Unicode related problems. In Bangla Unicode characters there are some confusions because a word can be written in completely different ways. We have to normalize the news so that we can easily work with those natural language texts. For example,

'সো' can be written as (স then ে then া) or as (স then ো)

'গৌ' can be written as (গ then ে then ৗ) or as (গ then ৌ)

2.4.2 Finding Original Date

Studying lightning news we find that normally reports are published few days after the actual incidents. So for our analysis, we consider up to 3-day differences.

At first, from news published date of each news, we find day name and convert this into numbers 1 to 7. Then, we take all the day names from the news text. If a Bangla day name like 'সোমবার' is found, corresponding English name "Monday" is taken. Then we convert these day names into numbers 1 to 7. Finally, we calculate the absolute difference between news published day name and each day name found from the text. This day difference helps us to find all the original dates from a news. If there is no day name is found, then keyword like 'আজ' (Today) or 'গতকাল' (Yesterday) is considered and used according to above procedure. If nothing is found in news text then the news date is taken as original date.

2.4.3 Finding Upazila Name and Keywords

Sometimes several news reports are clustered in a single report which creates anomalies in our news dataset. Furthermore, if it is only a lightning news there may be many places which are not lightning event places at all. So, we have to be selective in extracting place names from news texts. From our analysis of many news, we find that every news report is described in some sentences. Normally, it starts with a brief description of an event. One or two sentences tell about the place and time of the event. Then next few sentences describe this event in details. Using this technique, we try to find only lightning event places.

At first, we search for Bangla district names from our Upazila Database in a news text and corresponding English district names are taken. Then upazilas of those districts are taken in an Upazila Map where each Bangla upazila name is mapped by English equivalents. By extracting districts first we remove some ambiguities in finding upazilas, for instance, if a sentence contains a word like 'ফরিদপুর' (Faridpur) then 'ফরিদপুর উপজেলা' (Faridpur Upazila) can be found because we have no way to verify whether this is a district or upazila. Of course, we could check for 'উপজেলা' (Upazila) word after a place name. But this would be too inefficient. Natural language texts are not written like this all the time and lots of upazilas would be missed. If no district is found then all the upazilas are taken in the Upazila Map. We split the news text by '।' (Bangla Full Stop). We also add the title to news text for precision. So we have Bags of Sentences now from which we will make candidate sentences and look for upazilas and keywords in these sentences.

We start traversing from the last sentence because we find this efficient in finding place names. During our traversal, if keywords namely 'স্বাস্থ্য কমপ্লেক্স' (Health Complex) or 'হাসপাতাল' (Hospital) are found in a sentence, we skip those sentences because someone can be struck by lightning at one place and can be taken to a hospital at another place.

If a sentence contains accident keywords mentioned above in Table 3, we consider this as an accident event. In contrast, if this sentence contains ambigu-

Table 3 Keyword set

Lightning	Death	Injured
বজ্রপাত,বজ্রাঘাত,বজ্র	নিহত,মারা,মৃত্যু,প্রাণহানি,প্রাণ হারান,প্রাণ হারিয়েছে,প্রাণ হারায়	আহত, হতাহত, দগ্ধ,ঝলসে,পুড়ে, অসুস্থ

ous keywords such as ‘ঘর চাপা’ (pressed by house), ‘সাপের কামড় (snakebite)’, ‘গাছ চাপা’ ‘গাছ পড়ে’ ‘গাছ উপড়ে’ (pressed by tree), ‘চাপা পড়ে’ (by pressed), then we avoid this sentence from further processing. If these ambiguous keywords are not found, we further process it. We look for lightning keywords such as ‘বজ্রপাত’, ‘বজ্রাঘাত’, ‘বজ্র’ in this sentence or search two sentences forward for these three lightning keywords. If found then the accident sentence is considered as candidate lightning sentence. Now we search for upazila name in the candidate lightning sentence. We search for Bangla upazila names from our Upazila Map and add English equivalent to our Found Upazila Set. We also find keywords from this sentence and add this to our Keyword Set. We save this sentence in our Candidate Sentence List. If no upazila is found in this sentence, then we start from immediate next sentence and start traversing next sentences until any upazila from our Upazila Map is found in one of these sentences. During our traversal, if we ignore ambiguous sentences. We did so because when we are in news description, we have to traverse forward to find the place of the event. In the remaining next sentences, we apply the same procedure described above. When the entire text is traversed, only the candidate lightning sentences remain. If we cannot find any upazila after traversing all the sentences, we search for districts in our saved candidate sentences.

2.4.4 Finding Unique News

In our news database, we have lots of duplicate news because we have taken news from multiple newspapers. For finding unique news we try TF IDF and Cosine Similarity using upazila names and original dates. But we are unable to find an appropriate similarity score because of aggregation of multiple lightning news in a single report. Hence, whenever we find the same date and same place on two news report we remove one place from one of the two reports. This is how only the unique news remains.

2.5 Making Correlation Between LIS Data and Newspaper Data

From our processed Unique News Database and LIS Database, we start making the correlation. We take one news from our processed Unique News Database at a time

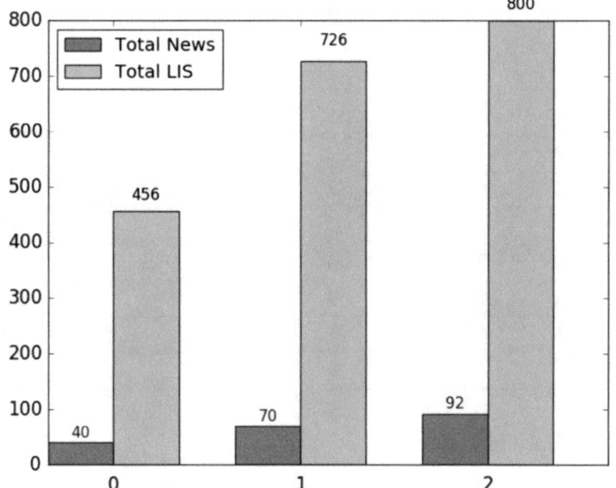

Fig. 2 Number of unique news matched with LIS data from 0 to 2 day difference

and match with LIS Database. We take each Original Date against each Upazila from that news and search in the LIS Database. If we find LIS for a date and place then this is added to total LIS list for this news.

3 Result Analysis

From January 1, 2007, to April 8, 2015, we get total of 541 news. Among this 434 is unique. We find that 118 news is reported multiple times in different newspapers. We have total 37644 LIS data in our time-space domain. We continue our further processing using unique news.

If we take up to two day differences for each news and take cumulative sum of matched news data with LIS data every time, then we can find a better correlation between news data and LIS data.

However, for our analysis, we only consider 0 day difference which means the same date as the original date found from news report. We see in Fig. 2 that, there are 40 news data match with LIS data on same date as the original date. In our news data, there are very few news from 2007 to 2009. Because, back then online newspapers were not very widespread in our country. But from 2010 there are lots of news reports regarding the lightning incidents. And our LIS data is taken up to April 8, 2015. The data in Table 4 indicates the total number of LIS data, the total number of news and number of news matched in every year from 2007 to April 8, 2015.

We plot a Gaussian Histogram of our 456 LIS values which are the exact match with our news data. From the graph (Fig. 3 shows the graph) we see that most LIS value lies in the range below 500000. There are also many radiance values around 1000000.

Table 4 News data match with 0-day difference

Year	Totale LIS	Total number of news	Number of news matched
2007	4959	03	00
2008	3901	04	00
2009	3874	12	00
2010	5575	57	08
2011	4046	60	06
2012	3920	96	05
2013	4663	80	11
2014	4512	83	06
2015	940	39	04

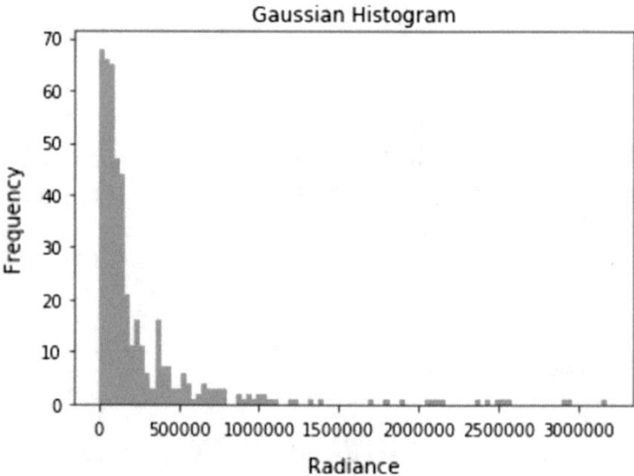

Fig. 3 Radiance value frequency

We draw three heatmap (Shows in Fig. 4) with 456 LIS data which are matched with news data. In this map, we see that most Lightning occurred during May to August. In Bangladesh, these 4 months is mainly monsoon season.

We find out top 10 upazila (Shows in Table 5) according to the number of the report and take the number of LIS data for individual news. We find out the range of the radiance value for each upazila.

4 LIS Data Processing Anomalies

Many news data do not match with LIS data due to LIS data processing anomalies [26]. We have found total 48 lightning reports on those anomalous dates. In our process, original dates and upazila names of those reports were found correctly. Here is a sample data of LIS anomalies depicted in Table 6.

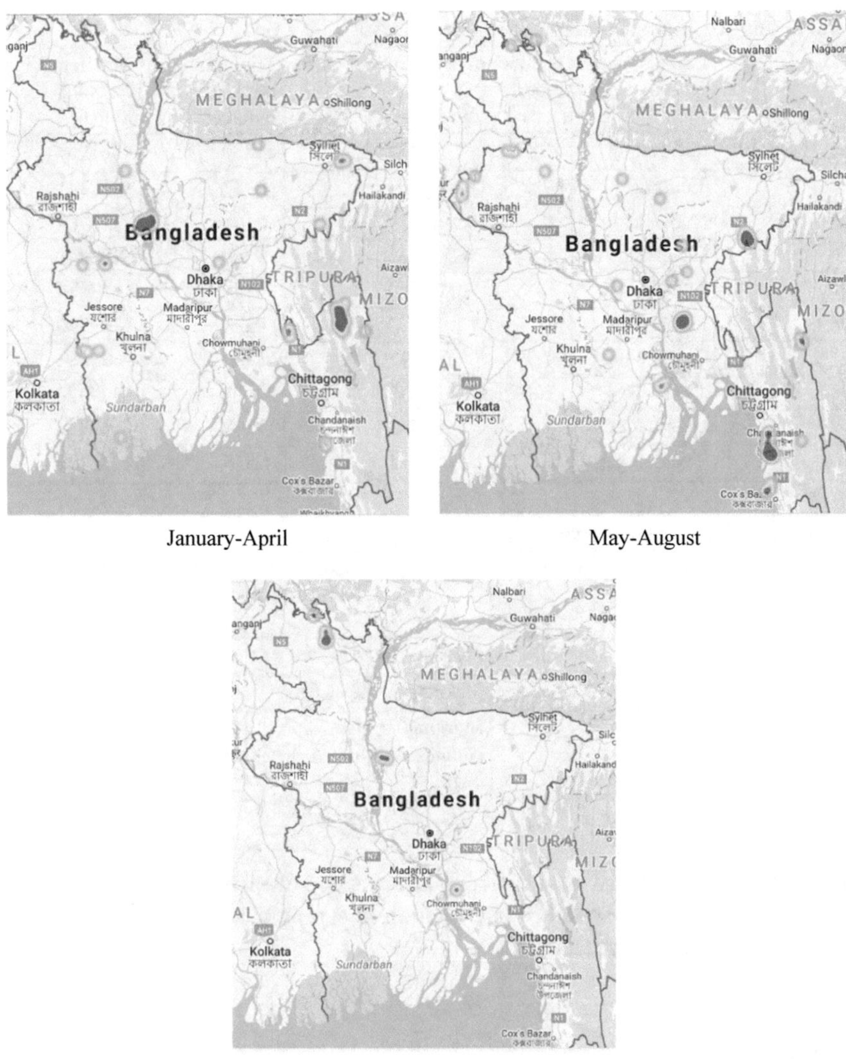

Fig. 4 Heatmap according to season

5 Conclusion

We have analyzed the lightning that occurred in Bangladesh in past several years. In our mapping of the local news with the LIS data, we only considered the news that was published shortly after a real event in LIS database. Using our analysis, we have identified those places and times where the severe damage is caused due to lightning. As the LIS data is no more updated from 2016, it became very important to analyze

Table 5 Top 10 Upazila according to newpaper

No	Upazila name	Number of report	Number of LIS	Minimum range of radiance	Maximum range of radiance
01	Shibganj	16	115	11730	4281909
02	Tahirpur	9	160	9987	5667741
03	Nikli	9	59	4870	115768
04	Chakaria	8	72	9740	193841
05	Kamalganj	8	243	4158	19488
06	Moheskhali	8	52	4274	290990
07	Gomastapur	7	97	5536	39088
08	Assasuni	7	37	22921	399079
09	Dharmapasha	7	262	4340	22239
10	Shyamnagar	6	267	5556	35478

Table 6 LIS data processing anomalies

Original date from news data	Dead	Injured	Time	Upazila extraction	Total LIS	Data processing anomalies
13/08/2014	7	10	Morning	Correct	0	Yes
01/04/2015	2	0	Morning	Correct	0	Yes
23/05/2011	32	51	Morning, Noon, Night	Correct	233	No

those data and understand how lightning affects people's lives in our country. This analysis would be very useful to take caution in certain districts of our country to save people's lives to a great extent.

References

1. Harrison, R.G.: Fair weather atmospheric electricity. J. Phys. Conf. Ser. **301**(1), 012001 (2011). ISSN 1742-6596. https://doi.org/10.1088/1742-6596/301/1/012001
2. Rakov, V.A., Uman, M.A.: Positive and bipolar lightning discharges to ground. In: Lightning: Physics and. Effects, pp. 214–240. Cambridge University Press (2003)
3. Lightning.gsu.edu. Archived from the original on Jan 15, 2016. Retrieved December 06, 2017
4. Bangladesh Bureau of Statistics (BBS): Bangladesh Disaster-Related Statistics 2015: Climate Change and Natural Disaster Perspectives. BBS, Ministry of Planning, Dhaka (2016)
5. The Daily Star, Lightning Strikes Most in May in Bangladesh. http://www.thedailystar.net/backpage/lightning-strikes-most-may-85924
6. The Daily Sun, Lightning Threat in Bangladesh. http://www.daily-sun.com/post/227889/Lightning-threat-in-Bangladesh

7. Ohsawa, T., Hayashi, T., Mitsuta, Y.: Intraseasonal variation of monsoon activities associated with the rainfall over Bangladesh during the 1995 summer monsoon season. J. Geophys. Res. **105**(D24), 29445–29459 (2000)
8. Salahuddin, A., Isaac, R.H., Curtis, S., Matsumoto, J.: Teleconnections between the sea surface temperature in the Bay of Bengal and monsoon rainfall in Bangladesh. Global Planet. Change **53**, 188–197 (2006)
9. Khatun, M.A., Rashid, M.B.: Climate of Bangladesh. Hygen HO. Norwegian Meteorological Institute and Bangladesh Meteorological Department, Dhaka (2016)
10. Christian, H.J., Blakeslee, R.J., Boccippio, D.J., Boeck, W.L., Buechler, D.E., Kevin, T.D., Goodman, S.J., Hall, J.M., Koshak, W.J., Mach, D.M., Stewart, M.F.: Global frequency and distribution of lightning as observed from space by the Optical Transient Detector. J. Geophys. Res. Atmos. **108**(D1), ACL 4-1–ACL 4-15 (2003)
11. Yuan, T., Di, Y., Qie, K.: Variability of lightning flash and thunder-storm over East/Southeast Asia on the ENSO time scales. Atmos. Res. **169**(Part A), 377–390 (2016)
12. Bond, D.W., Steiger, S., Zhang, R., Tie, X., Orville, R.E.: The importance of NOx production by lightning in the tropics. Atmos. Environ. **36**, 1509–1519 (2002)
13. Cecil, D.J., Buechler, D.E., Blakeslee, R.J.: Gridded lightning climatology from TRMM-LIS and OTD: dataset description. Atmos. Res. **135–136**, 404–414 (2014)
14. Qie, X., Toumi, R., Yuan, T.: Lightning activities on the Tibetan Plateau as observed by the lightning imaging sensor. J. Geophys. Res. **108**(D17), 4551 (2003)
15. Global Hydrology Resource Center (GHRC): LIS Space Time Domain Search. https://lightning.nsstc.nasa.gov/lisib/lissearch.pl?origin=ST&lat=23.5&lon=90.5&alat=7&alon=6&donob=both
16. Lightning & Atmospheric Electricity Research. https://lightning.nsstc.nasa.gov/lis/overview_lis_instrument.html
17. Google Maps API, Reverse Geocoding. https://developers.google.com/maps/documentation/javascript/examples/geocoding-reverse
18. Bangladesh.Gov.bd. http://www.bangladesh.gov.bd/site/view/upazila-list/
19. Wikipedia, Upazilas of Bangladesh. https://en.wikipedia.org/wiki/Upazilas_of_Bangladesh
20. Prothom Alo. http://www.prothomalo.com/
21. bdnews24. https://bangla.bdnews24.com/
22. Bhorer Kagoj. http://www.bhorerkagoj.net/
23. Jugantor. https://www.jugantor.com/
24. KalerKantho. http://www.kalerkantho.com/
25. Github, Crawler4j. https://github.com/yasserg/crawler4jh
26. Global Hydrology Resource Center (GHRC): TRMM-LIS data processing anomalies. https://lightning.nsstc.nasa.gov/data/lis_data_anomalies.html

A Computational Approach for Project Work Performance Analysis Based on Online Collaborative Data

Nikitaben P. Shelke and Poonam Gupta

Abstract The educational data mining plays an important role in understanding the behavior of students as well as analyzing, evaluating, and predicting the performance of the students at all level. The group work is also an inevitable part of the education domain. The tremendous rise in the use of online tools by students have created the opportunity to generate high-level views of information about the behavior of the students working on any particular project. This collaborative data can be useful for assessment of the project work on individual as well as group basis by using the data mining techniques. In this paper, we aim to present a framework which exploits this data to transform it into understandable format and use it for students' performance analysis and assessment. The result of k-means algorithm of clustering identifies the different levels of project groups as well as individual group members.

Keywords Clustering (k-means) · Educational data mining · Online collaboration · Performance analysis

1 Introduction

Data mining is used for Knowledge Discovery in Database that is used for extracting the meaningful data from large inconsistent, incoherent storage of data. The extracted patterns and relationships between data become helpful in decision making.

Data mining allows separating the disorganized and repetitive noise in data and meaningful data. It can understand what relevant data is and then make good use of that information to assess likely outcomes and helps in making informed decisions.

N. P. Shelke (✉) · P. Gupta
Computer Engineering Department, G.H. Raisoni College of Engineering and Management, Wagholi, Pune 412207, Maharashtra, India
e-mail: nikitashelke@gmail.com

P. Gupta
e-mail: poonam.gupta@raisoni.net

A. Abraham et al. (eds.), *Emerging Technologies in Data Mining and Information Security*, Advances in Intelligent Systems and Computing 813,
https://doi.org/10.1007/978-981-13-1498-8_5

The data mining techniques have been introduced into new fields of Statistics, Databases, Machine Learning, Pattern Recognition, Artificial Intelligence, Computational Capacities and so on [1]. The educational data mining (EDM) is one such field which holds high importance in the higher educational disciplines.

EDM has been defined by the International Educational Data Mining Society [2] as "an emerging discipline, concerned with developing methods for exploring the unique types of data that come from educational settings, and using those methods to better understand students, and the settings which they learn in".

The EDM is basically used for analyzing the huge amount of students' data and help the instructors to make the informed decision while making the assessments. There are different techniques for exploring data collected from different educational disciplines which can be either offline or online. Offline data can be collected from educational institution and their different departments. This offline data includes the data collected from the conventional system, i.e., data available on paper. The online data can be collected from online sources. These online sources consists of the online e-leaning systems developed by the institution itself, open-source learning platform MOODLE, MOOC (Massive open online course) and so on.

Some of the institutions may provide the online learning platform is being used to create virtual learning groups from different geographical locations that help in rapid growth of collaboration in learning. In this case, discovering the learning patterns of students by analyzing the system usage can play a most important role [3]. Traditionally, researchers applied DM methods to educational context. In modern times also, the EDM uses many data mining algorithms such as k-means (Clustering), C4.5, Naïve Bayes (Classification), GSP (Sequential Pattern Mining), ID3 (Decision Tree), Association Rule Mining, etc. [3].

There are multiple educational data mining stakeholders apart from students. Romero [4] have identified the stake holders as Learners/Students, Teachers/Instructors, Course Developers, Universities/Training Providers, and Administrators. These stakeholders play important and distinct roles in the education system. EDM techniques can be used for the specific objectives for all these stakeholders to analyze, predict and assist them for the performance improvement [4].

The various tools used in EDM have been discussed in [5]. RapidMiner, WEKA, KEEL, KNIME, Orange, and SPSS are the tools which are used in EDM process [5], Tableau, D3.js, and InfoVis tools along with some python packages can used for the data visualization of the EDM results [5].

The context of this research paper is that the electronic activity data generated by students while working in group for given project can be used for analysis of their performance. The performance analysis is divided in two parts: the group-wise performance and individual group member-wise performance. Also, this analysis will be helpful to the project coordinator/facilitator to make the informed assessment on the basis of results obtained from the data of online collaborative tools. Here, k-means algorithm of cluster analysis has been used to identify the similar behavior of groups as well as group members and then examine the framework of each group.

This paper road map is arranged as follows: Sect. 2 states survey of related work done in past describing the efficiency, drawbacks of work done till time. Section 3 describes the system architecture, input data and identification of parameters for clustering. In Sect. 4 results are discussed. Then, Sect. 5 concludes this paper with details of the future work.

2 Related Work

There has been numerous research works in the field of educational data mining (EDM). The International Educational Data Mining Society is supporting the research done in educational domain. The large numbers of publications have been published in Journal of Educational Data Mining (JEDM) as well as many other prominent publications. This section focuses on the related work done in this discipline.

Barros and Verdejo [6] whose DEGREE system was used to enable students to submit proposals in text format edit and improve them, until agreement was reached. In this paper, an approach has been presented to characterize group and individual member's behavior in computer-supported collaborative work with a particular set of attributes. The approach for this research by Barros and Verdejo [6] collaboration is conversation-based. They have not considered learner's belief, which may be required for a fine-grained study of collaborative interactions with more extensive modeling [6].

Soller [7, 8] analyzed knowledge sharing conversation data such that a candidate presents and explains their new understanding to colleagues or peers; colleagues or peers try to understand it. Their approach of Hidden Markov models and multidimensional scaling showed successful results to analyze the knowledge sharing interaction. Their analysis suggested that Hidden Markov modeling method is able to differentiate effective and ineffective knowledge sharing communication among the peers. It has also suggested the fact that online knowledge sharing behavior can be assessed by analyzing the sequence of conversation activity. However, Soller [7, 8] required group members to use a special interface using sentence starters, based on Speech Act Theory. This was the limitation of their research as requirement of special interface, which was limited to single collaboration medium [7, 8].

Merceron and Yacef [9] have applied the clustering to the data obtained from intelligent teaching assistant system Logic-ITA used by School of Information Technologies at the University of Sydney. This paper aims to characterize students having difficulties. The students are clustered in similar groups with regard to their ability by using k-means and hierarchical clustering. The future work proposed in this paper includes improving their Data Mining research tool, in particular its visualization facilities to help teachers in their interpretation [9].

Oyelade et al. [10] have implemented k-mean clustering algorithm with the Euclidean distance measure for analyzing students' result data. The data considered in [10] is the academic examination result data on which the clustering algorithm has been applied to monitor the student progress. They have concluded that this monitoring can be helpful for improving the future academic results in the subsequence academic semesters [10].

Kay and Yasef [11] have done the analysis for frequent patterns of collaborative learning activity on the interactive tabletop [11]. The electronic activity traces can be used to be a very good source to understand the strategies followed by groups which are working collaboratively. This paper presented the work based upon research in online collaborative learning and have proposed a methodology that can be used as a starting point to guide future research on the identification of patterns from educational tabletop settings [11]. This approach has ability to identify the further research on supporting online collaborative learning through the use of tabletops and machine learning techniques [11].

Amrieh et al. [12] have established student academic performance model which applies behavioral features and assesses the academic performance of the students. Three data mining techniques are used for developing this academic performance model: Decision Tree, Artificial Neural Network and Naïve Bayes [12]. In [12], they have considered data related to demographic, academic as well as behavioral features of students.

McBroom et al. [13, 14] have examined different techniques of following the development of student behavior during the semester using online submission system data, and different approaches to analyzing this development [13, 14]. K-means clustering is used to identify the students' behavior, the changes in the behavior over time as well as the impact of certain behavior on final exam grades. The results of this work has shown the relationship between these behaviors and final exam results along with how these behaviors might be visible early enough in the semester for instructors to mediate for better performance by students [13, 14].

Xie et al. [15] have proposed a framework for grouping students based on their behavioral sequences logged in system as well as online predicting mechanism, to effectively assign a student to the most matching group on a basis of the newly logged data and the historical models by using the clustering algorithm.

Choi and Lee [16] have integrated the theory- and data-driven approaches in order to understand student learning outcomes, activities, and patterns as they interact with course content using a popular Learning Management System. For theory-driven approach, K-means clustering and multiple regression analysis are applied and for data-driven approach, classifiers based on machine learning algorithms have been built [16].

To outline the related work, various researchers have explored the options for optimizing the analysis, prediction and evaluation of students based on educational data using data mining techniques. This proposed work focuses on the group work done collaboratively over the online tools available. The final analysis will help project guides and project coordinator to evaluate student's project work performance in group as well as individually. Additionally, the extended model of this research can help to discover the useful patterns indicating the better practices to be followed by students in order to improve the performance.

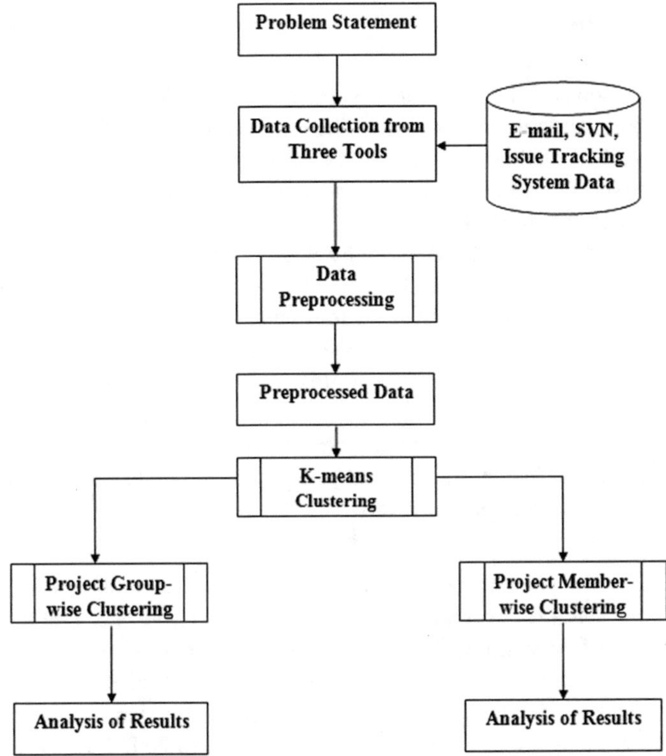

Fig. 1 System architecture

3 Proposed System

The proposed system details are given in following sub-sections. Figure 1 depicts the system architecture.

The objective of this system is to analyze the project work performed by different project groups after the integrating the data from different collaborative tools. Our system will then be used by project guide or project coordinator to make informed evaluation based on the work done on online collaborative data. This evaluation will show which project group is working effectively and it will also help to analyze the performance of each project group member with the overview of approach of team members' work and where to focus on their training efforts to develop the relevant skills.

3.1 Input Data

We have collected data from the open-source software projects, where the logs of online collaborative tools can be found. These projects have the processes/tools that can be defined by collaboration based processes/tools. The activity stream of tools from these projects will act as the electronic traces of the actions taken by the group and individual group member.

The tools we have considered in the proposed system consists of SVN (Subversion), Issue Tracking System, and E-mail. The data from these 3 tools- SVN, Issue Tracking, and E-mail- includes the electronic traces of activities performed by the participants of the project working on these 3 online collaborative tools. The logs of each of these actions performed by the group members are stored including time and group.

The input data considered in our work consists of seven project group (PG1, PG2, …, PG7) and each project group contains nine group members. The total numbers of events for three tools are approximately 175000. The details of the count of events for these three tools are given in next sub-section.

- **SVN**—SVN (Subversion) repository is one of the three tools considered here. As per the Version Control with Subversion book [17], SVN can be defined as a version control system for tracking the incremental versions of files and directories. It contains collection of files for any software creation or changes made over the period of time. SVN permits to discover the changes which resulted in each of those versions and helps in the arbitrary recall of the same [17].

Each project group has its individual SVN repository. Each project group has count of SVN committed (i.e. either the expected full change completed or rolled back). The data contains project group-wise SVN count and individual project group member SVN count also. The total number of project group-wise SVN count considered in our work is as shown in Fig. 2.

Initially, we have assumed that the extensive usage of SVN system will be done by the most effective project group. SVN_Count attribute has been considered for the k-means clustering applied on collective project group data as well as individual project group member data. The resultant patterns from both clustering algorithms can help determine the efficient ways to work with online collaborative tool SVN.

- **Issue Tracking System**—The issue tracking system is the second tool from which the data has been collected. The issue tracking system can also be referred to as the ticketing system. The idea behind this tool is to create an issue ticket for any task that the project group has to execute.

The issue tracking system can keep the track of status of the ticket created for any kind of job. This job can be the change request from client, any improvement or fixing of any bug reported. Figure 3 shows the number of issues (project group-wise) of the issue tracking system considered in the project.

Here, there is an assumption that the strongest project group should have maximum number of issue tickets. However, there are other parameters to be considered like

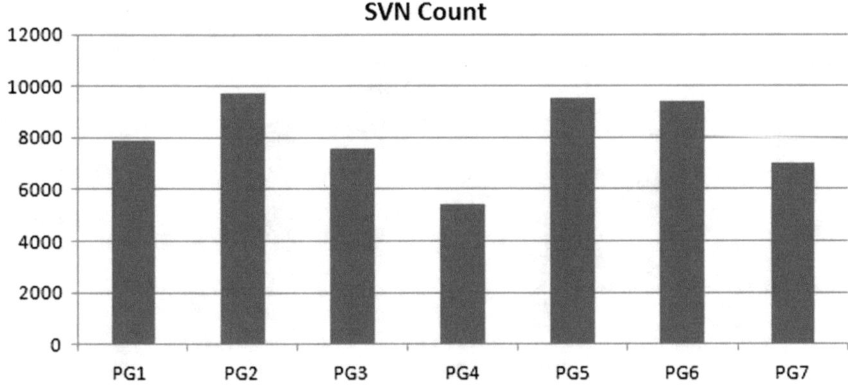

Fig. 2 Count of SVN events project group-wise. The total number of SVN committed for each project group

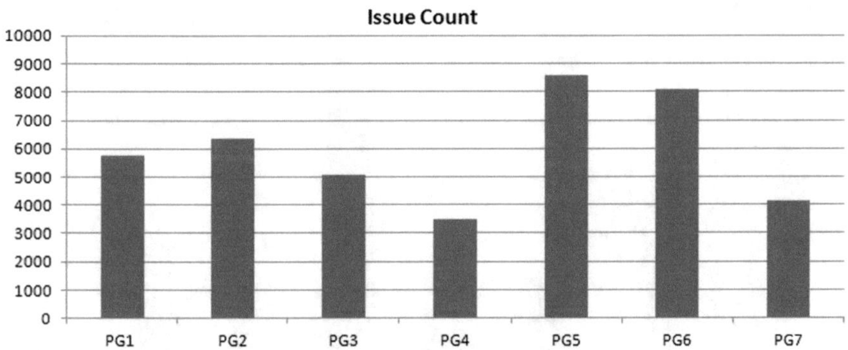

Fig. 3 Count of issue events project group-wise. Total number of issue events considered from issue tracking system

issue ticket type and issue ticket priority because the lower priority issue ticket may be finished in less time and if any project group work only on low priority issue ticket then they may have high number of issue ticket count.

Therefore, only issue ticket count should not be taken into account while doing the performance analysis. Hence, the issue tracking system data collected here has more attributes as well such as count of issue ticket, type, priority, views, and votes. These parameters are very significant for analysis of the performance of each project group as well as each project group member.

The total numbers of issues for range of different priorities for each project group are as shown in Fig. 4. The priority 5 is highest priority (critical priority) and priority 1 is the lowest one (minor priority).

The total numbers of issues for range of different types for each project group are as shown in Fig. 5. The type of issue tickets such as bug, improvement, new feature, sub-task, etc., are mapped while preprocessing the data Type 1,…, Type 5.

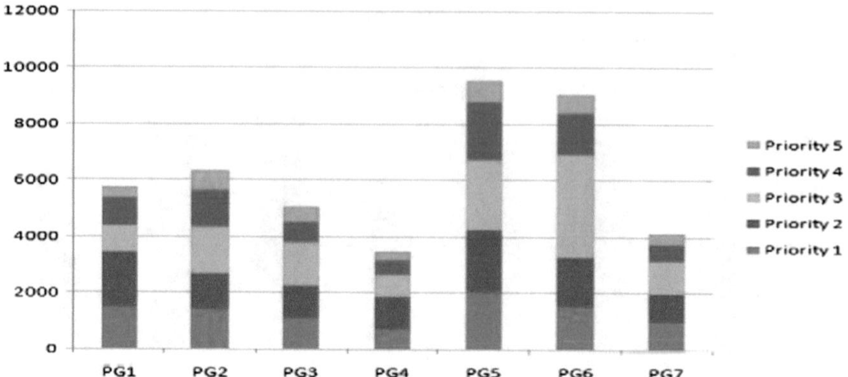

Fig. 4 Count of issues by priority and project group. Each bar represents the total number of issues with different priorities of the issue for each project group

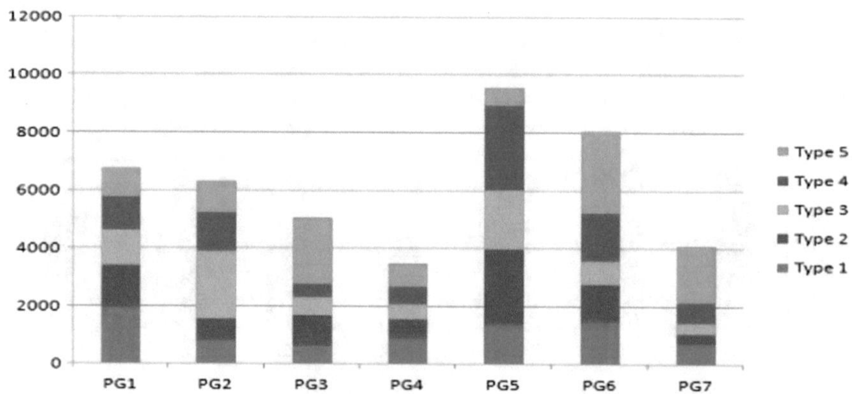

Fig. 5 Count of issues by type and project group. Each bar represents the total number of issues with different type of the issue for each project group

- **E-mail**—The next tool considered in this work is the e-mailing system. The data contains the total number of the e-mails sent by individual project group member as well as total number of the e-mails sent by the whole project group.

The total number of project group-wise e-mail counts considered in our work is as shown in Fig. 6.

3.2 Preprocessing of Data

Once the data has been collected from various online collaborative tools (SVN, Issue Tracking System, and E-mail Services), the next step required is data preprocessing.

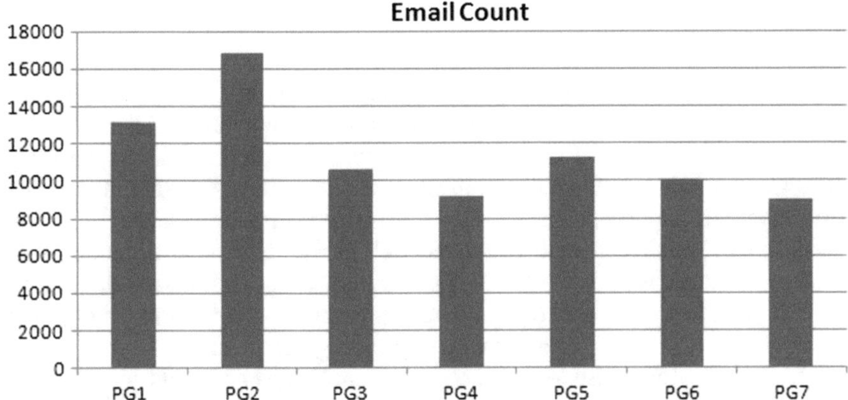

Fig. 6 Count of e-mail events project group-wise. Total number of e-mails considered from issue tracking system

The data collected is in the raw format. The data preprocessing is required to obtain the high quality data to fulfill the conditions enforced by the clustering algorithm. To obtain the high quality data for applying the data mining techniques few sub-tasks are required to be performed. In our case, data selection, data cleaning, data transformation, and data integration activities are performed.

The data selection has been done as per its usefulness in the proposed work based on the parameters considered on which the clustering will be applied. Here, we have eliminated the fields which are not useful for our work.

Once the data selection is done, data cleaning is the next step. It is applied on the data set to deal with noisy and missing data. The data records with missing values have been removed. Only complete cases have been considered for further data preprocessing.

The data transformation is done for the common data attributes such as project group names and individual project group members. This transformation is required in order to make sure that the data in different columns have the same format. For instance, in e-mail the individual project group member had the name format as 'Last Name, First Name'; however, the SVN and issue tracking system the name format was 'First Name, Last Name'. Thus, the data transformation step plays important role here.

The data integration step is needed as we have collected the data from three different tools for seven project groups. This step is performed to merge the data for further analysis.

3.3 Clustering

Clustering is an unsupervised way of analyzing the data in different fields such as educational data mining, statistics, machine learning, etc. [18]. Clustering is the process of grouping similar data items in one cluster. Clustering has been very popular algorithm in educational data mining, especially in analysis and prediction of performance of students [18].

In our work, we needed multiple parameters to analyze the data for accurate results, which is facilitated well by the clustering algorithm. By using k-means algorithm of clustering, we have been able to analyze and present meaningful patterns of project work done by the project groups as well as project group members.

The Project Group-wise k-means clustering and Project Member-wise k-means clustering has been implemented to get the results. The k-means algorithm is popular, uncomplicated, efficient clustering algorithm for EDM [18]. We have used R-language for k-means algorithm implementation (Group-wise and Member-wise) with Euclidean distance measure.

We have used the RStudio IDE for implementing the k-means algorithm in r-language. We decided on k-means algorithm implementation because of the less sensitivity to seed initialization and more efficient [19]. Hierarchical clustering may not have been good choice for the large data sets as it has high requirement of the time and memory capacity.

Project Group-wise Clustering—Project Group-wise Clustering is performed with seven parameters selected. These seven parameters included from the 3 collaborative online tools that have been considered. The k-means clustering has been implemented with $k = 3$ to $k = 7$. We have analyzed the results with $k = 3$ to $k = 7$, and we found that the results for $k = 3$ are more meaningful. The detailed results are discussed in Sect. 4. The results of Project Group-wise Clustering can be used by the project coordinator to understand which project group is performing well and which project groups have the performance gaps. They can make the informed decision for further evaluation of the project group based on this analysis.

Project Member-wise Clustering—Project Member-wise Clustering is also implemented with same seven parameters used in Project Group-wise Clustering. Even if these parameters are same as the project group-wise clustering, they exemplify individual project member activities and not project group activities. Here, k-means algorithm used with number of centers $k = 3$ to $k = 7$, and the results show $k = 4$ has more meaningful patterns. The detailed results are discussed in Sect. 4. The results of the Project Member-wise Clustering can be used by the project guide/facilitator for the further evaluation of an individual project group member. The individual skills can also be identified with this analysis such as which member has the leadership quality, which member is better at testing or coding. These results can also play an important role to suggest the useful patterns and help in modifying the behavior as required.

Table 1 Project group-wise clustering output with k = 3

Clusters	Project groups	Distinctive meaningful activity behaviors
Cluster 1	PG1, PG2, PG6	• Moderate count of issue tickets, SVN and E-mail • Moderate % of issue ticket types and priority • Infrequent votes and views
Cluster 2	PG5	• High count of issue tickets, SVN and e-mail • Moderate % of issue ticket types and priority • Frequent votes and views
Cluster 3	PG3, PG4, PG7	• Low count of issue tickets, SVN and e-mail • Low % of issue ticket types and priority • Very infrequent votes and views

4 Results

The result of our work is divided in two parts; the first part includes the result of project group-wise clustering and the second part includes the result of project member-wise clustering. As we have shown in previous section, we have selected the range of value of center k (k = 3 to k = 7) to check which result shows the most meaningful outcome of this analysis.

The first part of result, i.e., result of project group clustering has been considered with k = 3. The cluster 1 includes three project groups PG1, PG2, PG6 and is characterized by largely average events of SVN, e-mail and issue tracking system. The cluster 2 includes only one project group PG5 with overall high activities in all three tools. This project group shows the strongest characteristics and can be referred to as most effective project group. Cluster 3 again includes three project group PG3, PG4, PG7 with largely low activities of SVN, e-mail and issue tracking system. These project groups show the weakest group patterns.

Table 1 shows the result of project group clustering with three clusters and their distinctive meaningful activity behaviors.

The above results can help the project/course coordinator as it draws attention to the behavior of the strongest group against the behavior of average and weakest groups. This can also help to identify more effective ways to use the online collaborative tools by using these results.

The second part of result, i.e., result of project group member clustering has been considered with k = 4. Table 2 shows the result of project group member clustering with four clusters and their distinctive meaningful activity behaviors.

Based on these results, we have identified the cluster tags as well to all 4 clusters. For instance, the cluster 4 represents the developers, as the count of SVN is considerably higher here. It is a common knowledge that the coding is done and updated mostly by the developers and the files are updated in SVN. They also work on the tickets when there is any bug fix or new feature request has been raised in the issue

Table 2 Project member-wise clustering output with k = 4

Clusters	Number of students	Distinctive meaningful activity behaviors	Cluster tag
Cluster 1	18	• Moderate count of SVN, E-mail • High issue tickets count • Moderate % of issue ticket types and priority • Infrequent votes and views	Testers
Cluster 2	19	• Very low count of issue tickets, SVN and e-mail • Low % of issue ticket types and priority • Very infrequent votes and views	Others
Cluster 3	9	• High count of e-mail • Moderate count of issue tickets, SVN • Moderate % of issue ticket types and priority • Very frequent votes and views	Manager/technical lead
Cluster 4	17	• Moderate count of issue tickets, e-mail • High SVN count • Low % of issue ticket types and priority • Infrequent votes and views	Developers

system. Similarly, cluster 1 was tagged as 'tester', cluster 2 as 'others' and cluster 3 as 'managers/technical leaders'. Thus, clustering helped in finding the individual skills of the project group members, along with their activity levels in various tools.

The performance can also be analyzed for each member in the project group by its respective project guide, which in turn can help in further evaluation of the project group as a whole as well as single project group member. These results can also be the starting point of project guide/project coordinator discussion with the project groups about their behavioral aspects, which can be helpful for members in future project work in subsequent sessions.

5 Conclusion and Future Work

Our work helps in understanding the use of data from online collaborative tools for performance analysis of students while they are working on the project. This keeps the evaluation transparent also. This data mining technique helps in preparing the understandable presentation of data, to provide feedback to students on developing the skills they required to sustain in high pressure working environment of IT industry while they are employed.

This system can also help the facilitators to understand the level of students (strong/average/weak) and focus their teaching accordingly. The results obtained here can be the starting point to further evaluate the groups and individual members. It can assist the project groups and their guides, so that they can monitor the applicable aspects of the group's activities, present feedback and point out where are the issues.

The future work involved with this work is to find out the positive and negative outcomes of certain patterns. This involves further analysis, which can help to automate the discovery of the most significant rules and provide the remedy to the unproductive patterns of collaborative tools for more effective work patterns. The data can be enriched by collecting the data from other collaborative tools such as wiki, chat, forums, etc.

The extended work of this research can be done with the association rule mining algorithm; which may help to identify the effective rules. Support, confidence, and lift concepts can be used to find the rules which can be helpful to recommend the useful path to be followed in the future by same groups or members. This can also be helpful next batch of students who will be working on the project. These rules can be stored and used in the future, so that the users of system can explore them. The further explanation and the remedy to poor patterns can also be stored in order to avoid the ineffective rules followed by the project group members.

References

1. Baradwaj, B., Pal, S.: Mining educational data to analyze students' performance. Int. J. Adv. Comput. Sci. Appl. **2**, 63–69 (2011)
2. International Educational Data Mining Society. http://www.educationaldatamining.org/
3. Shelke, N., Gadage, S.: A survey of data mining approaches in performance analysis and evaluation. Int. J. Adv. Res. Comput. Sci. Softw. Eng. **5**, 456–459 (2015)
4. Romero, C., Ventura, S.: Educational data mining: a review of the state-of-the-art. IEEE Trans. Syst. Man Cybern. Part C: Appl. Rev. **40**(6), 601–618 (2010)
5. Slater, S., Joksimović, S., Kovanovic, V., Baker, R.S., Gasevic, D.: Tools for educational data mining a review. J. Educ. Behav. Stat. 85–106 (2016)
6. Barros, B., Verdejo, M.F.: Analyzing student interaction processes in order to improve collaboration: the DEGREE approach. Int. J. Artif. Intell. Educ. **11**, 221–241 (2000)
7. Soller, A., Lesgold, A.: A computational approach to analyzing online knowledge sharing interaction. In: Proceedings of 11th International Conference on Artificial Intelligence in Education, AI-ED, pp. 253–260 (2003)
8. Soller, A.: Computational modeling and analysis of knowledge sharing in collaborative distance learning. User Model. User-Adap. Inter. **14**, 351–381 (2004)
9. Merceron, A., Yacef, K.: Clustering students to Help Evaluate Learning. Technology Enhanced Learning, vol. 171, pp. 31–42. Springer (2005)
10. Oyelade, O.J., Oladipupo, O.O., Obagbuwa, I.C.: Application of K-means clustering algorithm for prediction of student's academic performance. Int. J. Comput. Sci. Inf. Secur. **7**, 292–295 (2010)

11. Martinez, R., Yacef, K., Kay, J., Al-Qaraghuli, A., Kharrufa, A.: Analyzing frequent sequential patterns of collaborative learning activity around an interactive tabletop. In: Proceedings of the 4th International Conference on Educational Data Mining (EDM) (2011)
12. Amrieh, E.A., Hamtini, T., Aljarah, I.: Preprocessing and analyzing educational data set using X-API for improving student's performance. In: IEEE Jordan Conference on Applied Electrical Engineering and Computing Technologies (AEECT), pp. 1–5 (2015)
13. McBroom, J., Jeffries, B., Koprinska, I., Yacef, K.: Mining behaviours of students in autograding submission system logs. In: Proceedings of 9th International Conference on Educational Data Mining, pp. 159–166 (2016)
14. McBroom, J., Jeffries, B., Koprinska, I., Yacef, K.: Exploring and following students' strategies when completing their weekly tasks. In: Proceedings of 9th International Conference on Educational Data Mining, pp. 609–610 (2016)
15. Xie, T., Zheng, Q., Zhang, W.: A behavioral sequence analyzing framework for grouping students in an e-learning system. Knowl.-Based Syst. **111**, 36–50 (2016)
16. Choi, H., Lee, J.E., Hong, W.-J., Lee, K., Recker, M., Walker, A.: Exploring learning management system interaction data: combining data-driven and theory-driven approaches. In: Proceedings of the 9th International Conference on Educational Data Mining, pp. 324–329 (2016)
17. Collins-Sussman, B., Fitzpatrick, B.W., Pilato, C.M.: Version Control with Subversion for Subversion 1.7. O'Reilly Media
18. A. Dutt, M.A. Ismail, T. Herawan: A systematic review on educational data mining, vol. 3536. IEEE Access (2017)
19. Data Mining Algorithms in R/Clustering/K-Means. https://en.wikibooks.org/wiki/Data_Mining_Algorithms_In_R/Clustering/K-Means

Design of Business Canvas Model for Social Media

Jan Hruska and Petra Maresova

Abstract Social media are a powerful instrument and companies use them every year with an increasing tendency. With the growing spread of social media, it can be confusing for companies which of the social media to use, how much to invest, and how to manage them. The main objective of this paper is to create a model for an effective use of social networks based on the business model canvas. Furthermore, the aim is to use analyzed data to create a general model of how not only companies should treat social media, what the important measures are, and, conversely, what is taboo on social media. Firstly, the authors analyzed the business model canvas and then with a combination of known principles effective on social networks created a model that will allow companies to use social networks more effectively.

1 Introduction

Social media are an important topic that impacts businesses, customers, coworkers, and everyone connected to them [1]. Lon Safko and David K. Brake identify three of the most important goals in business as to increase revenues, improve profitability, and ensure that you remain relevant, competitive, and alive in your industry. Safko also mentions that a systematic approach on social media can increase the company and brand value by engaging people in new forms of communication, collaboration, education, and entertainment [2]. Another generation of Internet-based applications where users control communication, promise to enhance promotional efforts with social marketing issues. Those applications can directly engage consumers in their motivation and creative process by producing and distributing information through collective content sharing, writing, social networking, social bookmarking, and syndication. It can also significantly enhance the power of viral marketing by increasing

J. Hruska (✉) · P. Maresova
University of Hradec Králové, Rokitanského 62, 500 03 Hradec Králové, Czech Republic
e-mail: jan.hruska.3@uhk.cz

P. Maresova
e-mail: petra.maresova@uhk.cz

A. Abraham et al. (eds.), *Emerging Technologies in Data Mining and Information Security*, Advances in Intelligent Systems and Computing 813,
https://doi.org/10.1007/978-981-13-1498-8_6

speed at which consumers share all kind of contributions and opinions with growing audiences. Strategic issues such as priority audience preferences, selection of appropriate applications, tracking and evaluation, and related costs are considered, Web 2.0 will continuously develop to allow promotion practitioners a more direct access to consumers with less dependency on traditional communication channels [3]. Success on social media is related to the effective use of management tools. Traditional management deals with the planning, organization, administration, monitoring, and control [4–6]. Other parts of management are concerned with the soft elements that relate to motivation, inspiration, vision and leadership [7]. The interactions among customers play an increasingly important role not only in maintaining brand image and enhancing customer loyalty, but also in affecting the formulation of marketing strategies for a certain brand or product [8, 9]. According to IBM research, there is a large perception gap between what the customers seek via social media and what companies offer [10]. Various authors have dealt with the area and its model expression. For example, Wu et al. [1] proposed theoretical visual framework to model in social network group decision making is put forward with following three main components: (1) trust relationship; (2) recommendation mechanism; and (3) visual mechanism. These models make it possible to analyze technological changes that alter the cost of communication [11]. The aim of this paper is to create model for effective use of social networks based on the business model canvas.

2 Used Methods and Tools

The model created in this paper is based on traditional business model canvas introduced by Alexander Osterwalder in 2010 and used by countless companies. For the needs of the social media model in this work, it does not take into account company logistics, factories, and is targeted only on social media work. The basis for the model is also the following questions: Why customer seek a company or brand via social media and interact? What would make a customer interact more and interact with optimistic attitude? And does social engagement influence the customer's feelings of loyalty toward a company as businesses hope it does? The model is created in Cmap. Cmap software empowers users to create, analyze, share, and criticize knowledge models represented as concept maps.

3 Design of Social Media Business Canvas Model

Generally, a business model is how a company creates value for itself while delivering products and services to customers [12]. This model was analyzed at the beginning and in this model, social media will be applied. Products for customers can be thought of here as solutions to satisfy a customer's problem or need. Those two things can be combined on social media, a company can fulfill an individual's need, which can be

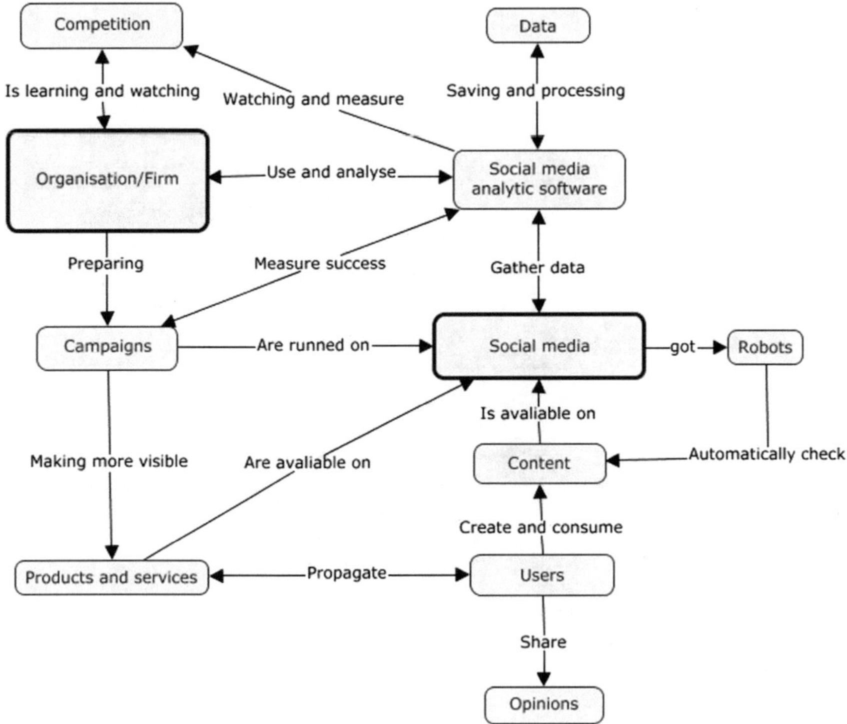

Fig. 1 Social media connections (*Source* own based on [3, 8, 9])

just to be entertained and solve a problem with a coherent advertisement (Fig. 1). We should not understand this model as a model to run an entire company, it is a model inside a model specially for social media. See the figure below to better understand the importance of the model described later.

This is what happens when a business model is combined and applied to social media (Fig. 2).

Value Proposition
The first big group is value proposition. It refers to your targeted audience. Facebook in particular has a very elaborate advertisement system and when a company correctly defines their audience, then it is easy to do so on the social media.

Segments
The second big part of the model represents customer segments. It needs to be said that customers do not exist to buy, but companies exist for them. It is important to have this in mind when doing any kind of business. In this section of the model the archetype of a customer is built. Who are they and why would they buy or use the company's service?

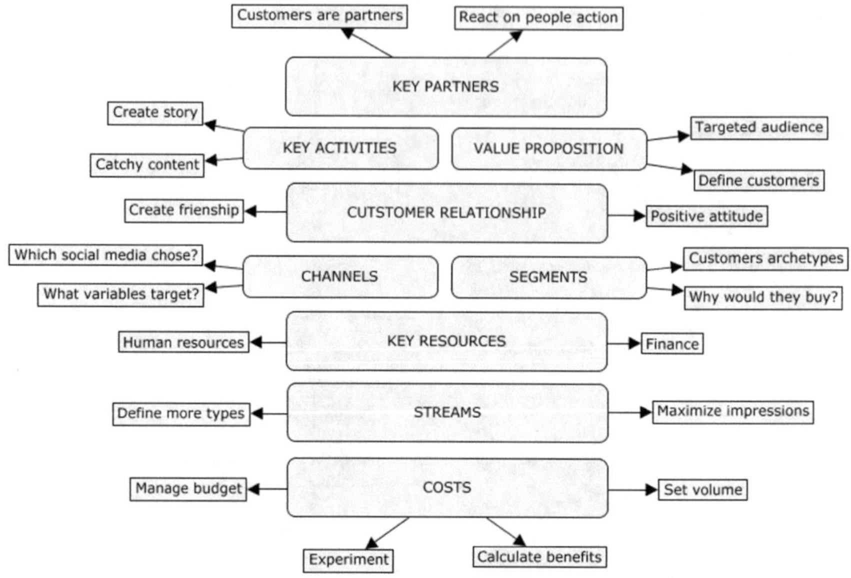

Fig. 2 Business model canvas for social media (*Source* own based on [10–15])

Channels
The third part of the model illustrates channels. Facebook and YouTube reached more than 1000 million users, that means that more than one-third of all internet users in the world, and more than one sixth of the global population, are active members of these networks [13]. Before 1990, the only channel was the physical channel, which contained sellers, stores. But in this age, we have virtual channels, web, mobile, and especially social media with a very complex and specialized system exactly for that case. The problem with the physical channel is that it is expensive and has to be on a very good location for people to see it, if the physical channel is supposed work. With social media, it is possible to target exactly which age category is needed, exactly which sex is needed etc. If the company wants to increase its revenue, it can target only people who already bought something; it can target people in specific groups, geographic locations, and much more. The big question in this part of the model is what social media a company should use. Facebook is necessary, but if it should add Instagram, Twitter, and others will depend on what the company does and how big it is.

Customer Relationship
The fourth part of the model is customer relationship; in a normal business model, it is how we get customers. When combining a business model with social media, what matters more is how we maintain a good relationship with customers, how we create entertaining content and react in comments and posts just like the customers want, an assertive, calm attitude always with a bit of humor in the case of social media. On social media, the important question is how to create content that would go viral,

because social media can get companies thousands of potential customers very fast with a minimum investment [14].

Streams

The next step in a normal business model is revenue streams. In other words, how you will make profit. However, this model does not aim to describe logistics principles, storing stocks, and similar issues. In the case of social media, companies should ask the question how to spend as little as possible to engage as many people as possible. In this part, companies should also determine if there are going to be multiple revenue streams. For example, having customers subscribe and later offer them interesting products and services via email, offer some giveaway for liking or sharing something on social media and expect income later on, or offer discounts when customers take some action. Key resources are the sixth part of the model. What important assets does the company need to do things? For simplification, in this model it is mainly finances and human resources with good ideas. The partnership in this model is again something a little bit different than in a normal business model. Without doubt, the most important partnership has to be with people who react to your posts, comment, and otherwise engage with your content. It is vital to create a strong positive partnership with potential customers. Key activities and routine are also a part of the model. To make sure that this model works, it is very important to tell a story and create content that is entertaining and interesting for targeted audience. It is not about pushing people to buy; it is about creating a good friendly relationship and showing them a product or service in an entertaining way, giving them more that they expect. Social media is from a large part about entertainment and companies should not use them to irritate customers by selling too hard.

Costs

Finally, all this is summed up by costs. On social media, and especially Facebook, it is very easy to plan costs strategically, depending on how big the company is, how many people it wants to target, and at what particular time. With every interaction with people on social media, the company gets a sense of what potential customers engage and who is the most active, and with this information, each interaction can be more specialized.

4 Discussion of Proposed Solution to Other Models

Many authors agree on the importance of visualizing and systematizing processes running on social networks [13–15]. Hanna et al. [15] describes a systematic way of understanding and conceptualizing online social media, as an ecosystem involving both digital and traditional media. Hofacker et al. [13] presented framework for social media challenges for marketing managers. One of the reasons why social media campaigns not working is because executives and managers are not investing enough time and thoughts on being actively engaged with users on social media [16]. Pradiptarini indicate that organizations should not only be involved with the online community, but also offline community to extend their relationship and customers loyalty [17].

Based on the mentioned business model canvas for social media can be recommended in general:

- building a good rapport with customers;
- timely and forthcoming responses to questions and posts by users;
- creating entertaining and catchy content.

Further research will be focused on simulation approaches enabling to determine more about what are the exactly key areas to focus on social media based on actual data and data from past and their outcomes.

5 Conclusion

Managers must cope with the specific challenges of social media such as more complex issues for marketers. The aim of this paper was to design model for an effective use of social networks based on the business model canvas. The main areas that need to be addressed are value proposition, segments, channels, customer relationship, streams and costs. This model is general and does not take into account any specifics of the companies (such as market segment, size of company) at this stage of proposal.

Acknowledgements This study was supported by the project Excellence 2016/17 (University of Hradec Kralove, Faculty of Informatics and Management) and internal research of Faculty of Informatics and Management, University of Hradec Kralove.

References

1. Jian, W., Francisco, C., Hamido, F., Enrique, H.-V.: A visual-interaction-consensus-model for social network group decision making with trust propagation. Knowl.-Based Syst. **122**, 39–50 (2017)
2. Safko, L., David, K.: Brake The Social Media Bible: Tactics, Tools, and Strategies for Business Success. Wiley, New Jersey (2010)
3. Rosemary, T., Neiger, B.L., Hanson, C.L., McKenzie, J.F.: Enhancing promotional strategies within social marketing programs: use of Web 2.0 social media. Health Promot. Pract. **9**, 338–343 (2008)
4. Mintzberg, H.: The Nature of Managerial Work. Harper & Row, New York (1973)
5. Morgan, G.: Images of Organization. Sage Publications, Newbury Park (1986)
6. Taylor, F.W.: The Principles of Scientific Management. Harper, New York (1911)
7. Kotter, J.P.: John Kotter on What Leaders Really Do. Harvard Business School Press, Boston (1999)
8. Rui, H., Jia-wen, W., Helen, D.S.: Customer social network affects marketing strategy: a simulation analysis based on competitive diffusion model. Physica A **469**, 644–653 (2016)

9. Glen, D.: Opinion piece: Social media: Should marketers engage and how can it be done effectively? J. Direct Data Digit. Market. Pract. **9**, 274–277 (2008)
10. Baird, C.H., Parasnic, G.: From social media to social customer relationship management. Strateg. Leadersh. 39:30–37 (2011)
11. Adalbert, M.: Online Social Networks Economics, vol. 47, pp. 169–184. Elsevier (2009)
12. Osterwalder, A., Pigneur, Y.: Business Model Generation: A Handbook for Visionaries, Game Changers, and Challengers. Wiley, New Jersey (2010)
13. Hofacker, C.F., Belanche, D.: Eight Social Media Challenges for Marketing Managers, vol. 20, pp. 73–80. Elsevier (2016)
14. Mangold, W.G., Faulds, J.D.: Social Media: The New Hybrid Element of the Promotion Mix, vol. 52, pp. 357–365. Elsevier(2009)
15. Richard, H., Andrew, R., Victoria, C.L.: We're all Connected: the Power of the Social Media Ecosystem, vol. 54, pp. 265–273. Elsevier (2011)
16. Vaynerchuk, G.: The Thank You Marketing. HarperCollins, New York, NY (2011)
17. Pradiptarini, Ch.: Social media marketing: measuring its effectiveness and identifying the target-market. J. Undergrad. Res. (2011)

Horticultural Approach for Detection, Categorization and Enumeration of On Plant Oval Shaped Fruits

Santi Kumari Behera, Junali Jasmine Jena, Amiya Kumar Rath and Prabira Kumar Sethy

Abstract The basic and primary step of any image processing approach, which classifies the similar areas in the image and helps in further analysis, is Segmentation. This paper reports a segmentation algorithm for automatic singulation, categorization and enumeration of on-plant oval shaped fruits for satisfying the purpose of automatic yield estimation. The algorithm is based on thresholding of color channels that are derived from specific color spaces. Thresholding of RGB color space has been used in the process of singulation and thresholding of YCbCr color space has been used in the process of categorization. In the process of enumeration, edge detection and dilation operations have been used. Results obtained were satisfactory basing upon various performance metrics.

Keywords Automated systems · Color space segmentation · Pre-harvesting techniques · Yield estimation · Wide range applications · Loss reduction

---

Industrial relevance: Automated systems have found wide range of applications in the field of horticulture. Its use has shown tremendous reduction in cost, human labor, and time and has added to the improvement of accuracy and precision. The proposed approach can be used for development of an automated system, which can estimate the amount of yield prior to harvesting. It can also estimate the number of mature fruits, those are needed to be harvested early to reduce loss and wastage. Harvesting at a correct maturity stage can help farmers in selling the fruits at a higher economic value. The system can prove highly beneficial for large fruit orchards by providing ease in monitoring and maintenance.

S. K. Behera (✉) · J. J. Jena · A. K. Rath
Veer Surendra Sai University of Technology, Sambalpur, Burla 768018, India
e-mail: b.santibehera@gmail.com

J. J. Jena
e-mail: junalijena@yahoo.com

A. K. Rath
e-mail: Akrath_cse@vssut.ac.in

P. K. Sethy
Sambalpur University, Sambalpur, Burla, India
e-mail: prabirasethy@suniv.ac.in

A. Abraham et al. (eds.), *Emerging Technologies in Data Mining and Information Security*, Advances in Intelligent Systems and Computing 813,
https://doi.org/10.1007/978-981-13-1498-8_7

1 Introduction

Automation of human visual system can be effectively accomplished by using image processing techniques. It is also known as machine vision. It can be categorized as one of the data mining approach as it retrieves information from images and analyzes them. Image processing has found a major application in the field of Agriculture by effectively reducing time, cost and labor as compared to manual techniques. Novell et al. (2012) stated Computer vision is a rapid, economic, consistent and objective inspection technique, which has expanded into many diverse industries [1].

Image segmentation, which separates the product region from background in the image, is one of the most important tasks in image processing since it is the first step in image analysis after the image capture to sub-divide an image into meaningful regions (Mizushima et al. 2013) [2]. After segmentation only, any subsequent analysis proceeds. It is applicable to both 2D and 3D imaging techniques. Segmentation is performed basing upon a certain characteristic such as shape, color, region, texture etc. Zheng et al. (2009) proposed a mean-shift based color segmentation method to separate green vegetations from background noises and shadows [3]. Pichel et al. (2006) proposed a heuristic segmentation algorithm based on merging of sub-optimal segmentation applicable to over-segmented images [4]. A segmentation approach based on adaptive local threshold proposed by Navon et al. (2005) aimed at segmenting color images [5]. Mery et al. (2005) proposed a hybrid algorithm for segmentation of color food images in which, optimal linear combination of RGB color components, global thresholding and morphological operations were performed [6]. Application of segmentation to 3D images can be observed in the work by Mitra et al. (2004) in which they have used active SVM to segment multi-spectral remote sensing images [7].

Features extraction also plays an important role in processing and analysis of images. Conversion of qualitative features of images into quantitative parameters enables mathematical operations to be applied on them followed by performance analysis and evaluation. Moreda et al. (2009, 2012) reviewed various techniques for features extraction of fruits and vegetables [8, 9].

Automated yield estimation, may be of preharvest or postharvest time, is also a major and ultimate application of image processing in the field of horticulture. It directly indicates the act of counting fruits or vegetables as on the objective background. The method proposed by Payne et al. (2013, 2014), estimated the mango crop yield prior to harvesting [10, 11]. A similar work proposed by chinchuluun et al. (2009), aimed at determining citrus count using Bayessian classifier, morphological operations and watershed segmentation [12]. Cubero et al. (2016) reviewed various automated systems based on machine vision for inspecting citrus fruits from the field to postharvest [13]. Qureshi et al. (2016) proposed a machine vision for counting mango fruits using KNN and SVM [14].

2 Materials and Methods

The primal recognition technique, that human visual system adopts to distinguish between mature and immature fruits and to identify fruits amidst background leaves, is the color recognition technique. Therefore, a novel approach is proposed using color space segmentation for performing three functions at once, i.e., singulation or detection, categorization, or labeling and enumeration or counting. The approach is applicable to oval-shaped fruits as wider fruit area gives good color contrast from the background areas. The algorithm successfully fulfilled the objective when applied to images of five species of oval-shaped fruits, i.e., mango, orange, apple, pomegranate and tomato. In this paper, segmentation technique based on a specific thresholding criteria, has been applied to color spaces to obtain the required result.

2.1 Color Space Segmentation

The proposed technique detects the presence of mature fruits in the objective image. After detection, it counts the number of partially mature or fully mature fruits. Basing upon the functionalities, algorithm can be divided into three parts as follows.

2.1.1 Singulation

Process of Singulation aims at detection of all single units of fruits those are either partially matured or fully matured. The key idea here is to detect the "blushy" or reddish part of the fruit that indicate its maturity. In the algorithm, this function is accomplished by extracting the "R" channel and "G" channel of the RGB color space and obtaining a binary image by using a thresholding criteria, "R>G". Those pixels of RGB image which do not satisfy the criteria, constitute white pixels in the binary image and vice versa.

2.1.2 Categorization

Categorization function can be performed by analyzing the binary image obtained from step of singulation. More number of white pixel indicates the fruits in the image are immature and vice versa. So, categorization can be done by calculating the ratio of white pixels to black pixels. Here, it is assumed that if twenty-percent or less portion of the image are black pixels, i.e., "ratio ≥5", then fruits in the image are immature and no counting is done. Only it will display that the fruits are immature. However, if the criterion is not satisfied then it is concluded that fruits in the image are mature and next categorization criterion is verified. In the next categorization criteria, it is

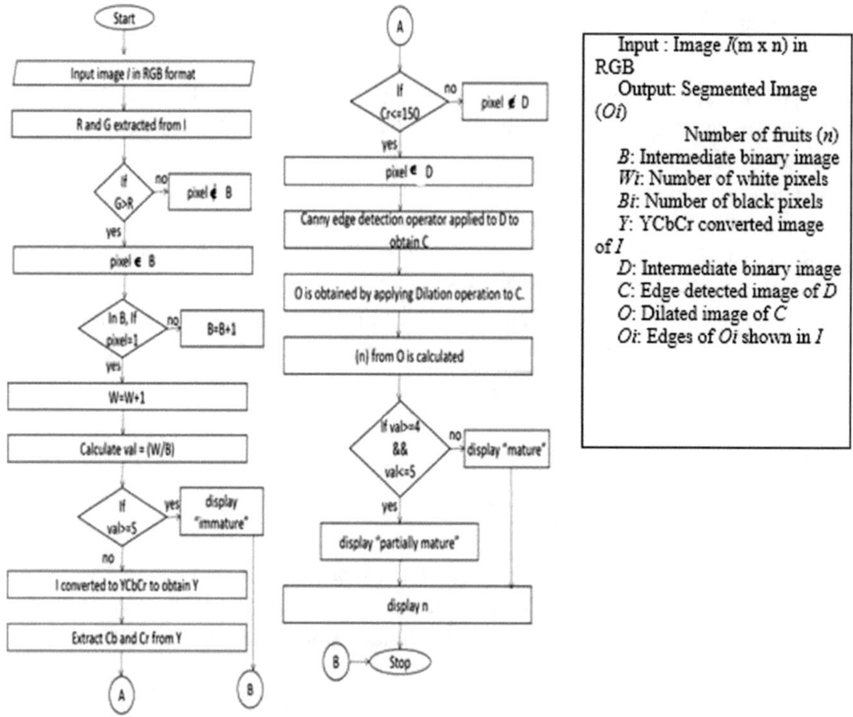

Fig. 1 Flowchart of proposed algorithm

assumed that, if the ratio lies between 4 and 5, then fruits are less mature, if not, then fruits are more mature. Then the algorithm proceeds to the function of enumeration.

2.1.3 Enumeration

This is the last functionality of our proposed approach. In this step, number of mature fruits is counted. For better contrast, RGB image is converted into YCbCr color space and "Cr" channel is extracted. A binary image is obtained by applying the thresholding criteria, "$\text{Cr} \leq 150$" (Payne et al. 2013). In the obtained binary image, black areas indicate the fruits in the image. Boundaries of black areas are detected by applying Canny edge detection operation (Canny 1986) followed by morphological operations [15]. Dilation, which is a morphological operation to bolden out the edges, is used to bold out the detected boundaries to reduce the cases of false detections due to presence of fine edges. Finally, automated count for detected objects is performed. Flowchart of the proposed algorithm is given in Fig. 1.

2.2 *Statistical Analysis*

Statistics is a mathematical tool for quantitative analysis of data and as such it serves as the means by which we extract useful information from data (Peters 2001) [16]. The basic statistical technique for error evaluation is calculating "Percentage of Accuracy". Its formula stated as follows.

$$\%\,\text{Accuracy} = \frac{\text{Obsered Value}}{\text{True Value}} \times 100\% \tag{1}$$

Standard deviation is the statistical technique which analyzes the extent up to which data is scattered from the mean value or the deviation of the data from the mean value, hence indicates the uniformity in data. Formula for standard deviation and standard error of the mean stated as follows.

$$S = \sqrt{\frac{\sum_{i=1}^{n}(x_i - \overline{x})^2}{n-1}} \tag{2}$$

$$S_{\overline{x}} = \frac{S}{\sqrt{n}}, \tag{3}$$

where, S is standard deviation, n is the number of samples in data set, x_i is the value of ith sample in data set and $\overline{x}$ is the mean value of all samples in data set.

3 Results and Discussion

A set of sample images were simulated for verification of assumed categorization criteria and satisfactory results were obtained. A set of twelve sample images of five varieties of fruits and their respective simulation output is given in Figs. 2, 3, 4, 5, 6, 7, 8, 9, 10, 11, 12 and 13.

On giving the input image of Figs. 2a–13a, we got the respective intermediate binary image, i.e., Figs. 2b–13b from which the number white pixel and black pixel were estimated. The ratio of white pixel to black pixel was calculated and categorized as mature, less mature and immature. The converted YCbCr image were shown in Figs. 2c–13c and the segmented output image is shown in Figs. 2d–13d. All observations are given below in a nutshell, in Table 1.

3.1 *Performance Analysis*

Performance of the proposed algorithm was analyzed basing upon various performance metrics and its behavior was studied.

Table 1 Observation table after simulation of images

Images	No. of white pixel (w)	No. of black pixel (b)	Ratio (w/b)	Output	Images	No. of white pixel (w)	No. of black pixel (b)	Ratio (w/b)	Output
Fig. 2	43859	21821	2.10	More mature	Fig. 8	63442	22461	2.78	More mature
Fig. 3	15202	21820	0.69	More mature	Fig. 9	36202	19891	1.82	More mature
Fig. 4	54024	32376	1.67	More mature	Fig. 10	182560	15432	11.83	Immature
Fig. 5	112173	20850	5.38	More mature	Fig. 11	193390	16890	11.45	Immature
Fig. 6	97665	20561	4.75	Less mature	Fig. 12	53678	36245	1.48	More mature
Fig. 7	61441	22756	2.70	More mature	Fig. 13	42952	28417	1.51	More mature

Fig. 2 **a** Original mango image **b** intermediate binary image **c** YCbCr image **d** output image

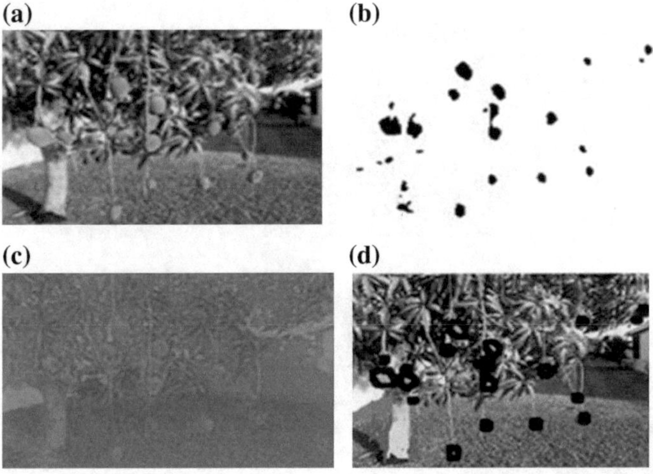

Fig. 3 **a** Original mango image **b** intermediate binary image **c** YCbCr image **d** output image

3.1.1 Accuracy

For measuring efficacy, sample images were tested using our proposed approach and satisfactory results were obtained in most of the cases. Accuracy was estimated by comparing manual count of fruits in the images against automated count resulted by the system and using Eq. (1). Table 2 gives the values of manual count, automated

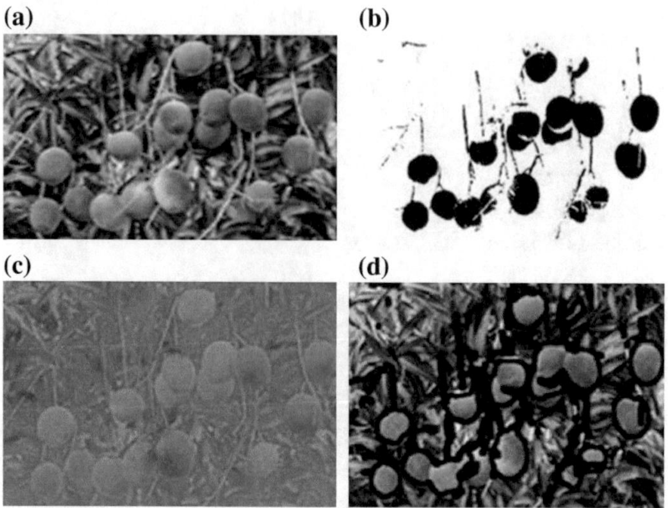

Fig. 4 **a** Original mango image **b** intermediate binary image **c** YCbCr image **d** output image

Fig. 5 **a** Original mango image **b** intermediate binary image **c** YCbCr image **d** output image

count and percentage of accuracy obtained after simulation of samples. Average percentage accuracy of the system was found out to be 60%.

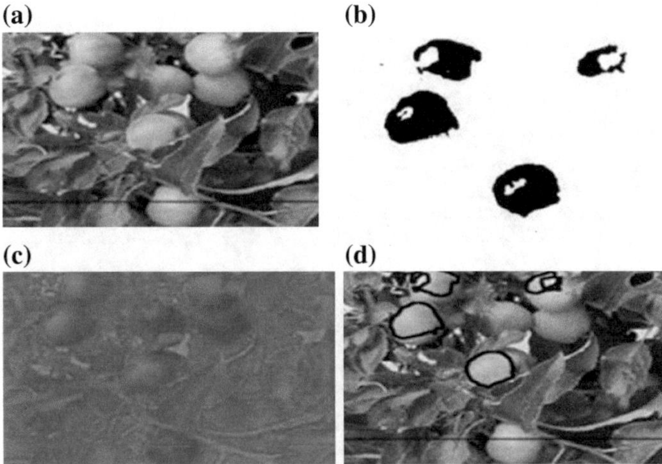

Fig. 6 **a** Original apple image **b** intermediate binary image **c** YCbCr image **d** output image

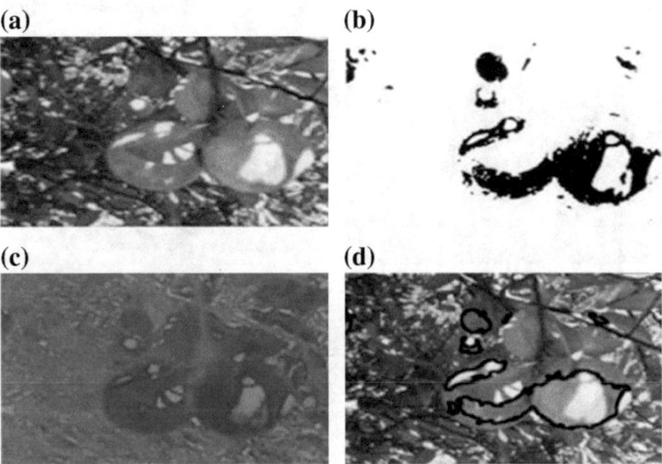

Fig. 7 **a** Original orange image **b** intermediate binary image **c** YCbCr image **d** output image

3.1.2 Applicability

The proposed work was verified to be successfully applicable to four types of oval shaped fruits, i.e., apple, orange, tomato, and mango.

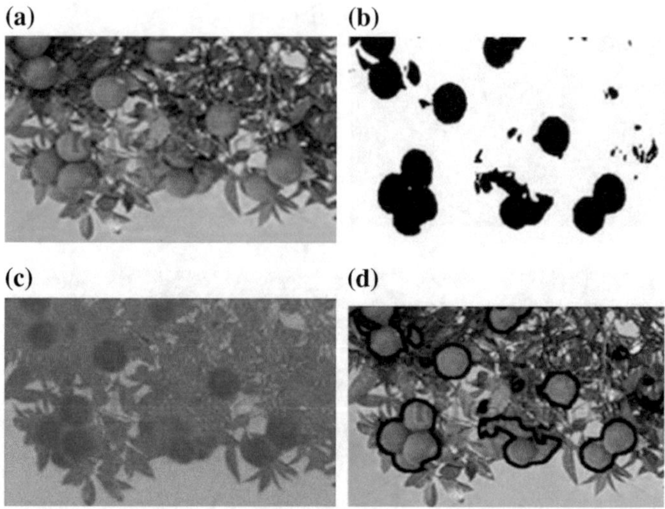

Fig. 8 **a** Original orange image **b** intermediate binary image **c** YCbCr image **d** output image

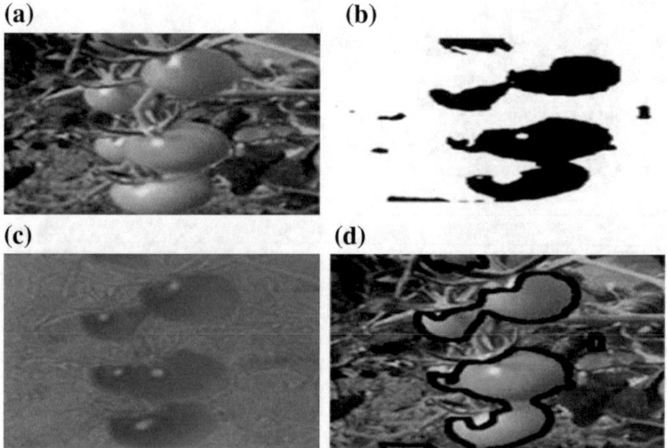

Fig. 9 **a** Original tomato image **b** intermediate binary image **c** YCbCr image **d** output image

3.1.3 Usability

The proposed algorithm was successfully used for performing three types of functions prior to harvesting in the field of fruit cultivation. The functions were given as follows.

i. Singulation (Detection)
ii. Categorization (Labeling)
iii. Enumeration (Counting).

Fig. 10 **a** Original immature mango image **b** intermediate binary image **c** YCbCr image **d** no output image

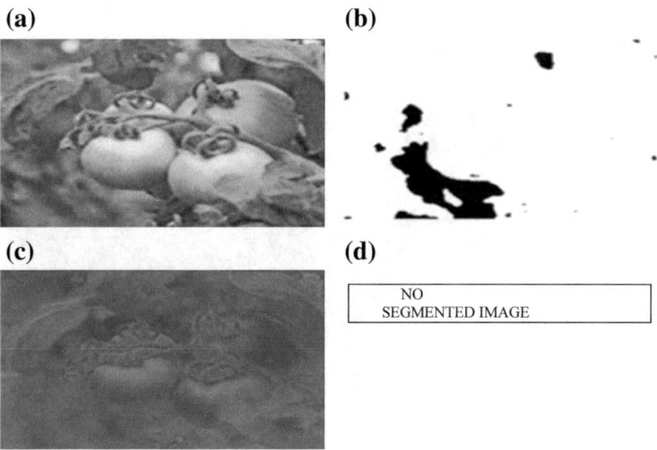

Fig. 11 **a** Original immature Tomato image **b** intermediate binary image **c** YCbCr image **d** no output image

3.2 Factors Affecting the Performance

After analysis of segmented images it was observed that climatic conditions and fruit position or alignment, greatly affect the performance of the system. Fruits in the diffused sunlight or shadow regions are not properly detected. An example of such type is shown in Fig. 14a. Overlapped fruits were also sometimes not detected properly. An example of such type is shown in Fig. 14b. Sometimes the fruits at the background trees got detected, which added to false detection. Similar example is

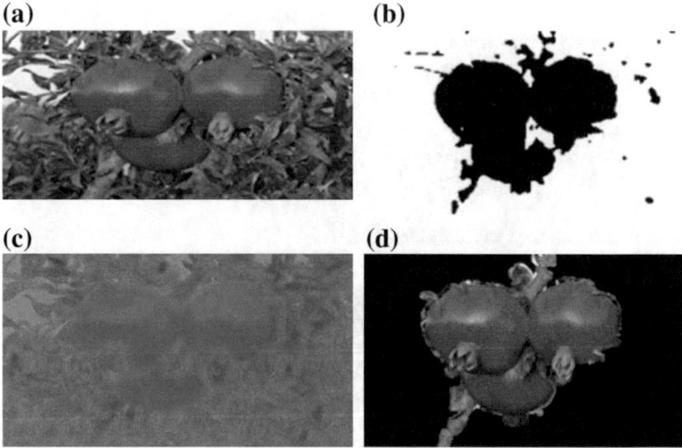

Fig. 12 **a** Original pomegranate image **b** intermediate binary image **c** YCbCr image **d** output image

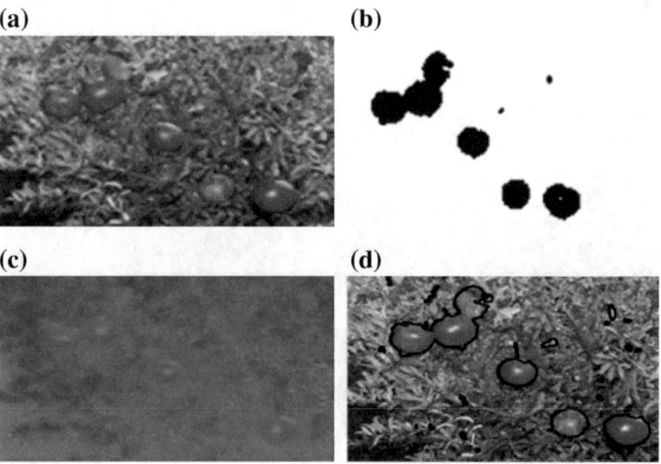

Fig. 13 **a** Original pomegranate image **b** intermediate binary image **c** YCbCr image **d** output image

shown in Fig. 14c. Fruits, which got covered by leaves and twigs, were sometimes detected in multiple segments. Such an example is shown in Fig. 14d. Algorithm was also unable to detect the immature green fruits as shown in Fig. 14e.

4 Conclusion and Future Scope

The proposed algorithm efficiently employed the concept of segmentation upon different color spaces to satisfy the purpose of singulation, categorization, and enumeration. However, further improvement in the algorithm may be done for detection and

Table 2 Values of automated count, manual count and %accuracy of each image

Images	Manual count	Automated count	Accuracy in %	Images	Manual count	Automated count	Accuracy in %
Fig. 2	30	18	60.00	Fig. 8	22	13	59.09
Fig. 3	17	9	52.94	Fig. 9	6	4	66.67
Fig. 4	17	8	47.06	Fig. 10	27	immature	No counting
Fig. 5	21	12	57.14	Fig. 11	4	immature	No counting
Fig. 6	7	4	57.14	Fig. 12	3	5	60
Fig. 7	9	5	55.56	Fig. 13	6	9	66.67

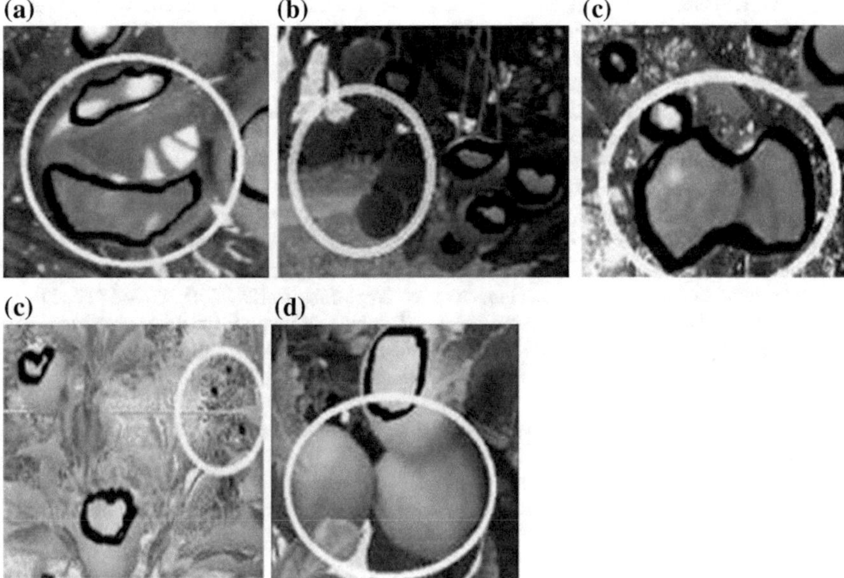

Fig. 14 Showing all the factors affecting the performance

counting of green or immature fruits and for improvement of the accuracy of the algorithm. The feature of fruit quality evaluation may also be added to increase its usability.

References

1. Garrido-Novell, C., et al.: Grading and color evolution of apples using RGB and hyperspectral imaging vision cameras. J. Food Eng. **113**(2), 281–288 (2012)
2. Mizushima, A., Renfu, L.: An image segmentation method for apple sorting and grading using support vector machine and Otsu's method. Comput. Electron. Agric. **94**, 29–37 (2013)
3. Zheng, L., Zhang, J., Wang, Q.: Mean-shift-based color segmentation of images containing green vegetation. Comput. Electron. Agric. **65**(1), 93–98 (2009)
4. Pichel, J.C., Singh, D.E., Rivera, F.F.: Image segmentation based on merging of sub-optimal segmentations. Pattern Recogn. Lett. **27**(10), 1105–1116 (2006)
5. Navon, E., Miller, O., Averbuch, A.: Color image segmentation based on adaptive local thresholds. Image Vis. Comput. **23**(1), 69–85 (2005)
6. Mery, D., Pedreschi, F.: Segmentation of colour food images using a robust algorithm. J. Food Eng. **66**(3), 353–360 (2005)
7. Mitra, P., Shankar, B.U., Pal, S.K.: Segmentation of multispectral remote sensing images using active support vector machines. Pattern Recogn. Lett. **25**(9), 1067–1074 (2004)
8. Moreda, G.P., et al.: Non-destructive technologies for fruit and vegetable size determination–a review. J. Food Eng. **92**(2), 119–136 (2009)
9. Moreda, G.P., et al.: Shape determination of horticultural produce using two-dimensional computer vision–a review. J. Food Eng. **108**(2), 245–261 (2012)
10. Payne, A.B., et al.: Estimation of mango crop yield using image analysis–segmentation method. Comput. Electron. Agric. **91**, 57–64 (2013)
11. Payne, A., et al.: Estimating mango crop yield using image analysis using fruit at 'stone hardening' stage and night time imaging. Comput. Electron. Agric. **100**, 160–167 (2014)
12. Chinchuluun, R., Lee, W.S., Ehsani, R.: Machine vision system for determining citrus count and size on a canopy shake and catch harvester. Appl. Eng. Agric. **25**(4), 451–458 (2009)
13. Cubero, S., et al.: Automated systems based on machine vision for inspecting citrus fruits from the field to postharvest—a review. Food Bioprocess Technol. **9**(10), 1623–1639 (2016)
14. Qureshi, W.S., et al.: Machine vision for counting fruit on mango tree canopies. Precis. Agric. 1–21 (2016)
15. Canny, J.: A computational approach to edge detection. IEEE Trans. Pattern Anal. Mach. Intell. **8**, 679–698 (1986)
16. Peters, C.A.: Statistics for analysis of experimental data. In: Powers, S.E., Bisogni Jr., J.J., Burken, J.G., Pagilla, K. (eds.) AEESP Environmental Engineering Processes Laboratory Manual. AEESP, Champaign, IL (2001)

A Novel Deterministic Framework for Non-probabilistic Recommender Systems

Avinash Bhat, Divya Madhav Kamath and C. Anitha

Abstract Recommendation is a technique which helps and suggests a user, any relevant item from a large information space. Current techniques for this purpose include non-probabilistic methods like content-based filtering and collaborative filtering (CF) and probabilistic methods like Bayesian inference and Case-based reasoning methods. CF algorithms use similarity measures for calculating similarity between users. In this paper, we propose a novel framework which deterministically switches between the CF algorithms based on sparsity to improve accuracy of recommendation.

Keywords Collaborative filtering · Non-probabilistic approach · Switching recommender systems · Sparsity · Similarity metrics · Genre independence

1 Introduction

Recommender systems are used for prediction of the behavior of a user to provide personalized content and services. A recommendation is a suggestion made to a user, keeping in mind his previous choices to improve his decision-making process. A recommender system carries this operation and predicts the user's rating to a product or a service without overwhelming them with sizeable data [1–3]. To provide relevant recommendations, a system must understand user's preferences, which can be done using Collaborative Filtering or Content-based Filtering or Hybrid methods [1, 3–5].

A. Bhat (✉) · C. Anitha
The National Institute of Engineering, Mananthavadi Road, Vidayaranya Puram, Mysuru, Karnataka, India
e-mail: avinashbhatneelavar@gmail.com

C. Anitha
e-mail: anithac.cse@nie.ac.in

D. M. Kamath
SJCE, JSS STU Campus, Manasagangotri, Mysuru, Karnataka, India
e-mail: divyamadhavkamath@gmail.com

A. Abraham et al. (eds.), *Emerging Technologies in Data Mining and Information Security*, Advances in Intelligent Systems and Computing 813,
https://doi.org/10.1007/978-981-13-1498-8_8

This paper aims to provide a framework for recommender systems which gives better accuracy, based on sparsity of the dataset. We focus on scope of switching recommender system, a type of hybrid recommender, which can switch between techniques based on a certain criterion [5].

Collaborative Filtering includes methods which make recommendations depending on behavioral analysis of other users in the system. These algorithms predict the preferences of the users and then rank the items based on the preferences to generate the recommendation list [1, 3, 6]. CF algorithms are of two types, User-based CF and Item-based CF. User based CF is also called user–user CF in which a group of users whose ratings similar to that of one user is listed and their behavior is analyzed collectively to predict preferences of the target user [1, 3, 7]. Item-based CF is also called item-item CF in which the recommendations are generated based on similarity between the rated items to other items which are not rated [1, 3, 7].

The concept of similarity metrics is used in data mining for determining the similarity between items or objects [1, 8]. Similarity is a value in the range [0, 1] that gives the measure of the degree to which the two items are alike [7]. Here, 0 corresponds to no similarity and 1 corresponds to total similarity. There are a number of measures used to calculate the similarity, few of which have been discussed in this paper.

- Euclidean Distance—Distance between two points in a Euclidean space is called Euclidean Distance. This notion is employed in recommender systems for similarity identification, more close two points are, more the similarity [8, 9]. The Euclidean distance E(x, y) between two points x = (x1, y1) and y = (x2, y2) is given by

$$E(x, y) = \sqrt{(x1 - x2)^2 + (y1 - y2)^2} \tag{1}$$

- Manhattan Distance—Collective summation of the extent from initial point to end point when traveled strictly vertical and/or horizontal. It returns the absolute distance between two data points [10]. Mathematical expression is,

$$M(x, y) = \sum_{k=1}^{n} |x_k - y_k| \tag{2}$$

- Pearson Correlation—The statistical correlation between common ratings of two users to calculate the similarity is called Pearson correlation [8, 9]. A threshold, which is the minimum number of correlated items that is necessary for performing the computation, is set for computation of similarities. Equation is as follows:

$$P(x, y) = \frac{N \sum xy - (\sum x)(\sum y)}{\sqrt{\left[N \sum x^2 - (\sum x)^2\right]\left[N \sum y^2 - (\sum y)^2\right]}} \tag{3}$$

- Spearman Rank Correlation—Spearman Correlation uses the same computation as the Pearson method. The difference lies in the fact that the items are represented with respect to their ranking as opposed to their ratings. The rating ranges from a rank of one, given to higher ratings and for lower ratings the rank reduces.
- Cosine Similarity—It uses a vector-space approach, where each user is represented as a vector. The similarity is obtained by calculating the cosine distance between the rating vectors making use of their dot product [8].

$$sim = \cos\theta = \frac{AB}{|A||B|} = \frac{\sum_{i=1}^{n} A_i B_i}{\sqrt{\sum_{i=1}^{n} A_i^2}\sqrt{\sum_{i=1}^{n} B_i^2}} \tag{4}$$

- Tanimoto Distance—It is a variation of the Jaccard Distance and is defined as ratio of intersecting values to the union of values [8]. This can be represented by the equation,

$$T_s(X, Y) = \frac{\sum_i X_i \bigwedge Y_i}{\sum_i X_i \bigvee Y_i} \tag{5}$$

The reason these measures were selected for this paper is because they represent different classes of calculating similarity, in essence, based on distance, based on correlation and based on vector. They are compared to identify the suitable measures for each kind of CF algorithms [11, 12].

The paper is organized as follows. Section 2 focuses on related work, followed by a description of proposed framework in Sect. 3. The recommender system built for the purpose of analysis is discussed in Sect. 4. Results and discussion of the analysis and conclusion in given in Sects. 5 and 6 respectively.

2 Related Work

Recommendation systems started off predominantly as offline systems rather than online real-time systems we use today. The capacity of computers to provide recommendations was recognized early in history of computing [7]. Elaine Rich, in his paper on User Modeling via Stereotypes [13], introduced a computer based librarian which grouped the books based on similarity of user input to hard coded information. Tapestry [14], by David Goldberg et al. proposed a model which would allow users to read any mail (by which it meant any online article), and rate them which would further call for recommendations, close to how recommenders function today. Further, John T. Riedl et al. proposed Grouplens architecture [6, 15] for collaborative filtering. During this period, collaborative filtering and recommenders became a trend leading to a number of recommenders for various domains, like Ringo [16] for music, Amazon for e-commerce systems, Netflix for online streaming. The Netflix Prize awarded 1 million who could beat Netflix recommender by ten percent, and was won by BellKor's Pragmatic Chaos [1, 17], by implementing matrix factorization methods to improve the density of the dataset enabling the recommender to perform better. Recommender techniques include probabilistic models based on information retrieval, Bayesian inference [3, 7, 18], and case-based reasoning methods [19]. A survey on these models was done by Bobadilla et al. [20]. A switching hybrid recommender system can switch between recommended techniques by applying certain criterion. One such approach combined Naive Bayes classification approach with the collaborative filtering [21]. An experimental study was done by Michael D. Ekstrand et al. which let the users choose the algorithm for recommendation [22]. Advantages of improving the similarity measures in recommenders was shown by Sarwar et al. and Massa et al. who implemented Cosine similarity, adjusted cosine similarity, Pearson and Spearman Correlation for recommenders [23, 24]. Best comparison is done by Chandelier et al. who implemented various similarity as well as error measures to validate the prediction [25]. This paper aims at improving the accuracy of non-probabilistic collaborative filtering using a switching recommender system which uses sparsity of the dataset to switch between the algorithms. This framework follows up on Ekstrand et al. model [22], but by automating the choice of algorithm, we aim to make the system faster.

3 Proposed System

3.1 Overview

Current recommendation systems make use of a single type of similarity measure, or in case of collaborative filtering, either user based or item-based. However, this can be optimized for better accuracy. We see that the sparsity of the dataset plays a vital role in the accuracy of these methods. The proposed system provides a way to automatically select the various stages of recommendation based on the dataset.

In this section, we present a framework which is able to determine the type of CF and the similarity measure which can provide the best possible prediction for a particular dataset. The framework is designed to diagnose the dataset based on its sparsity, and is able to pick better algorithms for the given data.

The framework can be thought of as a black box, to which the input is the dataset, and the output is a list of recommendations (Fig. 1b).

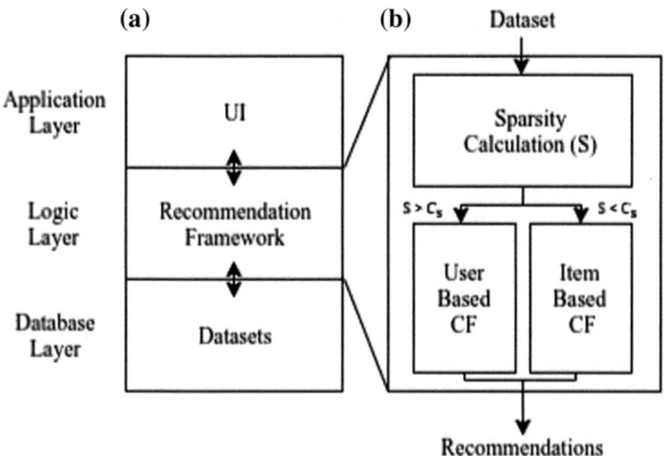

Fig. 1 **a** Three-tier architecture of the framework. The datasets are fed to the framework which outputs the results to the user. **b** Expansion of the framework

3.2 Layout

A dataset can be viewed in two different ways, a user-oriented view which is used by User-based CF and an item oriented view used by Item-based CF. The user oriented view looks like {*user*1: {*item*1: *rating*, *item*2: *rating*…}, *user*2: {*item*1: *rating*…}…}.

The item oriented view is {*item*1: {*user*1: *rating*…}, *item*2: {*user*1: *rating*… }… }.

The input to the framework is a user oriented view. In order to transform the dataset to item oriented format used by Item-based CF, a *transform Matrix* function is used.

Algorithm 1. The *transform Matrix* function.

```
begin
for user in user_ortn
      for item in user_ortn[user]
            item_ortn[item][user]= user_ortn[user][item]
return item_ortn
```

The backbone of the framework lies in the sparsity calculation. Sparsity is nothing but the population of the dataset. The algorithm for sparsity calculation is as follows.

Algorithm 2. The *calculate Sparsity* function.

```
begin
num_user = max_number_of_users
num_item = max_number_of_items
slots = num_user * num_item
for user in user_ortn:
      for item in user_ortn[user]:
            density = density + 1
      end for
end for
sparsity = density/slots
return sparsity
```

This function returns the sparsity S of the dataset. After a comparison of S with the sparsity coefficient C_S, the framework redirects the code to a User-based CF module or an Item-based CF Module. Sparsity Coefficient C_S is a specific percentage value, which if exceeded, the dataset can be said to be dense. Once the code is branched, the framework chooses the similarity measure which is suitable for the type of collaborative filtering being performed. The output of the framework is the list of possible recommendations.

4 Evaluation

The proposed framework is based upon the analysis done on two datasets: MovieLens and Jester. Two recommendation engines, one User-based and one Item-based, were built for the purpose.

4.1 Datasets

MovieLens was developed by the GroupLens project at the University of Minnesota [6, 26]. This data contains ratings on scale 1–5. Datasets used were as follows:

- *MovieLens Older Dataset Small* was a dataset of size 5 MB. It has 100,000 ratings from 1000 users on 1700 movies.
- *MovieLens* 1 *M Dataset* consists of 6 MB data holding about 1 million ratings from 6000 users on 4000 movies.
- *MovieLens Latest Dataset Small* of size 1 MB with 100,000 ratings applied to 9,000 movies by 700 users was used.

The Jester dataset is the dataset released from the Jester Online Joke Recommender System [27]. This data contains ratings on scale −10 to 10. Datasets used were,

- *Jester Dataset 1* is a dataset of size 3.9 MB which contains data from about 24,983 users.
- *Jester Dataset 2* is a dataset of size 3.6 MB which a collection of data about ratings of 23,500 users.
- *Jester Dataset 2+* is a dataset of size 5.1 MB. This dataset stores information of 500,000 ratings from 79,681 users.

The data from these datasets were loaded to the recommender system in the user oriented view. Based on the sparsity calculated by the sparsity algorithm (Algorithm 2), the datasets were classified as sparse or dense. Since the scale of rating of these datasets were different, they were normalized to the scale [−1, 1] during the evaluation.

4.2 System Specification

Analysis was performed on a 64-bit Ubuntu 16.04 LTS platform with a 4 GB memory, and a quad core Intel i5-5200U CPU with clock speed of 2.2 GHz.

4.3 Recommender Systems

Two recommender systems, one user-based and another item-based were written in Python language to compare the efficiency and accuracy of the similarity of efficiency measures [28]. The user-based recommender calculates the similarity between a person and other people and predicts the possible data of the person based on the most similar people. The item-based recommender calculates the similarity between two items based on the ratings received. Once the dataset was loaded to the recommender, the data was randomly split into training and test data. Prediction of possible rating of test data was made using the available training data which was further compared to actual data available. Lower error between the two data is a measure of better prediction accuracy. The random division of data for testing and training purpose was done again and the process is repeated, leading to calculation of average error value.

5 Results and Discussion

The calculated average error values are based on two accuracy metrics below.

5.1 Accuracy Metrics

The recommendations made have to be tested for accuracy as well as speed to understand which performs better in this particular dataset. The data was divided as 85% for training, and 15% for testing. The analysis is based on the validity or the accuracy of the results. Two measures of evaluating the accuracy of predictions are used to check the validity of the predictions [1, 3, 7].

- Root Mean Squared Error (RMSE)—The difference between predicted values and observed values (prediction error) when taken individually are called residuals. RMSE is the standard deviation of these residuals. It is a measure of how spread out these residuals are. It is a good error metric for numerical predictions [29]. Mathematically,

$$RMSE = \sqrt{\frac{1}{n}\sum_{i=1}^{n}\left(y_i - \hat{y}_i\right)^2} \tag{5}$$

- Mean Absolute Error (MAE)—The sum of the absolute value of the residuals is MAE. It indicates the margin of an error which can be expected from the forecast on an average [29]. It is formulated as,

$$MAE = \frac{1}{n}\sum_{i=1}^{n}\left|y_i - \hat{y}_i\right|^2 \tag{6}$$

5.2 Results

The sparsity of the dataset was calculated and the outcome is as represented in Table 1. The datasets were tested using the similarity measures mentioned above considering the User-based CF and Item-based CF (Figs. 2 and 3). The graphical representation of Figs. 2 and 3 is depicted in Figs. 4 and 5.

5.3 Discussion

In order to develop the proposed framework, two different genre of datasets were evaluated. Based on the sparsity of the datasets, following observations were made.

- Denser the dataset, higher the variation in the ratings and hence more possibility for error. This leads us to our first argument that efficiency of similarity measures is independent of the genre of the dataset and rather it only depends on the sparsity of the dataset.

Table 1 Classification of the datasets based on sparsity

Genre	Dataset	Volume	Sparsity	Dense/sparse
Movie	MovieLens 1 M	22384240	0.044684	Sparse
	MovicLcns 100 k older	6023136	0.016574	Sparse
	MovieLens 100 k latest	1569152	0.063533	Sparse
Jokes	Jester dataset 1	6500	0.991385	Dense
	Jester dataset 2	6500	0.986769	Dense
	Jester dataset 2+	17640	0.955272	Dense

ITEM-BASED COLLABORATIVE FILTERING

	SPARSE DATASETS					
SIMILARITY MEASURES	**1M Dataset**		**100k Dataset Latest**		**100k Dataset Older**	
	MAE	**RMSE**	**MAE**	**RMSE**	**MAE**	**RMSE**
Euclid Distance	0.2977	0.3436	0.3904	0.5064	0.5937	0.6894
Manhattan Distance	0.2978	0.3435	0.3905	0.5063	0.5938	0.6893
Pearson Correlation	0.4482	0.5146	0.6500	0.7986	0.8011	0.9431
Spearman Correlation	0.3185	0.3829	0.4061	0.4869	0.4579	0.5515
Cosine Similarity	0.3631	0.4190	0.4776	0.6122	0.5637	0.6840
Tanimoto Distance	0.7548	1.0070	0.6879	0.9147	0.7992	1.0549
	DENSE DATASETS					
SIMILARITY MEASURES	**Jester Dataset 2+**		**Jester Dataset 1**		**Jester Dataset 2**	
	MAE	**RMSE**	**MAE**	**RMSE**	**MAE**	**RMSE**
Euclid Distance	3.5855	4.7018	4.5583	5.6827	4.3113	5.4467
Manhattan Distance	3.8214	4.7928	4.8498	5.9823	4.3377	5.3231
Pearson Correlation	3.7232	4.8707	4.5674	5.7247	4.4863	5.5783
Spearman Correlation	4.7306	5.7467	4.9540	5.9838	4.8581	5.8347
Cosine Similarity	3.6502	4.8486	4.2797	5.4621	4.3655	5.4575
Tanimoto Distance	4.1897	5.3187	5.2042	5.8388	4.4991	5.4918

Fig. 2 MAE and RMSE values of different similarity measures for item-based CF

- Item-based algorithm calculates the similarity of the items based on the user's ratings. In a sparse matrix, it is feasible to calculate the relationship between the items rather than that between the users. Two users might rate two different sets of items making it impossible to calculate the similarity between them. Hence, Item-based CF has better accuracy for dense datasets.
- In a sparse dataset, since there are less ratings, the accuracy of calculation of the similarity between the users is less. Distance measures can be a better approach towards similarity calculation. Let's consider the MovieLens datasets, where the ratings are more mutually exclusive, which makes the distance calculation a viable approach. Hence, Euclidean Distance metric performs distinctively well (Fig. 3), making it a good choice for an Item-based recommender. In a dense matrix, since

USER-BASED COLLABORATIVE FILTERING						
	SPARSE DATASETS					
SIMILARITY MEASURES	1M Dataset		100k Dataset Latest		100k Dataset Older	
	MAE	RMSE	MAE	RMSE	MAE	RMSE
Euclid Distance	0.6671	0.8260	0.6064	0.7679	0.6996	0.8665
Manhattan Distance	0.7134	0.8833	0.6654	0.8400	0.7473	0.9240
Pearson Correlation	0.6748	0.8236	0.6970	0.8594	0.7019	0.8538
Spearman Correlation	0.7549	0.9232	0.7435	0.9139	0.7889	0.9626
Cosine Similarity	0.7861	0.9599	0.7783	0.9548	0.8297	1.0116
Tanimoto Distance	0.7844	0.9589	0.7851	0.9671	0.8356	1.0188
	DENSE DATASETS					
SIMILARITY MEASURES	Jester Dataset 2+		Jester Dataset 1		Jester Dataset 2	
	MAE	RMSE	MAE	RMSE	MAE	RMSE
Euclid Distance	4.0290	4.9961	4.6158	5.3930	4.1081	4.9523
Manhattan Distance	4.0640	5.0421	4.6837	5.4644	4.1350	4.9839
Pearson Correlation	3.8612	4.7787	4.2533	5.0416	3.8065	4.6107
Spearman Correlation	4.0111	4.9713	4.5724	5.3570	3.9642	4.7892
Cosine Similarity	3.8737	4.8127	4.0810	4.8777	3.7926	4.6388
Tanimoto Distance	4.1450	5.1252	4.8286	5.6070	4.1837	5.0346

Fig. 3 MAE and RMSE values of different similarity measures for user-based CF

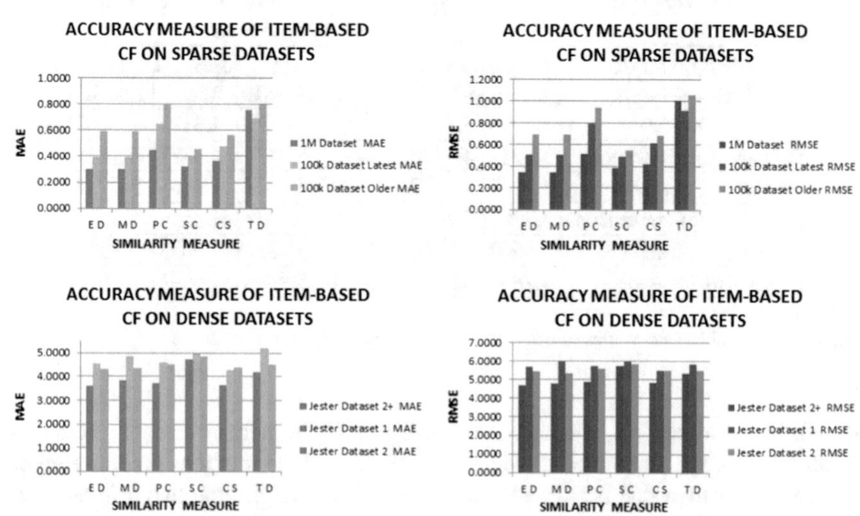

Fig. 4 Comparison of performance of similarity measures on the datasets in item-based CF

there are copious ratings, correlation and vector based approaches have a better accuracy. In Jester dataset, Pearson Correlation gives a low error (Fig. 4), making it appropriate for User-based recommender.

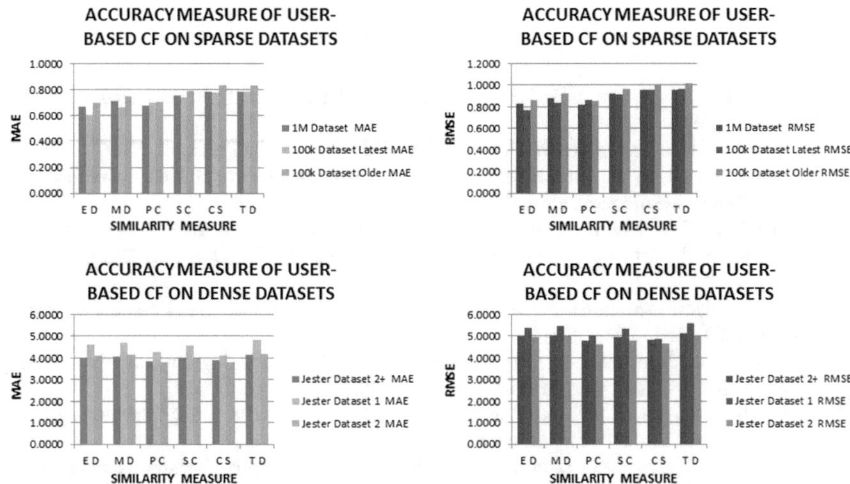

Fig. 5 Comparison of performance of similarity measures on the datasets in user-based CF

6 Conclusion

The proposed framework can be used as a recommendation system for any genre of dataset. The user can simply use the proposed system as a black box which takes any dataset as input and gives the best recommendation as output, making the user free from the complex logistics involved. The following conclusions formed the basis for the development of the framework. First, similarity measures depend only on the sparsity of the dataset. Second, User-based CF gives better accuracy for dense datasets, while Item-based CF gives better accuracy for sparse datasets. Finally, most suitable similarity measure for User-based CF was found to be Pearson correlation with an accuracy in the range of 91–94% for sparse datasets and in the range of 94–96% for dense datasets and Euclidean distance was found appropriate for Item-based CF with accuracy range of 99.4–99.7% for sparse datasets and 94–96% for dense datasets.

Acknowledgements The authors express their thanks to GroupLens and Jester team who compiled excellent datasets for research on recommender systems. We are grateful to the Principal of our college The National Institute of Engineering, Mysuru for encouraging us and providing the required facility.

References

1. Lü, L., Medo, M., Yeung, C.H., Zhang, Y.C., Zhang, Z.K., Zhou, T.: Recommender systems. Phys. Rep. **519**(1), 1–49 (2012)
2. Ullman, J.D.: Mining of Massive Datasets. Cambridge University Press, pp. 92–97 (2011)
3. Ricci, F., Rokach, L., Shapira, B.: Introduction to recommender systems handbook. Recommender Systems Handbook, pp. 1–35 (2011)
4. Adomavicius, G., Tuzhilin, A.: Toward the next generation of recommender systems: a survey of the state-of-the-art and possible extensions. IEEE Trans. Knowl. Data Eng. **17**(6), 734–749 (2005)
5. Burke, R.: Hybrid recommender systems: survey and experiments. User Model. User Adap. Interact. **12**(4), 331–370 (2002)
6. Herlocker, J.L., Konstan, J.A., Terveen, L.G., Riedl, J.T.: Evaluating collaborative filtering recommender systems. ACM Trans. Inf. Syst. (TOIS) **22**(1), 5–53 (2004)
7. Ekstrand, M.D., Riedl, J.T., Konstan, J.A.: Collaborative filtering recommender systems. Found. Trends® Human Comput. Interact. **4**(2), 81–173 (2011)
8. Amatriain, X., Pujol, J.M.: Data mining methods for recommender systems. Recommender Systems Handbook, pp. 227–262 (2015)
9. Shimodaira, H.: Similarity and recommender systems. School of Informatics, The University of Eidenburgh, 21 (2014)
10. How to build a recommendation system, Mickael Le Gal (2014)
11. Sondur, M.S.D., Chigadani, M.A.P., Nayak, S.: Similarity measures for recommender systems: a comparative study. J. Res. **2**(03) (2016)
12. Bagchi, S.: Performance and quality assessment of similarity measures in collaborative filtering using mahout. Procedia Comput. Sci. **50**, 229–234 (2015)
13. Rich, E.: User modeling via stereotypes. Cogn. Sci. **3**(4), 329–354 (1979)
14. Goldberg, D., Nichols, D., Oki, B.M., Terry, D.: Using collaborative filtering to weave an information tapestry. Commun. ACM **35**(12), 61–70 (1992)
15. Resnick, P., Iacovou, N., Suchak, M., Bergstrom, P., Riedl, J.: GroupLens: an open architecture for collaborative filtering of netnews. In: Proceedings of the 1994 ACM Conference on Computer Supported Cooperative Work, pp. 175–186 (1994)
16. Shardanand, U., Maes, P.: Social information filtering: algorithms for automating "word of mouth". In: Proceedings of the SIGCHI Conference on Human Factors in Computing Systems, pp. 210–217 (1995)
17. Koren, Y.: The bellkor solution to the netflix grand prize. Netflix Prize Doc. **81**, 1–10 (2009)
18. Condli, M.K., Lewis, D.D., Madigan, D., Posse, C.: Bayesian mixed-E ects models for recommender systems. In: Proceedings of ACM SIGIR, vol. 99 (1999)
19. Smyth, B.: Case-based recommendation. The Adaptive Web, pp. 342–376 (2007)
20. Bobadilla, J., Ortega, F., Hernando, A., Gutiérrez, A.: Recommender systems survey. Knowl. Based Syst. **46**, 109–132 (2013)
21. Ghazanfar, M., Prugel-Bennett, A.: An improved switching hybrid recommender system using naive Bayes classifier and collaborative filtering (2010)
22. Ekstrand, M.D., Kluver, D., Harper, F.M., Konstan, J.A.: Letting users choose recommender algorithms: an experimental study. In: Proceedings of the 9th ACM Conference on Recommender Systems, pp. 11–18 (2015)
23. Sarwar, B., Karypis, G., Konstan, J., Riedl, J.: Item-based collaborative filtering recommendation algorithms. In: Proceedings of the 10th International Conference on World Wide Web, pp. 285–295 (2001)
24. Massa, P., Avesani, P.: Trust-aware collaborative filtering for recommender systems. CoopIS/DOA/ODBASE **1**(3290), 492–508 (2004)
25. Candillier, L., Meyer, F., Boullé, M.: Comparing state-of-the-art collaborative filtering systems. In: International Workshop on Machine Learning and Data Mining in Pattern Recognition, pp. 548–562 (2007)

26. MovieLens Datasets. https://grouplens.org/datasets/movielens/
27. Jester Datasets. http://eigentaste.berkeley.edu/dataset/
28. Toby, S.: Programming Collective Intelligence. O'Reilly Media, pp. 7–27 (2007)
29. Wit, J.: Evaluating recommender systems: an evaluation framework to predict user satisfaction for recommender systems in an electronic programme guide context, Master's thesis, University of Twente (2008)

EEG Signal Analysis Using Different Clustering Techniques

Chinmaya Kumar Pradhan, Shariar Rahaman, Md. Abdul Alim Sheikh, Alok Kole and Tanmoy Maity

Abstract Electroencephalogram (EEG) is a test used to detect neurological disorders by checking the electrical activity in the brain. EEG is done to check problems related to the electrical activity of brain and its disorders such as epilepsy and types of seizures occurring, sleep disorders such as narcolepsy, encephalitis, brain tumor, Stroke, Dementia, etc. EEG may also be used to check out if a person is brain dead in persistent coma. The electrical activity of brain is checked with the help of electrodes attached to the scalp and recorded in computer. With increasing database performance of K-Means, and FCM is not efficient. In this paper we propose spatial clustering algorithm for EEG signals to improve the efficiency.

Keywords Electroencephalogram · Clustering techniques · K-Means · Fuzzy C-Means · Spatial clustering

C. K. Pradhan (✉) · S. Rahaman (✉)
Department of Electronics & Communication Engineering, Dream Institute of Technology, Kolkata, India
e-mail: chinmayapradhan07@gmail.com

S. Rahaman
e-mail: shaiar.rahaman13@gmail.com

Md. Abdul Alim Sheikh (✉)
Department of Electronics & Communication Engineering, Aliah University, Kolkata, India
e-mail: alim.sheikh16@gmail.com

A. Kole
Department of Electrical Engineering, RCC Institute of Information Technology, Kolkata, India
e-mail: alokkole123@yahoo.co.in

T. Maity
Department of Mining Machinery Engineering, Indian Institute of Technology (ISM) Dhanbad, Dhanbad, India
e-mail: tanmoy_maity@yahoo.co.in

A. Abraham et al. (eds.), *Emerging Technologies in Data Mining and Information Security*, Advances in Intelligent Systems and Computing 813,
https://doi.org/10.1007/978-981-13-1498-8_9

1 Introduction

The Electroencephalogram (EEG) is a test to check the voltage signal arising from neural activity in the brain. Due to activity of neutrons in the brain EEG is produced. It is measured by two ways. In one of the method electrodes are fixed to the patient's scalp and the variation in the electrical signals are measured in terms of voltage. To check the disorders in neural activity EEG signals are classified using classifiers. The information about the human brain and neurological disorders is found based on the output from the electrodes. The signal measured from the brain is contaminated with variety of noise sources which should be reduced to get obvious pictures about the brain functionality. The mental activities should be distinguished through the EEG test. If the EEG recordings are corrupted and composed with different attributes for noise levels, the identification of frequencies will be mismatched. When the frequencies are mismatched, automatically the mental activities are distinguished and lead to false diagnosis.

Different de-noising techniques are applied on the signals to reduce the levels of the noise. Classification of EEG signal is very time-consuming task, because a human expert is required to classify them properly. In this article, we have applied three different clustering algorithms, i.e., Fuzzy C-means, Spatial and K-Means to classify the EEG signals.

The further organization of this paper is as follow. Section 2 Reviews the approaches in order to classify EEG signals based on the movement of brain on Left-Right Limb. Section 3 Proposed methodology, Sect. 4 Demonstrate the results obtained from the comparison of different clustering techniques and Sect. 5 Conclusion.

2 Literature Review

The author of this paper has referred to the different articles on clustering techniques. Clustering is a traditional problem in data analysis. Where in the task is to divide the objects into similar groups based on criteria, typically distance between the objects. Many different clustering algorithms have been developed over the decades. More discussions are available about the clustering techniques, [1–4]. K-means is a well-known type of unsupervised learning clustering algorithm which is used when unlabeled data is available, i.e., requires no prior information about data points [5, 6]. In Clustering algorithms, fuzzy clustering is the mostly used algorithm. Fuzzy set theory was actually proposed by Zadeh in the year 1965 and gave an idea of uncertainty of belonging which was described by a member function. The Electroencephalogram, i.e., EEG signal is a voltage signal arising from the neural activities. EEG can be used to classify different mental states and to find out related disorders in neural activity. Different Neural activities are analyzed using EEG signal and are classified by different classifiers. The input data set is clustered using K-means clustering and fuzzy c-means (FCM) clustering and is input to neural network. In order to classify unknown data points a neural network tools is used called as Neuro Intelligence. Using K-means or FCM clustering techniques the nonlinear time series data into Normal or Abnormal categories. To train the neural network, clustered data set is used. First Nonlinear Time Series (NLTS) measurements are extracted from unknown EEG signal and then input to trained neural network to classify the EEG signal.

A. K-means clustering is a popular clustering method, initially designed for signal processing, but is popularly used in cluster analysis, data mining and pattern recognition applications [7, 8]. As its name itself suggest that it is having K number of clusters of input variables of equal no of groups. Every cluster belongs to an observation means no of data points and the average of data points is known as centroid. The clustering is done by reducing the Euclidean distance between data and the corresponding cluster center. The method follows a simple approach to classify a given data with no of observations into m number of clusters. First, k centroids are defined one for each cluster. Assuming centroids to be far distance from each other initially, points belonging to the given dataset is taken and associated with the nearest centroid. After these k new centroids are obtained, a new iteration is done between the same data set points and the nearest new cluster center and the process will continue so that no change of centroid location is possible.

Steps for k-means clustering algorithm:

(1) Select n cluster centers i.e. centroids.
(2) Euclidean distance method is used to calculate the distance between the objects and the centroid.
(3) Find the nearest centroid for each data point. Group the data point based on minimum distance from the centroid.
(4) Repeat steps 2 and 3 until no object is left.

B. Fuzzy c-means clustering:

Fuzzy clustering is a method of clustering, where data points can belong to one or more clusters. Based on the distance between data point and the centroid, membership is assigned to each data point. Membership values are always between ‘0’ and ‘1’. It is developed by Dunn in 1973 and improved by Bezdek in 1981. And it is frequently used in pattern recognition. The membership value of data point is large, whose distance is minimum, i.e., near to centroid. Summation of membership value of each data point should be equal to one [2, 8, 9].

$$J_m = \sum_{i=1}^{N}\sum_{j=1}^{C} u_{ij}^{m}\left\|X_i - C_j\right\|^2, \qquad (1)$$

where

m Real Number greater than 1.

Uij Degree of membership of xi in the cluster j.
Xi The i-th of dimensional measured data.
Cj The d-dimension center of the cluster.
||*|| Norm expressing the similarity between any measured data and the center [2, 9].

Fuzzy partitioning is carried out through an iterative optimization of the objective function shown above, with the update of membership uij and the cluster centers Cj by [9, 8]

$$u_{ij} = \frac{1}{\sum_{k=1}^{C} \frac{\|X_i - C_j\|}{\|X_i - C_k\|}^{\frac{2}{m-1}}} \tag{2}$$

This iteration will stop when

$$\max_{ij}\left\{\left|u_{ij}^{k+1} - u_{ij}^{k}\right|\right\} < \varepsilon, \tag{3}$$

where ε is a termination criteria between 0 and 1, whereas k are the iteration steps. This procedure converges to a local minimum or a saddle point of Jm. It is a tool used in Image processing, Plant and animal ecology, Transcriptomics.

3 Proposed Methodology

3.1 Spatial Clustering

Spatial data means data related to space. Spatial data consists of data that have a spatial component [1]. The spatial parameter combined with clustering is defined as

$$S_{pij} = \sum_{k \in w(xj)} u_{ik} \tag{4}$$

w(xj) is the window function having centered on pixel xj in spatial domain. The spatial function SPij is the probability of pixels xj belongs to j-th cluster. The objective function is below

$$J_m = \sum_{i=1}^{N} \sum_{j=1}^{C} u_{ij}^{m} \|X_i - C_j\|^2 \tag{5}$$

Spatial algorithm contains following steps as follows:

Step1. Select the S number of clusters manually.
Step2. Create clusters and determine the centroid.

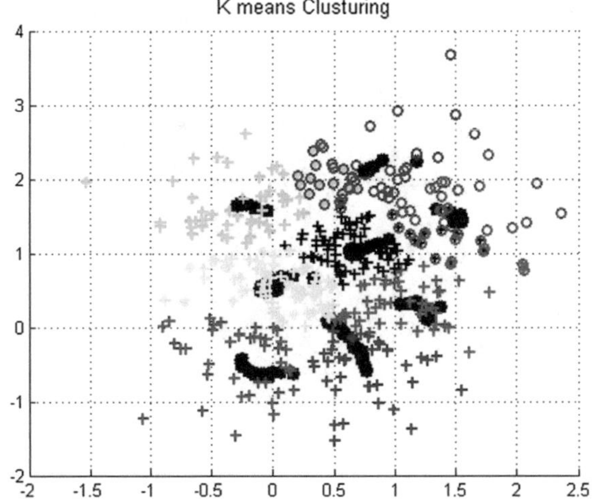

Fig. 1 K-Means clustering

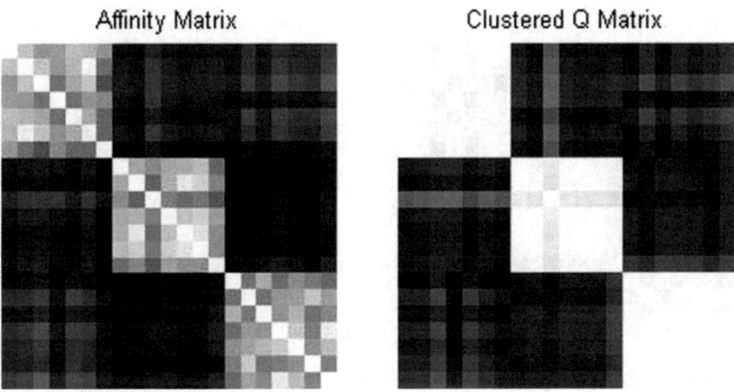

Fig. 2 Affinity matrix and clustered Q matrix

Step3. Assign every object of the cluster which minimizes the variance between the object and the centroid.
Step4. Recalculate centroids by taking average of all the objects in the cluster.
Step5. The process will continue to get the convergence.

4 Results and Discussion

The comparison table represents the difference between the K-Means, Spatial and FCM. The Spatial clustering method contains the better accuracy than other classification methods (Figs. 1, 2, 3 and 4) (Table 1).

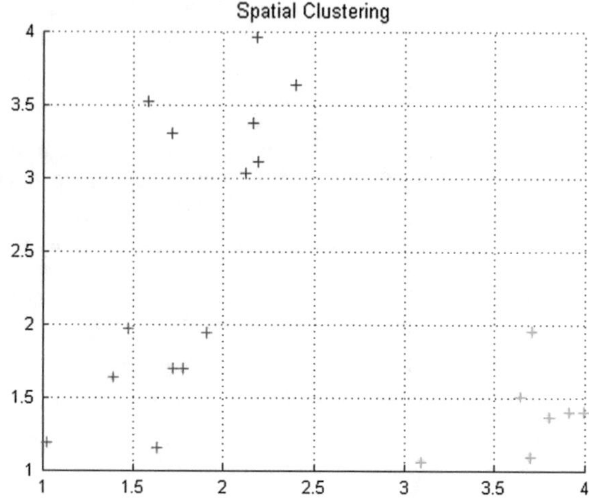

Fig. 3 Spatial clustering

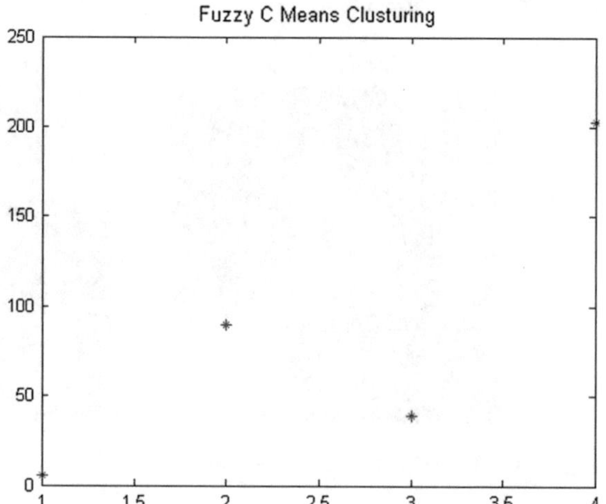

Fig. 4 FCM clustering

Table 1 Comparison table between TPR and FPR

Clustering method	TPR	FPR	Accuracy (%)
K-Means	1.4423	0.1580	90.1246
Spatial	1.4481	0.1507	90.5749
FCM	1.4388	0.1592	90.0388

5 Conclusion

Different clustering techniques are used to classify EEG signals obtained from the electrical activity of the brain. Here in this paper, we proposed spatial clustering method on EEG signal image database in comparison with K-means and Fuzzy

C-Means clustering algorithm. The results obtained from different clustering methods, i.e., K-means, FCM and Spatial are 90.12%, 90.03%, and 90.57% respectively. From the results it can be observed that spatial clustering method outperforms k-means and FCM clustering methods used on EEG signals analysis.

References

1. Smith, T.F., Waterman, M.S.: Identification of common molecular subsequences. J. Mol. Biol. **147**, 195–197 (1981)
2. Funmilola, A., Oke, O.A., Adedeji, T.O., Adewusi, E.A.: Fuzzy K-C-means clustering algorithm for medical image segmentation. J. Inf. Eng. Appl. **2** (2012)
3. Berkhin, P.: Chapter "A survey of clustering data mining techniques". In: Grouping Multidimensional Data, pp. 25–71. Springer (2006)
4. Xu, R., Wunsch, D.: Survey of clustering algorithms. IEEE Trans. Neural Netw. (2005)
5. Easwaramoorthy, D., Uthayakumar, R.: Estimating the complexity of biomedical signals by multifractal analysis. IEEE Conference (2010)
6. Gath, I., Geva, A.B.: Unsupervised optimal fuzzy clustering. IEEE Trans. Pattern Anal. Mach. Intell. (2002)
7. Faraoun, K.M., Boukelif, A.: Neural networks learning improvement using k-means clustering algorithm to detect network intrusions. Int. J. Comput. Intell. (2007)
8. Hegde, N.N., Nagananda, M.S., Harsha, M.: EEG signal classification using K-means and FCM clustering methods. IJSTE **2** (2015)
9. Hekim, M., Orhan, U.: Subtractive approach to fuzzy c-means clustering method. J. ITU-D, **10**(1)

Automatic Geospatial Objects Classification from Satellite Images

Shariar Rahaman, Md. Abdul Alim Sheikh, Alok Kole, Tanmoy Maity and Chinmaya Kumar Pradhan

Abstract LiDAR data has several advantages for classification of objects from satellite images. LiDAR data acquisition occurs in 24 h which contains height information of the objects. The morphological are used for extracting image features. As urban object detection is more difficult for shadow, bushes shrubs mixed with huts. This method gives an automatic approach for classification of the object from satellite images. It also presents an automatic approach for extraction of roads, vegetation with higher indexed and lowers indexed from the point clouds of LiDAR data. In the first step point Clouds from LiDAR data are preprocessed and then digital elevation model (DEM) are generated from that particular location. Then we have Created AOI using the normalized difference between DEM and DTM. Finally, the pixels of different objects are classified using spatial model. The experimental results are very promising. To identify terrain and non-terrain points from the raw LiDAR data an automated filtering algorithm is developed with the classification.

Keywords LiDAR · DEMS · Point clouds · Spatial model

S. Rahaman (✉) · C. K. Pradhan (✉)
Department of Electronics and Communication Engineering, Dream Institute of Technology, Kolkata, India
e-mail: shaiar.rahaman13@gmail.com

C. K. Pradhan
e-mail: chinmayapradhan07@gmail.com

Md. Abdul Alim Sheikh (✉)
Department of Electronics and Communication Engineering, Aliah University, Kolkata, India
e-mail: alim.sheikh16@gmail.com

A. Kole
Department of Electrical Engineering, RCC Institute of Information Technology, Kolkata, India
e-mail: alokkole123@yahoo.co.in

T. Maity
Department of Mining Machinery Engineering, Indian Institute of Technology (ISM) Dhanbad, Dhanbad, India
e-mail: tanmoy_maity@yahoo.co.in

A. Abraham et al. (eds.), *Emerging Technologies in Data Mining and Information Security*, Advances in Intelligent Systems and Computing 813,
https://doi.org/10.1007/978-981-13-1498-8_10

1 Introduction

The Automatic Geospatial model for Classification process includes the detection and restoration of bad lines in the satellite images, image registration, geometric rectification, atmospheric correction, radiometric calibration and, and topographic correction. Remote Sensing is becoming more and more important information and data source, like for GIS and for many applications. The quality evaluation and conversion of data among different formats are also necessary before they can be incorporated into a classification procedure. Practitioners and Scientists have made a new developed techniques for advanced classification and improved classification approaches for risk assessment early warning systems before the event, and the disaster-alerting systems moment the event occurs and impact assessment after the event. The idea behind the classification is a sample pixel in an image that represents all the specific classes which can be selected all other pixels in the image. The object-oriented classification of a segmented image is made different from performing a per-pixel classification. In the first analysis the spectral information is not constrained by using pixel information. He or she may choose different shape measures which associated with each object in image to measure the spectral information in conjunction with various in the mean dataset. This introduces flexibility and robustness. The classification algorithms is very efficient for analysis different spectral and spatial attributes of each objects pixels that can be input to a many different method like maximum likelihood, nearest-neighbor, etc.

2 Literature Review

Here we discussed about the different general approach and methods that is use for remote sensing images for extraction of man-made object. We introduce the proposed method for object classification and their applications in remote sensing.

A. General methods for different object Extraction from SAR images.

(1) Geometric Shapes of homogeneous regions [1] with marked contrast of roads and buildings to detect pixel-level differences are often composed of bilinear boundaries and nearby objects of boundaries in remote sensing images [2]. It is a set of efficient and robust method to develop, evaluate a building extraction system. This method employing LiDAR point clouds from airborne because LiDAR datasets are available for large-scale analysis purposes [3]. Thus the most commonly used methods for roads, vegetation, building classification lines detection [4]. In recent year spatial, spectral and intensity differences between the two images in the image segmentation phase. The result is to creation of intensity images recognize as segments or patches in the landscape objects defined as individual areas with shape and spectral homogeneity. It is utilized in many context for man-made object extraction, correlation analysis over the entire scene [5]. In a different context, by converting cloud points [6] into line

segment which employed for road and building extraction with a minimum loss of data.

(2) Morphological Filtering: A morphological tool for investigating geometric structure in binary and grayscale images by means of opening, erosion, closing dilation, where a region shrinks by opening and for closing expands the region [4–9]. Based on designing imaging geometry operator which satisfies desirable properties. The erosion used to calculate a degree of change of the minimum value of h-f in a 'f' neighborhood of 'f' defined by the domain of h function. It distinguishes meaningful adjacent change pixels from irrelevant one [4, 7, 10, 11], the vast majority of shape processing. Pomerleau (1998) is the pioneer of using height difference and iteratively window size for increasing the structure element has a gray scale distribution. The neighborhood can be defined either city-scale modeling along LiDAR profiles (Shan and Sampath 2008) [12, 13], through neighboring points within a certain distance.

(3) Progressive Densification Filtering: Triangulated irregular networks are more complex than rasters, however they are more efficient space wise. A Triangulated Irregular Network (TIN) is a digital data structure [14] is updated by adding the new terrain points used in a geographic information system (GIS) for polynomial surface [15]. The preliminary terrain surface updated by adding the lowest point is obtained as the Triangular Irregular Network of low point sets. Therefore flexible and fewer points need to be stored color-coded according to class with regularly distributed points [16]. It is representing the continuous spatial phenomenon known as elevation.

(4) Interpolation Filtering: A system of filtering incorporating multirate filters which process data at various rates like upsampling, downsampling [17]. The Haar filter based on the contextual information of a building using point-wise attributes and suitable on data filtering have been made to study more complex images. Sub-band coding for speech processing in vocoders [18, 19]. At each level the surface was iteratively interpolated towards accessing, interpreting, and analyzing multispectral satellite images. The sample rate is reduced by Decimation of a signal. It eliminates unnecessary information and compacts the data, allowing more information to be stored, processed, or transmitted to predict cell values in the data [20]. The sample rate of a signal increases by Interpolation.

3 Proposed Methodology

Many sensors can work in the mode of discrete returns (typically first, second, third, and last) or full waveform technology. In a three-dimensional point cloud system, these data points are usually intended to represent the external surface of an object. A large man-made homogenous structure requires adaptive window sizes to identify ground regions. Due to discrete and irregular distribution the intrinsic characteristics of point clouds and then a large-scale scene the LiDAR data very large data size in usually provide comprehensive results. The classification task (Fig. 1) can be

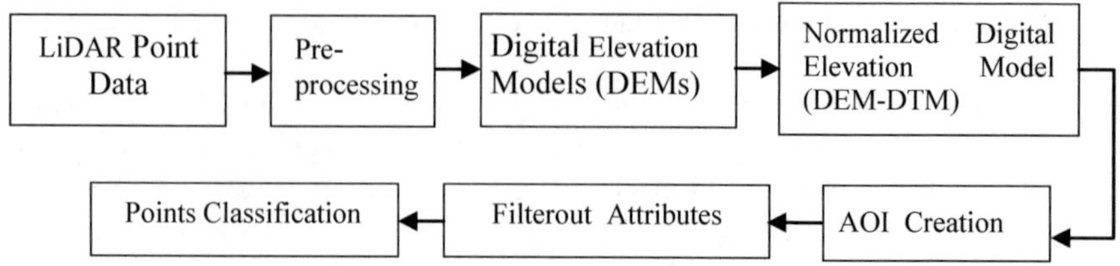

Fig. 1 The proposed approach for classification

strengthened by first aggregating points and then analyzing segment attributes applied to high-resolution images of point clouds in satellite images (Fig. 2).

Normally look at various types of land cover, bare earth, urban, forest, brush, high grasses. By filter out non-ground points like roads, grass and bridge we will get a smooth Digital Elevation Model (DEM). Statistical value is the average of the squares of the differences between known points and modeled points then square root of it. We will obtain a DEM by voiding, the vegetation points, roof top man-made objects, vegetation from DSM data. A useful model in the fields of study such as soils land use planning, safety, and hydrology landuse planning/safety. Modern LiDAR systems data used to find identifiable objects in the intensity images paint stripes, concreteasphalt edges, etc.

Determine precise 3D location with field techniques then determine elevation from the LiDAR surface at that XY, subtract, and statistically summarize. Mean of likelihood pixels given by

$$\mu\, ei = \frac{1}{n}\sum_{j=1}^{n} X_{ij} \ (i = 1,\ 2\ldots m) \tag{1}$$

Maximum likelihood estimation (MLE) could discover multiple parameters that maximize the likelihood function, it could discover that there is no maximum, or it could even discover that there is no closed form to the maximum.

Variance–covariance matrix

$$\sum\nolimits_{e} = \frac{1}{m}\sum_{i=1}^{m} (X_i - \mu_e)(X_i - \mu_e) \tag{2}$$

In Eq. (2)

m : number of bands
n : number of pixels

It should be checked before adopting the maximum likelihood classification to determine if the distribution of training data will fit the normal distribution or not (Figs. 2, 3, 4, 5, 6, 7) (Tables 1, 2, 3 and 4).

Fig. 2 LiDAR data from satellite images

Fig. 3 **a** Panchromatic images **b** Digital elevation model

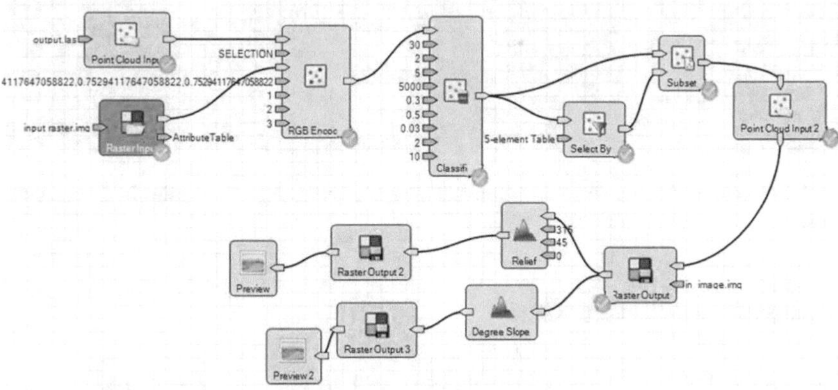

Fig. 4 Spatial model for classification

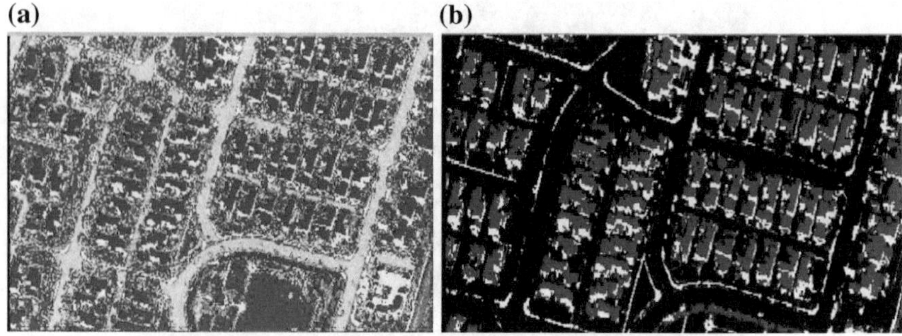

Fig. 5 **a** Shows combined image **b** Shows residential part with some noise

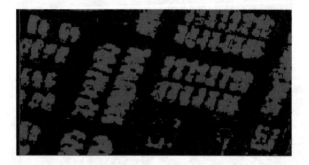

Fig. 6 Shows clear view of resident after morphological operation

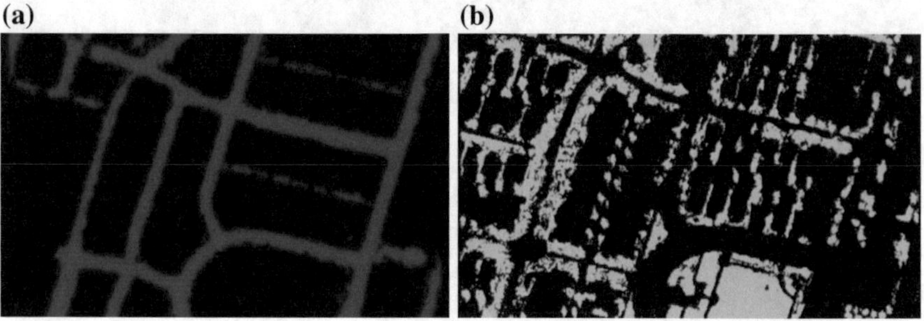

Fig. 7 **a** Shows road networks in the combined image. **b** Shows vegetation index in the combined image

4 Results and Discussion

A PC of CPU Intel (R) Core(TM) i3-4590 at 3.30 GHz and 6 GB RAM is used to perform the experiments. The method is realized in the ERDAS Imagine and MATLAB program. Accuracy assessment is done due to evaluate all images are normalized to thematic raster layer. The method has correctly classified all data with

Table 1 Gives point information abouts pixels

Point information	
Number of points	126319
Unknown return	0
Point format	1
Return 1	121041
Return 2	5246
Return 3	32

Table 2 Gives statistics information abouts pixels

Statistics information	
Min	343.131
Max	406.566
Mean	373.093
Median	372.371
Mode	373.61
Std. Dev.	12.774
File signature	LASF

Table 3 Gives the error matrix abouts pixels of everyobjects

Reference data				
Classified data	Unclassified data	Main road	Sub-road	Higher indexed
Unclassified data	0	0	0	0
Main road	0	5	0	0
Sub-road	0	0	15	0
HIG higher indexed	0	0	0	3
Lower indexed	0	0	0	0
Resident	0	0	0	0
Resident2	0	0	0	0
Wall	0	0	0	0
Column total	0	5	15	3

a accuracy of 97% as shown in Table 5. Total accuracy calculated as dividing the total number of correctly classified pixels by the total number of reference pixels. Error matrices compare the relationship between known ground truth reference data with the corresponding pixels in the lidar data of the classification results procedure. The Error Matrices are given in Tables 3 and 4.

Accuracy Assessment is an organized way of comparing classified objects with reference data, aerial photos, or other data. The data for Lidar points in the CellArray can be saved for further reference or to refine past evaluations.

Table 4 Gives the error matrix abouts pixels of everyobjects

Reference data table from RAW data				
Classified objects	Lower indexed	Resident	Resident2	Wall
Unclassified	0	0	0	0
Mainroad	0	0	0	0
Sub-road	0	0	0	0
Higher indexed	0	0	0	0
Lowerer indexed	8	0	0	0
Resident	0	26	0	0
Resident2	0	0	40	0
Wall	0	0	0	3
Column total	8	26	40	3

Table 5 Accuracy totals of classified objects overall classification accuracy 100%

Class type	Reference totals value	Classified totals value	Number correctly detected	Producer's accuracy (%)	User's accuracy (%)
Unclassified	0	0	0	–	–
Mainroad	5	5	5	100.00	100.00
Sub-road	15	15	15	100.00	100.00
Higher indexed	3	3	3	100.00	100.00
Lower indexed	8	8	8	100.00	100.00
Resident	26	26	26	100.00	100.00
Resident2	40	40	40	100.00	100.00
Wall	3	3	3	100.00	100.00
Totals	100	100	100		

The error matrix compares assigned pixels class value to the reference pixels class values in a matrix. The accuracy totals report calculates percentages of accuracy based upon the statistics results of the error matrix.

Kappa Statistics (K^) Given By

$$k = \frac{N \sum_{i=1}^{n} Xii \sum_{i=1}^{n} (Xi * X_{+i})}{N^2 \sum_{j=1}^{n} (X_{i_n} * X_{+j})}, \tag{3}$$

where Xin = the marginal totals of row i and column I, number of rows in the matrix is given by N, Xi = Total number of observations in the row j and column i,

N = total number of observations

Conditional Kappa for each Category.

Class Type	Kappa value
Unclassified	0.00
Mainroad	1.00
Sub-road	1.00
Higher indexed	1.00
1 Lowerer indexed	1.00
Resident	1.00
Resident2	1.00
Wall	1.00

Overall Kappa Statistics = 1.0000

5 Conclusion

The classification map is also called thematic map. From the observation, we can say unsupervised classification was noisier than that of supervised classification was not smooth. From the research observation we can say classification will be quite difficult if the same region or object contains different color. This would be good to use if all of the classes within the image are not known to us prior. It was hard to classify trees and some of the hardwood areas. A compact analysis of input RAW datasets from satellite image. Owing to the vast amount of satellite imagery data available with 155 increasing resolution. This method erroneously classifying a homogeneous pixel as a non-homogeneous pixel. The upper right corner will be changed so that the extents of the orthophoto are a full multiple of the new resolution. Also, there is a continuous change in the infrastructure with the development of different villages, towns, cities, states countries, etc. However, erdas classification using spatial model is very Versatile, Seamless, Complete, Flexible. This method have successfully perform objects extraction from LiDAR data.

References

1. Geman, S., Geman, D.: Stochastic relaxation, Gibbs distributions, and the Bayesian restoration of images. IEEE Trans. Pattern Anal. Mach. Intell. **6**, 721–741 (1984)
2. Im, J., Jensen, J.R., Tullis, J.A.: Object-based change detection using correlation image analysis and image segmentation. Int. J. Remote Sens. **29**(2), 399–423 (2008)
3. Borrmann, D., Elseberg, J., Lingemann, K., Nüchter, A.: The 3D hough transform for plane detection in point clouds: a review and a new accumulator design. 3D Res. **2**, 1 (2011)
4. Alpert, C., Yao, S.: Spectral partitioning: the more eigenvectors, the better. In: Conference on Design Automation, pp. 195–200 (1995)
5. Arikan, M., Schwärzler, M., Flöry, S.: O-snap: optimization-based snapping for modeling architecture. ACM Trans. Graph. **32**, 1–15 (2013)
6. Axelsson, P.: Processing of laser scanner data-algorithms and applications. ISPRS J. Photogramm. Remote Sens. **54**, 138–147 (1999)

7. Alharthy, A., Bethel, J.: Heuristic filtering and 3D feature extraction from LiDAR data. Int. Arch. Photogramm. Remote Sens. Spat. Inf. Sci. **34**, 23–28 (2002)
8. Arefi, H., Reinartz, P.: Building reconstruction using DSM and orthorectified images. Remote Sens. **5**, 1681–1703 (2013)
9. Axelsson, P.: DEM generation from laser scanner data using adaptive TIN models. Int. Arch. Photogramm. Remote Sens. Spat. Inf. Sci. **33**, 110–117 (2000)
10. Nevatia, R., Ramesh, B.: Linear feature extraction and description. Comput. Vis. Graph. Image Process. **14**, 257–269 (1980)
11. Abdullah, A., Rahman, A., Vojinovic, Z.: LiDAR filtering algorithms for urban flood application: review on current algorithms and filters test. Int. Arch. Photogramm. Remote Sens. Spat. Inf. Sci. **38**, 30–36 (2009)
12. Ortner, M., Descombes, X., Zerubia, J.: A marked point process of rectangles and segments for automatic analysis of digital elevation models. IEEE Trans. Pattern Anal. Mach. Intell. **30**(1), 105–119 (2008)
13. Gueguen, L., Soille, P., Pesaresi, M.: Change detection based on information measure. IEEE Trans. Geosci. Remote Sens. **49**(11), 4503–4515 (2011)
14. Duda, R.0., Hart, P.E.: Use of the Hough transformation to detect lines and curves in pictures. Comm. ACM, 15, 1972, pp. 11-15Nima Ekhtari, M.R. Sahebi, M.J. Valadan Zoej. Int. Arch. Photogramm. Remote Sens. Spat. Inf. Sci. **XXXVII**. Part B3
15. Baltsavias, E.P.: A comparison between photogrammetry and laser scanning. ISPRS J. Photogramm. Remote Sens. **54**, 83–94 (1999)
16. Boykov, Y., Jolly, M.: Interactive graph cuts for optimal boundary & region segmentation of objects in N-D images. Int. Conf. Comput. Vis. **1**, 105–112 (2001)
17. Davidson, J.: Stereo photogrammetry in geotechnical engineering research. Photogramm. Eng. Remote Sens. Eng. **51**, 1589–1596 (1985)
18. Peng, J., Zhang, D., Liu, Y.: An improved snake model for building detection from urban aerial images. Pattern Recogn. Lett. **26**(5), 587–595 (2005)
19. Kumar, S., Hebert, M.: Detection in natural images using a causal multiscale random field. Proc. IEEE Int. Conf. Comput. Vis. Pattern Recognit. **1**, 119–126 (2003)
20. d'Angelo, P., Lehner, M., Krauss, T.: Towards automated DEM generation from high resolution stereo satellite images. In: Proceedings of ISPRS Congress, Beijing, China, vol. 37, pp. 1137–1342, Part B4 (2008)
21. Awrangjeb, M., Ravanbakhsh, M., Fraser, C.S.: Automatic detection of residential buildings using lidar data and multispectral imagery. ISPRS J. Photogramm. Remote Sens. **65**(5), 457–467 (2010)
22. Landa, J., Procházka, D., Šťastný, J., 201. Point cloud processing for smart s stems. Acta Universitatis Agriculturae Et Silviculturae Mendelianae Brunensis 61, 2415–2421
23. Sampath, A., Shan, J.: Building boundary tracing and regularization from airborne LiDAR point clouds. Photogramm. Eng. Remote Sens. **73**, 805–812 (2007)
24. Fischler, M.A., Bolles, R.C.: Random sample consensus: a paradigm for model fitting with applications to image analysis and automated cartography. Commun. ACM **24**, 381–395 (1981)
25. Zhang, Z., Faugeras, O.: Finding clusters and planes from 3D line segments with application to 3D motion determination. In: European Conference on Computer Vision (ECCV), pp. 227–236 (1992)
26. Haala, N., Kada, M.: An update on automatic 3D building reconstruction. ISPRS J. Photogramm. Remote Sens. **65**, 570–580 (2010)
27. Huang, X., Zhang, L., Li, P.: Classification and extraction of spatial features in urban areas using high resolution multispectral imagery. IEEE Geosci. Remote Sens. Lett. **4**(2), 260–264 (2007)
28. Boykov, Y., Funka-Lea, G.: Graph cuts and efficient N-D image segmentation. Int. J. Comput. Vis. **70**, 109–131 (2006)
29. Jin, X., Davis, C.H.: Automated building extraction from high-resolution satellite imagery in urban areas using structural, contextual, and spectral information. EURASIP J. Appl. Signal Process. **2005**, 2198–2206 (2005)

30. Ren, Z., Zhou, G., Cen, M., Zhang, T., Zhang, Q.: A novel method for extracting building from lidar data-c-s method. Int. Arch. Photogramm. Remote Sens. Spat. Inf. Sci. **XXXVII**. Part B1
31. Blaschke, T.D., Evans, S., Haklay, M.: Visualizing the city: communicating urban design to planners and decision-makers. Planning support systems: integrating geographic information systems, models, and visualization tools, pp. 405–443 (2001)
32. Bernardini, F., Bajaj, C.: Sampling and reconstructing manifolds using alpha-shapes. In: 9th Canadian Conference on Computational Geometry, pp. 193–198 (1997)
33. Comaniciu, D., Meer, P.: Mean shift: a robust approach toward feature space analysis. Pattern Anal. Mach. Intell. **24**, 603–619 (2002)
34. Cook, W.J., Cunningham, W.H., Pulleyblank, W.R., Schrijver, A.: Combinatorial Optimization, 1st edn. Wiley, New York, NY, USA (1998)
35. Dorninger, P., Pfeifer, N.: A comprehensive automated 3D approach for building extraction, reconstruction, and regularization from airborne laser scanning point clouds. Sensors **8**, 7323–7343 (2008)
36. Song, Z., Pan, C., Yang, Q.: A region-based approach to building detection in densely build-up high resolution satellite image. In: Proceedings International Conference on Image Processing, pp. 3225–3228 (2006)
37. Karantzalos, K., Paragios, N.: Recognition-driven two-dimensional competing priors toward automatic and accurate building detection. IEEE Trans. Geosci. Remote Sens. **47**(1), 133–144 (2009)
38. Tsai, V.: A comparative study on shadow compensation of color aerial images in invariant color models. IEEE Trans. Geosci. Remote Sens. **44**(6), 1661–1671 (2006)
39. Hirschmüller, H.: Stereo processing by semiglobal matching and mutual information. IEEE Trans. Pattern Anal. Mach. Intell. **30**(2), 1–14 (2008)

RAiTA: Recommending Accepted Answer Using Textual Metadata

Md. Mofijul Islam, Sk. Shariful Islam Arafat, Md. Shakil Hossain, Md. Mahmudur Rahman, Md. Mahmudul Hasan, S. M. Al-Hossain Imam, Swakkhar Shatabda and Tamanna Islam Juthi

Abstract With the increasing software developer community, questions answering (QA) sites, such as StackOverflow, have been gaining its popularity. Hence, in recent years, millions of questions and answers are posted in StackOverflow. As a result, it takes an enormous amount of effort to find out the suitable answer to a question. Luckily, StackOverflow allows their community members to label an answer as an accepted answer. However, in the most of the questions, answers are not marked as accepted answers. Therefore, there is a need to build a recommender system which can accurately suggest the most suitable answers to the questions. Contrary to the existing systems, in this work, we have utilized the textual features of the answers' comments with the other metadata of the answers to building the

Md. Mofijul Islam (✉) · Md. Mahmudur Rahman
Department of Computer Science and Engineering, University of Dhaka, Ramna, Dhaka 1000, Bangladesh
e-mail: akash.cse.du@gmail.com

Md. Mahmudur Rahman
e-mail: mahmudur@cse.du.ac.bd

Sk. Shariful Islam Arafat · Md. Shakil Hossain · Md. Mahmudul Hasan
S. M. Al-Hossain Imam · S. Shatabda
Department of Computer Science and Engineering, United International University, Madani Avenue, Satarkul, Badda, Dhaka 1212, Bangladesh
e-mail: sharifultroublesome@gmail.com

Md. Shakil Hossain
e-mail: shakil.shaion@gmail.com

Md. Mahmudul Hasan
e-mail: ablaze.dip@gmail.com

S. M. Al-Hossain Imam
e-mail: hossainimam313@gmail.com

S. Shatabda
e-mail: swakkhar@cse.uiu.ac.bd

T. Islam Juthi
Department of Computer Science and Engineering, University of Asia Pacific, Green Road, Farmgate, Dhaka 1215, Bangladesh
e-mail: juthi.islam09@gmail.com

A. Abraham et al. (eds.), *Emerging Technologies in Data Mining and Information Security*, Advances in Intelligent Systems and Computing 813,
https://doi.org/10.1007/978-981-13-1498-8_11

recommender system for predicting the accepted answer. In the experimentation, our system has achieved 89.7% accuracy to predict the accepted answer by utilizing the textual metadata as a feature. We have also deployed our recommendation system web application, which is publicly accessible: http://210.4.73.237:8888/. We also deployed our system as Facebook Messenger Bot Application, which is accessible at https://www.facebook.com/RAiTABOT/, for helping the developers to easily find the suitable answer for a StackOverflow question.

Keywords Recommender system · Machine learning
Natural language processing · Crowdsourced knowledge

1 Introduction

Human life is full of questions. In addition, the Internet makes the world as a village to ease the communication and sharing of knowledge with the people. As a consequence, question answering (QA) sites are growing their popularity rapidly. A number of question answering sites, such as StackOverflow, Quora, and Yahoo Answer, have already established their reputation up to a limit that people are engaging more and more on those sites to get help with their questions' answer. In these sites, a person can ask questions or answer of any existing questions according. Moreover, they can also redirect the questions to the expert users. As a result, a question may have different answers from different persons.

With the increasing participation of people in those sites, a number of questions and answers are increasing very quickly. So, the searching of the most suitable answers is becoming strenuous day by day. At this point, it becomes annoying for a new user who needs urgent answer. Fortunately, some sites allow their community to mark the most suitable answer to a question. For example, till 2017, almost 15 million questions have asked and 23 million answers have given in StackOverflow; however, only 7 million questions' answers are marked as accepted answers by the StackOverflow community [1]. Hence, almost 50% questions' answers are not marked. Moreover, some questions', where the answers are marked as accepted, can get new suitable answers than the accepted answers which are not marked as accepted answer. Therefore, QA sites need a recommender system which can rank the answers to questions which help users to get the most suitable answers.

The main challenge to design an accepted answers recommender system is to select the appropriate set of features which enable us to accurately select the accepted answers. QA sites, such as StackOverflow, Quora, and Yahoo Answer, are providing some information (metadata) along with the answers. With the help of metadata associated with the questions and the answer, we can predict the accepted answer for a particular question. In this work, we have focused on the textual metadata of the answers with other factors provided by a question answering site. As there are a number of question answering sites, it is a tough decision to take one for our work. In this situation, we have taken StackOverflow as it is the largest and most trusted online community for developers throughout the recent years. Moreover, in StackOverflow, a lot of questions are not labeled as accepted answers. To overcome this situation, we have designed a recommender system to sort out the answers.

The major contributions of this work are summarized below:

- The novelty of this work is to utilize the textual metadata (sentiment) with the other factor of the answers to recommend the accepted answer.
- Rigorous experimentations have been performed to identify the most suitable feature groups which help to sort out the questions' answers.
- A web application and Facebook Messenger Bot application are developed so that user can use this recommendation system to identify the suitable answers.

The rest of the work is organized as follows: Sect. 2 shows the existing related works. In Sect. 3, we present the proposed accepted answer recommendation system methodologies. Performance evaluation of the proposed recommendation system as well as the deployed web application has been presented in Sect. 4. Finally, we conclude our work and present the future steps in Sect. 6.

2 Related Work

Predicting accepted answer for online question answering sites is not new in the town. There are works on it. A work to predict the score of an answer is presented in [2]. In this paper, the authors have examined different features of the content and the formatting of the questions. Mainly, the features contain the number of lines, links, tags, and paragraphs with some other factors on the polarity and the subjectivity of the questions. In addition, they have considered the number of answers, favorites, views, comments, and time to the accepted answer. Moreover, [3] presents a work to predict the quality characteristics for academic question answering sites using ResearchGate as a case study. In this work, the authors have focused on the peer judgments like votes. They have developed prediction model testing with naive Bayes, support vector machine (SVM), and multiple regression models. In [4], a work shows to predict accepted answer in massive open online course that is done. In this work, they present a machine learning model using historical forum data to predict the accepted answer. They have designed their model considering three types of features: *User features*—name and details (number of posted question, number of answers given, and number of accepted answers) of questioners and responders; *Thread features*—answers, comments, and timing metadata; and *Content features*—number of words of the question and answer, similarity measure (Jaccard similarity and KL Divergence) between question and answers. In [5], a work to find the best answer is described with the help of four categories of features—linguistic, vocabulary, meta, and thread. In this work, normalized log-likelihood and Flesch–Kincaid Grade are used. They have used age and rating score as the meta-features. Moreover, Blooma et al. [6] present to find the best answer in Community-driven Question Answering (CQA) services. They have used three different categories of features to design their model: *Social features*—asker/answerer authority and user endorsement; *Textual features*—answer length, question and answer length ratio, number of unique words, number of non-stop word overlap, and number of high-frequency words; and *Content-appraisal features*—accuracy, completeness, presentation, and reasonableness. In addition, Tian et al. [7] present another work to find the best answer in CQA services using the features like *answer context*, *question–answer relationship*, and *answer content*. In addition, [8] has presented a recent work to find the best answer in CQA. Moreover, [9] has discussed the prediction of closed questions based on the accepted answer findings using reputation.

A number of works have been done to predict the correct, accepted, and best answer in Question Answering (QA) services. But most of the works are based on questioners, answerers, contents, number of answers, and the similarity of answers with questions. In our knowledge, there is a lack of work on textual metadata, like sentiments of the answers' comments, with the other features did not explore for sorting the answer. So, we have designed a recommendation system based on answer features as well as textual metadata to find the suitable answer. Moreover, we performed an extensive experimentation to find out the best features group set which can help to improve the recommender systems' performance.

3 Prediction Model for Accepted Answer

Community Question Answering (CQA) service provides facilities to ask questions and get answers in online communities. StackOverflow (StO) is the largest and the most trusted online CQA service nowadays. A large number of people in the world have trust in this service. As a large community, it has a large number of users as well as questions. Around 62.8% of the questions in StO have marked the accepted answer. As a result, about 37.2% questions are remaining without any accepted answer marking. So, we have designed a prediction model (RAiTA) to predict the accepted answers.

3.1 Feature Selection

StackOverflow (StO) provides the comments of all the questions and answers with a number of attributes. The attributes contain question details with id, score, view count, comment count, etc. It also contains answer-related information as count, score, creation date, comment count, comment score, etc. From all the information, we have used two categories of features to design our prediction model.

1. *Thread features*—it contains question title, tags, details, number of responder with response details, question score with favorite count, answer rating, and number of commenter with comment score. It also contains view counts as well

as timing metadata like question creation time, response time, answer creation time, etc.

2. *Textual Metadata*—this is a finding from the textual data. It represents the sentiment of the responder as well as commenter whether they are positive, negative, or neutral with their comments and responses.

3.2 Feature Finalization

To design the prediction model, we have tested Random Forest (RF), Logistic Regression (LR), AdaBoost (AdaB), and Support Vector Machine (SVM). On the other hand, Recursive Feature Elimination (RFE) is used to finalize the features from our feature set. After a extensive experimentation, which is presented in Sect. 4, we have selected 23 attributes from the StO provided attributes' list to examine our work. Feature finalization is completed with the following steps.

- *Thread Features Extraction*—all the important attributes related to the thread according to our research are filtered out and added into the feature set to be tested.
- *Textual Metadata Analysis*—after analyzing the text using nltk library [10] for each of the comments, we get the sentiment results. The library analyzes the comments and returns us a sentiment of the text as positive, negative, or neutral. Then, it is used as a feature.
- *Feature Elimination*—in this step, we have used scikit-learn library to select the features according to the recursive feature elimination method. After the elimination, the feature set is updated.

With the finalized feature set, we have tested our prediction model. After that, the best result along with others has been shown in Sect. 4.

3.3 Different Measures for Model Evaluation

In our work, we have used different measures to evaluate the performances of the prediction model. Definitions of the measures are given here with respect to Positivity (P) and Negativity (N). Both the characteristics are divided into two parts—False (False Positive, FP and False Negative, FN) and True (True Positive, TP and True Negative, TN).

- *Accuracy (Acc)*—accuracy is the closeness of a measured value to a standard value. It is generally measured in percentages. Accuracy can be measured as

$$Acc = \frac{TP + TN}{P + N}$$

.

- *Sensitivity* (*Sen*)—sensitivity refers to the true positive rate. It can be calculated as

$$Sen = \frac{TP}{P}$$

- *Specificity* (*Spec*)—specificity refers to the true negative rate. It is defined as

$$Spec = \frac{TN}{N}$$

- *Area Under the Curve* (*AUC*)—Receiver operating characteristics are important to visualize the accuracy of a model. This is an FP–TP curve. The area under the curve is defined as AUC value of a model.
- *Matthews Correlation Coefficient* (*MCC*)—Matthews correlation coefficient is important to measure the quality of our binary classifier. It can be calculated directly using positive and negative values of the model. It is defined as

$$MCC = \frac{TP/\tau - \sigma \times \rho}{2}$$

where

$$\tau = TP + TN + FP + FN$$

$$\sigma = \frac{TP + FN}{\tau}$$

$$\rho = \frac{TP + FP}{\tau}$$

4 Performance Evaluations and Discussion

Experimenting with a number of different classifiers, our evaluation shows that we have better results. We have evaluated the work using different measures described in Sect. 3.3. The datasets and all the results are given below.

4.1 Dataset Description

The dataset contains all the comments as well as the metadata of the questions of StackOverflow. We extracted these questions' data from the StackOverflow Data Exchange system [1]. The information of the experiments dataset is presented in Table 1.

Table 1 Dataset details

No of instances	49992
No of unique questions	20449
No of unique answers	23191
No of accepted answers	12843
Unique questions without accepted answers	7606
Attributes	23

Table 2 Results of the sentimental analysis

	Accuracy	Sensitivity	Specificity	AUC	MC
Feature and sentiment	**0.897**	**0.848**	0.923	**0.886**	**0.774**
Only features	0.888	0.835	0.918	0.876	0.755
Only sentiment	0.65	0.024	**0.994**	0.509	0.079

*All the simulation is done under tenfold validation

4.2 *Effectiveness of Sentiments in Accepted Answer*

The textual metadata as sentiment is very effective in our work. We have tested the effect of sentiments a number of times in our prediction model. The summary of the experiment without feature optimization is shown in Table 2.

Analyzing the results in Table 2, we observed that when we incorporate the sentiment of answers' comment, our model works better with 89.7% accuracy, AUC and MCC values, 0.886 and 0.774, respectively. Moreover, we find these significant results without any optimization of feature set.

4.3 *Classifier and Feature Selection and Evaluation*

In this work, Random Forest (RF), Logistic Regression (LR), AdaBoost (AdB), and Support Vector Machine (SVM) are tested to find the best suited classifier as well as feature set. The experimental result is shown in Table 3.

From the result in Table 3, random forest gives the best result of 88.45% accuracy with a set of 10 features. So, we have selected random forest classifier for our prediction model. And the list of selected attributes are AnswerCommentCount, AnswerCount, AnswerScore, CommentScore, Neutral, Positive, QuestionCommentCount, QuestionFavoriteCount, QuestionScore, and QuestionViewCount.

Table 3 Results of different classifiers with different feature sets

Classifier		Random forest	Logistic regression	Ada boost	SVM
Accuracy (%)	No of features 11	88.17	66.03	69.69	83.77
	No of features 10	**88.45**	66.12	69.68	82.50
	No of features 9	88.09	66.11	69.60	81.31

*All the simulation is done under tenfold validation

Moreover, the final prediction model is tested with different classifiers and evaluated the quality and effectiveness according to various measures. The evaluation result is shown in Table 4.

After analyzing Table 4, we see that random forest classifier achieves the best accuracy, AUC and MCC value as 89.7%, 0.886 and 0.774, respectively. Moreover, the other classifiers achieve significant accuracy too. The achievement in AUC value is visually represented in Fig. 1.

Table 4 Evaluation results for different classifiers

Classifier	Accuracy	Sensitivity	Specificity	AUC	MCC
SVM (sigmoid)	0.651	0.048	0.981	0.514	0.03
SVM (rbf)	0.648	0.005	**1**	0.503	0.057
Logistic regression	0.652	0.051	0.981	0.516	0.09
Random forest	**0.897**	0.848	0.923	**0.886**	**0.774**
AdaBoost	0.692	0.436	0.832	0.634	0.291
Decision tree	0.890	0.851	0.911	0.881	0.76
Gaussian naive bayes	0.408	**0.949**	0.111	0.530	0.1

*All the simulation is done under tenfold validation

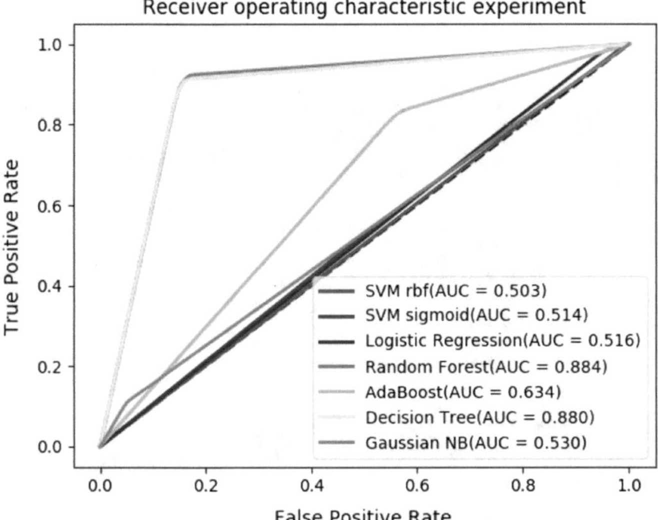

Fig. 1 ROC curves for different classifiers

5 Application Implementation

A web application for the proposed accepted answer recommendation system, RAiTA, is deployed and publicly accessible at http://210.4.73.237:8888/. We also deployed our system as Facebook Messenger Bot Application, which is accessible at https://www.facebook.com/RAiTABOT/.

5.1 Application System Architecture

We developed the RAiTA recommendation system as a three-layer modular architecture, which is depicted in Fig. 2, so that we can easily modify and enhance the system performance. This three-layer architecture enables us to separate the recommendation engine layer from the data and user layer. As a result, we can easily modify the recommendation engine layer without modifying the other layer.

In the StackOverflow data exchange layer, we utilize the StackOverflow data exchange API to extract the questions and answer data, such as question ID, list of answers, and their comments. After extracting the requested question data, it is fed into the recommendation system engine for generating the suitable answers.

In the recommendation engine layer, when recommendation API manager calls recommendation engine, feature extraction module is called to extract the necessary feature data from the StackOverflow data exchange. After extracting the feature data using question and answer feature extraction module, sentiment analysis of

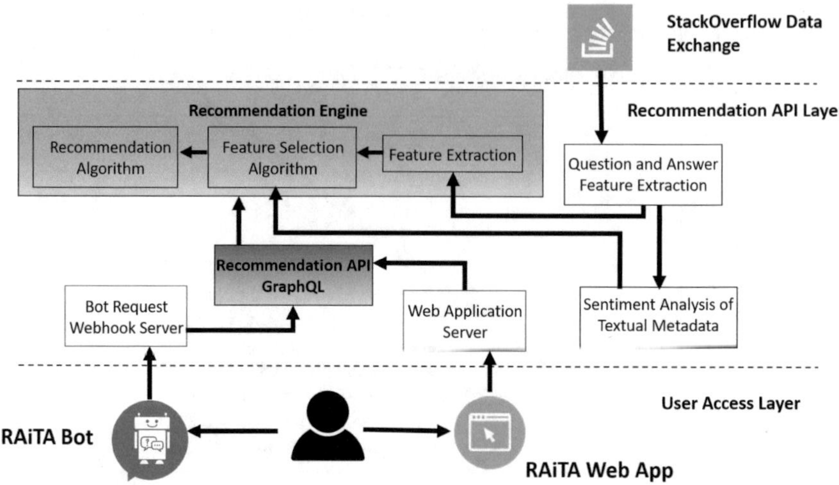

Fig. 2 RAiTA application system architecture

textual metadata is employed to identify the sentiment of the answers comments using NLTK library. Finally, feature extraction module fed the extract and processed sentiment data to the feature selection algorithm to find out the best possible feature from the data. After getting the selected features, recommendation engine employs the pretrained recommendation algorithm for identifying the best possible answer. Then, the list of accepted answer is returned through the GraphQL Recommendation API.

To simplify the user access, we develop two applications, RAiTA Bot and web application. Both the applications access the recommendation engine through their corresponding application server, while the application server only calls the recommendation API module to extract the recommendation engine output. As a result, user access layer is completely separate from the recommendation engine, which enables us to easily modify the recommendation approach without modifying the user application.

In this web app, there are basically two modules involved to predict the accepted answer. First, feature extraction modules extract the related features of the given question from the StackOverflow data exchange system [1]. Second, the recommendation engine uses these extracted features and employs the pretrained model to sort answers to that question. Moreover, in each answer, we showed the likelihood percentage of being an accepted answer. RAiTA's full web application is developed using the Python Django framework.

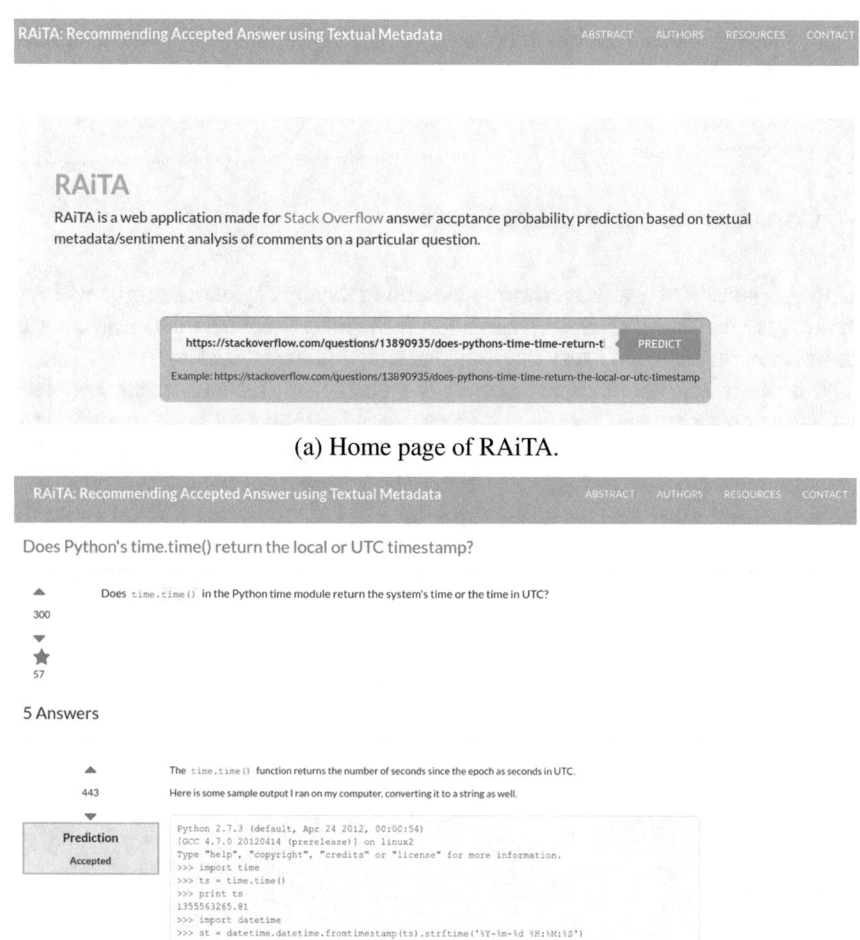

(a) Home page of RAiTA.

(b) Recommended answers page

Fig. 3 Web application of RAiTA

5.1.1 Web Application

The user interface of this recommendation system is very easy to use. Users just need to enter the StackOverflow question link in the search box, which is depicted in Fig. 3. After clicking the predict button, our system will sort the answers to this question based on the appropriateness, which is presented in Fig. 3.

In this web app, there are basically two modules involved to predict the accepted answer. First, feature extraction modules, which extract the related features of the given question from the StackOverflow data exchange system [1]. Second, the recommendation engine uses these extracted features and employs the pretrained model to sort answers to that question. Moreover, in each answer, we showed the likelihood percentage of being an accepted answer. RAiTA's full web application is developed using the Python Django framework.

6 Conclusions and Future Work

As the popularity of StackOverflow is increasing day by day, more people will rely on it to give and to get their answers. So, it is the time to confirm the acceptance of the answers for each and every question using an automated system.

In order to find the accepted answer, we design our recommendation system, RAiTA. It uses answers' metadata information with sentiment analysis to find the accepted answer for each answered question. After a vigorous experiment, the model is proven to be efficient and effective to find the accepted answers. The result shows that it can predict the accepted answer up to 89.7%. The AUC and MCC values for the model are also significant. Moreover, the model has fully functional web application "RAiTA" along with faster response. In addition, the web application "RAiTA" is very handy to use which has a user-friendly interface too.

The success of finding accepted answer in StackOverflow has interested us to extend the system to design an adaptive recommender system for other QA services like Quora, Yahoo, ResearchGate, etc., which can learn and transfer the knowledge. Furthermore, other textual features can be utilized to improve the prediction performance.

References

1. Stackexchange data explorer for stackoverflow. https://data.stackexchange.com/stackoverflow/. Accessed 28 Sept 2019
2. Alharthi, H., Outioua, D., Baysal, O.: Predicting questions' scores on stack overflow. In: Proceedings of the 3rd International Workshop on CrowdSourcing in Software Engineering, CSI-SE@ICSE 2016, Austin, Texas, USA, pp. 1–7, 16 May 2016
3. Li, L., He, D., Jeng, W., Goodwin, S., Zhang, C.: Answer quality characteristics and prediction on an academic q&a site: a case study on researchgate. In: Proceedings of the 24th International Conference on World Wide Web Companion, Companion Volume, WWW 2015, Florence, Italy, pp. 1453–1458, 18–22 May 2015
4. Jenders, M., Krestel, R., Naumann, F.: Which answer is best?: predicting accepted answers in MOOC forums. In: Proceedings of the 25th International Conference on World Wide Web, WWW 2016, Companion Volume, Montreal, Canada, pp. 679–684, 11–15 April 2016
5. Calefato, F., Lanubile, F., Novielli, N.: Moving to stackoverflow: best-answer prediction in legacy developer forums. In: Proceedings of the 10th ACM/IEEE International Symposium on Empirical Software Engineering and Measurement, ESEM 2016, vol. 13, pp. 1–10, Ciudad Real, Spain, 8–9 Sept 2016
6. Blooma, M.J., Chua, A.Y., Goh, D.H.: Selection of the best answer in CQA services. In: Seventh International Conference on Information Technology: New Generations, ITNG 2010, Las Vegas, Nevada, USA, pp. 534–539, 12–14 April 2010

7. Tian, Q., Zhang, P., Li, B.: Towards predicting the best answers in community-based question-answering services. In: Proceedings of the Seventh International Conference on Weblogs and Social Media, ICWSM 2013, Cambridge, Massachusetts, USA, 8–11 July 2013
8. Elalfy, D., Gad, W., Ismail, R.: A hybrid model to predict best answers in question answering communities. Egypt. Inform. J. (2017)
9. Lezina, G., Kuznetsov, A.: Predict closed questions on stack overflow. In: Proceedings of the Ninth Spring Researchers Colloquium on Databases and Information Systems, Kazan, Russia, pp. 10–14, 31 May 2013
10. NTLK, natural language toolkit. http://www.nltk.org/. Accessed 28 Sept 2019

LIUBoost: Locality Informed Under-Boosting for Imbalanced Data Classification

Sajid Ahmed, Farshid Rayhan, Asif Mahbub, Md. Rafsan Jani, Swakkhar Shatabda and Dewan Md. Farid

Abstract The problem of class imbalance along with class overlapping has become a major issue in the domain of supervised learning. Most classification algorithms assume equal cardinality of the classes under consideration while optimising the cost function, and this assumption does not hold true for imbalanced datasets, which results in suboptimal classification. Therefore, various approaches, such as undersampling, oversampling, cost-sensitive learning and ensemble-based methods, have been proposed for dealing with imbalanced datasets. However, undersampling suffers from information loss, oversampling suffers from increased runtime and potential overfitting, while cost-sensitive methods suffer due to inadequately defined cost assignment schemes. In this paper, we propose a novel boosting-based method called **L**ocality **I**nformed **U**nder-**Boost**ing (LIUBoost). LIUBoost uses undersampling for balancing the datasets in every boosting iteration like Random Undersampling with Boosting (RUSBoost), while incorporating a cost term for every instance based on their hardness into the weight update formula minimising the information loss introduced by undersampling. LIUBoost has been extensively evaluated on 18 imbalanced datasets, and the results indicate significant improvement over existing best performing method RUSBoost.

Keywords Boosting · Class imbalanced classification
Undersampling · RUSBoost · SMOTEBoost

S. Ahmed · F. Rayhan · A. Mahbub · Md. Rafsan Jani · S. Shatabda · D. Md. Farid (✉)
Department of Computer Science and Engineering, United International University, Dhaka, Bangladesh
e-mail: dewanfarid@cse.uiu.ac.bd
URL: http://cse.uiu.ac.bd

S. Ahmed
e-mail: sahmed133002@bscse.uiu.ac.bd

A. Abraham et al. (eds.), *Emerging Technologies in Data Mining and Information Security*, Advances in Intelligent Systems and Computing 813,
https://doi.org/10.1007/978-981-13-1498-8_12

1 Introduction

Class imbalance refers to the scenario where the number of instances from one class is significantly greater than that of another class. Traditional machine learning algorithms such as support vector machines [10], artificial neural networks [21], decision tree [13, 16], and random forests [7] exhibit suboptimal performance when the dataset under consideration is imbalanced. This happens due to the fact that these classifiers work under the assumption of equal cardinality between the underlying classes. However, many of the real-world problems such as anomaly detection [12, 15] and facial recognition [24] where supervised learning is used are imbalanced. This is why researchers came up with different methods that would make the existing classifiers competent in dealing with classification problems that exhibit class imbalance.

Most of the existing methods for dealing with class imbalanced problem can be categorised into sampling techniques, cost-sensitive methods and ensemble-based methods. The sampling techniques either increase the number of minority class instances (oversampling) or decrease the number of majority class instances (undersampling) so that imbalance ratio decreases and the training data fed to some classifier becomes somewhat balanced [3, 14]. The cost-sensitive methods assign higher misclassification cost to the minority class instances which is further incorporated into the cost function to be minimised by the underlying classifier. The integration of these cost terms minimises the classifiers' bias towards the majority class and puts greater emphasis on the appropriate learning of the minority concept [29]. Ensemble methods such as Bagging [6] and Boosting [1] employ multiple instances of the base classifier and combine their learning to predict the dependent variable. Sampling techniques or cost terms are incorporated into ensemble methods for dealing with the problem of class imbalance, and these methods have shown tremendous success [9, 28]. As a matter of fact, these ensemble methods turned out to be the most successful ones for dealing with imbalanced datasets [17].

In order to reduce the effect of class imbalance, the aforementioned methods usually attempt to increase the identification rate for the minority class and decrease the number of false negatives. In the process of doing so, they often end up decreasing the recognition rate of the majority class which results in a large number of false positives. This can be equally undesirable in many real-world problems such as fraud detection where identifying a genuine customer as fraud could result in loss of loyal clients. This increased false positive rate could be due to under-representation of the majority class (undersampling), over-emphasised representation of the minority class (oversampling), or over-optimistic cost assignment (cost-sensitive methods). The most successful ensemble-based methods also suffer from such problems because they use undersampling or oversampling for the purpose of data balancing, while the cost-sensitive methods suffer from over-optimistic cost assignment because the proposed assignment schemes only take into account the global between-class imbalance and do not consider the significant characteristics of the individual instance [30].

In this paper, we proposed a novel boosting-based approach called **L**ocality **I**nformed **U**nder-**Boost**ing (LIUBoost) for dealing with class imbalanced problems. The aforementioned methods have incorporated either sampling or cost terms into boosting for mitigating the effect of class imbalance and have fallen victim to either information loss or unstable cost assignment. However, LIUBoost uses undersampling for balancing the datasets while retaining significant information about the local characteristics of each of the instances and incorporates that information into the weight update equation of AdaBoost in the form of cost terms. These cost terms minimise the effect of information loss introduced by undersampling. We have used K-nearest-neighbour (KNN) algorithm [25] with small K value for locality analysis and weight calculation. These weights are not meant to mitigate the effect of class imbalance in any way. However, these weights are able to differentiate among safe, borderline and outlier instances of both majority and minority classes and provide the underlying base learners with a better representation of both majority and minority concepts. Additionally, LIUBoost takes into account problems such as class overlapping [18] and the curse of bad minority hubs [31] that occur together with the problem of class imbalance. The aim of this study is to show the effectiveness of our proposed LIUBoost both theoretically and experimentally. To do so, we have compared the performance of LIUBoost with that of RUSBoost on 18 standard benchmark imbalanced datasets, and the results show that LIUBoost significantly improves over RUSBoost.

The remainder of the paper has been arranged as follows. Section 2 presents related work and motivation behind our proposal, Sect. 3 presents our proposed method and Sect. 4 provides the experimental results. Finally, we conclude in Sect. 5.

2 Related Work

Seiffert et al. [28] proposed RUSBoost for the task of imbalanced classification. RUSBoost integrates random undersampling at each iteration of AdaBoost. In different studies, RUSBoost has stood out as one of the best performing boosting-based methods alongside SMOTEBoost for imbalanced data classification [17, 23]. A major key to the success of RUSBoost is its random undersampling technique which, in spite of being a simple non-heuristic approach, has been shown to outperform other intelligent ones. Due to the use of this time-efficient yet effective sampling strategy, RUSBoost is more suitable for practical use compared to SMOTEBoost [9] and other boosting-based imbalanced classification methods which employ intelligent undersampling or oversampling, thus making the whole classification process much more time-consuming. However, RUSBoost may fall victim to information loss when faced with highly imbalanced datasets. This happens due to its component random undersampling, which discards a large number of majority class instances at each iteration, and thus the majority class is often underrepresented in the modified training data fed to the base learners. Our proposed method incorporates significant information about each of the instances of the unmodified training set into the iterations of RUSBoost in the form of cost in order to mitigate the aforementioned information loss.

Fan et al. proposed AdaCost [11] which introduced misclassification costs, for instance, into the weight update equation of AdaBoost. They theoretically proved that introducing costs in this way does not break the conjecture of AdaBoost. However, they did not develop any generic weight assignment scheme that could be followed for different datasets. Their weight assignments were rather domain specific. Karakoulas et al. [22] proposed a weight assignment scheme for dealing with the problem of class imbalance where false negatives were assigned higher weights compared to false positives. Sun et al. proposed three cost-sensitive boosting methods for the classification of imbalanced datasets AdaC1, AdaC2 and AdaC3 [29]. These methods assign greater misclassification cost to the instances of the minority class. If an instance of the minority class is misclassified, its weight is increased more forcefully compared to a misclassified majority class instance. Furthermore, if a minority instance is correctly classified, its weight is decreased less forcefully compared to a correctly classified majority instance. As a result, appropriate learning of the minority instances is given greater emphasis in the training process of AdaBoost in order to mitigate the effect of class imbalance. All these methods assign an equal cost to all instances of the same class considering the between-class imbalance ratio. None of them take into account local characteristics of the data points.

Most of the methods proposed for classification of imbalanced datasets only take into account the difference between the number of instances from the majority and the minority class and try to mitigate the effects of this imbalance. However, this difference is only one of the several factors that make the task of classification extremely difficult. But these additional yet extremely significant factors are often overlooked while designing algorithms for imbalanced classification [26]. One of these factors is the overlapping of the majority and minority classes. Prati et al. [27] studied the effect of class overlapping combined with class imbalance by varying their respective degree and deduced that overlapping is even more detrimental to the classifier performance. Garcia et al. [18] examined the performance of six classifiers on datasets where class imbalance and overlapping were high and noticed that KNN [25] with a small value of **K** (local neighbourhood analysis) was the best performer under such circumstances. These observations point towards the feasibility of dealing with the problem of class overlapping in imbalanced datasets through incorporating information about the local neighbourhood of the instances into the training process. Another factor responsible for degrading the performance of classifiers in imbalanced datasets is the effect of bad minority hubs. These are instances of the minority class that are closely grouped together in the feature space. If such a group is close to a majority instance, that majority instance will have a high probability of being misclassified [31]. Such effects are not taken into account in the cost assignment scheme proposed by aforementioned cost-sensitive methods for imbalanced classification. However, our proposed method attempts to mitigate the effects of class overlapping and bad minority hubs by taking into account the local neighbourhood of each of the instances while assigning weights to them.

In some recent proposals, authors have incorporated locality information of the instances into their methods in different ways for dealing with imbalanced datasets. He et al. proposed ADASYN [20] oversampling technique which takes into account number of majority class instances around the existing minority instances and creates more synthetic samples for the ones with more majority neighbours so that the harder minority instances get more emphasis in the learning process. Blaszczynski et al. proposed Local-and-Over-All Balanced Bagging [4] which integrates locality information of the majority instances into UnderBagging. In this approach, the majority instances with less number of minority instances in their local neighbourhood are more likely to be selected in the bagging iterations. Bunkhumpornpat et al. proposed Safe-Level-SMOTE [8] which only uses the safe minority instances for generating synthetic minority samples. Han et al. proposed Borderline-SMOTE [19] which only uses the borderline minority instances for synthetic minority generation. Furthermore, Napierala et al. used locality information of the minority instances to divide them into aforementioned categories such as safe, borderline, rare and outlier [26]. All these aforementioned methods suggest that locality information of minority and majority instances is significant and can be used in the learning process of classifiers designed for imbalanced classification.

3 LIUBoost

The pseudocode of our proposed method LIUBoost is given in Algorithm 2. LIUBoost calls Weight_Assignment method given in Algorithm 1 before boosting iterations begin. This method returns two sets of weights $Weight^-$ and $Weight^+$ used, respectively, to decrease and increase the weights associated with an instance. $Weight^+$ are added inside the exponent term of the weight update equation for the misclassified instances at the iteration under consideration, while $Weight^-$ are added for the correctly classified instances. As a result, weight of the instances with greater $Weight^+$ grows rapidly if they are misclassified while weight of the instances with greater $Weight^-$ drops rapidly if they are correctly classified. Thus, LIUBoost puts greater emphasis on learning the important concepts rapidly. Additionally, LIUBoost performs undersampling at each boosting iteration for balancing the training set.

The alpha terms determine how significant the predictions of each of the individual base learners are in the final voted classification. These terms also play an important role in the weight update formula which ultimately minimises the combined error. Since LIUBoost has modified the original weight update equation of AdaBoost by adding cost terms, the alpha term needs to be updated accordingly in order to preserve coherence of the learning process. The alpha term has been updated according to the recommendations from [29]. One thing to notice here is that LIUBoost combines sampling method and cost-sensitive learning in a novel way. The proposed weight assignment method assigns greater $Increase\_Weight$ to borderline and rare instances while assigning less $Increase\_Weight$ to safe instances due to the way it analyses local neighbourhood. Napierala et al. [26] proposed a similar method

Algorithm 1 Weight_Assignment(dataset X,k)

for each instance $x_i \in$ in the training set, X **do**
  find **k**-nearest nearest neighbours for the instance
  $N_s \leftarrow$ number of neighbours with same class
  $N_o \leftarrow$ number of neighbours with opposite class
  if $N_s == 0$ **then**
    $Weight^{+}(i) \leftarrow \delta$
    $Weight^{-}(i) \leftarrow \frac{1}{N_o}$
  else if $N_o == 0$ **then**
    $Weight^{+}(i) \leftarrow \frac{1}{N_s}$
    $Weight^{-}(i) \leftarrow \delta$
  else
    $Weight^{+}(i) \leftarrow \frac{1}{N_s}$
    $Weight^{-}(i) \leftarrow \frac{1}{N_o}$
  end if
end for Return the $Weight^{+}$ and $Weight^{-}$

for grouping only the minority instances into four categories such as safe, borderline, rare and outlier. However, LIUBoost also distinguishes the majority instances through weight assignment. When the majority and minority classes are highly overlapped, which is often the case with highly imbalanced datasets [27], undersampling may discard a large number of borderline and rare majority instances which will increase their misclassification probability. LIUBoost overcomes this problem by keeping track of such majority instances through assigned weights and puts greater emphasis on their learning. This is its unique feature for minimising information loss.

4 Experimental Results

4.1 Evaluation Metrics

As evaluation metrics, we have used area under the Receiver Operator Curve (AUCROC) and area under the Precision–Recall Curve (AUPR). These curves use Precision, Recall (TPR) and False Positive Rate (FPR) as underlying metrics .

$$Precision = \frac{TP}{TP + FP} \tag{1}$$

$$TPR = \frac{TP}{TP + FN} \tag{2}$$

Algorithm 2 LIUBoost($dataset = (X, Y)$)

```
$m$ = number of instances
$T$ = number of boosting iterations
$(Weight^{+}, Weight^{-})$ = Weight_Assignment($dataset$,$k$)
for $i \leftarrow 1$ to $m$ do
    $D_i \leftarrow \frac{1}{m}$
end for
for $t \leftarrow 1$ to $T$ do
    $undersampled\_dataset$ = undersampling($dataset$)
    $h_t \leftarrow$ Decision_Tree($undersampled\_dataset$)
    $mis\_sum \leftarrow \sum_{h_t(x_i) \neq y_i} D_i^t \cdot Weight^{+}(i)$
    $cor\_sum \leftarrow \sum_{h_t(x_i) = y_i} D_i^t \cdot Weight^{-}(i)$
    update parameter $\alpha_t \leftarrow \frac{1}{2}\log\frac{1+cor\_sum-mis\_sum}{1-cor\_sum+mis\_sum}$
    if $\alpha_t \leq 0$ then
        $t \leftarrow t-1$
        return to statement 8
    end if
    for $i \leftarrow 1$ to $m$ do
        if $y_i \neq h_t(x_i)$ then
            $D_{t+1}(i) \leftarrow D_t e^{-\alpha_t \cdot y_i \cdot h_t(x_i) \cdot Weight^{+}(i)}$
        else
            $D_{t+1}(i) \leftarrow D_t e^{-\alpha_t \cdot y_i \cdot h_t(x_i) \cdot Weight^{-}(i)}$
        end if
    end for
    normalise $D$
end for
$g \leftarrow \sum_{t=1}^{T} \alpha_t h_t$
Return $h = sign(g)$
```

$$FPR = \frac{FP}{FP + TN} \tag{3}$$

Receiver Operating Characteristic (ROC) curve represents false positive rate (fpr) along the horizontal axis and true positive rate (tpr) along the vertical axis. A perfect classifier will have Area Under ROC Curve (AUROC) of 1 which means all instances of the positive class instances have been correctly classified and none of the negative class instances have been flagged as positive. AUROC provides an ideal summary of the classifier performance. For a not so good classifier, TPR and FPR increase proportionally which brings the AUROC down. A classifier which is able to correctly classify high number of both positive and negative class instances gets a high AUROC which is our goal in case of imbalanced datasets.

AUPR represents tpr down the horizontal axis and precision down the vertical axis. Precision and TPR are inversely related, i.e., as precision increases, TPR falls and vice versa. A balance between these two needs to be achieved by the classifier, and to achieve this and to compare performance, AUPR curve is used. Both of the aforementioned evaluation metrics are held as benchmarks for the assessment of classifier performance on imbalanced datasets. However, AUPR is more informative for cases of high-class imbalance AUROC. This is because a large change in false positive counts can result in a small change in the FPR represented in ROC.

4.2 Results

We have compared the performance of our proposed method LIUBoost against that of RUSBoost over 18 imbalanced datasets with varying imbalance ratio. All these datasets are from KEEL Dataset Repository [2]. Table 1 contains a brief description of these datasets. The algorithms have been run 30 times using tenfold cross-validation on each dataset, and the average AUROC and AUROC are presented in Tables 2 and 3, respectively. Decision tree estimator C4.5 has been used as base learner. Both RUSBoost and LIUBoost have been implemented in python.

From the results presented in Table 2, we can see that with respect to AUROC, LIUBoost outperformed RUSBoost over 15 datasets. However, with respect to AUPR, LIUBoost outperformed RUSBoost over 14 datasets out of 15 that is shown in Table 3.

Table 1 Description of datasets

Datasets	Instances	Features	IR
pima	768	8	1.87
glass5	214	9	22.78
yeast5	1484	8	38.73
yeast6	1484	8	41.4
ecoli-0-3-4_vs_5	200	7	9
abalone19	4174	8	129.44
pageblocks	548	10	164
led7digit-0-2-4-5-6-7-8-9_vs_1	443	7	10.97
glass-0-1-4-6_vs_2	205	9	11.06
glass2	214	9	11.59
glass6	214	9	6.38
yeast-1_vs_7	459	7	14.3
poker-8-9_vs_6	1485	10	58.4
haberman	306	3	2.78
winequality-red-8_vs_6	656	11	35.44
glass0	214	9	2.06
glass-0-1-5_vs_2	172	9	9.12
yeast-0-2-5-7-9_vs_3-6-8	1004	8	9.14

Table 2 Average AUROC comparison

Dataset	RUSBoost	LIUBoost
glass5	0.977	**0.987**
ycast5	0.984	**0.988**
yeast6	0.916	**0.921**
ecoli-0-3-4_vs_5	**0.987**	0.981
abalone19	0.784	**0.801**
glass-0-1-4-6_vs_2	0.701	**0.780**
glass2	0.697	**0.794**
page-blocks0	**0.988**	**0.988**
glass6	0.961	**0.966**
yeast-1_vs_7	0.785	**0.794**
poker-8-9_vs_6	0.791	**0.792**
haberman	0.599	**0.647**
winequality-red-8_vs_6	0.708	**0.727**
led7digit-0-2-4-5-6-7-8-9_vs_1	0.943	**0.953**
glass0	0.858	**0.869**
glass-0-1-5_vs_2	0.646	**0.725**
yeast-0-2-5-7-9_vs_3-6-8	**0.941**	0.938
pima	0.689	**0.704**

We have performed Wilcoxon pairwise signed rank test [32] in order to ensure that the improvements achieved by LIUBoost are statistically significant. The test results indicate that the performance improvements both with respect to AUPR and AUROC are significant since the null hypothesis of equal performance has been rejected at 5% level of significance in favour of LIUBoost. Wilcoxon test results can be found in Tables 4 and 5.

Table 3 Average AUPR comparison

Dataset	RUSBoost	LIUBoost
glass5	0.766	**0.835**
yeast5	0.690	**0.742**
yeast6	0.457	**0.548**
ecoli-0-3-4_vs_5	**0.930**	0.915
abalone19	**0.998**	**0.998**
glass-0-1-4-6_vs_2	0.209	**0.258**
glass2	0.257	**0.263**
page-blocks0	0.905	**0.907**
glass6	0.893	**0.923**
yeast-1_vs_7	**0.403**	0.344
poker-8-9_vs_6	0.188	**0.249**
haberman	0.344	**0.392**
winequality-red-8_vs_6	0.192	**0.242**
led7digit-0-2-4-5-6-7-8-9_vs_1	0.648	**0.759**
glass0	0.708	**0.753**
glass-0-1-5_vs_2	0.220	**0.263**
yeast-0-2-5-7-9_vs_3-6-8	**0.835**	0.824
pima	0.529	**0.544**

Table 4 Wilcoxon signed rank test based on average AUROC

RUSBoost	LIUBoost	Hypothesis (alpha = 0.05)	p-value
11.5	159.5	Rejected for LIUBoost	0.00068

Table 5 Wilcoxon signed rank test based on average AUPR

RUSBoost	LIUBoost	Hypothesis (alpha = 0.05)	p-value
23.5	146.5	Rejected for LIUBoost	0.0037

5 Conclusion

In this paper, we have proposed a novel boosting-based algorithm for dealing with the problem of class imbalance. Our method LIUBoost is the first one to combine both sampling technique and cost-sensitive learning. Although good number of methods have been proposed for dealing with imbalanced datasets, none of them have proposed such an approach. We have tried to design an ensemble method that would be cost-efficient just like RUSBoost but would not suffer from the resulting information loss and the results so far are satisfying. Additionally, recent research has indicated that dividing the minority class into categories is the right way to go for imbalanced datasets [5, 26]. In our opinion, both majority and minority instances should be divided into categories and the hard instances should be given special importance in imbalanced datasets. This becomes even more important when the underlying sampling technique discards some instances for data balancing.

Class imbalance is prevalent in many real-world classification problems. However, the proposed methods have their own deficits. Cost-sensitive methods suffer from domain-specific cost assignment schemes, while oversampling-based methods suffer from overfitting and increased runtime. Under such scenario, LIUBoost is cost-efficient, defines a generic cost assignment scheme, does not introduce any false structure and takes into account additional problems such as bad minority hubs and class overlapping. The results are also statistically significant. In future work, we would like to experiment with other cost assignment schemes.

References

1. Afza, A.A., Farid, D.M., Rahman, C.M.: A hybrid classifier using boosting, clustering, and naïve bayesian classifier. World Comput. Sci. Inf. Technol. J. (WCSIT) **1**, 105–109 (2011). ISSN: 2221-0741
2. Alcalá-Fdez, J., Fernández, A., Luengo, J., Derrac, J., García, S., Sánchez, L., Herrera, F.: Keel data-mining software tool: data set repository, integration of algorithms and experimental analysis framework. J. Mult. Valued Logic Soft Comput. **17** (2011)
3. Batista, G.E., Prati, R.C., Monard, M.C.: A study of the behavior of several methods for balancing machine learning training data. ACM SIDKDD Explor. Newsl. **6**(1), 20–29 (2004)
4. Błaszczyński, J., Stefanowski, J., Idkowiak, Ł.: Extending bagging for imbalanced data. In: Proceedings of the 8th International Conference on Computer Recognition Systems CORES 2013, pp. 269–278. Springer (2013)
5. Borowska, K., Stepaniuk, J.: Rough sets in imbalanced data problem: Improving re–sampling process. In: IFIP International Conference on Computer Information Systems and Industrial Management, pp. 459–469. Springer (2017)
6. Breiman, L.: Bagging predictors. Mach. Learn. **24**(2), 123–140 (1996)
7. Breiman, L.: Random forests. Mach. Learn. **45**(1), 5–32 (2001)
8. Bunkhumpornpat, C., Sinapiromsaran, K., Lursinsap, C.: Safe-level-smote: safe-level-synthetic minority over-sampling technique for handling the class imbalanced problem. Advances in Knowledge Discovery and Data Mining, pp. 475–482 (2009)
9. Chawla, N.V., Lazarevic, A., Hall, L.O., Bowyer, K.W.: Smoteboost: improving prediction of the minority class in boosting. In: European Conference on Principles of Data Mining and Knowledge Discovery, pp. 107–119. Springer (2003)
10. Cortes, C., Vapnik, V.: Support-vector networks. Mach. Learn. **20**(3), 273–297 (1995)
11. Fan, W., Stolfo, S.J., Zhang, J., Chan, P.K.: Adacost: misclassification cost-sensitive boosting. ICML **99**, 97–105 (1999)
12. Farid, D., Nguyen, H.H., Darmont, J., Harbi, N., Rahman, M.Z.: Scaling up detection rates and reducing false positives in intrusion detection using NBtree. In: International Conference on Data Mining and Knowledge Engineering (ICDMKE 2010), pp. 186–190 (2010)
13. Farid, D.M., Al-Mamun, M.A., Manderick, B., Nowe, A.: An adaptive rule-based classifier for mining big biological data. Expert Syst. Appl. **64**, 305–316 (2016)
14. Farid, D.M., Nowé, A., Manderick, B.: A new data balancing method for classifying multi-class imbalanced genomic data. In: Proceedings of 5th Belgian-Dutch Conference on Machine Learning (Benelearn), pp. 1–2 (2016)

15. Farid, D.M., Zhang, L., Hossain, A., Rahman, C.M., Strachan, R., Sexton, G., Dahal, K.: An adaptive ensemble classifier for mining concept drifting data streams. Expert Syst. Appl. **40**(15), 5895–5906 (2013)
16. Farid, D.M., Zhang, L., Rahman, C.M., Hossain, M., Strachan, R.: Hybrid decision tree and naïve bayes classifiers for multi-class classification tasks. Expert Syst. Appl. **41**(4), 1937–1946 (2014)
17. Galar, M., Fernandez, A., Barrenechea, E., Bustince, H., Herrera, F.: A review on ensembles for the class imbalance problem: bagging-, boosting-, and hybrid-based approaches. IEEE Trans. Syst. Man Cybern. Part C (Applications and Reviews) **42**(4), 463–484 (2012)
18. García, V., Sánchez, J., Mollineda, R.: An empirical study of the behavior of classifiers on imbalanced and overlapped data sets. In: Proceedings of Progress in Pattern Recognition, Image Analysis and Applications, pp. 397–406 (2007)
19. Han, H., Wang, W.Y., Mao, B.H.: Borderline-smote: a new over-sampling method in imbalanced data sets learning. Advances in Intelligent Computing, pp. 878–887 (2005)
20. He, H., Bai, Y., Garcia, E.A., Li, S.: ADASYN: adaptive synthetic sampling approach for imbalanced learning. In: IEEE World Congress on Computational Intelligence. IEEE International Joint Conference on Neural Networks, 2008, IJCNN 2008, pp. 1322–1328. IEEE (2008)
21. Hopfield, J.J.: Artificial neural networks. IEEE Circ. Devices Mag. **4**(5), 3–10 (1988)
22. Karakoulas, G.I., Shawe-Taylor, J.: Optimizing classifers for imbalanced training sets. In: Proceedings of Advances in Neural Information Processing Systems, pp. 253–259 (1999)
23. Khoshgoftaar, T.M., Van Hulse, J., Napolitano, A.: Comparing boosting and bagging techniques with noisy and imbalanced data. IEEE Trans. Syst. Man Cybern. Part A Syst. Hum. **41**(3), 552–568 (2011)
24. Liu, Y.H., Chen, Y.T.: Total margin based adaptive fuzzy support vector machines for multiview face recognition. In: 2005 IEEE International Conference on Systems, Man and Cybernetics, vol. 2, pp. 1704–1711. IEEE (2005)
25. Mani, I., Zhang, I.: KNN approach to unbalanced data distributions: a case study involving information extraction. In: Proceedings of Workshop on Learning from Imbalanced Datasets, vol. 126 (2003)
26. Napierala, K., Stefanowski, J.: Types of minority class examples and their influence on learning classifiers from imbalanced data. J. Intell. Inf. Syst. **46**(3), 563–597 (2016)
27. Prati, R.C., Batista, G., Monard, M.C., et al.: Class imbalances versus class overlapping: an analysis of a learning system behavior. In: Proceedings of MICAI, vol. 4, pp. 312–321. Springer (2004)
28. Seiffert, C., Khoshgoftaar, T.M., Van Hulse, J., Napolitano, A.: Rusboost: a hybrid approach to alleviating class imbalance. IEEE Trans. Syst. Man Cybern. Part A Syst. Hum. **40**(1), 185–197 (2010)
29. Sun, Y., Kamel, M.S., Wong, A.K., Wang, Y.: Cost-sensitive boosting for classification of imbalanced data. Pattern Recogn. **40**(12), 3358–3378 (2007)
30. Sun, Z., Song, Q., Zhu, X., Sun, H., Xu, B., Zhou, Y.: A novel ensemble method for classifying imbalanced data. Pattern Recogn. **48**(5), 1623–1637 (2015)
31. Tomašev, N., Mladenić, D.: Class imbalance and the curse of minority hubs. Knowl. Based Syst. **53**, 157–172 (2013)
32. Wilcoxon, F.: Individual comparisons by ranking methods. Biom. Bull. **1**(6), 80–83 (1945)

Comparative Study of Different Ensemble Compositions in EEG Signal Classification Problem

Ankita Datta and Rajdeep Chatterjee

Abstract The leading perspective of this paper is an introduction to three $(Type-I, Type-II, \text{ and } Type-III)$ types of ensemble architectures in Electroencephalogram (EEG) signal classification problem. Motor imagery EEG signal is filtered and subsequently used for three different types of feature extraction techniques: Wavelet-based Energy and Entropy $(EngEnt)$, Bandpower (BP), and Adaptive Autoregressive (AAR). Ensemble architectures have been used in various compositions with different classifiers as base learners along with majority voting as the combined method. This standard procedure is also compared with the mean accuracy method obtained from multiple base classifiers. The Type-I ensemble architecture with EngEnt and BP feature sets provides most consistent performance for both majority voting and mean accuracy combining techniques. Similarly, Type-II architecture with EngEnt and AAR feature sets provides most consistent performance for both majority voting and mean accuracy combining techniques. However, the Type-III ensemble architecture contributes highest result 82.86% with K-Nearest Neighbor (KNN) classifier among all three types.

Keywords BCI · EEG · Ensemble architecture · AAR · Bandpower · Wavelet energy entropy · Majority voting · Classification

1 Introduction

Brain–Computer Interface (BCI) is direct communication between the human brain and computer system without involving any muscles or motor neuron activities [1, 2]. BCI can replace, improve, and restore the nerves system by decoding the brain signal into control commands. Among all brain activity measurement techniques, the

A. Datta (✉) · R. Chatterjee
School of Computer Engineering, KIIT University, Bhubaneswar 751024, Odisha, India
e-mail: ankita.datta24@gmail.com

R. Chatterjee
e-mail: cse.rajdeep@gmail.com

A. Abraham et al. (eds.), *Emerging Technologies in Data Mining and Information Security*, Advances in Intelligent Systems and Computing 813,
https://doi.org/10.1007/978-981-13-1498-8_13

Electroencephalography (EEG) is the most widely used method. It is very popular among researchers due to its fine spatial and temporal resolution, cost-effectiveness, and portability. If the subject imagines that he/she is performing a mental task without performing it in reality such as limb movements, it is called motor imagery [3] in BCI domain. The application of motor imagery brain state classification can help the fully or partially paralyzed persons to do usual activities by their brain [4] only. There are many challenges in the EEG signal classification problem. Few important issues are eye movement, muscles movements, etc. These affect the recording as the signal is very weak in terms of amplitude (order of some microvolts) [5, 6]. Preprocessing techniques are employed to remove these unwanted signals before using appropriate feature extraction techniques.

An ensemble classifier is a supervised classification problem where each instance associated with the decision class [7–11]. The aim of an ensemble classifier is to learn from training sets using a group of different base classifiers and then the same built models are validated with test sets. The outputs (i.e., predicted decision class) of different base classifiers are combined in order to get a final output. There are many combining strategies available in literature; however, only majority voting and mean accuracy techniques are implemented in this paper. In our previous works [12], we have proved that in motor imagery brain state classification problem, an ensemble classifier outperforms individual base classifiers. Base classifiers are used to construct the ensemble classifier. Here, K-Nearest Neighbor (KNN) [13], Support Vector Machine (SVM) [14], and Naive Bayes (NB) [15] are used as base classifiers.

This paper is divided into four sections. Section 1 represents an introduction to BCI and ensemble classifier, and Sect. 2 describes methodology used in our experiments. Then, we discuss and analyze obtained results in Sect. 3. Finally, Sect. 4 provides us the conclusion of our study for this paper.

2 Method

2.1 Dataset Description

Here, the dataset (BCI competition II (2003) Dataset III) has been taken from the Department of Medical Informatics provided by Institute for Biomedical Engineering, University of technology, Graz [16]. The sampling rate of the dataset is recommended at 128 Hz. The EEG signal was recorded using the standard IEEE 10–20 electrode placement system (Fig. 1). Although EEG signal is available for three different electrodes $C3$, Cz, and $C4$, we have considered observations from $C3$ and $C4$ electrodes as they are dominant for human left–right-hand movements. According to the source, dataset contains left–right-hand movements of EEG signal of a healthy female subject for 6 s as shown in Fig. 2.

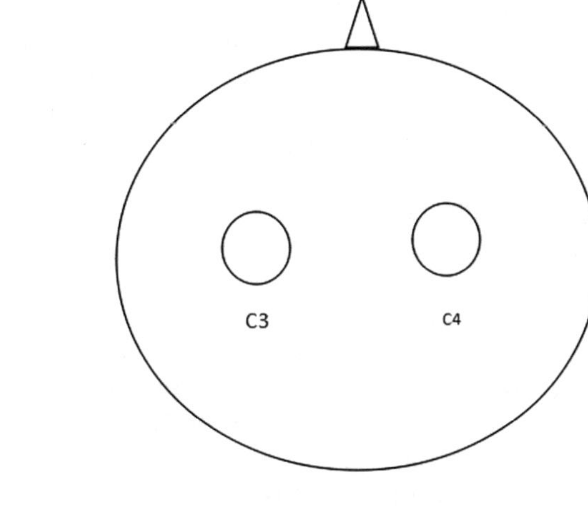

Fig. 1 IEEE 10–20 electrode placement for C3 and C4 electrodes

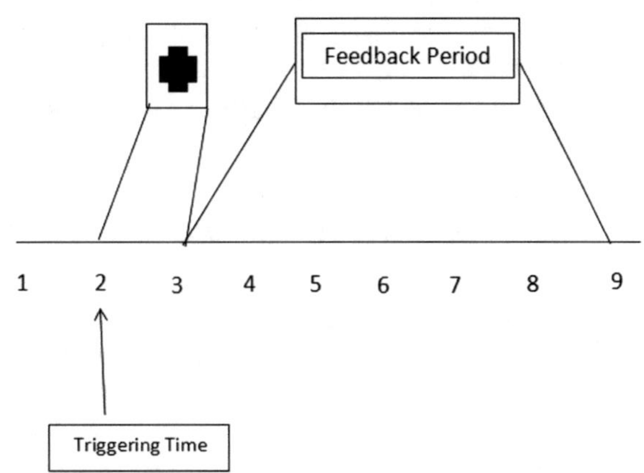

Fig. 2 Region of interest (6 s) extracted from raw EEG signal

2.2 Preprocessing

The whole dataset is of 280 trials (i.e., samples) and divided into 140 trials for training and 140 trials for test set. An elliptic bandpass filter has been used to extract the interested frequency bands of 0.5–50 Hz. Five frequency bands, known as delta (0.5–4 Hz), theta (4–7 Hz), alpha (7–13 Hz), beta (13–25 Hz), and gamma (25–50 Hz) waveforms, have been considered in computing bandpower feature set. Thus, frequency bands beyond this range are considered as noise.

2.3 Feature Extraction

Wavelet energy and entropy (*EngEnt*):- Wavelet is very appropriate for feature extraction of EEG signal and is proved empirically in literature as well as in our past works. It can distinguish between both temporal and spectral domain features of a signal. It

decomposes the signal into detail and coarse approximation coefficients. The down-sampled output of the first high-pass and low-pass filters provides the detail $D1$ and approximation $A1$. This derivation is repeated for another two times to its third level. The Daubechies (db) mother wavelet of order 4 and $D3$ level features has been used in this paper. Subsequently, energy and entropy computed from the $D3$ feature set instead of using the actual wavelet coefficients as features in order to reduce the dimensionality [17, 18].
Bandpower (*BP*):- It determines the percentage of total power present in a fixed frequency interval. As per the past related research articles, alpha and beta brain waveforms play a significant role in computing bandpower features for motor imagery in the said frequency intervals [19].
Adaptive Autoregressive Parameters (*AAR*):- AAR model parameters represent the signal characteristics. The Recursive Least Square (RLS) variant of AR model has been used for feature extraction. As per [20], the experimental parameter settings for RLS AAR are $p = 6$ (AR order), $aMode = 1$, $vMode = 2$, and $UC = 0$. UC is the model update coefficient to control the model adaptation ratio. It is implemented on the filtered EEG signal without considering any specific frequency band.

The feature set lengths for EngEnt, BP, and AAR are 12, 10, and 4, respectively.

2.4 Used Ensemble Architectures

We have used three different ensemble architectures to study the effect of ensemble classifiers composition. We basically are motivated from the paper [21] where similar architectures were used with different base classifiers. In type-I architecture (Fig. 3), the training sets are trained by same base classifier but with different parameter configurations. In type-II architecture (Fig. 4), the training sets are trained by the different base classifiers. However, in type-III architecture (Fig. 5), different feature sets are used to train by a classifier instead of varying the classifiers in ensemble with one feature set. In the abovementioned architectures, the decision generated by the classifiers is combined by the majority voting [8] technique. We have also computed the mean accuracy as another combining technique. The descriptions of all architectures are briefly shown in Table 1.

2.5 Experiment

In Type-I architecture, we have used KNN as the only base classifier with K values 5, 7, and 9. Table 2 shows the classifier parameters used by Type-I architecture. In Type-II architecture, three base classifiers SVM, KNN, and NB have been used. But in this experiment, the parameter configuration for the KNN base classifier is changed, whereas parameter setting for the remaining base classifiers kept unchanged. The detailed combination of different base classifiers is displayed in Table 3. The above-

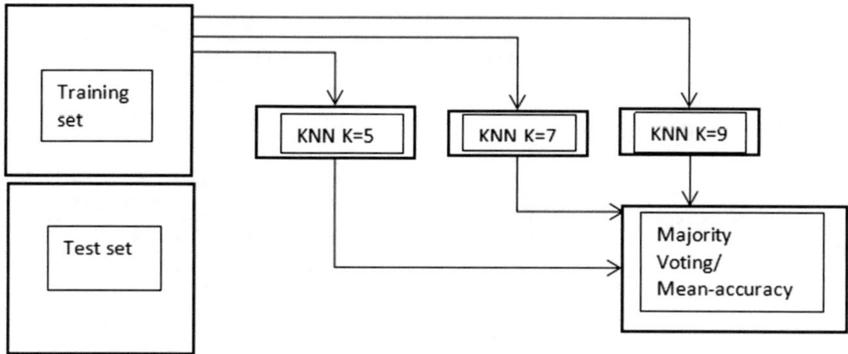

Fig. 3 Type-I ensemble architecture

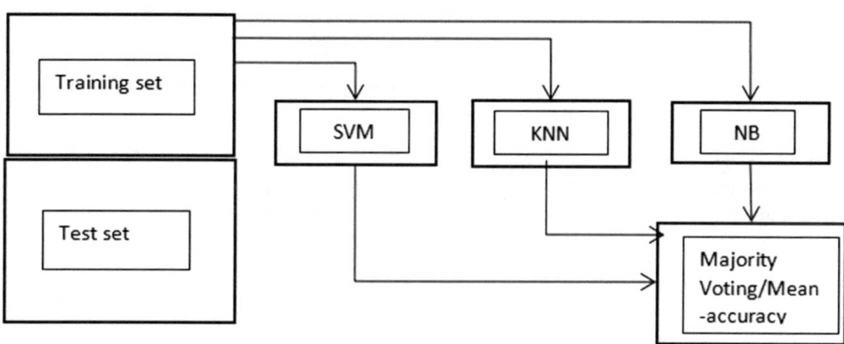

Fig. 4 Type-II ensemble architecture

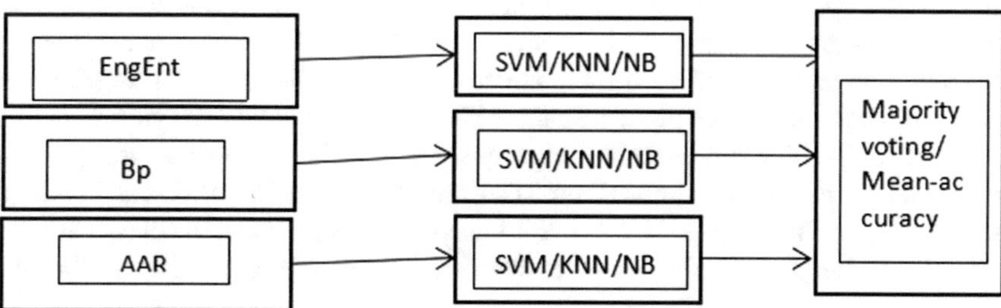

Fig. 5 Type-III ensemble architecture

Table 1 The description of each architecture

Architecture	Description
Type-I	A single feature set with same base classifier with varying parameter
Type-II	A single feature set with different base classifiers
Type-III	The different feature sets with same base classifier

Table 2 Configuration of the classifiers used in Type-I architecture

Base classifier	Parameter
KNN	K = 5
	K = 7
	K = 9

Table 3 Configuration of the classifiers used in Type-II architecture

Categories	Base classifier	Parameter
Combination 1	SVM	RBF
	KNN	K = 5
	NB	–
Combination 2	SVM	RBF
	KNN	K = 7
	NB	–
Combination 3	SVM	RBF
	KNN	K = 9
	NB	–

Table 4 Configuration of the classifiers used in Type-III architecture

Base classifier	Parameter
KNN	5
	7
	9
SVM	RBF
NB	–

mentioned Type-I and Type-II architectures are first applied on EngEnt feature set after that applied on BP and AAR feature sets. In Type-III architecture at once, we are using one base classifier on three already mentioned feature sets: EngEnt, BP, and AAR. In the first three phases, again we have used KNN with varying K values 5, 7, and 9. The K value was tuned rigorously and as the said values provide best and consistent results, we choose them. The kernel function for SVM is always kept as Radial Basis Function (RBF). The third classifier NB is used in its classical form without any alteration. Table 4 provides details of base classifiers used in the Type-III architecture.

3 Results and Discussion

Ensemble architectures are said to be efficient if its individual learners exhibit contradictions to one another, i.e., diversity among base classifiers. There are different methodologies for bringing diversity to the system: (a) using different training param-

Table 5 The results derived from Type-I architecture

Features	Majority voting	Mean accuracy
EngEnt	81.43	81.19
BP	82.14	81.19
AAR	74.29	76.67

Table 6 The results derived from Type-II architecture

Features	Combination 1		Combination 2		Combination 3	
	Majority voting	Mean	Majority voting	Mean	Majority voting	Mean
EngEnt	80.71	78.81	80.71	79.29	80.71	79.29
BP	71.43	60.71	71.43	61.19	71.43	60.95
AAR	78.57	78.81	77.86	79.29	78.57	77.14

Table 7 The results derived from Type-III architecture

Classifier	Parameter	Majority voting	Mean accuracy
KNN	K = 5	80.71	79.76
KNN	K = 7	82.86	79.52
KNN	K = 9	81.43	78.81
SVM	RBF	77.86	69.29
NB	–	81.43	76.43

eters for a single learner type, (b) using different types of classifiers, and (c) using different types of feature sets for building the prediction model [22–24].

The objective of this paper is to compare different ensemble architectures with different base classifiers and feature set combination. The results of Type-I architecture with majority voting combined technique and mean accuracy of base classifiers are shown in Table 5 [25]. Similarly, Table 6 displays the majority voting and mean accuracy (*obtained from individual base classifiers accuracies*) results of Type-II architecture with combination 1, 2, and 3, respectively. Initially, we focus on the Type-I architecture where BP gives the highest result with majority voting combining technique and $EngEnt$ provides the second best result. It is observed that the Type-II architecture with different base classifiers combination and $EngEnt$ feature set always give the highest result than any other feature set combinations. AAR also gives the second highest result in Type-II architecture. In case of Type-III architecture, KNN classifier with K value 7 provides the maximum accuracy 82.86% with majority voting combining technique (Table 7). The NB classifier comes up with the second best result 81.43% over the SVM variants. The graphical representations of Tables 5 and 6 are shown in Figs. 6 and 7.

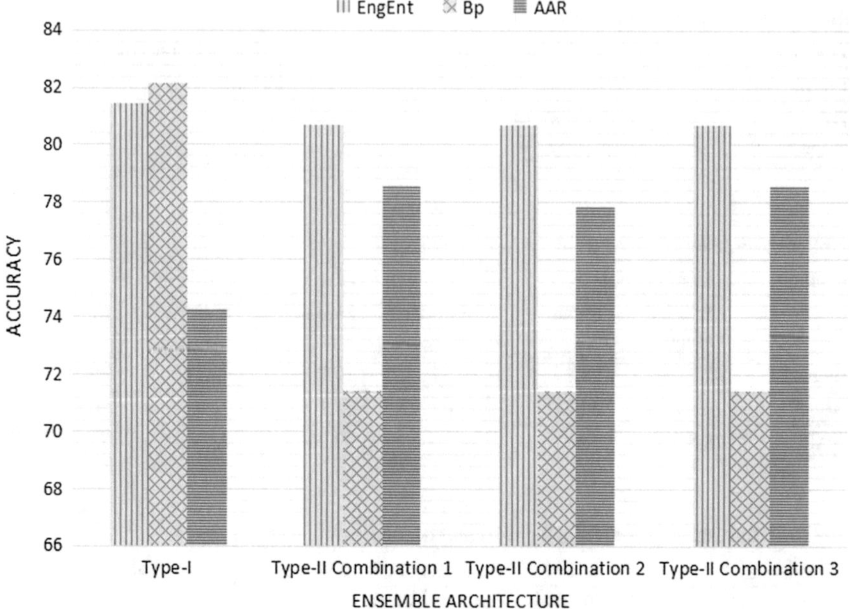

Fig. 6 Accuracies obtained from Type-I and Type-II architectures with majority voting

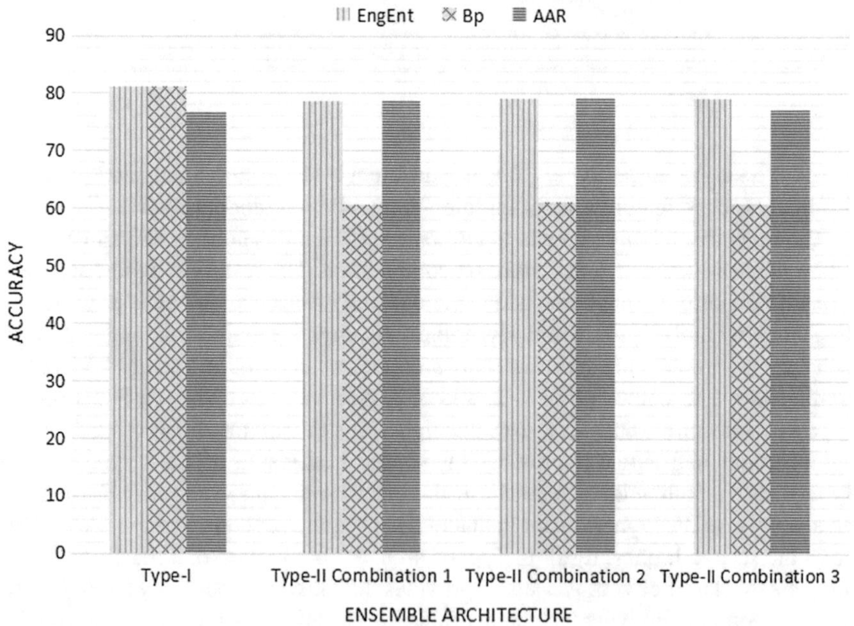

Fig. 7 Accuracies obtained from Type-I and Type-II architectures with mean accuracy

In our study, we have examined all three said types of ensemble composition on EEG signal classification. With varying K value, KNN generates different decision boundaries for each dataset, which brings higher diversity in predictions for the Type-I architecture. The Type-II architecture also demonstrates diversity for EngEnt and AAR but it is not as good as compared to Type-I for BP. The reason for overall inconsistent performance in Type-II is possibly due to the similar outputs from all the learners. The Type-III uses all three categories of datasets (EngEnt, BP, and AAR) for training. One of its variants provides the highest result among all the observations in this paper. In a broad sense, the three mentioned ensemble architectures reflect good competitiveness among themselves. However, wavelet-based EngEnt as feature set and Type-I as ensemble composition are concluded most reliable performer for motor imagery EEG signal classification.

4 Conclusion

This paper presents ensemble architectures with varying base classifiers and feature sets for motor imagery EEG signal classification. After discussing the results in the previous section, we have concluded that wavelet-based $EngEnt$ is most appropriate feature extraction technique because mostly it performs better than others. The Type-II architecture with AAR feature set composition perform better than Type-I architecture; otherwise, the Type-I ensemble architecture can be considered as most reliable type due to its consistent performance. However, it is noticed that out of all empirical observations, Type-III architecture with KNN base classifier and k value as 7 is the highest performing ensemble composition. The another observation is that the KNN classifier is the most stable classifier for ensemble approach to motor imagery signal classification. Our future work will be an extensive survey of different classifiers in the aforesaid ensemble composition and to examine their competitiveness in terms of performance.

References

1. Goel, M.K.: An overview of brain computer interface. In: Proceedings of Recent and Emerging Trends in Computer and Computational Sciences (RETCOMP), 2015, pp. 10–17. IEEE (2015)
2. Townsend, G., Graimann, B., Pfurtscheller, G.: Continuous EEG classification during motor imagery-simulation of an asynchronous BCI. IEEE Trans. Neural Syst. Rehabil. Eng. **12**(2), 258–265 (2004)
3. Chaudhari, R., Galiyawala, H.J.: A review on motor imagery signal classification for BCI. Sig. Process. Int. J. (SPIJ) **11**(2), 16 (2017)
4. Lotte, F.: Study of electroencephalographic signal processing and classification techniques towards the use of brain-computer interfaces in virtual reality applications. Ph.D. thesis, INSA de Rennes (2008)
5. Ilyas, M.Z., Saad, P., Ahmad, M.I.: A survey of analysis and classification of EEQ signals for brain-computer interfaces. In: 2015 2nd International Conference on Biomedical Engineering (ICoBE), pp. 1–6. IEEE (2015)

6. Vaid, S., Singh, P., Kaur, C.: EEQ signal analysis for BCI interface: a review. In: 2015 Fifth International Conference on Advanced Computing & Communication Technologies (ACCT), pp. 143–147. IEEE (2015)
7. Rahman, A., Tasnim, S.: Ensemble classifiers and their applications: a review (2014). arXiv:1404.4088
8. Rokach, L.: Ensemble-based classifiers. Artif. Intell. Rev. **33**(1), 1–39 (2010)
9. Lei, X., Yang, P., Xu, P., Liu, T.J., Yao, D.Z.: Common spatial pattern ensemble classifier and its application in brain-computer interface. J. Electron. Sci. Technol. **7**(1), 17–21 (2009)
10. Bhattacharyya, S., Konar, A., Tibarewala, D., Khasnobish, A., Janarthanan, R.: Performance analysis of ensemble methods for multi-class classification of motor imagery EEG signal. In: 2014 International Conference on Control, Instrumentation, Energy and Communication (CIEC), pp. 712–716. IEEE (2014)
11. Nicolas-Alonso, L.F., Corralejo, R., Gómez-Pilar, J., Álvarez, D., Hornero, R.: Ensemble learning for classification of motor imagery tasks in multiclass brain computer interfaces. In: 2014 6th Conference on Computer Science and Electronic Engineering Conference (CEEC), pp. 79–84. IEEE (2014)
12. Datta, A., et al.: An ensemble classification approach to motor-imagery brain state discrimination problem. In: International Conference on Infocom Technologies and Unmanned Systems (ICTUS). IEEE (2017). (In Press)
13. Hu, D., Li, W., Chen, X.: Feature extraction of motor imagery EEG signals based on wavelet packet decomposition. In: 2011 IEEE/ICME International Conference on Complex Medical Engineering (CME), pp. 694–697. IEEE (2011)
14. Siuly, S., Li, Y.: Improving the separability of motor imagery EEG signals using a cross correlation-based least square support vector machine for brain-computer interface. IEEE Trans. Neural Syst. Rehabil. Eng. **20**(4), 526–538 (2012)
15. Bhaduri, S., Khasnobish, A., Bose, R., Tibarewala, D.: Classification of lower limb motor imagery using k nearest neighbor and naïve-bayesian classifier. In: 2016 3rd International Conference on Recent Advances in Information Technology (RAIT), pp. 499–504. IEEE (2016)
16. BCI-Competition-II: Dataset III, Department of Medical Informatics, Institute for Biomedical Engineering, University of Technology Graz (2004). Accessed 6 June 2015
17. Chatterjee, R., Bandyopadhyay, T.: EEQ based motor imagery classification using SVM and MLP. In: 2016 2nd International Conference on Computational Intelligence and Networks (CINE), pp. 84–89 IEEE (2016)
18. Chatterjee, R., Bandyopadhyay, T., Sanyal, D.K.: Effects of wavelets on quality of features in motor-imagery EEQ signal classification. In: International Conference on Wireless Communications, Signal Processing and Networking (WiSPNET), pp. 1346–1350. IEEE (2016)
19. Brodu, N., Lotte, F., Lécuyer, A.: Comparative study of band-power extraction techniques for motor imagery classification. In: 2011 IEEE Symposium on Computational Intelligence, Cognitive Algorithms, Mind, and Brain (CCMB), pp. 1–6 IEEE (2011)
20. Schlögl, A.: The electroencephalogram and the adaptive autoregressive model: theory and applications. Shaker Aachen (2000)
21. Rahman, A., Verma, B.: Effect of ensemble classifier composition on offline cursive character recognition. Inf. Process. Manag. **49**(4), 852–864 (2013)
22. Alani, S.: Design of intelligent ensembled classifiers combination methods. Ph.D. thesis, Brunel University London (2015)
23. Nascimento, D.S., Canuto, A.M., Silva, L.M., Coelho, A.L.: Combining different ways to generate diversity in bagging models: an evolutionary approach. In: The 2011 International Joint Conference on Neural Networks (IJCNN), pp. 2235–2242. IEEE (2011)
24. Yu, L., Wang, S., Lai, K.K.: Investigation of diversity strategies in SVM ensemble learning. In: Fourth International Conference on Natural Computation, 2008. ICNC'08, vol. 7, pp. 39–42. IEEE (2008)
25. Sun, S., Zhang, C., Zhang, D.: An experimental evaluation of ensemble methods for EEG signal classification. Pattern Recogn. Lett. **28**(15), 2157–2163 (2007)

Improving Quality of Fruits Picked in Orchards Using Mahalanobis Distance Based Model

Kalaivanan Sandamurthy and Kalpana Ramanujam

Abstract A challenge in fruit orchards is the collection of fruits lying on the floor. They need manual collection which is time-consuming and labor-intensive. Automated collection is desired in such a scenario. A predefined collection path is required for automated collection using robots. A novel path planning methodology is introduced to perform nonuniform coverage, and paths are evaluated with the information available about fruit distribution probability in the environment. A fruit probability map was developed describing the potential of fruits at every location. A mathematical model defining the shortest path to a fruit distribution using Mahalanobis distance is evaluated. This path is compared with the paths used by a farmer for the collection of fruits. The amount of time spent by fruits lying in the ground has to be minimized. An even bigger challenge is the decay of fruits lying in the ground. Though optimality of the path is not guaranteed with respect to fruit distribution, usage of Mahalanobis distance model has significantly reduced the time spent by fruits lying in the floor. The presented results are promising for the usage of automated robots to collect fruits. This will lead to a reduction in the demand for manual labor. Extending the model with elevation data could further improve the results by increasing energy efficiency and decreasing time consumed.

Keywords Mahalanobis distance · Path planning · Fruit picking

K. Sandamurthy (✉) · K. Ramanujam
Department of Computer Science and Engineering, Pondicherry Engineering College, Pillaichavady 605014, Puducherry, India
e-mail: kalai_4390@yahoo.co.in

K. Ramanujam
e-mail: rkalpana@pec.edu

A. Abraham et al. (eds.), *Emerging Technologies in Data Mining and Information Security*, Advances in Intelligent Systems and Computing 813,
https://doi.org/10.1007/978-981-13-1498-8_14

1 Introduction

1.1 Cashew Fruits

India is the largest processor of cashew and second largest producer in the world. The export of cashew stood at 82,302 metric ton with a value of rupees 5168.78 crores [1] during 2016–2017. The production from India stood at 294,590 metric ton in the year 1990 and witnessed an increase to around 613,000 metric ton in the year 2010 [2]. The production is widely distributed around eight major states located across the country. The nuts along with the fruits fall to the ground on a daily basis. Fruits lying on the ground for an elongated period of time is a problem in cashew orchards. The quality of the fruits is deteriorated due to contamination from the soil and insects. Frequent collection of nuts is required to avoid this decrease in quality. This has paved the way for specific adaptations in the farmland and a number of management and control measures by the farmer. None of them has been fully efficient since frequent manual collection is required. This is a physically demanding job and with the decrease in availability of cheap labor for agricultural tasks, automation is the much desired solution for this problem.

1.2 Fruit Collection

Various automated systems have been tested for specific crops in literature. Trunk shaker [3, 4] was the earliest known system introduced for automation. It was designed for deciduous fruits like orange and mandarins. It clamps to the trunk of the tree and gives out a vigorous shake making the fruits to fall on the ground. It did less damage to the trees when compared to canopy shaker [5]. However, the fruits which fell on the ground were prone to bruises, affecting the quality of the fruits. Defoliation and damage to the bark were observed in these techniques. An automatic harvester was introduced in the MAGALI project [6]. A black and white camera attached to a grasping tool detected the presence of fruits and plucked them. Around 50% of harvesting was achieved. An optimized approach involving both human and machine was addressed in Agribot project [7]. Detection of fruits was done by human and the gripper equipped with laser telemetry plucked the fruits without causing any damage to the plants. Among the above methods, Agribot is considered to be the safest since there was less damage done to the plant and also the fruit. However, in a cashew orchard, this is not feasible since the quantity of fruits is huge. Hence, an automated system for collection of fruits from the ground remains the only viable solution.

1.3 Objectives

This paper focuses on addressing the path planning for collection of fruits from the farmland. The methodology must account for the fruits in the floor that are nonuniformly distributed with respect to time (the moment when fruits fall) and space (the location in the farmland). The main objectives of the path planning algorithm are

1. To prevent loss of quality, the time fruits spend lying on the floor must be minimized;
2. Locations with higher distribution of fruits must be visited initially;
3. The algorithm must be able to exploit the nonuniform distribution of fruits;
4. Maximize the number of fruits collected.

To satisfy the first constraint, a model using Mahalanobis distance is proposed. Objectives 2 and 3 require knowledge about the distribution prior to performing a search operation in the given environment. This suggests an offline path planning approach. Offline algorithms have a priori knowledge about the environment. This makes the computation of paths to be performed before the robot is introduced into the field. They tend to be cheap since the requirement of high-end sensors is eliminated when compared to online techniques.

2 A Priori Knowledge for Path Planning

The information used as a priori knowledge for the path planner is presented here. Based on this knowledge, a map with the information on location of fruits and obstacles is constructed. A fruit potential is assigned to the locations, which describes the expected number of fruits at each location. With this available information, a collection path is planned. During the process of collection, the actual location of fruits is collected. The map is updated with this available information, thereby for the following day the location with higher potential of fruits can be visited first. As shown in Fig. 1, the map is updated on a daily basis, scheduling the visits to locations with less fruits at the end of the collection cycle.

A novel model for planning of the path which satisfies the objective of minimizing the time spent by a fruit lying on the floor is evaluated. The performance of the model is evaluated with customary path planned and referenced to the collection behavior of a farmer. To have a reference for our model and compare with practical experience, a reference farm was selected, and data was collected with the help of researchers in agriculture domain.

Figure 1 details the flow of the path planning model. Data collected in the farm is used to form a map detailing the fruit distribution of the reference farm. Potential locations for fruits are analyzed using a priori knowledge of the farm obtained on a daily basis. A path is then planned to the locations and followed for collection of

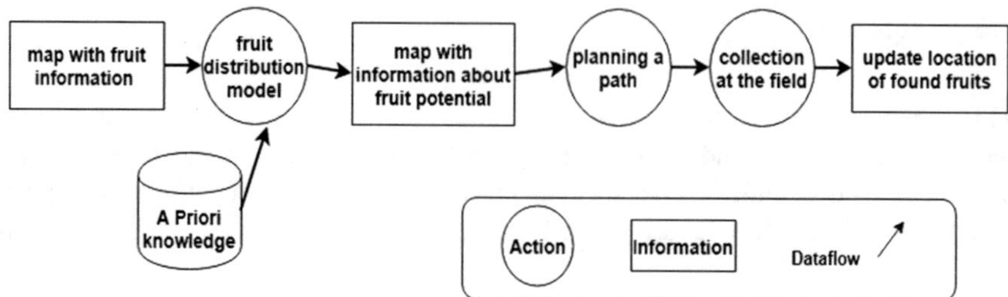

Fig. 1 Path planning model

Table 1 Parameters considered

N	Total number of locations
k	Index of location
Y	Length of the map
i	Location in length
j	Location in width
F_{time}	Time a fruit spends lying on the ground
t_{pickup}	Time of collection of fruit
t_{fall}	Time of fall to the ground
β	Rate of decay
$Y_{i,j}$	Yield at location *i,j*
MD	Mahalanobis distance
$D_{i,j}$	Measured distance at location *i,j*

fruits. The model employed for selection of which fruit to be picked next is the main focus on this work.

Table 1 lists the various parameters considered in evaluating the performance of the proposed model. Mahalanobis distance [8] is represented as

$$\Delta^2 = (\mathrm{X} - \overline{\mathrm{X}})\mathrm{S}^{-1}(\mathrm{X} - \overline{\mathrm{X}}), \quad (1)$$

where x is a row vector consisting of the measurements for observation. In this model, the vector consists of F_{time} and distance value. $\overline{x}$ is the measured mean of the sample. S denoted the covariance matrix of the given sample, the variance represented in the diagonal of the matrix, and covariance of each variable in the remaining entries of the matrix.

Objective 4 requires that maximum number of fruits to be collected per day. As indicated earlier, the yield is related to the potential of fruits present at a given location. The maximization of yield is represented as

$$\sum_{k=1}^{n} Yi(k), j(k). \quad (2)$$

Here, N denotes the total number of locations, and i(k), j(k) denote the locations visited at instance k. The potential $Y_{i,j}$ at location i,j is denoted by $Y_{i,j} = P_{i,j}$, where p_{ij} is the potential at location $l_{i,j}$ specified in the map. Areas containing obstacles cannot be accessed by the robot, and hence they are denoted by a potential value of $p_{i,j} = \infty$. A dynamic programming solution can be achieved for maximizing the objective of Eq. (2) resulting in a shortest path to the location with higher potential of fruits. It has the following limitations:

1. The occurrence of oscillatory motions between the current location and the next location with highest potential of fruits.
2. Noncompliance with objective 1 which is to minimize the time spent by a fruit lying on the floor.

Conventional path planning algorithms can plan shortest path to given locations, but cannot minimize the time spent by individual objects lying on the floor. This increases the decay function $\boldsymbol{\beta}$. To minimize this function, Eq. (3) has to be satisfied which is dependent on the values in Eq. (4).

$$\text{Min}(F_{time}) \tag{3}$$

$$F_{time} = t_{pickup} - t_{fall} \tag{4}$$

3 Algorithm to Determine Mahalanobis Distance

Conventional path planning models determine the shortest path from starting location to given location by calculating the Euclidean distance between the two given points. This holds well, when the objective is to plan only a shortest feasible path. The given problem has one more additional objective of minimizing the floor time and also maximizing the yield. Path planning using shortest distance cannot satisfy the remaining constraints. The behavior of the distribution is analyzed by its statistical properties. A novel correlation analysis employing Mahalanobis distance for determining the outliers within the available dataset is presented. The following algorithm calculates Mahalanobis distance for every individual fruit in the given location.

The algorithm for a given problem returns the best possible successive location from each location at a given moment.

4 Performance Evaluation Based on Simulation

The quality of the proposed strategy was compared with the pickup strategy used by a farmer. As the data about moment of fall and location was not available from practice, a realization of the potential map was performed. For construction of the above map, the following components were utilized:

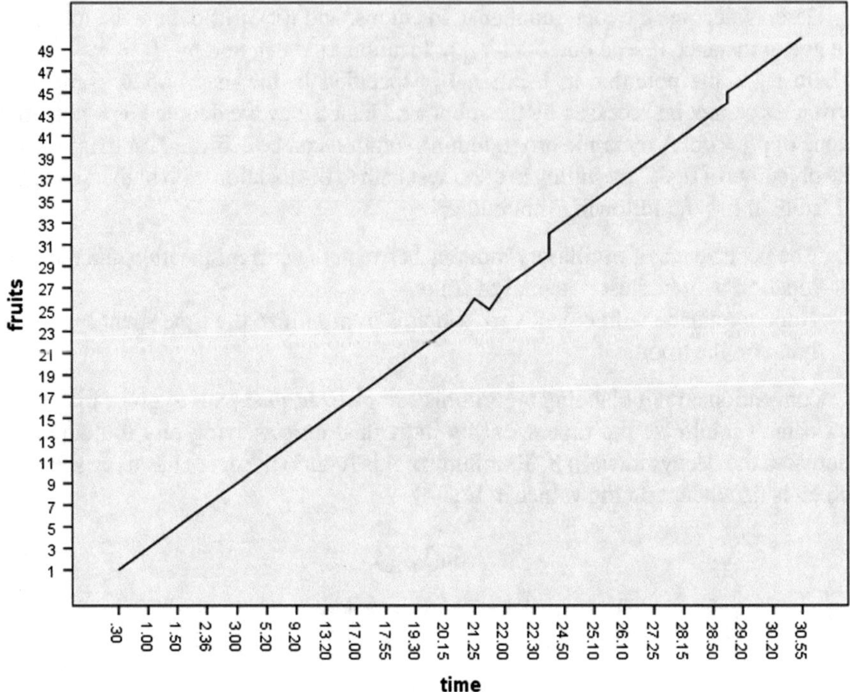

Fig. 2 Fruits at a given time interval

1. The number of fruits at a given time interval, and
2. The distribution of fruits for an interval.

It can be inferred from Fig. 2 that as time increases the number of fruits on the ground has a steep increase.

The data contained information about the number of fruits and the time of fall for each fruit. Conventional planning method would be to pick up the object which is nearer to the current location and then iteratively proceed with the same logic. This would lead toward a solution satisfying Eq. (1). Since the time of fall is not considered here, the decay function associated with each fruit grows exponentially as F_{time} increases. Since distance of an object and time of fall are correlated components, determination of Mahalanobis distance revealed the presence of outliers.

```
Algorithm for computing Mahalanobis distance
  k =N-1

  WHILE k > 1
    FOREACH F_time
      Determine all possible L_i,j
  FOREACH L_i,j
  Calculate MD from F_time and D_i,j
  END FOREACH F_time
  Select the location with highest MD value
  END FOREACH L_i,j
  k=k-1
  END WHILE
```

Utilizing this model ensured the minimization of decay component associated with every fruit. The dataset consisted of the distance of fruits from a given location and the t_{fall}. Assuming that the time of start of a picking operation is known, the F_{time} is calculated. Figure 3 represents the distance of object from a starting location. Figure 4 represents the Mahalanobis distance by considering the distance of object and also the F_{time} of every object. This results in a distribution as shown in Fig. 4. "A point that has a greater Mahalanobis distance from the rest of the sample population of points is said to have higher leverage since it has a greater influence on the slope or coefficients of the regression equation." Since the object having higher Mahalanobis distance has a greater influence on the coefficients of the regression equation, they tend to have higher influence on the performance of the proposed model, i.e., $\boldsymbol{\beta}$ the rate of decay.

To validate this, the paths adopted by a farmer and the simulated model were compared. A farmer's path inclined more toward picking up of nearest fruits from his location. The simulated path takes into account F_{time} adopted in the calculation of Mahalanobis distance. It resulted in additional time to complete the operation but reduced the decay time which was our objective in Eq. (2). 50 items were considered and the mean and standard deviation obtained were $\boldsymbol{\mu} = 1.958$, $\boldsymbol{\sigma} = 1.233$.

The time taken by humans and simulated time taken to visit all the locations in a given field with measurement of 100 square feet were noted. As the number of instances increase, the time taken by human increases due to biological constraints. Though simulation time takes double the time taken by farmer, it takes into account decay function and plans the visits. Hence, minimization of F_{time} is achieved.

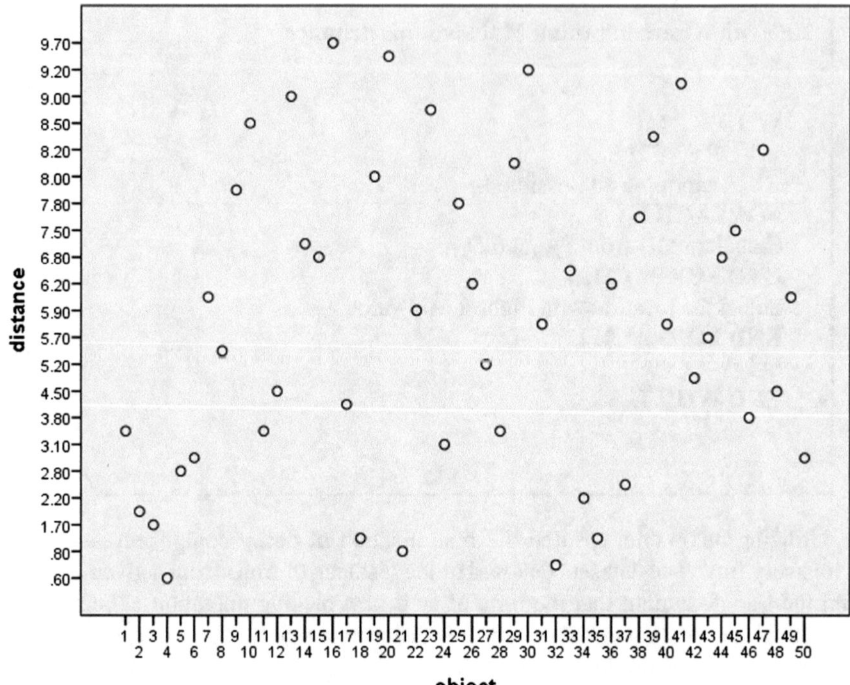

Fig. 3 Distance of object from starting location

Table 2 Results of ANOVA[a] test

Model		Sum of squares	df	Mean square	F	Sig.
1	Regression	8551.083	2	4275.542	291.110	.000
	Residual	660.917	45	14.687		
	Total	9212.000	47			

[a]Dependent Variable: object

The result of ANOVA test is presented in Table 2. The total number of fruits was considered as dependent variable and, F_{time}, measured distance was given as predictors for the test. With the given degrees of freedom, i.e., 2, the obtained F-value rejects the validity of null hypothesis.

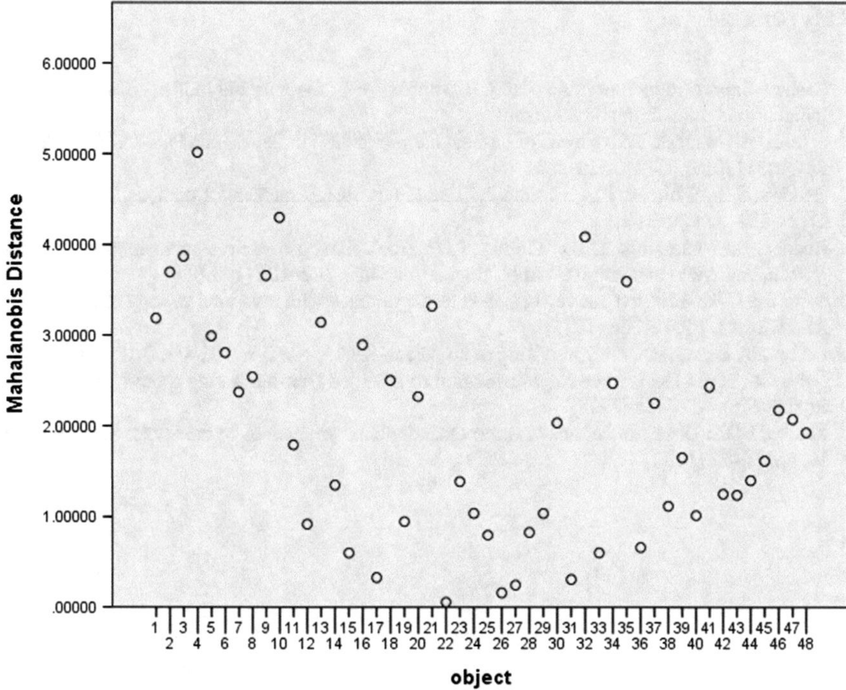

Fig. 4 Mahalanobis distance considering F_{time} and fruit distance

5 Conclusion

A novel model for path planning that takes into account the different independent variables defining a location has been presented and evaluated. The results obtained fulfilled the objectives laid out in Eqs. (3) and (4). Maximization of the objectives presented in Eq. (2) has not been fully satisfied since complete coverage of the given environment was not performed. This ignores the fruits falling during the time of collection which has not been addressed in this model. The speed of the robot employed for collection was not taken into consideration, and also the effect of sharp turns present in the computed path plays a major role in determining the total time taken. These are the limitations to be addressed in future work along with an implementation using a larger dataset of various orchards.

References

1. Cashew Export Council of India (CEPCI), Ministry of Commerce and Industry, Government of India. http://cashewindia.org/statistics
2. Senthil, A., Magesh, P.: Analysis of cashew nut production in India. Asia Pac. J. Mark. Manag. Rev. **2**(3) (2013). ISSN 2319-2836
3. Hedden, S.L., Whitney, J.D., Churchill, D.B.: Trunk shaker removal of oranges. Trans. ASAE **27**(2), 372–374 (1984)
4. Hedden, S.L., Churchill, D.B., Whitney, J.D.: Trunk shakers for citrus harvesting. Part II: tree growth, fruit yield and removal. Appl. Eng. Agric. **4**(2), 102–106 (1989)
5. Summer, H.R.: Selective harvesting of Valencia oranges with a vertical canopy shaker. Trans. ASAE **16**(6), 1024–1026 (1973)
6. D'EsnonA Grand, Robotic harvesting of apples, in ASAE; SME, no. 01 85, 210–214 (1985)
7. Ceres, R., et al.: Design and implementation of an aided fruit-harvesting robot (Agribot). Ind. Robot **25**(5), 337–346 (1998)
8. Klecka, W.R.: Discriminant analysis. In: Quantitative Applications in the Social Sciences, vol. 19, pp. 55–57 (1980)

Viable Crop Prediction Scenario in BigData Using a Novel Approach

Shriya Sahu, Meenu Chawla and Nilay Khare

Abstract Agriculture leads a vital role in the human surveillance where it lies as the initial step for the human civilization. Due to the excessive need for food, the agricultural practices are in large-scale production as a business which is termed as "Agribusiness". Modern agriculture allows both biological and technological developments such as plant breeding, agrochemicals, genetic breeding, remote sensing, crop monitoring, sensor nodes, and automatic maintenance system. The integration of sensor nodes in the farming field leads to the generation of huge data which can need a learning algorithm for analyses to determine a specific solution. There are various machine learning algorithms in practices which are not suitable for handling large datasets. In this paper, a novel algorithm "Agrifi-prediction algorithm" is created which has the functionality of loading the dataset in hdfs and comparing the previous dataset with the current processing dataset. The experimental process is carried out by using the Hadoop framework with MapReduce programming model by analyzing the meteorological and soil dataset and finally compared with the machine learning algorithm to evaluate the accuracy.

Keywords Agriculture · BigData · Prediction algorithm · Meteorological dataset · Hadoop framework

---

S. Sahu (✉)
Department of Computer Science & Application, Bilaspur University, Bilaspur 484886, India
e-mail: shriyasphd@gmail.com

M. Chawla · N. Khare
Department of Computer Science & Engineering, MANIT, Bhopal 462001, India
e-mail: meenuchawlamanit@gmail.com

N. Khare
e-mail: nilay.khare@rediffmail.com

A. Abraham et al. (eds.), *Emerging Technologies in Data Mining and Information Security*, Advances in Intelligent Systems and Computing 813,
https://doi.org/10.1007/978-981-13-1498-8_15

1 Introduction

BigData is the generally defined as a massive amount of data which is generated from different sources in structured, semi-structured, or unstructured format. The main functionality of BigData is mentioned as 4 V's such as volume, velocity, veracity, and variety. Volume denotes the size of the data which will be usually in terabytes, petabytes, and exabytes. Velocity is defined as the processing speed and the variety defines the various types of data which may lie in any of the following categories: structured, unstructured, and semi-structured data. Veracity defines the prediction accuracy which is processed by verifying the data [1, 2].

The major steps in information system are data collection, processing, reports generation, distribution, analysis/monitoring, prediction, and decision-making represented in Fig. 1. Data collection is an approach of collecting and measuring data from different sources to get accurate outcome. Data processing is the process of converting the raw data into useful or needed data. Analysis/monitoring, prediction, and decision-making are the final process, where the collected data are processed to find out a particular outcome.

The data generated in agricultural field are mainly from remote sensing, satellite, and farming technique. This is generally referred to as precision agricultural [3]. There are different types of data generated in agricultural field from various sources (Table 1).

In the current world, the agricultural practices are still made by assumptions of the well-versed farmers even though there are lots of technological changes and improvements. There is always a need for a perfect agricultural model where it supports predictive and assessment capability to help the private and the public sectors, i.e., both government and nongovernment [3, 4].

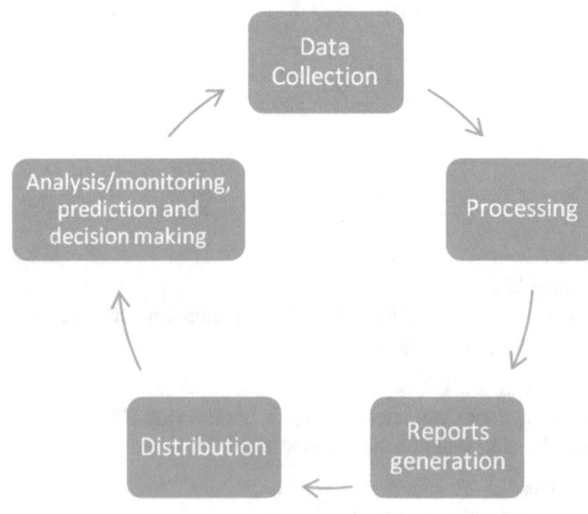

Fig. 1 Stages of data processing from the initial stage as collection to final stage as prediction

Table 1 Agricultural data generation source

S. No	Data type	Data sources
1	Sensor data	Remote sensing device, GPS integrated devices, fertilizers, moistness, farmer's queries, temperature, and equipment logs
2	Historical data	Monitoring farm and yield, weather conditions, GIS data, and labor data
3	Social and web data	Feedback and reviews, agro-advisor blogs, groups in social websites, and websites
4	Publications	Farmers and customers feedback, blogging sites like agro-advisor, agricultural blogs, social media groups, web pages, and data from search engines
5	Streamed data	Service usage (for metering/billing), server activity, website clicks, and geo-location of devices, people, and physical goods
6	Other data	Billing and scheduling systems, agricultural departments and other agricultural equipment manufacturing company

In today's era, the farmers are still estimating their agribusiness procedure physically even there are numerous innovations exist around the universe. Still, the procedure for agricultural business is static. With the progression of few horticulture assets, we are going to anticipate the harvest yield creation in a synchronous procedure. The traditional way is to monitor the crop field by a variety of embedded technique, but the gathered data has no proper organizing and analyzing methods. Agribusiness holders regularly neglect to get an outline of the present crop development stature since information is not spread over progressive systems. The expanding multifaceted nature of development procedures and for all time developing measure of data prompt to an over-burden of the agribusiness holders as for process checking, data analysis, and fault detection. In this manner, maintenance intervals are not chosen correctly and optimization potential with respect to throughput [5, 6].

Hadoop is an open-source programming framework, and data handling platform helps you to aggregate and apply agricultural intelligence anywhere around the natural world. More than midpoints and chronicled weather data can be examined with a less computational speed. After analyzing the crop development of a month, crop management should be done. Through the MapReduce, huge data are been investigated without challenges. By making a novel algorithm as "Agrifi algorithm", contingent upon the machine learning methods is utilized to deal with the crop development framework. The objective of this work is represented in Fig. 2 using the novel Agrifi-prediction algorithm.

2 Related Work

In paper [7], the evaluation takes place by NDVI technique which is calculated from the energy of electromagnetic waves obtained using the target crop. NDVI is calcu-

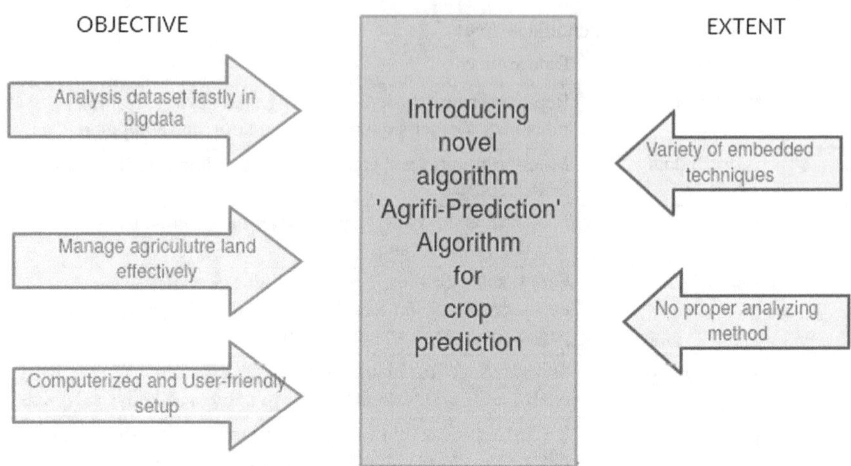

Fig. 2 The main objective of the research work for the crop prediction using the Agrifi-prediction algorithm

lated by the cosine function which is triply modulated with mean value, first stage of variables, and amplitude. The approach for determining the initial period of one crop at 8-day composite is obtained by moderate resolution imaging spectroradiometer. The estimator accuracy is not maintained, and random results are analyzed.

In paper [8], the oil consumption is proposed based on the amount of oil extracted and sold in markets with respect to demand by utilizing the support vector regression algorithm. The volume of oil extracted is calculated in association with many other factors. So, neural network algorithm is utilized to calculate the functionality between the social and economic inputs where the consumption of oil is the final solution. The error report of neural network algorithm is 0.0005 for the past 5 years, and support vector regression produces better accuracy than the other algorithms for the consumption of oil.

In paper [9], analysis and evaluation of water management have been taken place for the society living being and security. A novel technique is used to handle the large volume of fine-grained data where the robust probabilistic classifier is used. The data is in arbitrary dimensionality and in bulk volume. Water management field is taken and cost estimation of the work is high. Forecasting process has an offbeat procedure which is not a dependable one.

In paper [10], Eolic parks require remote sensing techniques where it generates huge amount of data and utilizes cloud computing techniques for storage of those data. The integration of hardware and software architecture allows processing this large, distributed, and heterogeneous datasets. The initial stage is the utilization of raspberry pi to save 80% of data that are used for communication and stored in the SD card. The second stage is the usage of classical database management for the local server. The final stage is the integration of cloud computing for data storage and processing with the visualization task. The demonstrating scalability of the system fails.

In paper [11], prediction is done on the wind and solar resource depends upon timescales. The meteorological variables are handled for the prediction where it requires multitude of disparate data, various levels of models that have specific time frame, and application of computational intelligence approaches for the integration of all the model and observed information and allows taking decision-making process. The requirement for enhanced conjectures of the renewable vitality factors absolutely is turning out to be progressively imperative as we look to mix more wind and solar energy into the grid.

In paper [12], the strategy initially starts by obtaining the entire data description about the dataset. Choose training sets to have the final result as the estimator in unbiased for the given data. The main objective is to obtain complete knowledge about the entity which helps for the statistical calculation. The proposed technique is used for increasing the efficiency without maximizing the monitoring cost, and the entire process is continued where it can be easily scalable for the monitoring of data obtained from distant sensor node system.

In paper [13], the risk happens in apple due to the apple scab in the Himalayan region that is identified by the prediction. For the prediction process, the beta-regression approach is used as classical formula for calculating the extreme index, and the implementation is carried out in Python language to forecast the apple scab in apple. The input dataset is collected from surveillance of pest dataset about apple scab.

In paper [14], a dynamic threshold method is proposed which is utilized for obtaining the freeze-up and break-up dates of big lakes from microwave brightness temperature data. The passive microwave brightness temperature data is obtained from the 35 biggest lakes from the northern hemisphere for past three decades. The distant sensing node systems discuss individually the issues in it. The main issues in handling the huge sensed data are how to manage and maintain the quality of data from various views. These freeze-up and break-up dates were validated from 18 stations' observations. The correlation coefficient in linear form calculated between observation and retrievals are 0.926 and 0.936.

In paper [15], a framework for the utilization of Information and Communication Technology (ICT) services is proposed in agricultural process to gather large data. BigData analysis for the agriculture data results in providing a different perspective for weather decisions, increase yield productivity, and reduce the expenses for harvesting, pesticide, and fertilizers. The entire process is carried out using the data processing technique and weather forecasting application.

In paper [16], a heuristic algorithm and multi-temporal technique are proposed for the detection of paddy growth using the huge data obtained from the remote sensing technique. The crop growth generally has a lot of data extraction phases which helps to find the water resource quantity, planting date, and required fertilizer for the improvement of crop yield. This paper focuses on the main challenges of how to determine the field for the growth of paddy from the result of previous paddy growth, with the utilization of distant sensing node technique. Sensing data collected from the non-irrigated farm field has low efficiency.

In paper [17], the proposed work is used to forecast the integration of different varieties of noises and mono-value of iron portion in series of sale price with improved accuracy. The prediction accuracy of this integration approach is more than that of any existing prediction model. The proposed integration approach is more feasible and reliable for large dataset. In particular, the integration approach is more suitable for mining and decision-making process. This effectively identifies the portion of iron in the dataset. Finally, the prediction has more accuracy in making decision and support enterprises.

In paper [18], they proposed the yield prediction from the input images generated by Radarsat 2. This work has a novel calculation technique from the relationship between the yield obtained from situ and backscatter. The landscape data collected from the Boussalem which is located in Tunisia. The data is collected at the edge of the season with the Radarsat2 images. Single data is integrated with the data obtained during acquisition stage and next data during flowering period. Final consequence shows that there is primary correlation coefficient obtained.

In paper [19], they proposed a BigData platform for a geospatial data obtained from the yield from the single farm level up to the continental scale. They identified the coupling of BigData approach with the integrated repository and services (PAIRS), for the decision support system for agro-technology transfer (DSSAT) crop model. This foresees the global scale of geospatial analytics, and PAIRS provides the solution for the heterogeneous data to integrate the crop models with the dataset size of hundreds of terabytes.

In paper [20], they proposed a framework for processing the live BigData with the help of load balancing, parallel processing, and filtration. The entire framework will be responsible for execution, storing the final solution, and making decision from the result obtained from the processing unit. The outcome is used for analyzing the real distant sensing system using geospatial data.

In paper [21], they presented U-Agriculture service based on cloud infrastructure. This may be utilized by farmers, services, and organization in management of BigData. The main process of U-Agriculture is to make a decision from the sensed data. Based on the following strategies, they presented architecture for U-Agriculture: 1. irrigation time, 2. seeding and yielding time, and 3. sowing fertilizer in the fields.

3 Problem Description

The main issues faced by the farmers in the agricultural land are the identification of disease in the crop. The crop diseases are difficult to identify in the earlier stage which may lead to serious issues in crop cultivation that spreads mainly due to the climate change. The production of crops is mainly damaged by diseases which lead to huge loss to the agricultural production. During the green revolution, the usage of insecticides and pesticides is increased to maximize the yield; however, the use of insecticides and pesticides depletes the soil nourishment [22]. The relationship between the weather and the possibility of occurrence of disease is quite complex and depends on many other conditions [23]. Apart from the crop disease, yield prediction plays a vital rolc in agricultural problem.

4 System Objective

- The principle target of this work is to foresee the event of hazard that considers the crop production management, which is vital for a government as well as for farmers or agriculture business holders or trading companies.
- Crop management assumes a noteworthy part in production and good impact in modern countries. A proper crop management dataset can spare the cost, the absence of interest, and costs of excess.
- This work will likewise be exceptionally helpful for the farmers to screen the harvest in an incessant premise utilizing this prediction steps.

5 System Design

The present research work utilizes a novel algorithm as "Agrifi algorithm" to foresee the stature of crop by deciding the agriculture land data and advises which area of crop would be healthy. Contingent on different farms, we are intending to regularize the dataset fields that will suit for all different crop managements. The data of soil can be dictated by the weather data in view of the weather-related datasets of earlier month containing the base temperature, most extreme temperature, sun sparkle (hours), rainfall (mm), moistness (percentage), and wind speed (kmph). By gathering the fields of the dataset, it is being investigated through the BigData innovation. The paper executes the Agrifi algorithm utilizing Hadoop stage.

5.1 Selection Model

In the selection model, a remote sensor hub is prototyped that actualized our proposed structure for detecting meteorological data. The sensor hub is a shrewd receptive gadget that spotlights on gathering soil dampness progression information regarding encompassing the alteration in atmosphere. The hub is built under open source which can be effectively reconstructed. It provides two clients characterized factors managing the level of data granularity and test interims. It is particularly intended for applications that need more duration.

5.2 Prediction Model

In this model, crop growth forecasting system is proposed to determine and conjecture crop growth stage after some period in view of dataset. The forecasting system depends on fundamental approach created by knowledge learning calculations. Building up an essential process, dataset is processed in the Hadoop systems, nourished to machine learning calculations for preparing and approving for further process. The attributes of meteorological data are temperature, sun sparkle, rainfall, moistness, and wind speed, together with the previous month values.

A stature loop has been built on the overall flow of the research work to validate the crop prediction process comparing with the predefined month of the crop stature. As a result, running a map job using HDFS (as represented in Fig. 3) in the iteration process for n times by processing the input obtained from the existing output, the job identifies the overall standardized data (Sd). This model creates the way for user data manipulation at any iteration. The input data can be obtained from the soil retrieval approaches or existing predicted data. The standardized data are then reduced using HDFS to classified data (Cd) by checking the dataset. Using a function called similar, the dataset has been predicted with the stature of the crop.

In the technological development in weather technology, the estimating of climatic data has turned out to be more precise and fine-grained, and it is accessible to people in general. Our framework can incorporate meteorological information from month to month from climate determining. On the input side, readings from meteorological information strategies can be coordinated to advance enhance precision. Physical estimation offers extraordinary exactness in the resources and manual work. The exact true value from remote sensor estimation can be utilized as the contribution

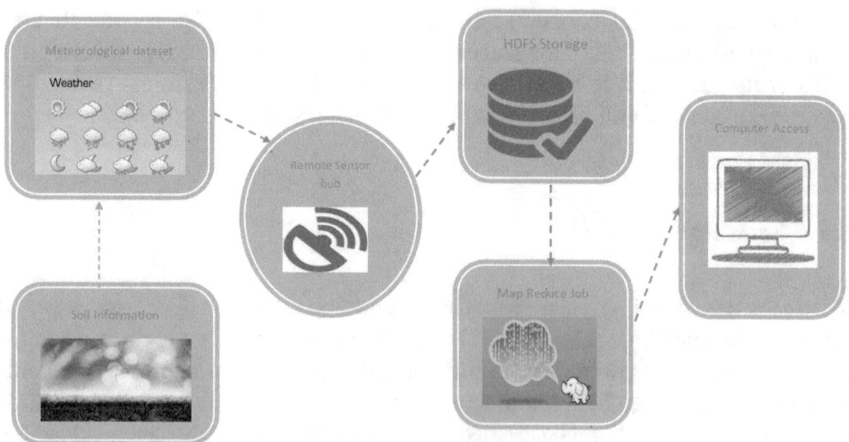

Fig. 3 Viable crop prediction architecture from the dataset collection to processing the dataset with MapReduce and HDFS storage

to amend the technique, as the final result tends to "float" initialize to earliest stage after a few execution.

```
  Input: Dataset (temp, wind, sunrise, rainfall stored in
HDFS)
  Output: Crop Stature Prediction
  For each item(i)
  Isum= ∑i; i• 0 to ∞
  Imean = Isum /N;
  Generate Sd using HDFS
  then
  For each range
  If (Sd == v)
    v++;
  Else
    Return sqrt((i2-((i2)/n))/(n-1))
  Generate Cd using HDFS
  Similar()
  If(Cd ==predefined Cd)
  crop healthy;
  Else
  Return crop unhealthy;
```

Algorithm: Agrifi-Prediction Algorithm

6 Experimental Result

Meteorological and soil datasets are collected from the field using the sensing hub from the field [24–31]. The meteorological dataset contains precipitation, temperature, cloud cover, vapour pressure, wet–day frequency, and reference crop evapotranspiration, and the soil dataset contains nitrogen(N), phosphorous(P), potassium(K), soil organic carbon(SOC), pH level, soil type, and moisture which are the main parameters [32–34].

The device is consummate in the Ubuntu 14.04 LTS operating system; with Hadoop 2.6.0, the dataset is composed from different sources to forecast the relevant output. The MapReduce implementation of the system takes 2.44 min in single-node Hadoop setup and represented in Ubuntu terminal as shown in Figs. 4 and 5.

6.1 Confusion Matrix

The confusion matrix is the predictive analysis which contains table with two rows and two columns. The negative and positive estimates are reported from both the predicted and actual cases as true and false for positive and negative case (Tables 2).

$$\text{Precision} = \frac{\text{TP}}{\text{predicted condition positive}} = 0.9721$$

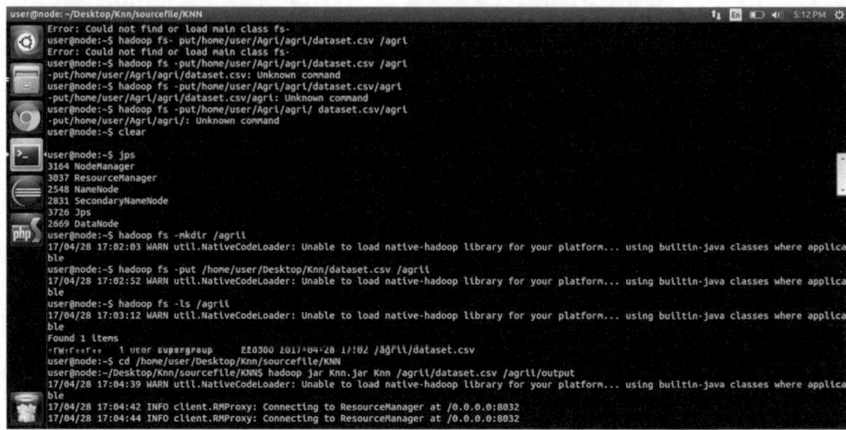

Fig. 4 On top of cipher denoted by the Hadoop framework starting process

```
user@node: ~/Desktop/Knn/sourcefile/KNN                                    5:11 PM
17/04/28 17:04:45 WARN mapreduce.JobSubmitter: Hadoop command-line option parsing not performed. Implement the Tool interface and execute your a
pplication with ToolRunner to remedy this.
17/04/28 17:04:46 INFO mapred.FileInputFormat: Total input paths to process : 1
17/04/28 17:04:46 INFO mapreduce.JobSubmitter: number of splits:2
17/04/28 17:04:47 INFO mapreduce.JobSubmitter: Submitting tokens for job: job_1493378575368_0001
17/04/28 17:04:49 INFO impl.YarnClientImpl: Submitted application application_1493378575368_0001
17/04/28 17:04:49 INFO mapreduce.Job: The url to track the job: http://node:8088/proxy/application_1493378575368_0001/
17/04/28 17:04:49 INFO mapreduce.Job: Running job: job_1493378575368_0001
17/04/28 17:05:24 INFO mapreduce.Job: Job job_1493378575368_0001 running in uber mode : false
17/04/28 17:05:24 INFO mapreduce.Job:  map 0% reduce 0%
17/04/28 17:06:24 INFO mapreduce.Job:  map 100% reduce 0%
17/04/28 17:07:08 INFO mapreduce.Job:  map 100% reduce 100%
17/04/28 17:07:10 INFO mapreduce.Job: Job job_1493378575368_0001 completed successfully
17/04/28 17:07:10 INFO mapreduce.Job: Counters: 49
        File System Counters
                FILE: Number of bytes read=12600
                FILE: Number of bytes written=343406
                FILE: Number of read operations=0
                FILE: Number of large read operations=0
                FILE: Number of write operations=0
                HDFS: Number of bytes read=228123
                HDFS: Number of bytes written=61
                HDFS: Number of read operations=9
                HDFS: Number of large read operations=0
                HDFS: Number of write operations=2
        Job Counters
                Launched map tasks=2
                Launched reduce tasks=1
                Data-local map tasks=2
                Total time spent by all maps in occupied slots (ms)=115021
                Total time spent by all reduces in occupied slots (ms)=31441
                Total time spent by all map tasks (ms)=115021
                Total time spent by all reduce tasks (ms)=31441
                Total vcore-seconds taken by all map tasks=115021
                Total vcore-seconds taken by all reduce tasks=31441
```

Fig. 5 The above figure is represented by MapReducing program that is completed successfully with 100% mapping and reducing process

Table 2 Confusion matrix table denoting positive and negative case

	Predicted negative	Predicted positive
Actual negative	TN 173	FP 27
Actual positive	FN 16	TP 941

$$\text{Recall} = \frac{\text{TP}}{\text{condition positive}} = 0.9832$$

$$\text{F1 score} = \frac{2}{\frac{1}{\text{recall}} + \frac{1}{\text{precision}}} = 0.9775$$

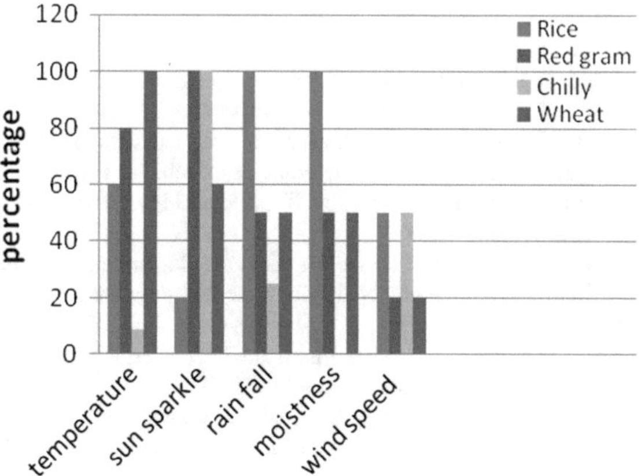

Fig. 6 The figure is represented by MapReducing program that is completed successfully with 100% mapping and reducing process

$$\text{Accuracy} = \frac{\text{TP} + \text{TN}}{\text{total population}} 0.9628$$

Four different crops are considered to compare the parameters such as temperature, sun sparkle, rainfall, moistness, and wind speed. The comparison of rice, red gram, chilly, and wheat is shown in Fig. 6.

7 Conclusion

Agriculture is the main source of the human surveillance with the increase in the food and daily activities. However, the growth in the instruments in the farming leads to the generation of large amount of data where this needs a proper system to maintain and analyze the data. This paper proposed a novel "Agrifi-prediction algorithm" to forecast the appropriate crop for the farm field from the meteorological and soil dataset. The experimental phase is implemented in Hadoop framework by loading both the meteorological and soil datasets as the input dataset with the source program MapReduce programming model. The proposed system obtains the accuracy of 0.963 derived from the confusion matrix. The proposed algorithm helps the agribusiness to predict the suitable crop from the loaded input dataset which enhances the agricultural business.

References

1. Mintert, J., Widmar, D., Langemeier, M., Boehlje M., Erickson, B.: The Challenges Of Precision Agriculture: Is Big Data The Answer. Presentation at the Southern Agricultural Economics

Association (2015)

2. Lamrhari, S., Elghazi, H.: A profile-based big data architecture for agricultural context. analysis of soil behaviour and prediction of crop yield using data mining approach. In: 2nd International Conference on Electrical and Information Technologies ICEIT' (2016)
3. Bendre, M.R., Thool, R.C., Thool, V.R.: Big data in precision agriculture through ICT: rainfall prediction using neural network approach. In: Proceedings of the International Congress on Information and Communication Technology. Advances in Intelligent Systems and Computing (2016)
4. Fan, W., Chong, C., Xiaoling, G., Hua, Y., Juyun, W.: Prediction of crop yield using big data. In: 8th International Symposium on Computational Intelligence and Design (2015)
5. Gandhi, N., Armstrong, L.J., Petkar, O., Kumar Tripathy, A.: Rice crop yield prediction in India using support vector machines. In: 13th International Joint Conference on Computer Science and Software Engineering (JCSSE) (2016)
6. Gandhi, N., Armstrong, L.: Applying data mining techniques to predict yield of rice in humid subtropical climatic zone of India. In: International Conference on Computing for Sustainable Global Development (INDIACom) (2016)
7. Suwantong, R., Srestasathiern, P., Lawawirojwong, S., Rakwatin, P.: Moving horizon estimator with pre-estimation for crop start date estimation in tropical area. In: American Control Conference (2016)
8. MahyaSeyedan, S., NazariAdli, N., Omid Mahdi Ebadati, E.: Performance analysis and forecasting on Crude Oil: novel Support Vector Regression application to market demand. In: International Conference on Soft Computing Techniques and Implementations (ICSCTI), 8–10 Oct 2015
9. Chatzigeorgakidis, G., Karagiorgou, S., Athanasiou, S., Skiadopoulos, S.: A MapReduce based k-NN joins probabilistic classifier. In: IEEE International Conference on Big Data (Big Data) (2015)
10. Moguel, E., Preciado, J.C., Sánchez-Figueroa, F., Preciado, M.A., Hernández, J.: Multilayer big data architecture for remote sensing in Eolic Parks. IEEE J. Sel. Top. Appl. Earth Observ. Remote Sens. (2015)
11. Ellen Haupt, S., Kosovic, B.: Big data and machine learning for applied weather forecasts. In: IEEE Symposium Series on Computational Intelligence (2015)
12. Xie, H., Tong, X., Meng, W., Liang, D., Wang, Z., Shi, W.: A multilevel stratified spatial sampling approach for the quality assessment of remote-sensing-derived products. IEEE J. Sel. Top. Appl. Earth Observ. Remote Sens. (2015)
13. Vhatkar, D., Kumar, S., Agrionics, G.C., Dogra, S., Mokheria, P., Kaur, P., Arora, D.: Low cost sensor based embedded system for plant protection and pest control. In: International Conference on Soft Computing Techniques and Implementations (ICSCTI) (2015)
14. Chi, M., AtliBenediktsson, J.: Big Data for Remote Sensing: Challenges and Opportunities. IEEE Publication, (2016)
15. Bendre, M.R., Thool, R.C., Thool, V.R.: Big data in precision agriculture: weather forecasting for future farming. In: First International Conference on Next Generation Computing Technologies (NGCT-2015) (2015)
16. Mulyono, S., Ivan Fanany, M.: Remote sensing big data utilization for paddy growth stages detection. In: IEEE International Conference on Aerospace Electronics and Remote Sensing Technology (ICARES) (2015)
17. Ming, J., Ru, N., Sun, J.: Study of iron concentrate price forecasting models based on data mining. In: IEEE International Conference on Cloud Computing and Big Data Analysis (2016)
18. Barbouchi, M., Chokmani, K.: Yield estimation of the winter wheat using Radarsat 2 Polarimetric SAR Reponse. In: 2nd International Conference on Advanced Technologies for Signal and Image Processing—ATSIP'2016 (2016)
19. Badr, G., Klein, L.J., Freitag, M., Albrecht, C.M., Marianno, F.J., Lu, S., Shao, X., Hinds, N., Hoogenboom, G., Hamann, H.F.: Toward large-scale crop production forecasts for global food security. Copyright by International Business Machines Corporation (2016)

20. Kalaiselvi, G., Priyanga, P., Priyanga, S., Sandhya, R.: Big data analytics for detecting crop cultivation using co-location mining algorithm. Int. J. Adv. Res. Biol. Eng. Sci. Technol. (IJARBEST) **2**(10) (2016)
21. SukanyaNandyala, C., Kim, H.: Big and meta data management for U-Agriculture mobile services. Int. J. Softw. Eng. Appl. **10**(2) (2016)
22. Paul, M., Vishwakarma, S.K., Verma, A.: Analysis of soil behaviour and prediction of crop yield using data mining approach. In: International Conference on Computational Intelligence and Communication Networks (2015)
23. Veenadhari, S., Misra, B., Singh, C.D.: Machine learning approach for forecasting crop yield based on climatic parameters. In: International Conference on Computer Communication and Informatics ICCCI (2014)
24. Global Temperature Time Series. http://data.okfn.org/data/core/global-temp
25. National Aeronautics and Space Administration. http://data.giss.nasa.gov/gistemp/
26. GLOBAL Land-Ocean Temperature Index. http://data.giss.nasa.gov/gistemp/tabledata_v3/GLB.Ts+dSST.txt
27. REMSS data. http://data.remss.com/msu/monthly_time_series/RSS_Monthly_MSU_AMSU_Channel_TLT_Anomalies_Land_and_Ocean_v03_3.txt
28. OpenEI. https://en.openei.org/datasets/dataset?sectors=water§ors=wind
29. Wind Integration National Dataset Toolkit. http://www.nrel.gov/grid/wind-toolkit.html
30. Extreme Wind Speed Data Sets: Directional Wind Speeds. http://www.itl.nist.gov/div898/winds/directional.htm
31. Climate: monthly and annual average relative humidity GIS. https://en.openei.org/datasets/dataset/climate-monthly-and-annual-average-relative-humidity-gis-data-at-one-degree-resolution-of-the-world
32. MazharRathore, M., Ahmad, A., Paul, A., Daniel, A.: Hadoop based real-time big data architecture for remote sensing earth observatory system. In: 6th International Conference on Computing, Communication and Networking Technologies (2015)
33. Alves, G.M., Cruvinel, P.E.: Big data environment for agricultural soil analysis from CT digital images. In: IEEE Tenth International Conference on Semantic Computing, (2016)
34. Bendre, M.R., Thool, R.C., Thool, V.R.: Big data in precision agriculture through ICT: rainfall prediction using neural network approach. In: Proceedings of the International Congress on Information and Communication Technology. Advances in Intelligent Systems and Computing. Springer Science+Business Media, Singapore (2016)

A Graph Based Approach on Extractive Summarization

Madhurima Dutta, Ajit Kumar Das, Chirantana Mallick, Apurba Sarkar and Asit K. Das

Abstract With the advent of Information technology and the Internet, the world is producing several terabytes of information every second. Several online news feeds have popped up in the past decade that reports an incident almost instantly. This has led to a dire need to reduce content and present the user only with what is necessary, called the summary. In this paper an Extractive Summarization technique based on graph theory is proposed. The method tries to create a representative summary or abstract of the entire document, by finding the most informative sentences by means of infomap clustering after a graphical representation of the entire document.

Summarization systems work to reduce full sized articles to produce fluent and concise summaries that convey the main idea or the central idea of the passage along with relevant information. A summary is a text that is produced from one or more texts, containing a significant portion of the information present in the original text, and the length of which is much lesser than that of the original text. The systems produce paragraph length summaries. Automatic document summarization dates back to Luhn's work [1]. Many methods [2–4] have been proposed to extract the important concepts from a source text and to build the intermediate representation. Early methods [5, 6] focussed on the frequency of words present in the document to determine the concepts and ideas being highlighted in the document. Linguistic approaches on the other hand attempts to achieve true "semantic understanding" of

M. Dutta (✉) · A. K. Das (✉) · C. Mallick · A. Sarkar · A. K. Das
Indian Institute of Engineering Science and Technology, Shibpur, Shibpur, India
e-mail: madhurima.pg2016@cs.iiests.ac.in

A. K. Das
e-mail: writetoajit@yahoo.com

C. Mallick
e-mail: chirantana9@gmail.com

A. Sarkar
e-mail: as.besu@gmail.com

A. K. Das
e-mail: akdas@cs.iiests.ac.in

A. Abraham et al. (eds.), *Emerging Technologies in Data Mining and Information Security*, Advances in Intelligent Systems and Computing 813,
https://doi.org/10.1007/978-981-13-1498-8_16

the source document. The use of deep semantic analysis offers to create a quality summary that is somewhat close to a summary that is made by a human. Such approaches work effectively with a detailed semantic representation of the document and a domain specific knowledge base of the specific language.

Though there are several kinds of summarization techniques in the paper, only the extractive summarization techniques are discussed. The Extractive Summarizers basically work in three relatively independent phases. All summarizers work to produce an intermediate representation of the text at first. Raw input is preprocessed to remove stopwords i.e. frequently used articles, conjunctions and prepositions. Punctuations are removed as well in some cases. From the remaining portion, often the "term-frequency" or the "inverse document frequency" or both are found and stored as key-value pairs. These metrics help to understand the crux of the information in the document and process it in a particular way. Next, weights are assigned to the sentences based on the intermediate representation. In the final phase, the sentences are selected with the highest scores in a greedy approach. The stages are thoroughly described in the subsequent sections.

A. Intermediate Representation

The main task of this step is to identify the information hidden in the original document. "Topic Representation" approaches are used widely where the topic word pertaining to the particular document are identified. Some of the most popular summarization methods focuses on topic representation so that the central idea of the document is not lost. These approaches include topic word or topic signature approaches [7], frequency, TF-IDF etc. Other approaches like lexical chains [8] use widely available resources like WordNet which helps to establish similarity between semantically related words. Another method of extractive summarization is with Latent Semantic Analysis [9] which helps to identify patterns of word co-occurrence and these roughly feature as topics. Suitable weights are assigned to each pattern subject to fulfillment of some conditions. In "Indicator representation" [10] approaches each sentence in the input is represented as a list of indicators of importance such as "sentence length", "location in the document", "presence of certain phrases", etc. In graph models, such as LexRank [11], the entire document is represented as a network of sentences by means of weighted graphs where the edge weights correspond to the similarity between the sentences they connect.

B. Score Sentences

The task of scoring sentences differs on the basis of the approach being used to create its intermediate representation. It is given that the sentence which is of more relevance will be given a higher weightage than the others. This is exceptionally followed in the topic representation approaches. The score is assigned after examining of how important a particular sentence is to a particular document. For indicator representation methods, the weight of each sentence is determined by taking the values of the indicators into account, most commonly by using machine learning techniques. In the graphical approaches, metrics like the measures of similarity between the vertices is determined and then further processing is carried on.

C. Selecting Summary Sentences

The last and final step is responsible for constructing the resultant summary. Generally it selects the best combination of sentences in the form of paragraph that gives up the key information of the original text in a concise and fluent manner. Care is taken to minimize redundancy so that similar sentences in the summary are avoided. The summary should be coherent to the original text. The genre of the document must control what sentences should go in the summary. A document can be in the form of a webpage, a news article, email, a chapter from a book and so on.

The rest of the paper is organized in the following manner. Section 1 talks about our background study which includes some previous works in the field of Extractive Summarization. In Sect. 2 the proposed algorithm is explained. The results and analysis are presented in Sect. 3. Finally, the paper is concluded with possible future direction in Sect. 4.

1 Background Study

Interest in automatic text summarization arose early as in 1950s. As mentioned before, extractive summarization chooses a subset of sentences from the original text. These sentences are supposedly the most important sentences containing the important information from the text. Over the years, a host of techniques have been applied to perform extractive summarization. Topic representation approaches like "Topic Word" and "Topic Signature" methods have been applied from the very beginning [12]. These approaches consider a frequency threshold to classify certain words as topic words. Frequency driven approaches measure the density of topic words. Metrics like *tf* (term frequency) and *idf* (inverse document frequency) often combined together to help in identifying topic words as well. Another well known approach is centroid summarization [13]. It computes salience of the sentences using a given set of features.

Latent Semantic Analysis [14] is an unsupervised technique used for deriving an implicit representation of text based on the observed co-occurrence of words. It works in lines of dimensionality reduction, initially filling a $n \times m$ matrix where each row corresponds to words from the input and the columns correspond to the sentences.

In graph based summarization methods, the sentences are represented as vertices of a graph and they are connected by means of weighted edges. In [15] a bipartite graph is created to represent the sentences and topics individually after PageRank algorithm is applied to rank the sentences. This topical graph is then used again to rank the sentences using HITS (Hyperlink Induced Topic Search) [16], another popular ranking algorithm used to rank websites initially. A coherence measure is used to find out the relevance of a sentence to a certain topic and then decide whether to accept or reject the sentence for summary after optimisation is applied on the scores.

In [17] the TextRank algorithm [18] is used to assign scores to each of the sentences. Distortion measure has been taken into account to find out the semantic difference of the sentences and thus a distortion graph is constructed. Based on this distortion graph, the sentences are ranked again. The higher ranked sentences are eligible to go into the summary.

In [19] the "tweet similarity graph" has been obtained to establish the similarity measures between the pairs of tweets. The similarity metrics used are "Levenshtein Distance", cosine similarity, semantic similarity. It also compares the URLs and the hashtags used in the tweets to compute the similarity measure. The tweets are treated as vertices and the vertices with the highest degree are selected to go into the final summary.

2 Proposed Algorithm

We have tested our method on various news articles obtained from BBC News feed using BeautifulSoup [20] library of Python [21]. BeautifulSoup is a web scraping tool of Python. It helps us in scraping only the text portion of the XML documents. Once the data is successfully read in, the articles are scraped of its stopwords. Stopwords are the frequently occurring words of any language like *the*, *is*, *are*, *have*, etc. and other articles, common verbs, prepositions which do not convey any special information about the sentence. Next, Proper Nouns which are generally the names of persons, animals, places, or specific organisations or the principal object on which the article is based are removed from the sentences after the document is part of speech tagged. Every pair of the remaining portion of the sentences are checked for cosine similarity according to the Eq. (1) to construct the similarity matrix.

$$cos_{similarity}(x, y) = \frac{x \cdot y}{|x||y|} \tag{1}$$

where x and y are sparse vectors.

The similarity measure is based on the content overlap between sentences. The sentences are represented as bag-of-words, so each of them will be a sparse vector and define measure of overlap as angle between vectors. Removal of the proper nouns removes any chances of biasing the computation of similarity matrix. Each of the sentences are treated as vertices of a similarity graph $SG = (V, E)$. Edges are established between vertices and the weights are assigned according to their similarity measure. The clustering coefficient (c_u) of the vertex ($u \in V$) is computed according to Eq. (2).

$$c_u = \frac{2T(u)}{deg(u)(deg(u) - 1)} \tag{2}$$

where $T(u)$ is the number of triangles through node u and $deg(u)$ is the degree of u.

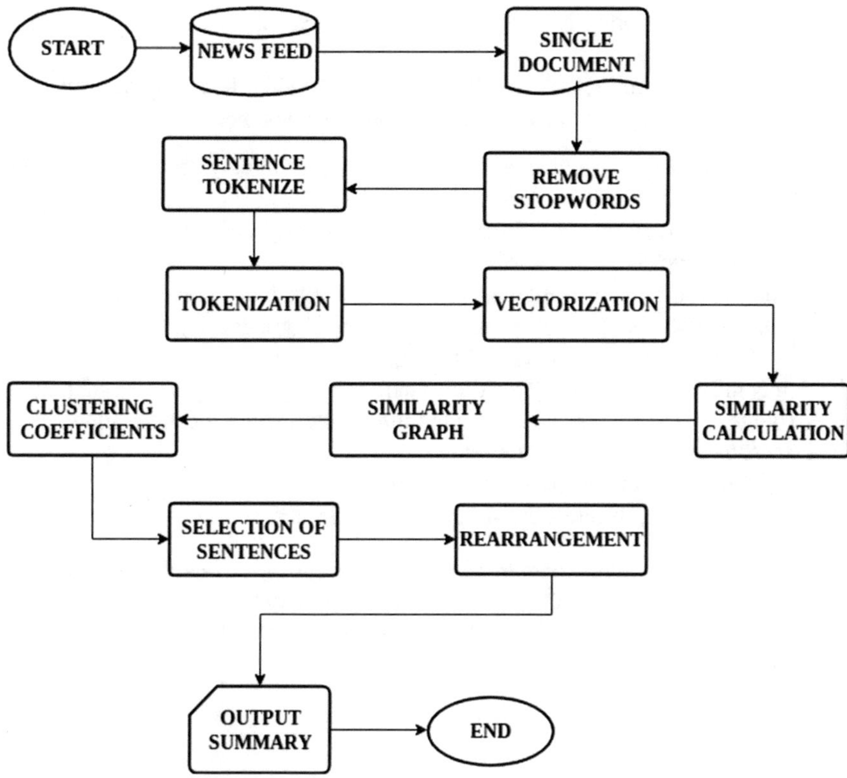

Fig. 1 Flowchart of the proposed system

With the similarity matrix and the graph as input the sentences are clustered by means of Infomap clustering. The average clustering coefficient (avg_c_u) is computed using the maximum c_u of each cluster. From each of the cluster, those sentences are selected whose $c_u > avg\_c_u$. This step is repeated until the summary length is reached. The flowchart of the algorithm is shown in Fig. 1.

Algorithm: Extractive Summarization

The outline of the algorithm is as follows:
Input: Text document from a web page scraped with the help of BeautifulSoup library.
Output: A summary of the input text.
Begin

1. Encode the text document in UTF-8 format.
2. Remove the stopwords from the encoded text
3. The text is tokenized into individual sentences using the NLTK library.
4. for each sentence

a. The NLTK library is used to tag the Parts of Speech of each of the words in the sentences.
b. The named entities, i.e. the Proper Nouns are identified and removed.
c. Remaining words in a sentence is treated as a text vector.

5. Cosine similarity is calculated between every pair of text vector using Eq. (1) and the similarity matrix S is obtained.
6. Construct a graph with sentences as nodes and weight of edges between every pair of sentence is the similarity between them.
7. Modify the graph removing edges with weight less than a predefined threshold.
8. Compute clustering coefficient of each node of the graph using Eq. (2) and compute the average clustering coefficient c_{avg}.
9. Apply infomap clustering algorithm to partition the given graphs into subgraphs.
10. Arrange subgraphs based on the maximum clustering coefficient computed.
11. For each subgraph of the graph
    if maximum clustering coefficient $< c_{avg}$
        remove the subgraph as redundant.
    else
        remove node with maximum clustering coefficient from the subgraph and store it into summary file.
12. If number of sentences in summary < predefined size, then go to step 10.
13. Return summary.

End

3 Results

The method has been thoroughly tested on the news articles generated by the BBC News feed. We have used the NLTK library [22] and the BeautifulSoup library of Python 2.7 [21] to help with our processing.

Evaluation using Rouge:

Since the early 2000s the ROUGE [23] metric is widely used for automatic evaluation of summaries. Lin introduced a set of metrics called Recall-Oriented Understudy for Gisting Evaluation (ROUGE) to automatically determine the quality of a summary by comparing it to reference summaries developed by humans which is generally considered as ground truth. There are several variations to the Rouge metric among which the popularly used metrics are explained below:

3.1 Rouge_N

The Rouge_N metric is a measure of overlapping words typically considered in N-grams between a candidate summary and a set of reference summaries. It is computed using Eq. (3)

$$Rouge\_N = \frac{\Sigma_{s \in refsum} \Sigma_{gram_n \in S} Count_{match}(gram_n)}{\Sigma_{s \in refsum} \Sigma_{gram_n \in S} Count(gram_n)} \tag{3}$$

where N stands for the length of the N-gram, $gram_n$, is the maximum number of N-grams co-occurring in a candidate summary and $Count_{match}(gram_n)$ is the same words occuring in the set of reference summaries in exactly the same sequence. Here value of N is considered as 1 and 2.

3.2 Rouge_L

Using this metric the longest common subsequence is sought between the reference summaries and the system generated summary. Although this metric is more flexible than the Rouge_N metric, it has a drawback because it requires all the N-grams in consecutive positions. ROUGE_L is a longest common subsequence (LCS) metric which seeks for the LCS between the reference and system summaries. Here, a sequence $A = [a_1, a_2, \ldots, a_n]$ is considered as a subsequence of another sequence $B = [b_1, b_2, \ldots, b_n]$ if there can be defined a strictly-increasing sequence of indices for B (i.e. $i = [i_1, i_2, \ldots, i_k]$ such that for all $j = 1, 2, \ldots k, a_{ij} = z_j$). The longest common subsequence for A and B can be considered as the sequence common to both B and A having sentence structure level similarity and identifies the largest occuring N-gram. In general it is assumed that ROUGE_L is based on the idea that pairs of summaries with higher LCS scores will be more similar than those with lower scores. LCS-based Recall, Precision and F-measure can be calculated to estimate the similarity between a reference summary X (of length m) and a candidate summary Y of length n according to Eqs. (4)–(6):

$$R_{lcs} = \frac{LCS(X, Y)}{m} \tag{4}$$

$$P_{lcs} = \frac{LCS(X, Y)}{n} \tag{5}$$

$$F_{lcs} = \frac{(1 + \beta^2) R_{lcs} P_{lcs}}{(R_{lcs} + \beta^2 P_{lcs})} \tag{6}$$

where, LCS(X, Y) is equal to the length of the LCS of X and Y and $\beta = \frac{P_{lcs}}{R_{lcs}}$.

Table 1 Rouge values of the proposed method with respect to ground truth

Metric	Recall	Precision	f-score	Recall	Precision	f-score
Length	10%			15%		
Rouge-1	0.69672	0.35417	0.55173	0.63030	0.43333	0.51358
Rouge-2	0.34711	0.17573	0.23333	0.29268	0.20084	0.23821
Rouge-L	0.55738	0.28332	0.37569	0.44848	0.30833	0.36543
Length	20%			25%		
Rouge-1	0.59538	0.46250	0.51991	0.57198	0.61250	0.59155
Rouge-2	0.26656	0.18410	0.20706	0.26172	0.28033	0.27071
Rouge-L	0.45989	0.35833	0.40281	0.42023	0.45000	0.43461

The results of our evaluation are given below in Table 1. It can be seen that for summaries which is 25% of the size of the original document gives better results than of shorter sizes.

The unbiased ground truth or the reference summaries of the news articles were obtained from our fellow peers and research scholars having different field expertise.

4 Conclusion and Future Work

In this paper we have tried to present a new graph based approach on achieving extractive summarization. By removing named entities from the sentences, the similarity measure for each pair of sentences becomes unbiased, hence important words are emphasized. This process gives us interesting results when evaluated on news articles. In the future we hope to refine the noun-pronoun resolution so that the results are further improved. Also instead of using Infomap clustering algorithms, several graph based and other clustering algorithms may be applied and a comparative study will be made as future work.

References

1. Nenkova, A., McKeown, K.: A Survey of Text Summarization Techniques. Springer Science+Business Media (2012)
2. Meena, Y.K., Gopalani, D.: Evolutionary algorithms for extractive automatic text summarization. In: Procedia Comput. Sci., 48(Suppl. C), 244 – 249 (2015). (International Conference on Computer, Communication and Convergence (ICCC 2015))
3. Saggion, H., Lapalme, G.: Generating indicative-informative summaries with sumum. Comput. Linguist. **28**(4), 497–526 (2002)
4. Gong, Y., Liu, X.: Generic text summarization using relevance measure and latent semantic analysis. In: Proceedings of the 24th Annual International ACM SIGIR Conference on Research and Development in Information Retrieval, pp. 19–25. ACM (2001)

5. Dunning, T.: Accurate methods for the statistics of surprise and coincidence. Comput. Linguist. **19**(1), 61–74 (1993)
6. Hovy, E., Lin, C.-Y.: Automated text summarization and the summarist system. In: Proceedings of a Workshop on Held at Baltimore, Maryland: 13–15 Oct 1998, TIPSTER'98, pp. 197–214, Stroudsburg, PA, USA, 1998. Association for Computational Linguistics
7. Lin, C.-Y., Hovy, E.: The automated acquisition of topic signatures for text summarization. In: COLING'00 Proceedings of the 18th conference on Computational linguistics, pp. 495–501. Association for Computational Linguistics Stroudsburg, PA, USA (2000)
8. Wei, T., Lu, Y., Chang, H., Zhou, Q., Bao, X.: A semantic approach for text clustering using wordnet and lexical chains. Expert Syst. Appl. **42**(4), 2264–2275 (2015)
9. Alpaslan, F.N., Cicekli, I.: Text summarization using latent semantic analysis. J. Inf. Sci. **37**(4), 405–417 (2011)
10. Kan, M.-Y., McKeown, K.R., Klavans, J.L.: Applying natural language generation to indicative summarization. In: Proceedings of the 8th European Workshop on Natural Language Generation, vol. 8, pp. 1–9. Association for Computational Linguistics (2001)
11. Erkan, G., Radev, D.R.: Lexrank: graph-based lexical centrality as salience in text summarization. J. Artif. Intell. Res. 457–479 (2004)
12. Harabagiu, S., Lacatusu, F.: Topic themes for multi-document summarization. In: Proceedings of the 28th Annual International ACM SIGIR Conference on Research and Development in Information Retrieval, pp. 202–209. ACM, New York, NY, USA (2005)
13. Radev, D.R., Jing, H., Stys, M., Tam, D.: Centroid-based summarization of multiple documents. Inf. Process. Manag. **40**, 919–938 (2003)
14. Landauer, T.K., Foltz, P.W., Laham, D.: An introduction to latent semantic analysis. Discourse Process. **25**(2–3), 259–284 (1998)
15. Parveen, D., Strube, M.: Integrating importance, non-redundancy and coherence in graph-based extractive summarization. In: Proceedings of the 24th International Conference on Artificial Intelligence, IJCAI'15, pp. 1298–1304. AAAI Press (2015)
16. Kleinberg, J.M.: Authoritative sources in a hyperlinked environment. J. ACM **46**(5), 604–632 (1999)
17. Agrawal, N., Sharma, S., Sinha, P., Bagai, S.: A graph based ranking strategy for automated text summarization. DU J. Undergrad. Res. Innov. **1**(1) (2015)
18. Mihalcea, R., Tarau, P.: TextRank: bringing order into texts. In: Proceedings of EMNLP-04 and the 2004 Conference on Empirical Methods in Natural Language Processing, July 2004
19. Dutta, S., Ghatak, S., Roy, M., Ghosh, S., Das, A.K.: A graph based clustering technique for tweet summarization. In: 2015 4th International Conference on Reliability, Infocom Technologies and Optimization (ICRITO) (Trends and Future Directions), pp. 1–6. IEEE (2015)
20. Beautifulsoup documentation. https://www.crummy.com/software/BeautifulSoup/bs4/doc/. Accessed 29 Nov 2017
21. Python 2.7.14 documentation. https://docs.python.org/2/index.html. Accessed 29 Nov 2017
22. Bird, S., Klein, E., Loper, E.: Natural Language Processing with Python. O'Reilly (2009)
23. Lin, C.-Y.: Rouge: a package for automatic evaluation of summaries. In: Proceedings of the ACL Workshop: Text Summarization Braches Out 2004, pp. 10, 01 2004

Targeted Marketing and Market Share Analysis on POS Payment Data Using DW and OLAP

Giridhar Maji, Lalmohan Dutta and Soumya Sen

Abstract Point of Sales (POS) terminals are provided by the banking and financial institutes to perform cashless transactions. Over the time due to different conveniences, use of digital money and online card transactions increased many folds. After each successful payment transaction at the POS terminals, a transaction log is sent to the POS terminal provider with payment-related financial data such as date and time, amount, authorization service provider, cardholder's bank, merchant identifier, store identifier, terminal number, etc. These data are useful for analytical processing which are useful for business. This paper proposes to process these huge transactional data using ETL process and thereafter construction of a data warehouse (DW) which enables the POS provider to employ certain analytical processing for business gain such as knowing own market share as well as position in market with respect to card payments, geographic location-wise business profiling, own as well as competitor's customer segmentation based on monthly card usage, monthly amount spent, etc.

Keywords EFTPOS · Card payments · DW schema model · OLAP · Market share · Customer segmentation

1 Introduction

Banking and financial establishments handle large amount of payments and transactions every day and gigantic quantity of raw data which is stored in the operational

---

G. Maji
Department of Electrical Engineering, Asansol Polytechnic, Asansol 713302, India
e-mail: giridhar.maji@gmail.com

L. Dutta · S. Sen (✉)
A.K.Choudhury School of I.T., University of Calcutta, Kolkata 700106, India
e-mail: iamsoumyasen@gmail.com

L. Dutta
e-mail: lalmohan.dutta@gmail.com

A. Abraham et al. (eds.), *Emerging Technologies in Data Mining and Information Security*, Advances in Intelligent Systems and Computing 813,
https://doi.org/10.1007/978-981-13-1498-8_17

systems. These data could be processed to identify hidden patterns and trends and even predict future scenarios. Large financial institutions store raw data in DWs for further analysis, helping top management decision-making. Generally, financial institutions collect competitor information from many different sources like surveys, official published results, from business partners, field executives, etc. Most of them are costly, time-consuming, contains large noise, and sometimes manipulated due to human intervention. Banks provide POS terminals to the merchant partners to be installed and used in retail outlets, small shops, ticket counters, bus rail counters, etc. The big merchants have large numbers of retail outlets across the countries in geographically and demographically diverse places. They get their POS terminal service from the bank. In a process, the concerned bank gets an opportunity to have large amount of POS payment data from many regions. POS terminal generated payment data are also commonly known as Electronic Funds Transfer at Point of Sale (EFTPOS) data. It is a common method of payment at POS or checkouts. Customers mainly use debit or credit cards at EFTPOS terminals. These payment data contain some amount of useful competitor data along with the other common data. Some of the useful data are customer card type and provider bank, promotional features given to the customer, total transaction amount, and customer demographic details also available if they subscribe to some loyalty program offered by the retailer and there is an agreement on data sharing among bank and retailer. Our aim is to use the data available from different POS terminals throughout the country and design a DW schema with properly identified dimensions and ETL process for OLAP analysis. This OLAP tool will be able to produce reports on competitor's performance in different geographies, demography as well as market position. It will be easier to get an estimate on what percentage of cardholders are using which bank's card; what is the average amount of transaction value for each of bank's customers in some city in some quarter of some year. This kind of insight on competitor's business with almost noise free reliable data will help management to decide on future strategy and focus area, promotional campaigning, etc. in a very short time frame with much less cost.

In the next subsection, a brief discussion is given about the payment workflow that happens at POS terminal and different data fields that are stored and finally touch upon the required basics on data warehousing and OLAP analysis.

1.1 Basics of POS Payment Workflow

Typical workflow of any POS payment activity is depicted in Fig. 1. Commonly followed steps are as follows:

1. Bank A (the Issuer) where customer has an account issues (debit) card to the customer. In response to customers' application for credit or debit card, bank does some background checks (creditworthiness, CIBIL score, repayment capability, etc.) and based on that issues the card to the customer with set limits on maximum transaction amount with others.

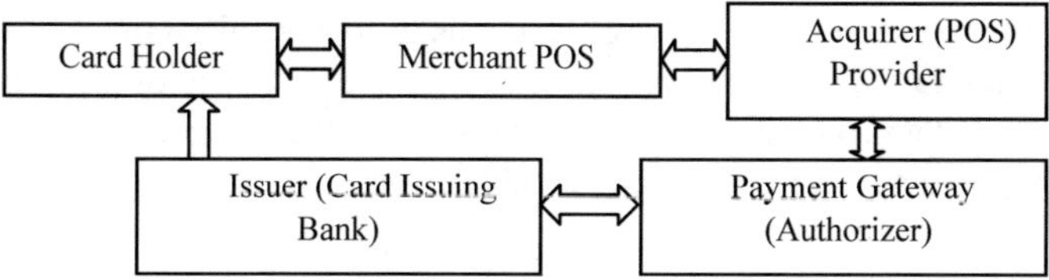

Fig. 1 Card authorization and process flow at a POS

2. Cardholder (Customer) swipes the card to pay his bill at POS terminal.
3. Card details along with payment data are sent to the POS provider bank (the acquirer) for further processing.
4. The acquirer sends authorization request to the authorizer.
5. Authorizer checks with the card issuing bank (Issuer) if the cardholder is eligible for the transaction, limit on amount, if it has sufficient balance (for debit cards), etc. Issuing bank agrees to pay the bill amount to the acquirer bank and authorizer makes them binding and validates the transaction.

6 and 7. Acknowledgement is returned from issuing bank to authorizer to the acquirer bank

8. Acquirer, i.e., the POS provider bank credits the amount into the merchant's current account maintained with the bank.
9. Invoice cum bill of items is generated at the POS and given to the customer. Now the customer is the legal owner of the items bought.

Acquirer bank normally offers many discounts and loyalty benefits to their customers such as cashbacks. These discounts and cashbacks are recovered from the merchant. Normally, the acquirer bank takes around 2% of the bill amount as the commission. Interested reader can refer to [1, 2] for more details about card payments and settlement issues.

1.2 POS Tender Payment Data Attributes

Any retail outlet point of sales (POS) system normally stores data in a normalized form (generally in 3-NF). It stores three basic types of data in at least 3 or more tables in operational databases:

Transaction header contains a row per transaction and constitutes of fields such as transaction timestamp, counter_id, store_id, employee_id (who has completed the transaction at the POS terminal), customer_id (if the customer is subscribed to some loyalty program), etc. **Transaction details** normally contain sold item details with each item a row. Data fields are item pricing information, discount, package details, checkout time, etc. **Transaction tender** entity contains payment information along with payment type, mode of payment, card details, bank details, link to unique customer identifier if subscribed for discounted price or loyalty programs. In our study, transaction tender data are considered as these data also contain information about total amount spent by a customer. With proper analysis of such data by the POS terminal providers, bank can get a true picture of its market share in different customer segments in various geographical regions.

1.3 Basics of Data Warehousing and OLAP

Online Analytical Processing (OLAP) tools feed on the huge amount of data stored in a DW and produce quick answers to the business queries that need to consolidate a great amount of detail low-level raw transaction data. OLAP tools along with the DW help to analyze hidden patterns and business trends. OLAP tools are used to manage and organize multidimensional data cubes where data are represented in the form of fact tables and dimension tables. In data warehouse, types of schema [3] are based on how dimension tables are connected to fact table. Dimensions identify the different business perspectives of trend analysis. Facts are the main events, whereas dimensions provide the context. Measures are used to quantify the fact values. Dimensions may have concept hierarchy. A concept hierarchy [4] is the abstraction of one dimension from detailed value to aggregated value.

Data cube or cuboid is the most fundamental way to represent data in data warehouse that help to model and visualize the data in multidimensional format. Base cuboid is the cuboid that corresponds to a given fact table. Repeated roll-up operations are applied on base cuboid to generate all possible combination of the cuboids which forms lattice of cuboids [4].

In this study, we shall model a DW schema and identify related dimensions to analyze the POS sales transaction data stored with the POS service provider bank and discuss different business intelligences that can be harvested from such data and used for the business growth.

2 Related Study

Traditionally, research on POS terminal data from retail stores focuses mainly on either market basket analysis or sales trend and seasonal variation analysis using data mining [5–8]. In some studies, it is used for customer behavior and buying habit analysis. For example, authors in [7] used POS transaction data for customer segmentation based on consumption data using clustering and classification techniques. Authors in [8] analyze consumer purchase behavior using the same data. Authors in [6] have used association rule mining to understand cross-product purchasing trends in retail stores by analyzing POS transaction data. Some studies focus on item placement on the floor and sales; they consider the attributes like shelve location for the items, combo offers, etc. [5]. Authors in [9] discuss mobile payment in e-commerce businesses as another upcoming way of paying bills using smartphones and mobile wallets. Data reduction and consolidation techniques for huge amount of time-stamped transactional data including POS data have been discussed in [10]. The authors in [11] discussed in detail the banking POS data, ETL processes, and use of different data mining techniques applied to that data with many business applications. But in almost all the studies, another less used part of the POS terminal data has been never used. Authors in [12] used Electronic Funds Transfer at Point of Sale (ETFPOS) data to cluster and classify merchants for market segmenting using the famous Recency, Frequency, and Monetary (RFM) analysis technique with a small set of data. In a recent study, authors in [13, 14] applied clustering and classification techniques to find customer groupings of a retail store using EFTPOS transaction data (quite a large volume of data) collected from a major bank in Australia who provided the POS services to retail merchants. The idea of segmenting market based on customer preferences or attributes has been first introduced by Smith [15]. Doyle in [16] defines market segmentation which talks about subdividing a market into different classes so that the customers of every class have similar behavior or same types of needs. Distinctive marketing strategy needs to be applied for every class.

POS terminals are provided to the retail merchants by some bank with whom the retailer maintains its current accounts, so after sales the payments done using card swapping goes directly to the business account of the merchant in acquirer bank. This offers a great opportunity for the acquirer bank to collect such data and analyze for its own benefit such as customer profiling. There may be a business collaboration between a POS service provider bank (Acquirer) and a retail chain; then, this customer data can also be utilized by the bank to identify potential clients to issue cards and also to sell other banking products with minimized risk of defaulting.

This type of DW-based OLAP analysis with telecom Call Detail Record (CDR) data has been studied in [17]. We have conceptualized this type of approach [17] on POS payment tender data to focus on market share of different banks in the cards and payments market segment with an additional aim to identify potential customer groups and places to venture in future along with effects of promotional discounts on average transaction amount.

3 Problem Statement

The objective of this study is to analyze this huge amount of POS payment data to develop an analytical model to leverage business intelligence. We have chosen to analyze the card transaction data used by customers of many different banks during sales at different merchant's retail outlets as well as transaction amounts along with promotional offers given with the cards in different defined regions. Existing business tie-ups with the big retail merchants and large POS provider banks can help in sharing customer data (available through different loyalty programs by retailers) and then customer segmentation based on their demography is possible. Promotional offers and discounts on concerned bank's cards can be offered to selected customer segments only in those regions where market share is less compared to other competitor card provider banks. Moreover, this analysis is helpful for both of these business organizations. A big advantage of this methodology is that the data are generated by the system itself, and hence no data cleaning method is required which saves money. Another important aspect of this work is that it could be carried out to any demographical location where business is carried out through POS terminals.

4 Proposed Methodology

Data warehouse is designed here to use in data mining process where the data could be readily used for mining valuable business information. The creation of a data warehouse is followed by ETL (Extract–Transform–Load) processing on the transactional data. As the outcome of the ETL process data is loaded into the appropriate data warehouse schema which is organized here in terms of the dimension tables, fact table and thereafter the suitable schema type are identified. We also identify the concept hierarchies of selected dimensions. Once the DW schema is constructed, we then work with data cubes and further form the lattice of cuboids for quick execution of user queries.

The problem is conceptualized in terms of the area/region, i.e., geographical location of the stores, card provider bank types, customer group based on demography, and the time of the year. Therefore, four dimensions are identified, namely, location, issuer (card provider bank), customer segment, and time. All of these dimensions have concept hierarchy which is shown in Fig. 2. Customer grouping based on demography and other attributes are not shown with concept hierarchy as customers, i.e., cardholders can be clustered based on age group, sex, monthly income, profession, etc. Again such detail customer information is only available for those cardholders who have subscribed to the loyalty program offered by the retail store; otherwise, only basic information limited to name, address, phone number, card type, card number, card provider bank, authorization service provider (VISA/MASTER/RUPAY, etc.), total transaction amount, etc. are accessible. Preprocessing is required to load the POS data into the data warehouse. Now the payment transaction data of 1-week duration is sourced from several operating databases of different branches, offices, as well as from data marts and then it is selected for unique customer ids (or card numbers while customer details not available). Data are aggregated based on different customer groups with the calculated measures such as average weekly transaction amount and average number of transaction by a group of customer.

Now to load data of store location dimension, the actual locations of the stores are mapped with the defined geographic region (locality or city) as multiple POS terminals map to a single store and multiple stores map to city unit of the retailer. Store locality is taken as the fundamental information of this dimension (A city may contain a few store locality). EFTPOS data contain the POS terminal ids which identify the store associated uniquely. Merchant master table provides the list of stores along with their address location and city. Store ids of a region (in this case a city) are identified at first and then data are aggregated by collecting and summing

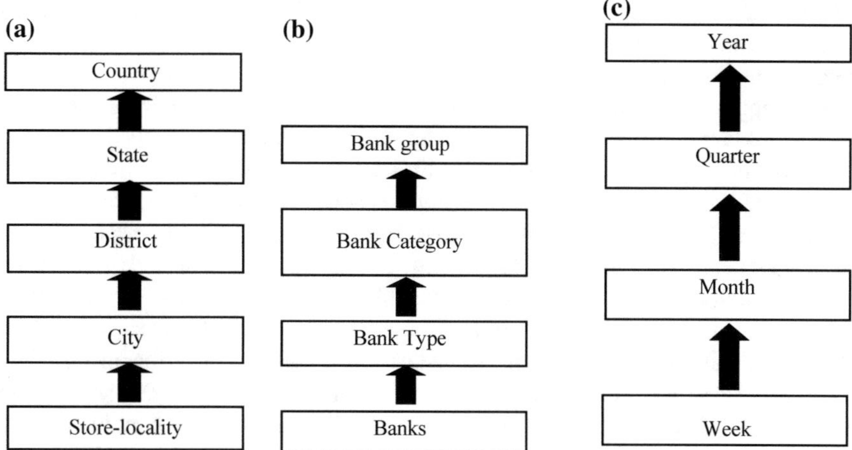

Fig. 2 Concept hierarchies of different dimensions used for analysis

all payment transactions (number of transaction and transaction amount) of all users. Different location abstractions for analysis are shown in the form of concept hierarchy in Fig. 2a.

Issuer bank dimension table provides the list of different card issuing banks with publicly available attributes. These banks can be categorized into different groups based on some attributes, thus establishing the concept hierarchy on this dimension. Again based on type of operation banks can be commercial, investment bank or cooperative banks, etc. The concept hierarchy of bank dimension is shown in Fig. 2b.

Time is an inevitable dimension for majority of the data warehouses. Management could analyze their business on different abstractions of time such as year, quarter, month, week, etc. So we can take most granular level as a week and start our study with any 1 week's data. Every EFTPOS transaction stores the timestamp entry. Therefore, timestamp values are aggregated and represented in different abstractions. Figure 2c represents this concept hierarchy.

The only fact table in this schema will represent data like customers of profession "IT" at City "Kolkata" have on average "18" number of payment transactions with average amount per transaction of "Rs. 1200" made with card issued by "HDFC bank" in the month of "May 2017" with a market share percentage of "13%".

4.1 Schema

The given data warehouse framework consists of four dimensions, namely, store location, issuer bank, time, and customer group. As all of the selected dimensions are directly connected to the fact table, the type of the schema is **Star Schema**. Hence, all the dimension tables make an entry in the fact table. The star schema is constructed with bare minimum requirement. We could have considered snowflake schema where some of the dimensions may be references to sub-dimension tables (We could have made a separate dimension table with different promotion details and linked that from issuer bank dimension using promotion key) but that would have made the analysis complex and warehouse performance is best when it is mostly de-normalized and number of joins are minimum and star schema is better than snowflake in that respect. So whenever possible we tend to construct star schema. If multiple business goals of management have common dimensions, in that case fact constellation can be considered. In our case, if customer relationship management unit as well as bank's own product sales unit (products like insurance, mutual fund, loans mobile wallet, etc.) data are also integrated into the same DW with some common dimensions, then it will be wise to adapt to fact constellation rather than creating separate DW or data marts. In [18], authors have considered fact constellation to deal with multiple types of business goals in a single warehouse schema. Finally, the star schema is shown in Fig. 3 along with the associated dimension tables and fact table. Location is a crucial dimension here; therefore, location of the retailer (i.e., the merchant for whom bank has provided POS services) is considered as a location dimension as it will help to analyze the business demographically. Again issuer bank (who has provided cards to the customer) is an essential dimension for our analysis. There exists some clustering of banks in India such as nationalized banks and privately owned banks. Depending on the market capitalization, banks may be categorized into small, medium, and large. POS service provider bank (acquirer) may be as well the issuer (i.e., Customer is having a credit or debit card issued by the same bank that has also provided the POS terminal to the retailer merchant). So from this dimension, top management can easily get hold of the big picture of cards and payment's business. What percent of card transactions are done by which bank; authorization is done by which authority, for every city individually or for a region, even for a whole country in almost real-time basis can be reported by proper roll-up aggregation on these dimensions. Time is crucial dimension for business analysis as every organization wants to view their share of business in different perspectives of time.

Three possible different measures are chosen in fact table. The organization could choose any number of measures which are significant for their business.

4.2 Using OLAP Cube

In order to analyze the business, the fact table could be extended to represent lattice of cuboids with the associated four dimensions. This will generate 16 cuboids. However, in Fig. 4, we show lattice of cuboids with three dimensions due to the scarcity of space. Lattice of cuboids is summarized at a level of store locality–week–issuer bank. As in the case of business, all combinations of dimensions are interesting at different levels of different dimensional cuboids presented for analysis. As every dimension

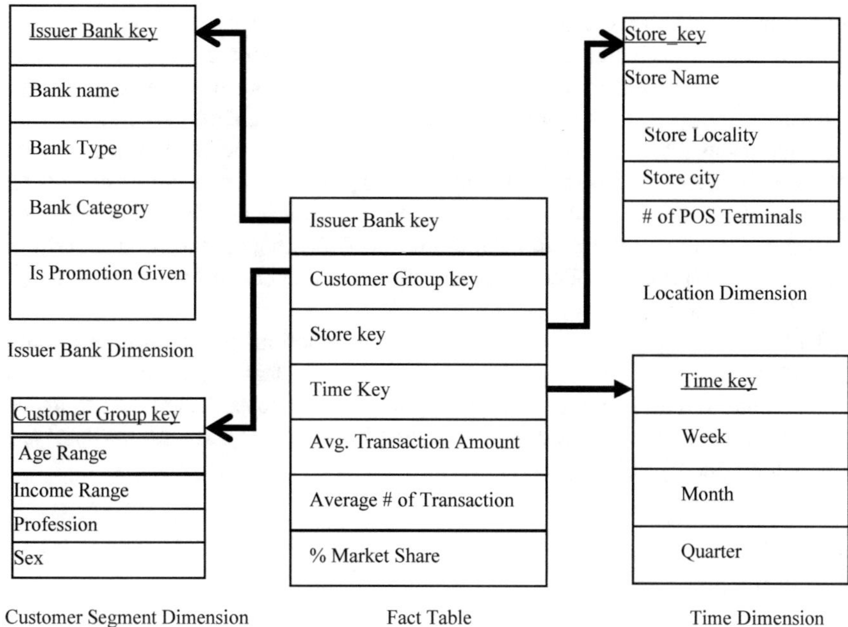

Fig. 3 Star schema proposed for market share analysis

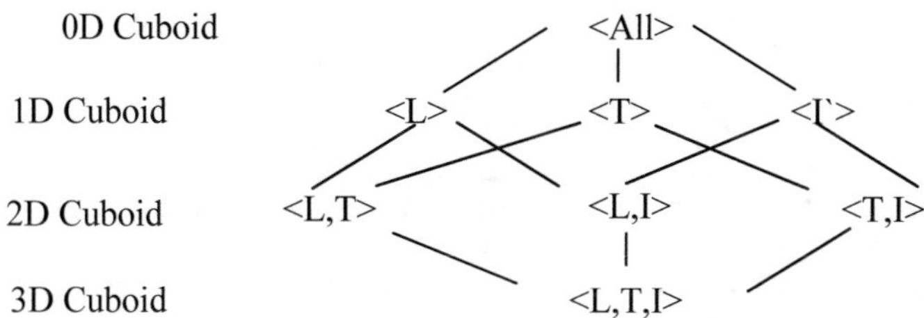

Fig. 4 Lattice of cuboids for location (L), time (T), issuer (I) dimension

of here has concept hierarchy, all the cuboids could be represented in different forms to meet the business requirements.

5 Conclusion and Future Work

This paper gives a new idea to utilize POS payment data to compute the market share of different card issuer banks and their card transaction amounts during different times of the year in different parts of the country. This vital accurate business knowledge about competitors' market share enables POS provider bank to plan future business strategy to increase market share and to identify potential places to foray into, customer segment to target to, time of the year to put forward some promotional campaigns, etc. For the above purpose suitable, DW schema has been identified and dimensions selected along with important measures. Concept hierarchies for the selected dimensions are discussed with lattice of cuboid structure for visual OLAP analysis. Main advantage of this system is that acquirer bank now can collect most precise and accurate information about other banks without incurring huge cost on manual market survey or collection of data through other un-trustworthy sources and market speculation.

This analytical processing could be extended to various other business applications (such as mutual fund, insurance, home loan, car loan, share trading, etc.) in future. Data mining techniques can be applied to such consolidated data to understand relationship between customer demography and other attributes like card use, card type, spending amount, etc.

References

1. FirstData, Payments 101: Credit and Debit Card Payments, A First Data White Paper, Oct 2010
2. Herbst-Murphy, S.: Clearing & settlement of interbank card transactions: a MasterCard tutorial for Federal Reserve payments (2013)
3. Sen, S., Ghosh, R., Paul, D., Chaki, N.: Integrating XML data into multiple Rolap data warehouse schemas. AIRCC Int. J. Softw. Eng. Appl. (IJSEA) **3**(1), 197–206 (2012)
4. Sen, S, Roy, S., Sarkar, A., Chaki, N., Debnath C.N.: Dynamic discovery of query path on the lattice of cuboids using hierarchical data granularity and storage hierarchy. Elsevier J. Comput. Sci. **5**(4), 675–683 (2014)
5. He, M., Li, J., Shao, B., Qin, T., Ren, C.: Transforming massive data to pragmatic target marketing practice. In: 2013 IEEE International Conference on Service Operations and Logistics, and Informatics (SOLI), pp. 408–412. IEEE, July 2013
6. Avcilar, M.Y., Yakut, E.: Association rules in data mining: an application on a clothing and accessory specialty store. Can. Soc. Sci. **10**(3), 75 (2014)
7. Xie, Y., Zhang, D., Fu, Y., Li, X., Li, H.: Applied research on customer's consumption behavior of bank POS machine based on data mining. In: 2014 IEEE 9th Conference on Industrial Electronics and Applications (ICIEA), pp. 1975–1979. IEEE, June 2014
8. Zuo, Y.: Prediction of consumer purchase behaviour using Bayesian network: an operational improvement and new results based on RFID data. Int. J. Knowl. Eng. Soft Data Paradig. **5**(2), 85–105 (2016)
9. Pousttchi, K., Hufenbach, Y.: Enabling evidence-based retail marketing with the use of payment data—the Mobile Payment Reference Model 2.0. Int. J. Bus. Intell. Data Min. **8**(1), 19–44 (2013)
10. Leonard, M., Wolfe, B.: Mining transactional and time series data. Abstract, presentation and paper, SUGI, 10-13 (2005)
11. Berhe, L.: The Role of Data Mining Technology in Electronic Transaction Expansion at Dashen Bank SC (Doctoral dissertation, AAU) (2011)
12. Bizhani, M., Tarokh, M.: Behavioral rules of bank's point-of-sale for segments description and scoring prediction. Int. J. Ind. Eng. Comput. **2**(2), 337–350 (2011)
13. Singh, A., Rumantir, G.W.: Two-tiered clustering classification experiments for market segmentation of EFTPOS retailers. Australas. J. Inf. Syst. **19** (2015)

14. Singh, A., Rumantir, G., South, A.: Market segmentation of EFTPOS retailers. In: The Proceedings of the Australasian Data Mining Conference (AusDM 2014), Brisbane (2014)
15. Smith, W.R.: Product differentiation and market segmentation as alternative marketing strategies. J. Mark. 21(1) (1956)
16. Doyle, C.: A Dictionary of Marketing. Oxford University Press (2011)
17. Maji, G., Sen, S.: A Data warehouse based analysis on CDR to depict market share of different mobile brands. In: 2015 Annual IEEE India Conference (INDICON). IEEE (2015)
18. Maji, G., Sen, S.: Data warehouse based analysis on CDR to retain and acquire customers by targeted marketing. In: 5th International Conference on Reliability, Infocom Technologies and Optimization (2016)

Promises and Challenges of Big Data in a Data-Driven World

Sonali Mukherjee, Manoj Kumar Mishra and Bhabani Shankar Prasad Mishra

Abstract Big Data has currently become a catchword and gained significant attention from academia and industry alike due to its innumerable applications in various fields, such as social media, marketing, public sector, healthcare, education, etc. Big Data is about dealing with enormous amount of structured, semi-structured, and unstructured data that is produced in a rapid manner which is difficult to handle using conventional systems. Big Data is creating different obstacles due to its immense volume and diverse types. Several tools and technological advances (e.g., Hadoop, Hive, Pig, Spark, etc.) are being developed to analyze and tackle Big Data for adding values and making radical changes in business, education, and society. In spite of having many challenges, Big Data is a boon to science. This article makes an attempt to collect and scrupulously document contributors of Big Data, pitfalls, benefits, tools, enabling technologies, and some prevailing applications as well.

Keywords Big Data · Challenges · Hadoop · MapReduce

1 Introduction

The burgeoning amount of data, generating from everywhere with ever known speed, volume, and variety, has become an important concern of today's world. Data analysts

---

S. Mukherjee (✉) · M. K. Mishra · B. S. P. Mishra
School of Computer Engineering, Kalinga Institute of Industrial Technology, Deemed to be University, Bhubaneswar 751024, Odisha, India
e-mail: mukherjee.sonali28@gmail.com

M. K. Mishra
e-mail: manojku.mishra05@gmail.com

B. S. P. Mishra
e-mail: mishra.bsp@gmail.com

A. Abraham et al. (eds.), *Emerging Technologies in Data Mining and Information Security*, Advances in Intelligent Systems and Computing 813,
https://doi.org/10.1007/978-981-13-1498-8_18

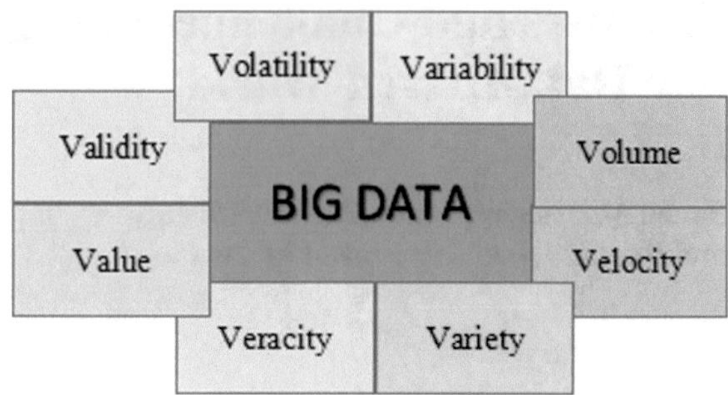

Fig. 1 Big Data and its characteristics

are facing tremendous challenges to handle and to gain full benefit of these data. A survey shows that Internet is one of the primary reasons for more than 60% data evolution [1]. According to IBM, data of 2.5 quintillion bytes are being created on daily basis; however, 90% of the data in the worldwide now have been created in the last 2 years alone. This huge amount of data is named Big Data. According to McKinsey, Big Data refers to datasets whose sizes are beyond the capacity of typical database software tools to capture, store, manage, and analyze [2]. Big Data is described in terms of 8 V's.

Volume: Is the quantity of data produced by any enterprise or organization.

Velocity: Defines the growth rate of data in today's world.

Variety: Defines that Big Data deals with homogeneous as well as heterogeneous data structure.

Veracity: Multiple sources of data make data unreliable and ambiguous.

Value: Is the importance of data that comes after mining it.

Validity: Holds the accuracy of data.

Volatility: Defines the lapse of time and data is worth storing in repository.

Variability: Means that data flow can be inconsistent and unpredictable with respect to time.

Figure 1 exhibits different characteristics of Big Data.

The rest of the paper is organized as follows: Sect. 2 narrates different contributors of data generation with diagram. Section 3 is about Big Data challenges versus benefits. Section 4 mentions various Big Data tools. Section 5 provides some conventional fields influenced by Big Data. Section 6 describes future with Big Data and Sect. 7 is conclusion.

2 Contributors of Data Generation Until Now

In 1944, IBM invented mainframe computer and initiated the era of storing and managing of data. At that time, not individuals but organizations and companies

Fig. 2 Contributors of data generation

were the only entities responsible for data generation. Traditional relational database system was enough to handle those data. Invention of world wide web (www) in 1989 has triggered the production of data. Before that data production rate was very low. The main impact of data generation from 1989 to 1996 was web content. Lou Rosenfeld and Peter Morville defined web content broadly as "the stuff in your website". The stuff includes data, documents, e-services, applications, audio, images, personal web pages, and video files. Web pages were available for reading only. In 1990, first ever website CERN was created, and a file was uploaded and published there. After that, https://SixDegrees.com, a social media site, was launched in 1997. In 2004, Mark Zuckerberg launched another social media site, popularly known as Facebook, which captures now 100 TB of data approximately per day [3]. It is apparent that this huge amount of data cannot be dealt with existing tools. At the present time, researchers are coming up with many solutions to handle Big Data. Nowadays, Internet of things (IOT) has become one major source of data generation. Figure 2 shows contributors of data generation from 1944 until now.

3 Big Data: Big Challenges Versus Big Benefits

Big Data benefits come at a cost of big challenges. Existing systems are not efficient to handle Big Data, so we need to bring changes to the traditional systems to cope up with Big Data. In this section, we will focus on subsets of these Big Data challenges and benefits.

3.1 Data Filtering

As data is coming from multiple sources, it is difficult to extract quality data from it. The worth of data lies in its value or quality. We need to stay away from dirty data and dark data as well. Dirty data is false data. Dark data is unused data.

3.2 Storing and Management

Big Data is data of enormous amount ranging from a petabyte (1 PB = 1000 TB) to an exabyte (1 EB = 1000 PB). Storing massive amount of heterogeneous data is one big challenge associated with Big Data. Existing architectures are incapable to handle these data. So both storing and managing these data are peril. Big Data and grid computing both focus on huge volume and diverse variety of data. These two technologies can be merged to store and manage Big Data effectively [4, 5].

3.3 Dealing with Hadoop

Hadoop ecosystem is a framework that incorporates many tools (like Flume, Zookeeper, HBase, etc.) to support Big Data handling. HDFS, one core component of Hadoop, is capable of managing huge amount of heterogeneous data. MapReduce another core component of Hadoop handles processing of that data in distributed manner. But Hadoop itself is not easy to manage. Most of the developers spend more time to understand Hadoop than to understand Big Data.

3.4 Security

Due to the vast amount of Big Data, keeping it safe is another troublesome matter. If Big Data has privacy risks, then organizations will not be convinced to use it. Every organization tries to preserve their sensitive data. Big contributors of Big Data, social media, gather information about individuals. So, privacy must be preserved while accessing those information [6]. Das et al. proposed framework to analyze tweets [7].

3.5 Adding Smartness to Cities

Cities are developing rapidly with digitization and modernization programs. Cities are turning into smart cities. The purposes of smart city are crime prohibition, accelerate energy usage, extend the rate of economic growth, reduce the consumption of resources and traffic management, and provide comfort to citizens using smart technologies. Every device connected to Internet (like sensors, CCTVs, GPS-enabled devices, etc.) is generating huge amount of divergent data. These bulk of data cannot be handled easily. These data must be stored as well as analyzed. Living in Big Data era is not researchers' choice. But it is a part of development. As a city full of smart gadgets produces huge amount of data, it has to adopt Big Data and Big Data tools.

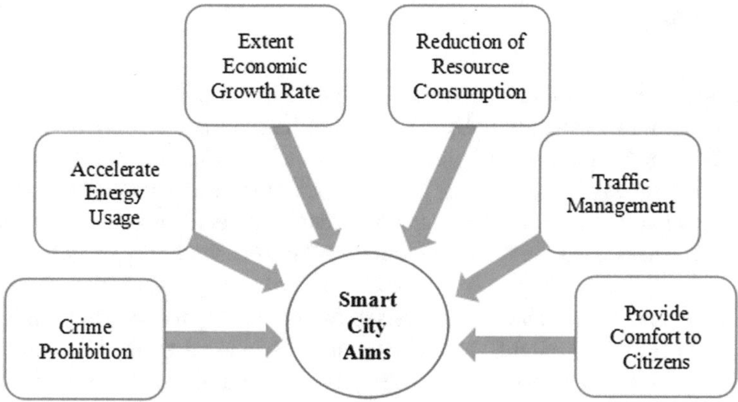

Fig. 3 Smart city goals

Big Data analytics can be helpful for smart grid and traffic congestion management [8]. Figure 3 displays different aims of smart city project.

3.6 *Affixing Value to Business*

People refer Big Data as huge amount of data that cannot be handled by traditional database system. But organizations see Big Data as treasure. Business companies are not satisfied with only present market analysis but also want to predict future to build stronger trade stratagem [9]. Big Data analytics can be useful for this. Big Data analytics integrate data from multiple sources and extract value, meaningful pattern from huge amount of data and help organizations to make decision. So despite having different challenges of Big Data, it comes with a time to empower business strategy.

4 Big Data Tools

Massive volume of heterogeneous data generated with high velocity is difficult to manage within the traditional systems termed as Big Data. It requires new tools and techniques that can easily capture, aggregate, store, and analyze data for enhancing its value. The tools can help us to process huge amounts of data to give an insight into the fast-changing situations. This section describes various tools to handle Big Data.

4.1 Hadoop

People often get confused Hadoop with Big Data. Hadoop is not Big Data although it plays a major role in handling Big Data. According to Apache, Hadoop framework processes large datasets in distributed environment. It provides scalability from single server to thousands of machines, each having local storage and computation power. It supports scalability (new machines can be added without decreasing performance), cost-effectiveness (Hadoop is an open-source platform), flexibility (handles unstructured data), and fault tolerance (if any device fails or any hardware error occurs, new device is allocated) [10]. Hadoop consists of two core components. They are Hadoop distributed file system (HDFS) and MapReduce. They are described below:

HDFS: HDFS file system stores huge (in terms of terabytes and petabytes in size) amount of data. Initially, data is stored in Hadoop cluster. HDFS divides data into multiple parts and loads them into different nodes. They are further replicated (usually three times) to guarantee fault tolerance. HDFS architecture follows master–slave format. The master node (NameNode) manages various slaves (DataNodes). NameNode stores the metadata. NameNode and DataNode communicate through signals usually known as Heartbeat.

MapReduce: Google first introduced MapReduce technique. Apache Hadoop used the technique of MapReduce. Hadoop MapReduce is an open-source framework performs both parallel and distributed data-intensive reckonings [11]. It is similar to a device that takes input, processes it, and generates the required output. MapReduce is based on master–slave approach. MapReduce has two components: Mapper and Reducer. When MapReduce master (JobTracker) receives a job from Client, it divides it among different slaves (TaskTracker), which is handled by mapper. After getting the solutions, reducer aggregates the solutions and returns it to the master. Figure 4a shows Hadoop master–slave architecture and 4(b) depicts Hadoop ecosystem, source: http://blog.agro-know.com/?p=3810ss. Table 1 summarizes different tools of Hadoop ecosystem. Figure 5 represents Hadoop MapReduce framework.

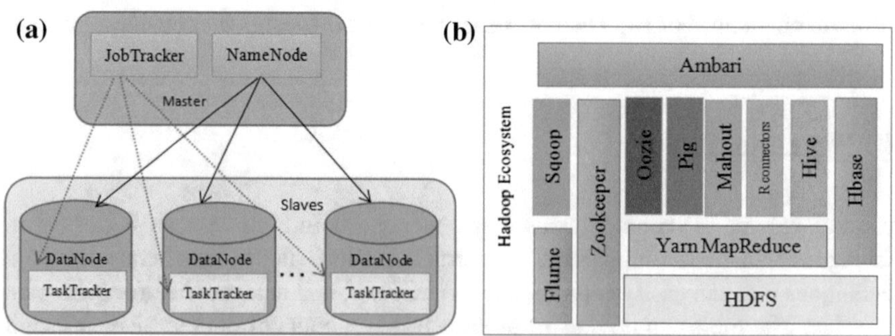

Fig. 4 **a** Hadoop master–slave architecture **b** Hadoop ecosystem

Table 1 Tools of Hadoop ecosystem

Developer	Tools	Application
Apache software foundation	HDFS	Stores large dataset in a distributed fashion
	Hbase	Acts like database
	Pig	High-level language makes program writing easier
	Sqoop	Transfers data between Hadoop and RDBMS
	Flume	Collects data from multiple sources and store it into single repository
	Zookeeper	Controls subtasks in a distributed environment
	Oozie	Controls the dependency among various jobs
	Mahout	Performs data mining on huge amount of data

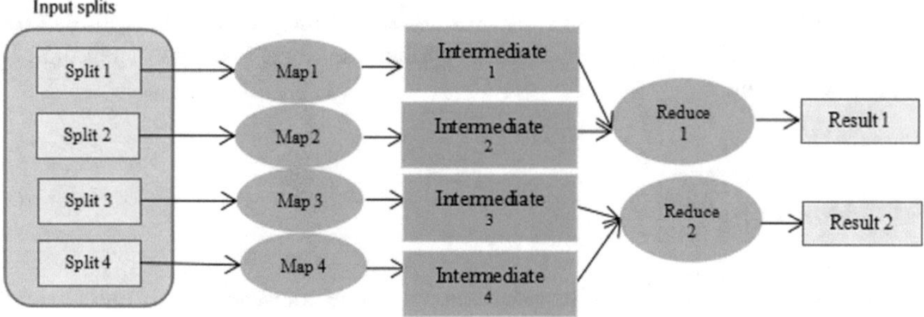

Fig. 5 Hadoop MapReduce framework

4.2 *Spark*

Spark is an open-source framework for parallel processing of Big Data. Spark was first developed in AMPLab in 2009. Then, Apache made it open sourced in 2010. It is more efficient than Hadoop MapReduce. Spark programs can be written in Java, Scala, or Python. The main difference between Spark and Hadoop is that Hadoop does not perform processing inside memory where Spark is an in-memory computing system. Spark guarantees fault tolerance by using RDD (Resilient Distributed Dataset), where Hadoop ensures fault tolerance using replication of dataset. Xu et al. used Spark for real-time data analytics [12]. Figure 6 shows Spark architecture.

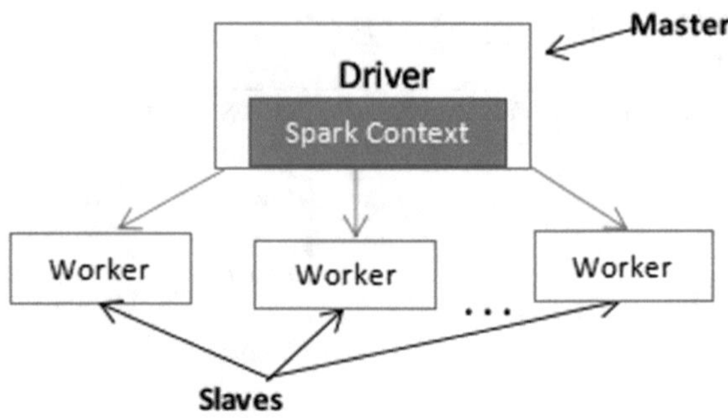

Fig. 6 Spark architecture

5 Fields Influenced by Big Data

It is incredible to witness how quickly Big Data is influencing many areas that deal with huge amount of heterogeneous data. One can see Big Data as an indispensable part of societal progress and innovation. It will make resounding impact on the way people learn and companies do business. In this section, we will take a look at the influence of Big Data on some prevalent application areas and how it is going to change everything.

5.1 Marketing

Marketing is based on buyer/seller relationship. Companies use various tools to analyze customers' behaviors, their likes, and dislikes. Market analysts use technologies to capture facial expressions of customers about a particular product advertisement. These facial expressions are analyzed using Big Data analytics [13]. Social media posts are also used to make decision regarding products. It helps companies in customer retention. These huge amount of data is mined using Big Data tools.

5.2 Image Processing

CCTV cameras and sensors are used rapidly at present days. Millions of images must be analyzed to detect one object. Existing models are not effective to satisfy the purpose. Hua et al. proposed a new method to detect the location of a missing child in a crowded space. This proposed model can analyze millions of captured images to identify the missing child [14].

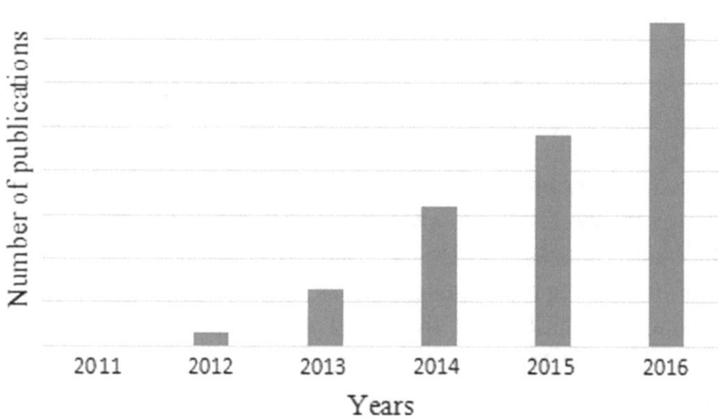

Fig. 7 Publications of various fields influenced by Big Data

5.3 *Healthcare*

The healthcare industry is not only generating huge amount of diverse data (medical imaging, pharmacy, laboratory reports, insurance, and other managerial data) but also the rate of data production is unmanageable by the existing systems. In 2011, U.S. healthcare related data reached 150 exabytes. So, Big Data is affecting healthcare industry.

Figure 7 shows pictorial representation of total number of publications from 2011 to 2016 of abovementioned application areas of Big Data. This graph is made using Publish or Perish software tool, version 5 for Windows 10. We measured number of publications considering "Title words only" feature of abovementioned software.

6 Future with Big Data

Big Data is in its inception phase. In spite of many challenges and the risks, today everyone is talking about and trying to embrace Big Data. Effective use of Big Data has endless potential that can open vast opportunities to the world at large. In future, Big Data can be explored for progress and well-being of the society. Let us point out some possible future gains from Big Data.

Big Data can make colossal impact on the medical care and healthcare industry. Effective analysis of medical data can give better outcomes for patients to receive cost-effective and finest medical care. Sophisticated data analytics may yield exciting outcomes and can revolutionize the way business operations are carried out and many more. It can make significant impact on the enterprise decision-making and maximize the development of the next-generation products and services. Prime Minister's office is also using Big Data to analyze citizen's ideas, thoughts, and sentiments about government action by using www.mygov.inplatform [15]. Big Data can transform education system. It can analyze huge amount of students' nature to make constructive decisions regarding their carrier [16]. It can be helpful in many other areas, such as providing precise response to disaster recovery, making optimal energy consumption decisions, genome sequencing, astronomical star surveys, identifying risk of structures like bridges and thus avoiding disasters, making accurate predictions of hurricanes, earthquakes, and thereby taking preventative actions.

Many of our expectations for the future may come true if we deal with Big Data judiciously. As we may see many promises in the future of Big Data, there are legitimate concerns that work with massive stores of data could lead to privacy threats and mistaken predictions.

7 Conclusion

This paper presented an interesting survey that includes contributors of Big Data, pitfalls, benefits, tools, enabling technologies, and some prevailing applications as well. This article has also discussed how helpful Big Data is and its scope of applying actionable knowledge driven by Big Data in practical environments.

References

1. Big Data Black Book. dreamtech press, authored by DT editorial Services, edition: 2015
2. McKinsey & Company. http://www.mckinsey.com/insights/Mgi/research/technology_and_innovationbig_data_the_next_frontier_for_innovation
3. Sruthika, S., Tajunisha, N.: A study on evolution of data analytics to Big Data analytics and its research scope. In: 2015 International Conference on Innovations in Information, Embedded and Communication Systems (ICIIECS), pp. 1–6. IEEE (2015)
4. Mishra, M.K., Patel, Y.S.: The role of grid technologies: a next level combat with big data. In Techniques and Environments for Big Data Analysis, pp. 181–191. Springer International Publishing (2016)
5. Mishra, M.K., Patel, Y.S., Ghosh, M., Mund, G.B.: A review and classification of grid computing systems. Int. J. Comput. Intell. Res. **13**(3), 369–402 (2017)
6. Madden, S.: From databases to big data. IEEE Internet Comput. **16**(3), 4–6 (2012)
7. Das, T.K., Acharjya, D.P., Patra, M.R.: Opinion mining about a product by analyzing public tweets in Twitter. In: 2014 International Conference on Computer Communication and Informatics (ICCCI), pp. 1–4. IEEE (2014)
8. Kumar, S., Prakash, A.: Role of big data and analytics in smart cities. Int. J. Sci. Res. **5**, 12–23 (2016)
9. Manyika, J., Chui, M., Brown, B., Bughin, J., Dobbs, R., Roxburgh, C., Byers, A.H.: Big Data: The Next Frontier for Innovation, Competition, and Productivity (2011)
10. Menon, S.P., Hegde, N.P.: A survey of tools and applications in big data. In: 2015 IEEE 9th International Conference on Intelligent Systems and Control (ISCO), pp. 1–7. IEEE (2015)

11. Lakshmi, C., Nagendra Kumar, V.V.: Survey paper on big data. Int. J. Adv. Res. Comput. Sci. Softw. Eng. **6**(8) (2016)
12. Yadranjiaghdam, B., Pool, N., Tabrizi, N.: A survey on real-time big data analytics: applications and tools. In: 2016 International Conference on Computational Science and Computational Intelligence (CSCI), pp. 404–409. IEEE (2016)
13. Sravanthi, K., Subba Reddy, T.: Applications of big data in various fields. Int. J. Comput. Sci. Inf. Technol. **4** (2015)
14. Hua, Y., Jiang, H., Feng, D.: FAST: near real-time searchable data analytics for the cloud. In: Proceedings of the International Conference for High Performance Computing, Networking, Storage and Analysis, pp. 754–765. IEEE Press (2014)
15. Mukherjee, S., Shaw, R.: Big dataconcepts, applications, challenges and future scope. Int. J. Adv. Res. Comput. Commun. Eng. **5**(2), 66–74 (2016)
16. West, D.M.: Big data for education: data mining, data analytics, and web dashboards. Gov. Studies Brook. **4**, 1–0 (2012)

Accelerating Pairwise Sequence Alignment Algorithm by MapReduce Techniquc for Next-Generation Sequencing (NGS) Data Analysis

Sudip Mondal and Sunirmal Khatua

Abstract Next-generation sequencing (NGS) technologies and different types of sequencing machines are introduced in an enormous volume of omics data. For analysis of NGS data, sequence alignment is always an essential step in finding relationships between sequences. Pairwise sequence alignment is a challenging task for reasonably large input sequences. Smith–Waterman (SW) is a popular centralized algorithm for sequence alignment. However, as data is spreading expeditiously, conventional centralized sequence alignment tools are inefficient in terms of computational time. In this paper, we propose a distributed pairwise sequence alignment technique using MapReduce implemented on Apache Spark framework, called MRaligner. We have compared the result of the proposed MRaligner with Jaligner, an open-source Java implementation of the Smith–Waterman algorithm for biological sequence alignment, and found significant improvement in terms of computational time.

Keywords Sequence alignment · Next-generation sequencing
Distribute and parallel algorithms · Apache spark · MapReduce

1 Introduction

The next-generation sequencing (NGS) machines typically produce million base pair (bp) reads on a single run [1]. Analysis of this NGS dataset is a big challenge in the field of bioinformatics. The researcher has developed many programs and workflows for analyzing large-scale biological sequence datasets. Pairwise sequence alignment is one of the fundamental problems. There are many existing pairwise sequence align-

S. Mondal (✉) · S. Khatua
Department of Computer Science & Engineering, University of Calcutta,
Kolkata 700106, India
e-mail: sudip.wbsu@gmail.com

S. Khatua
e-mail: enggnimu_ju@yahoo.com

A. Abraham et al. (eds.), *Emerging Technologies in Data Mining and Information Security*, Advances in Intelligent Systems and Computing 813,
https://doi.org/10.1007/978-981-13-1498-8_19

ment techniques available which are unable to map this large volume of data. There is always a need for high computation for database search. To find out sequences similar to a given query sequence from different sequence databases, the search programs need to compute for an alignment score for every sequence in the database. Popular nucleotide databases like National Center for Biotechnology Information (NCBI), European Molecular Biology Laboratory (EMBL), and DataBank of Japan (DDBJ) are increasing twofold in every 15 months [2]. An enormous amount of data is being produced on daily basis. Major sources of data include Bioinformatics labs, sequencing machines (NGS), healthcare industry, hospital, etc. To make this data fruitful, it needs to be accurately stored and analyzed. Traditional techniques are inadequate to store and analyze such large amount of data.

Biological sequence analysis is one of the important research fields in bioinformatics, mainly through sequence alignment which is finding nucleotide sequence in the reference genome. There are two types of the reads as short and long that need to be handled or processed for analysis. Various automated software tools are being developed for both short- and long-read alignments. Next-generation sequencing (NGS) technology is playing a vital role in development and advancements of various sequencing algorithms [3]. Over the past few years, a wide range of programs for aligning short sequencing reads to a reference genome have been developed. Most of the algorithms that have been developed for short reads with low sequencing error rate [4] are very efficient. However, they are not applicable to the reads over 200 base pair (bp). CUSHAW2, a parallelized, accurate, and memory-efficient long-read aligner, is based on the seed-and-extend technique and uses maximal exact matches as seeds. They have compared with other popular long-read aligners BWA-SW, Bowtie2, GASSST, and got a satisfactory outcome [5]. With the advancement in the field of biology and technology, the production of billions of short sequence reads (mainly DNA and RNA) by the next-generation sequencing machines have become less time-consuming. In upcoming years, the size of biological sequence data will continue to increase tremendously. Sequence alignment is the mandatory step for sequence data analysis. There are many existing sequence alignment tools like BLAST, RMAP [6], MAQ [7], SOAP [8], etc. However, these existing centralized alignment tools will not be able to handle such massive volume of sequence data, and therefore highly distributed computing machines will be required.

In recent years, a distributed computing framework called Hadoop can be used for the processing and storage of big data. It has two major components, namely, MapReduce used as the programming model and Hadoop distributed file system (HDFS) for storing the data. The distributed platforms that use MapReduce model, such as Apache Spark, provide highly parallel and distributed computing environment using thousands of computer. The humongous data generated by the next-generation sequencing machines can be efficiently stored and analyzed by using these platforms. Some initiative toward using platforms like Hadoop for sequence alignment has already been taken such as CloudBurst [9], CloudAligner [10], etc. Moreover, distributed architecture like cloud computing can boost alignment algorithms with respect to emerging long sequence reads compared to short-read alignment which is no longer the bottleneck of data analyses [11].

A dynamic programming algorithm for computing the optimal local alignment score was first described by Smith and Waterman [12], and later improved in [13] for linear gap penalty functions. Problem Statement: In local pairwise sequence matching, dynamic programming (DP) guarantees to find an optimal alignment given a particular scoring function. There are many recent developments on Heuristic-based approach due to its efficiency over DP. So to overcome this issue, we will implement this dynamic programming approach to solving local alignment problem in the MapReduce environment. An open-source Java tool, Jaligner [14], is an implementation of the Smith–Waterman algorithm with Gotoh's improvement for biological local pairwise sequence alignment using the affine gap penalty model. There are already lots of developments done in the field of Heuristic-based approach. So here we are focusing on DP because it guarantees to give the optimal solution. Moreover, we will try to present a general framework where we can run any alignment algorithm on the MapReduce environment.

The rest of the paper is organized as follows: Sect. 2 represents the background of the work; Sect. 3 describes the implementation of our method. Section 4 provides the experimental results discussion and Sect. 5 concludes the paper with future work.

2 Background

Pairwise sequence alignment is used to identify a part of similarity that may express different types of relationships between two sequences. This paper presents the famous past and recent work on both local and global pairwise sequence alignment algorithms [15]. The growing sequence database has been creating many new challenges including sequence database searching and aligning long sequences. When the two sequences are long or when the number of sequences in the database is large, we can use the heuristic algorithm to get better accuracy for computational time over Smith–Waterman algorithm. The sequence alignment tools like BLAST, FASTA, and Sim2 are used in this technique [16].

MapReduce is a programming model and software framework first developed by Google [17]. Hadoop composed of two parts: MapReduce and Hadoop distributed file system (HDFS). MapReduce is an implementation of MapReduce designed for large clusters, and the Hadoop distributed file system (HDFS) is a file system optimized for batch-oriented workloads such as MapReduce. HDFS is a block-structured file system managed by a single master node like Google's GFS. The processing unit in the Hadoop ecosystem is the MapReduce framework. There are two primitive functions in MapReduce which are Map() that takes key/value pairs as input and generates an intermediate set of key/value pairs and other function is Reduce() which merges all the intermediate values associated with the same intermediate key.

3 Proposed Methodology

A local alignment tool Jaligner is a Java implementation of Smith–Waterman algorithm [12]. We have select Jaligner as our tool and try to run it in Apache Spark. The motive for selecting Jaligner [14] is that it purely implements Smith–Waterman algorithm in Java Language with the Gotoh's [13] affine gap penalty model. As our main objective is to implement a dynamic programming based alignment algorithm with MapReduce, we try to keep the main Jaligner as it is, so that our approach can be treated as a generic.

3.1 Framework

For our experiment, all the sequence data are taken from the sequence database NCBI [18]. Target or reference sequence of whole human chromosome number 1 is S1 of 250 million characters and a query sequence is S2 of 180 character. The normal Jaligner tool is unable to solve this huge data. We have proposed a distributed framework for implementing Jaligner using MapReduce technique with a realistic example.

One of the major advantages of Apache Hadoop is to scale our hardware when required. New nodes can be added incrementally without having to worry about the change in data formats or the file system. We can run the same experiment on different sizes of the cluster to add data node only. In our experiment, we set up a pseudo-distributed Hadoop cluster running on 8-core processor machine and 16-Gb system memory. We use Apache Spark running on single data node Hadoop cluster for parallel execution of the alignment process. Spark provides a simple standalone deploy mode and we launch a standalone cluster manually.

The workflow of alignment process shown in Fig. 1, describes that long-read sequence first split and then uploaded to HDFS from local storage. Apache Spark calls Jaligner function for parallel execution. The Map function, f, executes the alignment process with each split of the sequence against another sequence. The Reducer function, g, collects the split score from all mapper f and generates a maximum score for the result.

3.2 Algorithm

Here is a simple algorithm for our approach.

Input: Reference Sequence S1 = a1,a2, ... an, Query sequence S2= b1,b2, ... bm and JAligner()
Output: An alignment score E and similarity S is found between two sequence elements.

```
L1 = length(S1) and
L2 = length(S2)
while i=1 do
    P(p1,p2, ... pi) = Partition(S1) where,
            length(pi) >= (2 x L2)
End while
While j=1 do
    Call Spark module for JAligner (S2 , pj) where
            pj is a reference split of each partition
    S [ j ] = JAligner(S2,pj)
    R(r1,r2,...rj) = score { P( j ) }
    E = Max [ r(j) ]
End while
```

4 Results

In order to evaluate the performance of our technique, we compared it with Jaligner's score and performance. For our experiment, we need two sequences to compute. Target or reference sequence is referred to as S1 and the query sequence is referred to as S2. For comparison purpose, we are taking S2 as a fixed length query string and we will increase the length of S1 to check our results. We ran the experiments many times and computed the average values. Figures 2 and 3 summarize the results achieved as score and time comparison.

As we can see from Fig. 2 that all the alignment scores are same, so now we can conclude that with respect to accuracy our solution is qualified for purpose.

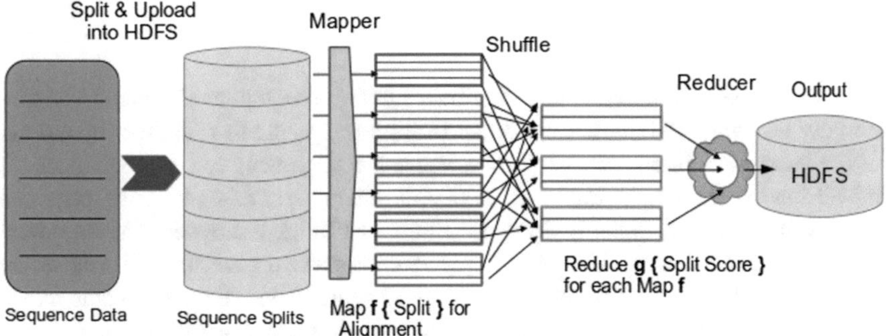

Fig. 1 Framework for the execution of our alignment workflow using MapReduce technique

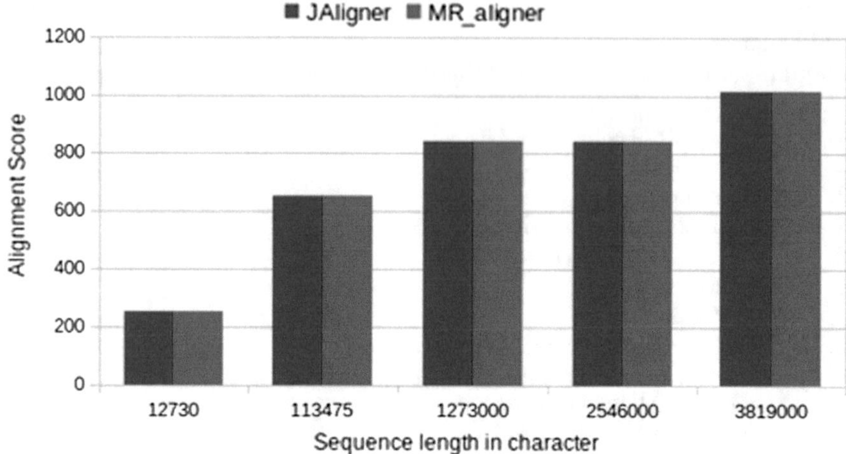

Fig. 2 Comparison of alignment score for Jaligner and MRaligner

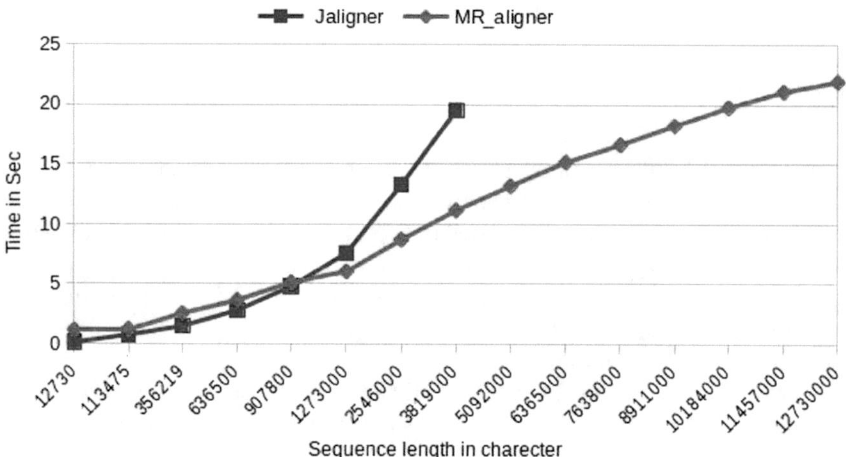

Fig. 3 Comparison of alignment time for Jaligner and MRaligner

As we can see from Fig. 3 that the time required to compute alignment is exponentially growing for Jaligner, whereas MRaligner is trying to straighten up nature.

The existing Jaligner can perform a sequence alignment of two sequences S1 and S2 having the length less than 5,000,000 characters. After this, the time and space complexity significantly dropped. But the MRaligner can perform alignment of two sequences, where S1 be the Reference or Target sequence and S2 be the query sequence. And S1 can be of any length and S2 must be less than 6000 characters. This gives MRaligner an upper hand advantage over the existing Jaligner. When S1 & S2 are small in length that is less 20,000 characters, Jaligner is the clear winner over MRaligner. Because starting the Spark Environment, it maintains that will incorporate extra overhead in MRaligner. And when the lengths of S1 & S2 are over 5,000,000 characters, Jaligner fails to perform the job as the size of the input string is unmanageable for it and the program fails to give the result. But on the other hand, MRaligner is able to execute the desired output.

5 Conclusion and Future Work

The framework that we provided to run any alignment algorithm in the MapReduce paradigm by using Apache Spark is generic, that is, we can replace any alignment algorithm with the Jaligner to perform alignment in Apache Spark environment. Moreover, an algorithm which is inadequate for pairwise long sequence alignment can be built for the alignment with this framework. As the framework is flexible and scalable, increased number of processor and a good distributed HDFS management will fasten up the processing. Moreover, data locality of processors will speed up the process further.

We have completed this work by applying a dynamic programming sequence alignment algorithm in MapReduce paradigm using Apache Spark and got satisfactory results. But there is a scope for improvement in the execution of the program in a more time-efficient manner. This can be achieved by applying more worker node in the cluster with good management of HDFS. Apart from this, we may implement the heuristic-based alignment algorithm for faster execution. We can further incorporate streaming techniques for receiving huge sequence data from any streaming service that is helpful for analyzing sequence data to predict better results.

References

1. Buermans, H.P.J., Dunnen, J.T.: Next generation sequencing technology: advances and applications. Biochim. Biophys. Acta **1842**, 1932–1941 (2014)
2. Benson, D.A.: GenBank. Nucleic Acids Res. **28**, 15–18 (2000)
3. Ekre, A.R., Mante, R.V.: Genome sequence alignment tools: a review. In: AEEICB16. 978-1-4673-9745-2 IEEE (2016)
4. Li, H., Durbin, R.: Fast and accurate long-read alignment with Burrows–Wheeler transform. Bioinformatics **26**(5), 589–595, 2010 (2009)
5. Liu, Y., Schmidt, B. Long read alignment based on maximal exact match seeds. In: Bioinformatics. ECCB 2012, vol. 28, pp. i318–i324 (2012)
6. Smith, A.D., Xuan, Z., Zhang, M.Q.: Using quality scores and longer reads improves accuracy of Solexa read mapping. BMC Bioinform. (2008)
7. Li, H., et al.: Mapping short DNA sequencing reads and calling variants using mapping quality scores. Genome Res. (2008)
8. Li, R., et al.: SOAP: short oligonucleotide alignment program. Bioinformatics (2008)
9. Schatz, M.C.: CloudBurst: highly sensitive read mapping with MapReduce. Bioinformatics **25**(11), 1363–1369 (2009)

10. Nguyen, T.: CloudAligner: a fast and full-featured MapReduce based tool for sequence mapping. BMC Res. Notes **4**, 171 (2011)
11. Li, W., Homer, N.: A survey of sequence alignment algorithms for next-generation sequencing. Brief. Bioinform. **11**(5), 473–483 (2010)
12. Smith, T.F., Waterman, M.S.: Identification of common molecular subsequences. J. Mol. Biol. **147**, 195–197 (1981)
13. Gotoh, O.: An improved algorithm for matching biological sequences. J. Mol. Biol. **162**, 705–708 (1982)
14. Moustafa, A.: JAligner: Open source Java implementation of Smith-Waterman. (2005)
15. Haque, W.: Pairwise sequence alignment algorithms: a survey. ISTA Information Science, Technology and Applications (2009)
16. Li, J.: Pairwise sequence alignment for very long sequences on GPUs. IEEE Int. Conf. Comput. Adv. Biol. Med. Sci. PMC (2013)
17. Dean, J., Ghemawat, S.: MapReduce: Simplified Data Processing on Large Clusters Google, Inc. (2004)
18. National Center for Biotechnology Information. http://www.ncbi.nlm.nih.gov

Entity Resolution in Online Multiple Social Networks (@Facebook and LinkedIn)

Ravita Mishra

Abstract The social network is a platform where the user can frame societal relations with people, friends, and colleagues who claim same significance, actions, and real-life connections. Social network is web-based services, and it grants individuals to build their personal profile (public/private) and also maintains the list of users with whom to share connections, post, updates, and view. Today, over 2.9 billion individuals across the globe use the Internet and about 42% of these users actively participate in OSNs. Recent statistics list about 226 active OSNs in 2017. Each online social network (OSN) offers a distinct set of innovative services that easily access to information. The main aim of Twitter is at particular instance; its retweet feature enables quick access to news, campaigns, and mess information, while pinboards of pinterest provide facility to reach the work of artists, photographers, and fashion designers to enjoy these services, accordingly and a user book herself/himself on multiple online social networks (OSNs). Research center shows that 91% users registered themselves on both Twitter and Facebook and 52% users on Twitter, LinkedIn, and Instagram. During registration on any OSN, a user creates an identity for herself/himself listing personal information (profile) and connections and sharing. Since varying policy and purpose of the identity creation on each OSN, quality, quantity, and correctness of her identity vary with the online social network. These give dissimilar existence of the same user and it is scattered across Internet; it has no accurate networks directing to one another. These unequal networks detach his/her from any privacy concerns. It emerges if the identities were implicative collated. However, desperate unlinked networks are a concern for various participants. The application of analysis/findings will be helpful in marketing and job recruitment, where the manager wants to check the employee profile on Facebook and LinkedIn. Facebook will give the detail of social activity and LinkedIn gives professional connection and their past experiences. The analysis/discovery can also be used in the security domain, recommendation system human resource management, and advertisement. The paper is organized in six parts: part I contains an introduction to the social network, part II contains literature

R. Mishra (✉)
Ramrao Adik Institute of Technology (RAIT), Nerul, Navi Mumbai 400706, India
e-mail: ravita.mishra@rait.ac.in

A. Abraham et al. (eds.), *Emerging Technologies in Data Mining and Information Security*, Advances in Intelligent Systems and Computing 813,
https://doi.org/10.1007/978-981-13-1498-8_20

survey, part III explains the limitation and research gap, part IV contains proposed methodology and block diagram, and part V contains conclusion and references.

Keywords Identity resolution · Identity search · Jaro distance · Online social networks (OSNs) · Finding nemo · Nudge nemo

1 Introduction

For information collection and making relationships, the user needs to make API calls, but each social network sets their limits on the number of calls permitted. This introduces the problem of limited information when the access to the full graph is limited and one can only access the graph through a limited number of API calls. These results are used by researchers and data scientists to usually repeat their analysis multiple times over time to improve accuracies [1, 2]. The main application of social network analysis is that users have been helping the industry in exciting ways to enrich user experience and service; it is also helpful for searching and linking user accounts that conform to the same individual in popular online social networks. A difficult task in social network analysis is finding the user behavior across multiple social networks [3]. The three important techniques to find the user behavior is self-identification, self-mention, and self-sensitive sharing. In self-identification, users explicitly mention their identities on other OSNs or on webpage using hyperlinks. In social network analysis, user not only accesses their account if the only username set available but also has to add some other parameters like basic information and other description and also used private profile for analysis [4]. When a search parameter is not supported by the API, then identity search fails and; it is challenging to retrieve candidate identities similar to a searched user identity on the mentioned parameter. The new methods need to be modified to start with a Facebook user identity and find the corresponding Twitter identity. The user identity linking method also fails sometimes.

2 Review of Research and Development in the Field

Entity/identity resolution issues can be categorized into different subways, namely, identity search and identity matching. Many kinds of survey have described multiple identity resolutions and matching techniques to connect various networks in the real world but they fail to analyze identity search techniques for finding identical identities to their possibility. It is a crucial issue, which has been addressed by many researchers in the past [5–8]

Benjelloun et al. [9] explained the D-Swoosh algorithm and its main task to distribute the entity resolution workload across many systems. In this system, genetic matching and merging functions are used and this algorithm ensures that the new merged records are distributed to all machines that may have matching records [10]. Here, author performs a detailed analysis on a test bed of 15 processors, where application expertise can eliminate some comparisons and where all records must be matched. The author also discovered that applying domain knowledge is very difficult, as it needs a good circulation of records across syntactic groups.

Peled et al. [11] introduced new methods for solving various entity resolution problems which are comparable user profiles across multiple OSNs [12]. Here, supervised learning techniques (classification) are used to match two user profiles from two different OSNs; this method uses extracted features from each one of the user profiles. Here, classifiers perform entity matching between two users' profiles for the following circumstances: in first, it matches entities across two online social networks; second, one searches user attribute by the same name; and third, de-anonymizing a user's network. The model was tested by collecting data from two popular OSNs, Facebook and Foursquare, and it evaluates the attainment of the model. Author develops classifiers which use multiple features: 1. Username-based features, 2. data-based features, and 3. topological-based features. The developed method was evaluated using real-life data collected from two OSNs, Facebook and Foursquare. In this research, logit boost algorithm seems to be the most appropriate algorithm and it solves the problem over all the categorical algorithms [11]. This method has different limitations, like the similarity between user finding, classification algorithm impacts problem, and accuracy of the algorithm which are not sufficient because of public profile.

Esfandyari et al. [13], suggested that the people open several accounts on diverse online social networks (OSNs) and enjoy different services, and people have lots of different types of profile information and these pieces of information from various sources can be achieved by identifying individual person's profile across social networks. Here, author addresses the problem because user identification process can be used as an allocation task. In this method, common public attribute features can be accessed by using application programming interface (API). For building negative instances, author proposes different methods and it uses the usual random selection to investigate the effectiveness of each method and the classifier requires training [14]. The effectiveness of this method is measured in real time by collecting different profiles from different OSNs like Google, Facebook, and Twitter.

Chung et al. [15] give the basic idea about linking problem of same users across different social networks and it also provides the basic solution for comparing two different profiles of users. This method is further enhanced by seeing some increased societal association, i.e., friends. The author also proposes a two-phase clustering (Hybrid Clustering) method and it generates a summary of each individual/person. In the first phase of the algorithm, it selects actively connected groups that considered as seeds. In the other step of the algorithm, it elects non-seeds to the aggregates based on the profiles' design of persons. The last dissemination comes over various groups that are regarded as the societal summary. The methods are tested on two real social networking datasets which are gathered by API; the analysis results are satisfactory in terms of feasibility of the recommended method, and the proposed approach also compares the two profiles, person–name attribute of different OSNs.

Goga [16], in this paper, focuses on the privacy issues and security concern of individual. For security point of view, in the social connections, the terms of service specify that users can create a single account. But maximum user forms more than accounts and by end malignant user often acts like honest users. In the social network, different users and their information also are correlated with different sites and accounts of single user. To solve this issue, entity matching techniques require to identify the accounts of single person. In this paper, author also considered different parameters for evaluating the profile attribute, namely, they are A—Availability, C—Consistency, I—Non-impersonality, and D—Discriminability. These parameters are used to find the quality of different profile attributes to match accounts. In this paper, author also used different matching schemes.

Classify the multiple accounts that belong to a one person: Mobile phones are good example of identification of single account of one single person because the sensors can log almost all user activities, and it gives a complete picture of persons. Sensors are used in many applications because sensors provide different monitoring services, such as applications to monitor sleep, heart rate, or the number of steps we do each day [17]. Sensors exist in early days but their use has increased widely because more sensors will be surrounding us [16].

Using graph structure of social connection to boost the account matching: The technique of de-anonymize graphs is used here, and the main idea behind such approaches is to start with a minor node that is considered as seed node; for this node, know the complete corresponding account information about the two graphs that are created by seed (u) and non-seed (v) node; lastly, the seed node propagates the matching techniques and it measures the similarity between the two graphs (u, v).

Sybil detection using cross-site information: The current research is focused on how to evaluate the identity of the honest users, and they verify the user's identity/detailed summary on different social networks like Xing, Facebook, LinkedIn, or Google+ [18, 19]. The available system detects only fake/untrustworthy identities that rely on evaluating information about the integrity that is available on a single site or domain.

Find which things accomplish a person popular across different sites: The person's popularity is an important parameter to correlate different social networks. In OSNs, Facebook, Twitter, and LinkedIn, users are mostly interested to build their accounts/profile more popular as compared to their friends [2, 20]. Users want to form profile more popular by investigating a number of questions. The answer to that questions will give a better understanding of what forms an account popular than other.

Interpret how to disseminate information from one social network to another: The main idea behind is to understand whether information distributed from one social network to another is authenticated or not [21].

Recommender with trigonous information: In this recommender, system plays important roles in recommending the content inside in a system using the information gathered inside the particular system [13]. This concept is possible to know the user's account details on different social networks of the user.

Campbell et al. [22], in this paper, provided a new technique of integrating information; from multiple sites, it creates a complete picture of user activities and their characteristics and trends. Here, author proposes a different category of search that is profile, content, and graph. In profile-based methods, it considers approximate string matching techniques and in content-based methods, it performs author identification/identity and finally, in graph-based methods, it applies new cross-domain community detection methods and it generates neighborhood-based features [23]. **Limitation**: One issue for training the fusion models is the problem of missing data. For example, the content features may be missing for some users due to an insufficient number of posted words.

Shuy et al. [24], in this paper, proposed the method that increases the reputation and dissimilarity of social media. The main purpose is to encourage many people who can enroll himself/herself on multiple online social networks and enjoy their benefit and services. At the time of registration, user can generate their entity and it represents her/his unique public picture in every OSN. The main purpose of identity linking techniques applied in different domains such as recommendation system and link prediction. The data in the social network are vast and complicated and face various challenges because the data have unique characteristics, and to solve these problems, new approaches are used, such as first extract different features and then create a predictive model for various perspectives.

Jain et al. [3, 25] explained how a single user can register himself/herself on multiple social networks to enjoy their different services. At the time of registration, user creates their identity and that identity establishes three main major dimensions; they are profile, content, and network [26]. User largely governs his/her identity framework on any social network and therefore can manipulate multiple aspects of it; and no one can mark his/her presence uniquely in the online social network. In this paper, literature has also proposed an identity search method on the basis of profile attributes but has left the other identity dimensions' content and network, which are not explored. Author also introduces two unique identity search algorithms based on content and network attributes and it is the improvement of the old identity search algorithm. A new identity resolution/search system called as Finding Nemo is also deployed, this system introduces identity search methods to find a Twitter user's identity on Facebook and also proposes a new identity search algorithms which access public as well as private profile information of users.

3 Limitation of Existing System

Limitation/Research Gap: Entity resolution in online multiple social networks has various research gaps, and existing systems have various limitations. Here, the problem can be solved by improved methods/techniques [13]. The various limitations/research gaps are listed below:

3.1 Limited Information

Social graph has limited or less information, the third-party individuals (researchers and data scientists) who would like to perform network analysis can only access the social graphs through a limited application programming interface (API) [27]. For example, in an intelligence agency, they are collecting information about some suspicious person or terrorists and then they want to merge the data for the same user across different online social networks to obtain a better picture of a possible person/terrorist. Social networks also allow the user to generate content, and user-generated content is usually messy (contains typos, malformed input, and bad GPS location for the check-in domain) and has lots of duplicates; it also causes the problem of graph tracking (updating).

3.2 Crowdsourcing for ER

In this scheme, machines first compute the probability that each pair of records refers to the same underlying real-world entity [28]. Next, ask humans to resolve record pairs and leverage transitivity to avoid asking humans to resolve every pair.

3.3 Imperfect Crowd

We can use simple techniques such as majority voting to reduce human errors. However, the intermediate results will contain some unresolved record pairs because we have not reached an agreement for these pairs [28]. It would be useful to utilize these intermediate answers and make some inferences about record pairs.

3.4 Improving Crowd ER

In the existing system, machine can update its model based on each answer from the crowd worker and recalculate the probabilities of the record pairs that will not ask humans. We can use each human's answer as a new training example and update the machine learning (ML) model based on the new example. We constantly recalculate the probabilities and reorder the pairs to ask humans; therefore, proving any theoretical bound on the number of questions would be nontrivial [28].

3.5 Limited API

Limited API introduces a new challenge where for each machine, we require an access token and to obtain an access token, it takes some time and social networking system usually prevents users from creating many tokens [19]. To obtain an access token on Twitter, one needs to first create an account, then create an application, and finally create an access token [27]. Twitter may suspend the account if he/she has created too many tokens.

3.6 Identity Linking Using Username Only

History of other attributes like profile picture and description, which change more frequently than username, can be used further to link user profiles [28]. We could not do so because of nonavailability of data of the respective attributes on other OSNs [3, 29].

3.7 Identity Search Dependency on API

If a search parameter is not supported by the API, it is challenging to retrieve candidate identities similar to a searched user identity on the mentioned parameter [30]. Therefore, the methods are asymmetric, i.e., the methods need to be modified to origin with a Facebook user integrity and find the identical Twitter identity.

3.8 Evaluation of Self-identified Users

Truth datasets of real-world users, used for evaluation of identity search and linking methods, contain those users who explicitly self-identify their identities on multiple social networks [16]. A validated dataset of users and their identities across OSNs who do not explicitly identify their own accounts, which is a challenge to gather. Therefore, applicability and performance of our methods on non-self-identified users are difficult to examine (Table 1).

Facebook always allows the user to first register themselves and then create a user profile, upload photographs, video, sending messages, and keeps in touch with friends, family, and colleagues [20, 31]. On the other hand, LinkedIn is totally analogous from the Facebook. The main task of LinkedIn is that it maintains professional figure, network, and brand; it is also helpful in finding a job, networking, endorsement, selecting new employees, getting sales leads, and even getting employability news, marketing, etc. [32–34]. The main aim is to collect common attributes of Facebook and LinkedIn which is helpful in identity matching.

Table 1 Data extracted (Attribute) from Facebook and LinkedIn

Facebook	LinkedIn
UserID	ID type (Passport, identity card, driving licence)
Username, name	Name
Profile picture	Profile picture
Location	Company name, company location
Gender	Gender
Birthday	Birthday
Hometown	Location
City	City
Languages known	Experience
Friends	Skill
Religion views	Past experience
Political views	Region preferences
Favorite quotes	Language
Relationship status	Professional skills
Websites/Webmails	Groups
Networks	Connections
Relatives	Recommendation
Primary education (Primary school/collage name, class, type of degree)	Educational qualification
Professional qualification/experience (Employer name, position, title, start date, end date, description)	Professional skills (Employer name, position, description)

4 Objective/Methodology of Proposed System

4.1 Objective/Methodology

In entity resolution, various problems are discussed above. It is difficult to find same entity on multiple social networks (Facebook and LinkedIn); our main objective is to find an identity on multiple social networks in a distributed system [12]. In available system, sharing budget and privacy policy is also a big challenge. We are trying to solve that problem using separate API; the main objectives of the proposed system are listed below [11, 15, 35].

4.1.1 ER on Multiple Social Networks

The problem of matching users is between two social networks (Google+ and Twitter) [18]. Our objective would be to obtain a holistic picture of each person's identity by unifying his/her information from multiple social networks. For example, say we have three social networks A, B, and C. After we have matched some users between A and B, we can define a unified graph D that aggregates information from A and B. We can use D to help us match users between A and C and between B and C [19].

We update D to reflect new matching pairs as we match more users between different networks. Then, we can go back and attempt to match more users between A and B again. We are also finding the person's online footprint (web search results, personal websites, and online profiles). For example, we could use DataMiner to scrape and parse LinkedIn people search results to extract full name, job title, company, and location. In addition, we could augment the online footprint to help resolve the person's identity [36].

4.1.2 Shared Budget

For each API operation, we have different budget values. For example, every 15 min, Twitter gives each user a budget of 15 API calls for each of the following operations: inlinks, outlinks, and friendships. In this case, the optimization is simply which node to make the API call on. For example, operation relationships require two nodes to be passed as parameters [18]. We have a budget of 15 relationships calls; we decide which node pairs to make these calls. Another type of budget is a single budget shared among API operations. In this case, split the budget among the operations. The user decides not only which node but which operation will make an API call that will yield the highest information gain as we probe the graph.

4.1.3 Distributed ER

In distributed ER, we have multiple machines used here; we assume that we have full access to the first graph (G), but limited access to the second graph (T). We will continue using this problem set. We will partition the first graph (G) and split the work for probing the second graph (T) to different machines. Each machine gets a partition of G and performs ER on the partition. We would like to find matching nodes for a number of G nodes in T [18, 36]. The node correspondences learned by a machine may be useful to another machine that shares some nodes and edges [36]. Thus, we can share this info among machines to help resolve more nodes. At the end, we combine the results from all machines. We can utilize and adapt graph partition techniques to split the ER task into different machines.

In the proposed system, the first block is the extraction of data from different social network sites (Facebook and LinkedIn), and next block preprocesses/cleans the data and stores in a flat file. The third block performs the identity search for the first site and it performs the attribute matching and creates candidate; next block extracts the shortlisted candidate. Next, it performs content matching of the shortlisted candidate. Final block gives the matched Facebook user identity [11].

4.2 The Architecture of Proposed System

The proposed system consists of the following blocks.

4.2.1 Data Extraction

The data will be gathered by the API from the online social network.

4.2.2 Data Cleaning/Preprocessing

The data obtained will be cleaned by removal of any missing or irrelevant data.

4.2.3 Data Storage

The data collected will be stored in a flat file.

4.2.4 Methodology

Data dictionary will contain the most used words on social media sites. The data dictionary can be used to get more keywords from the content of the user (Fig. 1).

4.3 Collective Clustering Algorithm

Collective clustering is used to match different attributes from the different social networks and solve identity resolution problem [30]. This method follows a greedy approach and agglomerative clustering algorithm, and its main purpose is to find the most matched/similar clusters and then merge that values. For finding cluster similarity, it requires clusters of reference [14, 23]. In the available system, each design cluster is expressed as the same real-world stuff [29, 38]. Comparing to other technique like transitive closure, use single-linkage clustering, but collective clustering uses an average linkage approach and it also defines the relationship between two

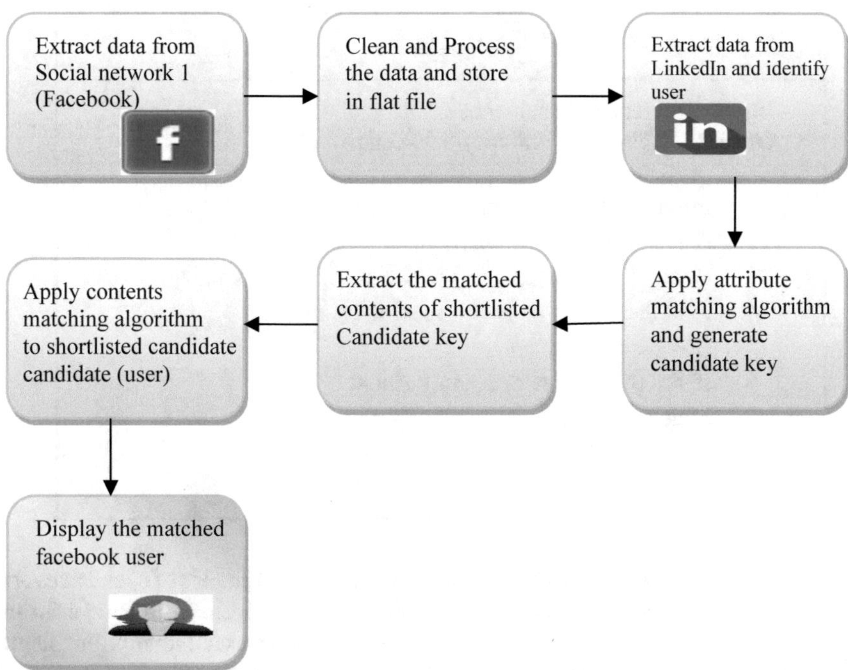

Fig. 1 Block diagram gives the brief idea of the proposed system (Extracting data and displaying matched entity)

different group's c_i and c_j as the average similarity between each reference in c_i and each reference in c_j: the similarities between these two clusters are expressed by the following equation:

$$sim(ci, cj) = 1|ci.R| \times |cj.R| \sum r \in cj.R.r' \in cj.Rsim\left(cj.r, cj.r'\right) \quad (1)$$

Here, $|c.R|$ represents the number of implication in cluster c.

Pseudo Code for Collective Clustering
Input: Reference of each cluster **Output**: display the similarity between two cluster 1. Initialize each reference as a cluster 2. Assign threshold value (t) 3. Compute Similarity(sim) between each cluster pair 4. Find the cluster pair with maximal sim(c_i ,c_j) 5. Compare sim(c_i ,c_j) 6. If sim (c_i ,c_j) greater than threshold 7. Merge c_i and c_j 8. Go to step 2 9. End

The above algorithm gives the similarity between two attributes of social network Facebook and LinkedIn [2, 29]. The attribute is taken from LinkedIn and performs matching algorithm, and it gives the matched Facebook user identity. For name matching, we used the character and length similarity and for company name and affiliation Jaccard similarity is used, and it compares two sets. Equation 1 computes the similarity between two clusters, and matched attributes are helpful for identifying the exactness of the accounts of the same individual or not. (Fig. 2).

Proposed Algorithm of Entity Resolution System: Some of the journal papers referred proposed an identity resolution system using clustering methods where the networks of the individual in the social group were being clustered together by the collective clustering algorithm. To optimize the performance of identity resolution process, we propose a system that utilizes a pairwise comparison string matching algorithm [29].

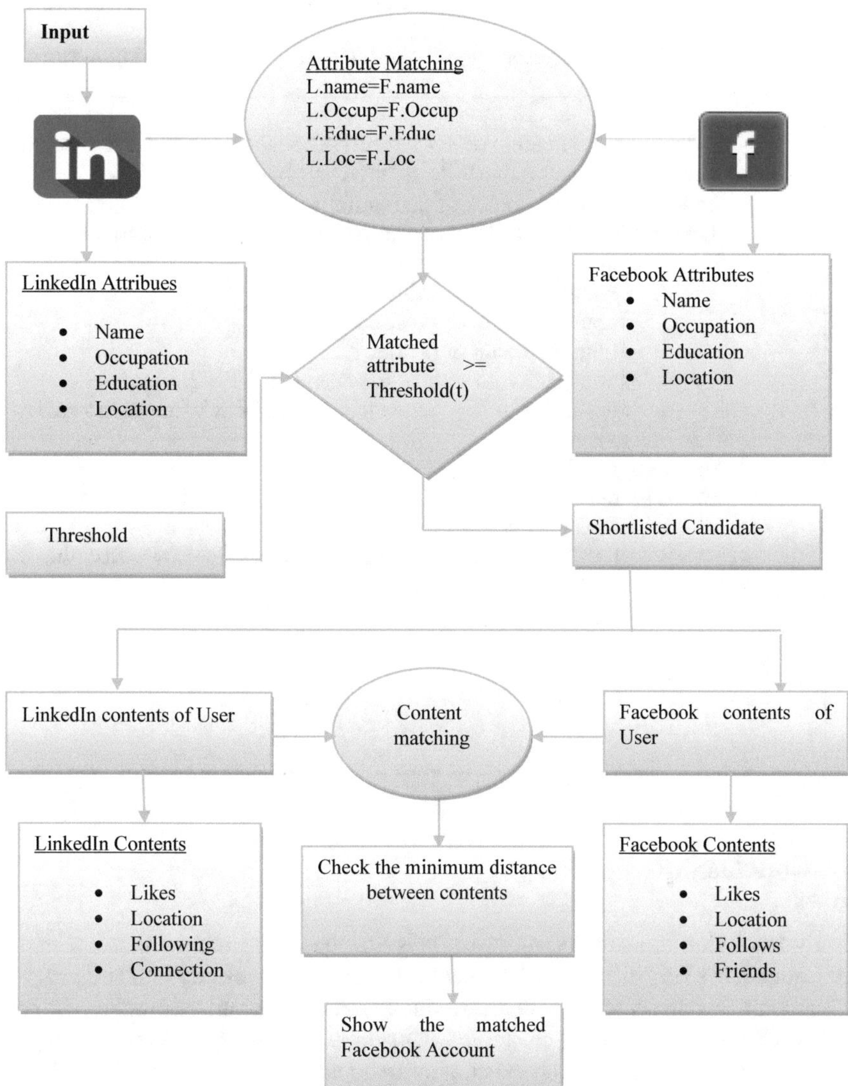

Fig. 2 Explains the different attributes of two social networking sites and its used proposed algorithm to find the matched entity

Pseudo code for Proposed algorithm (Attribute Matching Algorithm)

Let P1 and P2 are the two profiles(Facebook and LinkedIn), and having A1, A2,…An profile attributes of both profile
Input: Profile Attributes of one online social network
Output: Matched Identity on the second online social network

1. Start.
2. Assign threshold value.
3. Input the name of profile P1.
4. Input the name of profile P2.
5. Finding the Similarity between (P1.A1, P2.A1).
6. IF similarity is greater than a threshold value then predict it as a match,

then assign value 1.
ELSE assign value 0.

7. Repeat for all an attributes.
8. if values (total 1's) > threshold then it shows that the two profiles are the same person.
9. Profiles are similar.
10. End.

5 Conclusion

Entity resolution in social network analysis has the main problem that is identity matching. It is a very difficult task to match entity on Facebook and LinkedIn. Earlier research papers presented a solution by simply matching profile attributes of a user in two OSNs or performing clustering algorithms individually. Here, I am trying to resolve the drawback of the existing system and making an attempt to address the different problems of identity resolution in multiple OSNs. I am proposing a new approach for identity resolution technique that performs pairwise matching on identity attributes, content matching as well as self-mention search and used to access only public information and it finds the candidate identities. The proposed system is used in intelligence and law enforcement agencies, because they always suffer from the missing data problem. Second, the identity resolution system not only handles duplicates caused by entry errors but also handles data uncertainty. The proposed system also helps analysts to find user identities (e.g., spammers) across different OSNs. Even though this work has focused on two different social networks like LinkedIn and Facebook, it also tries to match few uncommon attributes of LinkedIn that will help to find the person's real identity. This work can also be extended to another social network to identify the real identity of person's Facebook and Instagram.

Proposed techniques helpful in the distributed environment and proposed identity search algorithms can access public information and some important attributes that will help to find candidate identities and other identities. The improved identity matching algorithms match candidate identities with the given identity and find the unique identity of the person. The method also tries to find the Facebook and LinkedIn sharing token value using API, and it will improve the searching and also define the cost of each attribute.

This paper gives an idea about the techniques that solve the problem of identifying user accounts across multiple online social networking sites which also increases the accuracy of finding correct identities of users. In this technique, merge the information of a single user having accounts at multiple online social networking sites. This aggregate information about a single user is useful in developing many applications such as follows:

Security domain: Malignant users create multiple accounts on different social networking sites to enhance reachability to the targets.

Recommendation domain: System is helpful in building friend recommendation feature. The recommendation feature can find user friend's identities on multiple social networking sites with their information on one social network, and it can suggest his/her to connect to the suggested friend's identities.

Enterprises: One of the problems of e-businesses face is not being able to realize exactly and their social audience to calculate their return on investment. Deduplicating their social audience by linking online social networking accounts can help calculate their ROI.

Human resource management: It can be used by the HR managers to check the profile of a candidate on different social platforms.

References

1. Wani, M.A., Jabin, S.: A sneak into the Devil's Colony—Fake Profiles in Online social networks. http://www.cps.gov.uk/legal/a_to_c/communications_sent_via_social_media/#a10
2. Dewan, P., Kumaraguru, P.: Towards automatic real-time identification of malicious posts on Facebook. In: 2015 13th Annual Conference on Privacy, Security and Trust (PST), July 2015, pp. 85–92. IEEE
3. Jain, P., Kumaraguru, P.: On the dynamics of username changing behavior on Twitter. In: Proceedings of the 3rd IKDD Conference on Data Science, 2016, CODS, New York, NY, USA, pp. 61–66. ACM (2016)
4. Xu 2, W., Esteva, M., Trelogan, J., Swinson, T.: A Case Study on Entity Resolution for Distant Processing of Big Humanities Data. IEEE. 978-1-4799-1293-3/13/$31.00 ©2013
5. Getoor, L., Dieh, C., P.: Link Analysis: a survey. In: ACM SIGKD Explorations Newsletter, **7**(2), December 2005, 3 https://doi.org/10.1145/1117454.1117456
6. Brizan, D. G., Tansel, A., U.: A Survey of Entity Resolution and Record Linkage Methodologies. Communications of the IIMA, **6**(2), 1–10 (2006)
7. Elmagarmid, A., K.: Duplicate Record Detection: a Survey. In: IEEE Transactions on knowledge and data engineering. IEEE (2007). **19**(1), January 2007

8. Benjelloun, A., Crainic, T. G, Bigras, Y.: Towards a taxonomy of City Logistics projects, **2**(3), 2010, pp. 6217–6228, https://doi.org/10.1016/j.sbspro.2010.04.032
9. Benjelloun, O., Garcia-Molina, H., Gong, H., Kawai, H., Larson, T.E., Menestrina, D., Thavisomboon, S.: D-swoosh: a family of algorithms for generic, distributed entity resolution. In: 27th International Conference on Distributed Computing Systems, 2007. ICDCS'07, pp. 37. IEEE (2007)
10. Bilgic, M., Licamele, L., Getoor, L., Shneiderman, B.: D-dupe: an interactive tool for entity resolution in social networks. In: Proceedings of the IEEE Symposium on Visual Analytics Science and Technology, 2006, pp. 43–50. IEEE (2006)
11. Peled, O., Fire, M., Rokach, L., Elovici, Y.: Entity matching in online Social networks. In: Social Com, pp. 339–344. IEEE (2013)
12. https://www.skullsecurity.org/blog/2010/return-of-the-facebook-snatchers
13. Esfandyari, A., Zignani, M., Gaito, S., Rossi, G.P.: User identification across online social networks in practice: pitfalls and solutions. J. Inf. Sci. (2016). https://doi.org/10.1177/0165551516673480
14. Xiao, C., Freeman, D.M., Hwa, T.: Detecting clusters of fake accounts in online social networks. In: 2015 ACM. ISBN 978-1-4503-3826-4/15/10. http://dx.doi.org/10.1145/2808769.2808779
15. Chung, C.-T., Lin, C.-J., Lin, C.-H., Cheng, P.-J.: Person identification between different online social networks. In: 2014 IEEE/WIC/ACM International Joint Conferences on Web Intelligence (WI) and Intelligent Agent Technologies (IAT). IEEE (2014). 978-1-4799-4143-8/14 $31.00 © 2014. https://doi.org/10.1109/wi-iat.2014.21
16. Goga, O.: Matching user accounts across online social networks: methods and applications. Computer science. LIP6—Laboratoire d' Informatique de Paris 6, 2014. HAL Id: tel-01103357. https://hal.archives-ouvertes.fr/tel-01103357
17. Ahmed, K., Panagiotis, G., Vassilios, S.: Duplicate record detection: a survey. IEEE Trans. Knowl. Data Eng. **19**(1) (2007)
18. Vesdapunt, N.: Entity resolution and tracking on social networks (2016). http://purl.stanford.edu/st867dy5990
19. Li, Y., Peng, Y., Xu, Q., Yin, H.: Understanding the user display names across social networks. In: 2017 International World Wide Web Conference Committee (IW3C2), published under Creative Commons CC-BY 4.0 License.WWW 2017 Companion, Apr. 3–7, 2017, Perth, Australia. ACM. 978-1-4503-4914-7/17/04
20. Adhikari, S., Dutta, K.: Identifying fake profiles on LinkedIn. In: PACIS 2014 Proceedings, vol. 278. http://aisel.aisnet.org/pacis2014/278
21. Shen, Y., Jin, H.: Controllable information sharing for user accounts linkage across multiple online social networks. In: CIKM '14, Nov. 3–7, 2014, Shanghai, China
22. Campbell, W.M., Li, L., Dagli, C., Priebe, C.: Cross-Domain Entity Resolution in Social Media, Aug. 2016
23. Zafarani, R., Tang, L., Liu, H.: User identification across social media. ACM Trans. Knowl. Discov. Data **10**(2), Article 16, 30 pp. (2015). http://dx.doi.org/10.1145/2747880
24. Shuy, K., Wangy, S., Tangz, J., Zafarani, R., Liuy, H.: User identity linkage across online social networks: a review. SIGKDD Explor. **18**(2), 5–17 (2015)
25. Jain, P., Kumaraguru, P., Joshi, A.: @I seek 'fb.me': identifying users across multiple online social networks. In: Proceedings of the 22nd International Conference on World Wide Web, Companion, New York, NY, USA, pp. 1259–1268. ACM (2013)
26. Jain, P., Rodrigues, T., Magno, G., Kumaraguru, P., Almeida, V.: Cross-pollination of information in online social media: a case study on popular social networks. In: Proceedings of the 2011 IEEE 3rd International Conference on Social Computing, SocialCom '11, pp. 477–482, Oct 2011
27. Vesdapunt, N., Garcia-Molina, H.: Identifying users in social networks with limited information. In: 31st IEEE International Conference on Data Engineering, ICDE 2015, Seoul, South Korea, Apr 13–17, pp. 627–638. IEEE (2015)
28. Saberi, M., Janjua, N.K., Chang, E., Hussain, O.K., Peiman, P.: In-house crowdsourcing-based entity resolution using argumentation. In: Proceedings of the 2016 International Conference

on Industrial Engineering and Operations Management Kuala Lumpur, Malaysia, Mar. 8–10, 2016
29. Bartunov, S., Korshunov, A.: Joint link-attribute user identity resolution in online social networks. In: The 6th SNA-KDD Workshop '12 (SNA-KDD '12) Aug 12, Beijing, China. Copyright 2012 ACM (2012). 978-1-4503-1544-9
30. Malhotra, A., Totti, L., Meira Jr, W., Kumaraguru, P., Almeida, V.: Studying user footprints in different online social networks. In: Proceedings of the 2012 International Conference on Advances in Social Networks Analysis and Mining, (ASONAM '12, pp. 1065–1070. IEEE Computer Society (2012)
31. Prieto, V.M., Álvarez, M., Cacheda, F.: Detecting LinkedIn spammers and its spam nets. Int. J. Adv. Comput. Sci. Appl. (IJACSA), **4**(9) (2013)
32. Bradbury, D.: Data mining with LinkedIn. Comput. Fraud Secur. 10, 5–8 (2011). Cao, Q., et al.: Aiding the detection of fake accounts in large-scale social online services. In: Proceedings of NSDI (2012)
33. Fire, M., et al.: Strangers intrusion detection-detecting spammers and fake profiles in social networks based on topology anomalies. Hum. J. **1**(1), 26–39 (2012)
34. Krombholz, K.: Fake identities in social media: a case study on the sustainability of the Facebook business model. J. Serv. Sci. Res. **4**(2), 175–212 (2012)
35. Zhang, H., Kan, M., Liu, Y., Shaoping: Online social network profile linkage-based on cost-sensitive feature acquisition. In: SMP 2014, CCIS 489, pp. 117–128. Springer, Berlin, Heidelberg (2014)
36. Kokkos1, A., Tzouramanis, T.: A hybrid model for linking multiple social identities across heterogeneous online social network. Springer International Publishing AG 2017 B. Steffen et al. (Eds.): SOFSEM 2017, LNCS 10139, pp. 423–435 (2017)
37. Dewan, P., Kumaraguru, P.: Facebook inspector: towards automatic real-time detection of malicious content on Facebook. Precog Indraprastha Institute of Information Technology—Delhi (IIITD), India Soc. Netw. Anal. Min. **7**(1) (2017)
38. Leskovec, J., Sosi'c, R.: SNAP: a general-purpose network analysis and graph-mining library. ACM Trans. Intell. Syst. Technol. **8**(1), Article 1, 20 pp. (2016). https://dx.doi.org/10.1145/2898361

Big Data Approach for Epidemiology and Prevention of HIV/AIDS

Nivedita Das, Sandeep Agarwal, Siddharth Swarup Rautaray and Manjusha Pandey

Abstract Nowadays, the disease is spreading and becoming noxious to the society inattentive of hospitalization that is present. Toxic diseases are the disorder by organisms, such as bacteria, viruses, fungi, or parasites, which happened in a normal body. Some toxic syndromes pass from one individual to another individual, some are transferred due to animals bite or insects, and others may happen by consuming contaminated water or food or by getting exposed to the organisms which are present in the environment. AIDS becomes a rapidly spreading and turning the life to death disease. HIV spreads from one individual to another individual in the population, in many different ways that may be due to semen and blood. The study of disease is called pathology, which includes the study of cause. This paper mainly focuses on the prediction of disease like HIV/AIDS using supervised learning system.

Keywords Predictive analysis · Training dataset · Test dataset
Proposed flowchart

1 Introduction

In computing analysis, datum is a fact that has been interpreted into a mode that is useful for progressing. In today's computing device and communication media, data are the fact converted into digital binary form. Collection of standardized data

N. Das (✉) · S. Agarwal · S. S. Rautaray · M. Pandey
School of Computer Engineering, KIIT, Deemed to be University,
Bhubaneswar 751024, Odisha, India
e-mail: niveditads26@gmail.com

S. Agarwal
e-mail: sandygarg65@gmail.com

S. S. Rautaray
e-mail: siddharthfcs@kiit.ac.in

M. Pandey
e-mail: manjushafcs@kiit.ac.in

A. Abraham et al. (eds.), *Emerging Technologies in Data Mining and Information Security*, Advances in Intelligent Systems and Computing 813,
https://doi.org/10.1007/978-981-13-1498-8_21

is required for a database [1]. In a relational database, it gathers schemas, tables, queries, reports, views, and other elements. Structured data are stored in database in sequential format. Data mining is the computing method of detecting designs in huge information sets' associating processes at the crossing of machine learning, statistics, and database system. Actually, prediction means forecast an uncertain event and is based upon a particular fact. The framework is actual layered structure indicating what kind of work done in thesis.

Big Data is actually an enormous size of data that cannot be gathered and handled using classical access within a given time limit [2]. There is a lot of misinterpretation; we are reversing the Big Data analytics [3]. We usually purpose the data to introduce to the data either in GB/TB/PB/EB/anything that is greater than size, which does not characterize the term "Big Data" entirely. A small number of data that can be indicated to as a Big Data provided by the context it is being used. Let us assume an example and try to explain this Big Data analytics to you. For instance, if you try to attach a record, i.e., 100 MB in size to an e-mail, we would not be capable to do so. As the e-mail system would not support the attached document with respect to e-mail can be revered to as Big Data analytics, dissimilar types of data are exploded as well as a way of machinery into the space obtainable for functioning with data. Big Data analytics is represented and appropriate objections linked to the operation along with data which are explained. The Institute of Medicine (IOM) in 2012 entitled "Excellent responsibility at low cost: The way to moderately learning health care in America" established that the American medical system has become far too complicated and too expensive to extend business as expected. The basic efficiency and ability to manage a rapidly depending upon a base and the world system focus on to keep patients in all the improvements, security and feature of concern, and the nation's economic balance and go comparative (Fig. 1).

Here, Big Data introduces a concept of 7 V's as shown in Fig. 2. Since the information is spreading immensely nowadays [4], Big Data defines both size and vision from unstructured, composite, noisy, mixed, representation, and the volume of data. These 7 V's are discussed in detail as **Value**: Importance of data or the value of information which includes data is called data value. The word value in Big Data plays an important role. It includes a massive volume and different varieties of data which are easy to access and [4] delivers quality analytics that helps in making decisions. It provides the actual technology. **Velocity**: it is acknowledged as how fast the data are constructed and stored. The data are very large and continuous in nature. **Volume**: it displays the total amount of data that may in Gigabytes or Terabytes or more than that [5]. Nowadays, data are generated by [3] digitally on systems such as media-related society, the volume of data to be evaluated is humongous. **Veracity**:

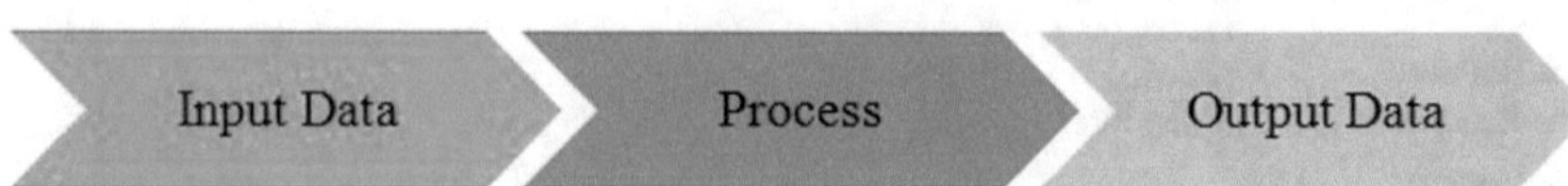

Fig. 1 Approach of data processing

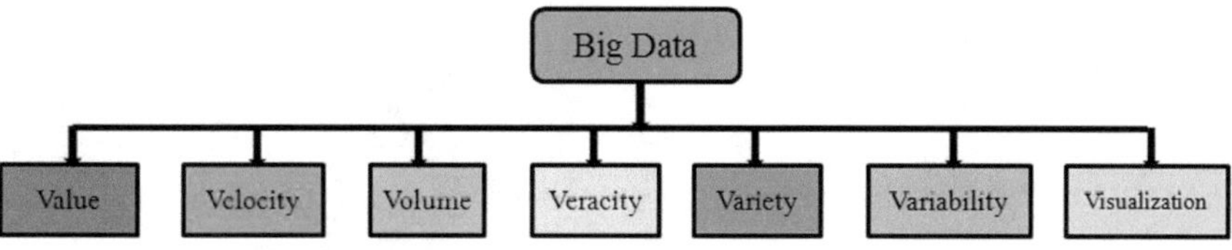

Fig. 2 Components of Big Data

the veracity concept in Big Data deals with bias, noise, and unstructured. Big Data feels that veracity in data analysis is the major issue when it compares with volume and velocity. **Variety**: different types of data being created are called data variety. This concept is to direct the attention to a lot of sources and different types of data which are structured and unstructured [6]. We have accustomed to collect data from sources like databases, file system, spreadsheets, etc. Nowadays data come in the mode of e-mails, pictures, videos, pdf, audio, etc. **Variability**: deviation of the text can hamper procedure to handle and manage it [7]. **Visualization**: It is the method which benefits to accept the definition of different data principles in rapid and detailed aspects.

1.1 Application of Big Data

Nowadays, Big Data plays a vital role in different fields. Big Data applications [8] are creating a new generation in all different sectors [9]. It also makes an easier lifestyle for human being. **Banking**: to detect fraud cases, Big Data is helpful. It identifies the misuse of credit cards and debit cards, etc. **Healthcare**: the Big Data continuously helps in the background of medical section. Actually, the newly introduced technology increases the cost of medical care. Big Data is the best solution for solving this problem.

Media, Communications, and Entertainment: Big Data is accumulating, considering, and handling customer vision. It influences mobile and social media willing. It also understands the patterns of real-time, media-willing usages. **Education**: Big Data shows an important role in higher education. It is used to measure lecture's potential to assure a better experience for both students and lectures. **Insurance**: it provides consumer observations for clear and straightforward commodities by considering and forecasting consumer action via data derived from media related to social, GPS-enabled devices and CCTV footage. Big Data also grants better consumer custody from insurance departments. **Energy and Utilization**: Big Data is a demanding element to determining key business problems for utility companies (Fig. 3).

Data are extracted from source systems in its original form, storing in a disorganized distributed file system. The storing of data across several data nodes in a simple hierarchical form of directories of files is known as Hadoop Distributed File

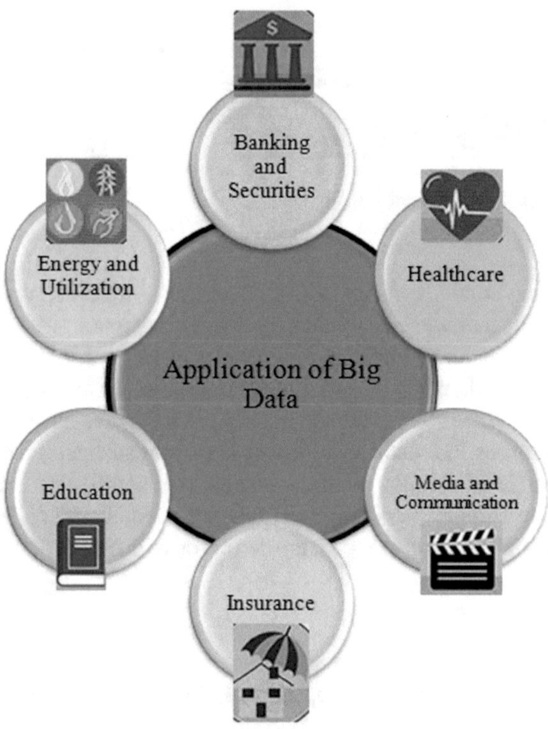

Fig. 3 Applications of Big Data

System (HDFS) [10]. Commonly, data are accumulated in 0.064 GB files in the data nodes with a high rank of compression. The HIV virus is a type of organism. Once someone infected with HIV, it stays in the body for the whole life. This virus damages the immune system. The immune system defects the body against the diseases. HIV virus is the source of AIDS. The full name of AIDS is Acquired Immune Deficiency Syndrome, which means deficient immune system. The difference between Big Data and relational databases is that the traditional table-and-column figure does not have Big Data but that relational databases have. In classical relational databases, a schema for the data is required [11].

1.2 Structure of HIV

Hospitalization for HIV means that most of the people stay together, with fewer people establishes AIDS [8]. Although there is presently no antidote for HIV with the correct medical care and backing, people with HIV can live long and active lives. To do this, it is specifically very important to take medical care perfectly and deal with any feasible reactions [12]. Before going to HIV life cycle, we should know the detailed about structure of HIV. Here, Fig. 4 indicates a virus structure [13].

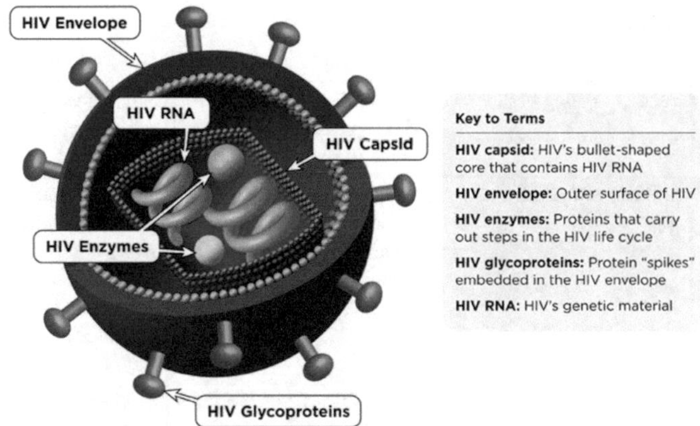

Fig. 4 A HIV virus structure

Figure 4 shows the structure of a Human Immunodeficiency Virus (HIV). The HIV enzymes represent two dot marks inside the cell, which aids the proteins, which achieve the steps of life cycle of HIV [14]. Next comes to HIV RNA, RNA means ribonucleic acid. It is a crucial molecule with extended chains of nucleotides. A nucleotide consists of a nitrogenous base, a ribose sugar, and a phosphate. Generally, RNA is a genetic material of HIV. The HIV capsid is the core of HIV which looks like bullet shaped and it contains both HIV RNA and HIV enzymes. HIV envelope is the outer of HIV. HIV glycoprotein is present at the outer surface of the HIV envelope in spike shape. It plays a vital role for the virus to get into the protein cell. Primarily, it attaches itself to proteins on the outer surface of the cell. Then, it acts like a spring-loaded mousetrap and snaps into a new conformation that drags the virus and cell close enough that the membranes fuse. Lastly, the HIV genome is released into the cell, where it immediately gets to work building a new virus [13].

2 Literature Survey

The primary objection of Big Data analytics in healthcare domain, and it includes secure, store, search, and share and analyze the healthcare data. Actually, Big Data refers to structured, semi-structured, unstructured, and the complex data generated by an automatic process. Health care is an example of acceleration, variation, and quantity, which are typical aspects of the data what it produces. These data are spread among different healthcare systems, health guarantee, investigator, etc. For adding these three V's, the veracity of healthcare data is also critical for its meaningful use toward developing untraditional research [15]. Actually, predictive analysis is used to predict the present conditions and future events. There are many research works going on for finding the drugs to get cured from this deadly disease but till now

Table 1 Literature survey

Year	Author	Paper title	Description
2015 [1]	Sean D. Young et al.	A “big data” approach to HIV epidemiology and prevention [18]	This paper addresses new tools and techniques for HIV infections
2016 [2]	S. Packiyam et al.	Big Data analysis for AIDS disease detection system [5] using cluster techniques.	The proposed path for gathering and supplying Big Data for analytics presented in this paper shows how important it is to prefer the technology migration
2014 [4]	Jokonya Osden	A Big Data framework for stop and regulation of HIV/AIDS, TB, and Silicosis in the mining industry	To predict and control the mining industry
2017 [8]	S. Packiyam et al.	AIDS detection system using Big Data analytics	In this paper, this tool is helped for predicting HIV/AIDS disease in very active manner than fresh access

there is no such type of success found. As playing a vital role in everywhere, here Big Data is enforced for improved conclusion which will have a good impact on the society [5]. So, in this chapter, HIV/AIDS is more focused and some measures are given for this disease. This chapter describes three types of measure; those are oral medicative, environmental predicative, and another one is an operational predicative measure [16, 13]. HIV and AIDS can be handled with proper treatment, but cannot be cured. Antiretroviral slows down the multiplication of the virus, but it does not wipe out. To support the medication, AIDS patients often get treatment to boost the immune system and fight against diseases [17]. A person can prevent AIDS by having safe sex and using clean needles. To make this HIV-relevant paper, the paper takes the help of literature survey (See Table 1).

In these, all paper in table representation is about “Big Data” approach to HIV epidemiology and prevention [18], Big Data analysis for AIDS disease detection system using cluster techniques, Big Data framework for the prevention and control of HIV/AIDS, TB, and Silicosis in the mining industry, and AIDS detection system using Big Data analytics.

The tools and their application [19] are presented in Table 2.

All the components are shown in Fig. 5, which represents the Hadoop ecosystem. It includes Zookeeper, Mahout, Hive, Pig, MapReduce, YARN, HDFS, Flume, HBase, and Cassandra [20].

Table 2 Big Data tools and its application

Sl. No	Big Data components	Application
1	Cloudera	It can help us in business to form an enterprise data hub to assign people in a group to better accessibility to those data are storing
2	Hadoop	It is an open-source framework. It is a technology which helps in the treatment of cancer and genomics. It is used to monitor the patient vital
3	HDFS	It provides high-throughput access to application data
4	Teradata	It includes implementation, business consulting, training, and support
5	MapReduce	It appears as a programming framework used for parallel computing application which proposed by Google. It provides a flexible and scalable foundation for analytics

Hadoop Ecosystem

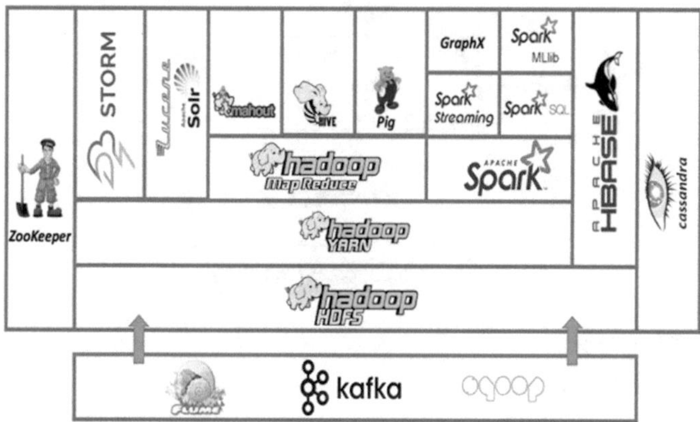

Fig. 5 Hadoop ecosystem

3 Methodology

This proposed algorithm in the form of flowchart gathers some datasets, i.e., train dataset and test dataset. This train dataset will follow data optimization technique. The optimized dataset will apply a supervised learning technique. Here, another dataset, i.e., test dataset, will apply as an input to this supervised learning technique. Now the condition will check as if all the datasets are identified, and then give the output; otherwise, applying another supervised learning technique and this supervised learning technique will take inputs from conditioned data and test dataset. After applying this supervised learning technique, this will give the final output, which will be going for predictive analysis (Fig. 6).

Then, the whole process is going to predictive analysis to give the result of the public awareness.

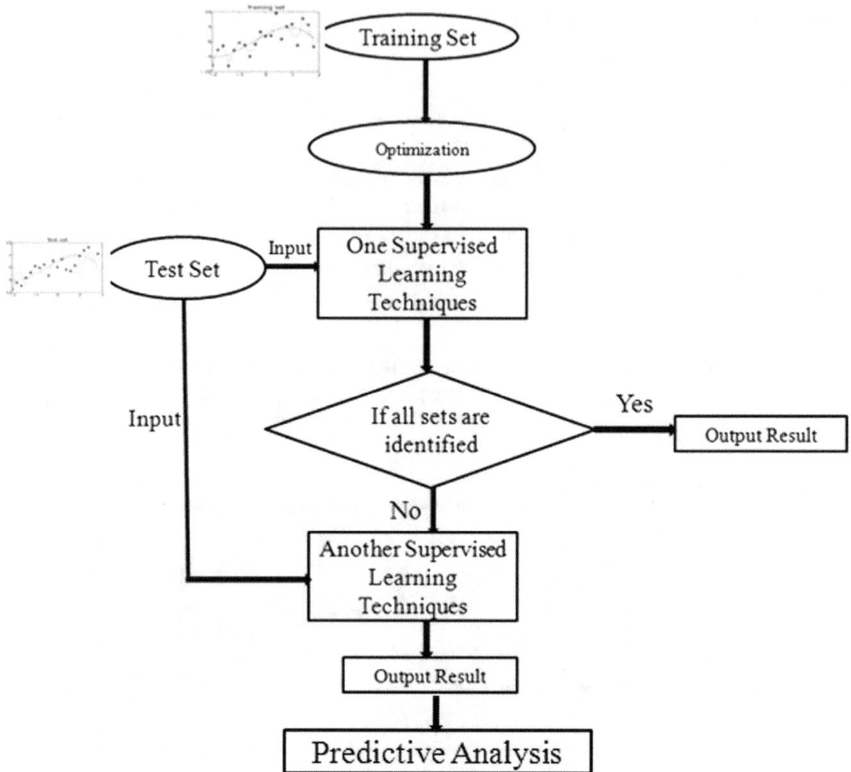

Fig. 6 Proposed flowchart

4 Discussion

To achieve the said healthcare facility that provides the coordination between patients and coordinators for gaining higher self-satisfactions is mandatory. As the clinical innovation is extending decision-making, delivery of the patient service to be also beneficial for improving their health has become inevitable. Ability to, access to, analyze, and document the patients' record anywhere, thus providing health care in any location has become the challenge of the day. Using this above-proposed flowchart, we can detect HIV/AIDS and hence cure at an early stage. So it will combat the spread of HIV. Previously, all known models or flowcharts were better but this flowchart will help to build a model for public to prevent HIV/AIDS. To improve the healthcare facility in future also, the research community must provide clinical delivery innovation, patient dialog, and collaboration, securing optimum operations of the healthcare system. The successful implementation of healthcare solutions is in high expectation.

A new future can be created for health care by using Big Data technologies. Today, medical costs are continually bruising worldwide. Using this Big Data technology, we can lower the increasing cost in medical examinations and differed results through disease prevention, early discovery of disease, thus improving the quality of life, predicting with a high degree of accuracy, and analyzing the interrelationship of examination value and diseases would help in the prevention of many fatal diseases similar to AIDS like various forms of cancer, infections, and other physical ailments.

References

1. Gandomi, A., Haider, M.: Beyond the hype: big data concepts, methods, and analytics. Int. J. Inf. Manage. **35**(2), 137–144 (2015)
2. Yadav, K., Pandey, M., Rautaray, S.S.: Feedback analysis using Big Data tools. In: 2016 International Conference on ICT in Business Industry & Government (ICTBIG), Indore, pp. 1–5 (2016)
3. Chen, C.L.P., Zhang, C.-Y.: Data-intensive applications, challenges, techniques and technologies: a survey on Big Data. Inf. Sci. **275**, 314–347 (2014)
4. Ward, J.S., Barker, A.: Undefined by data: a survey of Big Data definitions (2013). arXiv:1309.5821
5. Mohapatra, C., et al.: Map-reduce based modeling and dynamics of infectious disease. In: 2017 International Conference on I-SMAC (IoT in Social, Mobile, Analytics and Cloud) (I-SMAC). IEEE (2017)
6. Infographics—The four V's of Bog Data. IBM Big Data & Analytics Hub (2016). http://www.ibmbigdatahub.com/infographic/four-vsbig-data
7. Toshniwal, R., Dastidar, K.G., Nath, A.: Big Data security issues and challenges. Int. J. Innov. Res. Adv. Eng. (IJIRAE) **2**(2) (2015)
8. Murdoch, T.B., Detsky, A.S.: The inevitable application of big data to health care. JAMA **309**(13), 1351–1352 (2013)
9. Mohapatra, C., Pandey, M.: A review on current methods and application of digital image steganography. Int. J. Multidiscipl. Approach Stud. **2**(2) (2015)
10. Jokonya, O.: Towards a Big Data Framework for the prevention and control of HIV/AIDS, TB and Silicosis in the mining industry. Proc. Technol. **16**, 1533–1541 (2014)
11. Young, S.D.: A "big data" approach to HIV epidemiology and prevention. Prev. Med. **70**, 17–18 (2015)
12. Heredia, A., et al.: Targeting of mTOR catalytic site inhibits multiple steps of the HIV-1 lifecycle and suppresses HIV-1 viremia in humanized mice. Proc. Natl. Acad. Sci. **112**(30), 9412–9417 (2015)
13. Mohapatra, C., et al.: Usage of Big Data prediction techniques for predictive analysis in HIV/AIDS. In: Big Data Analytics in HIV/AIDS Research, pp. 54–80. IGI Global (2018)
14. Sahani, S.K.: Analysis of a delayed HIV infection model. In: International Workshop on Computational Intelligence (IWCI). IEEE (2016)
15. Wang, J., et al.: Regularity of herbal formulae for HIV/AIDS patients with syndromes based on complex networks. In: 2014 IEEE Symposium on Computational Intelligence in Big Data (CIBD). IEEE (2014)
16. Khalid, Z., Sezerman, O.U.: Prediction of HIV drug resistance by combining sequence and structural properties. IEEE/ACM Trans. Comput. Biol. Bioinform. (2016)
17. Zhang, X., Wang, J., Liang, B., Qi, H., Zhao, Y.: Mining the prescription-symptom regularity of TCM for HIV/AIDS based on complex network
18. Ansari, Md.T.J., Pandey, D.: Risks, security, and privacy for HIV/AIDS data: Big Data perspective. In: Big Data Analytics in HIV/AIDS Research, pp. 117–139. IGI Global (2018)

19. Katal, A., Wazid, M., Goudar, R.H.: Big data: issues, challenges, tools and good practices. In: 2013 Sixth International Conference on Contemporary Computing (IC3). IEEE (2013)
20. Landset, S., et al.: A survey of open source tools for machine learning with big data in the Hadoop ecosystem. J. Big Data **2**(1), 24 (2015)

A Survey on Local Market Analysis for a Successful Restaurant Yield

Bidisha Das Baksi, Varsha Rao and C. Anitha

Abstract Establishment of a new restaurant requires a paramount investment in it. Thus, a thorough analysis of different factors is important to the determination of the probable rate of success of the restaurant. Among the different factors, location plays a vital role in the determination of the success of the restaurant unit. The demographics of the location, existing cluster of restaurants established in the location, and the growth rate of the location need to be studied prior to selection of the optimal location. However, it is a cumbersome job to find the correct location, by analyzing each of these different aspects of the location. This paper studies the different aspects of a location which makes it a determining factor in the prediction of success of a restaurant. The proposed work aims at the creation of a web application that determines the locations suitable for the establishment of a new restaurant by using techniques of machine learning and data mining.

Keywords Restaurant market · Clustering algorithm
Location-based predictive analysis · Web application

1 Introduction

A new venture requires a thorough analysis of market conditions. Restaurant business is one of the most profitable businesses of today's time, and at the same time it involves immense risk. New food joints require large capital investment and operational costs. In [1], the author states that when the new outlet fails to break even, it closes in a

B. Das Baksi · V. Rao · C. Anitha (✉)
Department of Computer Science and Engineering, The National Institute of Engineering Mysuru, Mysuru 570008, Karnataka, India
e-mail: anithac.cse@nie.ac.in

B. Das Baksi
e-mail: bidisha.bksh@gmail.com

V. Rao
e-mail: varshar186@gmail.com

A. Abraham et al. (eds.), *Emerging Technologies in Data Mining and Information Security*, Advances in Intelligent Systems and Computing 813,
https://doi.org/10.1007/978-981-13-1498-8_22

short span of time, thereby incurring huge losses. Location of the new venture plays an important role in determining its success. Also, as in [2], it is demonstrated that the local socio-demographic condition has significant impact on the profitability of a restaurant. Emerging trends in the food industry play an important role while designing the business plan. Many other factors contribute to the type of menu and the price that can be charged.

It is imperative for restaurateurs to understand the likes and dislikes of the customers [3]. Since location is an important factor in determining success, a thorough analysis of the location needs to be performed before the investment is made. This analysis constitutes investigation of various parameters such as the demographic conditions, local preferences, tourist population, competition in the area, economy of the area, and so on [4, 5]. When an investor wants to open a new restaurant in a city, the investor is unsure of the area in the city that he/she should invest in. Opportunities for new restaurants in various regions can be established based on people's preferences.

To solve the above-stated problem, the proposed system intends to explore factors that determine the success rate of a restaurant by creating an application that accepts various attributes of the new venture as an input and provides the investor with a list of the areas that are suitable for his/her venture and also the approximate annual revenue that can be generated from each of these locations. We aim to design an automated system to determine locations suitable for the establishment of a restaurant business by applying concepts of K-means clustering. The primary objective of the proposed system is to help investors make an informed and optimal decision about selecting a location for their restaurant business before opening a new outlet in the city. This analysis helps investors take a calculated risk.

2 Related Work

This section summarizes the contributions by various researchers' for setting up and running a successful restaurant. There are many factors that determine the success of a restaurant. However, location is deemed to be the most significant factor among others. It is important that the restaurant is in the vicinity of that community, which has sufficient count of the ideal customers that the restaurant targets. Here, we focus on only those contributions that consider location as a major factor for analysis.

2.1 Analysis Based on Demographics and Psychographics

In [4], the author states that restaurant style and cuisine type helps in defining the demographics and psychographics of the customer, respectively.

For the identification of the ideal customer, the target market is selected and the ideal customer profile is defined. This involves the determination of three aspects of the ideal customer: the demographics, the psychographics, and the behavior. An approximate consumer profile would include demographics (age, gender, education, income, family, life, and geographic location), buying habits (role, frequency, online/mobile/offline, and social proof), psychographics (attitudes, values, lifestyle, personality, and interests/hobbies) and persona types (competitive, spontaneous, methodical, and humanistic).

The next step involves the determination of the area where the ideal customer crowd is in majority. For example, in America, a list of zip codes can be created, using Quick Fact Census. It will give information about the demographics identical to each of these zip codes.

Threshold value for population is decided on the basis of capital investment and also the size of business. Once the population is estimated, the percentage of crowd is calculated that can serve as the ideal customer for the restaurant. Based on this percentage of crowd, the list of zip codes can be ranked and ideal location can be confirmed.

Customer Activity Analysis
Customer activity plays an important factor, and it can be of two types: Foot traffic and vehicle traffic. The effect of traffic can be an additional factor to be considered in making decision regarding the best place for restaurant concept.
SAP digital consumer insight tool is a software which provides foot traffic report by leveraging location-enabled mobile phones. The data can be even broken down into age, gender, home zip code, etc. Thus, this tool can be used to estimate the foot traffic.
Vehicle traffic comprises vehicular traffic in the area near to the restaurant locality. Fortunately, all cities, countries, and states periodically conduct a traffic count.

Competitor Analysis
Competitor restaurants can help in better understanding the restaurant location demographics. However, competitor restaurant can pose threat to the restaurant based on similar concept. Complimentary restaurants are restaurants which have different restaurant concepts but similar price points. These complimentary restaurants can help in creating a market for the proposed restaurant.

Neighborhood
The infrastructure of the neighborhood area is a key aspect in location analysis. The future growth of the proposed restaurant can be predicted depending on the presence of educational institutions, movie hall, and many others in that area.

2.2 *Effect of Socio-demographic Factors on Location Pattern*

In [2], a number of socio-demographic factors are considered to understand their influence on restaurant location. According to the authors, factors such as cost of establishment, traffic flow, ingress, parking, visibility demographics, competition, and municipal regulations affect the selection of an ideal location for establishing a restaurant. Also, the assessment of current demand characteristics and how they are likely to evolve over a period of 10 or 20 years are also crucial for selecting the location. In [2], a geographic region is divided into subregions on the basis of zip codes. The probability of a zip code serving as an ideal region for the development of the business is determined. The restaurant location patterns across the US were expressed using the negative binomial (NB) regression model for each zip code. The likelihood of observing a count of restaurant number y_i in zip code i is considered as [2]

$$f(y_i) - \frac{e^{-\lambda_i}\lambda_i y_i}{y_i!} \quad (1)$$

The (conditional) expectation of y_i, λ_i, is specified as a log-linear function of a set of explanatory variables x_i, as [2]

$$\ln\ E(x_i) = \ln(\lambda_i) = x_i\beta \quad (2)$$

where β is the vector of coefficients of x_i.

In NB model, a gamma distributed term is incorporated in (2), such that

$$\ln\ E(x_i) = \ln\left(\lambda_i^*\right) = x_i\beta + v_i \quad (3)$$

The size of each zip code area varies significantly across the country; therefore, an exposure variable $\ln(AREA_i)$ is included in x_i with coefficient constraint as 1.

Therefore, if both sides of (3) are subtracted by the term $\ln(AREA_i)$, the left-hand side becomes $\ln E\left(\frac{y_i}{AREA_i/x_i}\right)$, which represents the restaurant density.

Thus, restaurant growth from t_1 to t_2 in a particular zip code i can be predicted as

$$\widehat{growth_i} = \frac{y_i(t_1)}{y_i(t_2)} = exp\left(\Delta x_i\hat{\beta}\right) \quad (4)$$

Several socio-demographic factors were considered, and the parameters considered for analysis were based on these factors. The parameters include the population count, median age, median income, educational attainment population, and renter occupation. Their analysis indicated that the higher the parameter values, the larger the supply of restaurants. On the other hand, the other factors which did not contribute were household size, male percent, and owner occupation. In [2], authors conclude that a prime location is directly linked customer in terms of intention, satisfaction, loyalty, and retention. According to [2], smaller cities amid rural areas appear more attractive than more highly urbanized areas to certain restaurant types. Some restaurant brands were directly affected by the housing tenure and owner occupancy percentage. Ethnic neighborhood characteristics also mattered for some brands.

2.3 Other Contributions

Many online un-authored information [5] and articles [4, 6] can be found to facilitate restaurateurs in considering important factors such as industry trends, location market area, competition, customer psychological, and potential revenue projection by providing questionnaires and survey forms. It can be difficult for a restaurateur to understand these factors and it is also time-consuming. According to [3], people rate a restaurant not only based on food but also on dine-scape factors such as facility aesthetics, lighting, ambience, layout, table setting, and servicing staff. These factors priority vary depending on location.

With the abundance of data available on the above factors, analysis on it can be done using various mathematical models and data analysis methods like multiple regression, neural networks, Bayesian network model, random forest, SVM, and many more can be used in predicting potential revenue of restaurant depending on various factors [1, 7]. Not only revenue prediction but also customer preference can be determined. Analysis provides more insights to the restaurateurs and can be used as reference for decision-making.

3 Proposed Work

The shortcomings of the referred papers were observed to be the following. In [4], the authors have categorized the income into three levels: Low, mid, and high. But the author has failed to elaborate on the range in which these three levels are decided. Further, the author has classified high income into high and high disposable income. This type of categorization leads to ambiguity and may not be suitable for all geographic locations around the globe.

In [2], the authors conclude the location of a restaurant and its type is heavily influenced by population density, ethnicity, and urbanization of an area. It highlights relationships between demand characteristics and total restaurant supply. Some market factors are universal in how they influence supply, such as higher population density, median household income, median age, and ethnicity. The factors such as ownership styles and local government policies did not find any weightage in the analysis. In [2], the model predicts business future growth of US-based restaurants. This model is not suitable for Indian-based restaurants. Also, prediction of business growth is not easy as the Indian market is volatile. A lot of other factors were not considered in the analysis such as the local government regulations and competition between different restaurant types.

Considering the Indian scenario, the GDP has grown rapidly in the last few years; accordingly, the lifestyle of people has changed. This transformation is visible with more and more people eating out frequently. Restaurants are cropping up in every corner of the city to meet this increased demand. In order for restaurants to survive in this highly competitive business, location plays a vital role. The other factors include

Table 1 Proposed dataset I

Area/Ward no.	Gender (M/F)	Annual income	Age (range)	Preferences (Chinese/Indian/Continental)	Amount usually spent on eating outside	Amount likely to be spent per meal	No. of members in the household

capital investment, operational costs, theme, and so on. Location of a new venture plays an important role in determining its success.

The government of India is also coming up with great initiatives such as the development of smart cities. We have proposed a system which provides the investor, a number of suitable locations for investment. The system determines the locations that possess the threshold characteristics required for a successful restaurant. Initially, data related to food preferences, type of cuisine, economy, type of population, ambience preferences, etc. are collected from the customers which are stored in the database and used for the market analysis. The proposed system takes as its input, the investment amount, targeted age group, type of cuisine, ambience, and related information from the investor and provides the suitable locations along with the approximate success rate at each location by using a suitable clustering algorithm on the collected data. The investor can choose among the given locations and take appropriate decision based on the analysis.

3.1 System Design and Datasets

This section describes the proposed system design and the datasets considered. The proposed system is a web application that any investor can use before localizing the restaurant location. Since the data considered need to be dynamically updated to reflect the changes in every part of the city, most of its content directly depends on the data available through various government and authorized websites. The rest of the data need to be collected through surveys. The data thus collected are proposed to be grouped on the basis of ward numbers as per the city corporation's information.

Dataset Representing User Demographics: Dataset I
This set consists of data related to user age, cuisine preferences, annual income, and expenditure habits. A survey form is created and distributed among the citizens of the restaurant to collect information from them (Table 1). The data so collected are cleaned and wrangled to form the desired dataset. This dataset helps to learn about the current industry trend and the market demands (Table 2).

Dataset Representing Areas of Importance: Dataset II
This set consists of areas or structures of economic interest. This can be collected through Google Maps API. This dataset helps to determine the relative importance of different regions (Table 3).

Table 2 Dataset I after data wrangling

Subregion group no.
Male (%)
Female (%)
Mean age range
Mean income
Mean no. of members in the household
Chinese cuisine (%)
Continental cuisine (%)
Indian cuisine (%)
Approximate population
Mean amount likely to be spent on a meal
Mean amount usually spent on food per month

Table 3 Proposed dataset II

Area/Ward no.	Region of interest (Mall/Park/Highway/Flyover/Theatre, Metro)

Table 4 Dataset II after data wrangling

Subregion group no.	Malls (%)	Airport (Yes/No)	Parks (%)	No. of flyovers	Movie halls (%)	Relative importance score

Table 5 Proposed dataset III

Restaurant location (Ward No.)	Specialty cuisine (Indian/Chinese/Continental)	Investment	Average cost for two	No. of customers visit per day	Popularity rating	Zomato rating

Dataset II is wrangled (Table 4). The areas are grouped into subregion and relative importance scores are assigned to each subregion. Based on the type of area considered, the weightage for each of the regions of interest is dynamically decided and considered. For example, the weightage for airport may differ from a Metropolitan city and a town city. The weightage given to each region of interest is stored in a database and updated in small intervals of time, depending on the current scenario of the city/town. For example, the percentage of malls in a particular region multiplied to its assigned weightage adds up to the relative importance score for that region.

Dataset Consisting of Existing Restaurant Data: Dataset III
This set consists of attributes of an existing restaurant business. This dataset can be formed by performing a survey and collecting information from manager of each restaurant. This dataset is used to train the model for predicting annual revenue of a restaurant (Table 5).

Table 6 User input for proposed work

Restaurant type (Indian/Chinese/Continental)
Annual Investment
Average cost for two
Crowd target per day
Points of interest (Mall/Movie Hall/Park/School/ etc.)

User Input
The investor gives input about the restaurant through the website. A survey form is put up on the website which is filled by the user and submitted to the server. The survey form is as shown in Table 6.

Design
The data collected in dataset I (Table 1) is subjected to k-means algorithm where the attributes chosen for the k-means algorithm are male percent, female percent, age range, cuisine preference, mean income, mean no. of members in the household, cuisine percentage, approximate population, mean amount to be spent on a meal, and the mean amount spent on dining outside per month. The optimal number of clusters that is to be formed for a region should be selected after visualization of data. The clusters thus formed have unique characteristics that differentiate them from one another and the members of the same cluster resemble one another in some way or the other. The idea of growth of restaurants on the basis of location factors, as studied in [2], is taken into consideration and each of the clusters is approximated to be suitable for certain types of restaurants. For example, as shown in [2], a higher male percentage in most areas leads to scarcity of all types of restaurants in that area except that of vegetarian restaurants which shows a higher rate of establishment around those areas, with higher male population. The input accepted from the user through the survey form is analyzed and converted into another data point, which is mapped against the clusters formed through k-means clustering algorithm. The appropriate cluster is selected and the subregions in that cluster are displayed as the preferable location for the setting up of the restaurant. The region of importance score is considered, and the subregions are ranked on the basis of that. The dataset III (Table 5) is used to train a machine learning model which is used to predict the annual revenue that can be generated by the restaurant to be established in the regions that are chosen to be suitable for its establishment. The result of this prediction can be used to re-rank the subregions in the chosen cluster.

4 Conclusion

This paper studies the different factors of a location which makes it an important aspect that is to be considered, prior to investing in a new venture at an unknown place. Guidelines for the optimal selection of location for a new venture are existing.

Some researches prove the correlation between the location factors and the growth of restaurant in that location. Based on these studies, our proposed system aims at creating a web application which recommends the appropriate location for the establishment of a new restaurant, thereby reducing the task of the investors in studying each individual factor. However, the proposed system is yet to be implemented and the accuracy is yet to be evaluated. The proposed system serves as the framework for the development of an application, which estimates suitability of locations for the development of new restaurant units.

References

1. Raul, N., Shah, Y., Devganiya, M.: Restaurant revenue prediction using machine learning. Int. J. Eng. Sci. **6**(4), 91–94 (2016)
2. Yang, Y., Roehl, W.S., Huang, J.-H.: Understanding and projecting the restaurantscape: the influence of neighborhood socio-demographic characteristics on restaurant location. Int. J. Hosp. Manag. **67**, 33–45 (2017)
3. Mahalingam, S., Jain, B., Sahay, M.: Role of physical environment (Dinescape factors) influencing customers' revisiting intention to restaurants. In: International Conference on Advances in Computing, Communications and Informatics (ICACCI) (2016)
4. Tarver, E.: How to Choose the Best Restaurant Location for Your Business. Online document
5. Unknown Author.: Restaurant market analysis—a guide. http://www.menutek.com/assets/restaurant-market-analysis.pdf
6. Karen, E., Spaeder.: Online documentation on How to Find the Best Location. https://www.entrepreneur.com/article/73784
7. Lasek, A., Cercone, N., Saunders, J.: Restaurant sales and customer demand forecasting: literature survey and categorization of methods. In: Leon-Garcia, A., et al. (Eds.) Smart City 360°. SmartCity 360 2016, SmartCity 360 2015. Lecture Notes of the Institute for Computer Sciences, Social Informatics and Telecommunications Engineering, vol. 166. Springer, Cham (2016)

Context Dependency Relation Extraction Using Modified Evolutionary Algorithm Based on Web Mining

Parinda Prajapati and Premkumar Sivakumar

Abstract A decisive region of natural language processing is semantic analysis, the study intensively of the significance of linguistic utterances. This theory puts forward algorithms that extract semantics from web sources using machine learning techniques. In semantic relation, finding relations between two entities is a process of element connection extraction. It is the chore of ruling semantic relations between two elements from given content when they have some interrelations. In this manuscript, we recommend a novel web mining-based English element connection extraction approach. With this approach procedure of web mining for different web pages, entities and connection highlight words for element connection extraction. Some existing methods can be evolved for the same. The method we have adopted is improved relational feature word extracting. The fallout gained signifies the initial steps on the mode to a comprehensive strategy for the analysis of web sources and feeds.

Keywords Relation extraction · Web mining · Feature extract · Information extraction

1 Introduction

Here, we focus on entity relation extraction which means determining the relation between entities. Nowadays, we have a bunch of online and offline data which is most of the time unstructured. To extract meaningful structured data and relation between entities from these documents is very important and complex task.

It is important to understand information extraction, feature extraction, and relation extraction between entities, methods of extraction like supervised and unsuper-

P. Prajapati (✉) · P. Sivakumar
Silver Oak College, Ahmedabad 382481, India
e-mail: parindaprajapati2011@gmail.com

P. Sivakumar
e-mail: premambal@gmail.com

A. Abraham et al. (eds.), *Emerging Technologies in Data Mining and Information Security*, Advances in Intelligent Systems and Computing 813,
https://doi.org/10.1007/978-981-13-1498-8_23

vised learning, types of supervised and unsupervised learning, and relation extraction techniques like RDF and OWL.

Feature extraction is the combined process of feature selection and feature extraction. In machine learning, feature extraction is the task to find out informative, nonredundant, and generalized steps mostly used to better human interpretations.

Dimensionality reduction is related to feature extraction where what actually gets done is to transform data from high-dimensional space to space of fewer dimension. In feature selection, there are mainly three tasks: information gain, search and feature selection to add or remove.

We have text or XML type of data or documents and all we have to do is detection and classification of semantic relation from this document. This process is known as relationship extraction.

As per the proposed flow, this paper is more focused about information extraction which is next level of relationship extraction where we extract structured form of data from unstructured or semi-structured form which is generally in machine-readable form. We can get IE form using classifier like generative or discriminative or using sequence model. It is also important to understand natural language processing (NL) which is interrelation between machine and human language.

Information extraction is having many methods of improvising for finding out relation between multiple relative words from different documents. It is also defining the relation internally within the documents. It can be used for strengthening the work of any search engine. The derived semantic relation will be helpful to reach to a particular sentence with efficiency.

There are different types of methods like supervised, semi-supervised, and unsupervised methods with their advantage and drawbacks. Supervised learning is where you have some prior knowledge about your problem or dataset, so that you can say that particular type of output can come for particular input [1].

Supervised learning difficulties can be further grouped into regression and classification problems. Classification: When your input falls into some specified category like "red" or "blue" or "disease" and "no disease". Regression: It contains predictions of value or any event and it has real values like "dollars" or "weight" [2].

Unsupervised learning is where you do not know that where your output data will get placed. The purpose of unsupervised learning is to model fundamental edifice or dispersal in the data in order to acquire more about data [1]. These are called unsupervised learning. There is no correct answer or any guide for this type of learning. Algorithms are meant to their own devices to determine and give the interesting structure in the data. Unsupervised learning difficulties can be further assembled into clustering and association problems.

Clustering: This is the technique of grouping data as per their behavior or property like grouping employees by their salary.

Association: An association rule learning problem is where you want to determine guidelines that define huge shares of your data, such as people that bargain X also tend to bargain Y. We are going to use supervised learning as it gives better results than unsupervised methods with more competences [3].

2 Related Background

Text mining is relatively new area of computer science, and its use has grown as unstructured data available continues to increase exponentially in both relevance and quantity. Through techniques such as categorization, entity extraction, sentiment analysis, and others, text mining extracts the useful information and knowledge hidden in text content.

Web mining is the submission of data mining methods to determine patterns from the World Wide Web. This data is collected via mining the web. It makes consumption of programmed gadgets to disclose and extract statistics as of servers. It gathered both ordered and formless data and data from browser, server logs, website and link structure, page content, and different sources.

Web structure mining uses diagram premise to examine the node and association arrangement of a website.

a. Take out pattern as of hyperlink in the network.
b. Mining the file formation: scrutiny of tree-like formation of page structures to express HTML or XML tag usage

Semantic World Wide Web has lots of web pages, and these pages and data are structured and designed in a way that computers can read it directly [4].

Naive Bayes classifier in machine learning, ***naive Bayes classifiers,*** is an ancestor of effortless probabilistic classifiers foundation on applying Bayes with sturdy (naive) liberty guess among the features [5].

Context information—Contex**t** is any information that can be used to characterize the circumstances of entities (i.e., whether a person, place, or object) that are measured significant to the dealings linking a user and application, counting the user, and the application themselves.

True negative case was negative and guess negative.
True positive case was positive and guess positive.
False negative case was positive and guess negative.
False positive case was negative and guess positive.

$$\text{Precision} - \text{TP}/(\text{TP} + \text{FP})$$

$$\text{Recall} - \text{TP}/(\text{TP} + \text{FN})$$

Moreover, this paper is going to use different types of relation extraction technique like RDF and OWL in a novel way. Resource description process (RDF) is a construction for recitation assets; on the web, it is intended to interpret and tacit by computers and not intended for being presented to people. RDF is printed in XML. Here, resource means it can be anything which is located by URL. Main application of this is data integration. It was originally designed to get metadata for a web resource, and later we started to use for other application. It is a data model. It gives subject, object, and predicate of sentence or data. It defines property value with property and resources. It works on triples. It is a statement about things.

OWL, web ontology language, is a family of facts depiction languages for novelist ontology. Ontology is ceremonial approach to illustrate catalog and classification set of connections, basically significant constitution of information for diverse domains: the nouns on behalf of classes of objects and the verbs representing kindred among the objects.

3 Literature Survey

Jianzhou LIU1, 2 et al. [1] introduced two tasks. First, build the relative feature words knowledge base. Other part assembles the framework of entity pairs and calculates the correspondence rate among those perspectives and the relation feature words of each relation type. Fully unsupervised method for relation extraction is based on web mining to resolve the above limitations, discriminative category matching (DCM) scheme.

Sanket S. Pawar et al. [6] introduced projected system that consists of subsequent modules.: 1. Query processing, 2. tuple set extraction, 3. IR engine, 4. candidate network generation, and 5. performance evaluation. The projected system gives comparable framework which demonstrates concert of IR system and finds out method with candidate network. Here, they have used IMDB for prototype 1 and reuters dataset for prototype 2. Prototype 1 has "Graph evaluation words execution time" which shows results of number of words vs execution time for IR style and determines the style. Another graph is for precision and recall. Prototype 2 has "Graph evaluation reminiscence contact with Time**".** Homogeneous appraisal constraint, high quantity of work on semi-structured data keyword search, and query workload have been taken seriously. Memory utilization issues for part A use hybrid algorithm, global pipeline algorithm, sparse algorithm, and naïve algorithm. For part B, layer 2 algorithm, part A of iterative search on range (ISR), and part B singular pass search (SPS) are used.

Sarika et al. [4] suggested latent semantic indexing (LSI) automatic mathematical technique. LSI starts with matrix of terms by credentials. This matrix is added to examine by singular value decomposition (SVD) two-factor analysis. This starts with one rectangular matrix and further rotten into three other matrixes of special form. This process is known as SVD. Offline component database preparation tasks use the LSI method, online component query recommendation engine, and the user interface. WORDNET dictionary here focuses on the synonymy and polysemy which were not available in the traditional search engine. Here, LSI and SVD give more relevant result over traditional system: Algorithm 1: how the page will be collected by crawler; Algorithm 2: for indexing the document; Algorithm 3: steps done by latent semantic indexing; Algorithm 4: extracting context from wordnet dictionary; and Algorithm 5: recommendation engine working.

Shu Tian et al. [5] reported three components on which they have worked. Text tracking is a tracking-based text detection. They have worked on embedded caption also. Framework for tracking-based text detection and recognition is Bayesian formulation of unified framework. Different types of methods are sliding window-based methods, connected component-based methods, and hybrid methods. A few botched belongings for anticipated move toward most likely because of text tracking. Frame average scheme has nastiest act. ROC graph is complicated for experiments. Generic Bayesian-based framework uses tracking-based text detection and recognition for surrounded description. For tracking-based text recognition term, polar over-segmentation method and an agglomerative hierarchical clustering algorithm are used.

Ryo Onuma et al. [7] proposed basically two methods: Way for producing context information set of connections and automatically accumulating context information. Extraction of relationship between context information and generation of context information network are carried out. Scheme is used for taking out imperative tip for small world construction. Main aim of their research was to levy the essential usefulness of the process for extracting important points for the small world structure and identifying problem. Effectiveness of the method has been evaluated by precision rate, recall, and F-measure. This research has resolved the following problems: load on the user in obtaining context information, intricacy in building up context information and administration, be short of lucidity as to whether concentration should be paying attention to fastidious information, monitoring program to acquire log, edge of graph to define relationship, small world network for important points, and shortest path to calculate.

4 Proposed Work

Step 1. Start
Step 2. Input file
Step 3. If input file is XML

Do
Read tag using parser;
Transform tags and values;
Acquire values for further process;

Else

Read text file;
Tokenize each sentence;
Cleansing the words (removal of extra symbol, white space, punctuation marks, etc.);

Step 4. Stemming process by porter's stemmer algorithm;

Step 5. Convert data to RDF;
Step 6. Process RDF

Do
Prefix for type, property, list, statement properties;

Step 7. Process OWL

Do
Calculate distance (computational logic);

Step 8. Sort distance and list words accordingly;
Step 9. Stop.

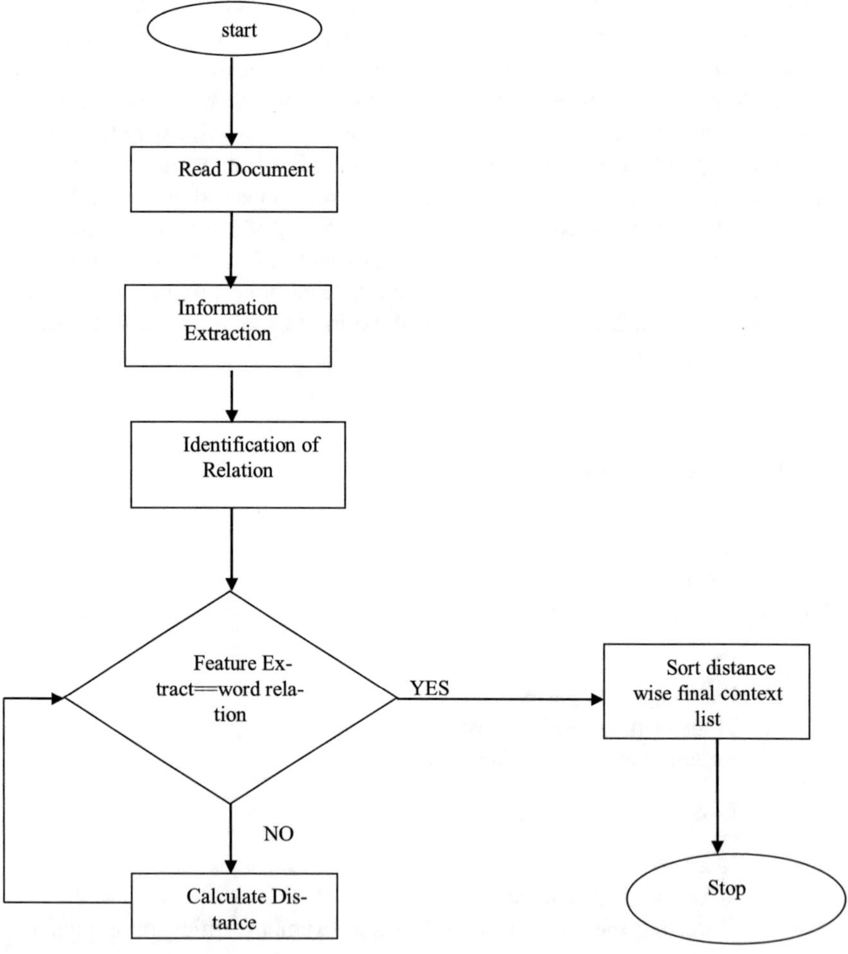

Table 1 ACE relation types and example

Type	Example
EMP-ORG	The CEO of Microsoft
PHYS	A military base in Germany
GPE-AFF	U.S. businessman
PER-SOC	A spokesman for the senator

Table 2 Comparison of accuracy between existing and proposed

Type	Base paper (%)	Proposed (%)
EMP-ORG	81	84.33
PHYS	76	81.24
GPE-AFF	66	80.16
PER-SOC	70	76.98

The data

ACE is an exploration and expansion program in information extraction paid for and supported by the U.S. Government. In 2004 (procedure of figuring out the value, quantity, or superiority of somewhat), four foremost kinds of kindred among seven kinds of things/businesses are defined. The thing/business types are PER (Person), ORG (Organization), FAC (Facility), GPE (Geo-Political Thing/business: countries, cities, etc.), LOC (Location), WEA (Weapon), and VEH (Vehicle). Apiece converse about/say of a thing/business has a talk about/say type: NAM1 (proper name), NOM (in name only/very small amount), or PRO (pronoun); for paradigm, George W.Bush, the head and he (bout up every brace of objects in array). The four relation types are EMP-ORG (Employment/Membership/Subsidiary), PHYS (Physical), PER-SOC (Personal/Social), and GPE-AFF (GPE-Association). It encloses 27 kin subtypes demarcated by ACE, but this paper solitary emphasizes on four thing/business relation types. Table 1 slants samples of separately kin type.

This dataset contains 348 credentials with 125 kilo vocabulary and 4400 relatives. There are two key credentials: newswire and broadcast news. Here, we have used the 150 sort out entity pairs for all relative form as of the ACR data. To construct relation, feature words of 50 entity duo have used and other are used to test.

The results

Here, methodology has tested over 150 entity brace. We pick the top 10 vocabulary as relative feature words as of all relation type. We accomplished 80.68% average precision. The exhaustive consequences are labeled in Table 2 followed by Fig. 1 of graph representation of the result.

As of the upstairs outcomes, it is not tough to realize that the EMP-ORG relative kind of abstraction verifies to have the utmost superlative results compared to other entities. The possible motive is that there are diverse dispersals as of web search outcomes for diverse relation-type entity pairs. For entirety up, the outcomes specify that our process is hopeful for entity relation extraction.

This research has tested method on other entities too. Results are shown in Table 3 and Fig. 2.

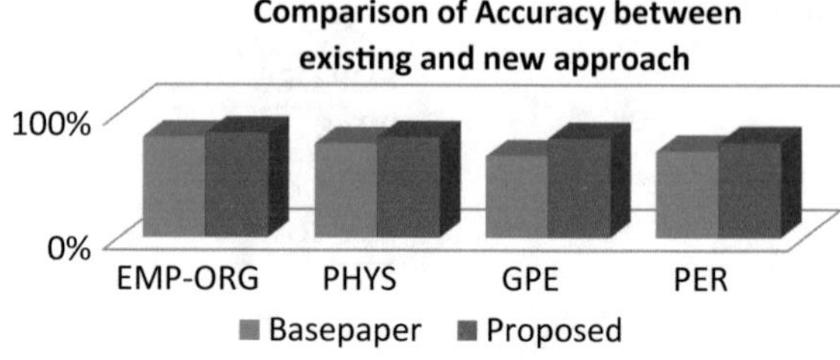

Fig. 1 Graph representation of comparison accuracy

Table Table 3 Accuracy of relation extraction

Type	Accuracy (%)
LOC	82
PRE	78
NOM	75
PRO	88
NAM	83

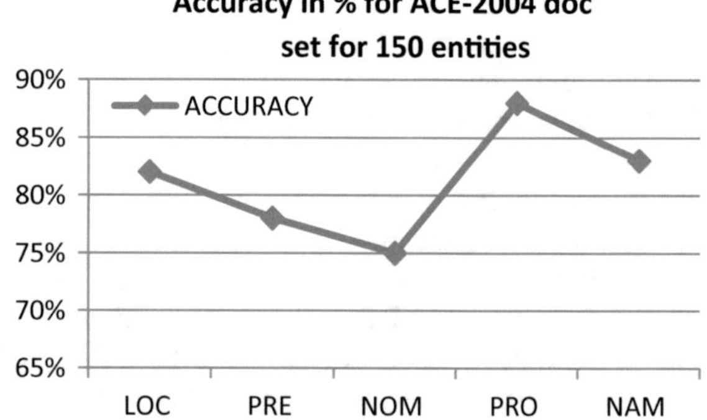

Fig. 2 Accuracy for ACE 2004 doc set

5 Conclusion

In this paper, we have introduced the means and approaches to signify relations and influences between words. We have proposed a method that permits a flexible and customizable way of generating connections between data sources. Methods reviewed from base paper and research papers are basically focusing on POS tagging, query expansion, and naïve Bayes classifier. The approach adopted for proposed flow is to strengthen the above-said method with hybrid approach. This allows an improved way of finding relations between words to access information about nearby links. We have defined the procedure of defining networks and used them to enlarge the connections and explore them for more meanings. We have assessed the upshot of the afresh-presented research algorithm with the ACE document set where we have got better results than the existing algorithm for the same entities.

References

1. Liu1,2, J., Xiao2, L., Shao2, X.: Context-dependency relation extraction based on web mining. In: 2016 Sixth International Conference on Instrumentation & Measurement, Computer, Communication and Control
2. Nasution, M.K.M., Sinulingga, E.P., Sitompul, O.S., Noah, S.A.: An Extracted Social Network Mining SAI Computing Conference 2016 July 13–15, 2016, London, UK
3. Soni, H.K.: Multi-objective association rule mining using evolutionary algorithm. Int. J. Adv. Res. Comput. Sci. Softw. Eng. **7**(5) (2017). ISSN: 2277 128X
4. Sarika, Chaudhary, M.: Context driven approach for extracting relevant documents from WWW. In: International Conference on Computing, Communication and Automation (ICCCA 2015)
5. Tian, S., Yin, X.-C., Su, Y., Hao, H.-W.: A Unified framework for tracking based text detection and recognition from web videos. IEEE Trans. Pattern Anal. Mach. Intell.
6. Pawar, S.S., Manepatil, A., Kadam, A., Jagtap, P.: Keyword search in information retrieval and relational database system: two class view. In: International Conference on Electrical, Electronics, and Optimization Techniques (ICEEOT) (2016)
7. Onuma, R., Nakayama, H., Abe, S., Kaminagat, H., Miyadera, Y., Nakamura, S.: Methods for Extracting the Polish Contexts in Research Document Creation

NoSQL Overview and Performance Testing of HBase Over Multiple Nodes with MySQL

Nabanita Das, Swagata Paul, Bidyut Biman Sarkar and Satyajit Chakrabarti

Abstract The escalating amount of web-based applications in the fields of social networks, media, biology, physics, and the Internet of things are continuously generating large volume of data or Bigdata in terabytes, petabytes, and zetabytes over a short period of time. Consequently, an immense amount of read and write requests is generated without much latency. It is an immediate concern to store and analyze such huge amount of mixed ASCII and non-ASCII data efficiently, economically, and in no time. The conventional database systems like MySQL are incapable to handle such large volume of data in real time. At this point, there is a claim that column-based NoSQL databases like Accumulo, Cassandra, HBase, or document-based Apache CouchDB, Couchbase, MongoDB are capable of handling such huge data volume efficiently. In this work, we focussed on column-based Apache HBase, a NoSQL distributed database management system developed in the Bigdata domain on distributed file system architecture provided by Hadoop (HDFS). Let us begin the discussion on NoSQL HBase and the association between HBase and Hadoop. Then some of the important features of HBase are explained. After that, we discussed the advantages and limitations of HBase in distributed data processing over the other NoSQL database management systems. Finally, we performed some experiments to compare the time performance of HBase with traditional database MySQL as data size increases.

Keywords Bigdata · Apache Hadoop · HDFS · NoSQL · Distributed DBMS HBase

---

N. Das · S. Paul (✉) · B. B. Sarkar
Techno India College of Technology, Kolkata, India
e-mail: swagatapaul@hotmail.com

N. Das
e-mail: das.nabanita2007@gmail.com

B. B. Sarkar
e-mail: bidyutbiman@gmail.com

S. Chakrabarti
University of Engineering and Management, Kolkata, India
e-mail: satyajit.chakrabarti@iemcal.com

A. Abraham et al. (eds.), *Emerging Technologies in Data Mining and Information Security*, Advances in Intelligent Systems and Computing 813,
https://doi.org/10.1007/978-981-13-1498-8_24

1 Introduction

We are living in an age of Internet and expect instantaneous results through web-based applications like Facebook, Amazon, Google, etc. This demand is resolved by many companies to focus on delivering targeted information, which is the key to their success. So a great amount of binary data needs to be efficiently stored and processed [1]. Unstructured data implies variable length field, and the records can be constituted of the subfields [2]. In today's data rain, multi-terabytes, petabytes, or even zetabytes of databases are common [3]. Facebook processes 2.5 billion content and 500+ terabytes of data per day [4]. Digital data growth will be 44–35 zetabytes on a yearly basis according to International Data Corporation [5]. Due to the increase in database size, some major problems arose. They are as follows: information and content management should be completed in a reasonable time, handling capabilities of business analysis [3]. There are some disadvantages of the traditional relational database management systems (RDBMS). Some of the disadvantages are resolved by a new scalable and distributed database system, which is different from the traditional RDBMS and is proposed as a NoSQL database. The strong features of these databases are simplicity in design, horizontal scaling, a finer control over availability, and a way of "scaling out" to automatically spread data across several servers; however, ACID property does not work [6]. An alternative to ACID is BASE [7]. NoSQL databases include four categories: Document-oriented (great for analysis—MongoDB), column-oriented (great for data access—HBase), key–value pair (great for speed in query—REDS, used for server-side customer history), and graph-oriented (great for graph-related queries—NEO4 J). It uses Hadoop distributed file system as storage [8]. Some of the salient features of HBase are schema-less, defined column families, no join, used sharding, horizontally scalable, no transactions, handle high growth of de-normalized data, good for semi-structured and structured data, NoSQL database, and disable a table and then drop [9].

High cost of the infrastructure and higher licensing costs are the problems today. Some of the examples of NoSQL databases are Bigtable by Google [10], Dynamo by Amazon [11], Apache Cassandra [12], Hypertable [13], CouchDB [14], Voldermort by LinkedIn [15], MemcacheDB [16], MongoDB [17], HBase, Accumulo, Amazon's SimpleDB, Github's CloudData, HPCC/DAS, Apache's Flink, Splice, RethinkDB, RavenDB, MarkLogic Server, etc. [3]. Rest of this paper contains the following sections: Sect. 2 performs literature survey. Section 3 describes Bigdata—Hadoop framework. Section 4 describes the elaborative study of HBase. Section 5 focuses on HBase architectural components. Section 6 illustrates the comparison of HBase with other leading NoSQL databases. Section 7 provides some experimental results with traditional RDBMS MySQL. Section 8 contains the advantages of HBase, and finally Sect. 9 contains the conclusion.

2 Related Work

Managing and processing huge amount of web-based data are becoming increasingly difficult via conventional database management tools. In this work, HBase is used to manage it to generate required results. It is a good choice to handle this kind of problem. HBase architecture search retrieves data at high speed [3]. Popular NoSQL database HBase is structured as two layers: (a) Layer-1: For distributed data storage and (b) Layer-2: indexing and elasticity for data distribution. This layered system often suffers from the problem of isolation between layers. An attempt is made to overcome this problem in this work. According to [18], HBase–BDB is designed, implemented, and evaluated over B+-tree file storage structure operating on HDFS. It shows a performance improvement of 30% in random read and 85% in random write. HBase is one of the databases to handle Bigdata. Data generated from geoscience provides much information. HBase is used to handle this data. Due to index level problem, many researchers have been going on to improve this problem [18]. According to STEHIX, a two-level index architecture is introduced [19]. The two indexes are index in the meta table and the region index. This two-level architecture helps HBase to process queries. Authors develop HConfig to optimize the performance of HBase. They consider bulk loading as their parameter. The HConfig enhances bulk loading job up to 2 ~ 3.7x speedup in throughput compared to default configuration [20].

This work focuses in the field of Bioinformatics (Biology + Information Technology); data are growing at a rapid rate. To store and manipulate this huge data to generate important patterns, it needs some solutions. The obvious solution is Bigdata. In the area of Bigdata, the most common distributed file system for storing and data manipulation is Hadoop–HBase. Many important projects are going on using Hadoop–HBase configuration; one such project is PNNL (Pacific Northwest National Laboratory) project [21]. Confidentiality, integrity, and availability (CIA) are the three very important parameters for data protection and privacy in database. According to [22], HBase is evolving at a great speed to dominate today's IT world. Though initial versions pay less attention in security, latest versions provide more security aspects to compete with RDBMS (has more inbuilt security features) in the area of transaction processing. In the area of Geographical Information System (GIS), the data are being increased as a variety of research works and data manipulation works are carried out to fulfill the demand of information. The Hadoop-based free open-source database HBase is used in this regard. In this paper, HBase is used to build a low-cost, flexible, and easy to maintain pyramid tile layer services—HBGIS (Hadoop-based geographical information system). Here, HBase is selected for its column-oriented capability which is better for massive data storage. HBGIS built with inexpensive hardware and provides large data warehouse and cluster configuration to enhance the quality and efficiency of the global image service in the area of remote sensing image management [23]. HBase is a kind of NoSQL database, which has only one index for the table and this index is the row key of the database. So to improve the data retrieval process, we need to concentrate on the design of row key.

A number of techniques have been modeled (Sequences, salting, padding, hashing, and modulo operation) to investigate four different designs: MD5 hashing function, presplitting HBase tables, random prefix, ID modulo prefix, ID MD5 modulo prefix, Line MD5 modulo prefix, and generalization of the primary key [24]. There are some tests conducted in HBase to find out the performance in single-machine and multi-machine cluster environment [25].

3 Bigdata—Hadoop

Five V's of Big Data (Volume): Google processes 100 PB data/day. Facebook processes 500 TB data/day, YouTube processes 1000 PB video storage, and 4 billion visits in a day [26]. Velocity: Data production will be 44 times bigger in the year 2020 than it was in the year 2009 [5]. Variety: Various types of data are added to Bigdata group, ranging from social media, feeds, multimedia, healthcare, sensor, etc. Veracity and Value: Uncertainty of data.

Hadoop: The framework enables petabytes or even zettabytes of data storage and processing them to generate new information patterns. Hadoop ecosystem comprises four stages: Data management tools are the uppermost, then data access tools, next comes data processing tools, and finally the data storage. Hadoop framework consists of three parts: (a) Hadoop core. (b) HDFS—when the storage capacity massively increased over time but the data access rate has not improved accordingly and the performance of the application reduces, HDFS retrieves the data from multiple disks in parallel to improve the access rate [3, 27]. (c) In MapReduce, the two operations are Map and Reduce—divide a large problem into smaller subproblems. **Hadoop Characteristics**: (a) Recovers quick hardware failure. (b) Streaming data access for batch processing. (c) Large datasets: HDFS is designed to support huge datasets of millions of rows and columns. (d) Simple coherency model: In HDFS applications, we need to follow write-once-read-many access model for files. (e) "Moving Computation is Cheaper than Moving Data" [27]. (f) Easily portable across heterogeneous hardware and software platforms [27]. (g) Robustness. and (h) Snapshot.

4 HBase: A NoSQL Database

eBay, Facebook, Twitter, Yahoo, and Adobe use HBase extensively. HBase tables have column family and column name to show data types of a given cell [3, 21]. A cell's value is uniquely identifiable [3, 1, 21]. Data retrieval results in MapReduce job [21, 9]. A set of column families is used to create a table [3]. The detailed architecture is given in Apache HBase documentation review [9].

Table 1 Difference between HBase and Bigtable

Characteristics	HBase	Bigtable
Storage	Region RegionServer	Tablet Tabler server
Timestamp	Milliseconds	Microseconds
File system	HDFS, Amazon S3, EBS	GFS
Data storage	HFile	SSTable
Coordination	ZooKeeper	Chuuy
Mapping	Cannot map storage files into memory	Possible
Locality	Locality groups and column family	Locality groups
Transactions	across multiple rows are not supported	Supports transactions
Storage management	Major compaction, Write-ahead log	Major compaction, Commit Log
Data processing	Hadoop MapReduce, MemStore	MapReduce, MemTables
Language used	Written with Java	Written with C++
Secondary index	Poor	Mature

5 HBase Architectural Components

HBase architecture has three main components [23]: (1) HMaster, (2) HRegion-Server, and (3) HBaseClient. It provides the interface to connect with underlying layers. The mapping information of regions to RegionServer is maintained in a special catalog table named .META [23, 28]. ZooKeeper maintains server state in the cluster and provides server status information.

6 Comparison of HBase with Leading NoSQL Databases

Hive is another component of Hadoop ecosystem and is a different technology than HBase [3]. Data can even be read and written from Hive to HBase and back again. Since Hive uses SQL-like language, HBase and Hive combined together for better data access from HBase tables with JDBC bindings [1, 23] (Tables 1, 2, 3, and 4).

Table 2 Difference between HBase and Hive

Characteristics	HBase	Hive
Real-time operations	Operations run in real time	Not to be used for real-time querying
Language used	Use custom language	Use SQL-like language HQL
ACID properties	Partially ACID compliant	Not ACID compliant
Updating process	Supports update statements	Do not support statement update
Queries used	Real-time queries	Analytical queries

Table 3 Difference between HBase and MongoDB

Characteristics	HBase	MongoDB
Data storage	Wide-column store	Document store
Implementation language	Java	C++
Indexing method	Secondary indexes	Geospatial indexing
Language used	JAVA API, Thrift, RESR API	Proprietary protocol using JSON
Trigger	Trigger available	Trigger not available
Replication method	Selectable replication factor	Master–slave replication
File system	HDFS	JSON/BSON
Consistency method	Immediate consistency	Eventual consistency
Searching method	No feature of searching	Good feature of searching

Table 4 Difference between HBase and MySQL

Characteristics	HBase	MySQL
Data storage	Wide-column store	Relational DBMS
Secondary indexes	No	Yes
Query language	No	Yes
Partitioning methods	Sharding	Horizontal, sharding with cluster or fabric
Replication methods	Selectable replication factor	Master–master, Master–slave replication
MapReduce	Yes	no
Consistency concept	Immediate Consistency	Immediate consistency
Foreign keys	No	yes not for MyISAM storage engine
Transaction concepts	No	ACID not for MyISAM storage engine

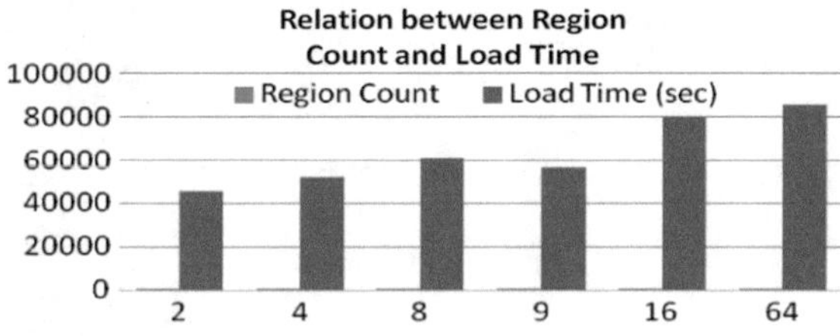

Fig. 1 Relation between region count and load\time

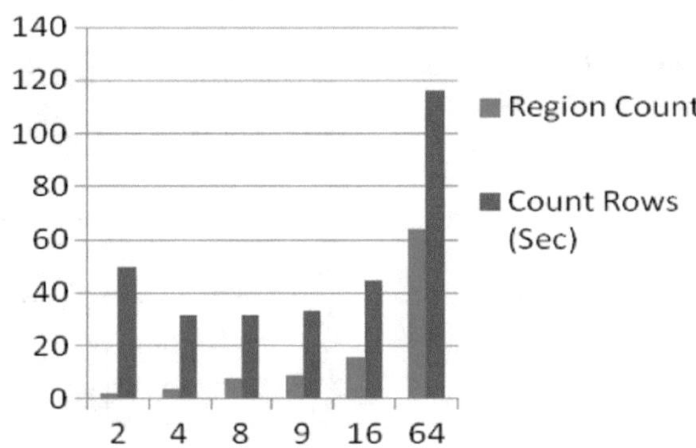

Fig. 2 Relation between region count and CountRow time

Table 5 Operating system, Hadoop, and HBase configuration

Items	Detailed description	Items	Detailed description
OS	Ubuntu 15.04 64 bits	Network interface	1Gbps Ethernet
JVM	64 bits	Storage	500 GB SATA
Hadoop	V-2.7.3	HBase	V-2.2.1
HDFS	V-2	No. of data nodes	30
MySQL	V-5.7	No. of master node	2
CPU	Intel Core i5	I/P file size	2 GB CSV to 128 GB CSV
Primary memory	DDR3, 4 GB		

7 Experimental Setup and Results

The following setup is considered for our experiment on Hadoop—HDFS–HBase (Figs. 1, 2 and Tables 5, 6).

We executed CountRows ten times for each of the region counts and we have taken the average of them (Tables 7 and 8).

Table 6 Results found on HBase for dataset 128gb

Region count	Load time (s)	Count rows (s)
Default	45625.65	49.670
4	52102.81	31.564
8	60565.04	31.420
9	56139.86	33.100
16	79563.94	44.750
64	85429.280	116.000

Table 7 Results found on HBase and MySQL

Data size (GB)	Load time HBase (min)	Load time MySQL (min)
2	20.4	13.8
4	45.5	27.1
8	67.3	56.2
10	83.9	69.2
16	124.7	151.7
32	267.6	330.7
64	384.8	804.8
128	568.1	1688.6

Table 8 Results found on HBase and MySQL

Data size (GB)	Scan time HBase (s)	Scan time MySQL (s)
2	0.354	5.55
4	1.065	13.7
8	1.137	21.5
10	1.568	23.9
16	2.208	35.47
32	6.522	57.2
64	16.432	82.4
128	47.281	113.6

8 HBase Advantages

HBase does not support SQL. Some of the attractive features of HBase make it special. However, HQL and HBQL are query languages in Hive and HBase, respectively [22, 23, 29]. In our experiments, we can see from the results that HBase is better than MySQL for large datasets. We can see the effect in the result when we reach dataset range more than 64 GB. The HBase database has a lot of advantages. The most important feature of HBase, which differentiates it from other leading NoSQL databases, is its column-oriented and column family storage. Due to this storage, scan operation will not read "not requested" columns at all (bloom filter), thereby improving processing speed [6]. Time range info skips the file if it is not in the time

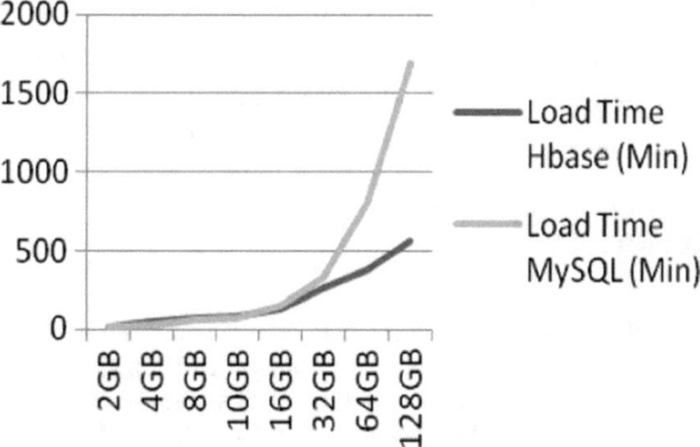

Fig. 3 Comparision between HBase and MySQL load time

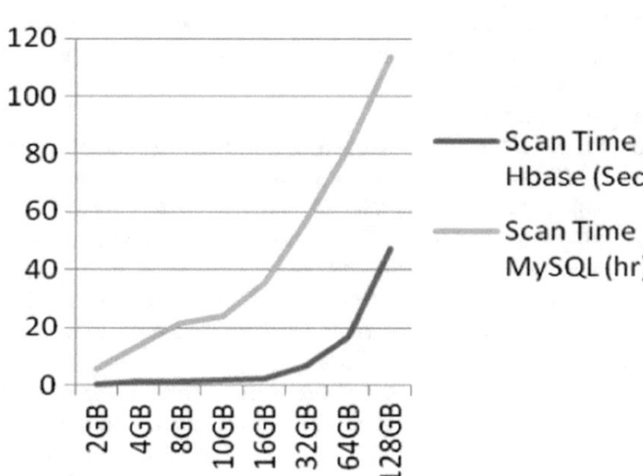

Fig. 4 Comparision between HBase and MySQL scan time

range, and the read is looking for. Minor compaction and major compaction improve read performance by reducing the number of storage files [28]. When a write returns, all readers will see the same value—strong consistency model, scaling automatically, and automatic sharding are very important features of distinction.

9 Conclusion

In this paper, we studied NoSQL, Bigdata, and some important features of Hadoop. Then, we looked into HBase. We made some comparative study between HBase and relational data store MySQL. The experimental results show that with smaller dataset, the performance of MySQL is better than HBase. When we increase the size of the datasets, the difference in performance is visualized. When the size of datasets increases, the performance of the NoSQL database HBase will also show its strength. Time to load and scan is both getting better in HBase. Figures 3 and 4 show these results. Therefore, with the increase of dataset, it is better to work with NoSQL databases such as HBase. We also reviewed the HBase physical architecture and logical data model. Then, we presented experimental results. Finally, some of the results show the importance of HBase in the NoSQL world which is unique and well suited for distributed data processing.

References

1. Bhupathiraju, V., Ravuri, R.P.: The dawn of big data-Hbase. In: 2014 Conference on IT in Business, Industry and Government (CSIBIG), pp. 1–4. IEEE, March 2014
2. Huang, S., et al.: Non-structure data storage technology: a discussion. In: 2012 IEEE/ACIS 11th International Conference on Computer and Information Science (ICIS), pp. 482–487. IEEE, May 2012
3. Vora, M.N.: Hadoop-HBase for large-scale data. In: 2011 International Conference on Computer Science and Network Technology (ICCSNT), vol. 1, pp. 601–605. IEEE, Dec 2011
4. www.techcrunch.com
5. www.emc.com
6. Yang, F., Cao, J., et al.: An evolutionary algorithm for column family schema optimization in HBase. In: 2015 IEEE First International Conference on Big Data Computing Service and Applications (BigDataService). IEEE (2015)
7. https://changeaas.com/2014/08/20/nosql-and-its-use-in-ceilometer/
8. www.mail-archives.apache.org
9. Naheman, W., et al.: Review of NoSQL databases and performance testing on HBase. In: Proceedings 2013 International Conference on Mechatronic Sciences, Electric Engineering and Computer (MEC). IEEE (2013)
10. Chang, F., Dean, J., et.al.: Bigtable: a distributed storage system for structured data. In: Seventh Symposium on Operating System Design and Implementation (OSDI), Seattle, WA, Usenix Association, Nov 2006
11. DeCandia, G., et al.: Dynamo: Amazon's highly available key-value store. In: The Proceedings of the 21st ACM Symposium on Operating Systems Principles, Stevenson, WA, Oct 2007
12. www.cassandra.apache.org
13. Judd, D.: Scale out with HyperTable. Linux Mag. (2008)
14. Anderson, J.C., et al.: CouchDB: The Definitive Guide, 1st edn. O'Reilly Media (2009). ISBN: 0596158165
15. www.project-voldemort.com
16. http://memcachedb.org/memcachedb-guide-1.0.pdf
17. Kristina, C., Michael, D.: MongoDB: The Definitive Guide, 1st edn. O'Reilly Media (2010). ISBN: 9781449381561
18. Saloustros, G., Magoutis, K.: Rethinking HBase: design and implementation of an elastic key-value store over log-structured local volumes. In: 2015 14th International Symposium on Parallel and Distributed Computing (ISPDC). IEEE (2015)
19. Chen, X., et al.: Spatio-temporal queries in HBase. In: 2015 IEEE International Conference on Big Data (Big Data). IEEE (2015)
20. Bao, X., et al.: HConfig: resource adaptive fast bulk loading in HBase. In: 2014 International Conference on Collaborative Computing: Networking, Applications and Work sharing (CollaborateCom). IEEE (2014)
21. Taylor, R.C.: An overview of the Hadoop/MapReduce/HBase framework and its current applications in bioinformatics. BMC Bioinform. **11**, Suppl. 12 (2010): S1
22. Srinivas, S., et al.: Security maturity in NoSQL databases-are they secure enough to haul the modern IT applications. In: 2015 International Conference on Advances in Computing, Communications and Informatics (ICACCI). IEEE (2015)
23. Xiao, Z., et al.: Remote sensing image database based on NOSQL database. In: 2011 19th International Conference on Geoinformatics. IEEE (2011)

24. Zdravevski, E., et al.: Row key designs of NoSQL database tables and their impact on write performance. In: 2016 24th Euromicro International Conference on Parallel, Distributed, and Network-Based Processing (PDP). IEEE (2016)
25. Naheman, W., et al.: Review of NoSQL databases and performance testing on HBase. In: Proceedings 2013 International Conference on Mechatronic Sciences, Electric Engineering and Computer (MEC), pp. 2304–2309. IEEE (2013)
26. www.slideshare.net/kmstechnology
27. www.hadoop.apache.org
28. www.mapr.com/blog/in-depth-look-hbase-architecture
29. Wei, D., et al.: Organizing and storing method for large-scale unstructured data set with complex content. In: 2014 Fifth International Conference on Computing for Geospatial Research and Application (COM. Geo). IEEE (2014)

NO SQL Approach for Handling Bioinformatics Data Using MongoDB

Swaroop Chigurupati, Kiran Vegesna, L. V. Siva Rama Krishna Boddu, Gopala Krishna Murthy Nookala and Suresh B. Mudunuri

Abstract Proliferation of genomic, diagnostic, medical, and other forms of biological data resulted in categorizing of biological data as bigdata. The low-cost sequencing machinery, even in small research labs, is generating large volumes of data which now needs to be mined for useful biological features and knowledge. In this paper, we have used a NoSQL approach to handle the repeat information of the entire human genome. A total of 12 million repeats have been extracted from the entire human genome and have been stored using MongoDB, a popular NoSQL database. A web application has been developed to query data from the database at ease. It is evident that bioinformaticians tend to shift their database development approach from traditional relational model to novel approaches like NoSQL in order to handle the massive amounts of biological data.

Keywords Bigdata · NoSQL · Bioinformatics · Microsatellites · Genomics
Human genome

S. Chigurupati · K. Vegesna · L. V. Siva Rama Krishna Boddu · G. K. M. Nookala
S. B. Mudunuri (✉)
Sagi Ramakrishnam Raju Engineering College, Bhimavaram 534204, India
e-mail: sureshverma@gmail.com

S. Chigurupati
e-mail: raviswaroop.chigurupati@gmail.com

K. Vegesna
e-mail: venkatkiran693@gmail.com

L. V. Siva Rama Krishna Boddu
e-mail: krishna2928@gmail.com

G. K. M. Nookala
e-mail: gopinukala@gmail.com

A. Abraham et al. (eds.), *Emerging Technologies in Data Mining and Information Security*, Advances in Intelligent Systems and Computing 813,
https://doi.org/10.1007/978-981-13-1498-8_25

1 Introduction

Biological data is being produced at a phenomenal rate even from small research laboratories and firms in the recent few years, as most of these labs are equipped with a variety of high-throughput sequencing machines and automated genome sequencers. On the other hand, the public data repositories like EBI and NCBI disseminate hundreds of petabytes of data every year [1]. Biologists till now are trying to analyze one or few genomes of their interest. But the recent advances in the field have given them new opportunities to compare and analyze several genomes at a time. The fields like single-cell genomics, microarrays, and NGS analysis are all generating huge volumes of data at a rapid pace. In a way, the biologists have already joined the bigdata club and have been facing challenges in storing and processing information [2].

Compared to fields like physics, information technology, business, astronomy, etc. that deal with enormous data over the decades, the field of modern biology has been submerged with bigdata in quick period [3]. This massive increase in data creates several opportunities for biologists to expand their horizons. At the same time, they also demand to adopt new techniques and methods that use bigdata technologies. Especially, the modern biology or the bioinformatics community has been developing several database-specific resources to mine useful information from this explosion of data. The underlying databases of many of these bioinformatics applications are based on a relational model.

Relational Database Management Systems (RDBMSs) like MySQL, ORACLE, etc. have been a popular choice for building several bioinformatic databases for the past few decades. RDBMS provides excellent storage efficiency, data integration, and good data retrieval speeds compared to file-based storage [4]. However, due to the increased rate of sequence data generation, the traditional relational databases can no longer handle such large data efficiently. In this context, non-relational databases (NoSQL databases) like CouchDB, MongoDB, HBase, Cassandra, etc. can be the ideal tools for bioinformatics application development [5]. Moreover, the current databases do not require updating of multiple rows and tables and do not rely on RDBMS features like concurrency, integrity, etc. In such cases, NoSQL works better than the relational databases.

A NoSQL database provides a mechanism for storage and retrieval of data that is modeled in a way other than the tabular relations used in relational databases. The internal data structures used by NoSQL databases are different from those used by RDBMS, making some of the operations to be executed faster in NoSQL.

In this paper, we have demonstrated the development of a simple bioinformatics application that uses a NoSQL database MongoDB to store and retrieve all the Simple Sequence Repeats (SSRS) also called as microsatellites of human genome that constitutes a size of around 12 million records. MongoDB is a cross-platform, document-oriented database that provides high performance, high availability, and easy scalability [6].

2 Materials and Methods

The complete DNA sequence of human genome (24 chromosomes) of size ~3.2 GB has been downloaded from National Center for Biotechnology Information (NCBI) genome server. The downloaded sequences were submitted to a DNA repeat detection software named Imperfect Microsatellite Extractor (IMEx) [7] to extract microsatellite repeats. Microsatellites are tandem repetitions of DNA motifs of size six or less that play a significant role in various fields and applications such as DNA fingerprinting, evolution, paternity studies, strain identification, genetic diseases, DNA forensics, etc. [8]. These repeats are also responsible for causing several neurological disorders and cancers [9]. Researchers show a keen interest in studying these repeats and their dynamics in various genomes. IMEx has been used to extract imperfect repeats (max imperfection of 10%) with a minimum size of 10 bp. A total of 1,17,97,339 (approximately 12 million) microsatellites have been extracted from the entire human genome. Table 1 shows the number of microsatellites detected in the entire human genome chromosome-wise.

The extracted repeat information is further converted to JSON/CSV format using python scripts and is imported to MongoDB, an open-source document database, and leading NoSQL database. MongoDB works on the concept of collections and documents. Database is a physical container for collections. Each database gets its own set of files on the file system. A single MongoDB server typically has multiple databases. Collection is a group of MongoDB documents and is the equivalent of an RDBMS table. A collection exists within a single database and they do not enforce a schema. Documents within a collection can have different fields in the form of key-value pairs. Documents have dynamic schema, which means that documents in the same collection do not need to have the same set of fields or structure and common fields in a collection. The MongoDB database has been installed in an Ubuntu Server, and a web-based query application has been developed on top of it.

3 Application Development

Several organism-specific microsatellite databases have been developed earlier using RDBMS such as MICdb [10], InSatDB [11], ChloroMitoSSRdb [12], EuMicroSatDB [13], CMD [14], etc. In this paper, we have created the application that uses a MongoDB database instead of traditional relational database. In MongoDB, we have created a database named "IMEx" and a collection named "chromosome" to which all the repeat information has been imported using the custom python scripts developed in-house.

A web-based query system has been developed to retrieve the desired information from the repeat data using HTML5, JavaScript, CSS, and PHP. The server-side scripting has been achieved with PHP code using Mongo-PHP drivers. An advanced search query form has been created using which the users can perform the following type of searches.

Table 1 Chromosome-wise microsatellite number detected by IMEx along with the sequence length of each chromosome

Chromosome	Sequence length (in mb)	Total repeats extracted
Chr 1	253.9	9,34,437
Chr 2	247.0	9,57,375
Chr 3	202.3	7,76,597
Chr 4	194.0	7,53,977
Chr 5	185.2	7,15,929
Chr 6	174.2	6,75,449
Chr 7	162.5	6,52,876
Chr 8	148.0	5,69,244
Chr 9	141.2	4,81,281
Chr 10	136.5	5,42,213
Chr 11	137.8	5,32,141
Chr 12	135.9	5,42,695
Chr 13	116.7	3,84,432
Chr 14	109.2	3,57,169
Chr 15	104.0	3,33,137
Chr 16	92.1	3,50,884
Chr 17	84.9	3,45,948
Chr 18	82.0	3,04,589
Chr 19	59.8	2,67,940
Chr 20	65.7	2,58,399
Chr 21	47.6	1,62,749
Chr 22	51.8	1,62,971
Chr X	159.2	6,10,823
Chr Y	58.4	1,24,079
Total:	3.2 GB	Total: 1,17,97,339

3.1 *Motif-Based Search*

The application allows the users to query repeats of a particular motif (GTA, AT, Poly A, etc.). User can select a particular chromosome and get all the repeats of a motif.

3.2 *Motif Size-Based Search*

The repeats are tandem repetitions of sizes 1–6 bp. Provision has been given in the application to extract repeats of a particular motif size (Mono, di, tri, etc.). For example, the user can extract all triplet repeats (motif size: 3) of X chromosome in order to study and analyze the microsatellites causing triplet repeat disorders in humans.

3.3 Repeats of a Specific Repeat Number

Some users might be looking for repeats that are of a particular length or more. Our application is provided with an option to query repeats that iterate more than n number of times. The thresholds can be given for each repeat size separately.

3.4 Repeats in a Specific Region

The application also allows you to fetch microsatellites from a specific region of DNA in a chromosome. User can give the start and end coordinates of the genomic location to get repeats from that particular region.

The complete architecture of the human genome repeat database application has been given in Fig 1.

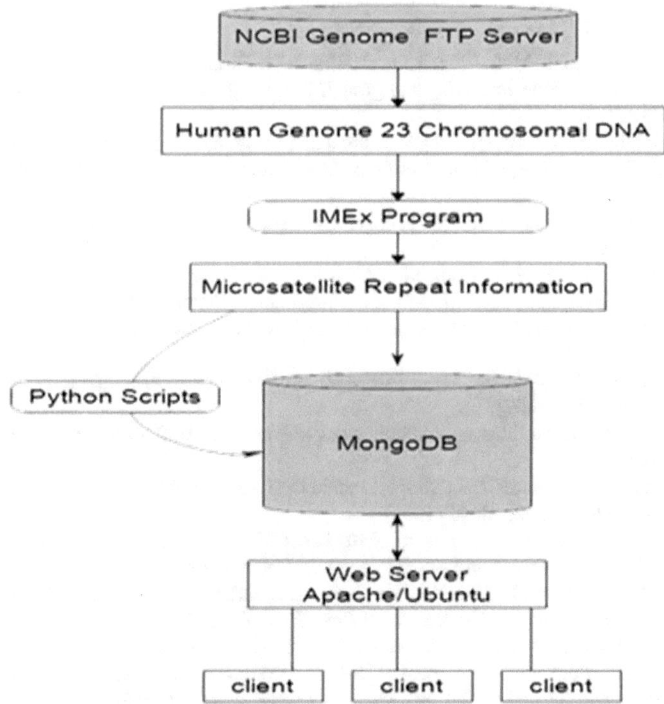

Fig. 1 Architecture diagram of the development of human microsatellite repeat database

The query form can be used with a combination of the above options. It has been observed that the usage of NoSQL approach has made the query retrieval very quick despite having 12 million records.

4 Conclusions

In this paper, we have demonstrated a model to develop biology application using MongoDB in place of traditional RDBMS. Moreover, we learned that bioinformatics is the primary field in which bigdata analytics are currently being applied, largely due to the massive volume and complexity of bioinformatics data. Using this application, biology researchers can query required data and could get the result faster than existing traditional SQL applications.

Acknowledgements The authors would like to thank Ms. Kranthi Chennamsetti, Centre for Bioinformatics Research and SRKR for her help in the extraction of microsatellites from human genome. This work is supported by SERB, Department of Science and Technology (DST), India (Grant ID://ECR/2016/000346).

References

1. Cook, C.E., Bergman, M.T., Finn, R.D., Cochrane, G., Birney, E., Apweiler, R.: The European bioinformatics institute in 2016: data growth and integration. Nucleic Acids Res. **44**(D1), D20–D26 (2015)
2. Marx, V.: Biology: the big challenges of big data. Nature **498**(7453), 255–260 (2013)
3. Singer, E.: Biology's big problem: there's too much data to handle. Quanta Mag. 2014. Accessed 26 Jan
4. Codd, E.F.: A relational model of data for large shared data banks. Commun. ACM **13**(6), 377–387 (1970)
5. Manyam, G., Payton, M.A., Roth, J.A., Abruzzo, L.V., Coombes, K.R.: Relax with CouchD-B—into the non-relational DBMS era of bioinformatics. Genomics **100**(1), 1–7 (2012)
6. Chodorow, K.: MongoDB: The Definitive Guide: Powerful and Scalable Data Storage. O'Reilly Media, Inc. (2013)
7. Mudunuri, S.B., Nagarajaram, H.A.: IMEx: imperfect microsatellite extractor. Bioinformatics **23**(10), 1181–1187 (2007)
8. Ellegren, H.: Microsatellites: simple sequences with complex evolution. Nat. Rev. Genet. **5**(6), 435–445 (2004)
9. Sutherland, G.R., Richards, R.I.: Simple tandem DNA repeats and human genetic disease. Proc Natl Acad Sci **92**(9) 3636–3641 (1995)
10. Mudunuri, S.B., Patnana, S., Nagarajaram, H.A.: MICdb3.0: a comprehensive resource of microsatellite repeats from prokaryotic genomes. Database (2014)
11. Archak, S., Meduri, E., Kumar, P.S., Nagaraju, J.: InSatDb: a microsatellite database of fully sequenced insect genomes. Nucleic Acids Res **35**(suppl_1), D36–D39 (2006)
12. Sablok, G., Padma Raju, G.V., Mudunuri, S.B., Prabha, R., Singh, D.P., Baev, V., Yahubyan, G., Ralph, P.J., Porta, N.L.: ChloroMitoSSRDB 2.00: more genomes, more repeats, unifying SSRs search patterns and on-the-fly repeat detection. Database (2015)

13. Aishwarya, V., Grover, A., Sharma, P.C.: EuMicroSat db: a database for microsatellites in the sequenced genomes of eukaryotes. BMC Genomics **8**(1), 225 (2007)
14. Blenda, A., Scheffler, J., Scheffler, B., Palmer, M., Lacape, J.M., John, Z.Y., Jesudurai, C., Jung, S., Muthukumar, S., Yellambalase, P., Ficklin, S.: CMD: a cotton microsatellite database resource for Gossypium genomics. BMC Genomics **7**(1), 132 (2006)

Thunderstorm Characteristics Over the Northeastern Region (NER) of India During the Pre-monsoon Season, 2011 Using Geosynchronous Satellite Data

Sandeep Thakur, Ismail Mondal, Phani Bhushan Ghosh and Tarun Kumar De

Abstract Northeastern Region (NER) of India commonly experiences the occurrence of rigorous convective squalls, commonly known as Thunderstorms (TS) or Nor'westers through the months of March to May. This study utilizes satellite images of Kalpana-1 satellite in Thermal Infrared (TIR) and Water Vapor (WV) channels to identify TS genesis, frequency, and track over the NER. The results obtained have been compared with Indian Meteorological Department (IMD) reports to check the accuracy of satellite image interpretations. 81 TS were detected over NER of India with 38 in the month of April and 43 in the month of May. The duration of TS follows an exponential distribution with maximum TS having lifetime of an hour or less and a few having lifetime of about 8–9 h. The TS occurred almost during any time domain of a day with maximum number of them occurring between 16 and 20 UTC. Most of the systems were formed and dissipated over the NER, while some systems were formed over the northwestern parts of Bihar, Jharkhand, and West Bengal and moved toward the NER under the influence of the Nor'westers collecting moisture, maturing, and dissipating over the NER. The dominant wind was from the northwest during the season. The precipitation from these TS accounts for 20–25% of annual rainfall and is very important for the agrarian economy of this region. The interpreted TS was found to be lesser (67%) than the number of events reported by IMD reports primarily because of the limitations in spatial resolution of the sensor.

Keywords Thunderstorm · Satellite images · Northeastern region (NER)

S. Thakur (✉) · I. Mondal · T. K. De
Department of Marine Science, University of Calcutta, 35 B.C. Road, Kolkata 700019, India
e-mail: sandeept.pu@gmail.com

P. B. Ghosh
Institute of Engineering & Management, Salt Lake, Sector-V, Kolkata 700091, India

A. Abraham et al. (eds.), *Emerging Technologies in Data Mining and Information Security*, Advances in Intelligent Systems and Computing 813,
https://doi.org/10.1007/978-981-13-1498-8_26

1 Introduction

A widespread characteristic above the Indo-Gangetic plain and NE India is the occurrence of rigorous local convective squalls, generally well known as thunderstorms throughout the pre-monsoon time of year (March–May). They usually approach a station from the northwest, and hence are called Nor'westers. Locally, they are also called *KalBaisakhi* in Bengali and *Bordoichila* in Assamese. These are greatly influenced by mesoscale convective systems [1].

The thunderstorms mostly form over the hot landmasses and rise up to converge with the air mass on top of it and start convection. These systems usually attain severity when continental air comes together with humid air from sea. Eastern and northeastern parts of India, i.e., West Bengal, Jharkhand, Orissa, North Eastern Region (NER), and fraction of Bihar, get affected by thunderstorm throughout pre-monsoon month of April–May [2]. Incessant heating of landmass through the day starts convection which intensifies by addition with humid air mass brought by the southeast winds from the Bay of Bengal [3]. Thunderstorms coupled with lightning, hailstones, and intense downpour inflicted widespread damage to the cultivated harvest, thatched huts, and tall poles, and cause inundations, ensuing in loss of life and home. Aviation is another sector that is severely affected by thunderstorms.

Predicting thunderstorms is a very complicated job in weather forecast, owing to their fairly miniature spatial and temporal size and the innate nonlinearity of their physical dynamics. In [4], he first put forward the idea for utilizing satellite images in weather prediction and forewarning. He put forward quite a few thoughts on how geostationary satellites might be implemented to improve forecasting and these are followed by many scientists such as [5–10], where they have studied various facets of thunderstorm remote sensing using geostationary satellites. Also, the large number of field experiments has been conducted toward thunderstorms in various parts of India. There have been several attempts to relate several of the early notions to the nowcasting of storms initiation and intensification. In [11], they proposed a method for the nowcasting of rainstorm commencement and expansion by means of both GOES-8 and Doppler radar data (WSR-88D). Reports of cumulonimbus cloud formations over Guwahati were put forward by [12], and many such studies about thunderstorm formation and their characteristics over eastern region have been reported [13, 14].

This study was carried out with an aim to observe TS from the Thermal Infrared (TIR) and Water Vapor (WV) image data obtained from Kalpana-1 satellite; to find the frequency and duration of TS affecting NER of India for 2011; to understand the occurrence of TS during different time domains of a day; to study the track followed by the TS; and to compare the TS occurrence as reported by IMD with that of satellite interpretation within NER of India to check the accuracy of satellite observation.

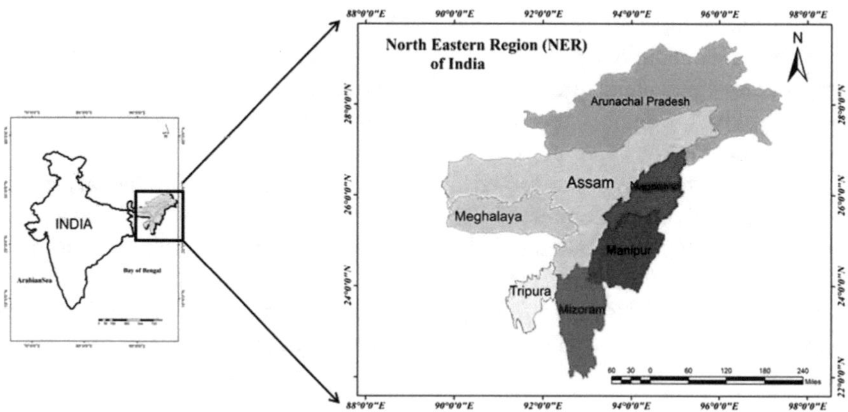

Fig. 1 Location map of the study area

2 Area of Study

The study area is the NER of India comprising the states of Assam, Meghalaya, Nagaland, Tripura, Arunachal Pradesh, Manipur, and Mizoram. It extends from 21°58′–29°30′ N latitude and 85°49′ E–96°30′ E longitude (Fig. 1). The region has distinctive climate disparities. The hasty transformation in landscape effects the climatic changes within short distances. Usually, the everyday temperature in the plains of Assam, Tripura, and in the western segment of Mizo Hills is about 15 °C in January, while in other parts of the region, it is between 10 and 15 °C. The upper stretches of Arunachal Pradesh experiences temperatures below freezing point. It is characterized by high rainfall brought about by monsoon.

Geographically, these states fall in the southeastern foothills of the mighty Himalayas. To the south lies the Bay of Bengal, responsible for bringing moisture rich air to the region. Also, the mighty Brahmaputra which runs through Assam influences the climate of the region.

3 Methodology

Hourly satellite images in TIR and WV channel from Kalpana-1 satellite for the months of April and May 2011 were taken for visual interpretation. The images were used for the identification of TS formations over the NER. The TIR data were analyzed, and many TS formations were identified over the region. Those were again studied in the WV images. The WV images helped in the differentiation of cumulonimbus clouds from cirrus clouds by the water vapor content. The paths followed by the TS were observed. The TS activity as reported by IMD during the period was also taken into account, and comparisons of both were made. A case study of one of the TS formed on May 27, 2011 was also carried out. Its duration, life cycle, and direction of motion were studied. The RH, wind direction, rainfall, air temperature, atmospheric pressure, and wind speed data were also analyzed during the life cycle of the thunderstorm. These data were collected from ISRO's Automated Weather Station (AWS) portal MOSDAC.

A comparative study of the TS activity reported by Indian Meteorological Department (IMD) and those interpreted by satellite imageries was done. The primary aim was to check the precision and effectiveness of satellite image interpretation. IMD reports on TS activity from April 14–May 30, 2011 were compared with satellite image interpretation during the same period. The comparison was done only for two categories, viz., total number of thunderstorms reported and the time of genesis.

4 Results and Discussions

4.1 Interpretation from WV and TIR Images

It was observed that the NER experienced several TS events during the period. The preferred region of development of convective systems was Eastern and sub-Himalayan West Bengal, Bihar Plateau, Jharkhand, and neighboring Bangladesh. Among the northeastern states, southern states such as Tripura, Mizoram, Meghalaya, and Assam saw extra development of thunderstorm systems. It was primarily due to greater insolation and a rich supply of moisture from the Bay of Bengal and the presence of the mighty Brahmaputra River. Most of the TS were formed locally and dissipated over the region, while some systems were formed over the northeastern parts of Bihar, Jharkhand, or eastern West Bengal and moved toward the NER under the influence of Nor'westers, collecting moisture on its way, and then maturing and dissipating over the NER [15].

The satellite images also give an idea about the wind pattern over the region throughout the pre-monsoon season. The dominant wind was from the northwest. Winds were also observed from the southeast. Sometimes, winds were also observed from the southwest. These winds bring in lots of moisture from the Bay of Bengal, which can help in the formation of large-scale TS under favorable conditions. Instances of merging of several independent systems and becoming a larger one, a "multi-cell cluster" on a few occasions were also observed [15]. There was also an instance of a TS developing into a depression, owing to favorable conditions on May 18, 2011. It moved south and lasted for 3 days before dissipating over coastal West Bengal and Orissa. The path followed by the system was independent of the normal wind pattern. As can be seen from Fig. 2, the TIR and WV images can be effectively used to identify the TS formation.

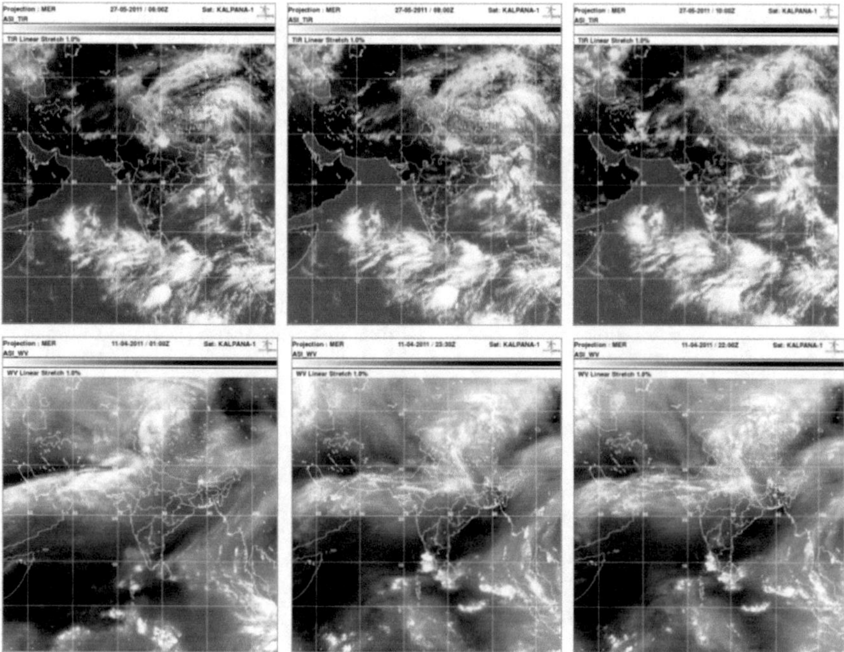

Fig. 2 TIR and WV images from Kalpana-1 satellite. Formation stage (left), maturation stage (middle), and dissipation stage (right)

4.2 Frequency and Distribution

In all, 81 systems were interpreted. Of these, 38 were formed in the month of April while 43 in the month of May. The maximum number of TS observed in a single day was 4. It was seen that some days were full of TS activities, while on some days, as many as four TS were formed. April had greater number of days without any such event. Of the 38 systems observed in the month of April, 4 were formed on April 13. 12 days saw one such activity. 4 days recorded three TS each, and 5 days had two such formations. In the month of May, two such systems were formed on 9 days, while 4 days saw three systems forming over the region on the same day. 13 days saw the formation of one system, while on 5 days there were no such TS (Fig. 3).

4.3 Duration

Duration of thunderstorms formed in April–May 2011 has been represented well in Fig. 4. Of the 81 systems interpreted, 26 systems developed, matured, and dissipated within an hour. 17 of them lasted up to 2 h. 11 systems lasted for 3 h, 9 systems lasted

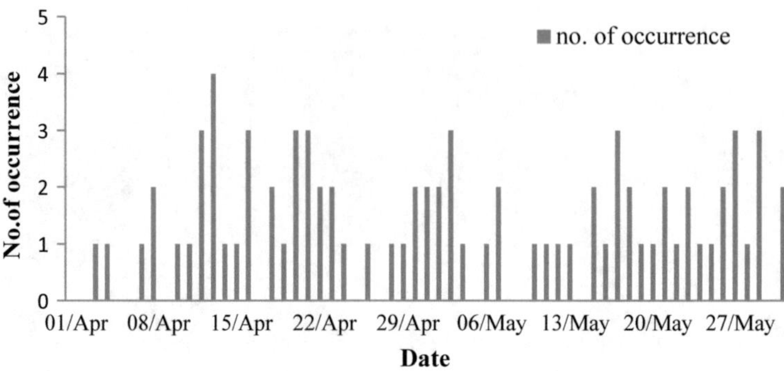

Fig. 3 Frequency of TS over NER during April–May 2011

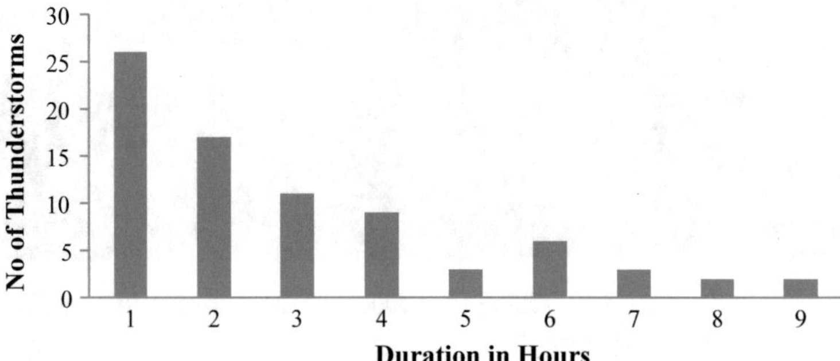

Fig. 4 Duration of TS formed in April–May 2011

4 h or less, 3 systems for 5 h or less, 6 systems for 6 h or less, 3 systems for 7 h or less, and 2 systems for 8 and 9 h each or less. While the duration of the systems is not fixed, yet the findings are in conformity of [15].

4.4 *Time of Genesis*

From the images, the time of genesis of TS was also plotted. The time domain was divided into six categories: 0–4 UTC, 4–8 UTC, 8–12 UTC, 12–16 UTC, 16–20 UTC, and 20–24 UTC. Figure 5 shows the number of formation of TS during the said time domain. From the figure, it can be concluded that maximum number of TS is formed between 16 and 20 UTC, i.e., 21:30–01:30 IST.

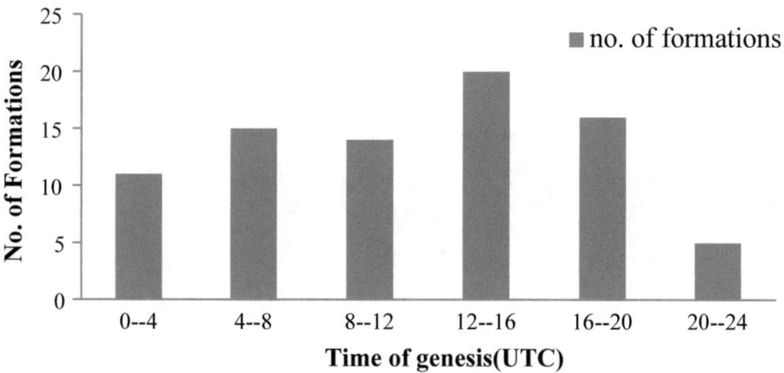

Fig. 5 Genesis of TS within different time domains

4.5 *Comparison with IMD Reports*

A comparison of the TS reported by IMD with those interpreted with the help of satellite images was done. Major differences were seen. All the TS activity interpreted by the help of TIR and WV data was plotted (Figs. 6, 7, 8 and 9) to find out the frequency of its occurrence over the region during the months of April and May. Total number of thunderstorms has been assessed (Fig. 6) and it was seen that, during the period, April 14–May 30, 2011, IMD reported 121 TS-related activities, while the total number of such formations observed in satellite images was only 65 for the period which is 67% of the events reported by IMD. IMD reported 40 such activities in the month of April and 81 in the month of May. While based on satellite images, only 24 (60%) could be interpreted in the month of April and 41 (50.6%) in the month of May (Fig. 6). It could be because of the spatial resolution of the satellite sensor used. Only large systems could be tracked by satellite images, while IMD uses a network of Doppler radars, and hence smaller events are also recorded.

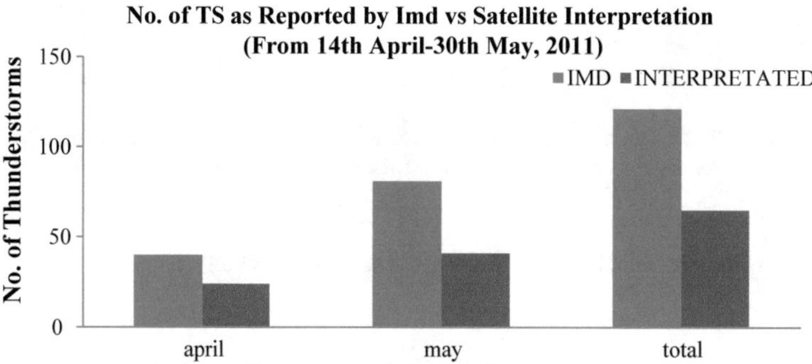

Fig. 6 No. of TS as reported by IMD versus satellite image interpreted

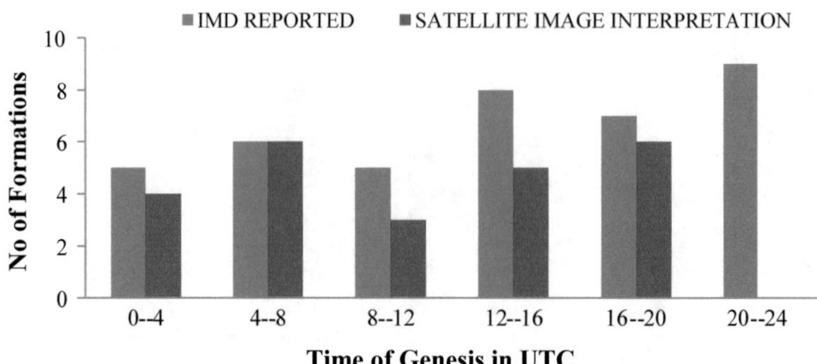

Fig. 7 Time of genesis (UTC) of TS as reported by IMD against satellite interpretation for the month of April 2011

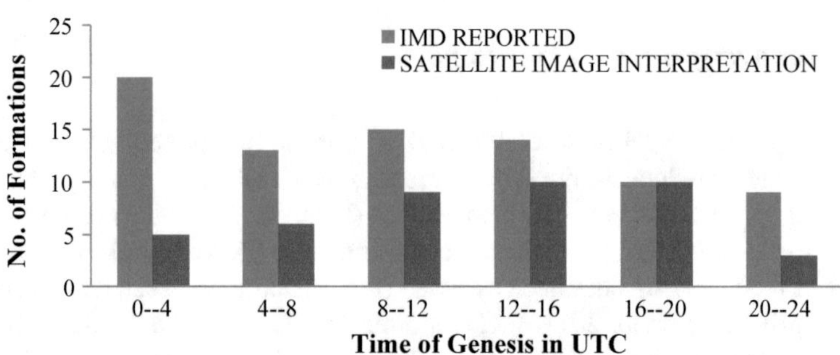

Fig. 8 Time of genesis (UTC) of TS as reported by IMD against satellite interpretation for the month of May 2011

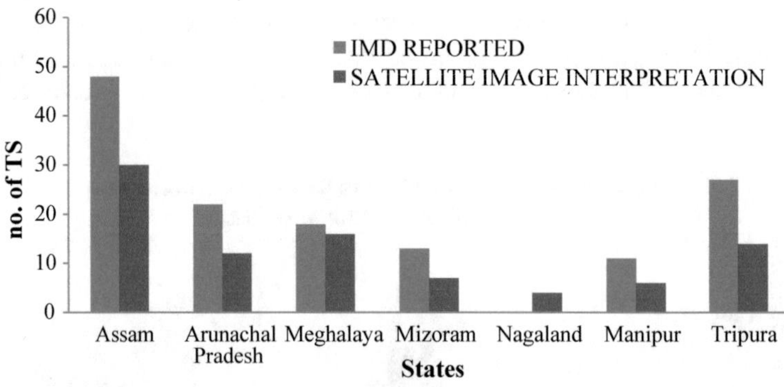

Fig. 9 TS in different states based on IMD reports and satellite interpretation

A comparison of the time of genesis (Figs. 7, 8) during different time domains as reported by IMD against those interpreted by the help of satellite images was done. For the month of April, IMD reported 9, 7, 8, 5, 6, and 5 such formations between

20–24 UTC, 16–20 UTC, 12–16 UTC, 08–12 UTC, 04–08 UTC, and 00–04 UTC, respectively, while the those interpreted during the same time period were 0, 6, 5, 3, 6, and 4, respectively. In the month of May, IMD reported 20, 13, 15, 14, 10, and 9 such formations between 00–04 UTC, 16–20 UTC, 08–12 UTC, 12–16 UTC, 16–20 UTC, and 20–24 UTC, respectively. In comparison 5, 6, 9, 10, 10, and 3 such formations were observed by satellite images.

The TS formation was differentiated state-wise. It was compared with IMD report for which also the same was done. As shown in Fig. 9, it can be summarized that the state of Assam was most affected by TS than any other states as per the IMD report. It was followed by Tripura, Arunachal Pradesh, Meghalaya, Mizoram, Manipur, and Nagaland, respectively.

According to the satellite images, Assam saw the formation of maximum TS, followed by Meghalaya, Tripura, Arunachal Pradesh, Mizoram, Manipur, and Nagaland, respectively. It could be because of the topography of Assam. It has more plains than other states, which allows for more insolation leading to increase in temperature. Also, two rivers, The Brahmaputra and Barak, play a big role in moisture supply in the region. The heated air is raised up by hill slopes that are found on all sides of Assam. The total number exceeds that of previous totals (considering each TS as unique) because some TS were reported in two or more states simultaneously.

4.6 Track

The tracks followed by the TS systems were mainly in three directions. It was deduced from Kalpana-1 satellite images. They are

- Northwest to Southeast,
- West to East, and
- Southeast to northeast.

Of these, northwest to southeast was the dominant track as most of the systems followed it, while some moved west to east or from southeast to northeast.

5 Conclusions

The TIR and WV image data can be used for observation of TS. Most of the TS were formed locally and dissipated over the region, while some systems were formed over the northwestern parts of West Bengal, Jharkhand, and Bihar and move toward the NER under the influence of Nor'westers, collecting moisture on its way, and then maturing and dissipating over the NER. The images also give an idea about the wind pattern over the region. The dominant wind was from the northwest. Winds were also observed from the southeast and southwest. Totally, 81 systems were interpreted. Of these, 38 were formed in the month of April, while 43 in the month of May. The frequency of TS formation was more in May. Of the 81 systems interpreted, 26 systems developed, matured, and dissipated within an hour. 17 of them lasted up to 2 h, while others lasted more than that. Maximum number of TS was formed between 16 and 20 UTC, i.e., 21:30–01:30 IST.

There were differences between the IMD reports and satellite image interpretation. During the period, April 14–May 30, 2011, IMD reported 121 TS-related activities, while the total number of such formations observed in satellite images was only 65 for the period. As per IMD report, maximum (25) number of TS were formed between 00:00 and 04:00 UTC, while the interpreted record shows that most (16) were formed between 16:00 and 20:00 UTC. The state of Assam was most affected by TS than any other states. Other states with notable formations are Meghalaya, Arunachal Pradesh, and Tripura. The differences between the IMD reports and satellite image interpretation are due to satellite resolution which cannot aid in observing small systems and wind gusts. Thus, only satellite observation is not enough to track thunderstorm activity. The tracks followed by the TS systems were mainly in three directions. They are northwest to southeast, West to East, and southeast to northeast.

Thus, it can be concluded that although coarse resolution satellite imageries can be used to track and monitor larger thunderstorm systems, yet it is not helpful in the detection of smaller systems. This calls for a more regular network of satellites and radar systems to act in unison for the exact forecasting of the storms.

References

1. Singh, S.: Physical Geography, Prayag Pustak Bhawan, pp. 521–528 (2009)
2. Basu, G.C., Mondal, D.K.: A forecasting aspect of thunder squall over Calcutta and its parameterization during premonsoon season. Mausam **53**, 271–280 (2002)
3. Williams, E.R., Mustak, V., Rosenfeld, D., Goodman, S., Boccippio, D.: Thermodynamic conditions favorable to superlative thunderstorm updraft, mixed phase microphysics and lightning flash rate. AtmosRes **76**, 288–306 (2005)
4. Purdom, J.F.W.: Satellite imagery and severe weather warnings. In: Preprints 7th Conference Severe Local Storms. Amer. Meteor. Soc., Kansas City, pp. 120–137 (1971)
5. Maddox, R.A., Hoxit, L.R., Chappell, C.F.: A study of tornadic thunderstorm interactions with thermal boundaries. MWR **108**, 322–336 (1980)
6. Anderson, C.E.: Anvil outflow patterns as indicators of Tornadic thunderstorms. In: Preprints 11th Conference Severe Local Storms. Amer. Meteor. Soc., Kansas City, pp. 481–485 (1979)
7. Adler, R.F., Fenn, D.D. Moore, D.A.: Spiral feature observed at top of rotating thunderstorm. Mon. Wea. Rev. **109**, 1124–1129 (1981)
8. Mccann, D.W.: The enhanced-V: a satellite observable severe storm signature. Mon. Wea. Rev. **111**, 888–894 (1983)
9. Heymsfield, G.M., Blackmer, R.H., Schotz, S.: Upper-level structure of Oklahoma Tornadic storms on 2 May 1979. I-Radar and satellite observations. J. Atmos. Sci. **40**, 1740–1755 (1983)
10. Heymsfield, G.M., Blackmer, R.H.: Satellite-observed characteristics of Midwest severe thunderstorm anvils. Mon. Wea. Rev. **116**, 2200–2223 (1988)
11. Roberts, R.D., Rutledge, S.: Nowcasting storm initiation and growth using GOES-8 and WSR-88D data. Weather Forecast. **18**, 562–584 (2003)

12. Sharma, G.N.: Cumulonimbus tops around Gauhatiairport. Ind. J. Met. Hydrol. Geophys. **29**, 705–716 (1978)
13. Sohoni, V.V.: Thunderstorms of Calcutta 1900–1926. IMD Scientific Note No. **3**(1), (1928)
14. Blanford, H.F.: Dirunal period of rainfall at Calcutta. Ind. Meteor. Memoirs **IV**, 39–46 (1886)
15. Litta, A.J., Mohanty, U.C.: Simulation of a severe thunderstorm event during the field experiment of STORM programme 2006, using WRF-NMM model. Current Sci. **95**, 204–215 (2008)

A Novel Approach to Spam Filtering Using Semantic Based Naive Bayesian Classifier in Text Analytics

Ritik Mallik and Abhaya Kumar Sahoo

Abstract Nowadays, big data analytics has taken a major role in different fields. In this way, text analytics is a challenging field which is a part of big data analytics. It is the process of converting understand text data into meaningful data for analysis to provide sentiment analysis. The main role of text analytics is to extract semantic meaning from contents of different text and to classify this text. We use text analytics in the field of spam filtering so that we can analyze text which is incorporated within spam mails. In the Internet era, as the number of email users is increased in exponential rate, spam filtering is a major issue for text classification. We propose semantic based Naïve Bayesian classification algorithm in spam filtering. We use R language which is open source and statistical analysis language, by which we can analyze spam mails for text analytics by calculating the accuracy of different subjects of spam mail.

Keywords Semantic based text analytics · Text classification · Spam filtering R language · Naïve bayesian classifier

1 Introduction

The objective of this literature review is to summarize the concepts of text analytics of big data while diving into the different tools and techniques used for analyzing text.

This literature review explains properly the concept of big data analytics as well as the types of analytics that can be done on big data. This review also sheds light on text analytics on big data. The various approaches to text analytics, the tools, and

R. Mallik (✉) · A. K. Sahoo (✉)
School of Computer Engineering, Kalinga Institute of Industrial Technology, Deemed to Be University, Bhubaneswar 751024, India
e-mail: ritikmallik@gmail.com

A. K. Sahoo
e-mail: abhayakumarsahoo2012@gmail.com

A. Abraham et al. (eds.), *Emerging Technologies in Data Mining and Information Security*, Advances in Intelligent Systems and Computing 813,
https://doi.org/10.1007/978-981-13-1498-8_27

techniques required to handle the processing of text are introduced and explained. A preview of the implementation of sample analysis performed is also contained in this review.

1.1 Big Data Analytics

Big data analytics is the process of examining large data sets to uncover hidden patterns, unknown correlations, market trends, customer preferences, and other useful business information. The analytical findings can lead to more effective marketing, new revenue opportunities, better customer service, improved operational efficiency, competitive advantages over rival organizations, and other business benefits [1].

The primary goal of big data analytics is to help companies make more informed business decisions by enabling data scientists predictive modelers and other analytics professionals to analyze large volumes of transaction data, as well as other forms of data that may be untapped by conventional business intelligence programs.

1.2 Diving into Text Analytics

The application of the concept of text analytics on big data is also known as text mining which is used to derive useful information from text. The useful information is usually derived by analyzing hidden patterns through statistical pattern learning. Text mining refers to collecting structured text as input, extracting hidden patterns from structured text, evaluating patterns, and interpreting the output. Text analytics usually involves text identification, text mining, text categorization, text clustering, entity modeling, link analysis, sentimental analysis, and visualization. The main objective is to convert text into data which requires analysis by the help of natural language processing concept and analytical methods [2].

In this paper, Sect. 2 explains about literature survey on big data analytics and different phases of text analytics. Section 3 tells about different phases of text analytics along with spam filtering process and Naive Bayesian classifier. Section 4 explains the proposed method which behaves as the combination of semantic analysis with Bayesian classification method. Section 5 shows experimental result analysis. Section 6 concludes the paper with future work.

2 Literature Survey

2.1 Big Data Analytics

Big data analytics is the ability to manage huge amount of dissimilar data (i.e., structured data, unstructured and semi-structured data) at a decent speed and within the required time frame to promote actionable decisions in real time.

Big data analytics can also be interpreted as those tools and datasets whose size is beyond the capability of conventional data acquisition and data management tools and techniques (ex. RDBMS, spreadsheets, etc.) to capture, store, analyze, and manage.

The purpose of big data analytics is to generate meaningful insight/value from otherwise raw data sets. The data size is interpreted as the potential to generate unparalleled value.

2.2 Characteristics of Big Data

Big data is not all about size. It has other dimensions by which it is evaluated. These are called as the 5Vs of big data. They include [2, 3]:

Volume: It implies the sheer size of the data set. Big data ranges between terabytes and yottabytes of data. The volume of the data is affected by the *time* and the *type* of data. What is considered as big data today might not meet the threshold of big data in the future as storage capacities might have been enhanced. The type of data depends on the industry in which it is being used.

Velocity: This is the rate at which data are being generated and analyzed in real time. Various types of data are streamed on a daily basis via electronic gadgets, text documents, etc. These data contain a plethora of information which can be used to generate some meaningful insights into the behavior of the individuals or machines using it. Such insights include the buying patterns/trends, demographics, etc.

Veracity: The trustworthiness of big data is being evaluated here. As big data does not originate from only one source, there might be an aura of unreliability present in one or more of these sources as these data tend to be imprecise and uncertain. Data based on situations that require human judgment are placed under this dimension, such as customer or user behavior on social media sites, etc.

Variety: This implies the structural differences in the data sets. The data could comprise unstructured, structured, semi-structured, or a combination of all. The heterogeneity of the data set is being evaluated here.

Value: The relevance of the insight generated is being evaluated here. Data in its raw/original form present very little value relative to its volume. A higher value can be obtained by analyzing larger volumes of data.

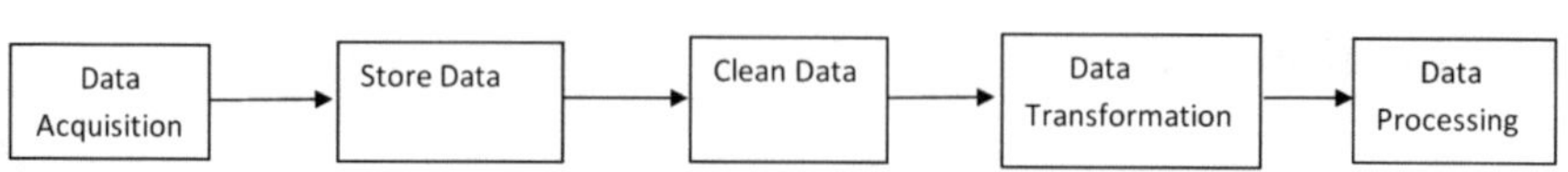

Fig. 1 Phases of analytics in big data

2.3 Phases of Analytics in Big Data

Big data analytics is done in the two following phases: Data acquisition, data storage, data cleaning, data transformation, and data processing [4] (Fig. 1).

3 Text Analytics

Text analytics is a subcategory under big data analytics which handles the extraction of information and facts from the unstructured data.

Text analytics is an emerging field for information recovery, statistics. Text mining can help us to obtain possibly valuable industry insights from text-based content.

Text analytics can also be interpreted as a gentle foray into the processing of text documents, typically unstructured data with the aim of generating valuable knowledge in a structured format [1]. The proximity, relationship, and association of the words within the text are examined by applying different tools and techniques. Examples of textual data include social network feeds, tweets, blogs, log files, etc. Text analytics is in some quarters perceived to be knowledge discovery from text. Text analytics is an integral part of any big data solution as 80% of the data available is in unstructured form (Fig. 2).

Text Mining Approaches: Certain approaches or techniques are utilized for text analytics [5]. They include:

Natural Language Processing: This technique makes use of linguistic concepts such as parts of speech and grammatical structures. The rationale behind this technique is to identify the *who, what, where, how, why*, etc., within the context. NLP is generally used for content analysis.

Using R: R is used to perform statistical analysis on text of various dimensions to ascertain the word/term frequency, term proximity, and document length, generate word clouds, plot statistical graphs, etc.

Sentiment Analysis: This involves the extraction of opinions, emotions, and sentiments from unstructured data. Opinion mining is used to track attitudes, feelings, etc. The polarity of the opinions is determined. Different techniques like natural language processing and text analytics are used to derive valuable information from source materials. Sentiment analysis refers to determining the attitude of speaker or writer with respect to any topic by measuring polarity of text document. The attitude of a person may be his or her evaluation or emotional communication. This sentiment may be neutral or positive or negative.

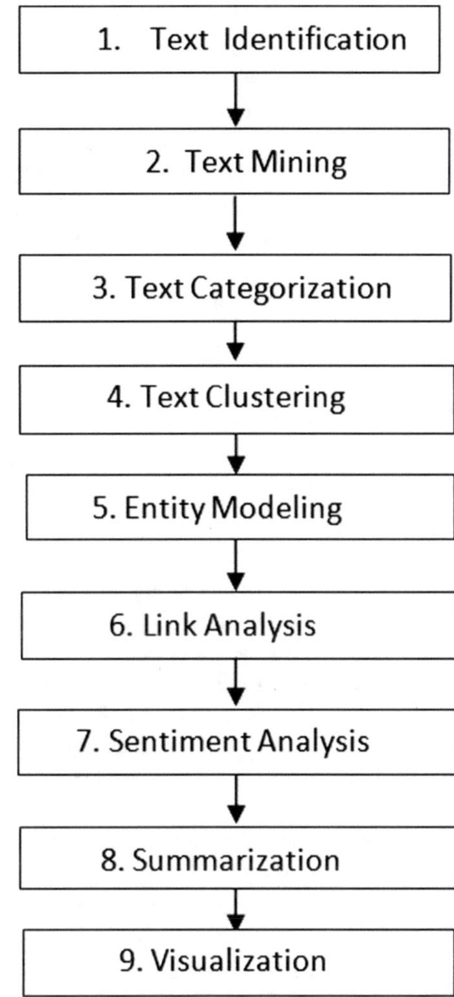

Fig. 2 Flow diagram of text analytics

Keyword/Term Based Approach: Terms that best describe a subject or text file are automatically identified. Key phrases or key words are used to represent the most relevant information contained in a document.

Information Extraction Approach: Structured data are extracted from unstructured text materials via the use of algorithms. There are two prominent subtasks in this approach.

Entity Recognizer: This subtask finds the names (entities) in text and classifies them into predefined categories, such as person, location, date, etc.

Relation Extraction: The semantic relationship among these entities is discovered. A good example could be: person "A" works at location "XYZ". The association between these two entities has been uncovered.

Text Summarization: This technique provides a short summary of a single document or more than one document by a computer program.

Methods of text summarization include:

Extractive Summarization: This sub-method requires the analyst to have a thorough understanding of the entire text and string along the salient units. The importance of the text units is evaluated based on their location within the text and frequency of occurrence. The text being analyzed is a subset of the original text.

Abstractive Summarization: The semantics are extracted from the text. The evaluation involves text units that may not be present within the original text document. The summary is generated by parsing the original text via NLP tools.

3.1 Text Analytics in Spam Filtering

Almost about 70% of the business emails are usually spam, they not only consume the time of the users but also they take up space of mailbox which causes wastage of network bandwidth and it also engulfs very important personalized emails. There are generally two types of methods which are classified into two categories, they are static method and dynamic method [6]. Static method includes spam mail identification which is based on the predefined address list whereas dynamic method considers email content and applies spam filtering decision with respect to this content. The filtering technique is applied on the list of words and phrases which define spam messages. Text categorization and data mining techniques are used for spam filtering by implementing machine learning methods. Naïve Bayesian classification method is based on Bayesian classifier which is used in manual spam email filtering. Bayesian filter is considered with high positive precision and low negative recall [7].

Hu et al. proposed a Chinese spam filtering approach with text classification method based on the semantic information and identified enormous potentiality in spam filtering with multiple classes and less feature terms. They used decision tree classification method in Chinese spam filtering [6]. Jialin Ma et al. proposed a message topic model (MTM) based on the probability theory of latent semantic analysis that uses latent Dirichlet allocation method [8] (Fig. 3).

3.2 Bayesian Classifier

Naïve Bayesian classification is based on Bayes' theorem which uses the conditional probability concept. This classification method is used for spam filtering which does not require preset rules and analysis of spam content. If computed conditional probability values are greater than the threshold value, then the message is classified as spam.

$$P(A|B) = [P(B|A) \cdot P(A)]/P(B) \tag{1}$$

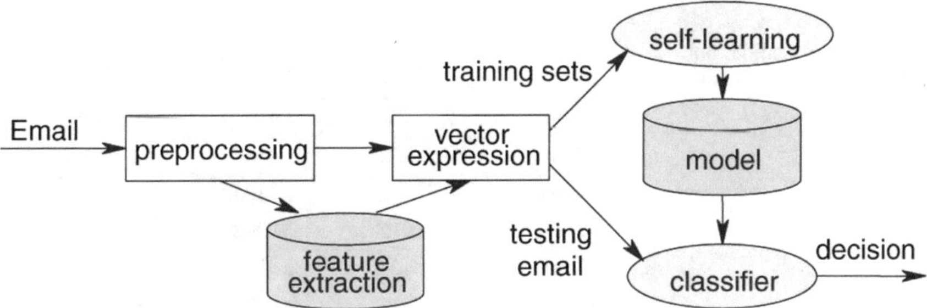

Fig. 3 The process of spam filtering

- P(*A*) is the unconditional probability of A. This means A does not take any information about B. However, the event B need not occur after event A.
- P(*A*|*B*) is the conditional probability of *A*, given *B*. This is also called posterior probability because it depends on the specified value of B.
- P(*B*|*A*) is the conditional probability of *B*, given *A*. It is also called the likelihood.
- P(*B*) is the marginal probability of *B* and acts as a normalizing constant.

4 Proposed Method

Steps:
In our approach, we first extract semantic information from texts, named as feature terms. The decision tree is built up by using feature terms and selected by pruning. The annotations are added with preprocessed text in natural language processing. The semantic information are the key element of classification. After building up decision tree, we apply Naïve Bayesian classifier to classify spam mail.

i. Databases of spam and non-spam messages are created.
ii. Appearance rate for each independent word of each database is calculated.
iii. Hash tables are created for each database consisting of independent words.
iv. Probability of each word in the hash tables is calculated for each incoming mail.
v. Probability of spam message is calculated using Bayes' method.

If the probability value is higher than the threshold value, then the message is classified as spam.

Table 1 Accuracy by considering classification performance of each subject

Subject	P	R	F1
Invoicing	0.97812	0.93712	0.95718
Training	0.90728	0.86614	0.88623
Recruiting	0.92752	0.88854	0.90761
Eroticism	0.88892	0.83980	0.86366
Website	0.95912	0.90985	0.93383
Selling	0.90823	0.87560	0.89162
Letter	0.86987	0.81750	0.84287
Defrauding	0.89656	0.84960	0.87245

5 Experimental Result and Analysis

The key of this semantics based text classification approach is extracting the semantic meanings accurately and precisely. As a result, the processing of annotations generating is very important and takes many efforts.

There are two kinds of error in text classification:

Texts that belong to one class are classified to other class. (false positives)

Texts that not belong to one class are classified into this class. (false negatives) (Table 1)

$$P = \frac{correct}{(correct + spurious)} = \frac{tp}{(tp + fp)} \tag{2}$$

$$R = \frac{correct}{(correct + missing)} = \frac{tp}{(tp + fn)} \tag{3}$$

$$F1 = \frac{2PR}{(P + R)} \tag{4}$$

By analyzing the value of P, R and F1, we classified our spam mail into different subjects with accuracy of correctness. By using semantic based Bayesian classifier, we classified spam mail and found that the accuracy of each subject related to spam mail. As compared to the existing work, this proposed algorithm was found to have the best accuracy of the proposed classifier for spam filtering.

6 Conclusion and Future Work

In this paper, we proposed semantic based Naive Bayesian classification spam filtering approach with text classification method based on semantic information. Semantic notations on the words and sentences are attached to achieve text semantic information. Spam mails unnecessarily cause loss of network bandwidth valuable memory spaces. Spam mails also contain viruses which could also damage the system and personal data. The proposed method identifies unused and used subject of spam mail. In future, we can use different types of classifier based on good performance in text classification domain which can be required for security purposes in a large organization.

References

1. Packiam, RM., Prakash, V.S.J.: An empirical study on text analytics in big data. In: Computational Intelligence and Computing Research (ICCIC),Cauvery College For Women, Tiruchirapalli, India, pp. 1–4 (2015)
2. Gandomi, A., Haider, M.: Beyond the hype: Big data concepts, methods, and analytics. Int. J. Inf. Manag. **35**(2), 137–144 (2015)
3. Ishwarappa, A.J.: A brief introduction on big data 5Vs characteristics and hadoop technology. In: International Conference on Computer, Communication and Convergence (ICCC 2015), vol. 48, pp. 319–324 (2015)
4. Vashisht, P., Gupta, V.: Big data analytics techniques: a survey. In: 2015 International Conference on Green Computing and Internet of Things (ICGCIoT), pp. 264–268 (2015)
5. Agnihotri, D., Keshri, V., Priyanka, T.: Pattern and cluster mining on text data. In: 2014 Fourth International Conference on Communication Systems and Network Technologies, pp. 428–432 (2014)
6. Hu, W., Du, J., Xing, Y.: Spam filtering by semantics-based text classification. Adv. Comput. Intell. (2016)
7. Yu, B., Xu, Z.: A comparative study for content-based dynamic spam classification using four machine learning algorithms. Knowl.-Based Syst. Elsevier (2008)
8. Jialin, M., Zhang, Y., Liu, J., Yu, K., Wang, X.A.: Intelligent SMS spam filtering using topic model. In: 2016 International Conference on Intelligent Networking and Collaborative Systems, pp. 380–383 (2016)

A Proposed Approach for Improving Hadoop Performance for Handling Small Files

Arnab Karan, Siddharth Swarup Rautaray and Manjusha Pandey

Abstract As the world is getting digitized, the speed in which the amount of data is overflowing from different sources in different formats, and it is not possible for the traditional system to compute and analyze this kind of data called big data. To properly analyze and process big data, tool like Hadoop is used which is open source software. It stores and computes the data in a distributed environment. Big data is important as it plays a big part in making big benefits for today's business It captures and analyzes the wealth of information of a company and quickly converts it into actionable insights. However, when it comes to storing and accessing of huge amount of small files, a bottleneck problem arises in the name node of Hadoop; so in this work, we propose a method to efficiently optimize the name node working by eradicating the bottleneck problem arising due to massive small files.

Keywords Big data · Hadoop · Name node · Small files · Bottleneck problem

1 Introduction

Big data came into existence when a huge amount of data (structured, unstructured and semi-structured) were unable to be handled by relational computing systems. In big data, huge amount of data are analyzed and processed in a short period of time as compared to the traditional database system. Big data is going to change the world completely, and to understand the phenomenon what big data is we should know the V's of big data namely. Big data is important as it plays a big part in

A. Karan (✉) · S. S. Rautaray · M. Pandey
School of Computer Engineering, KIIT, Deemed to Be University,
Bhubaneswar 751024, Odisha, India
e-mail: arnab.east@gmail.com

S. S. Rautaray
e-mail: siddharthfcs@kiit.ac.in

M. Pandey
e-mail: manjushafcs@kiit.ac.in

A. Abraham et al. (eds.), *Emerging Technologies in Data Mining and Information Security*, Advances in Intelligent Systems and Computing 813,
https://doi.org/10.1007/978-981-13-1498-8_28

making big benefits for today's business and it captures and analyzes the wealth of information of a company [1] and quickly converts it into actionable insights. It has improved the business quality in the fields of life sciences and health care, retail market, financial services and banking services, manufacturing industry, etc. Hadoop is the most efficient tool used to process big data. The problem arises when there is processing of huge amount of small files by HDFS [2]. HDFS follow a master/slave architecture in node. When a large number of small files are processed, the files are stored in the data node and metadata are stored in the name node. The name node resides in the main memory, and the size of the main memory is not large enough. So when a large number of small files are stored, a large number of metadata of each file are also stored in the name node. This gives rise to the bottleneck problem.

1.1 HDFS Architecture

An HDFS architecture follows master–slave architecture in which the name node is considered to be master and data node is considered to be slave. The management of file system name space and storing of file system metadata are done by name node.

As shown in Fig. 1, this single name node manages multiple data nodes in a concurrent manner. The data node stores the data given by the client in the form of file in HDFS. As this file size is big, it is broken down into blocks and these blocks are replicated independently among the data nodes (Default replication factor of HDFS is three). This replication mechanism is used because if there is a failure of data node in which the file is present, it can be retrieved from other data node. This block is represented in the form of two files. The first file contains the original data and the second file contains the block's metadata. The metadata generated is stored in the name node which resides in the primary using the handshake mechanism between name node and data node. The communication between name node and data node happens with the help of a heartbeat mechanism. In heartbeat mechanism, the name node sends heartbeat message every 3 s to check on the of data nodes. If any heartbeat is not responded by the data node, then it will be considered dead by the name node.

2 Literature Survey

Table 1 describes the objective of the research work done on the optimization area of big data tools, and the methods used to do optimization such as some techniques which have been needed in this research work such as extended name node with cache support, shadoop, and the application domain in which these have been implemented.

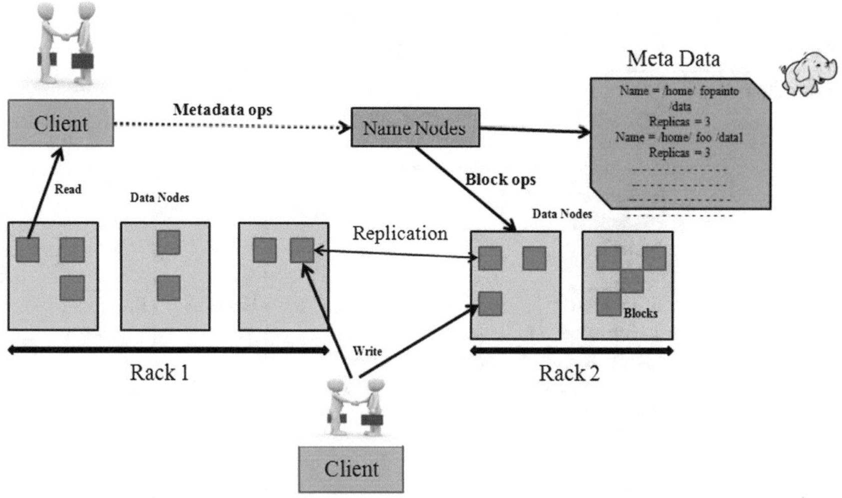

Fig. 1 HDFS architecture

3 Small Files Problem

The efficiency and performance of HDFS decrease drastically when a large amount of small files are stored in HDFS that result in serious access latency and high memory consumption [3]. The metadata is stored in the name node which resides in the main memory, and it is stored in the main memory for faster and efficient client service. For each block of file, the metadata is stored. When the file size is larger than the block size, the amount of metadata stored is justified. But when a large number of small files each less than the block size (default block size 64 mb) are stored, for each small file, a metadata [4] is created which in turn creates a large number of metadata thus putting a tremendous load on the name node. For example, assume that metadata in memory for each block of a file takes up about 150 bytes. Consequently, for a 1 GB file, divided into 16 64 MB blocks, 2.4 KB of metadata is stored. Whereas, for 10,500 files of size 100 KB each (total 1 GB), about 1.5 MB metadata is stored. This example shows that huge amount of small files takes a significant amount of space in name node and main memory. Again this huge amount of small files cause bottleneck problem while accessing these files in name node. This prevents optimal usage of HDFS for various applications.

3.1 Bottleneck Problem

The reasons for the bottleneck problem [5] that arises in the name node are:

Table 1 Survey on different approaches used in previous research work

Author	Objective	Technique/algorithm	Tools	Application
Xiayu hua et al.	Enhancing throughput of the Hadoop distributed file system for interaction-intensive tasks	Extended name node with cache support and storage allocation algorithm	Hadoop	File optimization
Peter p et al.	To increase performance and energy efficiency by obtaining the exact number of task for any workload running in Hadoop	Job profiling method for optimal resource provisioning for any MapReduce workload	Hadoop	Enterprise
PEI Shu-jen et al.	To increase the efficiency of the MapReduce by improving the scheduler	Task locality Improvement scheduler	Hadoop	Enterprise
Rong Gu et al.	Optimizing the underlying tasks and job execution and thereby reducing the time cost of each individual job and its tasks	Shadoop	Hadoop	MapReduce optimization

- A large amount of main memory space will be occupied by storing an enormous amount of metadata. Each metadata can occupy up to 150 bytes in the name node memory.
- The small files accessing speed also decrease as the load on name node increases while accessing [6] and maintaining large number of small files.

This work proposes an efficient method in which we can store and access massive small files of Hadoop without having the bottleneck problem. In the proposed approach, there will be a use of metadata table which will be residing on the secondary memory instead of main memory, at the same time, we will be performing a balanced binary search tree (AVL Tree) data structure on records/address stored on metadata table for faster access.

3.2 Research Gap

Hadoop uses a HDFS architecture, it works in a master–slave pattern where a name node is the master and many data nodes are the slave nodes. HDFS is designed in a way that there is a single and can have thousands of data node and ten thousands of client per cluster. The entire metadata [4] is stored in the main memory for efficient servicing and faster access. For each block of file, metadata is stored. Storing the metadata of larger files in name node whose size is greater than 64 mb (default block size used in Hadoop) is justified.

4 Proposed Framework

In general, metadata is stored in primary memory and the storage space of main memory is not vast. One of the functions of name node is it distributes the workload between the data nodes, and the information about the data or the data about the data (metadata) is stored in the name node. The problem arises when a vast number of small files are stored and processed by data nodes, for each file, a metadata is stored in a name node and a vast number of metadata are stored in name node and the master node has the responsibility [7] to distribute, keep track, and process data in data node. When the volume of metadata increases, the name node has to go through the bottleneck problem.

In this work, concept of metadata table is used. Instead of storing the metadata of small files in the name node which resides in the primary memory [8], it will be stored in the secondary memory which are nowadays enormous in memory capacity.

4.1 Advantages of Proposed Framework

In the previously used methods, where co-related small files are combined to form a large file and then that large file was processed. Combining of co-related files have certain disadvantage, if the small files are not co-related then it is not possible to combine them, hence giving rise to a small files are not co-related then it is not possible to combine them hence giving rise to small files problems again. Section 4.1 describes the proposed method used in this small files whether it is co-related or not it will be stored in secondary memory thus giving a ample amount of space for the name node to run efficiently in primary memory.

Another advantage is it helps in resolving in single point of failure (SPOF) in small files. If the name node containing the metadata crashes or becomes dysfunctional by any means, the metadata of small files is still safe as these are stored in the secondary memory.

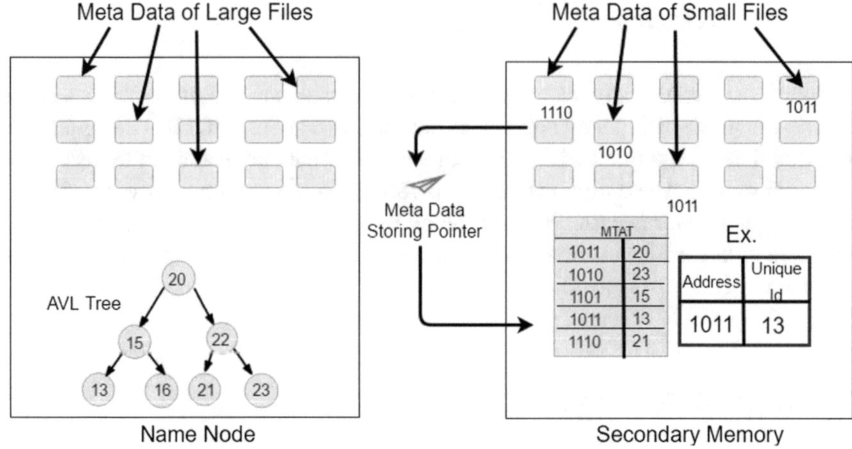

Fig. 2 Proposed model for diminishing the small file problem

As shown in Fig. 2, each metadata will be stored in some address space of secondary memory. A metadata address table(MDAT) will be maintained in the secondary [9] memory itself which maintains the address space of the stored metadata. By this, no metadata will be lost.

Algorithm 1: Meta Data storing Algorithim
Input:- S(f)-> Small Files
f(size)->File size
Step 1:- if (f(size)<threshold)
// Threshold value will be decided by the database Administrator
S_m -> $M(S_{(f)})$
/* Store the Meta Data of the small file in Secondary memory
S_m -> Secondary Memory
$M(S_{(f)})$ -> Meta Data of small files */
Else
N_N->$M(L_{(f)})$
// It is considered as large file and the Meta Data of large file $M(L_{(f)})$ will be stored in the Name Node(N_N)
Step 2:- S_P-> &Adr($M(S_f)$)
Store pointer (S_P) is used to refer to the address of the Meta Data of small file Adr($M(S_f)$) which reside in secondary memory.
S_P -> &MDAT
The address and unique id of small files that are stored in secondary memory are maintained in a table structure known as Meta Data Address Table (MDAT). MDAT resides in the secondary memory. The small file address is stored in the MDAT with the help of a pointer known as Meta data Store Pointer.

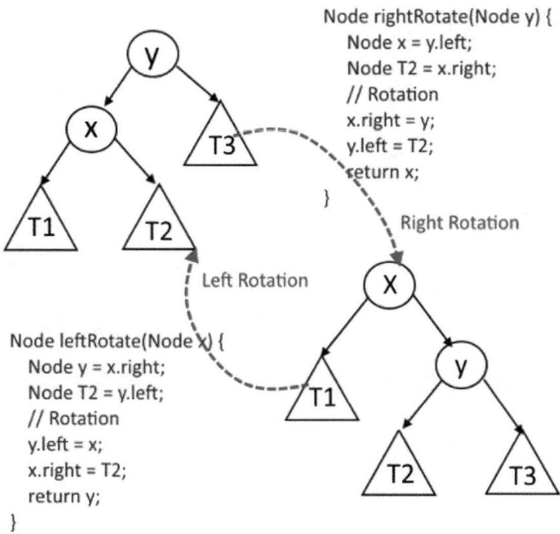
Node rightRotate(Node y) {
Node x = y.left;
Node T2 = x.right;
// Rotation
x.right = y;
y.left = T2;
return x;
}
Right Rotation
Left Rotation
Node leftRotate(Node x) {
Node y = x.right;
Node T2 = y.left;
// Rotation
y.left = x;
x.right = T2;
return y;
}
y
x
T1
T2
T3
X
y
T1
T2
T3

Fig. 3 AVL rotation

The address in the metadata table is stored in the form of binary balanced search tree in this work and AVL tree is used which will be stored in the main memory which makes it easier and faster at the time of accessing.

Algorithm 2:

An AVL tree data structure is designed on the basis of the file unique id present the metadata address table. The node in the AVL tree contains the file unique id and address of the metadata, when the client wants to access a file it directly goes to AVL tree after that the access pointer points it to the address of metadata present in the secondary memory. AVL tree is used so that the file can be accessed in O(log n) time. An AVL tree data structure is designed on the basis of the file unique id present in the metadata address (Fig. 3).

Then there will be four possibilities

1. **Left–Left Case**: x is the left child of y and y is the left child of z.

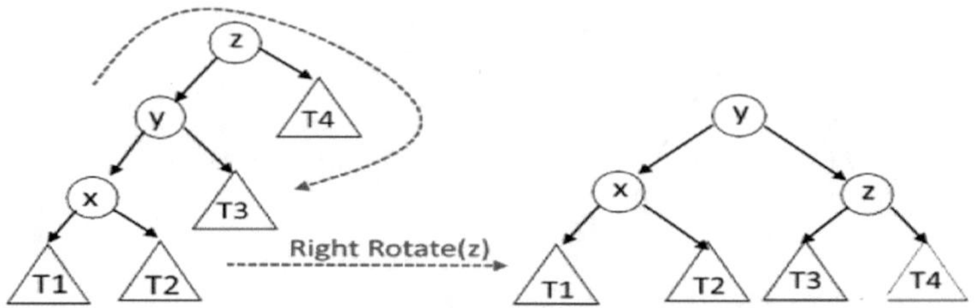

2. **Left–Right Case**: x is the right child of y and y is the left child of z.

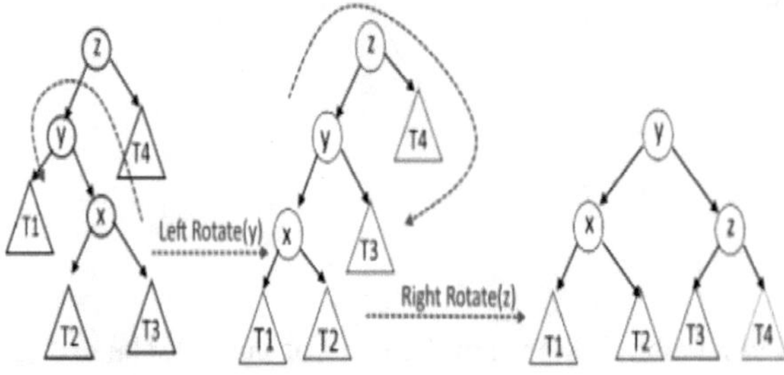

3. **Right–Left Case**: x is the left child of y and y is the right child of z.

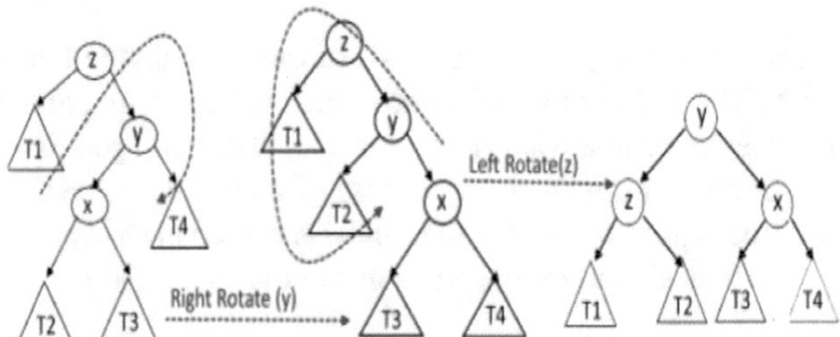

4. **Right–Right Case**: x is the right child of y and y is the right child of z.

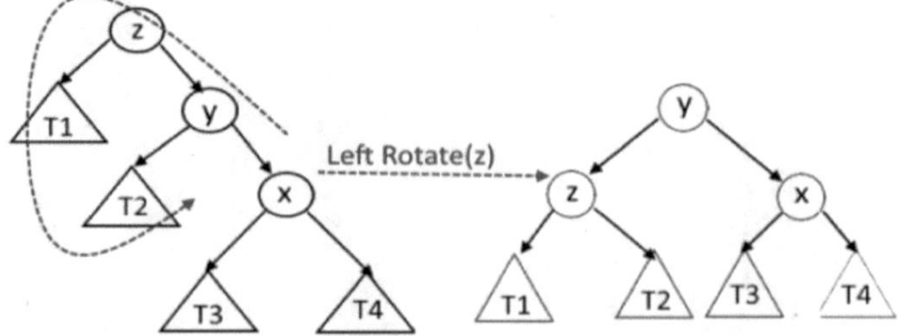

5 Conclusion

Hadoop file system is used to process large files very efficiently, but when it comes to small files, the computation power and efficiency of the HDFS decrease drastically. This happens because a large number of small files produce a large number of metadata which resides in the primary memory which accounts for the bottleneck problem of name node. This paper proposes a model in which instead of storing the metadata in the name node, it is stored in the secondary memory. A metadata address table (MDAT) is maintained in the secondary memory which contains the address of metadata present in secondary memory. A binary balanced search tree (AVL tree) is designed in name node which contains the address from the metadata address table (MDAT). This binary tree structure is constructed so that accessing if metadata will be faster and more efficient as we can access a metadata in O(log n) time.

References

1. Dong, B., et al.: A novel approach to improving the efficiency of storing and accessing small files on hadoop: a case study by powerpoint files. In: 2010 IEEE International Conference on Services Computing (SCC). IEEE (2010)
2. Jena, B., et al.: A survey work on optimization techniques utilizing map reduce framework in hadoop cluster. Int. J. Intell. Syst. Appl. **9**(4), 61 (2017)
3. Mackey, G., Saba S., Jun W.: Improving metadata management for small files in HDFS. In: 2009 CLUSTER'09. IEEE International Conference on Cluster Computing and Workshops. IEEE (2009)
4. Zhang, Y., et al.: Small files storing and computing optimization in Hadoop parallel rendering. Concurr. Comput. Pract. Exp. **29**(20) (2017)
5. Shu-Jun, P., Xi-Min, Z., Da-Ming, H., Shu-Hui, L., Yuan-Xu, Z. Optimization and research of hadoop platform based on _fo scheduler. In: 2015 Seventh International Conference on Measuring Technology and Mechatronics Automation (ICMTMA). IEEE, pp. 727–730 (2015)
6. Gupta, T., Handa, S.S.: An extended HDFS with an AVATAR NODE to handle both small files and to eliminate single point of failure. In: 2015 International Conference on Soft Computing Techniques and Implementations (ICSCTI). IEEE (2015)
7. Jena, Bibhudutta, et al.: Name node performance enlarging by aggregator based HADOOP framework. In: 2017 International Conference on I-SMAC (IoT in Social, Mobile, Analytics and Cloud) (I-SMAC). IEEE (2017)
8. Jena, B., et al.: Improvising name node performance by aggregator aided HADOOP framework. In: 2016 International Conference on Control, Instrumentation, Communication and Computational Technologies (ICCICCT). IEEE (2016)
9. Ganger, G.R., Kaashoek, M.F.: Embedded inodes and explicit grouping: exploiting disk bandwidth for small files. In: USENIX Annual Technical Conference (1997)

To Ameliorate Classification Accuracy Using Ensemble Vote Approach and Base Classifiers

Mudasir Ashraf, Majid Zaman and Muheet Ahmed

Abstract Ensemble methods in the realm of educational data mining are embryonic trends deployed with the intent, to ameliorate the classification accuracy of a classifier while predicting the performance of students. In this study, primarily miscellaneous mining classifiers were applied on our real academic dataset to foretell the performance of students and later on, ensemble vote (4) method was employed wherein the hybridization of predicted output was carried out, with majority vote as consolidation rule. The various classification techniques used in our study vis-à-vis Naive Bayes, k nearest neighbours, conjuctive rules and Hoeffding tree. The empirical results attained corroborate that there is a paramount significance in performance after application of ensemble vote (4) method. Furthermore, a novel attempt was made wherein, the pedagogical dataset was subjected to filtering process, viz., synthetic minority oversampling technique (SMOTE), with the intention to verify whether there is a considerable improvement in the output or not. The findings by and large have clearly confirmed that after application of SMOTE, the classifier achieved high accuracy of 98.30% in predicting the resultant class of students. Therefore, it calls upon the researchers to widen the canvas of literature by utilising the similar methods to unearth the different patterns hidden in datasets.

Keywords Data mining · Ensemble · K nearest neighbour · Naive bayes
Conjective rules · Hoeffding tree

M. Ashraf (✉) · M. Ahmed
Department of Computer Science, University of Kashmir, Srinagar 190006, J&K, India
e-mail: mudasir04@gmail.com

M. Ahmed
e-mail: ermuheet@gmail.com

M. Zaman
Directorate of IT & SS, University of Kashmir, Srinagar 190006, J&K, India
e-mail: zamanmajid@gmail.com

A. Abraham et al. (eds.), *Emerging Technologies in Data Mining and Information Security*, Advances in Intelligent Systems and Computing 813,
https://doi.org/10.1007/978-981-13-1498-8_29

1 Introduction

Data mining plays a predominant role in predicting the performance of students. There are numerous mining techniques such as regression, Naive Bayes, k means, neural networks, association rule, decision tree and so on and so forth that have been applied on various datasets pertaining to different areas to anticipate the performance of students. Using these methods, researchers can analyse, explore hidden information and obtain greater insights and potentials about data.

The principal focus of this paper is to predict the outcome performance of students by analyzing the various attributes of academic dataset associated to University of Kashmir. Primarily, base class classifiers such as Naive Bayes, k nearest neighbours, conjuctive rules and Hoeffding tree were independently applied on academic dataset to investigate the performance of each classifier and, subsequently performance of meta-classifier (ensemble vote) was also evaluated in contrast to base classifiers. Furthermore, from the past academic studies, it is quite evident that there has been significant work done in various domains of education data mining. Although, very few research studies have been executed using various ensemble techniques. Moreover, it becomes imperative for the researchers to use ensemble methods for predicting student performance. Therefore, in this study, a deliberate attempt has been made through the application of ensemble method such as vote (4). The ensemble vote (4) method has been used for classification of academic dataset associated with the University of Kashmir. This method combines four base classifiers namely Naive Bayes, KNN, conjuctive rules and Hoeffding tree, and their prediction output for improved classification of student results; thereby, comparing the classification accuracy, true positive rate (TP rate), false positive rate (FP rate), precision, error rate and other performance metrics, on ensemble vote (4) and individual classifiers while forecasting the performance of students.

2 Literature Review

Several comprehensive researches have been conceded that have highlighted the importance of data mining techniques in the realm of educational data mining. In this section, we will try to put forth an overview regarding the contribution of various researchers on the subject of academic data mining.

Data mining techniques have been exhaustively applied by researchers for predicting and improvising the performance of students. Dorina (2012) applied multiple techniques such as neural networks, KNN and decision tree on the educational dataset to forecast the performance of students with surpassing accuracy [1]. Among these three methods, neural network was found to be the best performer in disparity to other techniques. Ahmed and Elaraby (2014),used neuro-fuzzy method to identify students with low grades and significant empirical results were exhibited on the data [2]. Sonia and Laya (2015) employed a weighted modified algorithm to forecast the performance of students [3]. Kumar and pal (2011), investigated various attributes of dataset to identify candidates with greater potential and students who underperformed using Bayes classification [4].

Moreover, from the analytical assessment, viz., regression model applied by researchers namely Jothi and Venkatalakshmi (2014) made new strides and used clustering for examining and forecasting student's success rate [5]. Sheik and Gadage (2015) made some noteworthy efforts to observe the learning behavior of various models with the support of diverse open source tools to get an understanding of how these models evaluate, train and predict the performance of students [6].

Stafford et al. (2014) observed various wiki activity indicators and the final grades of the students [7]. The results demonstrated that there was considerable correspondence among the two factors and students who were engaged with wiki activities accomplished exceptional results. Junco et al. (2011) applied the analysis of variance (ANOVA) to determine the impact on learning outcome and student selection using twitter [8]. Giovanella et al. (2013) also examined effective collaboration of students in various social media applications such as blog, wiki, Delicious and Twitter as favorable learning performance indicators [9].

Suyal and Mohod (2014) examined students who required rigorous attention, by studying the association among various attributes using association rule mining [10]. Baderwah and pal (2011) used mining methods specifically decision tree on fields such as attendance, class test, assignment and semester marks for premature detection of students who are at menace [11]. A number of efforts have been put forth in this direction and a research faction comprising of Jeevalatha, Ananthi and Saravana (2016) applied decision tree on undergraduate students dataset covering a set of aspects such as higher secondary marks, communication skills and undergraduate marks for performance evaluation and placement assortment [12].

From prior review studies, it is further noticeable that the principal focus has been on forecasting the students performance based on several attributes visa-a-vis mid-term score, family income, attendance, assignment and so on while using various data mining techniques such as neural network, decision tree, Naive Bayes, association rule mining, SVM and so on. However, there has been substantial space in the sphere of studying the relationship between student's individual subjects and the overall outcome of a student using ensemble approach. Hitherto, there have been only self-effacing efforts to utilise ensemble vote method in EDM and there is tranquil deficiency of narrative/research demeanor in this background. Furthermore, novel methods are necessitated to be applied in the course of educational datasets to determine constructive and essential knowledge and details from educational backdrop.

3 Results and Discussion

Essentially in this section, we demonstrate empirical results obtained after applicability of SMOTE on real academic dataset collected from University of Kashmir. Subsequently, the dataset is analysed using classification techniques and ensemble

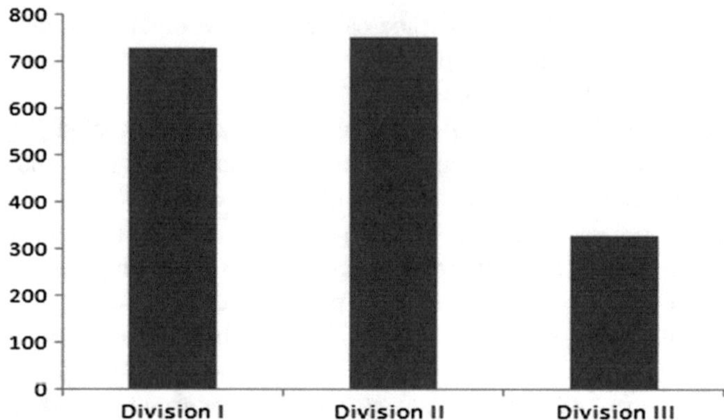

Fig. 1 Shows the imbalance in class distribution

vote (4) method. Eventually, performance of both classification and vote (4) ensemble method are then compared to estimate which methods can be significant and productive for students in making better predictions in the realm of educational data mining.

3.1 Balancing of Dataset

The dataset in our case has certain nonuniformity in classes as can be clearly seen in Fig. 1, wherein the percentage of students belonging to different classes is exemplified. Although, two classes, Division I and Division II, have a small variation in student's results. However, class Division III has a large discrepancy in output which can lead to biased outcome in determining the resultant class of a student. Therefore, as a matter of consequence, the possibility of students belonging to both classes Division I and Division II is by and large augmented and can guide towards the erroneous prediction of students.

Taking into deliberation the imbalance nature of the output class, viz., Overall_Grade, a supervised filtering technique namely SMOTE without replacement is applied, to ensure minimum biasing while training and testing the instances of a student on a particular technique. Therefore, this procedure can assist in predicting the right class of a student without being biased towards the majority class. After running SMOTE on a minority class, oversampled with 100 percent and with nearest neighbors of 5, across classification variables that includes three classes Division I, Division II and Division III, the size of instances is increased from 314 to 628. Figure 2 highlights the class distribution after application of SMOTE method.

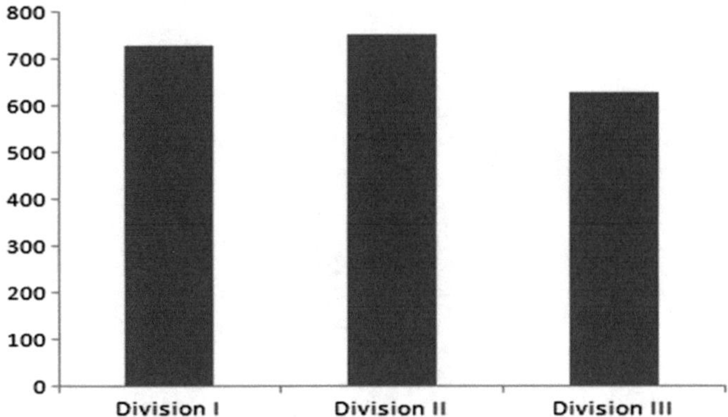

Fig. 2 Shows the balance in class distribution

3.2 Formulating Classifier's

One of the indispensable aims of this research is to examine various attributes of students through the application of mining techniques, viz., Naive Bayes, KNN, conjuctive rules and Hoeffding tree to make better advancements in the direction of forecasting student's success. Moreover, the experiments have been initially carried on individual classifiers and then later on contrasted with the ensemble vote (4) method. It was discovered that meta-classifier "ensemble vote (4)" with majority voting as consolidation rule, performed well with an accuracy of 98% on classifying students result. Furthermore, it was mainly done to obtain an insight for identifying which technique has shown prominent and remarkable performance on our data and results corroborate.

- Naive Bayes Classifier
  This classifier corresponds to Bayesian classifiers and various performance metrics associated with the classifier were taken into deliberation. The TP rate, FP rate and F-measure were recorded as 0.95, 0.03 and 0.95, respectively, which are documented in Table 1. The other parametric related with the classifier such as correctly classified, incorrectly classified, receiver operating characteristics (ROC) and other values demonstrated in Table 1 also witnessed significant results. Furthermore, the overall performance of this classifier with a 10-fold cross validation was found to be 95.50% which is largely satisfactory.
- K Nearest Neighbour
  This classification technique pertains to lazy category of classifiers. The classifier was run with a 10-fold cross validation on real academic dataset. The various performance measures corresponding to this specific technique including TP rate (0.91), FP rate (0.05) and F-measure (0.91) revealed that these values are noteworthy. Moreover, alternate parameters from Table 1 such as correctly classified (91.80%), incorrectly classified (8.18%), ROC (0.93) and so on and so forth,

Table 1 Exemplifies performance of diverse classifiers prior to filtering procedures

Classifier name	Correctly classified (%)	Incorrectly classified (%)	TP rate	FP rate	Precision	Recall	F-measure	ROC area	Rel. Abs. Err. (%)
Naïve Bayes	95.50	4.45	0.95	0.03	0.96	0.95	0.95	0.99	7.94
KNN	91.80	8.18	0.91	0.05	0.91	0.91	0.91	0.93	13.19
Conjuctive rule	82.50	17.49	0.82	0.12	0.70	0.82	0.75	0.89	39.22
Hoeffding tree	93.87	6.12	0.93	0.04	0.93	0.93	0.93	0.96	10.84

unveiled considerable results while operating with 10- fold cross validation. The all-encompassing results associated with the classifier were found equitable with an accuracy of 91.80%.

- ConjuctiveRules
  The third classifier in this study that we have applied on the current dataset belongs to another distinct category of classifiers videlicet rules. The measurements accomplished with this classifier after run on a 10-fold cross validation specifically TP rate (0.82), FP rate (0.12) and F-measure (0.75) were found to be realistic. Furthermore, Table 1 advocates that factors like correctly classified (82.50%), incorrectly classified (17.49) and F-measure (0.75) are also significant. Therefore, the overall performance achieved by the classifier in predicting the correct class of a student was found to be 82.50%.
- Hoeffding Tree
  The final classifier that we employed in the present study associates to class of trees. The performance metrics, viz., TP rate, FP rate and F-measure in this case are recorded as 0.93, 0.04 and 0.93, respectively. In addition, correctly classified (82.50%), incorrectly classified (17.49) and F-measure (0.75) associated with this classifier are conspicuous and are demonstrated in Table 1. Furthermore, this classifier achieved an accuracy of 93.87% in determining the actual class of a student.

After applying various classifiers on real academic dataset, while undertaking different attributes of dataset into revision, empirical results certified that Naive Bayes classifier performed outstanding against remaining classifiers with an accuracy of 95.50% in predicting the correct class of a student. Nevertheless, the classifier demonstrated minimum classification error of 4.4% in distinction to 8.18% (KNN), 17.49% (Conjuctive rules) and 6.12% (Hoeffding tree). Furthermore, it is worth noticing that various estimates displayed in Fig. 3 have been achieved without the application of SMOTE method.

The histograms furnished in Fig. 3 represent various performance metrics against different classification techniques. It is quite evident from Fig. 3 that conjuctive rules have rendered elementary results, and have been ranked lowest in assessment to other techniques. However, Naive Bayes on same evaluation estimates, viz., correctly classified, incorrectly classified, TP rate, FP rate, precision, recall and so on, has exhibited significant results in classifying the suitable category.

3.3 Classification After SMOTE

The dataset is subjected to filtering process to corroborate whether there has been a considerable transition after application of SMOTE to each classifier in predicting the final class of students. However, it was examined that there was substantial increase in predicting the conclusive class pertaining to student except conjuctive rule, wherein no transformation was replicated. The Navie Bayes classifier after passing through filtering process exhibited a significant rise in accuracy from 95.50% to 97.15% while

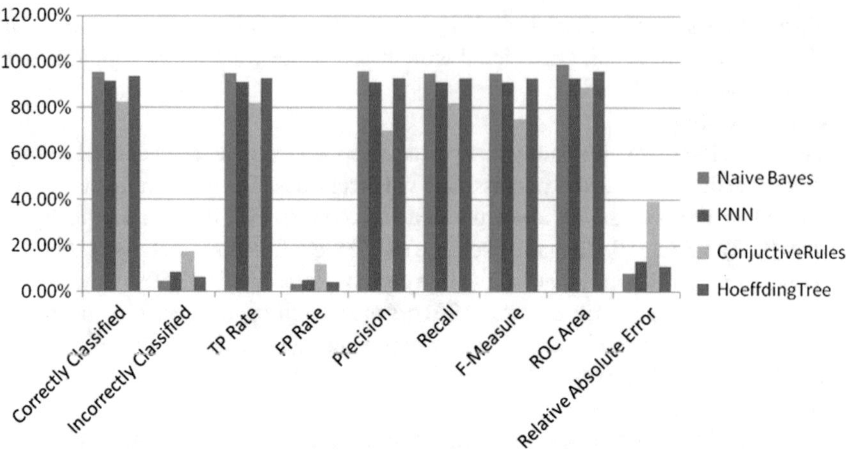

Fig. 3 Shows classification prior SMOTE

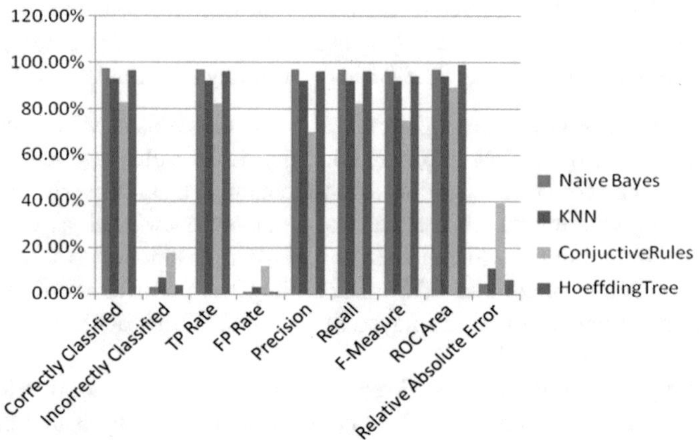

Fig. 4 Displays improvements after SMOTE

predicting the end class of student. Whereas, the accuracy of KNN and Hoeffding tree gets amplified from 91.80% to 92.79%, and 93.87% to 96.34% respectively. Table 2 exemplifies the various estimates after compliance of filtering process on above-mentioned classifiers.

The histograms of four classifiers after employing SMOTE indicate high relative absolute error in case of conjuctive rule. However, minimum relative absolute error has been recorded in Naive Bayes and then followed by Hoeffding tree and KNN classifiers. Besides, Fig. 4 demonstrates a significant improvement among correctly classified and incorrectly classified instances of a dataset than earlier corroborative results.

Table 2 Shows results after applying SMOTE method

Classifier name	Correctly classified (%)	Incorrectly classified (%)	TP rate	FP rate	Precision	Recall	F-Measure	ROC area	Rel. Abs. Err. (%)
Naïve Bayes	97.15	2.84	0.97	0.01	0.97	0.97	0.96	0.97	4.60
KNN	92.79	7.20	0.92	0.03	0.92	0.92	0.92	0.94	10.98
Conjuctive rule	82.50	17.49	0.82	0.12	0.70	0.82	0.75	0.89	39.22
Hoeffding tree	96.34	3.65	0.96	0.01	0.96	0.96	0.94	0.99	6.17

3.4 Using Ensemble Vote Without SMOTE

To make further strides in the direction of the contemporary problem of improving classification accuracy on academic dataset, primarily ensemble vote (4) without SMOTE was applied across variables of our dataset, and then later on, with the deployment of ensemble vote (4) with SMOTE, to certify whether there has been a considerable precision in classifying the students resultant class. The ensemble vote (4) comprising of classifiers Naive Bayes, KNN, conjuctive rules and Hoeffding tree were pooled with majority voting as combination rule and thereby, accomplished a significant classification accuracy of 97.30% as revealed in Table 3. Furthermore, the results acknowledged in Table 3 have not been subjected to filtering technique, viz., SMOTE. Therefore, the ensemble vote (4) with estimates in Table 3 such as correctly classified (97.30%), incorrectly classified (2.60%), TP rate(0.97), FP rate(0.01), F-measure (0.96), ROC (0.99) and so on, have publicised significant results than individual classifiers, viz., Naive Bayes, KNN, conjuctive rules and Hoeffding tree whose results have been stated earlier.

3.5 Using Ensemble Vote with SMOTE

In this method, ensemble vote (4) comprising of Naive Bayes, KNN, conjuctive rules and Hoeffding tree is applied, after the dataset has undergone through the process of SMOTE. This filtering technique is deployed with the intention to eliminate nonuniformity and inconsistency present in the dataset. So the classifier is not biased towards the majority class and consistency is achieved in our results. After exploitation of ensemble vote (4) with SMOTE, the results accomplished were quite remarkable than both individual classifiers as well as ensemble vote (4) without SMOTE. Furthermore, this classifier belonging to the category of meta-classifiers has shown outstanding classification accuracy of 98.30% as shown in Table 4, which is highly significant than any other classifiers discussed earlier.

The performance metrics associated with this classifier as displayed in Table 4 are also highly admirable, viz., correctly classified (98.30%), incorrectly classified (1.60%), TP rate (0.98), FP rate (0.01), F-measure (0.96), ROC (0.99) and so on.

Table 3 Demonstrates performance of classifiers using ensemble vote approach

Classifier name	Correctly classified	Incorrectly classified	TP rate	FP rate	Precision	Recall	F-measure	ROC area	Rel. Abs. Err.
Ensemble vote	97.30%	2.60%	0.97	0.01	0.97	0.97	0.96	0.99	18.03%

Table 4 Illustrates results of ensemble vote methods after application of SMOTE method

Classifier Name	Correctly Classified	Incorrectly Classified	TP Rate	FP Rate	Precision	Recall	F-Measure	ROC Area	Rel. Abs. Err.
Ensemble vote	98.30%	1.60%	0.98	0.01	0.98	0.98	0.98	0.99	22.08%

4 Conclusion

Primarily, the main objective of this study was to determine a comparison among meta-classifiers and individual classifiers, and as a consequence, identify whether there is any discrepancy in results after the application of meta- and base-classifiers. However, it was observed that after utilization of ensemble vote (4) on our real pedagogical dataset, the results revealed significant improvement of 97.30% while predicting the performance of students, in contrast to individual base classifiers. Furthermore, the academic dataset was subjected to filtering process wherein SMOTE was employed across both meta- and base-classifiers, to certify whether there is any substantial improvement in predicting the output class of students. The results exemplified that there was a significant development and, certified that SMOTE plays a pivotal role in enhancing the performance of classifiers.

The subject of Educational Data Mining and its association with various ensemble techniques has suffered from phenomenological drawbacks and there have been only meager attempts in this direction. Therefore, it calls for the attention of researchers to exploit various ensemble and other uncovered methods, which can further aid in the realm of education data mining.

References

1. Kabakchieva, D.: Student performance prediction by using data mining classification algorithms. Int. J. Comput. Sci. Manag. Res. **1**(4) (2012)
2. Ahmed, A.B.E.D., Elaraby, I.S.: Data mining: a prediction for student's performance using classification method. World J. Comput. Appl. Technol. **2**(2), 43–47 (2014)
3. Joseph, S., Devadas, L.: Student's performance prediction using weighted modified ID3 algorithm. Int. J. Sci. Res. Eng. Technol. **4**(5), 57–575 (2015)
4. Pandey, U.K., Pal, S.: Data Mining: A prediction of performer or underperformer using classification. Int. J. Sci. Information Technol. **2**(2), 686–690 (2011)
5. Jothi, J.K., Venkatalakshmi, K.: Intellectual performance analysis of students by using data mining techniques. Int. J. Innov. Res. Sci. Eng. Technol. **3**(3) (2014)
6. Nikitaben, S., Shriniwas, G .: A survey of data mining approaches in performance analysis and evaluation. Int. J. Adv. Res. Comput. Sci. Softw. Eng. **5** (2015)
7. T. Stafford, H. Elgueta, H. Cameron, Students' engagement with a collaborative wiki tool predicts enhanced written exam performance. Res. Learn. Technol. **22** (2014)
8. Junco, R., Heiberger, G., Loken, E.: The effect of Twitter on college student engagement and grades. J. Comput. Assist. Learn. **27**(2), 119–132 (2011)
9. Giovannella, C., Popescu, E., Scaccia, F.: A PCA study of student performance indicators in a Web 2.0-based learning environment. In: Proceedings of the ICALT 2013 (13th IEEE International Conference on Advanced Learning Technologies), pp. 33–35 (2013)
10. Suyal, S.R., Mohod, M.M.: Quality improvisation of student performance using data mining techniques. Int. J. Sci. Res. Publ. **4**(4) (2014)
11. Baradwaj, B.K., Pal S.: Mining educational data to analyze Ssudents' performance. (IJACSA) Int. J. Adv. Comput. Sci. Appl. **2**(6) (2011)
12. Devasia, T., Vinushree, T. P., Hegde, V.: Prediction of students performance using educational data mining. In: International Conference on Data mining and advanced computing (SAPIENCE), pp. 91–95. IEEE, Mar 2016

13. Kumar, V., et.al.: Knowledge discovery from database using an integration of clustering and classification. Int. J. Adv. Comput. Sci. Appl. **2**(3), Mar 2011

Big Data Ecosystem: Review on Architectural Evolution

Kamakhya Narain Singh, Rajat Kumar Behera and Jibendu Kumar Mantri

Abstract Big data is the collection of large datasets of different varieties, generated at an alarming speed with noises and abnormality. It is primarily popular for its five Vs namely volume, variety, velocity, value, and veracity. In an extremely disruptive world of open source systems that are playing a dominant role, big data can be referred to as the technology that can address the storage, processing, and visualization of such data which is too diverse and fast-changing. It has led to a complex ecosystem of new frameworks, tools, and libraries that are being released almost every day, which creates confusion as the technologists tussle with a swamp. This survey paper discusses the big data architecture and its subtleties that might help in applying the appropriate technology as a use case.

Keywords Big data · Big data ecosystem architecture
Big data processing and big data storage

1 Introduction

The concept big data has evolved due to an outburst of data from various sources like data centers, cloud, internet, Internet of things (IoT), mobile, sensors and other spheres [1]. Data have entered into every industry, all business operations and currently thought about a major factor in production [2]. Big data is broad and encompasses new technology developments and trends [3]. It represents the enormous volume of structured, semi-structured and unstructured data and usually in terms

K. N. Singh (✉) · J. K. Mantri
Department of Computer Application, North Odisha University, Baripada 757003, India
e-mail: kamakhya.vphcu@gmail.com

J. K. Mantri
e-mail: jkmantri@gmail.com

R. K. Behera
Kalinga Institute of Industrial Technology, Deemed to be University, Bhubaneswar 751024, India
e-mail: rajat_behera@yahoo.com

A. Abraham et al. (eds.), *Emerging Technologies in Data Mining and Information Security*, Advances in Intelligent Systems and Computing 813,
https://doi.org/10.1007/978-981-13-1498-8_30

of petabytes (PB) or exabytes (EB). The data are collected at an unparalleled scale which creates difficulty in making intelligent decisions. For instance, in the process of data acquirement, when the sourced data require decisions on cleanness, i.e., what data to discard and what data to keep, remains a challenging task and how to store the reliable data with the right metadata becomes a major point of concern. Though the decisions can be based on the data itself, greatly data are still not in a structured format. Blog content, Tweeter, and Instagram feeds are imperceptibly structured pieces of text while machine-generated data, such as satellite images, photographs, and videos are structured well for storage and visualization but not for semantic content and search. Transforming such content into a structured format for data analysis tends to be a problem. Undoubtedly, big data has the potential to help the industries in improving their operations and to make intelligent decisions.

1.1 The Five Vs of Big Data

Every day, 2500 petabytes (PB) of data are created from digital pictures, videos, transaction records, posts from social media websites, intelligent sensors, etc. [4]. Thus, big data is described as massive and complex data sets which are unfeasible to manage with traditional software tools, statistical analysis, and machine learning algorithms. Big data can be therefore characterized by the creation of data and its storage, analysis, and retrieval as defined by 5 V [5].

1. Volume: It denotes the enormous quantity generated in no time and determines the value and potential of it under consideration and requires special computational platforms in order to analyze it.
2. Velocity: It refers to the speed at which data is created and processed to meet the challenges and demands that lie in the path of development and growth.
3. Variety: It can be defined as the type of the content of data analysis. Big data is not just the acquisition of strings, dates, and numbers. It is also the data collected from various sources like sensors, audio, video, structured, semi-structured, and unstructured texts.
4. Veracity: It is added by some organizations which focus on the quality of the variability in the captured data. It refers to the trustworthiness of the data and the reputation of the data sources.
5. Value: It refers to the significance of the data being collated for analysis. The proposition of the value is easy to access and produces various quality analytics like descriptive, diagnosis, and prescriptive to produce insightful action in time-bound manner.

Additionally, two more Vs which represent visualization and variability (i.e. constantly changing data) are commonly used to make it to 7Vs but it fails to address additional requirements such as usability and privacy which are equally important.

1.2 Big Data Ecosystem Characteristics

A big data ecosystem must perform vigorously and be resource-efficient. Following are the desired characteristics of the big data ecosystem.

- Robustness and fault tolerance: Robustness is the ability of the system to cope with the error and during the execution and also to cope with erroneous input. Systems need to work correctly and efficiently despite the machine failures. Fault tolerance is the ability of the system to continue operating correctly in the event of the failure of some of its components. The system must be robust enough to handle machine failures and human errors. The systems must be human fault tolerance [6].
- Low-latency reads and updates: It is the measurement of delay time or waiting time experienced by a system. As far as possible, the big data system has to deliver low read time and low update time [7].
- Scalability: It is the ability of a system to manage a growing amount of work or its potential to be enlarged to accommodate the growth. The big data system has to promise for the highly scalable, i.e., in the event of increasing data and load, computing resources should be plugged-in easily.
- Generalization: The big data system to support a wide spectrum of applications with the operational functions of all dataset [6].
- Extensibility: When needed, the big data system provision to add functionalities with minimized cost.
- Ad hoc queries: Big data system should facilitate for ad hoc queries. As the need arises, the ad hoc queries can be created to obtain required information.
- Minimal Maintenance: Maintenance is defined as the work required in keeping the system runs smoothly. Big data system with modest complexity should be prioritized [6], i.e., the maintenance of the system should be kept as minimal as possible.
- Debuggability: Debuggability is defined as the capability of being easily debugged. When required, a big data ecosystem must present the necessary granular information to debug [6] and also facilitate for the required extent to which something can be debugged.

2 Big Data Ecosystem Architecture

The architecture presented below is representing the evolution of big data architecture.

1. Lambda (λ) architecture: In the earlier days, big data systems were constructed to handle three Vs of big data, namely volume, velocity, and variety to discover insights and make timely better business decisions. Nathan Marz coined lambda architecture (LA) to describe fault-tolerant (both against hardware failures and human mistakes), scalable, and generic data processing architecture. LA aims to

Table 1 λ architecture open source technology stack

Area	Technology stack
Data ingestion	Apache kafka, apache flume and apache samza
Batch layer	Apache hadoop, apache MapReduce, apache spark and apache pig
Batch views	Apache HBase, ElephantDB, and apache impala
Speed layer	Apache storm, apache spark streaming
Real-time view	Apache cassandra, apache HBase
Manual merge	Apache impala
Query	Apache hive, apache pig, apache spark SQL and apache impala

satisfy the needs for a strong, robust, and healthy system by serving a wide range of workloads and use cases with low-latency reads and updates. LA also aims that resulting system should be linearly scalable, and should scale out rather than up [8]. The architecture uses a combination of batch and real-time processing paradigm in parallel and runs on a real-time computational system [9]. λ has three layers namely:

Batch Layer: The batch layer is aimed to serve twofold purpose. The first purpose is to store the constantly growing immutable dataset into data sink and the second is to pre-compute batch views from the in-housed dataset. Computing the views is an ongoing operation, i.e., when new data arrives, it will be combined into the earlier existing views. These views may be computed from the entire dataset and therefore this layer is not expected to update the views frequently. Depending on the size of the in-housed dataset and cluster configuration, pre-computation could take longer time [10]. Batch layer produces a set of flat files containing the pre-computed views.

Speed Layer: While the batch layer continuously recompute the batch views, the speed layer uses an incremental approach whereby the real-time views are incremented as and when new data is sourced [10], i.e., it manages only the recent data and computes the real-time views.

Serving Layer: Performs indexing and exposes views for querying.

Incoming data are transmitted to both batch and speed layers for processing. The batch layer manages the immutable append-only raw data and then pre-computes the batch views [9]. The speed layer deals with recent data and computes the real-time views. At the other end, queries are answered by assimilation of both batch and real-time views. Both layers execute the same processing logic and output results in a service layer. Batch views are batch write and random read wherein real-time views are random write and random read [8]. Queries from back-end systems are executed based on the data in the service layer, reconciling the results produced by the batch and real-time views. The three-layer architecture is outlined in Fig. 1.

Open source technology stacks for λ architecture are presented in Table 1.

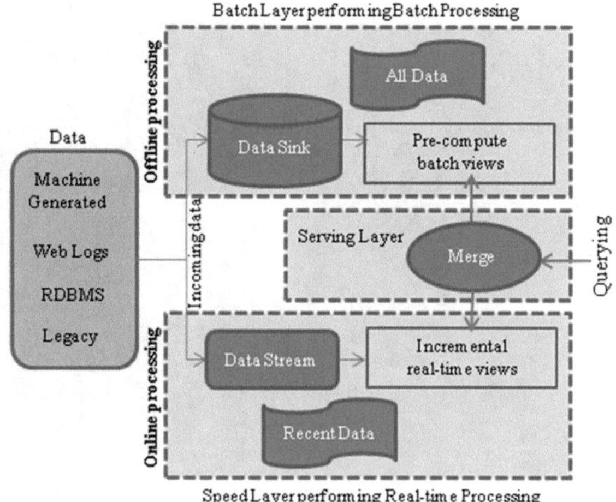

Fig. 1 Three layers of λ architecture, namely batch, speed, and serving layer

Error rectification of λ architecture is performed by allowing the views to be recomputed [11]. If error rectification is time consuming, the solution is to revert to the non-corrupted previous version of the data. This leads to a human fault-tolerant system where toxic data can be completely removed and recomputation can be done easily. The disadvantage of λ architecture is its complexity and its limiting influence. The batch and streaming processing pipeline requires a different codebase that must be properly versioned and kept in sync so that processed data produces the same result from both paths [12]. Keeping in sync the two complex distributed processing pipeline is quite maintenance and implementation, it would bring the same benefits and handle the problem. So the essence is to develop λ architectures using a "unified" framework which makes the same codebase available to both the speed and batch layer and combines the results of both layers transparently, which leads to the development of unified λ architecture.

2. Unified λ architecture: It combines both batch and real-time pipeline, which runs concurrently and the results merged automatically [13]. From processing paradigms, the architecture integrates batch and real-time processing pipeline by offering a single API. The architecture is outlined in Fig. 2.
   With a unified framework, there would be only one codebase to maintain. Open source technology stacks for unified λ architecture are the replica of λ architecture except the "Auto Merge" area. Spring "XD", and Summingbird which are the commonly used open source technology tool.

3. Kappa architecture: It is the simplification of λ architecture [14]. Kappa architecture is similar to λ architecture with the removal of batch processing paradigm. In summer 2014, Jay Kreps posted an article addressing pitfalls associated with λ architecture [15]. Kappa architecture is avoiding maintaining two separate codebases of batch layer and speed layer. It is handling real-time data processing and continuous data reprocessing using a single stream processing computation

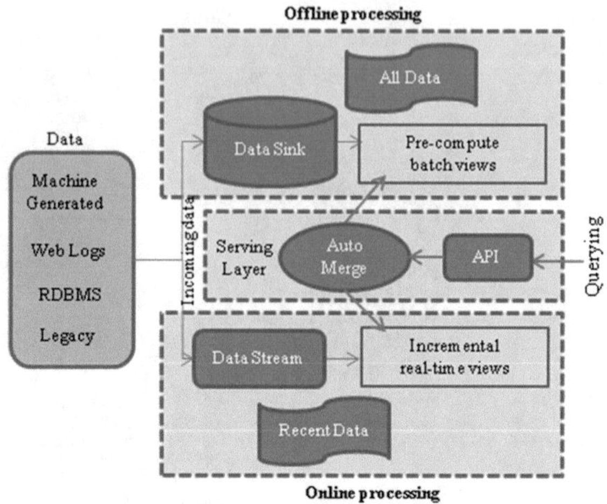

Fig. 2 Unified λ architecture

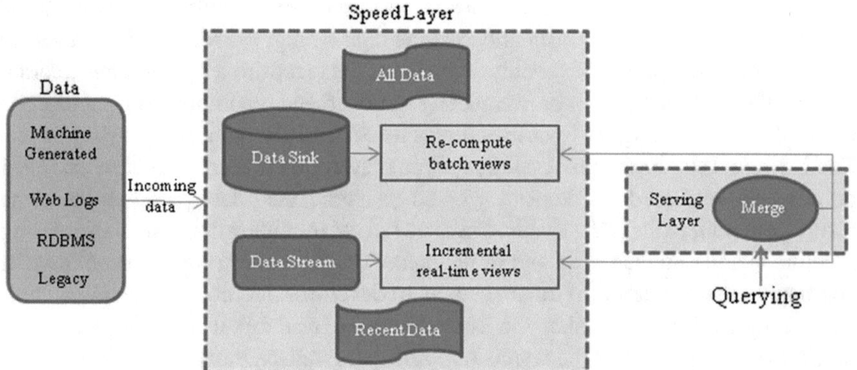

Fig. 3 Kappa architecture

model. Hence, it consists of two layers, namely speed and serving layer. The speed processing layer runs the stream/online processing jobs. Usually, a single online processing job is run to enable real-time data processing. Data reprocessing is done when some code of the online processing job needs to be tweaked. This is accomplished by running another changed online processing job and replaying all previous housed data. Finally, the serving layer is used for querying [15]. The architecture is outlined in Fig. 3.

Open source technology stacks for Kappa architecture are presented in Table 2.

4. Microservices architecture: It divides big data system into many undersized services called microservices that can run independently. This allows every service to run its own process and communicate in a self-ruling way without having

Table 2 Kappa architecture open source technology stack

Area	Technology stack
Data ingestion	Similar to λ architecture open source technology stack
Batch views	Apache HBase, ElephantDB
Speed layer	Apache storm, apache spark streaming
Real-time view	Apache cassandra
Queries	Apache hive, apache pig and apache spark SQL

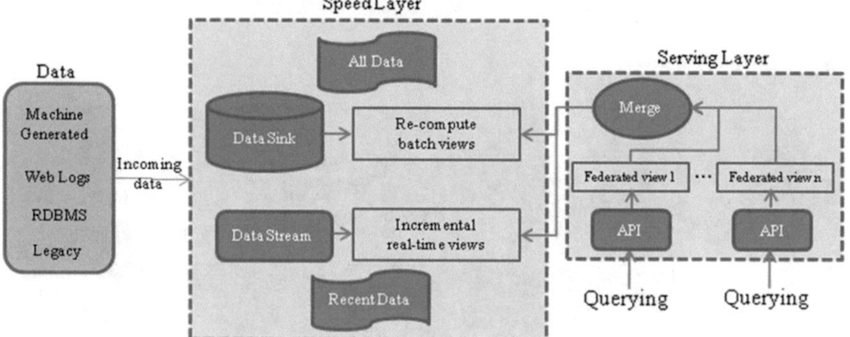

Fig. 4 Microservices architecture

to depend on other services or the application as a whole [16]. Martin Fowler defines that the services must adhere to the common architectural principles, including single responsibility, separation of concerns, do not repeat yourself (DRY), composability, encapsulation, loose coupling, use of consistent, standardized interfaces, etc. [17]. The architecture is outlined in Fig. 4.

Open source technology stacks for microservice architecture are identical to the λ architecture except for real-time view and federated view. Apache Cassandra is the commonly used open source technology tool for real-time view and Apache Phoenix is the commonly used open source technology tool for federated view.

5. Mu architecture: It ingests data into both batch and streaming process, but not for reliability viewpoint, because some work is better done in batches and some are in streaming paradigm [18]. The architecture is outlined in Fig. 5.

Open source technology stacks for Mu architecture remains same as Kappa architecture.

6. Zeta architecture: It is built on pluggable components and all together, it produces a holistic architecture [19]. Zeta is characterized by seven components, namely:
    - Distributed File System (DFS): It is the common data location for all needs and is reliable and scalable.

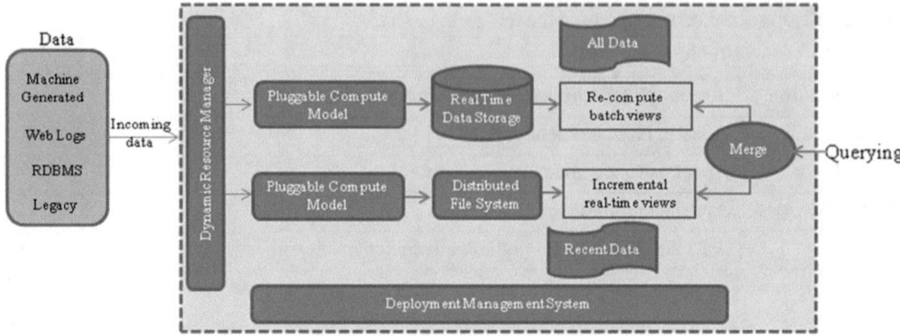

Fig. 5 Mu architecture

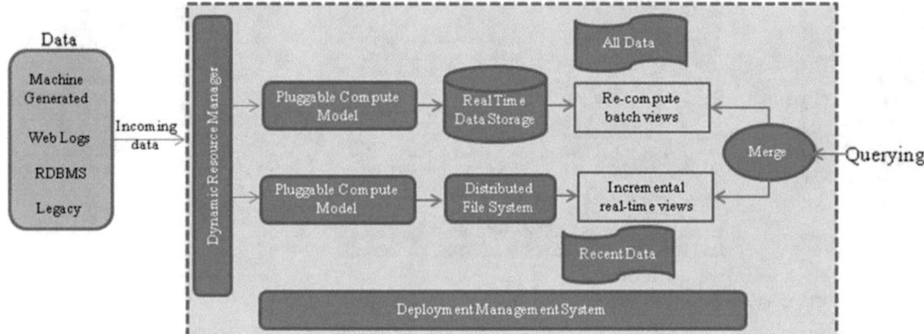

Fig. 6 Zeta architecture

- Real-time Data Storage: It is based on real-time distributed technologies, especially NoSQL solutions and is meant for delivering user supplied responses promptly and quickly.
- Enterprise Applications (EA): EA focuses to comprehend all business goals of the system. The examples of this layer are web servers or business applications.
- Solution Architecture (SA): SA spotlight is on a specific business problem. Unlike EA, it concerns a more specific problem. Different solutions can be combined to construct the solution for the more global problem.
- Pluggable Compute Model (PCM): It implements all analytic computations and are pluggable in nature as it has to cater to different needs.
- Dynamic Global Resource Management (DGRM): It allows dynamic allocation of resources that enables business to easily accommodate for priority tasks.
- Deployment/Container Management System: This guarantees a single, standardized method of deployment and implies that deployed resources do not concern about any environment changing, i.e., deployment in the local environment is identical with prod environment.

The architecture is outlined in Fig. 6.

Table 3 Zeta architecture open source technology stack

Area	Technology stack
Data ingestion	Similar to λ architecture open source technology stack
DGRM	Apache mesos and apache YARN
PCM	Apache spark and apache drill
Real-time data storage	Apache HBase and couchbase
DFS	HDFS
Batch view	Similar to kappa architecture open source technology stack
Real-time view	Apache cassandra
Queries	Similar to kappa architecture open source technology stack

Open source technology stacks for Zeta architecture are presented in Table 3.

7. IoT architecture (IoT-a): The Internet of Things (IoT) is an ecosystem of connected devices and possibly human, accessible through Internet. The IP address is assigned to the devices and it collates, and transfers data over the Internet without manual and human intervention. Examples of such devices are RFID sensors and sophisticated devices like smartphones. Data from such IoT devices transmitted to big data ecosystem to produce a continuous stream of information. In simplicity, IoT devices generate the data and are delivered to big data ecosystem to generate insight in time-bound manner [20]. The big data ecosystem uses scaling-out approach on commodity hardware to overcome the challenges posed by such devices. IoT-a is composed of three primary building blocks namely:

    - Ad hoc queries: Message Queue/Stream Processing (MQ/SP): It receives data from upstream system and depending on the business mandate, and performs buffering, filtering and complex online operations.
    - Database: It receives data from MQ/SP and provides structured and low-latency access to the data points. Typically, the database is a NoSQL solution with auto-shading and horizontal scale-out properties. The database output is of interactive nature, with an interface provided either through a store-specific API or through the standard interface SQL [20].
    - Distributed File System (DFS): In general, it receives data from either the DB and performs batch jobs over the entire dataset. If required, it can also receive data directly from MQ/SP block. This might include merging data from IoT devices with other data sources [20].

    IoT-a architecture is outlined in Fig. 7.
    Open source technology stacks for IoT-a architecture are presented in Table 4.

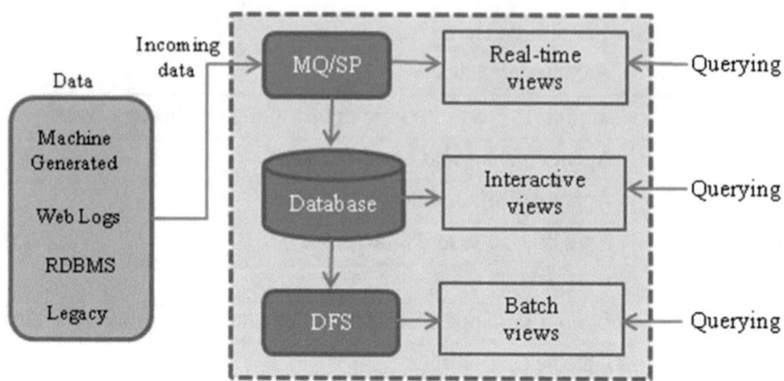

Fig. 7 IoT-a architecture

Table 4 IoT-a architecture open source technology stack

Area	Technology stack
Data Ingestion	Similar to λ architecture open source technology stack
Real-time views	Apache cassandra
Interactive processing	Apache spark
Interactive views	Apache drill
Batch processing	Apache mahout
Batch views	Apache hive
Queries	Apache hive, apache cassandra and apache drill

3 Discussion

This paper briefly reviews big data ecosystem architecture to the best of the knowledge, for discussions and usages in research, academia, and industry. The information presented discusses some research papers in the literature and a bunch of systems, but when it comes to the discussion of a small fraction of the existing big data technology and architecture, there are many different attributes that carry equal importance, weight and a rationale for comparison.

4 Conclusion

A theoretical study or a survey of various tools, libraries, languages, file systems, resource managers, schedulers, search engines, SQL and NoSQL frameworks, operational and monitoring frameworks had been highlighted in order to provide the researcher with the information for understanding big data ecosystem which are on a rapid growth raising concerns in terms of business intelligence and scalable management that includes fault tolerance and optimal performance. In this paper, big data architecture has been discussed though not in an elaborate manner, but hope this paper will serve as a helpful introduction to readers interested in big data technologies.

References

1. Borodo, S.M., Shamsuddin, S.M., Hasan, S.: Big data platforms and techniques. **17**(1), 191–200 (2016)
2. James, M., Michael, C., Brad, B., Jacques, B., Richard, D., Charles, R.: Big Data: The Next Frontier for Innovation, Competition, and Productivity. McKinsey Glob Inst. (2011)
3. Emerging technologies for Big Data—TechRepublic (2012). http://www.techrepublic.com/blog/big-data-analytics/10-emerging-technologies-for-big-data/
4. Every Day Big Data Statistics—2.5 Quintillion Bytes of Data Created Daily. http://www.vcloudnews.com/every-day-big-data-statistics-2-5-quintillion-bytes-of-data-created-daily/ (2015)
5. Big Data: The 5 Vs Everyone Must Know. https://www.linkedin.com/pulse/20140306073407-64875646-big-data-the-5-vs-everyone-must-know (2014)
6. Notes from Marz' Big Data—principles and best practices of scalable real-time data systems—chapter 1. https://markobigdata.com/2017/01/08/notes-from-marz-big-data-principles-and-best-practices-of-scalable-real-time-data-systems-chapter-1/ (2017)
7. The Secrets of Building Realtime Big Data Systems. https://www.slideshare.net/nathanmarz/the-secrets-of-building-realtime-big-data-systems/15-2_Low_latency_reads_and (2011)
8. Lambda architecture. http://lambda-architecture.net/ (2017)
9. Jay, K.: Questioning the lambda architecture. www.radar.oreilly.com 2014
10. The Lambda architecture: principles for architecting realtime Big Data systems. http://jameskinley.tumblr.com/post/37398560534/the-lambda-architecture-principles-for
11. Big Data Using Lambda Architecture. http://www.talentica.com/pdf/Big-Data-Using-Lambda-Architecture.pdf (2015)
12. Wikipedia lambda architecture. https://en.wikipedia.org/wiki/Lambda_architecture
13. Lambda Architecture for Big Data by Tony Siciliani. https://dzone.com/articles/lambda-architecture-big-data (2015)
14. Kappa architecture. http://milinda.pathirage.org/kappa-architecture.com/
15. Data processing architectures—Lambda and Kappa. https://www.ericsson.com/research-blog/data-processing-architectures-lambda-and-kappa/ (2015)
16. Microservices Architecture: An Introduction to Microservices. http://www.bmc.com/blogs/microservices-architecture-introduction-microservices/ (2017)
17. Data Integration Design Patterns With Microservices by Mike Davison. https://blogs.technet.microsoft.com/cansql/2016/12/05/data-integration-design-patterns-with-microservices/ (2016)
18. Real Time Big Data #TD3PI, 2015, http://jtonedm.com/2015/06/04/real-time-big-data-td3pi/
19. Zeta architecture. http://www.waitingforcode.com/general-big-data/zeta-architecture/read (2017)
20. Iot-a: the ınternet of thıngs archıtecture. http://iot-a.info

A Comparative Study of Local Outlier Factor Algorithms for Outliers Detection in Data Streams

Supriya Mishra and Meenu Chawla

Abstract Outlier detection analyzes data, finds out anomalies, and helps to discover unforeseen activities in safety crucial systems. Outlier detection helps in early prediction of various fraudulent activities like credit card theft, fake insurance claim, tax stealing, real-time monitoring, medical systems, online transactions, and many more. Detection of outliers in the data streams is extremely challenging as compared to static data since data streams are continuous, highly changing and unending in nature. Density-based outlier detection using local outlier factor (LOF) is the prominent method for detecting the outliers in the data streams. In this paper, we provide an insight on outlier detection and various challenges involved while detection of outliers in the data streams. We concentrate on density-based outlier detection and review major local outlier factor (LOF) based outlier detection algorithms in details. We also perform a comparative study of existing LOF algorithms to evaluate the performance in term of several parameters.

1 Introduction

Outlier detection is a vastly researched area and becomes an important aspect due to the rapid growth in data, storage, preprocessing, curation, analysis, and security issues concerned with data streams. Outlier observations are often found to be exceptionally dissimilar from other observations in its neighborhood. Outlier is a data point or a pattern which is not inline to the normal characteristics of remaining dataset, therefore it creates a suspicion that it is produced from a different or unintended source or mechanism. Outliers are often confused with errors generated due to technical or mechanical error, fraudulent activity, human mistake, noise, etc. However, if

S. Mishra (✉) · M. Chawla
Department of Computer Science and Engineering, Maulana Azad National Institute of Technology Bhopal, Bhopal 462 003, India
e-mail: m.supriya@outlook.com

M. Chawla
e-mail: chawlam@manit.ac.in

A. Abraham et al. (eds.), *Emerging Technologies in Data Mining and Information Security*, Advances in Intelligent Systems and Computing 813,
https://doi.org/10.1007/978-981-13-1498-8_31

outliers are correctly distinguished from such type of errors then the system can be rescued from several potential risks. For example, early detection of activities such as credit card fraud, insurance claim fraud, insider fraud monitoring, public health, and Internet traffic monitoring, etc. may prevent the system from several potential risks. A number of real-world applications exploit outlier detection in practice. Majority of research on this topic deals with detecting fraud, particularly financial fraud. Outlier detection can spot such activities and after that necessary actions can be taken.

These days applications produce copious data at high speeds continuously which are known as data streams. Voluminous data streams bring new challenges in outlier detection field. Outlier detection on static data is relatively easier, however, detection of outliers in data streams is nontrivial since data stream is continuous, highly changing, and unending in nature. And with the increase in the complexity, variety, speed, and size, the storage and mining of data streams become even more complex. Moreover, if the data stream is high dimensional and distributed in nature then the data points are very sparse, which in turn makes detection of outliers a troublesome task. Volume and speed of data stream further increase additional costs to the outlier detection. Thus, commonly used distance measure techniques such as Euclidean distance and the standard concept of nearest neighbors are less feasible to detect outliers. Hence, it is very difficult to design efficient algorithms to detect outlier in the data stream.

Outlier detection includes the approaches that are used for the purpose of discovering and removing outliers present in the datasets. Several approaches have been proposed and implemented for outlier detection for static as well as streaming data sets. These approaches can be classified into five major categories namely distance-based, distribution-based, nearest neighbor-based, clustering-based, and density-based. Among these approaches, density-based outlier detection is widely utilized as it detects outlier on the basis of the density of neighborhood. One category of density-based outlier detection approach utilizes the local outlier factor (LOF) [1] algorithm which assigns a real number to each data point, termed as outlier factor and determines the extent of being an outlier. For data points which lie deep inside a cluster, LOF value comes out to be 1 (approximately), and for other points, it comes greater than 1.

In this paper, we review state-of-the-art LOF algorithms and perform a comparative study in term of several parameters such as complexity, accuracy, scalability, etc. Rest of the paper includes the following sections. Section 2 summarizes past research in the area of outlier detection. Section 3 explains the working of prominent LOF-based algorithms. Section 4 presents a comparison study of LOF algorithms and at the last, we conclude the paper.

2 Related Work

Outlier detection has been an area of emphasis and hence sufficient literature is available on this topic. Initial researches conducted on outlier detection were majorly based on statistics [2]. Selection of a particular technique depends on distribution model of data, types of the attribute, environment, and the application. A thorough study of outlier detection is done by Hodge et al. [3]. A comprehensive survey of previously used techniques in this domain is done by Chandola et al. [4]. Detection methods for temporal data have been reviewed by Gupta et al. [5]. Whereas Rajasegarar et al. [6] provided a summary of existing anomaly detection algorithms for wireless sensor networks. In case of streaming environments where ample amount of data are generated at a high rate, it is essential to detect outliers in one scan and in an acceptable time limit. Approaches proposed for the streaming environment can be majorly categorized based on parameters like distribution model, clustering technique, and the distance between data points. In distribution-based approaches, the probability study is performed on data. The knowledge of underlying data is required in advance. Yamanishi et al. [7] have proposed a framework for data stream which includes two parts: data modeling and scoring. Sequential discounting estimation model (SD-EM) and sequential discounting for auto-regressive estimation model (SDAR-EM) [8] are used for learning, after which scoring method is applied for change point detection.

Clustering-based algorithms create clusters of data which represent the underlying distribution of data. Data points which fall apart from the clusters or a set of very less number of data points forming a small cluster are considered to be outliers. Clustering algorithms emphasize on finding out clusters rather than detecting outliers [9]. Differentiating outliers from clusters is more difficult in presence of noise. To eliminate this problem, an algorithm named DenStream is proposed in [10]. The framework proposed by Agarwal et al. [11] determines evolving clusters continuously and redefines dimensions included in each cluster. Distance-based approaches compute the distance of a point w.r.t to neighboring points and are of two types depending on distances, i.e., local and global. Angiulli et al. [12] and Yang et al. [13] exploited sliding window of data points in their experiments to detect global distance-based outliers. This approach was modified by Kontaki et al. [14] to optimize time complexity and memory consumption. For detection of local outliers, a density-based approach, which computes LOF was proposed in [1]. LOF-based subsequent approaches are proposed in [15–17].

3 LOF-based Outlier Detection

LOF-based algorithms compute a value for every data point and assign this value to it. This value is called LOF of the point. Depending on the LOF value, we can determine whether a point is an outlier or not. In this section, we have provided detailed working of major LOF algorithms. Figure 1 shows the general flow of all LOF-based algorithms [18].

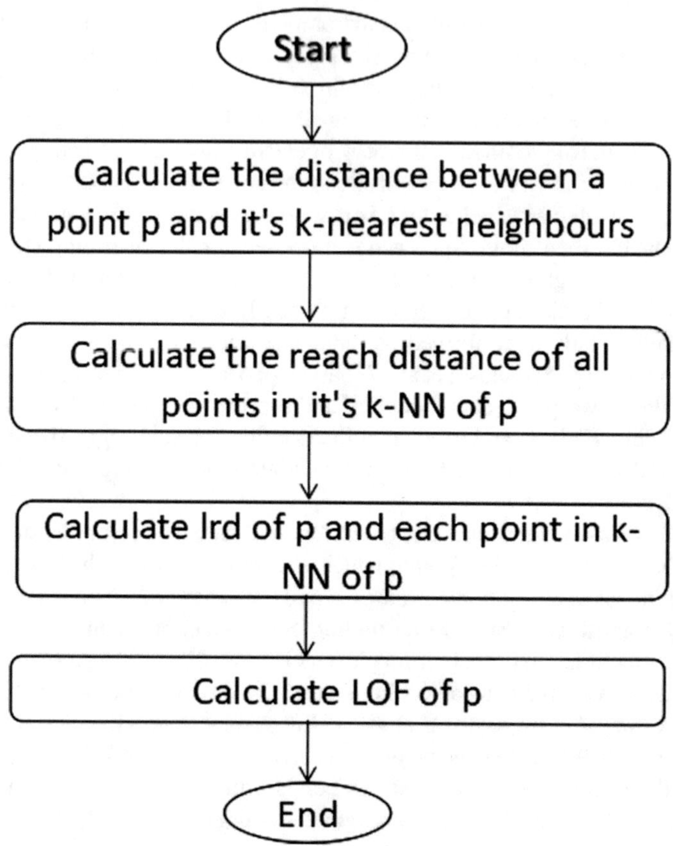

Fig. 1 Flow diagram of basic LOF algorithm [18]

3.1 Original LOF Algorithm

Original LOF technique designed by Breunig et al. [1] computes a value (LOF) for all data points existing in the dataset. LOF computation needs three values: k-distance, reach-distance, and local reachability density (LRD). These terms are defined as follows:

k-dist(p). It is the distance between a data point and it is kth nearest neighbor.

reach-dist. Reach-dist of point q with respect to point p is the maximum of two values: k-dist(q) and the distance between p and q, represented as $d(p, q)$. It is calculated as

$$\text{reach-dist}(q, p) = \max\{\text{k-dist}(q), d(p, q)\} \tag{1}$$

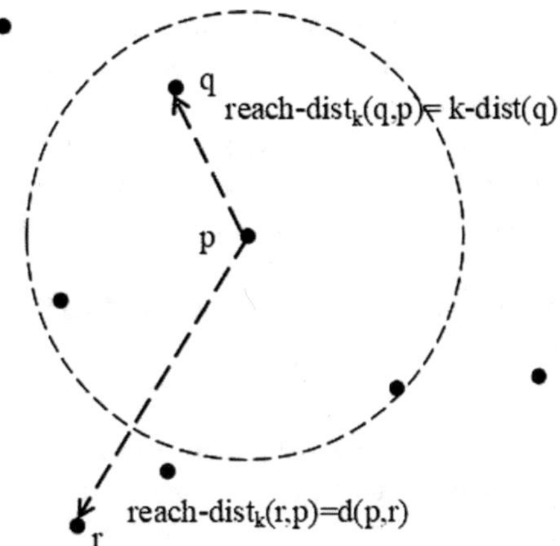

Fig. 2 Example of reach-dist(q, p) and reach-dist(r, p) for $k = 4$ [1]

In Fig. 2, reach-dist(q, p) is equal to k-dist(p) because k-dist(q) is greater than $d(p, q)$ and reach-dist(r, p) is equal to $d(p, r)$ because $d(p, r)$ is greater than k-dist(r) [1].

Local reachability density (LRD). LRD of point p can be defined as the inverse of average reach distances (reach-dist) of all the points in the k-neighborhood of point p. It can be calculated as given in Eq. 2.

$$LRD_k(p) = \frac{1}{k}\left(\sum_{q \in N(p,k)} \text{reach-dist}_k(p, q)\right)^{-1} \tag{2}$$

where $N(p, k)$ is set of points of k nearest neighbors of p.

Local outlier factor (LOF). $LOF(p)$ of a data point p is the average of ratios of $LRD(o)$ to $LRD(p)$ for all points o in the k-nearest neighborhood of point p. It is calculated as-

$$LOF_k(p) = \frac{1}{k}\sum_{o \in N(p,k)} \frac{LRD_k(o)}{LRD_k(p)} \tag{3}$$

Thus, it can be said that LOF of any objects depends on a single parameter which is the minimum number of points in the nearest neighborhood.

3.2 *Incremental LOF Algorithm (ILOF)*

Original LOF algorithm discussed in Sect. 3.1 starts working once the complete dataset is available in memory, i.e., static datasets. So, there was a need to modify the original algorithm to work for data streams. Pokrajac et al. [15] presented a modified algorithm called as incremental LOF (ILOF) algorithm. ILOF algorithm works well with data streams as well. It is able to detect an outlier when it arrives. This algorithm computes LOF of point p as soon as it arrives. Let $T = [t_1, t_n]$ be the time interval, and let $p_t = [p_1^D, \ldots, p_t^D] \in \mathbb{R}^D$ be the data points arrived in time $t \in T$ where D is the data dimensionality, ILOF computes LOF value of point p_t at time t. ILOF involves two main functions:

Insertion. In this step, a new point is added to the dataset and its reach-dist, LRD and LOF values are computed.

Maintenance. In this step, updation of k-distance, reach-dist, LRD and LOF values of existing affected points is done.

The steps for the incremental LOF algorithm are as follows:

1. Find out the k-nearest neighbors of any incoming point p, let it be set KNN. Then compute its k-distance (k-dist).
2. Now for all points $q \in KNN$ compute reach-dist(p, q) using Eq. 1.
3. Let S be the set of all data point who has a point p as their one of k-nearest neighbor.
4. For all points $o \in S$ and $q \in KNN$, update the k-dist(o), reach-dist(q, o).
5. If $o \in S$ then update set S, as $S \leftarrow S \cup \{q\}$.
6. For all $o \in S$ update $LRD(o)$ and $LOF(o)$.
7. Finally compute LRD and LOF values of point p, $LRD(p)$ and $LOF(p)$.
8. Return LOF.

Hence, implementing abovementioned steps ILOF algorithm computes LOF values at the time of data arrival.

3.3 Memory Efficient ILOF Algorithm (MILOF)

Algorithm discussed in Sect. 3.2 works well to detect outliers in data streams. It stores complete dataset in memory but in many real-time systems, often there is a restriction on storage. Like in sensor networks where each sensor unit has a very limited memory, it is not feasible to store the whole dataset. Actually, it is impractical and unnecessary to store the whole dataset in memory for the calculation of LOF since LOF value is affected by only k nearest neighbors and their corresponding LOFs. In [16], authors proposed an algorithm called as MILOF. This algorithm works in environments where the available memory allows to store only a subset $m \subset n$ data points from all the data points $n \in \mathbb{R}^D$ which have been arrived at time $t \in T$ and here $m \ll n$. Particularly, MILOF aims to find out outliers for the whole stream of data and not just for the m last data points. MILOF algorithm involves three steps: Summarization, merging, and revised insertion. These three steps are explained below .

Summarization. In MILOF algorithm, total available memory m is divided into two parts b and c, such that $m = b + c$. When a new data point arrives, it is inserted into memory b. If b becomes full then the first half of memory b (i.e., first $b/2$ points) is summarized and clusters are created using c-means (or k-means, here c is used since c

clusters are formed) clustering algorithm and kept in memory c. Let us say at step i of MILOF algorithm $b/2$ points are called as set $C^i = \{P_{i \cdot b/2+m} | m \in \{1, \ldots, b/2\}\}$. The c-means algorithm partitions this set into c clusters as $C^i = \{C_1^i \cup C_2^i \cup \ldots \cup C_c^i\}$ where cluster centers are represented by set $V^i = \{v_1^i, v_2^i, \ldots, v_c^i\}$. Now k-dist, LRD and LOF of each cluster center are calculated using Eqs. 4–7 as given below.

$$\text{k-dist}(v_j^i) = \frac{\sum_{p \in C_j^i} \text{k-dist}(p)}{|C_j^i|} \tag{4}$$

$$\text{reach-dist}(p, v_j^i) = \max\{\text{k-dist}(v_j^i), d(p, v_j^i)\} \tag{5}$$

$$LRD_k(v_j^i) = \frac{\sum_{p \in C_j^i} LRD_k(p)}{|C_j^i|} \tag{6}$$

$$LOF_k(v_j^i) = \frac{\sum_{p \in C_j^i} LOF_k(p)}{|C_j^i|} \tag{7}$$

Merging. After summarization step when $b/2$ new data points are inserted then again new cluster centers are generated. Let cluster centers generated at step i be V^i, these are merged with previous $(i-1)^{th}$ step cluster centers V^{i-1}. Hence, only single set of clusters centers $x = V^i \cup V^{i-1}$ are stored in memory at any instance. For set X, let W be set of their corresponding weights, weighted c-means algorithm is applied to X and set is partitioned into c' clusters such that $X = \{X_1 \cup X_2 \cup \ldots \cup X_{c'}\}$ and corresponding weight set $W = \{W_1 \cup W_2 \cup \ldots \cup W_{c'}\}$, having new cluster centers $Z = \{z_1, z_2, \ldots, z_{c'}\}$. Initially, Z is the same as the first set of cluster centers V^0. And thereafter this merging step is applied at each subsequent iteration.

Revised insertion. When a new data point comes, it is inserted as ILOF algorithm with slight modification. The difference is that in ILOF values are computed using pre-previous data points, whereas in this algorithm, values are computed using recent points as well as summarized cluster centers. A new data point insertion is done in two steps:

1. Compute LOF of the incoming data point. Find its k-nearest neighbors, and if any of the ith $(i < k)$ nearest neighbor is a cluster center, rest of neighbors are not needed because points in this cluster are eventually nearest neighbors. For all KNNs find k-dist, LRD, and LOF using Eqs. 1–3. k-dist of a cluster center z_j is calculated using Eq. 8. reach-dist of point p w.r.t a cluster center is calculated using Eq. 5.
2. Now k-dist, reach-dist, LRD and LOF of KNNs are updated as ILOF algorithm, if necessary. But in case of a cluster center updation is postponed to next merging step.

$$\text{k-dist}(z_j) = \frac{\sum\limits_{x_i \in X_j, w_i \in W_j} w_i \text{k-dist}(x_i)}{\sum\limits_{w_i \in W_j} w_i} \tag{8}$$

3.4 *Grid-LOF Algorithm (GLOF)*

LOF algorithms are highly complex, hence are not suitable for large data with high dimensions since the increase in the data dimensionality results in the increase in computation complexity. Besides this in original LOF [1], KNNs are determined by computing distances from the complete dataset, which is also time-consuming. Therefore in [17], a grid-based method (GLOF) is presented which enables us to reduce the time taken to compute KNN in LOF algorithms. In this method, complete dataset is divided into small regions termed as grids, then LOF of each grid is computed. Consider a data set S of n data points each having d dimensions as $S \in \mathbb{R}^D$. The complete algorithms can be summarized as following steps:

1. Data points in each dimension are divided into k equal distance intervals, called grids. So, there are $k \cdot d$ grids in data set.
2. Each data point $x_i \in S$ is associated to one of the grids $\{1, 2, \ldots, k^d\}$. If no point belongs to a grid that grid is not considered.
3. For every grid j of the dataset, a grid centroid C_j is calculated.
4. For every grid centroid C_j, $LOF(C_j)$ is calculated using original algorithm [1].
5. GLOFs for each data point is determined as if x_i belongs to grid j, then $LOFG(x_i) = LOF(C_j)$.

When the results of this algorithm were compared with original algorithm, GLOF had a better performance time [17].

4 Comparative Study

After thorough study, we have compared algorithms on parameters like complexity, accuracy, application, data dimensionality, etc. Complexity is a measure of runtime requirement and depends on size and type of dataset. For streaming data, computation complexity is higher. Accuracy is a measure of correction of the algorithm. Important factors affecting accuracy are dataset size, number of data dimensions, number of clusters, and noise. The basic algorithm is efficient for the static dataset, whereas ILOF, MILOF, and GLOF work are suitable for the streaming environment. Choice of the algorithm is decided on the basis of application areas. As data dimensions increase, the complexity of algorithm also rises accordingly. A number of data dimensions also impact complexity, accuracy, and memory consumption.

Table 1 Comparison of LOF algorithms

Algorithm	Complexity	Accuracy	Scalability	Dataset	Dimensionality
LOF	$O(n^2)$	High	Less	Static and small	Very low
ILOF	$O(n \log n)$	High	Less	Streaming and small	Low
MILOF	$O(n \log(b + c))$	High	High	Streaming and large	Medium
GLOF	$O(kd)$	Moderate	High	Large	High

The comparison results are summarized in Table 1. From our study, we have inferred that there is no single algorithm which works best for every scenario. Each algorithm has its pros and cons as well. Suitability and usage of algorithm vary with environment, dataset characteristics, application, etc.

5 Conclusion

From our study, we came to the conclusion that MILOF algorithm outperforms LOF, ILOF, and GLOF in terms of accuracy, performance, and scalability. MILOF also proved to be relatively stable in terms of computation time and accuracy on changing dataset size, dimensionality, etc. GLOF showed improved efficiency for high-dimensional datasets but at the cost of accuracy. We observed that parameter k is very important in GLOF algorithm. Algorithm's accuracy and computation time largely depend on parameter k. Hence in our future study, we aim to further analyze the parameter k and optimize it enhances the accuracy of GLOF algorithm.

References

1. Breunig, MM., Kriegel, HP., Ng R.T., Sander, J.: LOF: identifying density-based local outliers. In: Proceedings of the ACM SIGMOD International Conference on Management of Data (2000). https://doi.org/10.1.1.35.8948
2. Barnett, V., Lewis, T.: Outliers in Statistical Data, 3rd edn. Wiley (1994)
3. Hodge, V.J., Austin, J.: A survey of outlier detection methodologies. Artif. Intell. Rev. **22**, 85–126 (2004). https://doi.org/10.1023/B:AIRE.0000045502.10941
4. Chandola, V., Banerjee, A., Kumar, V.: Anomaly detection: A survey. ACM Comput. Surv. **41**(3), 58(2009), Article 15. https://doi.org/10.1145/1541880.1541882
5. Gupta, M., Gao, J., Aggarwal, C.C., Han, J.: Outlier detection for temporal data: a survey. TKDE **25**(1), 1–20 (2013)
6. Rajasegarar, S., Leckie, C., Palaniswami, M.: Anomaly detection in wireless sensor networks. IEEE Wirel. Commun. **15**(4), 34–40 (2008). https://doi.org/10.1109/MWC.2008.4599219
7. Yamanishi, K., Takeuchi, J.I.: A unifying framework for detecting outliers and change points from non-stationary time series data. In: SIGKDD, pp. 676–681 (2002). https://doi.org/10.1145/775047.775148

8. Yamanishi, K., Takeuchi, J.I., Williams, G., Milne, P.: On-line unsupervised outlier detection using finite mixtures with discounting learning algorithms. In: SIGKDD, pp. 320–324 (2000)
9. Aggarwal, C.C., Han, J., Wang, J., Yu ,P.S.: A framework for clustering evolving data streams. In VLDB, pp. 81–92 (2003)
10. Cao, F., Ester, M., Qian, W., Zhou, A.: Density-based clustering over an evolving data stream with noise. In: SIAM Conference on Data Mining, pp. 328–339 (2006). https://doi.org/10.1.1.104.3793
11. Aggarwal, C.C., Han, J., Wang, J., Yu, P.S.: A framework for projected clustering of high dimensional data streams. VLDB **30**, 852–863 (2004)
12. Angiulli, F., Fassetti, F.: Detecting distance-based outliers in streams of data. In: CIKM, pp. 811–820 (2007). https://doi.org/10.1145/1321440.1321552
13. Yang, D., Rundensteiner, E.A., Ward, M.O.: Neighbor-based pattern detection for windows over streaming data. In: Advances in Database Technology, pp. 529–540 (2009). https://doi.org/10.1145/1516360.1516422
14. Kontaki, M., Gounaris, A., Papadopoulos, A.N., Tsichlas, K, Manolopoulos, Y.: Continuous monitoring of distance-based outliers over data streams. In: ICDE, pp. 135–16 (2011). https://doi.org/10.14778/2994509.2994526
15. Pokrajac, D., Lazarevic, A., Latecki, L.J.: Incremental local outlier detection for data streams. In: CIDM, pp. 504–515 (2007)
16. Salehi, M., Leckie, C., Bezdek, J.C., Vaithianathan, T., Zhang, X.: Fast memory efficient local outlier detection in data streams. IEEE Trans. Knowl. Data Eng. **2812** (2016). https://doi.org/10.1109/TKDE.2016.2597833
17. Lee, J., Cho, N.W.: PLoS One. **11**(11), e0165972, 10 Nov 2016. https://doi.org/10.1371/journal.pone.0165972
18. LOF algorithm flow chart. https://www.researchgate.net/figure/284161916_fig3_Figure-5-LOF-algorithm-flow-chart

Context Level Entity Extraction Using Text Analytics with Big Data Tools

Papiya Das, Kashyap Barua, Manjusha Pandey and Siddharth Swarup Routaray

Abstract Private organizations like offices, libraries, and hospitals makes use of computers for computerized database, when computers became a most cost-effective device. After that E.F Codd introduced relational database model, i.e., conventional database. Conventional database can be enhanced to temporal database. Conventional or traditional databases are structured in nature. But always we do not have the pre-organized data. We have to deal with different types of data. That data is huge and in large amount, i.e., big data. Big data mostly emphasized into internal data sources like transaction, log data, emails, etc. From these sources, high-enriched information is extracted by the means of process text data mining or text analytics. In this research work, we will briefly discuss text analytics and its different types and tasks.

Keywords Computerized database · Conventional database · Temporal database Text data mining · Text analytics

1 Introduction

First of all data can be facts that can be recorded or stored. For example, any text can be considered as data like the name of any person, number, address, age, height, weight, etc. A picture, image file, and PDF can also be considered data. Database is a collection of related data. It can also be text and numbered data. Database is of

P. Das (✉) · K. Barua · M. Pandey · S. S. Routaray
School of Computer Engineering, Kalinga Institute of Industrial Technology (KIIT), Deemed to be University, Bhubaneswar 751024, Odisha, India
e-mail: papiyanita895@gmail.com

K. Barua
e-mail: kashyapbarua@gmail.com

M. Pandey
e-mail: manjushafcs@kiit.ac.in

S. S. Routaray
e-mail: siddharthfcs@kiit.ac.in

A. Abraham et al. (eds.), *Emerging Technologies in Data Mining and Information Security*, Advances in Intelligent Systems and Computing 813,
https://doi.org/10.1007/978-981-13-1498-8_32

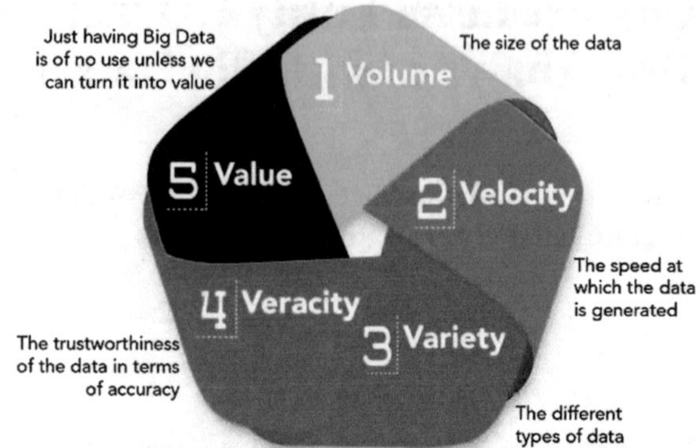

Fig. 1 V's of big data

different types. As we have already discussed text and numbered database, another type of database [2] is traditional database. Recently, numerous number of videos are uploaded and downloaded everyday. This task is done in the database of video which is created. One more database is Multimedia DB. Recently, NASA has geographic information system (GIS) for analyzing the spatial data. They are sending a lot of satellites;satellites are taking image of earths and send them back. Now, when database is going to be huge, then it is called data warehouse. Data mining is based on semi-structured data. Mining is of different types. One is text mining. Text mining is the process of distilling actionable insights from text. When a satellite is taking image for social media pictures and traffic information system. There is too much information too handle and organize. These are the bunch of texts which are nearly impossible to organize quickly. Text mining is all about organizing the unorganized text. The data scientists of IBM splits big data into [1] 4 Vs; Volume, Variety, Velocity, and Veracity (Fig. 1).

Sometimes, a fifth V value [1] is also considered. The only main aim of this division is to extract business value and to improve their organizational power. As business value is more focused in an organization. Volume the word itself is describing its nature. As the worlds records every year data generation graph is increasing. So it merely thinks about small storage capacities. Data's increasing nature starts from bits to bytes, bytes to kilobytes, and stops in exabytes till now. Many more may come in future. Variety deals with both structured and unstructured data. Traditional structured data includes users bank account statements (Table 1).

Audio images, video images, and logs are structured data mounted by unstructured data. Veracity is the biggest hurdle in big data analytics. It maintains the cleanliness of data and removes the dirty data. Velocity is the data generating rate. It is a tremendous and extensive method.

Table 1 Data measurement charts with its units and values

Data measuring units	Values
Bit	0, 1
Byte	8 bits
Kilobyte	1000^1 bytes
Megabyte	1000^2 bytes
Gigabyte	1000^3 bytes
Terabyte	1000^4 bytes
Petabyte	1000^5 bytes
Exabyte	1000^6 bytes
Zettabyte	1000^7 bytes
Yottabyte	1000^8 bytes

2 Application Domains

(i) **Banking and Securities**: Applications of big data in banking are very crucial. The Securities Exchange Commission (SEC) use big data for monitoring stock and financial market.

(ii) **Communications, Media and Entertainment**: By this, we collect and an analyze consumer insights. It tries to understand patterns of real-time contents of data.

(iii) **Healthcare Providers**: Hospitals use data from mobile for millions of patients for detecting several lab tests.

(iv) **Education**: Big universities and institutions use big data tools for managing student information according to a specific key or primary key, i.e., roll number. Tools may be SQL method.

(v) **Manufacturing and Natural Resources**: In this approach, big data uses predictive modeling for integrating large amounts of data. The types of data may vary like it can be graphical data, text data, or may be temporal data.

(vi) **Government**: Government uses big data in various fields like in Social Security Administration (SSA) and The Food and Drug Administration (FDA).

(vii) **Insurance**: In the insurance industry, they predict the customers insight and predictive behavior from different social media, GPS-enabled device, CCTV footage, etc.

(viii) **Finance**: Big data uses technical analysis in the financial market.

(ix) **Internet of Things**: Big data and IoT works concurrently. The targeted data is [12] extracted from the IoT device for preprocessing and it provides mapping of device interconnectivity.

(x) **Sports**: Big data uses prediction modeling in sport sensors for performance improvement of players. It also can predict winners in a match using [4] big data analytics.

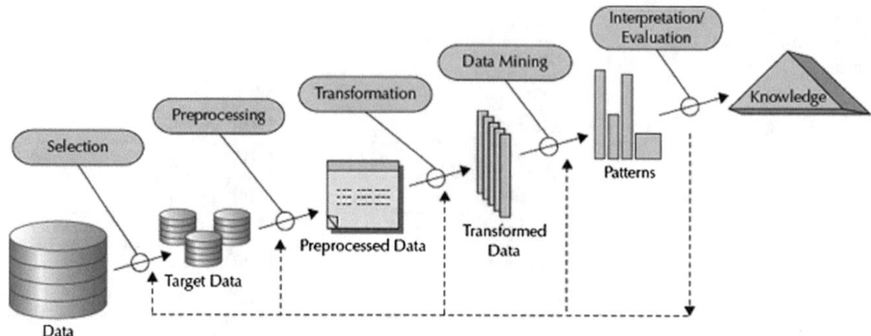

Fig. 2 Knowledge discovery process

3 Literature Survey

Text analytics is all about conversion of unstructured data into meaningful data for analysis. Different process and techniques of text analytics have been briefly described below. How text analytics has evolved and what is its techniques have been discussed depending upon some previous year papers.

3.1 Knowledge Discovery from Database

Organized text uses the knowledge discovery of database (KDD) process to extract useful information from the huge amounts of data. Knowledge discovery process includes different steps for gaining useful information. For extraction target data set is selected for preprocessing. Depending on the goal of KDD process technique for data mining is decided whether it is classification, regression, etc. After this mined knowledge or pattern is discovered as in Fig. 2.

3.2 Text Analytics

Text analytics [4] is all about gathering of meaningful information and data from dirty data for the analysis. It is also a complete package of mining, categorization, clustering, etc. It also involves process of tagging, annotation, pattern recognition, etc. Text analytics is based on application enriched analytical algorithms for drawing out outputs which are in XML format or in data warehousing applications. It is also a predictive analysis method. When the training data sets or texts come user categorized the texts into different portions or texts for classification. This classification process includes the text preprocessing steps. Steps are cleaning, tokenization, POS

tagging, transformation, evaluation, selection of attributes, insights identification, etc. Preprocessing includes different steps.
Text cleaning: Text cleaning involves removal of unwanted or dirty data. Example: ads in webpages, popups coming in websites, etc.
Tokenization: Tokenization is breaking of textual contents into words, symbols named as tokens.
Part Of Speech Tagging: POS tagging includes transformation, selection, mining, and evaluation. After the tokenization process, tagging is assigned to each token.
Text Transformation: Text transformation is also attribute generation. Its main approaches are bag of words and vector space.
Feature Selection: Feature selection is also attribute selection. Main aspect of this method is to remove redundant and irrelevant features.

3.3 Analytics Techniques

Text analytics techniques are based on different applications of text analysis. When the training data sets or texts come, user categorized the texts into different portions or texts for classification. This classification process includes the text preprocessing steps [5]. Steps are cleaning, tokenization, POS tagging, transformation, evaluation, selection of attributes, insights identification, etc. When the raw text or raw data comes, statistical or linguistic techniques applied for the text analysis. Then the texts are categorized according to their taxonomical behavior. Then the concepts and patterns are extracted for getting relationships in large amounts of text. After that accuracy level of the model is being checked. When there is unstructured, ambiguous data which is difficult to process, text preprocessing method is used as shown in Fig. 3.

Preprocessing includes different steps as in Fig. 4. First step is text cleaning which involves removal of unwanted or inconsistent data, e.g., ads in webpages, popups coming in websites, etc. Tokenization is breaking of textual contents into words and

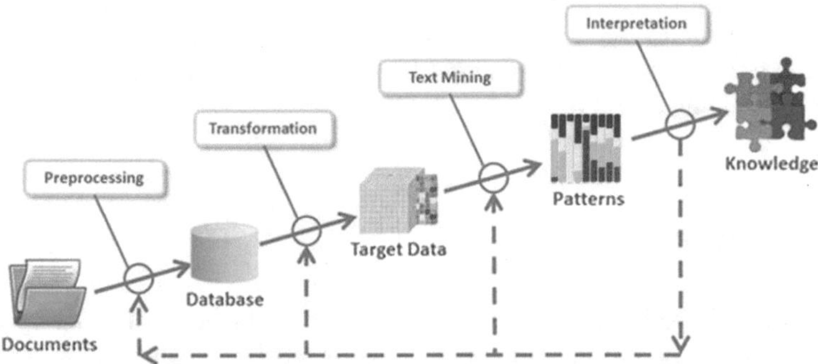

Fig. 3 Text mining Process

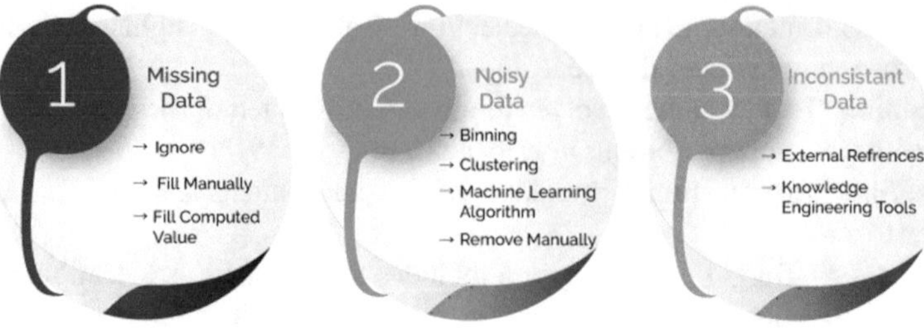

Fig. 4 Text preprocessing method

symbols named as tokens. Part of speech tagging includes transformation, selection, mining, and evaluation. After the tokenization process, tagging is assigned to each token [5]. Text transformation is also attribute generation. Its main approaches are bag of words and vector space. Feature selection is also attribute selection. Main aspect of this method is to remove redundant and irrelevant features.

3.4 Text Analytics Applications

Different applications of text mining are sentiment analysis, feedback analysis, competitive and computing intelligence, national security issues, accuracy and precision measurement of data, monitoring of social media, management of e-record and data. Many top free text analytics [5] software are there for analysis of text. Topic modeling and TM are the most wanted platforms for evaluation of data and getting targeted data set. Text analysis is also known as text mining. Mining is of different types: Data mining, text mining, web mining, etc. Data mining is extraction of valuable information from huge amount of data collected from data warehouse. Text mining is nothing other than text analytics.We have discussed text analytics [4] in the introduction part. Information gathered by Web Mining is the collection of data from traditional data mining methodologies and information covered over World Wide Web (WWW). Other mining techniques are sequence mining, graph mining, temporal data mining, spatial data mining (SDM), distributed data mining (DDM), and multimedia mining (Fig. 5).

3.5 State of the Art

Garcia et al. [5] proposed a technique for achieving highest performance in big data projects without preprocessing of data with tools like parsing and querying. In this paper, Hadoop and other open source parsers used in RDF raw data set. NX parser and Jena parser are used. But Jena parser is slowest. Because it has a large

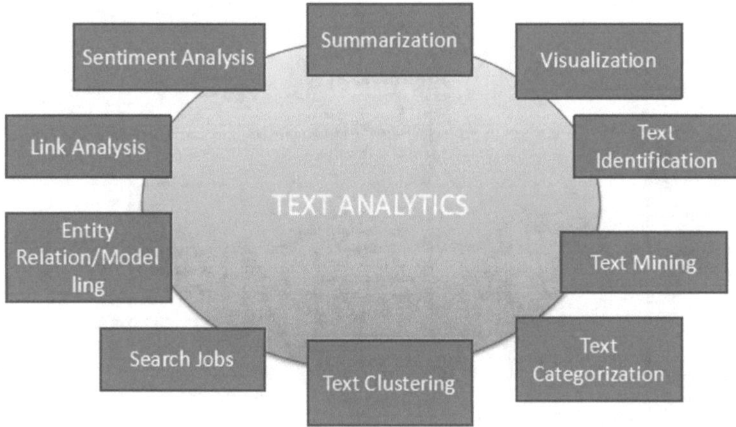

Fig. 5 Text analytics applications

code database and perform validation while parsing, which may cause data loss. Celik et al. [3] proposed a method about resumes. Job seekers upload resumes in many formats. In this work, kariyer.net and TUBITAK jointly worked to introduce free structured resumes. They have used ORP (ontological based resume parser) for extracting information in efficient way. English and Turkish resumes are converted in ontological format for concept matching tasks. But we can convert only English and Turkish resumes. Jose et al. [10] proposed a technique about the analysis of association of academic performance and age. Predictive model is used for forecasting student performance in higher level modules based on the contextual factors. But the limitation is they used MEC data set for predictive modeling, which has limited student records. Javed et al. [8] discussed job title classification based system. They used machine learning system which is very important in production of career building. Hua et al. [6] proposed a method on short text harvesting and analyzing the semantic knowledge. Short text is crucial to understand. It always do not justify the syntax of a written language. In short texts, dependency parsing is hard to apply. Due to short texts and equivocal property, it expands and generates data in large content, which is tough to handle. They derive their frameworks in three subcategories of text segmentation, type detection, concept labeling. Jianqiang et al. [9] proposed a research technique on text preprocessing methods on Twitter Sentiment Analysis. It analyzes public feeling towards different events for elicitation of sentiment features. They divide the classification performance in two types of tasks and epitomize the performance of classification using classifiers. Mandal et al. [11] discussed the use of Hadoop Distributed File System (HDFS) with MapReduced model for counting the number of consecutive words in Word processor. They showed use of MapReducing for generating a set of key-value pairs as output. Figure 6 describes the MapReduce framework.

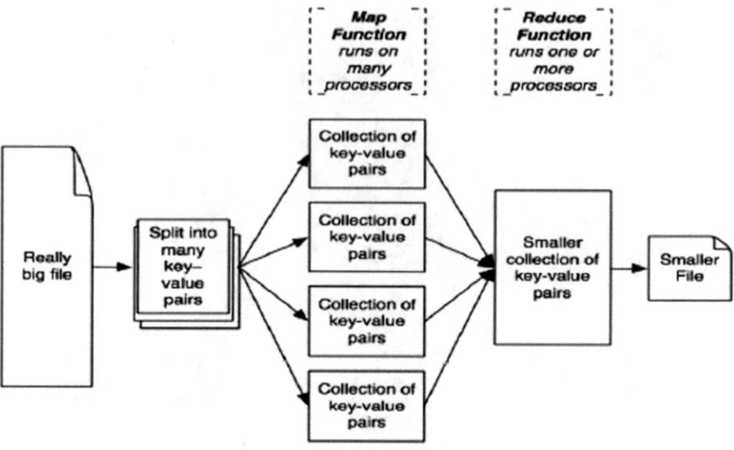

Fig. 6 MapReduce framework

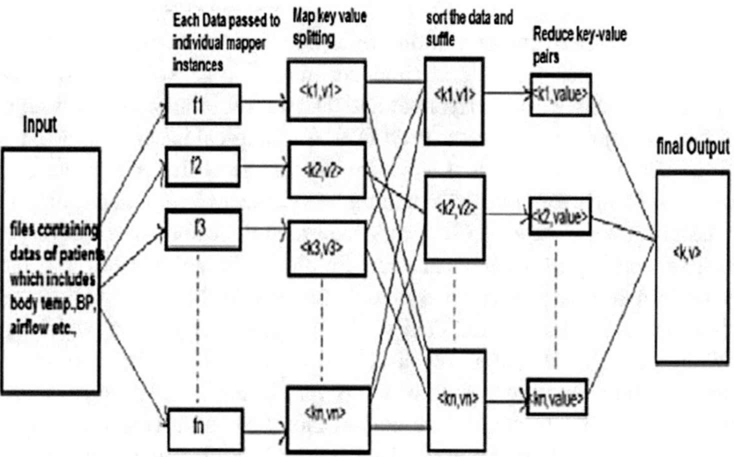

Fig. 7 Word processing using hadoop MapReduced with one input system

In Fig. 7, the input system is MapReduced for efficient processing of word [11] using Word Count method. It reads out text input files and counts the occurrence number of each word. The mapper function takes each line as input and divides it into words, producing a key-value pair of each word. The output produced by the mapper is then shuffled and sorted by value as per the occurrences of each word, which then becomes input to the reducer. Then the reducer sums the counts of every word and generates a single key-value with the word. Table 2 provides the summary of work carried out (Fig. 8).

Table 2 Survey papers

Year, Author	Paper title	Proposed mechanism	Advantages	Discussion
2013, Garcia et al. [5]	Analysis of Big data technologies and methods	Hadoop, open source parser	For providing rich set of information about tackling big data	Using of different CPU model for optimal approach
2013, Celik et al. [3]	Towards an information extraction system based on ontology to match resumes and jobs	Ontology based resume parser for finding resume	Plain text resume into ontology form by Ontology Knowledge Base (OKB)	System calculates percentage completeness depend upon work experience, education etc.
2016, Jose et al. [10]	Progression analysis of students in a higher education institution using big data open source predictive modeling tool	Predictive analysis depends upon academic performance and age	Research on student data set whose age is more than 30 to isolate factors in academic performance	Correlation analysis, Predictive modeling for determining data sets
2015, Javed et al. [8]	Carotene: A Job Title Classification System for the online recruitment domain	Carotene classification system for online recruitment	Job title classification by carotene architecture by SVM-KNN method	Used SVM and KNN method in Carotene architecture
2017, Hua et al. [6]	Understand Short Texts by Harvesting and Analyzing Semantic Knowledge	Chain model,pair wise model, Monte Carlo method	For maintaining accuracy and efficiency in short texts to extract semantic knowledge	For removal of ambiguity in short texts
2017, Jianqiang et al. [9]	Comparison research on text preprocessing methods on Twitter sentiment analysis	N-grams model, prior polarity model, Nave Bayes Classifier	To identify opinion expression in piece of text	Large Volumes of data
2015, Mandal et al. [11]	Architecture of efficient word processing using Hadoop MapReduce for big Data applications	Hadoop Map Reduce, Hadoop Distributed File System	To count the number of consecutive words and repeating lines	Time consuming method

(continued)

Table 2 (continued)

Year, Author	Paper title	Proposed mechanism	Advantages	Discussion
1995, Ulusoy et al. [13]	Research Issues in Real-Time Database Systems	Transaction/query processing,data buffering, CPU and IO scheduling	For satisfying timing constraints in real time database applications	Replacement of conventional database systems to eliminate disk access delays
2013, Joshi et al. [2]	Distributed database	Distributed Data mining	Partitioning method to store large amount of data on different site or server	Use of distributed data mining on distributed database
2016, Janani et al. [14]	Text mining research	Natural Language Processing (NLP), clustering	Efficiency increased with text mining tools in the extraction point	Text Mining in different languages
2014, Jadhav et al. [7]	Survey on Text Mining and its Techniques	Text Mining Process, NLP	Useful relevant information from dirty data	Ambiguity, Time consuming for handling lots of unstructured text

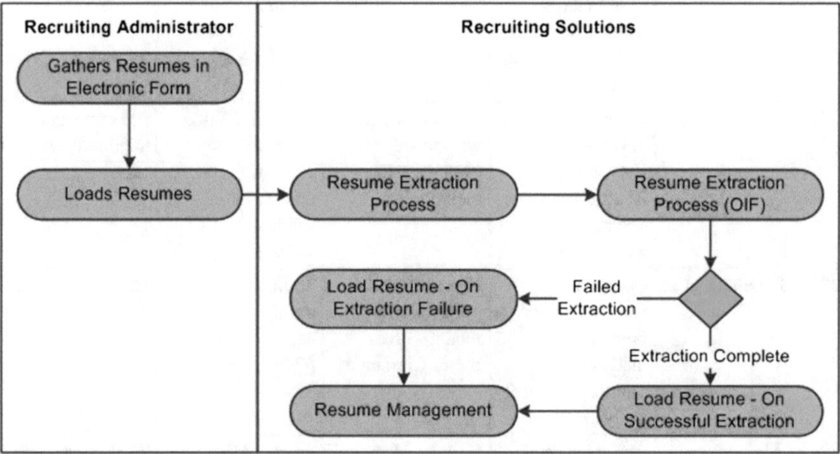

Fig. 8 CV parser model

4 Conclusion and Industry Applications

Text analytics is all about conversion of unstructured data into meaningful data for analysis. Different process and techniques of text analytics have been briefly described above. How text analytics has evolved and what is its techniques have been discussed depending upon some previous year papers. Text analysis techniques can be used in resume filtering in recruitment process in the companies. Our future scope will reflect on developing a CV parser model using text analytics process. The data sets of big data are large and complex in nature. So, many software have been introduced to tackle such large databases. CV parsing is such a technique for collecting CVs. CV Parser supports multiple languages, semantic mapping for skills, job boards, recruiter, and ease of customization. Recruiter companies use CV parser technique for selection of resumes. As resumes are in different formats and they have different types of data like structured and unstructured data, meta data, etc. The CV parser technique will provide the entity extraction method from the uploaded CVs.

References

1. Anuradha, J., et al.: A brief introduction on big data 5Vs characteristics and hadoop technology. Procedia Comput. Sci. **48**, 319–324 (2015)
2. Bamnote, G., Joshi, H.: Distributed database: a survey. Int. J. Comput. Sci. Appl. **6**(2), 09741011 (2013)
3. Çelik, D., Karakas, A., Bal, G., Gultunca, C., Elçi, A., Buluz, B., Alevli, M.C.: Towards an information extraction system based on ontology to match resumes and jobs. In: 2013 IEEE 37th Annual Computer Software and Applications Conference Workshops (COMPSACW), pp. 333–338. IEEE (2013)
4. Ferguson, M.: Architecting a big data platform for analytics. A Whitepaper prepared for IBM 30 (2012)
5. Garcia, T., Wang, T.: Analysis of big data technologies and method-query large web public RDF datasets on amazon cloud using hadoop and open source parsers. In: 2013 IEEE Seventh International Conference on Semantic Computing (ICSC), pp. 244–251. IEEE (2013)
6. Hua, W., Wang, Z., Wang, H., Zheng, K., Zhou, X.: Understand short texts by harvesting and analyzing semantic knowledge. IEEE Trans. Knowl. Data Eng. **29**(3), 499–512 (2017)
7. Jadhav, A.M., Gadekar, D.P.: A survey on text mining and its techniques. Int. J. Sci. Res. (IJSR) **3**(11) (2014)
8. Javed, F., Luo, Q., McNair, M., Jacob, F., Zhao, M., Kang, T.S.: Carotene: a job title classification system for the online recruitment domain. In: 2015 IEEE First International Conference on Big Data Computing Service and Applications (BigDataService), pp. 286–293. IEEE (2015)
9. Jianqiang, Z., Xiaolin, G.: Comparison research on text preprocessing methods on twitter sentiment analysis. IEEE Access **5**, 2870–2879 (2017)
10. Jose, M., Kurian, P.S., Biju, V.: Progression analysis of students in a higher education institution using big data open source predictive modeling tool. In: 2016 3rd MEC International Conference on Big Data and Smart City (ICBDSC), pp. 1–5. IEEE (2016)
11. Mandal, B., Sethi, S., Sahoo, R.K.: Architecture of efficient word processing using hadoop mapreduce for big data applications. In: 2015 International Conference on Man and Machine Interfacing (MAMI), pp. 1–6. IEEE (2015)
12. Narasimhan, R., Bhuvaneshwari, T.: Big data brief study. Int. J. Sci. Eng. Res. **5**(9), 350–353 (2014)
13. Ulusoy, O.: Research issues in real-time database systems: survey paper. Inf. Sci. **87**(1–3), 123–151 (1995)
14. Vijayarani, S., Janani, M.R.: Text mining: open source tokenization tools—an analysis. Adv. Comput. Intell. **3**(1), 37–47 (2016)

A Novel Hybrid Approach for Time Series Data Forecasting Using Moving Average Filter and ARIMA-SVM

Gurudev Aradhye, A. C. S. Rao and M. D. Mastan Mohammed

Abstract Time series data consists of a variety of information in its patterns. It is composed of both linear and nonlinear parts. Depending on nature of time series data, either linear model or nonlinear model can be applied. Instead of applying linear time series model like Auto Regressive Integrated Moving Average (ARIMA) and nonlinear time series model like Support Vector Machine (SVM) and Artificial Neural Network (ANN) individually on time series data, the proposed hybrid model decomposes time series data into two parts using Moving Average Filter and applies ARIMA on the linear part of time series data and SVM on nonlinear part of time series data. The performance of the proposed hybrid model is compared using Mean Absolute Error (MAE) and Mean Squared Error (MSE) with the performances obtained by the conventional models like ARIMA, SVM, and ANN individually. The proposed model has shown efficient prediction results when compared with the results given by the conventional models of time series data having trended patterns.

Keywords ARIMA · Artificial neural network · Support vector machine Moving average filter

1 Introduction

Time series forecasting techniques are data analytical methods that are widely used in different areas such as marketing, business process, financial sector, weather forecasting, and climate change variations. Time series data can be volatile or nonvolatile.

G. Aradhye (✉) · A. C. S. Rao
IIT(ISM) Dhanbad, Dhanbad 826004, India
e-mail: gurudevaradhye2015@gmail.com

A. C. S. Rao
e-mail: acrao232@yahoo.co.in

M. D. Mastan Mohammed
V. R. Siddhartha Engineering College, Vijayawada 520007, India
e-mail: mastan559@gmail.com

A. Abraham et al. (eds.), *Emerging Technologies in Data Mining and Information Security*, Advances in Intelligent Systems and Computing 813,
https://doi.org/10.1007/978-981-13-1498-8_33

For example, wind speed is volatile time series data while the global temperature is nonvolatile data. Some of the time series data are in a linear form which includes "growth of the body parts" and some are in nonlinear form includes "moving speed of vehicle". For time series forecasting, various techniques are mentioned in literature depending on nature of data. For linear time series data, Auto Regressive Integrated Moving Average (ARIMA) and for nonlinear time series data, models like Support Vector Machine (SVM) and Artificial Neural Network (ANN) are used. ARIMA model makes the prediction based on past observations and past errors. This model is widely used for the prediction of data, such as sugarcane yield [1], stock index [2], next day electricity price [3], and wind speed forecasting [4]. ANN forecasting model is used for nonlinear data. This prediction model is widely used in many applications such as electric demand [5], electricity price [6], river flow [7], and finical distress [8]. Support Vector Machine (SVM) model is also used for wide range of forecasting applications such as crude oil price [9], wind power [10], electricity load [11].

1.1 Introduction to Time Series Data

Time series data is a collection of data observation over a periodic time interval such as an hour, day, week, month, and year. It is used as data analytical tool for making predictions based on historical data. Depending on nature of data, various time series modelings can be applied. The models such as ARIMA, Simple Exponential Smoothing (SES) are used for modeling linear data. Whereas SVM, ANN approaches are used for modeling nonlinear data. As the data is not always a linear or nonlinear data, hybrid models are used for time series forecasting. The predicted values and real values of time series data are used to decide the performance criteria of models.

1.2 Introduction to ARIMA Model

This model was introduced by Box and Jenkins and it integrates Auto Regressive (AR) and Moving Average (MA) model, abbreviated as ARIMA [15]. Auto Regressive (AR) model is mathematically represented as

$$y_t = c + \phi_1 y_{t-1} + \phi_2 y_{t-2} + \cdots + \phi_p y_{t-p} + e_t \tag{1}$$

where c is constant and e_t is white noise. Above equation is of form regression with lagged values, which are used as predictors.
Moving Average (MA) model is mathematically represented as

$$y_t = c + e_t + \Theta_1 e_{t-1} + \Theta_2 e_{t-2} + \cdots + \Theta_q e_{t-q} \tag{2}$$

where e_t is white noise. The Moving Average is a regression with lagged errors. Combination of both AR and MA is said to be Auto Regressive and Moving Average (ARMA) model. It can be mathematically represented as

$$y_t = c + \phi_1 y_{t-1} + \phi_2 y_{t-2} + \cdots + \phi_p y_{t-p} + \Theta_1 e_{t-1} + \Theta_2 e_{t-2} + \cdots + \Theta_q e_{t-q} + e_t \tag{3}$$

The ARMA model with differencing parameter is represented as $ARIMA(p, d, q)$ [13].
Where
p: AR model parameter
q: MA model parameter
d: Number of differences made to make time series stationary.

1.3 Artificial Neural Network (ANN)

The ANN is widely used for forecasting the nonlinear time series data. Neural network can be of two layers or more layers. Neurons are handling units which are connected to each other. Feedforward ANN is mostly utilized for the time series forecasting. Feedforward ANN architecture [14] is shown in Fig. 1. At time t, the observation is y_t, which is the function of past observations. It is given by

$$y_t = g(y_{t-1}, y_{t-2}, \ldots . y_{t-n}) + e_t \tag{4}$$

where e_t is the error at time t.
The Sigmoid function is a bias function and it can be represented as

$$sigmoid(x) = \frac{1}{1 + e^{-x}} \tag{5}$$

w_{ij} represents the weight value, and b represents the bias value. To obtain weight coefficients, known data sequence is given as input to the model. Training sequence is given to model, which results in minimizing the global error function. Levenberg–Marquit (LM) is used for the training purpose. The results obtained from ANN are better than the ARIMA model for nonlinear time series data.

1.4 Support Vector Machine

Support vector machine regression is machine learning technique which is based on supervised learning. It limits the problem of local minima and over fitting as compared with ANN. Reason for that is Support Vector Machine Work on structural risk minimization rather than empirical risk minimization. SVM has its learning technique capable to work with high-dimensional space even with small number of

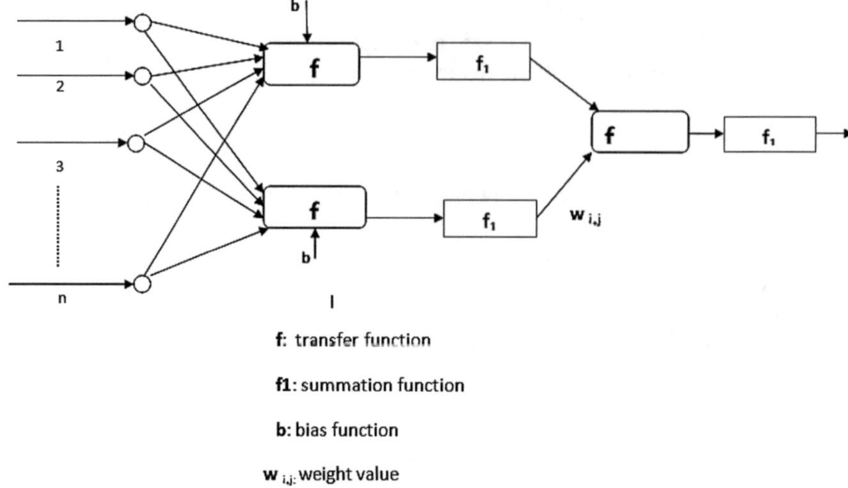

Fig. 1 Artificial neural network architecture

training set. In SVM, the small subset of data points is called as support vectors. There are four basic types of kernels Linear, Polynomial, Radial Basic Function (RBF), and sigmoid kernel. The equation for RBF is represented as

$$K(x, y) = exp(-\mu||x - y||^2), \mu > 0 \tag{6}$$

$$f(x) = \omega(\phi(x)) + b \tag{7}$$

The RBF has two parameters μ and b, which is cost parameter of error. In case of time series sliding window SVM is useful. In case of sliding window SVM, the next value of time stamp V_{t+1} is used as target for the input variable V_t. Similarly V_{t+2} is target for the input variable V_{t+1} and so on.

1.5 Time Series Decomposition

Time series data contains a large amount of information in form of patterns. This pattern information is helpful to categorize behavior of data observed over a time interval [15]. Three different types of time series patterns are given as

1. Trend: In given data trend pattern, if there exists long-term increase or decrease then such pattern is called a trend.
2. Seasonal: In given data, the seasonal pattern exists for seasonal factors such as a month, quarter, or a year.
3. Cycle: In given data, if there is a rise or falls that are not fixed period then cycle exits.

1.6 *Moving Average Filter*

Moving Average Filter is a classical method for time series decomposition. Moving Average Filter of order m is given by

$$T_1 = \frac{1}{m} \sum_{j=-k}^{k} y_{t+j} \tag{8}$$

where m=2k+1. Trend cycle is obtained by averaging the time series with k periods of t. The observations obtained are close in value and averages remove randomness in data. T_2 is Detrend part obtained by subtracting the trend part of time series from the original time series [12].
$T_2 = T - T_1$
T_1: Trend part of Time series.
T_2: Detrend(Residual) part of Time series.
T: Original Time series.

1.7 *Performance Criteria*

The sequence of actual values is represented by $X_{obs,i}$ whereas $X_{model,i}$ represents the forecasted values. Mean Square Error (MSE) is given by

$$MSE = \frac{\sum_{i=1}^{n} (X_{obs,i} - X_{model,i})^2}{n} \tag{9}$$

and Mean Absolute Error (MAE) is given by

$$MAE = \frac{\sum_{i=1}^{n} (X_{obs,i} - X_{model,i})}{n} \tag{10}$$

2 Proposed Architecture and Methodology

2.1 *Architecture*

The ARIMA model used for forecasting the linear time series data. Whereas SVM and ANN models are used for the nonlinear time series data. Both the models have its own advantages and disadvantages. The proposed model works on parallel computation. In this model, The Moving Average filter decomposes time series data into two parts namely trend data part and detrend data part. On trend data part, ARIMA model is applied for forecasting the values and on detrend data part SVM model is applied for forecasting the values. The results obtained from each part are combined to form the final forecasted values. The proposed model architecture is shown in Fig. 2.

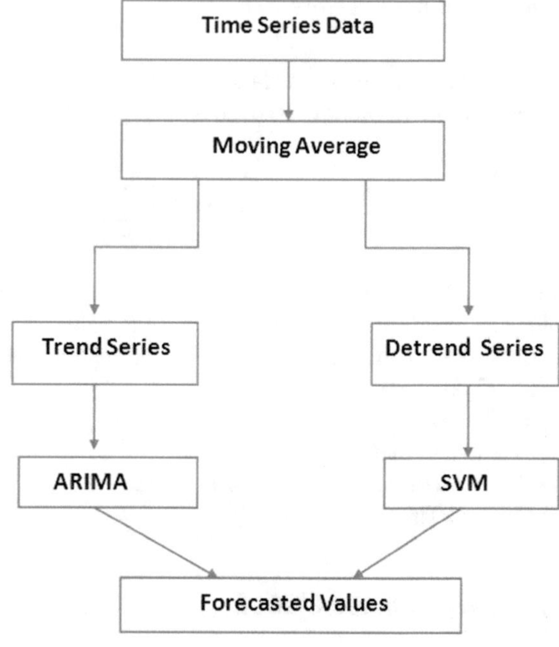

Fig. 2 Proposed model architecture

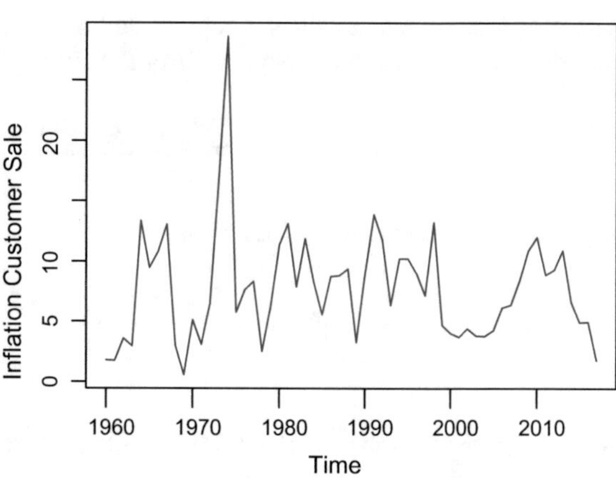

Fig. 3 Inflation consumer sale

2.2 *Datasets and Methodology*

The evaluation of the performance of the proposed model is done using two time series datasets. First, the historical data for inflation consumer sale price. Second, crude oil production in India. These datasets are used to compute performance measure. Time series plots of inflation consumer sale index from 1960–2016 of India [16] the important economic indicator and crude oil production in India from historical information from 1965–2014 [16] are shown in Fig. 3 and Fig. 5, respectively (Fig. 4).

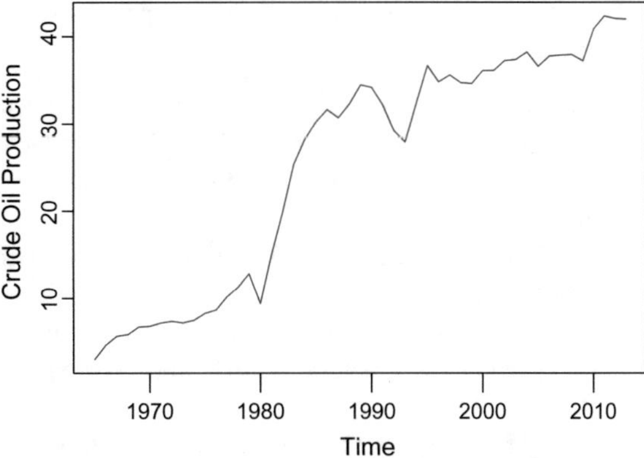

Fig. 4 Crude oil production

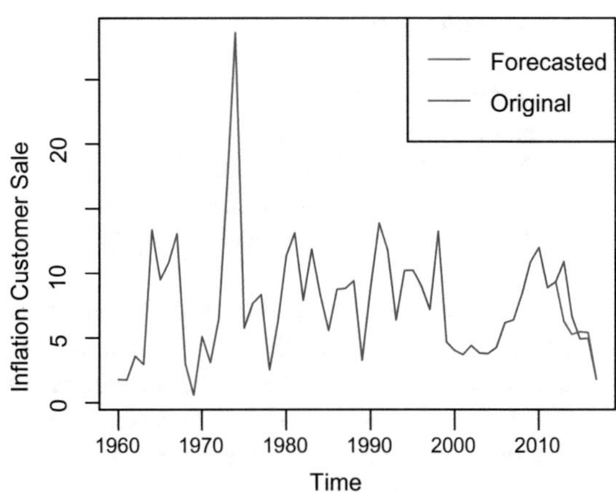

Fig. 5 Compare inflation customer sale ARIMA and original

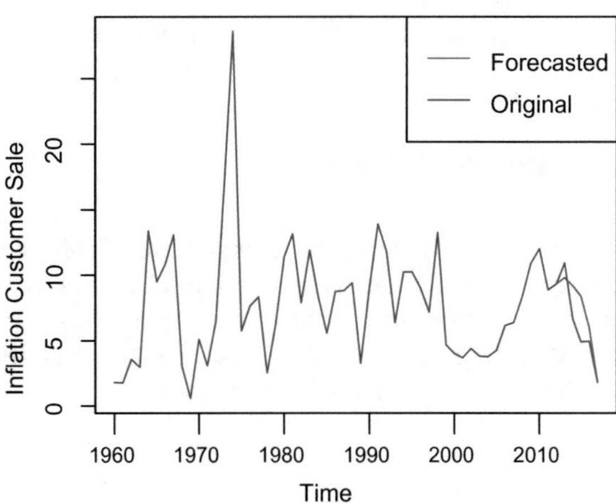

Fig. 6 Compare inflation customer sale SVM and original

In order to compare performance of proposed model, the following steps have been performed:

1. Apply ARIMA model: Eqs. (1) and (2) are used for finding AR(p), MA(q) parameters of ARIMA model. Apply ARIMA model for the mentioned time horizon and find predicted values and calculate MSE and MAE using Eqs. (8) and (9).
2. Apply ANN model: Eqs. (3) and (4) are used to find predicted values from ANN model. Find the values for the MSE and MAE using Eqs. (8) and (9).
3. Apply SVM model: Eqs. (5) and (6) is used to apply Support Vector Machine to find predicted values and Eqs. (8) and (9) are used to find the values of MSE and MAE.
4. Apply Moving Average Filter: Eq. (7) is used to find the trend series from given time series data. The Detrend part is obtained by subtracting trend series from original time series.
   $z = x + y$.
   z: Original Time series.
   x: Trend Part of Time series.
   y: Detrend Part of Time series.
5. Apply ARIAM for Trend part: Eqs. (1) and (2) are used for finding AR(p), MA(q) parameters of ARIMA model. x_1 represents the forecasted values of the trend series.
6. Apply SVM for Detrend Part: Eqs. (5) and (6) are used to apply Support Vector Machine. y_1 represents the forecasted value for Detrend Series.
7. Combine Results: Add the forecasted results obtained from ARIMA model and SVM model, which are x_1 and y_1 respectively, such that $z_1 = x_1 + y_1$.
8. Performance criteria: In our case, z_1 represents the forecasted values and z represents the original values. Use Eqs. (8) and (9) to find the MSE and MAE of the proposed model.
9. Compare the obtained results with the results of ARIMA, SVM, ANN models.

3 Numerical Results

In order to apply numerical computation, divide the dataset into training set and testing set. The model using training set is built and performance is checked with testing set. Performance criteria used for comparison are Mean Absolute Error(MAE) and Mean Squared Error(MSE). The experimental details of the proposed model by using the datasets are described as follows.

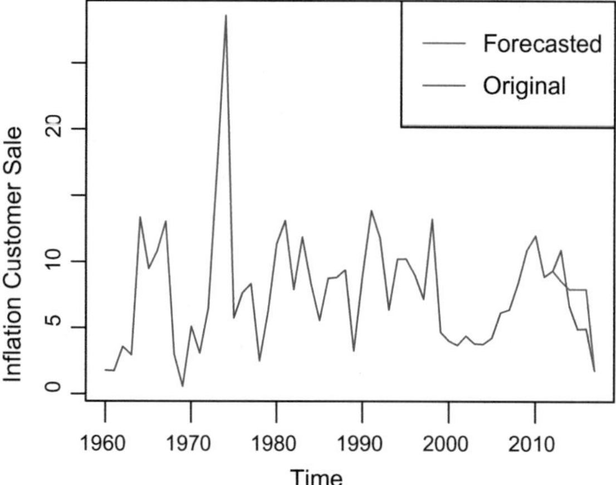

Fig. 7 Compare inflation customer sale ANN and original

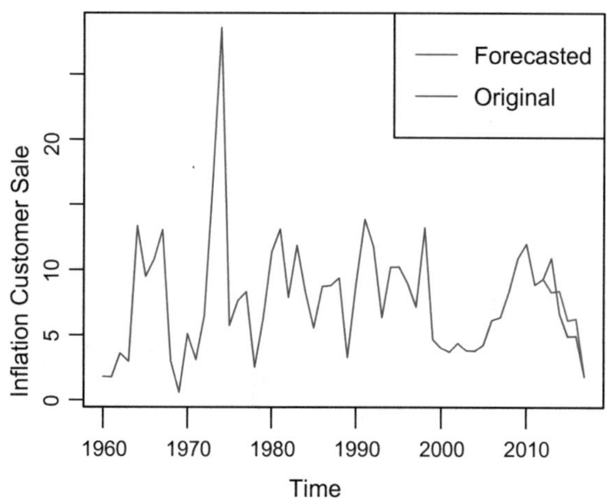

Fig. 8 Compare inflation customer sale proposed and original

3.1 Inflation Customer Sale Index

The dataset Inflation Consumer Sale Index for India is one of the important economical indicator. It contains historical data from 1960 to 2016 [16]. The corresponding results of ARIMA, SVM, ANN, and the proposed model are shown from Figs. 5, 6, 7 and 8. The original time series plot is shown in blue color while the forecasted is displayed in red. The results of MAE and MSE are described in Table 1. The proposed model shows better performance of MAE and MSE when compared with ARIMA, SVM, and ANN models on given dataset of Inflation Consumer Sale Index time series data.

Table 1 Result for inflation consumer index

Model	MAE	MSE
ARIMA	1.759	5.998
SVM	2.039	5.185
ANN	2.404	6.277
Proposed	1.722	3.265

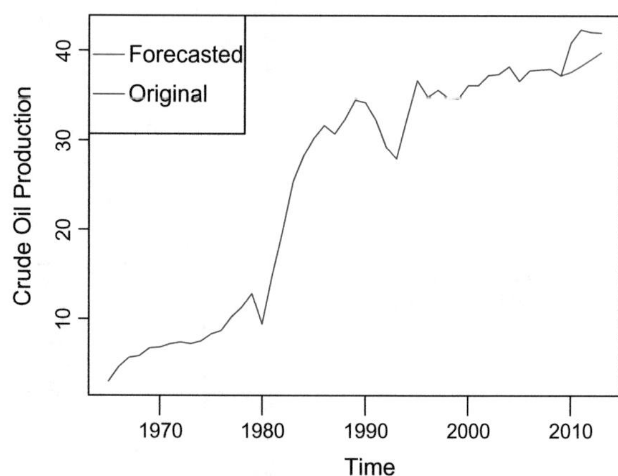

Fig. 9 Crude oil production for ARIMA and original

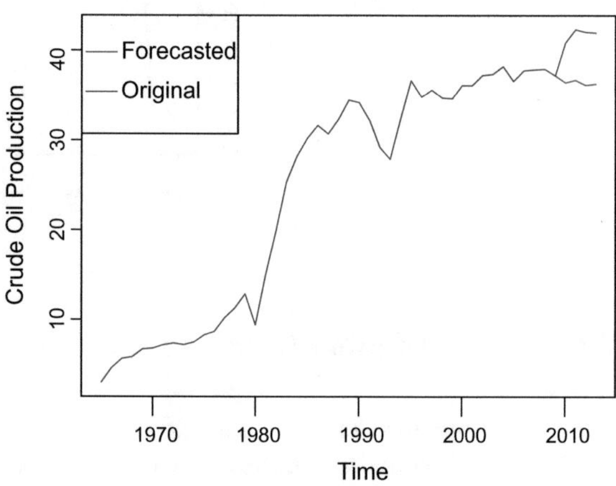

Fig. 10 Crude oil production for SVM and original

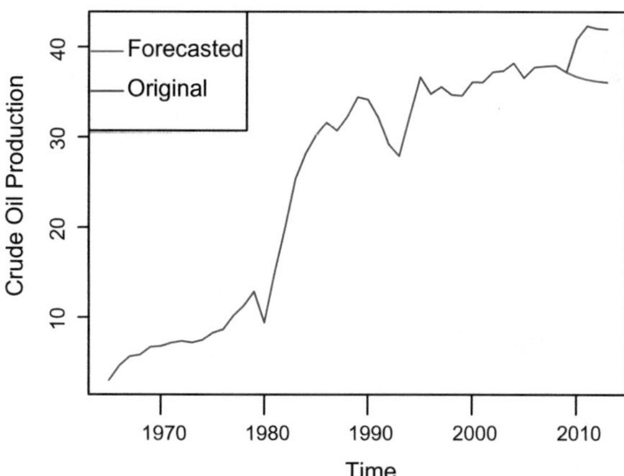

Fig. 11 Crude oil production for ANN and original

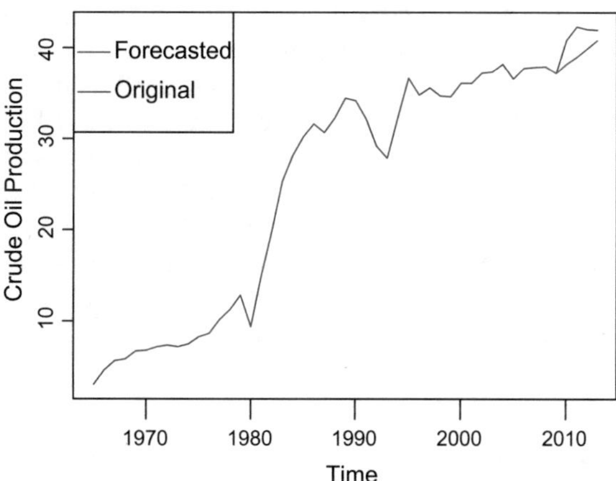

Fig. 12 Crude oil production for proposed and original

3.2 Crude Oil Production in India

The dataset Crude oil production in billion tones gallon for India is historical data from year 1965–2013 [16]. The corresponding results of ARIMA, SVM, ANN, and the proposed model are visualized from Figs. 9, 10, 11 and 12. Original time series plot is shown in blue color while the forecasted is displayed in red. The results of MAE and MSE are described in Table 2. The proposed model shows better performance of MAE and MSE when compared with ARIMA, SVM, and ANN models on given dataset of Crude oil production for India time series data.

Table 2 Result for crude oil production

Model	MAE	MSE
ARIMA	3.138	10.290
SVM	5.400	29.493
ANN	5.443	30.189
Proposed	2.314	5.991

4 Conclusion

In this paper, the time series forecasting models such as ARIMA, SVM, and ANN are used to find the performance criteria of time series data. Further, in the proposed model, the moving average filter is used to decompose the time series data into trend data and residual data. In the next step, the ARIMA model on trend component of time series data is used to obtain forecasted values. The residual component is obtained by subtracting the original time series and trend series. The residual component is found to be highly nonlinear in nature. In this scheme, SVM is used for nonlinear residual part modeling because SVM is prone towards mapping nonlinear time series data. The summation of forecasted values of trend component and residual component has shown better prediction values with higher accuracy when compared with ARIMA, SVM, and ANN models individually. Thus, the proposed model can be used in many time series forecasting areas for better prediction. In future, the accuracy can be improved by pre-considering the nature of time series data.

References

1. Suresh, K.K., Priya, S.: Forecasting sugarcane yield of Tamilnadu using ARIMA models. Sugar Tech. **13**(1), 23–26 (2011)
2. Wang, J.-J., Wang, J.-Z., Zhang, Z.-G., Guo, S.-P.: Stock index forecasting based on a hybrid model. Omega **40**(6), 758–766 (2012)
3. Contreras, J., Espinola, R., Nogales, F., Conejo, A.: ARIMA models to predict next-day electricity prices. IEEE Trans. Power Syst. **18**(3), 1014–1020 (2003)
4. Cadenas, E., Rivera, W.: Wind speed forecasting in three different regions of Mexico, using a hybrid ARIMA-ANN model. Renew. Energy **35**(12), 2732–2738 (2010)
5. Gonzalez-Romera, E., Jaramillo-Moran, M., Carmona-Fernandez, D.: Monthly electric energy demand forecasting based on trend extraction. IEEE Trans. Power Syst. **21**(4), 1946–1953 (2006)
6. Szkuta, B.R., Sanabria, A.L., Dillon, T.S.: Electricity price short-term forecasting using artificial neural networks. IEEE Trans. Power Syst. **14**(3), 851–857 (1999)
7. Kisi, O., Cigizoglu, H.K.: Comparison of different ANN techniques in river flow prediction. Civil Eng. Environ. Syst. **24**(3), 211–231 (2007)
8. Chen, W.-S., Du, Y.-K.: Using neural networks and data mining techniques for the financial distress prediction model. Expert Syst. Appl. **36**(22), 4075–4086 (2009)
9. Xie, W., Yu, L., Xu, S., Wang, S.: A new method for crude oil price forecasting based on support vector machines. Comput. Sci. ICCS **2006**, 444–451 (2006)

10. Wen, J., Wang, X., Li, L., Zheng, Y., Zhou, L., Shao, F.: Short-term wind power forecasting based on lifting wavelet, SVM and error forecasting. In: Unifying Electrical Engineering and Electronics Engineering. Springer, New York, pp. 1037–1045 (2014)
11. Li, X., Gong, D., Li, L., Sun, C.: Next day load forecasting using SVM. In: International Symposium on Neural Networks, pp. 634–639. Springer, Berlin, Heidelberg (2005)
12. Narendra, B.C., Eswara Reddy, B.: A moving-average filter based hybrid ARIMA–ANN model for forecasting time series data. Appl. Soft Comput. **23**, 27–38 (2014)
13. Yao, Q., Brockwell, P.J.: Gaussian maximum likelihood estimation for ARMA models. I. Time Series. J. Time Ser. Anal. **27**(6), 857–875 (2006)
14. Zhang, G.P.: Neural networks for time-series forecasting. In: Handbook of Natural Computing, pp. 461–477. Springer, Berlin, Heidelberg (2012)
15. Hyndman, R.J., George A.: Forecasting: principles and practice. OTexts (2014)
16. Crude Oil Production and Inflation Consumer Sale. http://www.datamarket.com

Social Network Analysis with Collaborative Learning: A Review

Abhishek Kashyap and Sri Khetwat Saritha

Abstract When the information is needed to be categorized in the view of user conduct, the approach of social network analysis is considered. It is the process of studying social structures through the use of networks and graph theory. It has come up as a vital procedure in modern sociology. Among them, one of the most up-to-date methods of using SNA is the study of Computer-Aided Collaborative Learning (CSCL). SNA is used to recognize how learners work in area of quality, length, frequency, and volume and communication schemes when applied to CSCL. Alongside, SNA can focus on detailed features of the network connection or the complete network as a unit. It utilizes graphical representation, written illustration, and information outline to consider the links inside a CSCL network. While applying SNA to a CSCL domain, the individual discussions are managed as a social network. This article presents a systematic overview of the basic aspects of social network analysis and is based on their attributes and relationships. It also assists to measure social processes in collaborative learning experiences which is shown through a sample case study.

Keywords Computer-aided collaborative learning · Sociometrics
Social media learning

1 Introduction

A social network is a network of social entities and interaction between these entities. The entities can be people or gatherings (groups) or associations (organization). Here, the people or gatherings can be dealt with as nodes of the social network have different social familiarities like easy-going acquaintance.

A. Kashyap · S. K. Saritha (✉)
Computer Science and Engineering Department, Maulana Azad National Institute of Technology, Bhopal 462003, Madhya Pradesh, India
e-mail: sarithakishan@gmail.com

A. Kashyap
e-mail: abhishekkashyap.coer@gmail.com

A. Abraham et al. (eds.), *Emerging Technologies in Data Mining and Information Security*, Advances in Intelligent Systems and Computing 813,
https://doi.org/10.1007/978-981-13-1498-8_34

Social network analysis is the calibrating and representing of flows and relationships [1]. The relationship is between the information processing entities. The entities can be organizations, or groups, or people, or computers. It provides both statistical and pictorial analysis of network and the relationships between humans. The number, size, and connections among subgroups of the network help in analyzing behavior of the network. Social network data consist of various elements. Social network analysis can be utilized as a part of solving or detecting cybercrimes utilizing the forensic analysis of social networking applications. SNA can be applied for management and consultancy services to business clients. This is called as Organizational Network Analysis. SNA also helps in analyzing different features of the network which can be used for various applications. For example in fraud detection, SNA normalizes the network features between zero and one which predicts whether the result obtained is legit or fraudulent. In order to predict the more accurate results in fraud detection, the output is categorized in on the basis of resource used in producing result. In order to have a more basic understanding of SNA methodology, some of methods belong to SNA is discussed.

1.1 Social Network Models

There are four network models used for SNA.

1.1.1 Model Representing Formal Methods

Formal methods include the use of graphical and mathematical techniques in social network analysis. There are three principle motivations to do as such and they are: one reason to represent the description of network minimally and systematically. Another reason is that mathematical representations enable us to apply the computer examination/analysis in the network information. Last reason is formal strategy representation of social network information is that the methods for graph processing and the standards of mathematics themselves propose different things [2]. The complete analysis of social network includes the distinction between the descriptive methods, analysis procedure (based in the decomposition of adjacency matrix), and statistical model.

1.1.2 Model Using Graphs for Representing Social Relations

When only one kind of graphic display is used in social network analysis, where nodes are present to represent entities and lines to represent relation among them. This kind of graphing things is borrowed from the mathematicians and they are known as sociograms. All the sociograms share regular component of utilizing a labeled circle for every actor in the population we describe. Envisioning a sociogram and furthermore its outline presented in the graphical representation give first depiction of social network information. When small graph is considered, this may suffice, for this simple relatively approach data and research questions are too complex.

1.1.3 Model Using Matrices for Representing Social Relations

In social network analysis, an extremely basic and most regular type of matrix where the number of rows and columns are the same as there are entities in our gathered information, and representation of ties between the actors is analyzed by the value presented in that corresponding shell, because of that most widely recognized matrix is binary. The presence of tie is represented by setting one in a cell, likewise zero speak to the nonappearance of tie. The matrix discussed above is the beginning stage for all network analysis. This sort of a matrix is termed as adjacency matrix in social network analysis because it speaks to who is alongside, or nearby whom in the "social space" mapped by the relations.

1.1.4 Statistical Models for SNA

Factual investigation of social networks traverses more than 60 years. Since the 1970s, one of the real headings in the field was to show probabilities of social ties between communicating units (social performers/actor), however before all else just little gatherings of actors were considered. Broad prologue to prior strategies is given by Wasserman and Faust [3]. The ERGM has been as of late reached out by Snijders et al. [4] keeping in mind the end goal to accomplish robustness in the evaluated parameters. The statistical literature on displaying social networks accept that there are n actors and data about binary relations between them. An n n matrix Y is there to represent the all binary relations. For instance, Yij = 1 on the off chance that i views j as friend. The entities are generally represented as nodes and the relations as ties between the nodes. At a point when the represented of relations is presented as undirected ties, then matrix Y is considered to be symmetric. Yij can be esteemed and not just binary, which imply the representation of strength of tie between actor i and actor j [5]. Moreover, every entity can have characteristics xi, for example, their statistic data. At that point, the n-dimensional vector X = x1, …, xn is a completely watched covariate information that is considered in the model (e.g., [6]). There are a few valuable properties of the stochastic models. Some of them are

1. The capacity to clarify imperative properties between entities that regularly happen, all things considered, for example, reciprocity, on the off chance that i is related to j then j will probably be by one means or another related with i; and transitivity, in the event that i knows j and j knows k, it is likely that i knows k.
2. Inference techniques for dealing with orderly occurred errors in the estimation of connections [7].

3. General approaches for parameter estimation and model examination utilizing Markov Chain Monte Carlo techniques (e.g., [8]).
4. Individual variability [9] along with properties of various entities [10] are considered.
5. Ability to deal with gatherings of nodes with comparable statistical properties [11].

There are a few issues with existing models, for example, degeneracy examined by [12] and adaptability specified by a few sources [4, 12]. The new particulars for the Exponential Random Graph Models are proposed in attempt to discover an answer for the temperamental probability by proposing marginally different parameterization of the models was then utilized.

1.2 Social Network Properties

There are a few properties of social networks that are exceptionally essential, for example, size, density, degree, reachability, distance, diameter, and geodesic separation. Here, we depict some more complicated properties which might be utilized as a part of social network analysis. The accompanying properties are taken from [1].

1.2.1 Property of Maximum Flow

Two actors are said to be totally connected by accounting how many different actors in the neighborhood of a source leads to the target.

For instance, suppose there is only single actor to whom message could be sent for retransmission despite the fact that the actor has many ways of reaching the target, and the connection will be considered a weak connection. On the other hand, if there are say four actors in the scene and each of whom possesses one or more ways of retransmitting the message then it is a strong connection. The suggestion made by the observed flow is that the weakest link in the connection will determine strength of connection.

1.2.2 Property of Hubbell and Katz Union

The most maximum flow approach concentrates on the weakness or then again repetition of connection between sets of actors—sort of a "quality of the weakest connection" argument. As an alternative approach, we should need to think about the quality of all connections as characterizing the network. On the off chance that we are occupied with how much two actors may impact on each other, or offer a feeling of basic position, the full scope of their connection ought to likely be considered. Regardless of whether we need to incorporate all connections between two actors, it may not make a great deal (in most cases) to consider a way of length 10 as critical as a way of length 1. These approaches match the aggregate connections between various actors (ties for undirected information, both sending and accepting ties for coordinated information). Every connection, be that as it may, is given a weight, as per its length. The more prominent the length, the weaker the connection.

1.2.3 Property of Taylor's Influence

The Hubbell and Katz approach is more accurate when it is applied to symmetric information, because more consideration is focused on the directions of connections (i.e., A's ties directed to B are similarly as essential as B's connections to A in characterizing the separation or on the other hand solidarity closeness between them). In the event that we are more particularly inspired by the impact of A on B in a directed graph, the Taylor impact approach gives an interesting elective. The Taylor measure, similar to the others, utilizes all connections, furthermore, applies an attenuation factor. Instead of standardizing on the entire resultant matrix, however, an alternate approach is applied. The column marginal for every actor is subtracted from the row marginal, and the outcome is then normed. It is necessary to maintain the balance between each present actors both side (sending and receiving) connections. Positive values then reflect a dominance of sending over accepting to the next performer of the combine—or adjust of impact between the two.

1.2.4 Property of Centrality and Power

All social scientist would concur that power is a basic property of social structures. There is substantially less agreement about what power is, and how we can depict and analyze its causes and results. There are three primary aspects of sociograms named as degree, closeness, and betweenness centrality. Degree centrality is thought to be the total number of ties for an actor and its persuasions are having more opportunities and choices. Closeness centrality is characterized as path length to other actors and its impact is immediate bargaining and exchange with different actors. Betweenness centrality is characterized as lying between each different sets of actor and its impact is handling contacts among actors to detach them or prevent connections. Above discussed are the primary methodologies that social network analysis has created to think about power, and the firmly related idea of centrality.

2 Collaborative Learning

Collaborative Learning approach is based on collective efforts of learners that achieve goals working in a group as a single entity and proper communication and contribute their work to increase the overall performance of group [13, 14]. About the collaborative learning, there is a number of research literature discussed and their conclusion primarily expressed that collaborative learning upheld the web-based learning. The interactions between the members and cohesive actions help in achieving the objective with ease [15]. New collaborative tools and reinforcing the learning in members and seniors collaborative training help learners to get more of a hands-on approach. When there is social interaction between groups and gatherings, then they become aware of their environment and shared their goals and also the members of gatherings build the knowledge necessary to solve the tasks [16–18]. The motivation of learners along with their skills and time spent on exchanges govern the effectiveness of collaborative learning. In order to measure and improve the accuracy of collaborative learning approach, the consideration of efficiency of collaborative tools should not be neglected. Collaborative learning also helps individuals (learners) in developing critical thinking, communication skills, and team spirit. It promotes effective social integration of learners from diverse groups. It brings out opportunities for learners to express their ideas and experiences.

The process of collaborative learning can be summarized in three steps. First step is to develop resources that empower learners. This can be achieved by encouraging interactions between learners in a playful, but purposeful, way. Forming groups leads to abundant information on topics, as each individual with different background contributes to the knowledge. Second step is to access complex ideas using concrete, visual, and tactile ways to present learning concepts of the learners. This helps in better understanding of objective and develop solutions to complex problems. Third step is to have exploratory discussion in the group where new meanings are developed by thinking collaboratively in a group. With the help of new meanings explored in the discussion, new alternatives can be developed and make the solutions more effective. Collaborative learning advancement empowers developers of learning frameworks to work as a network. Particularly significant to e-learning where developers can share and incorporate information with courses in a collaborative environment. Knowledge of a solitary subject can be pulled together from remote areas utilizing software frameworks. Collaborative techniques are processes, behaviors, and conversation that identify with collaboration between people. Structures, rubrics, outlines, and graphs are helpful in these circumstances to objectively document personal traits with the goal improving performance in current and future projects [19].

Collaborative environment: Mentioned stages are focused on social components, e.g., to make remarks, to share a link, or like/dislike mechanisms, while interactivity mechanisms are outer hyperlinks that make possible into the contents. Moreover, there is an absence of media formats that support aspects of interactivity which restricts the accessibility of interactive content.

In order to overcome these mentioned limitations, an environment allows to create multimedia-interactive entities supported by collaborative learning methodology guideline and a platform known as social media learning (SML). SML has three vital premises:

1. social interaction as the key for a fruitful collaborative effort,
2. interactivity inside and outside the interactive media content [20] and
3. reuses of mixed media assets.

SML accepts the elements of a social media platform and extends its features to education context.

Social media is, for the most part, portrayed by enabling the learners to receive critical and constructive positions in regards to the content while interacting with a high social dynamic environment. Following models present the important stages presented in the collaborative learning process [20, 21].

2.1 Analysis

The aim of this stage is the building of a common resources database. For this, each author utilizes seeking services to discover media resources, self-evaluates its significance, and offers it with the group (Divergence pattern). Evaluators assessed the shared resource, also there exist a regular communication through various communication services among evaluator and authors (e.g., remarks) which conclude the understanding about content and also the relevance of information (Building Consensus design).

2.2 Synthesis

The point of this stage is the organizing of ideas contained in the database. This progression is upheld by communication services where evaluators and authors contend about content in the resources. Various level structures of subjects are recommended for organizing relevant material, but a folksonomy (tagging systems) is a natural way to associate complex concepts.

2.3 Composition

The point of this stage is the organization of a learning object in light of reuse of the organized multimedia resources and interactive systems. Authoring tool has supported the abovementioned undertaking where a Scriptwriter makes some animation sequences with interactivity aspects. This procedure includes the selection of the best fragments of content and the consideration of focuses for intuitiveness.

2.4 Reproduction

In interactive mechanisms, a learner performs the necessary operations on the resource and interacts with the mechanisms. The learner's interactions are reported to the supervisor, who could take this data to enhance the material. The learners value the MLS object, and communications services, for example, discussions, talks or notice board can be used to argue about the resource and presentation of material (Building Consensus design).

3 Case Study

For analyzing interaction process in a collaborative learning scenario on the basis of SNA, there is a case study [13, 22]. There were 18 undergraduate students at Universidad Del Cauca University in Colombia. They all are between 19 and 21 years old. This learning activity is based on peer assessment and asynchronous social interaction process with the help of comments. It was supported by CSCL platform also called SML learning [22]. First, the process starts from the instructor where instructor gives a topic related to the course which was divided into three subtopics. A subtopic was chosen by the students and they have to make six teams of three members each. They were free to join any team. The whole process was planned for 7 days. While some student worked until 9th day.

3.1 Environment Setting

After created, their teams all the teams have to select a coordinator who takes all the responsibilities about the achievement of goals and the social interaction among peers.

3.2 Reference Material Delivery

Reference resources are delivered to teams by SML learning system which supports the resources visualization and some asynchronous social interaction mechanism like comments.

3.3 Making Essay

Writing an essay about the topics by the help of collaborative technique and this technique is also used for visualization. At most three pages were needed for the document in which achievement, abstract view of concept, and a quick review were written.

3.4 Peer Reviewing

Essays are reviewed by the teams in the following order: Team n reviewed the essay of team n+1, except the last team who access the 1st team essay. In the process, they all have to rate over a 5 point scale and also make at least 3 comments.

3.5 Making Essay II Version

From the comments and rating made by their peers, they are able to improve their essays.

3.6 Peer Reviewing

All the teams reviewed the new essays. In this way, assessment process was repeated.

The contextual investigation examined and all activities enlisted amid the experience. In the assessment procedure, the last score came to by the understudies in this action was 4.17 [13] out of a scale 0–5 (5 is the most extreme esteem). The analysis appears that assets in video arrange are more impactful than ones with text configuration. This perception is coherent with multimedia learning hypothesis which asserts that the multimedia is an impactful and flexible approach to introduce content in a learning strategies [23].

4 Conclusion and Future Work

This article presents an approach to analyze social interaction processes in collaborative learning scenarios based on timely calculated SNA sociometrics. To underpin collaborative learning approach, a successful collaborative learning experience with some of these approaches was efficiently accurate. As the results found that social interaction processes tend to be symmetric. Such a conclusion is based on the temporal behavior of the reciprocity metric and a strong correlation between in and out-grade metrics. Finally, the behaviors observed in SNA metrics indicate that measuring the final state of the interaction in a CSCL experience. It is not considered as an indication of gathering dynamics as observed in high variation in data analysis. Therefore, a temporal dimension of SNA metrics seems to provide valuable information to analyze CSCL scenarios. However, SNA theories must be brought closer to the concepts of such scenarios. As observed in the reviewed literature, content analysis techniques have been used as a relevant part of social interaction analysis processes. However, such results are handled independently of the SNA approach. We believe new categories can enrich or create alternative networks that can better represent interaction processes. As a future work, this approach will be applied to other experiences and scenarios and incorporate metrics related to clusters and interaction patterns in order to efficiently acquiring the output.

References

1. Jamali, M., Abolhassani, H.: Different aspects of social network analysis. In: IEEE/WC ACM International Conference (2006)
2. Hanneman, A., Riddle, M.: Introduction to social network methods. http://www.faculty.ucr.edu/hanneman/nettext/ (2005)
3. Wasserman, S., Faust, K.: Social Network Analysis: Methods and Applications. Cambridge University Press, Novemver (1994)
4. Snijders, T.A., Pattison, P.E., Robins, G.L., Handcock, M.S.: New specifications for exponential random graph models (2004)
5. Robins, G., Pattison, P., Wasserman, S.: Logit models and logistic regressions for social networks. Psychometrika **64**, 371394 (November 1999)
6. Hoff, P., Raftery, A., Handcock, M.: Latent space approaches to social network analysis. J. Am. Stat. Assoc. **97**, 10901098 (2002)
7. Butts, C.: Network inference, error, and informant (in)accuracy: a bayesian approach. Soc. Netw. **25**(2), 103140 (2003)
8. Snijders, T.: Markov chain monte carlo estimation of exponential random graph models. J. Soc. Struct. (2002)
9. Hoff, P.: Random Effects Models for Network Data, Irvine, California (2003)
10. Wang, Y., Wong, G.: Stochastic blockmodels for directed graphs, vol. 82, no. 8–19 (1987)
11. Handcock, M.: Assessing degeneracy in statistical models of social networks, working paper, University of Washington, Dec 2003
12. Smyth, P.: Statistical modeling of graph and network data. In: Proceedings of IJCAI Workshop on Learning Statistical Models from Relational Data, Acapulco, Mexico, Aug 2003
13. Claros, I., Cobos, R.: An Approach based on Social Network Analysis applied to a collaborative learning experience. IEEE Trans. Learn. Technol. **9**(2) (2016)
14. Vygotsky, L.S.: Mind in Society: The Development of Higher Psychological Processes. Harvard university University Press, Cambridge, MA, USA (1980)
15. Johnson, D.W., Johnson, R.T.: Cooperation and the use of technology. Handbook of Research for Educational Communications and Technology, vol. 1, p. 10171044. Springer, New York, NY, USA (1996)
16. Dillenbourg, P.: What do you mean by collaborative learning? In: Dillenbourg, P. (ed.) Collaborative Learning: Cognitive and Computational Approaches, p. 116. Elsevier, Amsterdam, The Netherlands (1999)
17. Anderson, C., Wasserman, S., Crouch, B.: A p! primer: logit models for social networks. Soc. Netw. **21**, 3766 (1999)
18. Stahl, G.: Contributions to a theoretical framework for CSCL. In: Proceedings of the Conference Computer Support Collaborative Learning: Found. CSCL Community, pp. 6271 (2002)
19. Moreno, R., Mayer, R.: Interactive multimodal learning environments. Educ. Psychol. Rev. **19**, 309–326 (2007)
20. Soller, A.L.: Supporting social interaction in an intelligent collaborative learning system. Int. J. Artif. Intell. Educ. **12**, 4062 (2001)
21. Claros, I., Cobos, R.: Social Media Learning:An approach for composition of multimedia interactive object in a collaborative learning experience. In: Conference Paper, June 2013
22. Claros, I., Cobos, R.: Social media learning: An approach for composition of multimedia interactive object in a collaborative learning environment. In: Proceedings of the IEEE 17th International Conference Computer Supported Cooperative Work Dec 2013, pp. 570575
23. Mayer, R.E.: Cognitive theory of multimedia learning, in The Cambridge Handbook of Multimedia Learning. Cambridge University Press, Cambridge, UK, pp. 3148 (2005)

Frequent Subpatterns Distribution in Social Network Analysis

Rahul Rane

Abstract Discovering the frequencies of a small subgraph in social networks graph is the most challenging part. Because social networking graphs have very few structural constraints which appear mostly in frequent patterns only. Many graph-based mining algorithms are available to solve this problem like Frequent Subgraph Mining and Motif Detection. Both algorithms find a small set of subgraph that repeats themselves in a large network for a specific amount of time. But motif detection gives the deep understanding of network functionality and also uncovers the structural property of complex network. The problem with analyzing the frequent patterns is their detection comes under computationally challenging problems. We can solve this problem by using sampling method which is not bound to the size of the graph and can compute millions of nodes. The full distribution of frequent pattern gives the statistical information about the underlying network which helps to understand network better and also use in analysis of the network. By using pattern distribution, we can uncover and identify the different aspect of Social Network which can be used in many real-life applications.

Keywords Frequent subgraph mining · Motif · Graphlets · Sampling algorithms

1 Introduction

Due to the rapid growth of social networks, a rich body of literature studies has focused on user behaviors, social graphs, information diffusions, trending topics, community detections, security and privacy in social networks. The large volume of users and contents in social networks, a variety of analytical metrics such as user relationships, interactions, influence, and topics, the streaming nature and high velocity of user-generated posts, contents and tweets has produced big data in social

R. Rane (✉)
Computer Science and Engineering Department, Maulana Azad National Institute of Technology Bhopal, Bhopal 462003, MP, India
e-mail: rahulrane165@gmail.com

A. Abraham et al. (eds.), *Emerging Technologies in Data Mining and Information Security*, Advances in Intelligent Systems and Computing 813,
https://doi.org/10.1007/978-981-13-1498-8_35

networks. Nowadays, large financial firm, telecom industries, and social network sites produce large data which create substantial research opportunity in data mining field [1]. Graphs are currently becoming more important in modeling and representing the information.

Social network analysis is significantly used in a wide range of applications and disciplines. Main applications include data aggregation and mining, network modeling and sampling, user attribute and behavior analysis, link prediction. It helps us to uncover, identify, and classify different aspects with the help of graph theory and networks. Many graph-based algorithms are available to calculate things like density, clusters, vertex centrality score, and modularity.

The graph patterns are the small subgraphs which repeat on a network for the significant amount of time. The existing notions of graph patterns are graphlet [2], Motif [3]. The graphlets are a small connected non-isomorphic induced subgraph of a large network. The induced subgraph must contain all edges between its nodes that present in a network. The motifs are partial subgraphs which may contain some of the edges in the large network.

The main task of Frequent Subgraph Mining (FSM) [4] algorithm is to identify frequent subgraph that appears in the graph network. The appearance of a subgraph in a network is supposed to be at a certain threshold. But the problem is it does not find subgraph in a single large network but the large database of small networks. By considering both Motif Detection and FSM, one can analyze them one by one and understand their statistical significance in the context of a network. However, the full distribution of patterns gives much more statistical information about the underlying network.

The distribution of embedded subgraph pattern in a large input network is based on a local and global measure of the networks. The local measures like a degree centrality which is directly focused on the vertices and edges and their expansion of edge happen in the neighborhoods only. The global measures such as density need to be considered full input graph to calculate one value.

Patterns are the small subgraphs. By considering the small pattern having few vertices, it creates huge amount of different non-isomorphic subgraphs. To enumerate all pattern in a single large graph is a computationally challenging part. Due to the exponential number of the pattern, the interpretation of mining result is limited. Even considering label graph adds tremendous information overhead to the graph database [5]. We can apply the following strategies to reduced information overload:

1. Use some aggregation method (e.g., entropy),
2. Concentrate on only small area of graph,
3. Applying various constraints that pattern needs to fulfill.

Discovering network motifs and frequent subgraph is computationally hard. Because the counting a topologies frequency is not an easy task, it requires solving subgraph isomorphism, which is also known as an NP-complete problem. There is no polynomial algorithm available to compute the canonical form of all possible arbitrary input pattern. Because exponentially, many subgraphs can get generated by considering fewer vertices which is very complex to sample. For many computationally hard counting problems, one has no choice but use approximation algorithm for practical purpose. In this situation, the approximation through uniform sampling is the most probable approach. The sampling approach does not depend on the size of nodes so millions of node can be computed which is the requirement of current social networking graphs.

2 Related Work

Recently, graph is used to modeling the complex structures like protein structure, biological network, chemical compound, and social network. Graphs are the most powerful data structure for the representation of object and concepts and their relationship. Graph mining has gained substantial importance in data mining. Many researchers propose novel methods and algorithm for graph mining. The classic frequent subgraph mining algorithm is to discover repeated subgraphs that appear to be as often as in the input graph with a certain threshold. There are many algorithms available for frequent subgraph mining which are based on apriori approach like AGM [6], FSG [7], SPIN [8], etc. and pattern growth approach like gSpan [4] and SUBDUE [9]. But all these algorithms only care about the candidate subgraph that appears in as many graphs different graph database and does not care about pattern appeared in a single graph. Recently, there has been some focus on the pattern that appears frequently on the single large graph [10].

Network with the similar global property may have different local structures to the local properties of any network that are the building block for the global properties. The local structure can be considered as the local motif and they become the building blocks for the big network. The distribution of local and global structure is important for the pattern distribution for many applications. The work of Milo et al. [3] is the pioneer for the network motif as per the definition motifs are the patterns which appears more in the real network than found in the randomly generated graph network so the frequency of small pattern found in the network is particularly higher than those found in the randomly generated graphs network. Later, the researcher has developed the network motifs in the significantly biological field. Apply it to various problem and application in the field of bioinformatics and improved its efficiency and performance for the large network. Kim et al. [11] use network motif to find the subgraph in the large biological network. The motif further uses to study different biological problems like cellular biology, gene structure, and cell behavior. Firana et al. [12] use network motif to understand the molecular mechanism in the cell behavior and proposed the new algorithm to detect motif efficiently with the high speed.

Many researchers use network motif to study the large graph and their relationship and infer many impotent results which may use in the decision-making. Network motifs are compared with random graph model also called as the null model and natural complex graph network which helps to get a deep insight of network and understand their functional significance and uncover the structural design of the complex network. Various methods and tools have been proposed to detect the motif in the network and give the solution to the computationally challenging problem like MFinder [13], FANMOD [14], Kavosh [15] and NETMODE [16].

Recently, we found some research focus on applying Motifs for Social Network Analysis [17, 18]. However, by applying this method on Social Network Analysis from bioinformatics, it creates some problems. As the genetic networks are more bound in the rule of physics or biology and representations are complex, every edge and vertices play the important role. But social networking graph contains lots of randomnesses and random noise. For example, the ways in which chemical molecules react with each other are a lot more restricted. But the social network graph cannot restrict because possible friendship connection can explore by meeting the different group of people so the expansion of graph in unpredictable. This problem can be solved using stochastic motif by introducing certain probability on the existence of edge [18].

Counting a subgraph pattern frequency is the difficult task because of subgraph isomorphism. To overcome this problem, uniform random sampling is the feasible approach. The excellence of sampling is based on the many performance metrics which are accuracy, convergence, and execution time. There are many sampling algorithms whose performance depend on the above constraints. For example, MFinder method is computationally expensive and also scales poorly with increasing number of motif size. Another problem with sampling is that they required the random access to the network with vertices as well as edges, which leads to create the problem with the restricted access data in the network like web network or hidden network.

Two random walk based sampling methods, namely MHRW (Metropolis–Hastings random walk) and SRW-RW (Simple Random Walk with Reweighting) for approximating the concentration of arbitrary-sized pattern graphs in a large network. The basic mechanism of both the methods is a Monte Carlo Markov Chain (MCMC) sampling over the candidate subgraph space, which is guaranteed to compute an unbiased estimate of the concentration of all the candidate subgraph of a given size simultaneously. These methods are scalable and significantly faster than existing methods [10].

3 Overview of Frequent Subgraph Mining

This section contains some graph definitions used in the process of FSM. Any FSM process could be contained in the following aspects.

3.1 Graph Representations

Usually, graph structure represented by adjacency matrix and adjacency list depends on the requirement of problem. Adjacency matrix represented by filling into the rows value and columns value of vertices and their intersection value represent a possible edge in between two vertices. The value at intersection point shows the total number of links between two vertices. Isomorphism testing required reliable labeling approach which ensures that two same looking graphs are lexicography labeled representation in the similar way irrespective of the order they appeared. The canonical lexicography labeling approach defines exclusive label codes for the input graph. Canonical labeling helps to isomorphism examining because it guarantees that a couple of graphs is isomorphic and their canonical labels guarantees to be similar [7]. One modest method of generating canonical labeling is associated with an adjacency matrix, the rows or column formed code with a list containing the value of integers having minimum value or maximum value lexicographic order. Another method based on the DFS lexicographic order is M-DFSC canonical labeling [10]. The adjacency matrix which helps to form canonical labeled representation form is also called the Canonical Adjacency Matrix.

3.2 Subgraph Enumeration

The method used for enumeration of all subgraph depends on link operation implemented by FSG, AGM, and different other methods are the extension operation. Join operation faced two problems: one is single join may produced by many candidates and another one is unnecessarily many join operations proposed. Extension operation faced the problem like it hidden nodes which recently introduced an edge may want to attach too.

3.3 Frequency Counting

The most important methods for graph frequency counting include Embedding Lists (EL) [7]. It helps to calculate the candidate frequency efficiently and is also responsible for the construction of candidate generation process. The other method is Recomputed Embedding (RE) which helps to improve the counting accuracy for candidate graph. The graph with single nodes proceeds to store his embedding label representation list of all of its incidence rate with the label into the memory. And other graphs with different node proceed to store in the embedding label representation form of tuples in the memory. The tuple includes an index of an embedding label list of the ancestor candidate graph with the embedding of a graph with its n-ode. The frequency of candidate graph is calculated from the large count of different candidate graphs present with its embedding list in the memory or database. The matching and counting of embedding label lists are faster but the problem is they create the problem for big graph and a large number of candidate subgraphs present in memory. To solve this problem, other approach can apply which keep log over the active candidate graph appear more frequently in a graph and restart the process and recomputed it using the log.

3.4 Frequent Subgraph Mining Algorithm

FSM algorithm roughly divide in the different approached one with the apriori-based approach and other is the pattern growth based approach. Although these algorithms differ in size of the graph, type of graph, search strategy they use, and the method they use to represent the graph, etc., there exist many different algorithms for different approaches. The technique used by the apriori-based FSM algorithm shares some alike characteristics with frequent item sets. The apriori-based algorithm has significant overloaded when a different size of repeated subgraphs needs to combine to produce the next level size graph candidates. In a pattern growth mining technique, graph extends by adding a new edge to the frequent subgraph and also searches every potential position to add edge. Mainly, the algorithm uses the depth-first search to count the candidate generated by repeated subgraph. The advantage of the depth-first algorithm is over the levelwise search like memory utilization is better and efficient subgraph testing is decent. Following are the different FSM algorithms mainly based on depth-first search.

3.4.1 gSpan

The gSpan is a graph-based pattern mining algorithm. A basic aim of this algorithm is discovering a frequent subgraph by no generation of candidate graph and also false pruning [4]. The gSpan used depth-first search and also be the first ever to use this method in FSM. This algorithm uses a canonical representation of graphs which is also called DFS Code. In this algorithm, DFS traversal is using to traverse the edges in a graph in which edges are visited in the order they come. The concatenation of edge is represented in which the order they come in a graph called DFS Code. The refinement in the graph is limited by gSpan because of the following two ways first, the pieces of a subgraph can be protracted at every node which stayed to the rightmost part of a path generated by DFS tree. Second, the piece of a subgraph generation is directed by the rate of the graph that appears in the lists. Though above pruning strategy cannot completely avoid the isomorphic piece generated by subgraph. So other methods like computing the canonical graph by considering lexicographically least DFS Code for every improvement use the permutations.

3.4.2 FFSM (Fast Frequent Subgraph Mining)

FFSM algorithm uses triangle matrix to represent graph whose node value is added to the diagonal and edge value in another place. The matrix code is represented by all the records of a matrix, which is all lines. Lexicographic order of candidate graph and isomorphic representation of same graphs will be the identical canonical form code called as Canonical Adjacency Matrix. In FFSM, to generate the refinement of metrics, we need to join them and at most two new structure comes as a result. In FFSM a newly added pair of edge-node might be added to the last node of the matrix. When the refinement is done, FFSM uses permutation of Canonical Adjacency Matrix to check that the produced adjacency matrix is in the canonical form. For some reason, if not found then refinement pruned the matrix. FFSM stores embedding of all subgraph to evade precise testing of isomorphism. The advantage of using FFSM is that it only stores the nodes which match and also ignored the edges, which helps to speed up the process of the link and postponed operations in the meantime embedding lists helps to calculate set operation on the nodes. FFSM also needs to check the occurrence of all extra set of subgraphs which are not in their canonical form [19].

3.4.3 SPIN

Large graph databases like Social Networking graph entire count of frequent subgraphs which may be too huge to permit a complete enumeration by using reasonable computational assets. This mining algorithm mainly focused on the maximal frequent subgraphs found in the large graph [8]. The main purpose of this algorithm is to find subgraph that is not being the part of any other frequent subgraphs in the large graph. This algorithm helps to reduce the magnitude generated by the output set at their finest case and also reduced the whole number of patterns which are mined into some magnitude. At first, it will mine all the repeated tree from the input graph database and then it reconstructs all the maximal subgraph from the mined tree. Also, SPIN has combined several other optimization techniques to accelerate the mining process for FSM.

4 Network Motif

Network motifs are subgraph pattern that occurs in a graph significantly more often. Each of this subgraph described as an interaction between the vertices in a specific pattern and also reflect a particular function [3]. Also, network motif provides a profound insight of networks and their purposeful ability to infer the statistical result and their discovery is computationally stimulating. Various notions have been proposed to overcome this challenging discovery of motifs. Network motif detection algorithm is mainly divided into numerous paradigm likes counting the frequent subgraph methods, sampling from big network methods, pattern growth using prune with constraints methods, and so on. Two steps are included in motif detection problems, first is computing the frequency of repetition of the subgraphs and second is assessing the subgraph importance. Following are the review of the different major algorithm on computational aspect of motif detection and respective benefits and drawbacks.

4.1 Mfinder

Mfinder is the first motif mining tool used to detect the motif in big networks. Its based on the full enumeration and sampling methods algorithms. Using this algorithm we can discover small motif which is below of size 5 or 6 motifs but if we exceed the size of motif it may not computationally viable. Kashtan et al. [13] presented Network Motifs discovery algorithm using the sampling which includes the edge based sampling on the complete network. This algorithm mainly focused on the induced subgraphs which help to discover motifs in both directed and undirected networks. When algorithm used sampling algorithm, one of the most important things is that it should take an unbiased sample. But many times, sampling creates problem to select the samples edge in a biased way and so Kashtan et al. suggested another method which uses weighting scheme which assigns dissimilar weights to the different subgraphs inside the network. The advantage of this algorithm is its computational time is asymptotically independents of network size.

4.2 Kavosh

Kavosh is able to discover network notifs in all type of network include directed or undirected graph networks [15]. The basic aim of the algorithm is improved in main memory usage. The basic idea for the algorithm is, first find all possible k-size subgraphs from the participated node, and then eliminate the particular node after that repeat this process till the last residual node. The most important subtask of this algorithm is its enumeration part. This subtask makes algorithm different from other. The enumeration process includes building a tree rendering to the restrictions which lead to improved memory usage and computation time. The tree structure helps to detect the individual subgraph which leads to an efficient solution.

5 Sampling Methods

In terms of the graph, sampling means in the large input graph, main task is to take a small graph satisfying certain properties of interest with particular domain those presented in the input graph. Many traditional sampling is considered the global properties of the graph. But local properties of graph play the main role in the sampling because they are the building blocks of the graph and satisfied the functional abilities. Traditionally, sampling is calculated by comparing the global metrics with the sample input to some other global metrics of the target output graph y considering an average degree. But we are interested mostly in the subgraph frequency and its counting. Both local and global metrics need to consider for accuracy and performance as per the requirement. Some of the different sampling methods discussed below fit perfectly for our problem definition.

5.1 Random Walk

The method of sampling requires to random walk into the graph. It generates a sample graph by using the random selection of node technique from the large set of input nodes by using uniform probability. As per the graph definition, the generated sample graph from large graph is called the induced subgraph. The process of moving for selection of the node in the large input set is basically the random walk process. In this case, the randomly selected next node w is uniform at the currently selected node v. Therefore, the probability to traverse a node w from node v is

$$P_{v,w}^{RW} = \begin{cases} \frac{1}{k_v}, & if\ w\ is\ a\ neighbor\ of\ v \\ 0 & otherwise \end{cases} \tag{1}$$

Lovsz et al. describe whole method in detail [20]. This method of the walk is very simple and also gives the prominent result in the analysis. The result shows the distribution for any applications is promising and the convergence time it takes is considerable. Problem with the method is that it is biased toward the inherited node. Consider the certain case like connected with directed graph, the probability of node is getting connected with the certain node v meets with time too

$$\pi_v^{RW} = \frac{k_v}{2.|E|} \tag{2}$$

This is classic RW samples nodes with respect to $\pi_v^{RW} \cong k_v$. We can say this process is undoubtedly biased towards the selection of high degree nodes.

5.2 Metropolis–Hastings Random Walk (MHRW)

The problem with Random Walk is the bias way of selecting the node. Instead of correcting the bias way of selecting the node after the walk, we can appropriately alter the transition metrics probabilities of the node so it meets the preferred uniform distribution. The Metropolis–Hastings algorithm use Markov Chain Monte Carlo technique. Because MCMC is based on the probability and most of the time gets very difficult to sample the data using the probability and its distribution [21]. To achieve the uniform distribution, we need to calculate transition probability.

6 Conclusion and Future Work

Discovering frequent subgraph in a large input graph is one of the most challenging problems for data mining community. This paper explores various appropriate frequent subgraph discovery algorithms suitable for the problem definition. This algorithm provides a wide range of description to some very noticeable and efficient subgraph discovery problems. We discussed some probable approaches available to use the distribution of frequent pattern to characterize a large input graph. We have also mentioned different approximation algorithms for computing these frequent subpattern distributions. However, they are based on sampling method we explore the relevant sampling methods to uncover the problem. Sampling method helps to scale very large graphs that present social networking produced. We are using pattern distribution to uncover many social network properties which help to understand the network better.

References

1. Vijayalakshmi, R., et al.: FP-GraphMiner-A fast frequent pattern mining algorithm for network graphs. J. Graph Algorithm. Appl. **15.6**, 753–776 (2011)
2. Milenkovi, T., Prulj, N.: Uncovering biological network function via graphlet degree signatures. Cancer Inf. **6**, 257 (2008)
3. Milo, R., et al.: Network motifs: simple building blocks of complex networks. Science **298.5594**, 824–827 (2002)
4. Yan, X., Han, J.: gspan: graph-based substructure pattern mining. In: Proceedings of 2002 IEEE International Conference on Data Mining, 2002. ICDM 2003. IEEE (2002)
5. Benjamin, C.: Towards using subpattern distributions in social network analysis. In: 2016 IEEE/ACM International Conference on Advances in Social Networks Analysis and Mining (ASONAM). IEEE (2016)
6. Inokuchi, A., Washio, T., Motoda, H.: An apriori-based algorithm for mining frequent substructures from graph data. Princ. Data Min. Knowl. Discov. 13–23 (2000)
7. Kuramochi, M., Karypis, G.: Frequent subgraph discovery. In: Proceedings IEEE International Conference on Data Mining, 2001. ICDM 2001. IEEE (2001)
8. Huan, J., et al.: Spin: mining maximal frequent subgraphs from graph databases. In: Proceedings of the Tenth ACM SIGKDD International Conference on Knowledge Discovery and Data Mining. ACM (2004)
9. Ketkar, N.S., Holder, L.B., Cook, D.J.: Subdue: compression-based frequent pattern discovery in graph data. In: Proceedings of the 1st International Workshop on Open Source Data Mining: Frequent Pattern Mining Implementations. ACM (2005)
10. Saha, T.K., Hasan, M.A.: FS3: a sampling based method for topk frequent subgraph mining. Stat. Anal. Data Min. ASA Data Sci. J. **8**(4), 245–261 (2015)
11. Kim, W., et al.: Biological network motif detection and evaluation. BMC Syst. Biol. **5.3**, S5 (2011)
12. Farina, L., et al.: Identification of regulatory network motifs from gene expression data. J. Math. Model. Algorithm **9.3**, 233–245 (2010)
13. Kashtan, N., et al.: Efficient sampling algorithm for estimating subgraph concentrations and detecting network motifs. Bioinformatics **20.11**, 1746–1758 (2004)
14. Wernicke, S., Rasche, F.: FANMOD: a tool for fast network motif detection. Bioinformatics **22**(9), 1152–1153 (2006)

15. Kashani, Z.R.M., et al.: Kavosh: a new algorithm for finding network motifs. BMC Bioinform. **10.1**, 318 (2009)
16. Li, X., et al.: Netmode: Network motif detection without nauty. PloS one **7.12**, e50093 (2012)
17. Hong-lin, X., et al.: Social network analysis based on network motifs. J. Appl. Math. **2014** (2014)
18. Liu, K., Cheung, W.K., Liu, J.: Stochastic network motif detection in social media. In: 2011 IEEE 11th International Conference on Data Mining Workshops (ICDMW). IEEE (2011)
19. Huan, J., Wang, W., Prins, J.: Efficient mining of frequent subgraphs in the presence of isomorphism. In: Third IEEE International Conference on Data Mining, 2003. ICDM 2003. IEEE (2003)
20. Lovsz, L.: Random walks on graphs. Combinatorics, Paul Erdos is Eighty **2**, 1–46 (1993)
21. Saha, T.K., Hasan, M.A.: Finding network motifs using MCMC sampling. CompleNet (2015)

Big Data Real-Time Storytelling with Self-service Visualization

Rajat Kumar Behera and Anil Kumar Swain

Abstract Stories help to communicate information and interpret knowledge. Once the data is collected, analyzed, cleansed, and transformed, the subsequent step is to extract potential value from it. Realization of value will happen, only when business-centric insights are discovered and translated to time-bound actionable outcome. To maximize the potential value, data should be decoded into a storytelling medium via visualization, which can be either static or dynamic. Big data visualization is to reveal stories from data tsunami, generated at an alarming speed with diversified formats. The stories tend to represent vital characteristics to enlarge users. Self-service visualization empowers users to uncover unique patterns, interesting facts, and relationships from the underlying data by building their own stories without the in-depth technical knowledge, possibly little handhold by IT department. In this survey paper, we first get familiar with big data storytelling with visualization and its related concepts, and then will look through general approaches to do the visualization. To get deeper about it, we will have discussion about truthful data visualization in self-service mode representing real view of the business. This paper also presents the challenges and available technological solution, covering open source for representing real-time view of the story.

1 Introduction

Big data is exemplified by sheer volume, soaring velocity, diversified variety, and/or inherent discrepancies, i.e., veracity (4 Vs) datasets that use a new paradigm of processing for insight discovery and time-bound decision-making [1]. Such data is generating unparalleled options for the businesses to culminate deeper insights to

R. K. Behera (✉) · A. K. Swain
KIIT Deemed to be University, Bhubaneswar, India
e-mail: rajat_behera@yahoo.com

A. K. Swain
e-mail: anilkumarswain@gmail.com

A. Abraham et al. (eds.), *Emerging Technologies in Data Mining and Information Security*, Advances in Intelligent Systems and Computing 813,
https://doi.org/10.1007/978-981-13-1498-8_36

Table 1 Benefits and time requirement by industry

Industry	Time Req	Expected benefits
Clinical care	Seconds	Reduce life risks and saves lives
Financial and stock market	Milliseconds	Enhance performance and business profit
Military decision-making	Seconds	Saves lives and enhance better performance
Intelligent transportation	Seconds	Save times and enhance living quality
Natural disasters	Minutes	Reduces life risks
Festivals/Crowd control	Seconds	Efficient crowd handling
Daily resources	Minutes	Efficient resource administration

Table 2 Benefits of data visualization technique

Sr#	Benefits	Percentages
1	Enhanced decision-making	77
2	Enhanced ad hoc data analysis	43
3	Enhanced collaboration/information sharing	41
4	Afford self-service capabilities to end users	36
5	Better return on investment	34
6	Time savings	20

reinforce the decision-making process and leads the struggle for not only to store and process the data in a significant way, but to present it in meaningful ways.

Different applications in real-time big data analytic can be used in many aspects of human life like quality of life, minimization of risks of lives, resource management efficiency and profitability in business, etc. So, we need to analyze and execute big data in real-time analytic field as soon as possible to get a fast response [2]. Real-time big data application related to industry like financial and stock market, intelligent transportations, early warning of natural disasters, etc., are important applications whose operations help in enhancing quality of life, reducing human risks and saving the lives of people. Due to real-time requirements, many challenges give more attention to collecting, transferring, processing and visualizing big data [3]. Table 1 is depicting the benefits and time requirement by respective industry [2].

To compete more efficiently and effectively, businesses are increasingly turning to data visualization technique that allows the decision maker to visualize a custom analytic view. Data visualization is the technique to represent the data, including variables and attributes in an orderly form for the unit of information visualization [4]. Visualization can be thought of as the "face" of big data. According to the respondent percentage of survey [5], Table 2 represents the benefits of data visualization.

The self-service aspect of the visualization allows analyst to create visualizations at their own, in their own time, while still matching the functionality and capabilities of non-self-service aspect. It is not merely the cost and efficiency savings that come as a by-service of self-service methods. With the right self-service data visualization tools and software, analysts are able to uncover interesting and unique patterns and relationships from the underlying data, and create striking visualizations to express those patterns through powerful, memorable visuals.

In an age where big data along with social, mobile and the cloud are all converging in new and exciting ways, data storytelling has become more essential than ever. Big data storytelling is a technique of delivering information resulting from multifaceted data analysis process in a way that allows the decision-makers to easily and quickly understand the context, understand its meaning and draw conclusions from it [6]. The vital step in constructing any storytelling is structuring easy-to-follow data stories which are the sequences of causally related events. Foremost, storytelling takes time to unfurl and the tempo matches the audience's aptitude to follow. Next, it holds the audience's attention by encompassing attention-grabbing background, typeset, and intrigue. Finally, it leaves a eternal perception, either by stimulating the audience's curiosity and making them want to study or observe more.

2 Preliminaries and Basic Concepts

In this section, we present the definitions and basic concepts of big data real-time analysis, real-time self-service visualization, and real-time data storytelling.

2.1 Big Data Analysis in Real Time

To perform real-time big data analysis, three major steps are required and is presented in Fig. 1.

Each of the steps is described below covering technical challenges:

1. Real-Time Data Collection—It is the method of collecting data from diversified sources and then integrates and stores in as-is form. The challenge involved in establishing trusted connection to various data sources and extracting and storing in real-time.
2. Real-Time Data Processing—The process essentially gives shape to stored raw data. Precisely, it cleanses and transforms for visualization or analysis consumption. Additionally, it produces smaller datasets (with aggregation/summarization

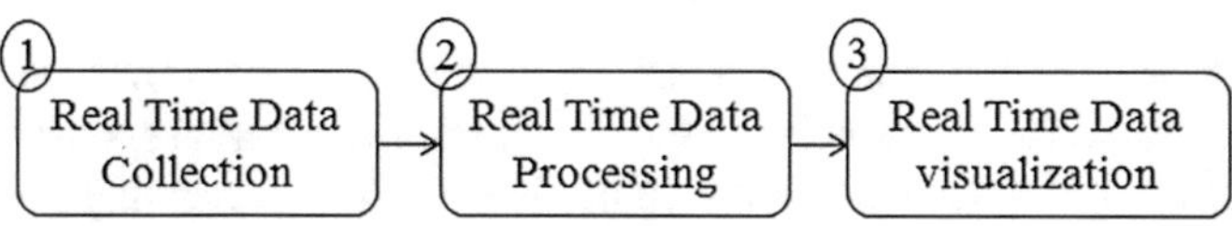

Fig. 1 Real-time data big data analysis steps

operation) which are generally consumed for visualization. It is also designed to detect real-time business in the processed data and to raise early warning for immediate action. Detection of exception is generally predefined business rules. The challenges surfaced while querying the large amount of data for abnormality detection.

3. Real-Time Data Visualization—This process involves monitoring the actions defined by the real-time decision-making process. The challenges surfaced when the visualization are explicitly performed by human in manually rather relying on available visualization tools.

2.2 *Big Data Self-Service Visualization*

Data visualizations are the method for expressing the insights found within big data. Self-service tools let the analyst without technical experience or know-how can make use of the tool without the need for external assistance to represent the insights from the underlying massive volume of data. It is added level of control, flexibility and customization over the work. Real-time self-service visualization allows making data-driven decisions with no gaps in information.

Traditional big data visualization development was confined to few technical professionals and few decision-makers were able to access the data. This approach was analogous to the conventional top-down business ladder where only a few decision-takers were making the decisions. This classical decision-making process does not fit modern era. A culture of bottom-up is getting traction for decision-making and enables everyone to visualize the data and analysis and has the potential to make even better data-driven decisions. In traditional visualization, the analyst and end user typically have no ability to create their own data visualizations or established data sources without the assistance of technical professionals. Self-service visualization empowers the analyst to explore any reachable dataset and to easily create personalized data visualization. Typically, it does not require a traditional deployment methodology such as build, test, and publish, since every analyst can simply build or extend the visualization with graceful methods such as drag-n-drop.

The difference between traditional visualization and real-time self-service visualization is presented in Table 3.

2.3 *Real-Time Data Storytelling*

Traditional visualization results in protracted visualization with over-eventful text, tables, graphs, and other visualization components. Decision-takers then regularly kick off exercise to reduce the mountain of visualization, demanding for summary visualization or exception visualization. The essence is to ensure that all the pertinent

Table 3 Difference between visualization and self-service visualization

Traditional visualization	Self-service visualization
Concept is old and exists around last 15 years	New concept, getting traction
IT department create visualization for each analyst/end user	Analyst creates visualization and answer questions for themselves, rather than having an IT department create visualization for each question/end user request
For each new question/end user ad hoc request, it places a burden on IT and goes through traditional build-n-deploy and it consumes time	For each new question/end user ad hoc request, analyst can build of its own and leads to quick turnaround time
Has a predefined view of the data	Operations like aggregation, summing are performed on the fly and hence the real-time view
Dimensions are decided at the time of designing the view by the IT department	Each end user may be asking different questions, and looking for different answers. So on-demand dimensions can be designed seamlessly
Generally not exploit In-Memory	In-memory instead of disk for any operation
Minimal support on "what-if" analysis	Enrich with real-time "what-if" analyses and leads to smarter decision

information still persists in the condensed version and the message is easily understandable. Frequently, the condensed version does not answer the vital question.

Real-time storytelling promptly provides the context, relevance, expectations and enables the decision-makers to grapple a vast quantity of facts swiftly and is combined with the power of efficient and effective real-time visualization.

3 Literature Review

In this section, we present the literature review of "big data solution stack to overcome technical challenge" and "history of storytelling in big data visualization".

3.1 Big Data Solution Stack

Big data analysis requires the use of best technologies at every step be at collating data, processing, and finally deriving the final conclusion in the form of visualization. The best technology helps in improving the processing power of the overall system for analyzing in real time. So each of the steps needs the implementation is not just only efficient but also economical. One of the ways to achieve the combination of efficiency and economy is open-stack technology with the adoption of parallel

Table 4 Big data analysis tool

Tool	Purpose
Apache Mahout	Machine learning and Data Mining
R	Predictive/Descriptive Analytics

and distributed approach. The solution stack is addressing the following technical challenges:

1. Real-Time Data Collection—Hadoop Distributed File System (HDFS) is a distributed file system and runs on commodity hardware [7] and is based on master–slave architecture. HDFS cluster comprises a single name node and more no of data nodes. Name node manages the file system namespace and regulates access to files by client requests whereas data nodes, usually one per node in the cluster, manage storage attached to the nodes that they run on. In HDFS, a large single file is split into sizable no. of blocks, which are stored in a set of data nodes. The name node executes different operations like opening, closing, and renaming files and directories related to file system and also determines the mapping of blocks to data nodes. The data nodes are used to execute when a client raises read and write requests. The data nodes perform different operations like block creation, deletion, and replication after getting instruction from name node [7]. Apache Flume and Sqoop are both eco-component of Hadoop which pull the data and load them into Hadoop cluster. Flume is responsible for collecting and aggregating large amount of log data whereas Sqoop is responsible for collecting, storing, and processing large amount of data. It scales data horizontally and multiple Flume tools cluster should put into action to collect large amount of data parallely and storing to staging area parallely as well [8].
2. Real-Time Data Processing and Data Storage—Spark is Apache Software Foundation tool for speeding up Hadoop data processing. The main feature of Spark is its in-memory computation that guarantees increase in the processing speed by caching the data and results in real-time processing. It provides fault tolerance through RDD (Resilient Distributed Dataset) which transparent data storage on memory and persists to disc only when needed. This helps to reduce most of the disc read and write cycle [9]. A collection of spark tool does parallel read and on the fly, dirty data is discarded and only the meaningful data to be transformed and loaded to Data Lake (No SQL).
3. Real-Time Data Visualization—Following tools are adequate for real-time analysis.

   3.1 Big Data Analysis Tool—The purpose of the big data analysis tool is to detect patterns, identify trends, and collect other valuable findings from the ocean of information. Table 4 depicts the tools to perform various analytics.

   3.2 BI Tool—BI (Business intelligence) tools are the technology that enables business/operational people to visualize the information to help/better the business/operation. Tableau is an in-memory BI tool for visually analyzing the data in real time. Tableau also enables users to share interactive and shareable dashboards depicting trends, variations, and density of the data in form of graphs and charts [10].

3.2 *History of Storytelling in Big Data Visualization*

In 1977, John Tukey wrote in Exploratory Data Analysis that Visualization is a means to extend storytelling and communicate patterns and trends in the visual realm. The role of visualization in data science and more so in big data science is inestimable [11].

In 1997, Behrens commented on Tukey's work saying "often likened EDA to detective work". The role of the data analyst is to listen to the data in as many ways as possible until a plausible story of the data is apparent [12]. Tukey's influence on the data science community remains to this day. Ultimately, the goal of visualization is to communicate answers to a question. The authors developed a generalized data science pipeline paralleling the elements of a story with a beginning, middle, and an end.

In 2014, Blue Hill Research has adapted Campbell's approach to analytics as shown in Fig. 2 [11].

In 2014, Mico Yuk and Stephanie Diamond defined an easy-to-follow storyboard. They defined storyboarding is to translate business requirements into a four segments that states the goal, measurements, and data visualization types [13]. The segments are

- Current State: What is happening now?
- Trends: How did it happen and what are the relationships?
- Forecast: What will happen in the future?
- What-if: How should we act in future to achieve or exceed the goals?

There are many variations to this approach. In the simplest case, a story has a beginning, middle, and an end.

In 2015, Linderman formulates five rules of storytelling [14] and is as follows.

- Make Opening Count—Explain the problem and hint the solution in first few lines.
- Be Vulnerable—Represent the challenges.
- Build Tension—Pitches are the subtle changes in the emotion. It is vital to identify the arc of the pitch and engage the audience through heightened tension.
- Revisit Value Proposition—Answer the value proposition and segue into the vision for the future.
- Call to action—The solution should be very specific and understandable.

Commonalities between the classic Hero's Journey and Linderman's approach to storytelling are twofold, i.e., the goal is to have an objective in mind and communicate a clear message.

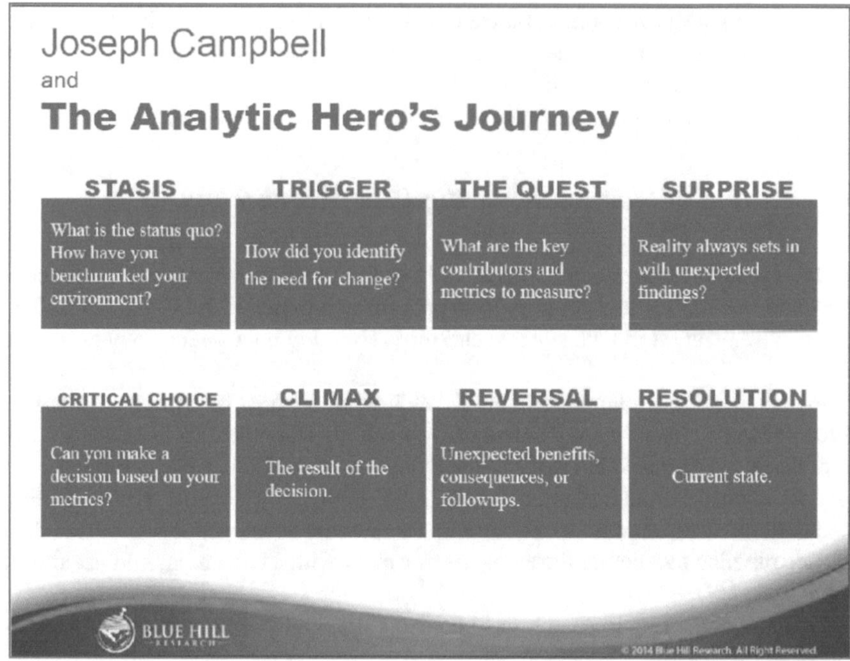

Fig. 2 The Hero's Journey. Eight steps of a story adapted to big data (Park and Haight 2014)

There has been renaissance in storytelling and analysts are approaching many industry segments with the recipes for successful storytelling and have rejuvenated interest in storytelling techniques.

In 2015, Tableau Software presented a white paper, keeping in mind the detective story. The elements of the story—from problem to tension to resolution—appear and are communicated from within the visualization are highlighted [15].

4 Proposed Approach

When data visualization is applied to big data, real-time storytelling is to engage decision-makers. Due to the continuous evolution of social media, complex visualizations require lots of deep analysis and is losing their relevance. Hence, it is critical to develop the accurate story and to reduce the time-bound attention from minutes to seconds.

The vital step in building any real-time self-service data visualization is to develop an easy-to-follow real-time storyboard. Real-time storyboarding is the visual representation of four-part diagram and is shown in Fig. 3 [13].

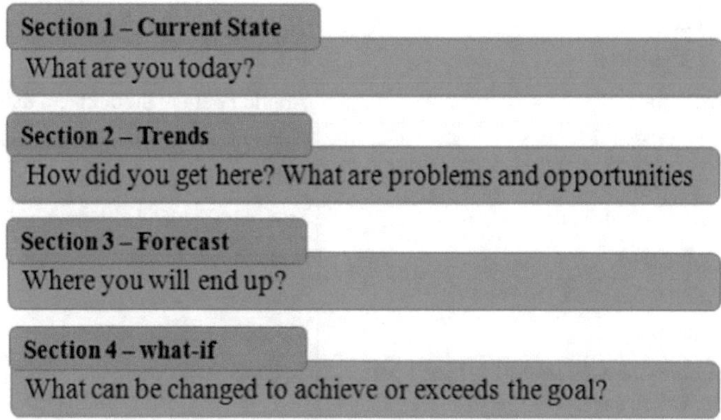

Fig. 3 Storyboard section

Easy-to-follow real-time storyboard is the collection and processing of actionable and particular real-time data that outlines a clear story. To create real-time storyboard, following steps can be followed.

1. Identification of the viewers—Knowing the viewers helps to quickly establish the manner of storyboard to construct and understand the approach for gathering helpful data and visual requirements. If the viewers are C-level executives or senior managers, expect to have little time to inspect immense detail. So data visualization for those viewers must be summarized views that give a past, present, and future of the business with drill down to further required details.
2. Document viewer's goals—A clear and unambiguous understanding of viewer's goals and existing pain points would assist to determine the scope of storyboard.
3. Define KPIs—Understanding the key performance indicator (KPI) that the viewers must view, monitor, or track. It is recommended to keep KPI count to fewer than 10 items combined [13].
4. Dig Deeper to Identify the Sole Purpose of the Story—It is the last step in developing the real-time storyboard. Identify closely on what visualization method is the ideal for better decision-making or analytics. Table 5 is depicting the commonly used data visualization methods. Additionally, identify the KPI which are impactful and are "unknown".

Figure 4 is depicting the KPI fitment chart.

5 Discussion

The art of real-time storytelling is simple and complex at the same time. Stories provoke thought and bring out insights. It is often overlooked in data-driven operations,

Table 5 Big data analysis tool

Sr#	Method name	Big data class
1	Treemap	Applied only to hierarchical data
2	Circle Packing	Applied only to hierarchical data
3	Sunburst	Volume + Velocity
4	Parallel Coordinates	Volume + Velocity + Variety
5	Steamgraph	Volume + Velocity
6	Circular Network Diagram	Volume + Variety
7	Table	Volume + Velocity + Variety
8	Bubble Chart, Scatter Plot, Line Chart, Bar Chart, Pie Chart	Volume + Velocity
9	Word Cloud (Textual data)	Volume
10	Candlestick Chart (Time Series data)	Volume + Velocity + Veracity
11	Map (Geographic data)	Volume + Velocity + Variety

Fig. 4 KPI fitment chart

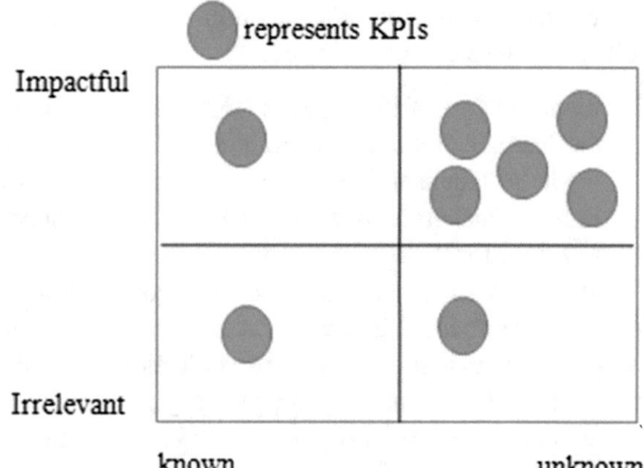

as it is believed to be a trivial task. What we fail to understand is that the best stories, when not visualized well in real time, end up being useless.

6 Conclusion

In practice, there are a lot of challenges for big data processing and visualization in real time. As all the data is currently visualized by computers, it leads to difficulties in the extraction, followed by its processing and visualization in real time. Those tasks are time-consuming and do not always provide correct or acceptable results.

The main focus of this paper is to give a brief survey of real-time storytelling with self-service visualization techniques that are used in big data. The approach on building any real-time story-telling with self-service data visualization is presented. Hope this paper will serve as a helpful opening to readers interested in self-service visualization and real-time storytelling in big data technologies.

References

1. Chen, C.L.P., Zhang, C.Y.: Data-intensive applications, challenges, techniques and technologies: a survey on big data. Inf. Sci. **275**(10), 314–347 (2014)
2. Jony, A.I.: Applications of real-time big data analytics. Int. J. Comput. Appl. **144**(5) (2016)
3. Mohamed, N., Al-Jaroodi, J.: Real-time big data analytics: applications and challenges. In: International Conference on High Performance Computing and Simulation (HPCS) (2014)
4. Khan, M., Khan, S.S.: Data and information visualization methods and interactive mechanisms: a survey. Int. J. Comput. Appl. **34**(1), 1–14 (2011)
5. Sucharitha, V., Subash, S.R., Prakash, P.: Visualization of big data: its tools and challenges. Int. J. Appl. Eng. Res. **9**(18), 5277–5290 (2014)
6. Data stories—how to combine the power storytelling with effective data visualization, https://www.slideshare.net/miriamgilbert08/data-stories-workshop-34390209
7. HDFS, https://hadoop.apache.org/docs/r1.2.1/hdfs_user_guide.html
8. Hurwitz, J., Nugent, A., Halper, F., Kaufman, M.: Big Data for Dummies. ISBN: 978-1-118-50422-2
9. Spark, https://www.spark.apache.org/
10. Empowering Insight with Tableau: A Case for Self-Service Analytics, https://www.ironsidegroup.com/2016/09/07/tableau-self-service-analytics/
11. Visualizing Big Data: Telling a Better Story—MODSIM World 2018, www.modsimworld.org/papers/2016/Visualizing_Big_Data.pdf
12. Behrens, J.T.: Principles and procedures of exploratory data analysis. Psychol. Methods **2**(2), 131–160 (1997)
13. Data Visualization for Dummies, http://pdf.th7.cn/down/files/1603/DataVisualizationForDummies.pdf
14. 5 Rules for Telling Stories with Your Pitch, https://andrewlinderman.com/2015/08/07/theres-an-easy-fix-for-a-dull-pitch-tell-a-story/
15. The 5 most influential data visualizations of all time, http://www.tableau.com/top-5-most-influential-data-visualizations

Informative Gene Selection Using Clustering and Gene Ontology

Soumen K. Pati, Subhankar Mallick, Aruna Chakraborty and Ankur Das

Abstract The main aspect of bioinformatics is to make an understanding between microarray data with biological processes as much as possible to ensure the development and application of data mining techniques. Microarray dataset is high voluminous containing huge genes, most of these are irrelevant regarding cancer classification. These irrelevant genes should be filtered out from the dataset before applying it in cancer classification system. In this paper, a clustering algorithm is used to group the genes whose similar expressions suggest that they may be co-regulated. Once the clusters are obtained, the biological knowledge is investigated for the genes associated with the clusters. A quality-based partition is determined by the co-expressed genes that have been incorporated with similar biological knowledge. Gene Ontology (*GO*) annotations are used to link the clusters to identify the biologically meaningful genes within the clusters. In the next phase, the fold-change method is used to pick up the differentially expressed genes from selected biologically meaningful genes within the clusters. These selected genes are termed as informative genes. The efficiency of the method is investigated on publicly accessible microarray data with the help of some popular classifiers.

Keywords Bioinformatics · Gene selection · Gene ontology · Clustering
Cancer identification · Microarray dataset

S. K. Pati · S. Mallick (✉) · A. Chakraborty
St. Thomas' College of Engineering & Technology,
Kolkata 700023, West Bengal, India
e-mail: subhan.mallick@gmail.com

S. K. Pati
e-mail: soumenkrpati@gmail.com

A. Chakraborty
e-mail: aruna.stcet@gmail.com

A. Das
Calcutta Institute of Engineering and Management,
Tollygunge, Kolkata 700040, West Bengal, India
e-mail: ankurdas8017@gmail.com

A. Abraham et al. (eds.), *Emerging Technologies in Data Mining and Information Security*, Advances in Intelligent Systems and Computing 813,
https://doi.org/10.1007/978-981-13-1498-8_37

1 Introduction

Microarray dataset is classically known to acquire big sets of annotations (genes), characterized by thousands or even additional of coordinates with ostensibly unknown correlations. This high cardinality regarding dimensionality has offered many challenges in analyzing the data, particularly when correlations among the genes are so complex. An investigation system considered for using the large set of genes will have high computational cost, slow learning process and poor classification accuracy due to the presence of high dimensionality. So, the selection of the most significant, informative, discriminative, and compact subset of genes from a high volume of genes is the target of a gene selection process for true diagnosis [1]. Essentially, gene selection [2] is a combinatorial optimization problem, which explores an optimal gene subset from a pool of 2^N competing candidate subset in a dataset, where the total number of gene is N.

In the paper, a biologically meaningful gene subset selection method is presented which gives maximum classification accuracy for classifying unknown samples. First, a hamming distance based *CHS* clustering technique [3] is applied on the experimental dataset. The *CHS* method is totally based on expression values and uses hamming distance to overcome the demerits (computational complexity and gene-to-gene interaction) of Euclidian distance in high dimensional space. Then, the *SAM* [4] is used to remove the redundant and irrelevant genes from each of the clusters and collect only the differentially expressed genes. Then Gene Ontology (*GO*) tree [5] is incorporated into the proposed method to gather the biological information. A GO tree is created using GO database. Involving genes in the expression data are mapped to GO tree and the nodes that are unmapped in GO tree are discarded accordingly from each cluster. So, biological knowledge is associated to each cluster and meaningful clusters are selected from initial set of clusters using *GO* tree. Then fold-change method [6] is used based on a cut-off value to select differentially expressed genes and form the informative gene subset. Some base classifiers (*SVM*, *MLP*, Naive Bayes, etc. [7]) are used to demonstrate the usefulness of the proposed methodology measuring the classification accuracy. The overall flow diagram of the methodology is shown in Fig. 1.

The paper is reported in four sections. Section 2 describes the clustering of the dataset and selection of important informative genes from biologically meaningful clusters with the help of gene ontology. Section 3 describes the experimental result and performance evaluation of the proposed method. The conclusions and future scope of this method are presented in Sect. 4.

2 Literature Review

In human beings, genes, made up of DNA, differ in volume from few hundred bases of DNA to more than two million DNA bases. Most genes are the same in all the

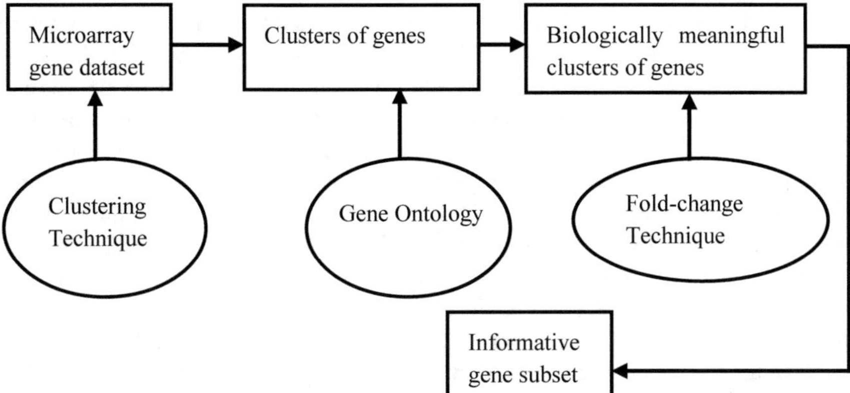

Fig. 1 Flow diagram of the proposed method

living organism of same species, but a small amount of genes (not greater than 1% of the whole) are slightly changed among the living organisms of same species. These small numbers of genes contribute unique physical features to each living organisms. One of the important features of DNA is that it can reproduce as well as create own copies [8]. Each strand DNA in the double helix structure that can help as a pattern for replicating the sequence of bases. It is vital when cells break up to new cell needs to have an identical copy of the DNA exist in the previous cell.

Generally, gene dataset is high dimensional and these datasets are required for suitable data mining technologies (for example clustering, gene selection, and classification) for further investigation [9]. Cluster analysis [3] is the assignment of grouping of a set of genes in such a mode that the genes in a cluster are more identical to each other than those are in different clusters. It is a primary objective for statistical dataset analysis including bioinformatics [10]. Cluster analysis can be done by different efficient clustering algorithms to make clusters using groups with similarities among members of that group, intervals or statistical distributions, dense area of input domain. So, clustering can be treated as a multi-objective problem of optimization [11]. The clustering is typically based on some similarity measures [12] described on the gene expression data (e.g., Euclidean distance, Manhattan distance, Hamming distance, and Pearson correlation).

The most significant problem in removing and investigating information from big databases is the high computational complexity. Dimension reduction [13] is the advantageous preprocessing step to select only important features for improving accuracy of the mining process. Feature selection can be done in either supervised or unsupervised way. Clustering as well as concurrent feature selection has been described in literature [14]. Selection of feature (subset) using supervised learning method is not always better than unsupervised method and vice versa. Performance of feature selection is satisfactory for both learning method [15].

The Gene Ontology (*GO*) scheme [5] has established three structured controlled vocabularies, namely biological processes (the functions of integrated living components, such as cells, organs, tissues and organisms at molecular level), cellular components (the cell parts or cellular surroundings), and molecular functions (the molecular level fundamental activities of a gene, such as catalysis or binding), which explain the relationship among the genes and gene products [16]. There are three disconnected parts related to this topic: (i) the development and maintenance of the ontology themselves (ii) the annotation of genes, which is involved with the associations between the ontology and the genes in the databases and (iii) development of tools that help the formation, maintenance, and utilization of ontology. The *GO* tree is prepared as an acyclic graph with direction, and every term has described relationships to one or more other terms in the same domain or other domains. The *GO* terminology is considered to be unbiased class, and includes terms applicable to prokaryotes and eukaryotes, single and multicellular organisms [16].

3 Gene Subset Selection Method

The biologically meaningful genes are selected for getting maximum classification accuracy during sample classification of microarray dataset. First, the proposed method uses a clustering technique [3] which gives an optimal number of clusters where the genes are grouped according to the similar expression values. Subsequently, a *GO* tree [5] will be used to find biologically meaningful clusters among all the clusters. Finally, the Fold-change method [6] is employed to select only the informative genes providing maximum classification accuracy.

3.1 Clustering of Genes

The Euclidean distance is familiar metric utilized for similarity measurement among two different objects. Though it is very helpful in low dimensional dataset but does not work sound in high dimensional microarray dataset. Rather, hamming distance is more effective for measuring similarity among genes with respective expression values in discrete domain [12]. Here, a Merging-Splitting based clustering method (*CHS*) [3] is utilized on initial set of clusters (n), which authenticates newly created clusters using validity indices and lastly, the set of optimal clusters is obtained by measuring the hamming distance [12] based similarity among genes.

3.2 Initial Clusters Selection

Initially, the clusters number (n) are obtained using the *CHS* methodology [3], where each cluster contains genes that have similar expression values. The statistical method *SAM* [4] is used to find differentially expressed (*DE*) genes among n clusters. This method is related to another statistical method popularly known as t-test method [4]. The test statistic of *SAM* is defined in Eq. (1).

$$d(i) = \frac{\bar{x}_i - \bar{y}_i}{s(i) - S_0} : i = 1, 2, \ldots, k \tag{1}$$

where S_0 is a small constant, x_i is the i-th gene expression value under one experimental condition, y_i is the i-th gene expression value under different experimental condition, and k is the number of genes in a cluster. Furthermore, $\bar{x}$ and $\bar{y}$ are the mean expression values under different conditions for i-th gene and S_0 is signified a percentile of the standard deviation values of all the genes. The standard deviation (detailed scatter of gene) $s(i)$ is defined using Eq. (2).

$$s(i) = \sqrt{\frac{\frac{1}{I} + \frac{1}{J}}{I + J - 2}\left[\sum_{i=1}^{I}\left(x_{ij} - \bar{x}_i\right)^2 + \sum_{j=1}^{J}\left(y_{ij} - \bar{y}_i\right)^2\right]} \tag{2}$$

where, I is the number of replicates in one experimental condition and J is the number of replicates in different experimental condition.

After finding differentially expressed genes from each cluster using *SAM* [4], the k number of clusters ($k \leq n$) is selected which contain only *DE* genes.

3.3 GO Tree Construction

The gene ontology is a very useful to all organisms [5] as it is a controlled vocabulary for all organisms. Experimental studies [5] among three types of GO (cellular component, biological process, and molecular function) depict a biological relation within the genes of the data. The biological process is best suited among all the rest categories of *GO* for this purpose.

Originally, *GO* hierarchical structure is represented by a directed acyclic graph (*DAG*). GO flat file format is parsed and converted it into an ordered tree where the order is specified by the number of children of each internal node. A GO-term may have two or more parents as well as it may have more than one path from the root also. To construct an ordered tree, GO-term's frequencies in several paths are regarded as distinctly. Thus the tree which is constructed itself an ordered tree whose root is a biological process and nodes are the GO-terms. The resultant graph (ordered tree) having *GO*-terms as its nodes and the root is the biological process term. The information accumulated in every node includes term identifier, term name, and relationship with parent term [5].

3.4 Biologically Meaningful Cluster Selection

The biologically enrich clusters are obtained by using *GO* [5]. The *GO* tree consists of GO-terms of particular genes which belong to the expression data. The *GO*-term of corresponding gene can be found by investigating the related database (like Saccharomyces Genome Data repository for yeast data). All those found *GO* terms are noticeable in the *GO* tree where one gene may have many corresponding *GO*-terms and all are marked. In time of the finding the relationship among the genes which are in a cluster, only the marking nodes will be under consideration while rest will be rejected.

Therefore, after forming *GO* tree for a dataset, only functionally related genes are retained for each cluster according to biological process and the l number of clusters are selected ($l \leq k$) which are biologically meaningful.

3.5 Fold-Change

The inclusion of biological knowledge (*GO* tree) in gene selection method enables us to select gene with better biological interpretation. Now, the fold-change concept [6] is used to select significant differentially expressed genes from l number of biologically meaningful clusters, which are considered as the informative genes in the proposed method. Because of its computational simplicity and interpretability, this method is preferred to biologist for gene selection purpose. It is frequently used to measure the variation in expression value of a respective gene. The fold-change for i-th gene is defined in Eq. (3).

$$FC_i = \frac{max(\bar{x}_\imath, \bar{y}_\imath)}{min(\bar{x}_\imath, \bar{y}_\imath)} \tag{3}$$

where $\bar{x}_i$ and $\bar{y}_i$ are the mean expression values of i-th gene for two different experimental conditions. The fold-change value is measured from the mean normalized values between two groups. The biologists usually think the cut-off value for Fold-change as $FC_i \geq 2$, as a important standard for identifying the differentially expressed genes.

Thus, after performing fold-change on a dataset, only p number of genes is identified whose fold-change value is greater than 2 and fully related to biological knowledge.

3.6 Gene Subset Selection Procedure

Suppose, the experimental microarray dataset is represented by a multidimensional array $(M)_{g\times s}$, where the rows are containing $|g|$ number of genes and columns are symbolized $|s|$ number of samples. The following procedure selects the important biologically meaningful genes which are responsible to give the maximum classification accuracy regarding the class of samples identification.

```
Algorithm: Gene Subset Selection (g,s)
Input: Gene expression dataset (M) g×s
      Output:Biologically meaningful p number of genes,
      (M) p×s
Begin
      Perform CHS algorithm [5] and obtain optimal-
      number(n) of clusters.
      Apply SAM algorithm [14]to reduce redundant genes
      from each cluster using equation (1) and obtain k
      number of clusters, where (k ≤ n).
      For each of k clusters
            For each gene
                  Find GO-term using biological pro-
                  cess.
                  Obtain functionally related genes us-
                  ing biological knowledge.
            End
      End
      Obtain biologically meaningful l number of clus-
      ters, where (l ≤ k).
      Apply Fold-change method using equation (3).
      Obtain desire gene subset which contain |p| num-
      ber of genes, where |p|<<|g|.
      Use some base classifiers to measure the effec-
      tiveness of the gene subset.
End
```

4 Experimental Results

Performance evaluation of the proposed methodology is discussed in this section based on publicly available benchmark microarray dataset. Here, *Yeast* [17], *Gastric cancer* [18], and *Leukemia* dataset [19] are used for experiment purpose and WEKA [20] data mining tool that contains some base classifiers is used for validation of the method.

Table 1 The outcomes of *CHS* method

Dataset	#Cluster	DB	Dunn	H	SC
Yeast	49	0.3483	0.8462	0.0028	0.7302
Gastric	64	0.4005	0.7649	0.0652	0.7754
Leukemia	69	0.5119	0.7708	0.0442	0.6340

The *Yeast* dataset is a set of 2884 genes under 17 conditions (based on time points). The *Gastric cancer* dataset contains 4522 genes and 30 samples (8 normal and 22 cancerous). The *Leukemia* dataset is a set of 7129 genes and 72 leukemia samples. Among these samples, 47 are acute lymphoblastic leukemia (ALL) and rest 25 are acute myeloblastic leukemia (AML).

4.1 The CHS Clustering Results

The *CHS* method [3] iteratively finds the clusters and lastly, regarding the validity indices optimal number of clusters is achieved and the output of the *CHS* algorithm is shown in Table 1 regarding the optimal number of clusters (#Cluster) and validity indices [21] (such as, *DB*, *Dunn*, Hartigan (*H*), and Silhouette (*SC*)).

4.2 Gene Subset Selection

SAM method [4] identifies the differential expression genes from each cluster and the unnecessary genes are removed from each cluster. For example, in Yeast dataset, the 30-th cluster contains 628 genes and after removing redundant genes by *SAM*, it contains 206 genes. The *GO* tree is constructed for each dataset and *GO*-term is defined for each gene of clusters. It is identified that a gene may be mapped with two or more GO-term. These terms are treated as separate nodes on the GO tree. Table 2 shows some levels of the tree, *GO*-term, and corresponding *GO*-id for Yeast dataset.

After applying gene ontology concept, the proposed method selects some biologically meaningful clusters according to *GO*-terms of each gene. Then fold-change method [6] is applied on biologically meaningful clusters to obtain the informative genes (shown in Table 3) (Table 4).

Table 5 describes the performance of the proposed method, where some classifiers [7], namely, Naive Bayes (*NB*), Multilayer Perceptron (*MLP*), Decision Tree (*J4.8* Random Forest (*RF*), and Rule-Based (*Bagging*) classifiers are used to measure the accuracy using selected gene subsets.

From Table 5, it observed that the *MLP*, *RF*, and *Bagging* classifiers provide 100% accuracy for *Gastric* cancer dataset. Similarly, for others dataset, the highest accuracies are marked by bold font. It is also noted that the average classification

Table 2 Example of *GO*-terms and GO-id on various levels for yeast dataset

Level	*GO*-term	*GO*-id
0	Gene ontology	GO: 0003673
1	Biological process	GO: 0008150
2	Development	GO: 0007275
2	Cellular process	GO: 0009987
2	Regulation	GO: 0050789
2	Physiological process	GO: 0007582
2	Viral life cycle	GO: 0016032
3	Cell differentiation	GO: 0030154
3	Cell communication	GO: 0007154
3	Cellular physiological process	GO: 0050875

Table 3 Biologically meaningful clusters

Dataset	#Initial cluster by *CHS* method	#Biologically meaningful clusters	#Selected genes
Yeast	49	14	7
Gastric	64	15	10
Leukemia	69	17	12

Table 4 The gene-id and gene name of the selected genes for considered datasets

Yeast dataset		Gastric dataset		Leukemia dataset	
Gene-id	Gene name	Gene-id	Gene name	Gene-id	Gene name
527	YDL164C	1470	MT1G	45	AFFX-HSAC07/X00351_5_at
1659	YKL190 W	2945	GIF	6606	X00437_s_at
2087	YMR076C	2967	ATP4A	322	D26598_at
942	YER018C	6375	HMGCS2	6967	M15395_at
1185	YGR152C	6681	CA2	2393	M95627_at
1913	YLR325C	1407	SST	6209	Z19554_s_at
789	YDR385 W	717	FCGBP	1779	M19507_at
–	–	1824	REG3A	6510	U23852_s_at
–	–	3068	ATP4B	1674	M11147_at
–	–	5779	LTF	4378	X62691_at
–	–	–	–	6584	Z54367_s_at
–	–	–	–	1868	M26602_at

Table 5 Description of selected genes for *Leukemia* dataset

Dataset	Classifier name	#Correctly classified samples	Accuracy (%)	Average accuracy
Yeast	NB	15	88.24	90.59
	MLP	**16**	**94.12**	
	J4.8	14	82.35	
	RF	**16**	**94.12**	
	Bagging	**16**	**94.12**	
Gastric	NB	28	93.33	98
	MLP	**30**	**100**	
	J4.8	29	96.67	
	RF	**30**	**100**	
	Bagging	**30**	**100**	
Leukemia	NB	63	87.50	88.33
	MLP	**65**	**90.28**	
	J4.8	61	84.72	
	RF	64	88.89	
	Bagging	**65**	**90.28**	

accuracy measuring by mentioned classifiers is above 88% for all dataset, which gives the effectiveness of the proposed method.

5 Conclusions

Methodical and neutral method for classification of samples is a vital treatment procedure of the disease and drug discovery. On the basis of biologist focal point, a small number of genes are desirable that controls the products before performing in strength analysis and costly experiments with a big set of genes. Thus, automatic detection of the informative gene subset which is biologically significant is highly advantageous. Here, a gene subset selection procedure is presented involving an effective clustering method, differentially expressed gene concept conjunction with biological knowledge. Gene ontology is a suitable biological resource to obtain informative genes. The algorithm *CHM* is employed for gene clustering based on gene expression values and obtained an optimal number of clusters for handling ambiguity and vagueness in high dimensional space. To reduce the computational complexity with respect to sample classification, the redundant genes are removed from each cluster using *SAM*. A *GO* tree is also attached with the proposed method to select biologically meaningful clusters among all clusters. Finally, the representative genes of every such cluster are considered by calculating greatest fold-change. External validation of the proposed method involving different classifiers regarding maximum sample classification accuracy has demonstrated the effectiveness of the proposed gene selection method.

References

1. Xu, X., Zhang, A.: Selecting informative genes from microarray dataset by incorporating gene ontology. In: Fifth IEEE symposium on bioinformatics and bioengineering. BIBE2005, pp. 241–245 (2005)
2. Das, A.K., Pati, S.K.: Rough set and statistical method for both way reduction of microarray cancer dataset. Int. J. Inf. Process. **6**(3), 55–66 (2012)
3. Pati, S.K., Das, A.K.: Missing value estimation for microarray data through cluster analysis. Knowl. Inf. Syst. Springer **52**(3), 709–750 (2017)
4. Zhang, Z.H., Jhaveri, D.J., Marshall, V.M., et al.: A comparative study of techniques for differential expression analysis on RNA-seq data. PLOS **9**(8), 1–11 (2014)
5. Rhee, S.Y., Wood, V., Dolinski, K., Draghici, S.: Use and misuse of the gene ontology annotations. Nat. Rev. Genet. **9**, 509–515 (2008)
6. Love, M.I., Huber, W., Anders, S.: Moderated estimation of fold change and dispersion for RNA-seq data with DESeq2. Genome Biol. **15**(12), 550 (2014)
7. Singh, U., Hasan, S.: Survey paper on document classification and classifiers. Int. J. Comput. Sci. Trends Technol. **3**(2), 83–87 (2015)
8. Kuzminov, A.: DNA replication meets genetic exchange: Chromosomal damage and its repair by homologous recombination. Proc. Natl. Acad. Sci. U.S.A. **98**(15), 8461–8468 (2001)
9. Wu, X., Zhu, X., Wu, G.Q., Ding, W.: Data mining with big data. IEEE Trans. Knowl. Data Eng. **26**(1), 97–107 (2014)
10. Wu, B.: Differential gene expression detection and sample classification using penalized linear regression models. Bioinformatics **22**(4), 472–476 (2006)
11. Liu, R., Liu, Y., Li, Y.: An improved method for multi-objective clustering ensemble algorithm. IEEE Congr. Evolut. Comput. 1–8 (2012)
12. Alamuri, M., Surampudi, B.R., Negi, A.: A survey of distance/similarity measures for categorical data. In: International joint conference on neural networks (IJCNN), pp. 1907–1914 (2014)
13. Rehman, M.H., Liew, C.S., Abbas, A., Jayaraman, P.P., Wah, T.Y., Khan, S.U.: Big data reduction methods, a survey. Data Sci. Eng. Springer **1**, 265–284 (2016)
14. Law, M.H.C., Figueiredo, M.A.T., Jain, A.K.: Simultaneous feature selection and clustering using mixture models. IEEE Trans. Pattern Anal. Mach. Intell. **26**(9) (2004)
15. Wolf, L., Shashua, A.: Feature selection for unsupervised and supervised inference: the emergence of sparsity in a weight-based approach. J. Mach. Learn. Res. **6**, 1855–1887 (2005)
16. Shivakumar, B.L., Porkodi, R.: Finding relationships among gene ontology terms in biological documents using association rule mining and GO annotations. Int. J. Comput. Sci. Inf. Technol. Secur. **2**(3), 542–550 (2012)
17. http://arep.med.harvard.edu
18. http://ailab.si/supp/bi-cancer/
19. http://www.genome.wi.mit.edu/MPR
20. http://www.cs.waikato.ac.nz/ml/weka/
21. Xu, R., Wunsch, D.: Survey of clustering algorithms. IEEE Trans. Neural Netw. **16**(3), 645–678 (2005)

Part II
Data Mining

Identifying Categorical Terms Based on Latent Dirichlet Allocation for Email Categorization

Aakanksha Sharaff and Naresh Kumar Nagwani

Abstract Being one of the most indispensable tools, managing email is a tedious task. There exist various email management tasks. Some of them are identifying email threads, clustering of emails, email categorization, email summarization, etc. Among these tasks, email categorization has emerged as an interesting research area. This paper presents a topic modeling approach to cluster emails and generating categorical terms. Categorical terms plays an important role in defining the categories. To identify the categorical terms, latent Dirichlet allocation (LDA) is done after cluster formation of emails.

Keywords Email categorization · Categorical terms · Latent Dirichlet allocation

1 Introduction

As the number of incoming email messages increases, it becomes very difficult to handle these emails. There are different tools for facilitating management of incoming emails, e.g., use of threads and use of folders or labels for categorizing incoming emails. Email categorization is a process of classifying text to a discrete set of predefined categories. Categorization of emails becomes difficult as different topics may be discussed in an email. Categorizing emails by identifying categorical terms is an important issue. It adds semantics to email management. In this paper, we propose an approach to identify (discover) categorical terms in a different way by adapting LDA as topic modeling approach. Categorical terms are used to define the category based on topic terms.

A. Sharaff (✉) · N. K. Nagwani
Department of Computer Science & Engineering,
National Institute of Technology Raipur, Raipur 492010, India
e-mail: asharaff.cs@nitrr.ac.in

N. K. Nagwani
e-mail: nknagwani.cs@nitrr.ac.in

A. Abraham et al. (eds.), *Emerging Technologies in Data Mining and Information Security*, Advances in Intelligent Systems and Computing 813,
https://doi.org/10.1007/978-981-13-1498-8_38

1.1 Email as Database

Email database consists of records of several users. Each user consists of several folders related to all_documents, sent_items, inbox, etc. Each file in folder consists of header section as well as content section. Hence, in this way, email files stored in all folders can be treated as email database.

1.2 Email Mining

Email Mining is a process of discovering useful and interesting patterns from email database using text mining approaches. There are various applications of Email Mining such as email clustering, email classification, email summarization, email visualization, email spam filtering, etc.

1.3 Email Categorization

Email categorization is also referred to as email filing and email classification. Due to the enormous volume of emails (sent/received), categorizing emails manually becomes a heavy burden for users. It is a process of assigning emails into predefined categories. Email categorization can be binary or multiple. In binary categorization, emails are categorized into two classes, i.e., spam and non-spam. In multiple categorization, emails are categorized into multiple classes, e.g., sports, entertainment, business, job opportunities, etc. Numerous researches have been reported in the field of email categorization [1–3]. The major techniques used in email categorization are Naïve Bayes, Support Vector Machine, Artificial Neural Networks, Rule learning, Decision Tree Based Algorithms, Logistics Regression, Lazy Learning, etc. [4].

1.4 Latent Dirichlet Allocation

Latent Dirichlet allocation (LDA) is a generative probabilistic model used to perform basic tasks such as classification, summarization, novelty detection, and similarity judgments over discrete corpora [5]. The basic idea of LDA is to represent document as random mixtures over latent topics where each topic is characterized by distribution of words [5]. According to Blei et al., LDA can be viewed as dimensionality reduction technique in spirit of Latent Semantic Indexing (LSI) and probabilistic Latent Semantic Indexing (pLSI). LDA is more feasible and effective in Information Retrieval tasks but gaining citations as an area of interest in machine learning field [6].

2 Related Work

Different strategies for managing email have been studied by Whittaker et al. [7]. They have shown the values for threading and search based tools for refinding emails. Managing emails by labeling incoming emails based on user preferences has been proposed by Armentano et al. [8]. Raghuveer presents a supervised approach for labeling of emails based on attachment and content of emails [9]. Alsmadi and Alhami [2] have proposed classification of emails based on N-grams.

An unsupervised learning framework has been proposed by Dredze et al. for generating a gist of email by selecting summary keywords [10]. Sayed et al. develops a three-phase tournament-based approach to classify emails by using World Cup final rules as heuristic approach [11]. An email answering recommendation has been developed using text classification technique by using multiple concept [12].

Dehgani et al. present various approaches for organizing emails in a better and an efficient way. Some of his approaches are finding email threads in a linear and tree structure conversations, categorizing emails by using its structural aspects, etc. [2, 13]. A brief survey of various text mining tasks performed over email has been discussed [14]. The techniques and tools used for performing email mining tasks are also presented by Tang et al. Email Mining tasks include email categorization, email contact analysis, email network property, email visualization, and email spam detection.

3 Methodology

The methodology of proposed work consists of five steps. The first step is to remove the terms, which are not required for our research, i.e., preprocessing of emails. Preprocessing of emails includes the task of removing stopwords and finding the root word by applying the process of stemming. In the second step, clusters are formed by applying textual similarity technique on email subject and content. After formation of cluster, LDA is applied and categorical terms associated with each topic have been generated. Next step is to calculate the frequent terms associated with each topic and filter out the most dominant terms. The overall research plan is depicted in Fig. 1 and explained below.

3.1 Preprocessing of Emails

Preprocessing of emails is performed to remove stopwords (e.g., is, am, are,) and to extract root words by applying stemming (e.g., converts covers, covered, covering to root word cover). The goal of preprocessing technique is to reduce the size of email.

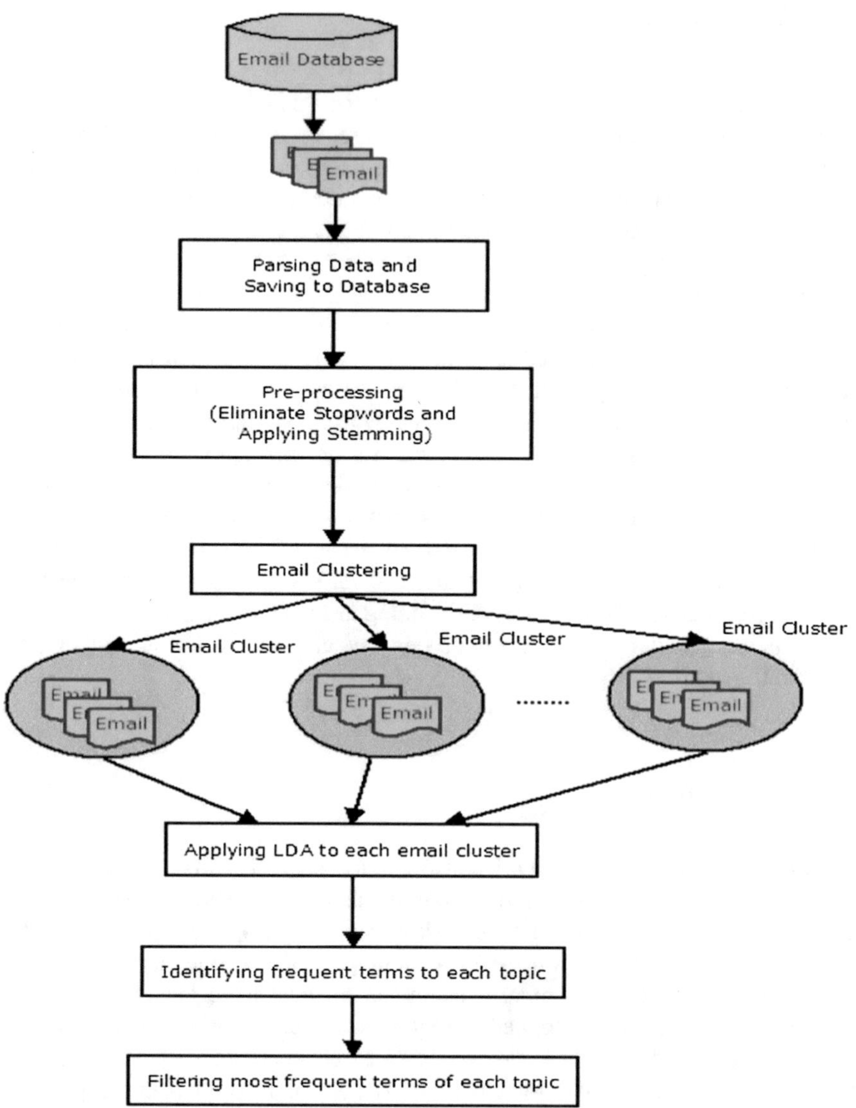

Fig. 1 The research methodology of the proposed work

3.2 *Creating Email Clusters Using Textual Similarity*

After performing preprocessing operations, clustering is applied to form clusters of preprocessed emails. Clustering of emails yields group of similar emails. Topic modeling approach will be applied to individual clusters to generate categorical terms.

3.3 Applying LDA on Email Clusters to Discover Topics

Once the email clusters are created, LDA is applied to individual email clusters by taking collection of email attributes (subject and body) of particular cluster. Nine topics are generated using ten terms present in each topic.

3.4 Identifying Frequent Terms for Each Topic

Capturing important terms and their relationship in any textual document are a critical part of text mining approach. Many researchers proposed approaches that form a group of terms called "topics". Individual terms are captured for each cluster once the clusters are created using LDA and occurrences of each term in all clusters are computed.

3.5 Removing (Filtering) Most Dominant Terms from Each Topic

Now based on the frequency of topic terms, filter out those terms with maximum frequency. These topic terms are considered as categorical terms.

4 Experiments and Results

The experiments are performed over publically available Enron dataset [15]. The dataset contains the folder information for each employee. Each message present in the folders contains the senders and the receiver email address, date and time, subject, body, text, and some other email specific technical details [15]. Implementation of proposed work is performed using Java programming language and LDA is implemented using Mallet API [16]. After preprocessing of emails, clusters are formed and based on LDA terms associated with each topic has been identified. These terms are mentioned in Table 1 and based on categorical terms, we provide category name. These categorical terms help in labeling (defining) categories.

5 Conclusion and Future Work

A methodology for identifying categorical terms and clustering emails based on LDA has presented. Proper categorical terms are required for effective categorization of emails. The goal of this work is to manage the emails systematically when there

Table 1 Generating categorical terms using LDA

Category name	Categorical terms
Resources	Gas, natural, oil, pant, tree, power, internet, web, electricity
Information technology	Information, message, email, news, communication, contact, research, data, transmission, project, images, question, request, call, to/from, review, list, spam, services, click, program, system, link, server, online, Internet, web, e-trade, file
Interface	Color, changes, font, size, center, left, class, align, font size, image, width, height, position, list, table, click, intended, updated, attached, link
Finance	Credit, rate, market, business, trading, deal, stock, order, seal, price, change, bill, financial, buy, fare, transaction, tax, billion, total, e-trade

is a vague idea of email topic. The future outlook for this research is to apply this technique for multi-topic identification of emails.

References

1. Youn, S., McLeod, D.: A comparative study for email classification. In: Advances and Innovations in Systems, Computing Sciences and Software Engineering, pp. 387–391. Springer Netherlands (2007)
2. Alsmadi, I., Alhami, I.: Clustering and classification of email contents. J. King Saud Univ.-Comput. Inf. Sci. **27**, 46–57 (2015)
3. Dehghani, M., Shakery, A., Mirian, M.S.: Alecsa: attentive learning for email categorization using structural aspects. Knowl.-Based Syst. **98**, 44–54 (2016)
4. Guzella, T.S., Caminhas, W.M.: A review of machine learning approaches to spam filtering. Expert Syst. Appl. **36**(7), 10206–10222 (2009)
5. Blei, D.M., Ng, A.Y., Jordan, M.I.: Latent Dirichlet allocation. J. Mach. Learn. Res. 993–1022 (2003)
6. Wei, X., Croft, W. B.: LDA-based document models for ad-hoc retrieval. In: Proceedings of the 29th Annual International ACM SIGIR Conference on Research and Development in Information Retrieval, pp. 178–185 (2006)
7. Whittaker, S., Matthews, T., Cerruti, J., Badenes, H., Tang, J.: Am I wasting my time organizing email? In: Proceedings of Conference Human Factors in Computing Systems—CHI, pp. 34–49 (2011)
8. Armentano, M.G., Amandi, A.A.: Enhancing the experience of users regarding the email classification task using labels. Knowl.-Based Syst. **71**, 227–237 (2014)
9. Raghuveer, A.: Automated content labeling using context in email. In: Proceedings of the 17th International Conference on Management of Data. Computer Society of India (2011)
10. Dredze, M., Wallach, H.M., Puller, D., Pereira, F.: Generating summary keywords for emails using topics. In: Proceedings of the 13th International Conference on Intelligent User Interfaces, pp. 199–206. ACM (2008)
11. Sayed, S., AbdelRahman, S., Farag, I.: Three-phase tournament-based method for better email classification. Int. J. Artif. Intell. Appl. **3**(6), 49 (2012)
12. Weng, S.S., Liu, C.K.: Using text classification and multiple concepts to answer e-mails. Expert Syst. Appl. **26**(4), 529–543 (2004)
13. Dehghani, M., Shakery, A., Asadpour, M., Koushkestani, A.: A learning approach for email conversation thread reconstruction. J. Inf. Sci. 0165551513494638 (2013)

14. Tang, G., Pei, J., Luk, W.S.: Email mining: tasks, common techniques, and tools. Knowl. Inf. Syst. **41**, 1–31 (2014)
15. Klimt, B., Yang, Y.: The enron corpus: a new dataset for email classification research. In: Proceedings of the European Conference on Machine Learning, Pisa, pp. 1–5 (2004)
16. Mallet, M.K.: Machine learning for language toolkit(2002). http://mallet.cs.umass.edu/. Accessed Feb 2014

The Role of Social Media in Crisis Situation Management: A Survey

Akanksha Goel, Manomita Chakraborty and Saroj Kumar Biswas

Abstract This paper aims at presenting a survey of the ongoing research on use of various social media platforms to manage various phases of a crisis situation. A thorough research has been carried out on the existing works and their limitations and scope of future have been highlighted to understand the progress made so far. The aim is to identify the areas that still require attention and to help build a system which aims at higher accuracy in information filtering and is robust to meet the real-time information demands to counter the crisis using social media platforms. The research has been categorized into two parts: Analysis of situational awareness information propagated during a crisis situation and second, information sharing among various communities and understanding community's role during disasters.

Keywords Disaster management · Sentiment analysis · Situational awareness Information sharing

1 Introduction

With the advent of a plethora of social media platforms, sharing any kind of information, opinion, or factual can be done in a fraction of a second [1]. While some of these platforms cater to one-to-one communication or a limited set of audience, platforms like Twitter broadcast this information to a much larger audience worldwide. Twitter houses a wide range of features such as geo-tagging, retweet, and hashtags that make it a subject of interest for researchers, despite Twitter's constraints such as unstructured language, sarcasm, and slangs, limitation of 140 characters per tweet [2, 3].

A. Goel (✉) · M. Chakraborty · S. K. Biswas
NIT Silchar, Silchar 788010, Assam, India
e-mail: akanksha.goel7@gmail.com

M. Chakraborty
e-mail: mou.look@gmail.com

S. K. Biswas
e-mail: bissaroj@yahoo.com

A. Abraham et al. (eds.), *Emerging Technologies in Data Mining and Information Security*, Advances in Intelligent Systems and Computing 813,
https://doi.org/10.1007/978-981-13-1498-8_39

During a crisis situation, the social media plays a crucial role in dissemination of information. Irrespective of the relationship shared among users, whatever updates are shared online are useful for building trust among the users and user communities [4–7]. In recent years, a lot of results have been drawn based on such research. The aim is to highlight those areas in these researches which are yet to be acknowledged and understanding the limitations so that assistance can be provided in the development of a robust and wholesome model for better management of crisis situation based on the data extracted from prominent social media platforms.

2 The Sentiment Analysis Framework for Crisis Management

The supervised sentiment analysis framework for crisis management is shown in Fig. 1. It is done in three phases: Initial processing, Training, and Evaluation [8]. The process of crisis management starts with collection of those tweets or social media posts which are in reference with a particular disaster. For this, many tools as well as Twitter API are available. Next step is initial processing of the collected data. This is segregated into two stages: Preprocessing or cleaning the dataset by removing redundant, personal messages or opinion based tweets. Preprocessing also includes removal of special characters, retweet symbol, hashtags, slangs, and URL. This step can be customized based upon requirements of each model. The second part is feature extraction. Feature extraction effectively means extracting those attributes from the dataset which are most useful. Various methods such as n-grams, bag-of-words [9] can be used for this step. Further, training step includes training machine learning algorithms such as Naïve Bayes, Support Vector Machine, and Maximum Entropy with a part of preprocessed dataset marked as training data. Evaluation step is the final step in the process in which the trained algorithms are tested using testing data and the results are analyzed using parameters such as Accuracy, Precision and Recall, and so on.

3 Social Media for Disaster Management

Sakaki et al. [4] group the study of social media for Disaster Management into two categories based on the type of data extracted: Situational Awareness and Information Sharing. Tweets pertaining to the occurrence of a disaster can be of two types, factual and subjective. Factual tweets which give the ground reality information such as infrastructural damages, evacuation routes, missing persons, and safety confirmations are grouped under Situational Awareness. This category of information plays a crucial role in decision making by first responders [10–12]. On the other hand, Information Sharing is a study of information exchange and behavioral analysis of

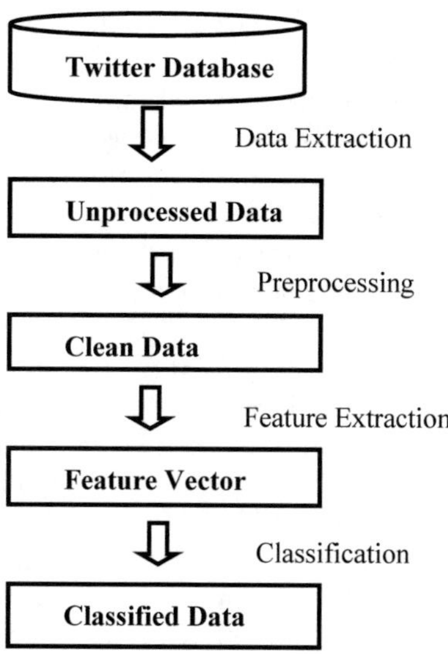

Fig. 1 Basic framework of sentiment analysis

various communities which surface during the occurrence of a crisis situation. Analyzing information of such kinds is helpful for identification of efficient ways of using social media as information sharing tools and to develop some strategies to contain risks and damages during a calamity.

3.1 Situational Awareness

Endsley [13] defines Situational Awareness as "perception of the elements in the environment within a volume of time and space, the comprehension of their meaning and projection of their status in near future." In simple words, he divides situational awareness into three consecutive activities, i.e., gathering information, derivation of clear understanding of this information and then projecting future based upon this understanding. As rightfully noted by Shekhar et al. [14] in their research, the tweets which are collected as a part of dataset contain a large number of redundant, rumor-initiating and unwanted tweets. These are categorized as noise in their research.

Yin et al. [15] proposed a model Emergency Situational Awareness-Automated Web Text Mining (ESA-AWTM). This model demonstrates real-time situational awareness by gathering data as an emergency situation unfolds. The project is guided by The Australian Government Crisis Coordination Centre (CCC). The information containing reports about the occurrence of the calamity is gathered and then it is processed and compiled and validated by authorized personnel. Logs of such information are maintained by the watch officer who is responsible for the information processing during an emergency event. These logs are useful for post-occurrence analysis. The model is distributed into five different phases and for each phase, a method is employed. The phases and the method employed are: Detect unexpected or unusual incidents, condensation and summarization of messages, classification and review of high-value messages, identification, tracking and management of issues, and forensic analysis. The basic challenge as assumed by the authors Yin et al. [16] is to make their model adaptive to various sorts of natural disasters such as cyclones, tornado, earthquakes, and not only floods.

Kebabcı et al. [16] developed a model for extracting and summarizing high profile tweets by analysis of post-occurrence Twitter data. Twitter API is used for data collection and only English language tweets are considered. The model was applied on the dataset of 2000 tweets collected after the Nagano Earthquake of November 22, 2014 in Japan. Support Vector Machine when applied 10-fold cross validation gives highest accuracy of 98% while Random Forest gives 97%. Naïve Bayes gives 93%. For the summarization, Hybrid TF-IDF yields best results in comparison to SumBasic and TF-ISF.

Vieweg et al. [17] studied the Twitter communication that took place during two events that occurred in the USA: The Red River Floods of 2009 and The Oklahoma Grassfires of 2009 to examine the situational features communicated during the events' occurrence. In case of Red River Floods, a 51-day collection period is taken right from the threat of the disaster to the time when maximum danger had passed while in case of Oklahoma Grassfire, this collection period was of 6 days, again, from threat time to end of fires. Initial analysis realized that 18% of Red River Floods on-topic tweets contained geo-location information while this number was 40% in case of the latter. The percentage of situational updates tweets was noted to be 49% and 56% for both the disasters respectively. This difference can be explained to be based upon: (1) Difference in the features of the disaster and (2) Change in the behavior of people over various stages of a disaster starting from threat to recovery. It was concluded that Geo-location feature would vary for various emergency situations depending upon a range of factors, some of which are mentioned above. In addition to geo-location, situational update was the feature which was observed to be given maximum weightage. Situational update has been categorized in the following categories: "Warning, Preparatory Activity, Fire Line/Hazard Location, Flood Level, Weather, Wind, Visibility, Road Conditions, Advice, Evacuation Information, Volunteer Information, Animal Management, and Damage/Injury reports". It was observed that the situational updates maximized as the disaster approached its peak. Another feature which was found to be worth noting is location referencing. Location referencing is defined as an ambiguous reference to a particular location. In case of Red River Floods, 8% of tweets contained this ambiguous location referencing while Oklahoma dataset contained only 6% of such tweets.

To et al. [18] proposed a framework in view of their two-point agenda, first, comparison of matching-based and learning-based techniques to identify relevant tweets from large volume of unfiltered dataset and second, the proposal of a modified matching-based technique as an attempt to improve the process of classification of relevant tweets. Their dataset comprised primarily of geo-tagged tweets. A step-by-step guide to their approach is: Removing spam, mapping of areas requiring assistance, identification of relevant tweets through matching-based and learning-based approaches, and sentiment analysis of relevant tweets. To et al. [18] aim to refine the collected hashtags and keywords by refining the collection through crowdsourcing and deleting irrelevant, though similar looking hashtags. This refinement modifies matching-based approach lexicon for better performance. A couple of points was noted by the authors, first, learning-based approach identifies relevant tweets better than matching-based approach. The reasons are identified as: Learning-based approach includes a higher number of relevant tweets which is not the case in matching-based approach. Another point noted is that recall for matching-based is higher that of learning based. Essentially, this means that the part of dataset selected by matching-based approach is smaller in size and better in quality than learning-based approach. After further analysis, it was concluded [18] that matching-based method yields better results. The results of the findings [18] are that Recall reduces in case of learning-based approach as it takes into consideration those hashtags and words which are not of any relevance to the topic. This reduces the quality of the dataset as a result.

Imran et al. [18] perform their experimentations on the datasets of tornado at Joplin and Hurricane Sandy. They used classification-extraction approach, Twitter-specific POS tagger ArkNLP. They separated the messages into two main classes: Personal and Informative. Informative class was further subclassified into classes such as Caution and Advice, Casualties and Damages, Donations, People, Information Sources, and Others. A set of multi-label classifiers is trained to automatically classify a tweet into one or more of these mentioned classes. Further analysis of these classified tweets was done through sequence labeling task. The features used for classification include words unigram, bigrams, POS tagging and others. In addition, other features such as binary and scalar features of tweets were also used. The results of the experiments done [18] indicated that class-specific models may lead to improvements in performance for some classes but not for others. It was observed that there was no consistent gain for tweets related to certain classes on performing experiments on several configurations of training and testing datasets.

Win et al. [19] propose a tweet monitoring system to identify disaster-related messages updated by people into a set of categories and then provide desired target information automatically. This system exploits a wide range of features ranging from Word Ngram, Word Cluster, Emotions, Lexicon-based Features, POS, Hashtags, and URLs. This system then creates a lexicon from the annotated tweets using PMI based score and frequency distribution of terms based upon the disaster dataset. The system aims at automatically creating different disaster datasets with annotated tweets. The annotations used at tweet level are of three types: Related and Informative, Related and Non-informative, and Not related. In the experiments conducted [9] a cross validation is done between features and classifiers by comparing various combinations. Some of the results are as tabulated in Tables 1 and 2.

Training and testing were carried out using 10-fold validation techniques. Naïve Bayes was observed to be the worst performer while LibLinear was always observed to be the fastest. The reduction in features was done by application of gain information theory based method and top 300 features were selected for analysis. Schulz et al. [20] adopt an approach based on the comparison of different classifiers and an

Table 1 Proposed features by LibLinear classifier [19]

Dataset	Precision	Recall	F1	Accuracy (%)
2013_Aus Bushfire (D1)	0.897	0.895	0.895	89.45
2013_Typh Yolanda (D2)	0.912	0.92	0.913	92.02
2014_Iceland Volcano (D3)	0.908	0.908	0.908	90.82
2015_Nepal Earthquake (D4)	0.748	0.751	0.749	75.05

Table 2 Performance of feature selection analysis on different datasets [19]

Dataset	Feature set		Accuracy of classifiers		
	Original features	Reduced features	LibLinear	SMO	Random forest
D1	6727	300	87.28	88.28	85.45
D2	11048	300	92.71	93.31	92.32
D3	4374	300	92.71	93.31	92.32
D4	37471	300	76.90	76.43	72.13

analysis of various features useful for detecting emotions in microposts. After initial preprocessing, several features such as Word Unigram, POS tagging, Character trigram and fourgram, syntactic and semantic features are extracted. Three datasets were used for evaluation and classification approaches were distributed into 3-class and 7-class problem. The results were that identification of 7-class sentiments was done with accuracy of 65.79% while three classes had an accuracy of 64.07%. While NBB outperforms all the classifiers in Set 1, NBM performs better for Set 2 because of increase in the size of vocabulary. Using NBM with POS tagging in Set 2 provides better results, however, syntactic and sentiment features are not of any use in these cases. They concluded their research by pointing out two points, first, a very large labeled dataset is required to train the classifiers and secondly, postprocessing the classified micropost can further reduce the set to extract relevant tweets. Studies discussed so far address the real-time classification of crisis situation related tweets for timely analysis of the data.

3.2 Information Sharing

Under highly uncertain events, people tend to share whatever information they get. Due to this communication, varied communities emerge. These communities are mainly constituted of those directly affected by unexpected or sudden events [21]. Each community has two constituent elements: actors and network among these actors. Certain networks are found to be denser while certain are negligibly smaller in comparison. Similarly, certain individuals can be identified as influencers within communities and the others are merely observers. In this section, an attempt is made to understand these communities and their intercommunity and intracommunity networks as well as the behavior of the actors of these communities in case of crisis situation particularly, through a survey of various studies conducted on the topic [22, 23].

In his study, "Organized Behavior in Disaster", Dynes [24] described eight socio-temporal stages of disaster showing how an individual shifts from one's self-assessment to concern about larger impact of the event and its effects on the community.

Disaster sociologists have described disaster to be "a unifying force that creates mutual helpfulness and emotional solidarity among those affected by it" [25]. Shklovski et al. [25] elaborated how the information is communicated and used and how a community responds during a crisis situation in two ways: first, what role a computer plays as a communication media for information exchange and connection with the locales during a calamity and second, they analyzed two community-based environments over the Internet to examine individual behavior and behavioral capabilities collectively. The main observations which emerged from the study explained irrespective of the impact of the disaster over global platforms such as news media, the effect is extremely local and thus the first responders in such situations are the locales. For this, Information Communication Technology (ICT) plays a crucial role. During a disaster, it becomes very essential to identify these local communities which are active in sharing information and provide a helping hand in the times of need and this serves as a motivation to develop social connections. The use of ICT [25] to find community can help facilitate social cohesion in geographical communities even post disaster, which is crucial for laying the base for the recovery operation.

The economist Arrow [26] identified trust as "a lubricant of the social system", while computer scientists such as Massa et al. [5] defined it as a user's opinion on another user's characteristics. Vedula et al. [10] base their research on the traces of trust amongst the network of users and identify influencers or those people who are highly trusted within a network. They use three psychological ideas to conduct these studies. For discovering influencers, they employ Influence-Passive Algorithm. Cohesiveness relates to the members of a group who share emotional and behavioral characteristics with one another and the group as a whole. Jaccard Similarity Metric method was employed for this. The authors concluded their studies by summarizing the observations that influence is a principle factor towards trustworthiness of users within a network, however, the presence of valence or sentiment ensures the stability of performance and response surface. Structural cohesion, while still useful has less of an impact than the other two factors, especially in the case of ephemeral (Mumbai attacks, Nice attacks, Hurricane Irene and Houston Floods) and political movement events (India Anti-Corruption).

In the paper "Disaster Analysis through Tweets" [14], Shekhar et al. analyzed three different cases: (i) Occurrence frequency of a disaster over a range of regions in given time (ii) Distribution of disasters over different geographical locations and (iii) Emotional analysis of people's response to disaster over Twitter. Sentiment

Table 3 Average accuracy (with standard error) and micro-averaged precision, recall, and F1 for the three sentiment classifiers, using their best configurations. The difference in accuracy between MaxEnt and the other classifiers is statistically significant (paired t-test, $p < 0.01$) [21]

Classifier	Accuracy	Precision	Recall	F1-score
Max Ent.	84.27±2.0	90.15	70.00	78.81
Dec. Tree	81.35±1.8	79.72	67.06	72.84
Naïve Bayes	78.63±2.2	75.78	71.16	73.72
Worry Lex.	79.41	95.74	52.94	68.18

Rating is given based upon the level of pessimism or negativity in the tweets posted by the users, and these emotions for different disasters can be subcategorized as negative sentiments, unhappy, depressed and angry. Together, this framework [14] measures and analyzes the emotional states of public in the event of natural calamity occurrence through the means of crisis analysis. This [14] provides us with the perception of real-time emotions of the public during the occurrence of a calamity. This framework [14] primarily focuses on Twitter and the authors [14] note the fact that other social platforms such as Facebook and not only microblogging platforms contain information in much detailed format about the emotions and the framework needs to be flexible enough to accommodate such data from these platforms.

Sakaki et al. [4] analyzed the behavior of Twitter users after the occurrence of the disaster. The study consists of two parts: (i) Posting Behavior Analysis: The change in posting activities of users during the occurrence of disaster and (ii) User Interaction Behavior Analysis: This research analyzes the replies and retweets that were made during the calamity. Goswami et al. [1] divided the users into two groups: One includes pre-retweeters: users who re-tweeted before the disaster. The other includes non-retweeters: users who did not retweet before the disaster. The authors also observe certain features of the replies and retweets that surface. In conclusion of their studies, they note that in a nutshell: (a) Behavior of Twitter users does not display any significant change of habits during occurrence of a disaster and (b) Rumors are heavily subdued by anti-rumor tweets.

Benjamin et al. [22] assess the feasibility of measuring public risk perception using social media analysis. In particular, they observed the message volume and sentiment variation over time, location, and gender. The results are derived by the application of three classifiers, i.e., MaxEnt (logistic regression), Naive Bayes, and a Decision Tree (ID3) classifier. It is found that MaxEnt exhibits the best accuracy and precision whereas Naive Bayes has a slightly better recall. Though the results obtained by performing experiments on Hurricane Irene dataset were not very conclusive, it was observed that the regions experiencing hurricanes on frequent basis expressed lower level of concerns. The dataset used in [22] was not apt to inject gender bias into the classifier due to gender neutrality of the tweets in general. The authors concluded that a great variety of demographic attributes need to be adjusted in order to adjust to the demographic bias inherent in social media. The results of their experiments are as depicted in Table 3.

4 Conclusion

Various studies have been conducted for both, Situational Awareness and Information Sharing aspects to understand how disasters can be better managed by depending upon the participation of those directly affected in various stages of a disaster. The primary and most prominent conclusion that can be drawn from this survey is that there is requirement for a model which accommodates not just one but all of the social media platforms which are highly popular among the users worldwide and a way to handle this massive volume of data [23, 24] that is generated on this platform each minute. The information which is shared on public platforms is extremely crucial and can play a game changer in disaster management when used effectively. A wholesome approach to filter out information which is of relevance and also takes into account various factors which influence this information exchange such as location, accessibility to information, the validity of this information, and degree of impact of the occurrence on the individual, and so on is what is needed. It is also noted that the models which have been proposed till now are largely depended upon the quality and quantity of training data thus putting their credibility and robustness to question. Lastly, the most important attribute of the model is evolution in real time [25]. The model that is the need of the hour should evolve in a way that information available through it is as per the current situation at ground level. The issues which have died should be removed and it should be able to accommodate issues as they arise over the cycle of crisis situation [26]. The stakes involved with each decision during a calamity or a disaster are huge. Thus, decisions taken with even a little amount of accurate information at hand make an unimaginable difference.

References

1. Goswami, S., Chakraborty, S., Ghosh, S., Chakraborty, A., Chakraborty, B.: A review on application of data mining techniques to combat natural disasters. Ain Shams Eng. J. (2016) (in press)
2. Stollberg, B.M., De Groeve, T.: The use of social media within Global Disaster Alert and Coordinations System (GDACS). In: 21st International Conference on World Wide Web (WWW '12 Companion). pp. 703–706. ACM, New York (2012)
3. Drakopoulos, G., Kanavos, A., Mylonas, P., Sioutas, S.: Defining and evaluating Twitter influence metrics: a higher-order approach in Neo4j. Social Network Analysis and Mining, pp. 7–52 (2017)
4. Sakaki, T., Matsuo, Y., Kurihara, S., Toriumi, F., Shinoda, K., Noda, I., Uchiyama, K., Kazama, K.: The possibility of social media analysis for disaster management. In: 2013 IEEE Region 10 Humanitarian Technology Conference, pp. 238–243, Sendai (2013)
5. Massa, P., Avesani, P.: Controversial users demand local trust metrics: an experimental study on epinions.com community. In: 20th National Conference on Artificial Intelligence, vol. 1, pp. 121–126. AAAI (2005)
6. Sen, A., Rudra, K., Ghosh, S.: Extracting situational awareness from microblogs during disaster events. In: 7th International Conference on Communication Systems and Networks (COMSNETS), pp. 1–6, Bangalore (2015)
7. Liu, B.: Sentiment Analysis and Opinion Mining. Claypool Publishers, Morgan (2012)

8. Bonchi, F., Castillo, C., Gionis, A., Jaimes, A.: Social network analysis and mining for business applications. ACM Trans. Intell. Syst. Technol. N.Y. **2**(3), Article 22, (2011)
9. Zimmer, M., Proferes, N.J.: A topology of twitter research: disciplines, methods, and ethics. Aslib J. Inf. Manag. **66**(3), 250–261 (2014)
10. Vedula, N., Parthasarathy, S., Shalin, V.L.: Predicting trust relations within a social network: a case study on emergency response. In: ACM on Web Science Conference. pp. 53–62. ACM, New York (2017)
11. Huang, B., Yu, G.: Research on the mining of opinion community for social media based on sentiment analysis and regional distribution. In: Chinese Control and Decision Conference, pp. 6900–6905, Yinchuan (2016)
12. Huang, B., Yu, G.: Research on the mining of opinion community for social media based on sentiment analysis and regional distribution. In: Chinese Control and Decision Conference, pp. 6900–6905, Yinchuan (2016)
13. Endsley, M.R.: Towards a theory of situational awareness in dynamic systems. Hum. Factors J. Vol. 37, issue 1, pp. 32–64 (1995)
14. Shekhar, H., Setty, S.: Disaster analysis through tweets. In: International Conference on Advances in Computing, Communications and Informatics (ICACCI), pp. 1719–1723, Kochi (2015)
15. Yin, J., Karimi, S., Robinson, B., Cameron, M.: ESA: emergency situation awareness via microbloggers. In: the 21st ACM International Conference on Information and Knowledge Management, pp. 2701–2703. ACM, New York (2012)
16. Kebabki, K., Karshgil, M.E.: High priority tweet detection and summarization in natural disasters. In: 23nd Signal Processing and Communications Applications Conference (SIU), pp. 1280–1283, Malatya (2015)
17. Vieweg, S., Hughes, A.L., Starbird, K., Palen, L.: Microblogging during two natural hazards events: what twitter may contribute to situational awareness. In: the SIGCHI Conference on Human Factors in Computing Systems, pp. 1079–1088. ACM, New York (2010)
18. Imran, M., Elbassuoni, S., Castillo, C., Diaz, F., Meier, P.: Practical extraction of disaster-relevant information from social media. (WWW 2013 Companion). In: 22nd International Conference on World Wide Web (2013)
19. Si Mar Win, S., Nwe Aung, Than.: Target Oriented Tweets Monitoring System during Natural Disasters. Int. J. Netw. Distrib. Comput. (2017)
20. Paulheim, H., Schulz, A., Schweizer, I., Thanh, T.D.: A fine-grained sentiment analysis approach for detecting crisis related microposts. In: ISCRAM (2013)
21. Pasarate, S., Shedge, R.: Comparative study of featureextraction techniques used in sentiment analysis. In: International Conference on Innovation and Challenges inCyber Security (ICICCS-INBUSH), pp. 182–186, Noida (2016)
22. Mandel, B., Culotta, A., Boulahanis, J., Stark, D., Lewis, B., Rodrigue, J.: A demographic analysis of online sentiment during hurricane irene. In: Proceedings of NAACLHLT, pp. 27–36, Stroudsburg (2012)
23. Abedin, B., Babar, A., Abbasi, A.:Characterization of the use of social media in natural disasters: a systematic review. In: IEEE Fourth International Conference on Big Data and Cloud Computing, pp. 449–454, Sydney, NSW (2014)
24. Dynes, R.R.: Behavior in Disaster: Analysis and Conceptualization, p. 265 (1969)
25. Shklovski, I., Palen, L., Sutton, J.: Finding community through information and communication technology in disaster response. In: 2008 ACM Conference on Computer Supported Cooperative work (CSCW '08), pp. 127–136. ACM, New York (2008)
26. Arrow, K.J.: The Limits of Organization. Norton, N-Y (1974)

Subset Significance Threshold: An Effective Constraint Variable for Mining Significant Closed Frequent Itemsets

S. Pavitra Bai and G. K. Ravi Kumar

Abstract Frequent Itemset (FIs) Mining is the most popular data mining technique and fundamental task for association rule and correlation mining. Closed Frequent Itemsets (CFIs) provide a lossless compact representation of Frequent Itemsets. Although many algorithms exist for mining Frequent Itemsets and Closed Frequent Itemsets, they lack human involvement for better guidance and control. Constraint-based data mining enables users to add their constraints along with the standard rules of the algorithm to suit their need. In this paper, we propose an effective constraint variable named Subset Significance Threshold (SST) which can be used along with minimum support threshold to mine significant CFIs. The CFIs which do not satisfy the constraint are considered insignificant and eliminated. Experiment analysis on various representational datasets proved the proposed constraint variable is effective in identifying insignificant CFIs.

Keywords Constraint mining · Significant itemset · Closed frequent itemset
Data mining · Data condensation · Approximate frequent itemset mining

1 Introduction

Frequent Itemset Mining plays an essential role in many important data mining tasks such as association rules mining, finding correlations between variables, etc. However, for a given minimum support, frequent itemset mining may identify a large number of itemsets as frequent. It is time-consuming for the user or any process to parse through these itemsets and find most useful itemsets. The alternative is to mine

S. Pavitra Bai (✉)
SJB Institute of Technology, BGS Health and Education City,
Dr. Vishnuvardhan Road, Kengeri, Bengaluru 560060, India
e-mail: s.pavitra@gmail.com

G. K. Ravi Kumar
BGS Institute of Technology, BG Nagar, Nagamangala, Mandya 571448, India
e-mail: ravikumargk@yahoo.com

A. Abraham et al. (eds.), *Emerging Technologies in Data Mining and Information Security*, Advances in Intelligent Systems and Computing 813,
https://doi.org/10.1007/978-981-13-1498-8_40

Closed Frequent Itemsets (CFIs). The CFIs mining techniques eliminate redundant frequent itemsets and substantially reduce the number of itemsets to process without loss of information. The CFIs are used as data model to generate frequent itemsets and association rules [1].

For any large and high dimensional data, even the number of CFIs generated can be enormous which fail to convey quality information. Also, it increases query processing time as there are more number of itemsets to parse [2]. In many situations, the user may want to apply additional constraints to filter patterns while mining. Constraint-based mining allows the user to mine patterns restricted by some constraints [3]. Similarly, constraints can be applied to mine only most significant CFIs so that the data can be analyzed quickly. Since the significance of an itemset is subjective, different users shall be able to set this parameter according to their needs. This parameter shall be used along with minimum support threshold to eliminate CFIs which does not satisfy both of these parameters and retain only significant CFIs in the mining process. In this paper, we define a new constraint variable called Subset Significance Threshold (SST) based on the closure property of the CFIs and rule confidence which is found effective in identifying significant CFIs.

The organization of this paper is as follows: Sect. 2 gives an overview of the related work. Section 3 defines the problem. Section 4 illustrates proposed constraint variable SST and the algorithm, which is used along with minimum support threshold to identify significant CFIs. Section 5 provides the experimental study of SST on different representational datasets. Finally, the paper is concluded in Sect. 6.

2 Related Work

Constraint-based pattern mining focuses on finding interesting patterns and reducing the number of patterns to analyze. The constraints are applied during frequent itemset mining or as a postprocessing step. Tran et al. [4] proposed to mine association rules restricted on constraint. The closed itemsets lattice is generated. The lattice is then partitioned based on the disjoint equivalence class and constraint mentioned. Then the association rules are mined quickly which inherently satisfy the given constraint. Duong et al. [5] proposed an efficient algorithm for mining frequent itemsets with a single constraint which is a set of itemsets. The algorithm mines only those frequent itemsets which contain the itemset of constraint set. Duong et al. [6] proposed mining of frequent itemset using double constraints. The closed itemset lattice for the dataset is first created. The itemsets in the lattice are arranged to group itemset having same equivalence class and support. Then the lattice is partitioned into disjoint equivalence classes. From the partitions, the frequent itemsets satisfying the two constraints are mined. The constraints include two sets, one to test monotone and another for anti-monotone. Tran et al. [7] proposed a method to mine association rules which intersect with given constraint itemsets. The closed frequent itemset lattice is generated. The association rule set is partitioned with constraints having same equivalence class. Then, these partitioned are mined independently for rules with confidence 1, and less than 1. Lazaar et al. [8] proposed a theory for mining closed frequent itemsets with any global constraints. The mining process shall consider a given query which may contain one or more constraints while mining the frequent patterns. Vo et al. [9] proposed mining of erasable itemsets using the constraint on subset and superset itemsets. An itemset is combined with other itemsets, and their profit gain is measured. If the gain of the combined superset is less than a given threshold, then the itemset is considered to be erasable. The algorithm used eliminate items that are part of a product which does not add much value.

In the most of the constraint-based pattern mining, the CFIs are mined first, and the constraints are applied while mining frequent patterns or association rules. For large and dense datasets, the number of CFIs generated is large and will have insignificant itemsets which can be ignored with little loss of support information but reduces large CFIs to parse.

3 Problem Definition

Consider Itemsets X and Y as CFIs for a given dataset and minimum support threshold. Let X be the proper superset of Y. Let S_X is the support count of X and S_Y is the support count of Y. As per the definition of Closed Itemsets, itemset Y is said to be closed if there exists no proper superset that has the same support as Y. So even if the support of Y is just 1 count greater than the support of its superset then Y is said to be closed, and it is retained in CFI list, i.e., Y is said to be CFI if and only if,

$$S_Y \geq S_X + 1 \quad (1)$$

For datasets of large size, the support count of CFIs is also huge. The difference of one or few support counts between superset X and subset Y CFIs does not add much value. If this difference is negligible to the user, then these subsets can be merged with their superset, thus reducing the number of CFIs generated. Of course, the accuracy of the support count of subset CFIs which are merged in this manner is affected but reduces the number of CFIs generated, which makes further processing of CFIs simple and faster.

4 Definition of Subset Significance Threshold (SST)

Let Y be a CFI and (X_1, X_2, X_3, …, X_n) be its set of immediate superset CFIs. Let X_{MAX} be the superset of Y which has highest support count among its immediate supersets and S_{MAX} be its support count.

The Subset Significance Threshold (SST) is a user-defined constraint variable. It is a variant of rule confidence used to mine association rules. As rule confidence is calculated between any two frequent itemsets, the SST uses only those itemsets which are immediate subset and superset pair. It is defined as the minimum percentage of support by which the subset must be more than the support of its immediate superset having highest support count among its supersets to consider it as significant. The user inputs the SST in percentage and its value ranges from 0% to more than 100%. So, calculation of the minimum support count by which the subset must be higher than its superset to become significant, i.e., Minimum Support for Subset Significance (MSSS) is given in Eq. 2.

$$MSSS = \frac{S_{MAX} \times SST}{100} \tag{2}$$

By applying the SST constraint variable on CFIs, a CFI is said to be significant if its support count is greater by MSSS of its superset with highest support count, i.e., Y is said to be significant CFI if,

$$S_Y > S_{MAX} + MSSS \tag{3}$$

CFIs that do not satisfy Eq. 3 is considered insignificant and eliminated. Using the closure property of CFIs, support of the eliminated CFIs or FIs is found from their immediate superset having highest support.

Note that the SST is the percentage threshold with respect to the support of its immediate superset and does not depend on the size of the dataset. So, this constraint variable applies to all type of datasets, i.e., small to large, dense to sparse, static and streaming data.

The identification of possible significant CFI using proposed Subset Significance Threshold (SST) is used as a postprocessing step after mining all CFIs for a given dataset. The algorithm for finding significant CFIs using SST is shown in Algorithm 1.

From the list of CFIs mined for a given dataset and minimum support threshold, each CFI is checked for its significance based on the proposed SST. The significant CFIs are identified, and the number of insignificant CFIs is counted. CFI from the mined CFI list is fetched one by one in line 2. The superset with highest support count is found from line 3 to 5. The minimum support count difference needed for the CFI to be identified as significant is calculated in line 6. From line 7 to 12, the CFI is checked to satisfy Eq. 3 to consider it as significant. If satisfied then it is added to the significant CFI list else the count of insignificant CFI is incremented. When all CFIs are processed, the significant CFI list and count of insignificant CFIs are returned in line 13.

5 Experimental Analysis

To verify the usability of the proposed SST constraint variable on mining significant CFIs, the frequency distribution of insignificant CFIs with respect to their supersets as mentioned in SST definition is studied on popular representation datasets. The

Algorithm 1 Finding Significant CFIs using SST

Require: Closed Frequent Itemsets (CFI_{List}) , SST%
Ensure: Significant CFIs, Count of Insignificant CFIs

```
InsignificantCFICount = 0

for each CFI from CFI_List do
  X_1...n = GetImmediateSupersetsOf(CFI)
  X_MAX = HighestSupportSuperset(X_1...n)
  S_MAX = X_MAX.Support
  MSSS = (S_MAX * SST)/100

  if CFI.Support > (S_MAX + MSSS) then
    SignificantCFIList.Add(CFI)
  else
    InsignificantCFICount = InsignificantCFICount + 1
  end if
end for

return SignificantCFIList, InsignificantCFICount
```

source of these datasets can be found in [10]. The characteristics of the datasets are given in Table 1.

The modified version of CHARM [11] closed frequent itemset mining algorithm in SPMF [12] open source data mining library is used to implement the proposed algorithm. The experiment is conducted by mining CFIs with an arbitrary minimum support and grouping them into different ranges of their significance calculated based on proposed Subset Significance Threshold (SST), i.e., their significance with respect to their immediate supersets. The distribution (histogram) of the CFIs in representative datasets is plotted from Figs. 1, 2, 3, 4, 5 and 6. It is observed that a large number of CFIs are skewed into very less significance value range. So this less significance range is further expanded to study the distribution of these large numbers of CFIs in more detail.

From the experiments conducted, it is observed that a large number of CFIs accumulate in lowest significance range for high and medium density data as shown

Table 1 Characteristics of representative datasets

Dataset	No. of transactions	No. of unique items	Minimum transaction length	Maximum transaction length	Average transaction length	Data density
Accidents	340183	468	18	51	33	High
Chess	3196	75	37	37	37	High
Mushroom	8124	119	23	23	23	Medium
T40I10D100K	100000	942	4	77	39	Medium
T10I4D100K	100000	870	1	29	10	Low
Retail	88162	16470	1	76	10	Low

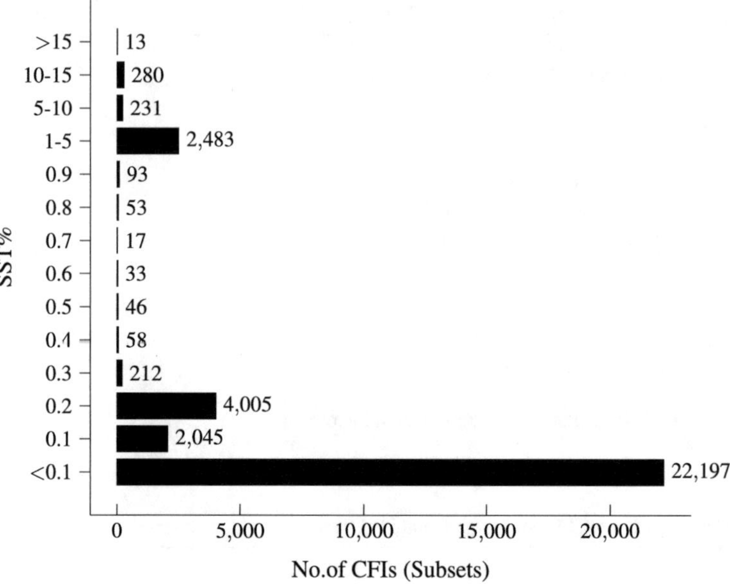

Fig. 1 Frequency distribution of CFIs (Subsets) on the range of significance threshold (SST) of accidents dataset with minimum support of 0.4

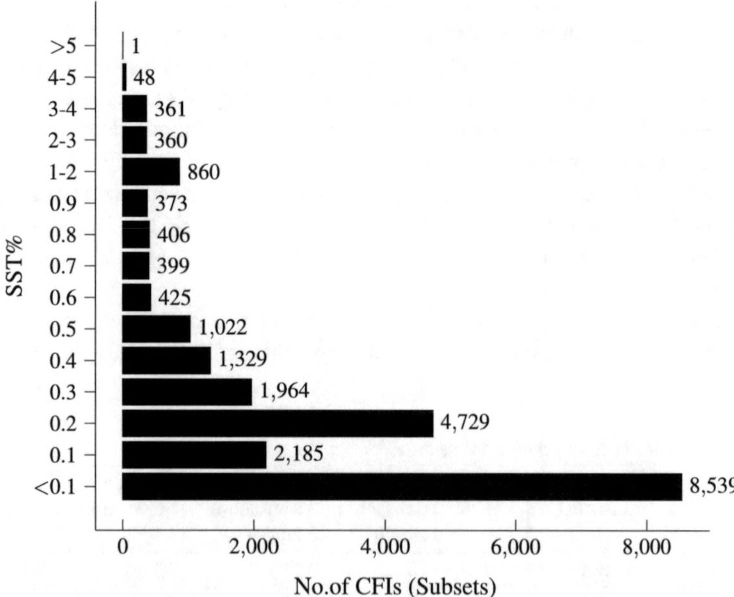

Fig. 2 Frequency distribution of CFIs (Subsets) on the range of significance threshold (SST) of chess dataset with minimum support of 0.7

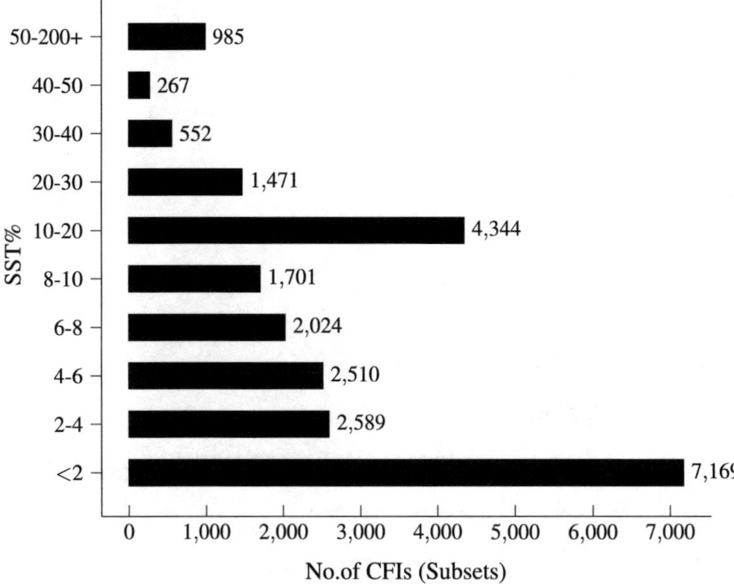

Fig. 3 Frequency distribution of CFIs (Subsets) on the range of significance threshold (SST) of mushroom dataset with minimum support of 0.03

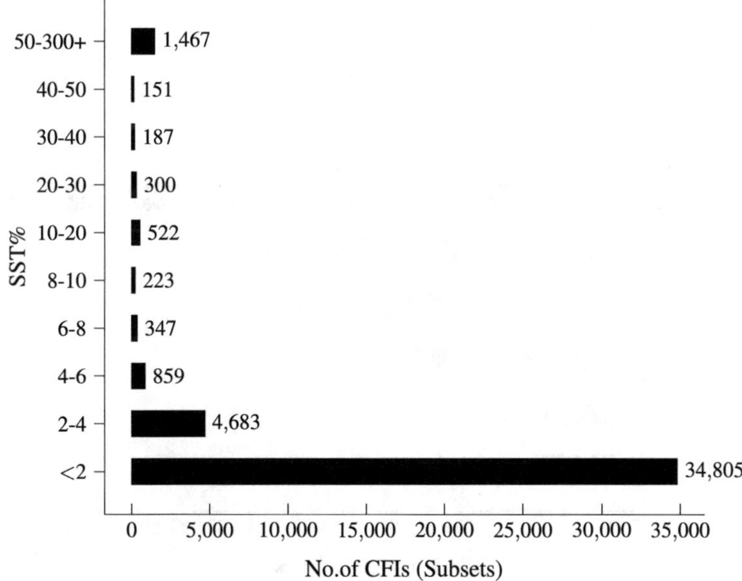

Fig. 4 Frequency distribution of CFIs (Subsets) on the range of significance threshold (SST) of T40I10D100K dataset with minimum support of 0.01

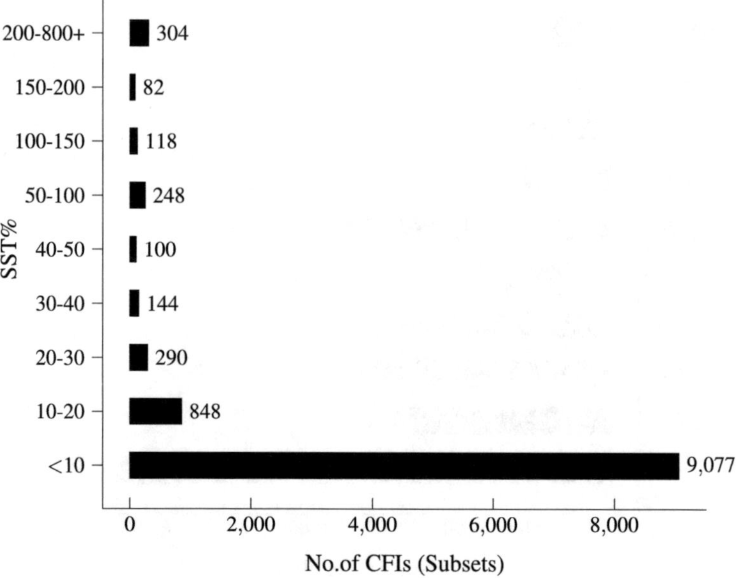

Fig. 5 Frequency distribution of CFIs (Subsets) on the range of significance threshold (SST) of T10I4D100K dataset with minimum support of 0.002

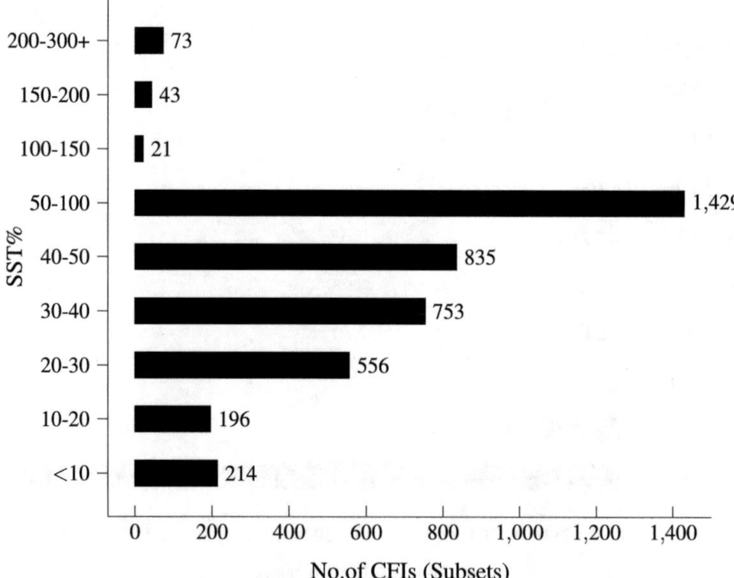

Fig. 6 Frequency distribution of CFIs (Subsets) on the range of significance threshold (SST) of retail dataset with minimum support of 0.001

in Figs. 1, 2, 3 and 4. These CFIs can be eliminated with very minimal effect on the accuracy of their support but yields a very high degree of compression of frequent itemsets.

For low-density data, the CFIs show mixed distribution, i.e., a large number of CFIs are distributed in the lowest significance range similar to high-density data as shown in Fig. 5 for the T10I4D100K dataset and also distributed across a couple of lower significance ranges as shown in Fig. 6 for retail dataset. So, in case of high dimensional sparse data, the proposed technique yields less compression for lower significance threshold. A higher degree of compression is achieved at the cost of considerable accuracy loss. An optimal Subset Significance Threshold can balance compression and accuracy.

The proposed Subset Significance Threshold (SST) is effective in eliminating insignificant CFIs and yields very high degree of compression with minimal loss of accuracy except for high dimensional sparse data. With a very less value of SST, most of the insignificant CFIs can be eliminated, and only significant CFIs are retained. These significant CFIs can be queried to find frequent itemsets and association rules. The percentage loss of accuracy for the itemsets eliminated is limited to the SST.

6 Conclusion

We have presented a new constraint variable named Subset Significance Threshold (SST) for mining significant closed frequent itemsets for any given dataset. The distribution of CFI subsets for their significance with respect to their immediate superset as defined by the SST is studied. The experimental result showed that a large number of the CFIs are distributed at lesser significance range. These CFIs are considered insignificant and eliminated with very minimal loss of accuracy in their support information. SST is a very effective mechanism for pruning insignificant itemsets and greatly reduce the number of CFIs generated for any given dataset. However, it is rarely observed that, in very dense data, the loss of accuracy for some eliminated CFIs is more than specified SST. It is due to the chaining effect in its superset hierarchy during the elimination process. In the future work, this rare condition is treated with an improved algorithm for pruning insignificant CFIs based on the SST and validated for its efficiency concerning information loss and any possible side effects.

References

1. Pasquier, N., Bastide, Y., Taouil, R., Lakhal, L.: Discovering frequent closed itemsets for association rules. In: Beeri, C., Buneman, P. (eds.) Database Theory ICDT99. Lecture Notes in Computer Science, vol. 1540, pp. 398–416. Springer, Berlin, Heidelberg (1999)
2. Han, Jiawei, Cheng, Hong, Xin, Dong, Yan, Xifeng: Frequent pattern mining: current status and future directions. Data Min. Knowl. Discov. Springer **15**(1), 55–86 (2007)

3. Nijssen, S.: Constraint-based mining. Encyclopedia of Machine Learning, pp. 221–255. Springer, US (2010)
4. Tran, A.N., Truong, T.C., Le, B.H., Duong, H.V.: Mining association rules restricted on constraint. In: Proceeding of IEEE International Conference on Computing and Communication Technologies, Research, Innovation and Vision for the Future (2012)
5. Duong, H., Truong, T., Le, B.: An efficient algorithm for mining frequent itemsets with single constraint. In: Proceeding of International Conference on Advanced Computational Methods for Knowledge Engineering, pp. 367–378. Springer (2013)
6. Duong, H., Truong, T., Vo, B.: An efficient method for mining frequent itemsets with double constraints. Int. J. Eng. Appl. Artif. Intell. **27**, 148–154 (2014)
7. Tran, A.N., Truong, T.C., Le, B.H.: An approach for mining association rules intersected with constraint itemsets. In: proceeding of International Conference on Knowledge and Systems Engineering, pp. 351–363 (2014)
8. Lazaar, N., Lebbah, Y., Loudni, S., Maamar1, M., Lemire, V., Bessiere, C., Boizumault, P.: A global constraint for closed frequent pattern mining. In: Proceeding of International Conference on Principles and Practice of Constraint Programming, pp. 333–349. Springer (2016)
9. Vo, B., Le, T., Pedrycz, W., Nguyen, G., Baik, S.W.: Mining erasable itemsets with subset and superset itemset constraints. Int. J. Expert Syst. Appl. Elsevier, 50–61 (2017)
10. Frequent Itemset Mining Implementations Repository. http://fimi.ua.ac.be/data/. Accessed 3 June 2017
11. Zaki, M.J., Hsiao, C.-J.: CHARM: an efficient algorithm for closed itemset mining. In: Proceeding of International Conference on Data Mining, pp. 457–473 (2002)
12. SPMF—An Open-Source Data Mining Library. http://www.philippe-fournier-viger.com/spmf/. Accessed 15 May 2017

An Automated Method for Power Line Points Detection from Terrestrial LiDAR Data

Arshad Husain and Rakesh Chandra Vaishya

Abstract Monitoring and maintenance of power lines are very essential because they are the elementary component of the power sector and play significant roles in various areas including electric network upgradation, designing, and the analysis of electricity by optimizing the routes. Mapping of power lines is an easier task with the help of laser scanner compared to traditional techniques because the power lines are located above the ground in particular to higher and lower voltage power line networks. Detection of power lines in densely situated cities in urban areas is a tedious task. Typically, in such complex environment, power lines are positioned as parallel to the rooftops of buildings. In the present study, an automated approach for the detection of power line points has been proposed using terrestrial LiDAR (Light Detection and Ranging) point cloud dataset. The proposed method includes four basic steps namely vertical slicing, neighborhood analysis, distance-based clustering, and analysis of individual clusters. The proposed method has been tested at a terrestrial LiDAR dataset. The completeness and correctness of proposed method are 98.54% and 96.89%, respectively.

1 Introduction

Power lines are the vital and rudimentary component of power sector and play significant roles in numerous application areas, i.e., electricity distribution monitoring [1], electric network upgradation, planning and designing [2], and investigation of power supply by adjusting the optimal routes [3, 4]. Airborne, Terrestrial, and Mobile Laser Scanning (ALS, MLS and TLS) are efficient, robust, highly precise, and widely used technology for the spatial data acquisition while each one of them has their own

A. Husain (✉)
GIS Cell, MNNIT Allahabad, Allahabad 211004, India
e-mail: rgi1501@mnnit.ac.in

R. C. Vaishya
CED, MNNIT Allahabad, Allahabad 211004, India
e-mail: rcvaishya@mnnit.ac.in

A. Abraham et al. (eds.), *Emerging Technologies in Data Mining and Information Security*, Advances in Intelligent Systems and Computing 813,
https://doi.org/10.1007/978-981-13-1498-8_41

advantages and disadvantages over each other. In order to calculate the range of the target point, time of travel of laser beam is calculated in pulse laser scanners while phase difference is used in continuous laser scanners. Along with the range of target point, scan angle is also determined. Further, these two calculated parameters, i.e., range and scan angle are converted in the X, Y, and Z values with the help of mathematical transformation called geo-location. Global Positioning System (GPS) and Inertial Navigation System (INS) are used for the transformation from the local coordinate values into global coordinate values, and generate the geo-referenced point cloud [5]. Along with the geometric information (X, Y, Z coordinate value), laser scanning system calculates some other important information such as intensity, number of returns, return numbers, etc. Laser scanning technology provides a swift and accurate alternative for mapping large areas at high resolution with less human and atmosphere interaction compared to traditional surveying and mapping systems, including photogrammetric systems [6]. The bulk data size is the only disadvantage of laser scanning system, and captured data size is very huge in size. Therefore, performing any kind of analysis at particular geospatial feature becomes very tedious and time consuming to use whole point cloud dataset at once. Geospatial feature extraction using laser scanning data facilitates the solution for this issue but the feature extraction method should be time efficient.

Initially, airborne laser scanning was widely accepted as a prominent technique for mapping power lines [1, 7]. Various methods had been proposed for the extraction of power lines from ALS point cloud data. Extraction of power lines using a bottom-up approach and Hough transform were applied iteratively to identify each single power line [8]. Hough transformation along with slope and peak values was applied for detection of power line points [9] also Hough transformation with geospatial features eigenvalues and eigenvectors was used for power lines points detection [10]. Gaussian mixture model was used for the classification of power line points [11]. The concept of parabolic curve fitting along with K-Dimensional (K-D) tree was used to identify the power line points [12]. Good results with ALS data can be obtained by using low-flying drones because it increases the point density of captured dataset. Data acquisition with ALS is constrained by the complex environment in urban areas. Typically, the buildings are extremely high in urban areas which make the working environment of the drone more complicated and dangerous [13].

Terrestrial and mobile mapping system is an emerging technology that can capture the road surface features rapidly and accurately [14]. Nowadays, terrestrial and mobile laser scanning is used widely for the extraction and modeling of roads [15], building facades [16], trees [17], and pole-like objects. Power lines are often distributed and attached with pole-like objects along the roadside, which make them accessible easily by terrestrial LiDAR. Point density of captured data is relatively high as compare to ALS point cloud data because TLS system is typically placed in proximity of power lines (range become shorter). Terrestrial LiDAR is capable of capturing data from a side view, which facilitates the acquisition of relatively complete point cloud dataset from multiple parallel line views and it can also be used to map power lines at any time (active remote sensing) therefore, TLS system provides the capacity for efficient mapping and managing the power lines.

Table 1 Statistical specification of the captured dataset

File size	No. of points	Street length (m)	Area (m^2)	Point Density (per m^2)
1.89 GB	42071105	218.16	15426.6547	2727.169

In this research, a facile automated method for detection of power line points from terrestrial laser scanning data has been proposed. The proposed method includes four basic steps, first step, i.e., vertical slicing, segment the dataset in a layer in which the power lines are situated. Further, neighborhood analysis is performed to identify the horizontal linear features in the second step of the proposed method. Distance-based clustering and analysis of individual clusters are performed in the third and last step of proposed method respectively in order to identify the power line clusters.

2 Test Data

The test data is captured from the Mahatma Gandhi Marg, Civil Lines, Allahabad city, Uttar Pradesh, India (25°26′55.5″N, 81°50′43.5″E) with the help of FARO Fo-cus3D X 330 TLS. It offers −60° to +90° vertical and 360° horizontal field of view, ranging from 0.6 m up to 330 m with distance accuracy up to ±2 mm and delivering high precision performance and coverage. It is easy to use as it has a dedicated touchscreen Liquid Crystal Display (LCD) display that shows the status information and allows the user to adjust the data capture parameters. The maximum elevation difference within the dataset is 21.86 m. There are very low slopes along the horizontal streets, and in some areas, streets and power lines are heavily blocked by trees therefore, a large number of power line points are missing in the captured dataset. Overall, the data set has an urban as well as nonurban behavior. The length of the street is 218.16 m. Figure 1 shows the Google Earth image of the corresponding location along with two different perspective view of the captured dataset. Red rectangle of Google earth image (Fig. 1) shows the particular test site. Statistical specification of the captured dataset is shown in Table 1.

File size is in gigabytes (GB), point density is in per meter square (/m^2) and total covered area by the captured dataset is in meter square (m^2).

3 Proposed Method

The proposed method for detecting power lines from TLS point cloud data is divided into four steps (Fig. 2). All these steps are discussed in detail in the following sections.

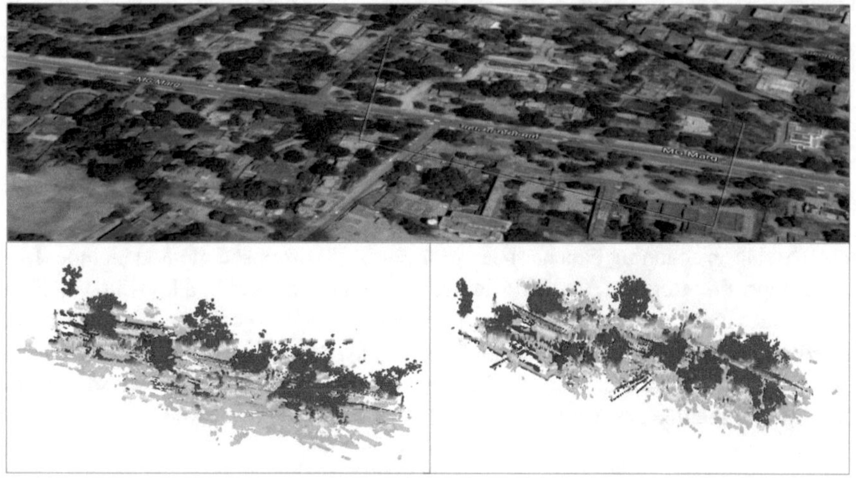

Fig. 1 Google Earth image of test site along with the captured dataset

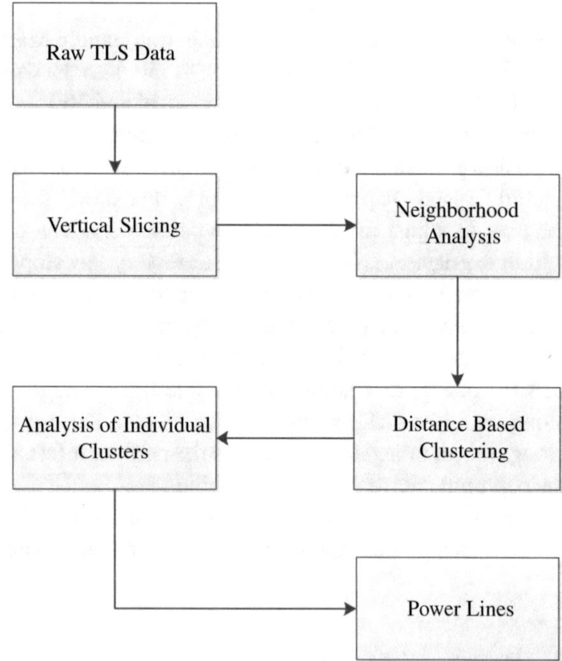

Fig. 2 Process sequence of proposed method

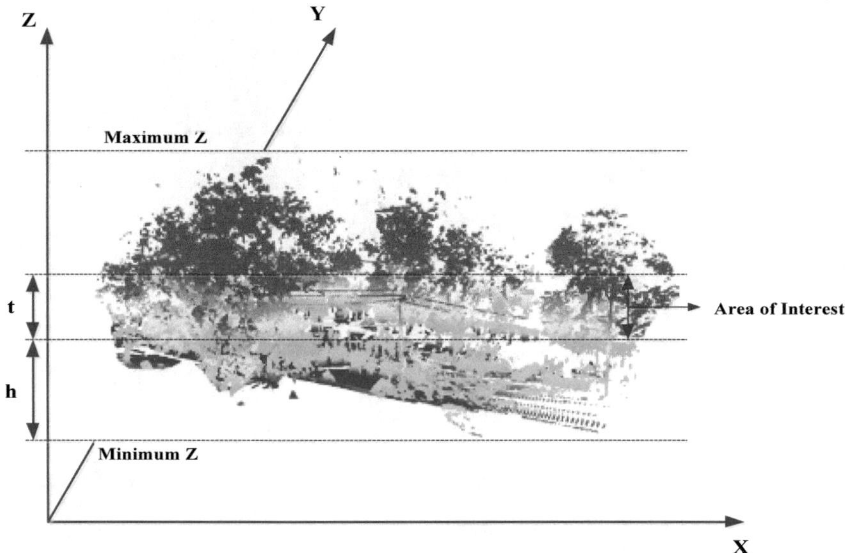

Fig. 3 Vertical slicing

3.1 Vertical Slicing

Initially, X, Y, and Z coordinate values of each point are taken as input and vertical slicing has been performed. For doing so, some (n) points having the lowest Z value are selected and their mean Z (z_{mean}) value is calculated. Calculated z_{mean} is used as minimum Z value of input dataset for further calculation. The reason of this calculation is to remove the effect of outliers. There might be a possibility that point having the actual minimum Z value may belong from outlier. Now the point of vertical slice heightened from (z_{mean} + h) to (z_{mean} + h+t) are taken for further processing because they belong to the area of interest (power line points) (Fig. 3). The objective of vertical slicing is to remove the unnecessary points which are not belonging to the area of interest and it improves the run-time performance of proposed method. Figure 4 shows the vertically sliced layer (area of interest) points.

The points belong to the area of interest are processed further, and three more successive steps have been performed at vertical sliced points for power line points detection.

3.2 Neighborhood Analysis

Points of various geospatial features such as tree crown and building facades along with the power line also persist in the vertically sliced layer. In order to suppress

Fig. 4 Vertically sliced layer (area of interest)

these points, Principal Component Analysis (PCA) based neighborhood analysis is performed. For doing so, the first point of the vertically sliced layer is chosen as a seed point and corresponding *k* neighbor points are selected with the help K-D tree. PCA has been performed at these selected points. In order to apply the PCA, maximum normalized eigenvalue (4) and corresponding eigenvector (5) are calculated with the help of variance–covariance matrix (1) of selected neighbor points (2, 3). Power lines are the horizontal linear feature therefore, to classify the horizontal linear features, the value of maximum normalized eigenvalue (α_{max}) is analyzed and it should be near to one, also the angle (θ) between Z axis to eigenvector corresponds to the maximum normalized eigenvalue (6) should be near to 90°. Likewise, all the points of the vertically sliced layer are chosen iteratively and PCA has been performed at corresponding neighborhood points. The reason of using K-D tree for neighborhood searching is because of its less time complexity.

$$Cov = \begin{pmatrix} \sigma_x^2 & \sigma_{xy} & \sigma_{xz} \\ \sigma_{xy} & \sigma_y^2 & \sigma_{yz} \\ \sigma_{xz} & \sigma_{yz} & \sigma_z^2 \end{pmatrix} \tag{1}$$

$$(Cov)_{3\times3} - (\lambda_{3\times1} \times I_{3\times3}) = 0 \tag{2}$$

$$\lambda = \begin{pmatrix} \lambda_{\max} \\ \lambda_{\text{mod}} \\ \lambda_{\min} \end{pmatrix} \tag{3}$$

$$\alpha_{max} = \frac{\lambda_{\max}}{\lambda_{\max} + \lambda_{\text{mod}} + \lambda_{\min}} \tag{4}$$

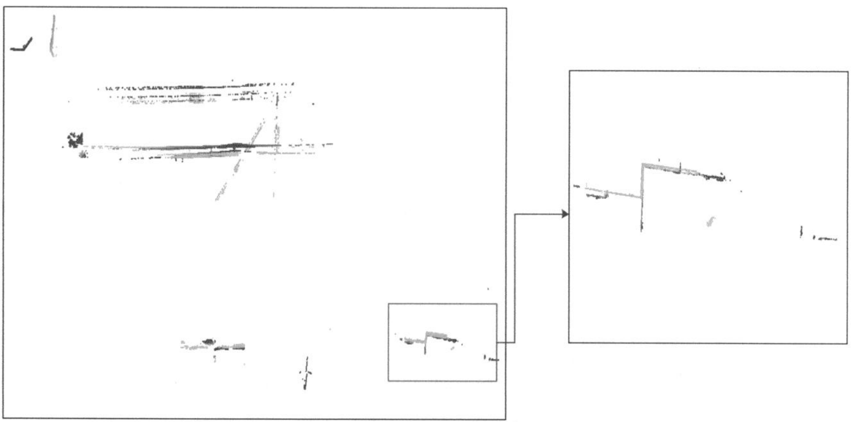

Fig. 5 Horizontal linear features obtained after applying neighborhood analysis

$$V\alpha_{max} = a\hat{i} + b\hat{j} + c\hat{k} \tag{5}$$

$$\theta = \cos^{-1}\left(\frac{c}{\sqrt{(a^2 + b^2 + c^2)}}\right) \tag{6}$$

where σ_x^2, σ_y^2 and σ_z^2 are the variances of X, Y, and Z coordinate values, respectively. Likewise,σ_{xy}, σ_{yz}, and σ_{xz} are the covariance between X-Y, Y-Z, and Z-X coordinates values respectively. Coefficients of unit vectors in X, Y and Z direction are *a*, *b*, and *c* respectively. Maximum, moderate, and minimum eigenvalues are $\lambda_{\max}$, λ_{mod} and $\lambda_{\min}$, respectively. Figure 5 shows the vertical linear features contain some other linear structures apart from the power lines.

3.3 Distance-Based Clustering

To swiftly group horizontal linear features into clusters, Euclidean distance based clustering approach is applied, which clusters discrete points on the basis of their Euclidean distances. An un-clustered point is grouped into a specific cluster if and only if its shortest Euclidean distance to the points in this cluster lies below a threshold (d_{th}). The objective of this step is to rapidly group the horizontal linear features into individual clusters. Further, each individual cluster is analyzed one by one to determine whether the selected cluster belongs to the power lines.

In order to perform the distance based clustering Queue data structure is used and the first point of horizontal linear feature dataset is inserted in queue. Further, Euclidian distance is calculated with rest of the points, if calculated Euclidian distance is less than the particular threshold value (7) then, all such points are inserted into the queue. Further, the first point is removed (dequeue operation) from the queue

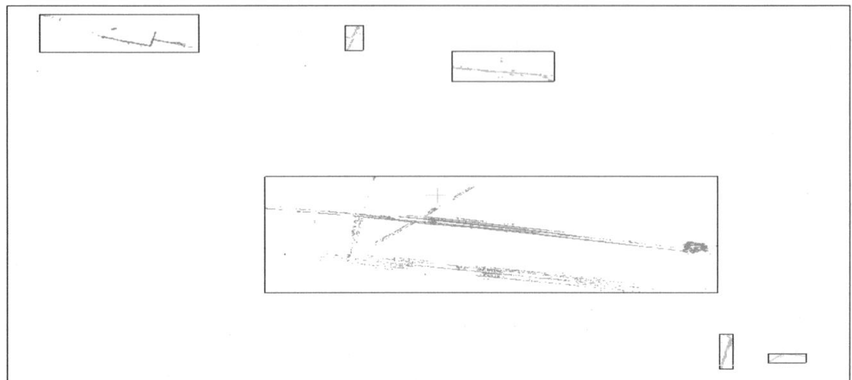

Fig. 6 Individual segmented cluster of vertical linear features

and put into the cluster, and similar procedure is applied for all remaining points of the queue, process is iteratively repeated until the queue becomes empty and all the points are grouped in single cluster.

$$\forall Point(i) : if(|P_s - P_i|) \leq d_{th} \tag{7}$$

where P_s shows the seed point and P_i shows i^{th} point. If the distance between them is less or equal to the d_{th}, then the seed point and selected i^{th} belong to the same cluster. Figure 6 shows the bounding box of individual segmented cluster of vertical linear features.

3.4 Analysis of Individual Clusters

In this step, false-positive clusters (apart from the power line clusters) are identified and removed from the rest horizontal linear features. For doing so, diagonal length of bounding box of each cluster is calculated by finding the coordinates of bottom left and top right corners (Fig. 7) and a curve is fitted using least square method for each cluster points (Fig. 8), based upon these two criteria each cluster is analyzed individually. If the calculated length of cluster is greater or equal to a threshold (l_{th}) and the nature of fitted curve should be linear or hyperbolic having face towards the positive Z direction then the selected cluster belongs from the power lines cluster.

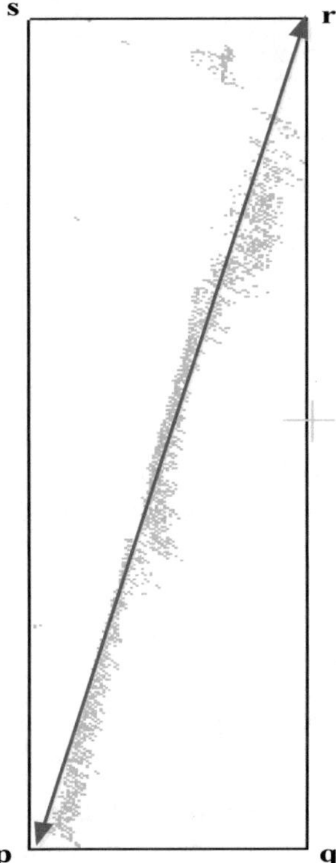

Fig. 7 Diagonal length calculation of individual cluster

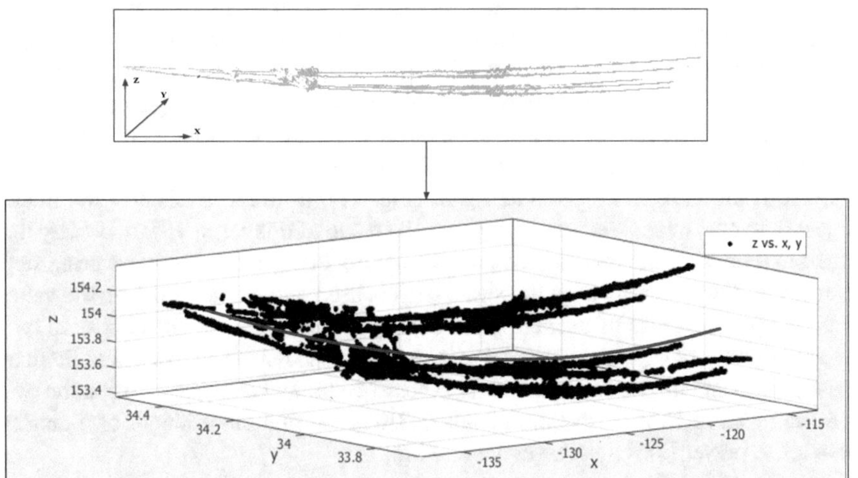

Fig. 8 Fitting of curve at individual cluster

Table 2 Parameters and their used values in various steps of the proposed method

Steps	Parameters	Used value
Vertical slicing	n, h, t	100, 5 m, 8 m
Neighborhood analysis	k, α_{max}, θ	50, 0.8, 15°
Distance-based clustering	d_{th}	0.2 m
Analysis of individual clusters	l_{th}	40 m

4 Results

Proposed methodology has been tested at captured terrestrial LiDAR point cloud dataset (Fig. 1). Statistical specification of dataset is given in Table 1. The method uses various parameters in different steps; used values of these parameters for the present study are listed in Table 2. These values may differ from one point cloud dataset to another.

4.1 Method Parameters Setting

In vertical slicing step (Sect. 3.1) three parameters n, h, and t are used for calculation of Z_{mean}, lower limit of vertical slice and upper limit of vertical slice respectively. It is assumed that not more than 100 points will belong to outlier in the captured dataset; therefore the value of n is set to 100. The height of electric poles present in dataset is height of 8 meters and the layers of power lines are bonded at poles from 5 meters from the bottom. Therefore, the values of h and k are set to 5 and 8 meters, respectively. Three parameters k, α_{max}, and θ, are used in the second step of the proposed method (neighborhood analysis) for number of neighbors, value of maximum normalized eigenvalue and angle between the eigenvector corresponding to maximum normalized eigenvalue and Z axis ($\hat{k}$). The value of k is set to the 50, because it provides more accurate result (Fig. 11). In order to identify the linear object the value of α_{max} is set to 0.8 [18]. The value of θ is set to 15° to identify the horizontal linear structure from the linear structure. In distance-based clustering step of method, Euclidian distance threshold (d_{th}) is used with value of 0.2 m, the value is chosen on the basis of average point density and average distance between two neighbor points of captured dataset. Parameter l_{th} is used in the analysis of individual clusters step and the value of parameter is set to 40 m. The distance between the two electric poles is 40 m in captured dataset. Thus, the minimum length of a cluster belongs to power lines must be equal to 40 m.

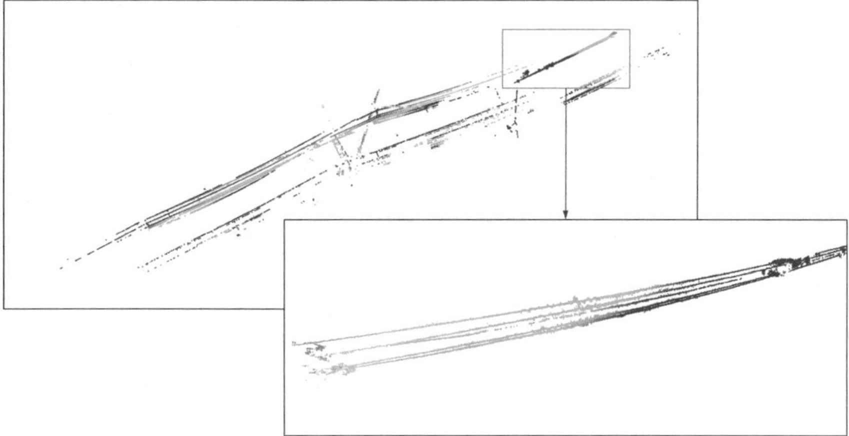

Fig. 9 Perspective view of reference power line points

4.2 Reference Data

Visual inspection has been performed at captured dataset in order to manually generate the reference data. Cloud Compare, v2.6.0 has been used to visualize and manual generation of reference data from the captured dataset. Figure 9 shows the manually generated dataset by the visual inspection from the captured dataset.

4.3 Output Dataset

After applying the proposed method at captured dataset, power lines are successfully detected having the completeness (8) and correctness (9) of 98.54% and 96.89%, respectively. Figure 10 shows the detected power line points at parameter values shown in Table 2. Statistical analysis at output dataset is performed with the help of generated reference dataset (Table 3).

$$Completeness = \frac{Power\ line\ points\ in\ output\ data}{Total\ number\ of\ points\ in\ reference\ data} \times 100 \tag{8}$$

$$Correcteness = \frac{Power\ line\ points\ in\ output\ data}{Total\ number\ of\ points\ in\ output\ data} \times 100 \tag{9}$$

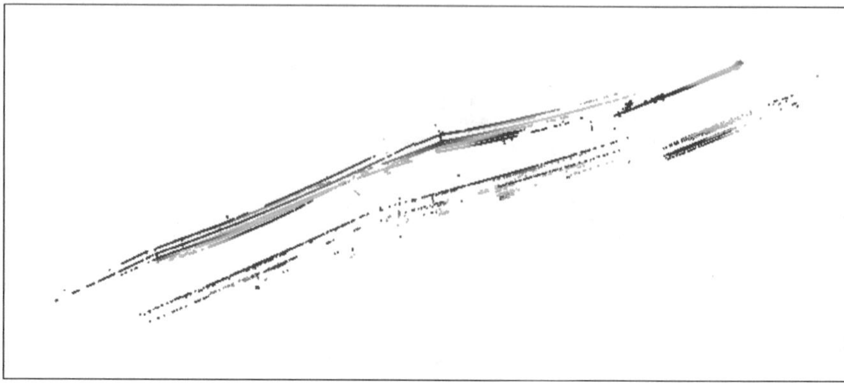

Fig. 10 Perspective view detected power line points

Table 3 Completeness and Correctness of output dataset

Total number of points in output data	726470
Power line points in output data	693599
Non-power line points in output file	32871
Total number of points in reference data	703876
Completeness	98.54%
Correctness	96.89%

5 Discussion

The proposed method uses only the X, Y, and Z coordinate values (geometric information) of each TLS point. The method does not require additional information such as intensity, scanning geometry, number of returns, return numbers, etc. The proposed method is dependent at point density because distance-based clustering is based on point density and used to individually segment the horizontal linear feature.

5.1 Execution Time Analysis

The proposed method is coded at Matlab2013a installed on Sony Vaio E Series notebook (OS: Windows7 64bit, CPU: Intel Core i3@2.4 GHz, RAM: 3 GB). The execution time of proposed method at parameter values, shown in Table 2 is 416.52 s. The execution time of the proposed method depends on the number of nearest neighbors (k) and height interval (p) of power line distribution (t-h). If the value of k increases, the more number of neighbors are originated in account corresponding to the selected seed point. Therefore, it increases the execution time of proposed method because it will take more time for finding additional number of neighbors

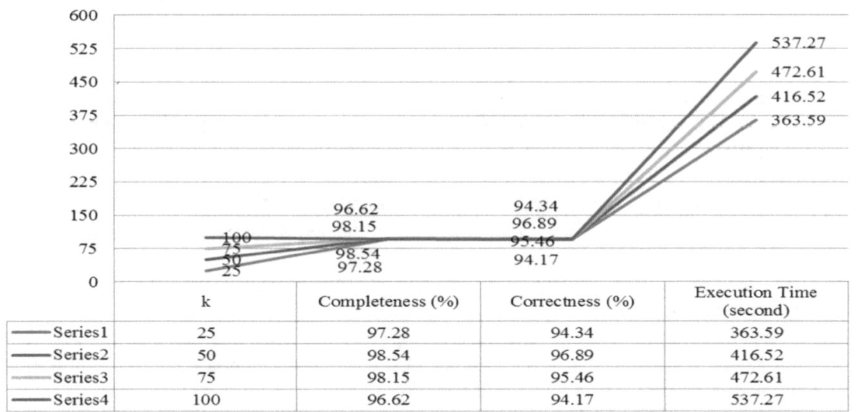

	k	Completeness (%)	Correctness (%)	Execution Time (second)
Series1	25	97.28	94.34	363.59
Series2	50	98.54	96.89	416.52
Series3	75	98.15	95.46	472.61
Series4	100	96.62	94.17	537.27

Fig. 11 Execution time, completeness and correctness of dataset at different values of parameter k

and vice versa (Fig. 11). Second parameter at which the execution time of proposed method depends is the difference (p) between the parameters t and h. If the value of p increases, the height of vertically sliced layer will increase and it will lead toward to the processing of more number of points and vice versa.

6 Conclusions and Future Recommendations

A novel, automated, and time efficient method for the detection of power line points from TLS point cloud data is proposed in this study. The proposed method includes four steps in order to detect the power lines. Only the geometric information (X, Y, and Z coordinate values) of each MLS point has been used by the method. K-D tree and PCA has been used for time-efficient nearest neighbor search and identification of horizontal linear features, respectively. The proposed method is dependent at point density. An initial assumption that is not more than 100 points will belong to outliers is established. The distance between two electric poles (knowledge base information) is used as threshold for filtering the power line clusters. The proposed method is implemented at Matlab2013a and run-time analysis is also performed. The method has been tested on a captured TLS dataset and corresponding power lines are detected. Completeness and correctness of method are 98.54% and 96.89%, respectively.

Future work will be focused on addressing the limitation of the proposed method such as dependency at point density and achieve the maximum level of completeness and correctness. Automated determination of optimized value of employed parameters is also an important future aspect.

References

1. Ussyshkin, R.V., Theriault, L., Sitar, M., Kou, T.: Advantages of Airborne Lidar Technology in Power Line Asset Management. In: Proceedings of the International Workshop on Multi-Platform/Multi-Sensor Remote Sensing and Mapping (M2RSM), Xiamen, China, pp. 1–5. (2011)
2. Lu, M.L., Pfrimmer, G., Kieloch, Z.: Upgrading an existing 138 kV transmission line in Manitoba. In: Proceedings of IEEE Power Engineering Society General Meeting, Montreal, QC, Canada, pp. 964–970. (2006)
3. Koop, J.E.: Advanced technology for transmission line modeling. Transm. Distrib. World **54**, 90–93 (2002)
4. Ashidate, S., Murashima, S., Fujii, N.: Development of a helicopter-mounted eye-safe laser radar system for distance measurement between power transmission lines and nearby trees. IEEE Trans. Power Deliv. **17**, 644–648 (2002)
5. Kukko, Hyyppä: J.: Small-footprint laser scanning simulator for system validation, error assessment, and algorithm development. Photogram. Eng. Remote Sens. **75**, 1177–1189 (2009)
6. Liu, X.: Airborne Lidar for DEM generation: some critical issues". Prog. Phys. Geog. **32**, 31–49 (2008)
7. Zhu, L., Hyyppä, J.: Fully-automated power line extraction from airborne laser scanning point clouds in forest areas. Remote Sens. **6**, 11267–11282 (2014). https://doi.org/10.3390/rs61111267
8. Melzer, T., Briese, C.: Extraction and modeling of power lines from als point clouds. In: Proceedings of 28th Workshop of the Austrian Association for Pattern Recognition (ÖAGM), Hagenberg, Austria, pp. 47–54 (2004)
9. Liu, Y., Li, Z., Hayward, R., Walker, R., Jin, H.: Classification of airborne lidar intensity data using statistical analysis and hough transform with application to power line corridors. In: Proceedings of the Digital Image Computing: Techniques and Applications, Melbourne, Australia, pp. 462–467 (2009)
10. Jwa, Y., Sohn, G., Kim, H.B.: Automatic 3D powerline reconstruction using airborne Lidar data. Int. Arch. Photogram. Remote Sens. **38**, 105–110 (2009)
11. McLaughlin, R.A.: Extracting transmission lines from airborne Lidar data. IEEE Geosci. Remote Sens. Lett. **3**, 222–226 (2006)
12. Liang, J., Zhang, J., Liu, Z.: On extracting power-line from airborne Lidar point cloud data. Bull. Surv. Mapp. **7**, 17–20 (2012)
13. Ou, T., Geng, X., Yang, B.: Application of vehicle-borne data acquisition system to power line detection. J. Geod. Geodyn. **29**, 149–151 (2009)
14. Lehtomäki, M., Jaakkola, A., Hyyppä, J., Kukko, A., Kaartinen, H.: Detection of vertical pole-like objects in a road environment using vehicle-based laser scanning data. Remote Sens. **2**, 641–664 (2010)
15. Jaakkola, A., Hyyppä, J., Hyyppä, H., Kukko, A.: Retrieval algorithms for road surface modelling using laser-based Mobile mapping. Sensors **8**, 5238–5249 (2008)
16. Zhu, L., Hyyppä, J., Kukko, A., Kaartinen, H., Chen, R.: Photorealistic building reconstruction from Mobile laser scanning data. Remote Sens. **3**, 1406–1426 (2011)
17. Jaakkola, A., Hyyppä, J., Kukko, A., Yu, X., Kaartinen, H., Lehtomäki, M., Lin, Y.: A low-cost multi-sensoral Mobile mapping system and its feasibility for tree measurements. ISPRS J. Photogramm. Remote Sens. **65**, 514–522 (2010)
18. El-Halawany, S.I., Lichti, D.: Detection of Road Poles from Mobile Terrestrial Laser Scanner Point Cloud, International Workshop on Multi-Platform/Multi-Sensor Remote Sensing and Mapping (M2RSM), Xiamen, China, Art. No. 5697364 0, (2011)

Stochastic Petri Net Based Modeling for Analyzing Dependability of Big Data Storage System

Durbadal Chattaraj, Monalisa Sarma and Debasis Samanta

Abstract With the exponential growth of information technology, the amount of data to be processed is also increased enormously. Managing such a huge data has emerged as the "Big Data storage issue", which can only be addressed with new computing paradigms and platforms. Hadoop Distributed File System (HDFS), the principal component of Hadoop, has been evolved to provide the storage service in the vicinity of Big Data paradigm. Although, several studies have been conducted on HDFS few works focus on the storage service dependability analysis of HDFS. This work aims to develop a mathematical model to represent the storage service activities of HDFS and formulates its dependability attributes. To achieve this, a stochastic Petri net (SPN) based modeling technique is put forward. The proposed model accurately quantify two important dependability metrics namely storage service reliability and availability of HDFS.

1 Introduction

Of late, the expansion of the methods for gathering information and the speed at which this information is produced has demonstrated the requisite for finding an approach to process them in a reasonable time. This is the so-called "Big Data" issue. Firms and organizations having huge storage infrastructure with quite an average usability require continuous updates of the technologies used in a data center to maintain high

D. Chattaraj (✉) · M. Sarma
Subir Chowdhury School of Quality and Reliability, Indian Institute of Technology Kharagpur, Kharagpur 721302, India
e-mail: dchattaraj@iitkgp.ac.in

M. Sarma
e-mail: monalisa@iitkgp.ac.in

D. Samanta
Department of Computer Science and Engineering, Indian Institute of Technology Kharagpur, Kharagpur 721302, India
e-mail: dsamanta@sit.iitkgp.ernet.in

A. Abraham et al. (eds.), *Emerging Technologies in Data Mining and Information Security*, Advances in Intelligent Systems and Computing 813,
https://doi.org/10.1007/978-981-13-1498-8_42

levels of service. As of late, the technology division began assessing technologies for storing Big Data. In this synergy, various distributed file systems (e.g., HDFS [2], GFS [9], MooseFS [15], zFS [16], etc.) have emerged. Hadoop Distributed File System (HDFS), the principal component of Hadoop, has been evolved to provide the Big Data storage service, where data is reliably kept in a distributed fashion into different servers. HDFS as it stands today aims to fulfill the following: (1) reliable data storage by avoiding data loss, (2) quick retrieval of data on demand, (3) capacity management to scale effortlessly including nodes for extending storage capability and enhancing general performance and fault tolerance of the framework and (4) storage of an enormous amount of data of any size, beginning from little too huge unstructured files. However to fulfill the aforementioned criteria, a precise storage service dependability measures of HDFS is necessary.

With the best of our knowledge, very few works were reported in the literature [4, 7, 11, 13, 17, 20] to address the aforesaid issues. M. K. McKusick and S. Quinlan in the conversation on "GFS: Evolution on fast-forward" [14] discussed about Namenode faults and utilization of the snapshot technology to reduce maintenance time. They also quoted that "Datanode fault is considered typical: HDFS sets up replicas to ensure data safety: The setup of checkpoints or replicas results in increased performance overhead for HDFS". K. V. Shvachko [18] examined the overhead of typical fault-tolerant technology and estimated the number of Datanodes limit from the viewpoint of the tasks. In addition to this, very few literatures [5, 6, 10] are reported recently towards the investigation of availability modeling for cloud storage infrastructure. In this synergy, authors were proposed different availability models for cloud storage cluster and virtualized system. Nonetheless, without the loss of generality, they did not focus on Big Data storage system and its underlying distributed file system (i.e., HDFS) modeling. Li et al. [11] introduced the stochastic Petri net based modeling of HDFS to quantify its storage service reliability and availability. However, the model that the authors were analyzed is now the outdated version of HDFS.

In order to address the aforementioned research gap of the existing state of the art, this paper focuses on the following research objectives: (1) To design a succinct mathematical model of HDFS to quantify two important dependability attributes namely reliability and availability. (2) To compute reliability and availability of HDFS using a concise numerical formulation based on Stochastic Petri Nets (SPNs) [19] together with Continuous Time Markov Chain (CTMC) [8]. (3) The distributed storage of data and its replication provides fault tolerance but increases the complexity of the system model resulting in massive state-space generation of computationally intractable size. This is known as "state-space explosion" problem [11]. This work addresses this issue. (4) To propose a general model of HDFS by considering the current HDFS federation architecture and formulate the expression for both reliability and availability parameters.

The remainder of this paper is organized as follows. A brief insight of HDFS federation architecture and the preliminaries of SPNs are discussed in Sect. 2. Section 3 presents the SPNs based system modeling of HDFS. Section 4 discusses the formulation of reliability and availability parameters of the proposed model. The simulated result of the proposed model is shown in Sect. 5. Finally, Sect. 6 concludes the paper.

2 Preliminaries

This section discusses the basic knowledge about HDFS Federation Architecture (HFA) and SPNs as follows.

2.1 HDFS Federation Architecture

The architecture of HDFS federation is available in [3]. For scaling the namespace service horizontally, HDFS federation utilizes multiple independent Namenodes or namespaces. As the Namenodes are configured in a federated environment, it works independently and does not require any coordination with other Namenodes. The Datanodes are used as a shared physical storage for blocks by all the Namenodes. In addition to this setting, individual Datanode enrolls with all the Namenodes in the Hadoop cluster and it sends periodic heartbeats and block reports signals to all the Namenodes. Moreover, a Block Pool (BP) is a set of blocks that belong to a single namespace or Namenodes. Datanodes physically store blocks for all the BPs in the cluster and independently manages each BPs. This enables a namespace to assign block identities independently for new blocks without coordinating with other namespaces. As a result, the failure of a Namenode does not prevent the Datanodes access with the help of other Namenodes in the cluster. It improvises the system reliability by eliminating the single point of failure issue. In order to model the aforesaid HDFS federation architecture and quantify its dependability attributes, in this paper, we utilize a mathematical tool namely SPNs.

2.2 Stochastic Petri Nets

Stochastic Petri Nets (SPNs) are a widely accepted mathematical tool for analyzing the performance and dependability of various distributed systems. The basic definition of SPN is given as follows.

Definition 1 SPN can be represented mathematically as $SPN = (\mathscr{P}, \mathscr{T}, \mathscr{F}, \mathscr{W}, \mathscr{M}_0, \lambda)$. First tuple $\mathscr{P} = (p_1, p_2, \ldots, p_p)$ is termed as a place set and it describes the state variables of the network: second tuple namely $\mathscr{T} = (t_1, t_2, \ldots, t_T)$ represents a finite set of transitions and the execution of each transition can change the value of the state variables of the network: third tuple $\mathscr{F}$ namely an arc set represents the flow relationship between places and transitions and vise versa where $\mathscr{F} \subseteq \mathscr{P} \times \mathscr{T} \cup \mathscr{T} \times \mathscr{P}$: fourth tuple describes a weight function between place and transition and vise versa, and it is denoted as $\mathscr{W} : \mathscr{F} \rightarrow \mathbb{N}$: Fifth and sixth tuples depict the initial state of the network and the firing rates allied with the transitions. Both of these tuples are expressed as $\mathscr{M}_0 = (m_{01}, m_{02}, \ldots, m_{0p})$ and $\lambda = (\lambda_1, \lambda_2, \ldots, \lambda_T)$ respectively.

Several extensions of the fundamental SPNs are reported in the literature [21], such as Generalized Stochastic Petri Nets (GSPNs) [1], Stochastic High-level Petri Nets (SHPNs) [12], etc. For analyzing dependability of HDFS data storage system, in this article, we adopt GSPNs.

3 Proposed Methodology

HDFS federation incorporates master–slave architecture for distributed storage so the model is segmented into two layers, Namenode layer and Datanode layer. In this connection, both Namenode layer and Datanode layer consist of n numbers of Namenodes and m number of Datanodes respectively. The failure-repair model of both the Namenode and Datanode is developed using GSPNs and they are discussed in the following subsections.

3.1 The Failure-Repair Model of Namenode

In the HDFS cluster, the Namenode is a single point of failure. The HDFS HA (high availability) addresses the problem by providing the option of running n number of Namenodes in the same cluster, wherein an active or standby configuration referred to as functioning Namenode (i.e., single number) and the standby Namenodes (i.e., $(n-1)$ numbers). The $(n-1)$ standby Namenodes are in hot standby allowing fast fail over to the failed Namenode. The failure-repair model of a Namenode is shown in Fig. 1a. In this model, the normal functioning of the Namenode is represented by the presence of token in place NN_up. During the operation, the Namenode may suffer from various failures and attacks, which leads to the failure of the Namenode. The failure of Namenode is denoted by the place NN_dwn and the action resulting in this failure is depicted by the transition T_NN_f. Immediate transition T0 has an inhibitor arc and so gets enabled only when the active Namenode (i.e., token in NN_up) fails and the place NN_up does not have any token. In case of failure of the functioning Namenode, the T0 gets enabled and the standby Namenode becomes the active Namenode. This action is represented by flow of token from place STNN_up to NN_up. Note that, here, due to design simplicity, we have shown only one token out of $(n-1)$ token in the place of STNN_up. The failed Namenode is repaired via transition T_NN_r and put in the standby mode. In addition to this,

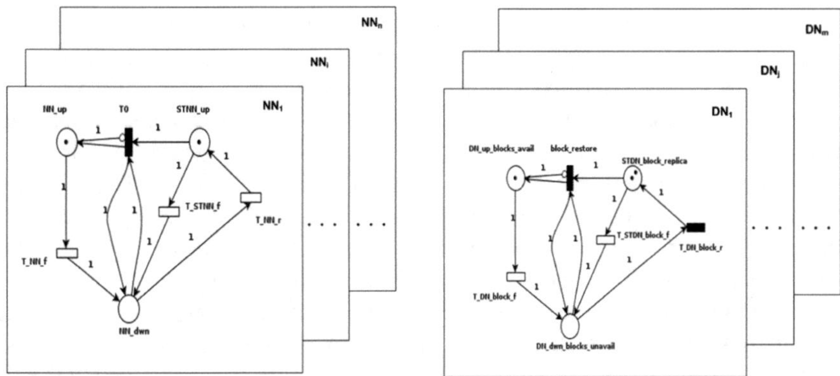

(a) Failure-repair model of a single Namenode **(b)** Failure-repair model of a single Datanode

Fig. 1 GSPN models of both Namenode and Datanode layers in HDFS federation

any standby Namenode can fail during hot standby mode and it is represented as transition T_STNN_f. The transition rates conform to exponential distribution for T_NN_f, T_NN_r and T_STNN_f transitions respectively.

3.2 *The Failure-Repair Model of Datanode*

The failure-repair model of a Datanode considering replication factor (RF) is equal to three is shown in Fig. 1b. In this model, the normal functioning of the Datanode vis-a-vis the availability of the blocks is represented by the presence of token in place DN_up_blocks_avail. During its operation, the Datanode may suffer from various failures (i.e., power failure, abnormal shutdown, etc.) and attacks, which leads to the failure of the Datanode and unavailability of the blocks. The failure of Datanode is denoted by the place DN_dwn_blocks_unavail and the action resulting in this failure is depicted by the transition T_DN_f. Immediate transition block_restore has an inhibitor arc and it gets enabled only when the active Datanode (i.e., token in DN_up_blocks_avail) fails and the place DN_up_blocks_avail does not have any token. In case of failure of the functioning Datanode, the blocks are unavailable and the transition block_restore gets enabled. It will restore the lost blocks from the hot standby Datanodes (i.e., $(m-1)$ Datanodes). This action is represented by flow of token from place STDN_block_replica to DN_up_blocks_avail. Note that, here, we consider the replication factor is equal to three where DN_1 is having one block and other two blocks are kept into different hot standby Datanodes say in DN_j and DN_m, respectively. In addition to this, the lost blocks are immediately repaired via transition T_DN_block_r by replicating them into other hot standby Datanodes. In addition to this, any standby Datanode can fail during hot standby mode and it is represented as transition T_STDN_block_f. The transition rates conform to exponential distribution for T_DN_f, T_DN_block_r and T_STDN_block_f transitions respectively.

In order to reduce the design complexity, m number of Datanodes is decomposed into a single layer as it has a large number of identical structures. The single decomposed layer of Datanodes is named as Datanode module. The Datanodes in the Datanode module are all assumed to be identical, and their failures are independent of each other. The failure-repair model for Datanode module (see Fig. 1b) is quite comparable to the practical application of the HDFS. Files that are broken into blocks are not a constraint to be stored as a block per Datanode but depending upon the storage capacity of the Datanode it can keep multiple blocks of different files. The one constraint that this configuration follows is—no two replicas of the identical block are to be stored in the same Datanode. This is done to keep the fault-tolerant property of the system intact. Therefore, the failure condition for the Datanode module with k Datanodes ($m = k$, see Fig. 1b) is defined as $F_{datanode\_module}$ = {(failure of 3 Datanodes) ∪ (failure of 4 Datanodes) ∪ (failure of 5 Datanodes) ∪ ... ∪ (failure of k Datanodes).}

The failure condition of the Datanode module starts only after the failure of 3 or more Datanodes at a time. This is because the replication factor is three as mentioned in the assumption. The failure of 3 or more Datanodes at a time follows the binomial distribution with parameters $k \in \mathbb{N}$ and $p, q \in [0, 1]$ and a Cumulative Distribution Function (CDF) of the module can be expressed as $F_{datanode\_module} = Prob[X \geq 3] = 1 - Prob[X \leq 2]$. Here, X is the random variable denoting the failure of individual Datanodes. So,

$$F_{datanode\_module} = 1 - \sum_{r=0}^{2} \binom{k}{r} p^r q^{k-r}. \tag{1}$$

Since, each Datanode is assumed to follow exponential distribution with constant failure rate λ then from Eq. 1 we get

$$F_{datanode\_module} = 1 - \sum_{r=0}^{2} \binom{k}{r} (e^{-\lambda \cdot t})^r (1 - e^{-\lambda \cdot t})^{k-r}. \tag{2}$$

Assume the Datanode module follows the exponential distribution with constant failure rate λ_m, for the mission time of one year. Therefore, the failure rate of the Datanode module is derived from Eq. 2 as

$$\begin{aligned}\lambda_m = (-ln[(1 - e^{-\lambda \cdot t})^k + k \cdot e^{-\lambda \cdot t} \cdot (1 - e^{-\lambda \cdot t})^{k-1} \\ + \frac{k \cdot (k-1)}{2} e^{-2\lambda \cdot t} (1 - e^{-\lambda \cdot t})^{k-2}])/(t).\end{aligned} \tag{3}$$

Here, the instantaneous time at which the failure rate of the Datanode module (see Eq. 3) has been evaluated is assumed to be one year or 8760 hours. In addition to this, for the shake of calculation simplicity, we assume that the mean time to repair (MTTR) of the Datanode module is one hour.

3.3 Basic GSPN Model of HDFS

After obtaining the failure-repair model of both the Datanode module and Namenode, a simple GSPN model of the HDFS storage system is being proposed and it is shown in Fig. 2. Note that, here we consider only two Namenodes (i.e., $n = 2$) with simple failure-repair strategy for reducing the design complexity of the overall system.

The presence of token in place HDFS_up represents the normal functioning of the HDFS with active Namenode functioning routinely. NN_dwn_HDFS_up represents the failure of functioning Namenode, but due to the presence of the standby Namenode (here $n = 2$), HDFS still functions normally, while the token in place STNN_dwn denotes the failure of both Namenode. The failure of a block with all its replicas is represented by presence of token in place DN_module_dwn. Datanode module can fail in two circumstances: (1) HDFS is up and running with active Namenode working normally and (2) Active Namenode suffers a failure, while standby Namenode takes control of the HDFS. A token in the place HDFS_up represents the success state of the system when active Namenode is working. Datanode module may suffer failure when the system is in this state. This action is represented by the firing of transition T_DNmod_f and the presence of token in place DN_module_dwn. Presence of token in place NN_dwn_HDFS_up represents success state of the system when active Namenode fails, and standby Namenode provides with fail over capabilities. HDFS in this state is still up and running, and Datanode module has the probability of failure. If Datanode module fails this is represented by firing of the transition T_DNmod_f2 and presence of token in place DN_module_dwn. The reachability matrix obtained from the aforesaid model is shown in Table 1.

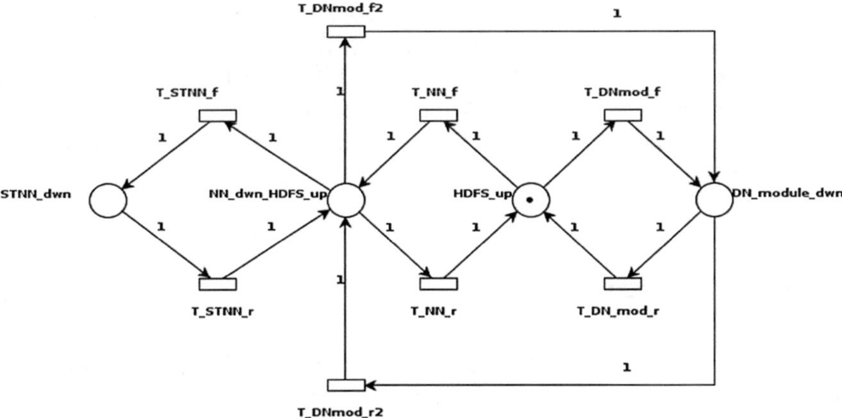

Fig. 2 Basic GSPNs model of HDFS federation architecture

Table 1 Reachability matrix of the proposed GSPNs model

Markings	HDFS_up	DN_module_dwn	NN_dwn_HDFS_UP	STNN_dwn
M_0	1	0	0	0
M_1	0	0	1	0
M_2	0	1	0	0
M_3	0	0	0	1

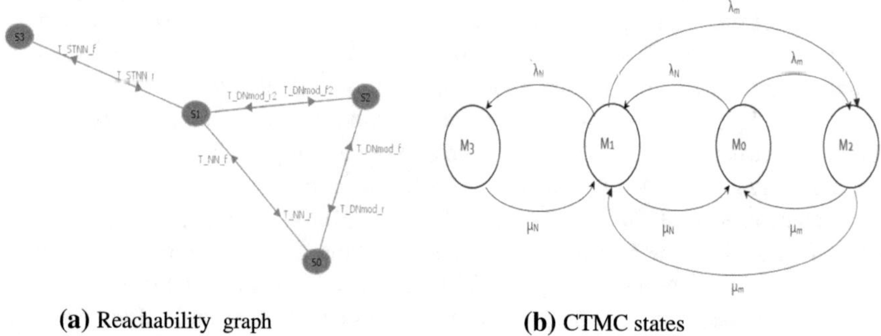

(a) Reachability graph **(b)** CTMC states

Fig. 3 Analysis of the proposed GSPNs model for HDFS federation

The Reachability Graph (RG)[1] obtained from the same model shown in Fig. 3a is mapped to Continuous Time Markov Chain (CTMC) to assist the dependability analysis of the HDFS model. The CTMC states derived from Fig. 3a are shown in Fig. 3b. A case based decomposition of Datanode Layer (see Sect. 3.2) in HDFS framework eliminates the possibility of the state-space explosion. This makes the analysis of the system feasible and tractable. The failure rate and repair rate of the Namenode is denoted by λ_N and μ_N respectively. This value remains the same for the failure and repair of the standby Namenode as both active node and hot standby node are identical devices with similar repair conditions. The proposed HDFS model is evaluated as follows.

4 Model Evaluation

From the CTMC obtained in Fig. 3b, the time-domain differential equations are obtained as follows:

$$\frac{dM_0(t)}{dt} = -(\lambda_N + \lambda_m) \cdot M_0(t) \tag{4}$$

[1]To obtain the RG, we simulate the proposed GSPNs model in PIPEv4.3.0 tool, for detail visit: http://pipe2.sourceforge.net/.

$$\frac{dM_1(t)}{dt} = \lambda_N \cdot M_0(t) - (\lambda_m + \lambda_N) \cdot M_1(t) \tag{5}$$

$$\frac{dM_2(t)}{dt} = \lambda_m \cdot M_0(t) + \lambda_m \cdot M_1(t) \tag{6}$$

$$\frac{dM_3(t)}{dt} = \lambda_N \cdot M_1(t) \tag{7}$$

The system initially is in the working condition and so,

$$M_{i\,(t=0)} = \begin{cases} 1 & \text{if } i = 0 \\ 0 & \text{otherwise} \end{cases} \tag{8}$$

To obtain the reliability of the system, we discard the set of failed states (i.e., states M_2 and M_3 is considered as the absorbing states). The solution of HDFS system reliability at time t is obtained as

$$R(t) = M_0(t) + M_1(t) \tag{9}$$

After taking the Laplace Transformation (LT) of Eqs. 4–7, incorporating the initial conditions (see Eq. 8) and taking the Inverse Laplace Transform (ILT), we get the reliability expression from Eq. 9 as follows:

$$R(t) = e^{-(\lambda_N+\lambda_m)\cdot t}[1 + \lambda_N \cdot t] \tag{10}$$

The availability of the HDFS is the steady state probability of states, which represent the normal functioning of the system. Steady state availability (SSA) is the system availability when the system has arrived at a steady state, i.e., $t \rightarrow \infty$ So, the equations for the availability calculation are presented as follows.

$$-(\lambda_N + \lambda_m) \cdot M_0 + \mu_N \cdot M_1 + \mu_m \cdot M_2 = 0 \tag{11}$$

$$\lambda_N \cdot M_0 - (\lambda_m + \mu_N + \lambda_N) \cdot M_1 + \mu_m \cdot M_2 + \mu_N \cdot M_3 = 0 \tag{12}$$

$$\lambda_m \cdot M_0 + \lambda_m \cdot M_1 - 2\mu_m \cdot M_2 = 0 \tag{13}$$

$$\lambda_N \cdot M_1 - \mu_N \cdot M_3 = 0 \tag{14}$$

$$M_0 + M_1 + M_2 + M_3 = 1 \tag{15}$$

hr The states M_2 and M_3 represent the failure states of the HDFS framework. So, after solving the steady state probability for states M_0 and M_1 in Eqs. 11–15, the following result is obtained.

$$A_S = M_0 + M_1 = \frac{(1 + \frac{2\lambda_N + \lambda_m}{2\mu_N + \lambda_m})}{1 + (\frac{2\lambda_N + \lambda_m}{2\mu_N + \lambda_m}) + \frac{\lambda_N}{\mu_N} \cdot (\frac{2\lambda_N + \lambda_m}{2\mu_N + \lambda_m}) + \frac{\lambda_m}{2\mu_m}(1 + \frac{2\lambda_N + \lambda_m}{2\mu_N + \lambda_m})} \quad (16)$$

Steady state availability of HDFS directly depends on the number of replicas of the block. Therefore, increasing the replication factor raises the availability.

5 Experimental Result

We assume the values of the proposed model configuration parameters namely mean time to failure (MTTF) and MTTR of individual Namenode or Datanode are 1000 hours and 1hour respectively. These configuration parameters are mostly obtained from the literature and the log files that are maintained by our Institute's computer maintenance team members. From Fig. 4a, it is observed that the storage service

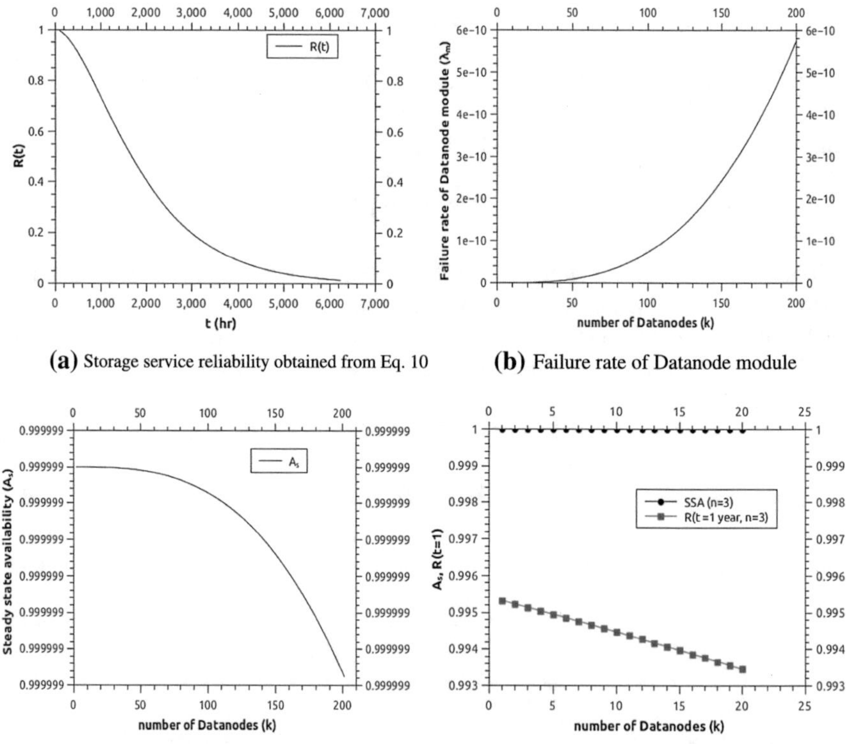

(a) Storage service reliability obtained from Eq. 10 (b) Failure rate of Datanode module

(c) Storage service availability obtained from Eq. 16 (d) Storage service reliability and availability with RF = 3

Fig. 4 Summary of the results obtained by evaluating the proposed HDFS model

reliability of HDFS has reduced over time. More precisely, the longer the HDFS system is put into operation the failure probability of the Datanodes will increase. As a result, the failure probability of Datanode module is also increased. Therefore, the storage service reliability is reduced over time. In addition to this, the number of blocks into which the client file is divided also has a great influence on storage service reliability. Increasing the number of blocks and its replicas needs more number of Datanodes (scale up the storage capacity) for storage, and as a result, increases the failure rate of the Datanode module (see Fig. 4b). Therefore, it decreases the storage service reliability (see Fig. 4a). Moreover, block replications have a positive effect on the storage service reliability and availability (see Fig. 4c, d) and it makes HDFS more fault-tolerant. Nonetheless, it can be inferred from the experimental results that the storage service reliability and availability of HDFS framework mostly depend on the failure rate of the Datanode module.

6 Conclusion

A stochastic Petri net based mathematical model for the federated Hadoop distributed file system architecture has been explored first time in this work. Further, this is the first time where the storage service dependability attributes of the same architecture have been analyzed accurately. Moreover, two important dependability metrics namely storage service reliability and availability have been formulated mathematically. Based on the current block replication policy of Hadoop distributed file system, it is being recommended that the default replication factor is equal to three as the right choice for higher storage service reliability and availability. The proposed stochastic Petri net based model analyses are able to substantiate the aforesaid claim. Further, the proposed approach of dependability attributes analyses is scalable and it can be applied for analyzing the dependability of any system without its limiting boundary. More significantly, the proposed methodology can be extended to analyze other dependability attributes such as maintainability, integrity, safety, etc. In the future work, we will extend the proposed model to measure the dependability metrics for large-scale Hadoop clusters with its sophisticated real-time settings.

References

1. Ajmone Marsan, M., Conte, G., Balbo, G.: A class of generalized stochastic Petri nets for the performance evaluation of multiprocessor systems. ACM Trans. Comput. Syst. (TOCS) **2**(2), 93–122 (1984)
2. Apache Hadoop: Hadoop Distributed File System (HDFS) (2017). http://hadoop.apache.org/. Available via Internet Source. Cited 22 Mar 2018
3. Apache Hadoop: HDFS Federation (2017). https://hadoop.apache.org/docs/stable/hadoop-project-dist/hadoop-hdfs/Federation.html. Available via Internet Source. Cited 22 Mar 2018

4. Bruneo, D.: A stochastic model to investigate data center performance and QoS in IaaS cloud computing systems. IEEE Trans. Parallel Distrib. Syst. **25**(3), 560–569 (2014)
5. Bruneo, D., Longo, F., Hadas, D., Kolodner, E.K.: Analytical investigation of availability in a vision cloud storage cluster. Scalable Comput. Pract. Exp. **14**(4), 279–290 (2014)
6. Bruneo, D., Longo, F., Hadas, D., Kolodner, H.: Availability assessment of a vision cloud storage cluster. In: European Conference on Service-Oriented and Cloud Computing, pp. 71–82. Springer (2013)
7. Dantas, J., Matos, R., Araujo, J., Maciel, P.: Models for dependability analysis of cloud computing architectures for eucalyptus platform. Int. Trans. Syst. Sci. Appl. **8**, 13–25 (2012)
8. Elsayed, E.A.: Reliability Engineering, vol. 88. Wiley (2012)
9. Ghemawat, S., Gobioff, H., Leung, S.T.: The google file system. In: Proceedings of 19th ACM symposium on Operating systems principles (SIGOPS), vol. 37. ACM (2003)
10. Ghosh, R., Longo, F., Frattini, F., Russo, S., Trivedi, K.S.: Scalable analytics for IaaS cloud availability. IEEE Trans. Cloud Comput. **2**(1), 57–70 (2014)
11. Li, H., Zhao, Z., He, L., Zheng, X.: Model and analysis of cloud storage service reliability based on stochastic petri nets. J. Inf. Comput. Sci. **11**(7), 2341–2354 (2014)
12. Lin, C., Marinescu, D.C.: Stochastic high-level Petri nets and applications. In: High-Level Petri Nets, pp. 459–469. Springer (1988)
13. Longo, F., Ghosh, R., Naik, V.K., Trivedi, K.S.: A scalable availability model for infrastructure-as-a-service cloud. In: 41st International Conference on Dependable Systems & Networks (DSN), pp. 335–346. IEEE (2011)
14. McKusick, K., Quinlan, S.: GFS: evolution on fast-forward. Commun. ACM **53**(3), 42–49 (2010)
15. MooseFS: Can Petabyte Storage be super efficient (2017). https://moosefs.com/. Available via Internet Source. Cited 22 Mar 2018
16. Rodeh, O., Teperman, A.: zFS-a scalable distributed file system using object disks. In: Proceedings of 20th Conference on Mass Storage Systems and Technologies (MSST), pp. 207–218. IEEE (2003)
17. Shafer, J., Rixner, S., Cox, A.L.: The hadoop distributed filesystem: Balancing portability and performance. In: International Symposium on Performance Analysis of Systems & Software (ISPASS), pp. 122–133. IEEE (2010)
18. Shvachko, K., Kuang, H., Radia, S., Chansler, R.: The hadoop distributed file system. In: 26th Symposium on Mass Storage Systems and Technologies (MSST), pp. 1–10. IEEE (2010)
19. Wang, J.: Petri nets for dynamic event-driven system modeling. Handb. Dyn. Syst. Model. **1** (2007)
20. Wu, X., Liu, Y., Gorton, I.: Exploring performance models of hadoop applications on cloud architecture. In: 11th International ACM SIGSOFT Conference on Quality of Software Architectures (QoSA), pp. 93–101. IEEE (2015)
21. Zeng, R., Jiang, Y., Lin, C., Shen, X.: Dependability analysis of control center networks in smart grid using stochastic petri nets. IEEE Trans. Parallel Distrib. Syst. **23**(9), 1721–1730 (2012)

Sachetan: A Crowdsource-Based Personal Safety Application

Sheikh Nabil Mohammad, Sakhawat Hossain, Md. Mustafijur Rahman, Mithun Das and Md. R. Amin

Abstract In developing countries such as Bangladesh, social and economic development is often hindered by lack of civilian safety due to increasing crime rate. In this paper, we have attempted to devise a solution to help citizens avoid potentially dangerous locations using data analytics. We designed and developed a system named "Sachetan", which is a mobile application that enables its users to privately share their own experience of incident occurrences with precise spatial and temporal information. Two data sources are combined and statistical insight and graphical representation of patterns have been presented to the users. In our research, we tried to find the pattern and determine the movements of high-frequency incident areas by analyzing the data.

1 Introduction

In recent years, Bangladesh has exhibited so much potential for development that this country is heralded as a role model of development for other third world countries. However, it is plagued by a high crime rate, and law enforcement often proves to be incapable of dealing with a high number of petty crimes such as mugging, robbery, road accidents, sexual harassment, theft, and so on. This is a problem with

---

S. N. Mohammad (✉) · S. Hossain · Md. M. Rahman · M. Das · Md. R. Amin
Shahjalal University of Science and Technology, Sylhet 3114, Bangladesh
e-mail: nabil-cse@sust.edu

S. Hossain
e-mail: saimon.sust@gmail.com

Md. M. Rahman
e-mail: nebir1993@gmail.com

M. Das
e-mail: mithun.das227@gmail.com

Md. R. Amin
e-mail: shajib-cse@sust.edu

A. Abraham et al. (eds.), *Emerging Technologies in Data Mining and Information Security*, Advances in Intelligent Systems and Computing 813,
https://doi.org/10.1007/978-981-13-1498-8_43

many developing or underdeveloped countries, and it becomes an important task to identify potential geographical locations where such incidents take place. Often such incidents are connected to poor infrastructure. For example, places that are not properly illuminated after dark or poor road conditions that make the place less frequented create a breeding ground for antisocial activities. We propose that if people can be informed of patterns regarding certain petty crime occurrence temporal data, they can alert themselves and increase their chances of avoiding such incidents. Most of the time, these incidents remain unreported leaving the potential victim(s) unaware of possible contingencies. Here, we therefore introduce a system where people share their own experience of incident occurrences anonymously with precise location information, incident time, and severity. It also gives users access to key insight and patterns of crime activity. We attempted to perform statistical data analysis to figure out the pattern and determine the movements of incidents by analyzing the data. We hope that long-term data of this platform and our analyzing results will help the governmental authorities to take proper steps to handle these problems across different geographical regions and prioritize developmental strategies.

2 Literature Review

Several recent efforts have been made to tackle these crime incidents. Research into the harms of harassment and legal remedies [1] have addressed the threats women face in the streets. Mobile applications like SOS Response [2], Not Your Baby App [3], Circleof6 [4], Safetipin [5], Abhaya [6], OnWatch [7], and SheSecure [8] were developed, which helped the victim in crime spot by providing signal or images to the security monitoring center. Some of them allow sending out GPS location of the user. Scholars were interested in analyzing crime hot spots [9] and burning times, which are becoming a major component of the work of criminologists, crime analysts, and crime prevention practitioners from the past decade. As social networks are the common platform of people to freely provide private information, for instance, on their current situation, opinions, etc., others try to use this opportunity to improve the prediction of crimes [10].

With the help of crowdsourcing [11], some attempts were made so that people can both provide crime-related information and be aware about their locations. Hollaback! [12] is a social movement organization that mainly works for raising awareness about street harassments against women. Protibadi [13] is also developed for sharing street harassment experiences for Bangladeshi women. It allows a user to report the location along with a description. HarassMap [14] collects and summarizes different types of harassments on a map mainly in Egypt. Another Bangladeshi movement, SafeStreet [15] is mentionable which empowers women in public places against sexual harassment. They also enable a woman to find a safe path that has less harassment hazard, at any point of time via their app.

3 Related Study

We evaluated the studies discussed above and identified some key issues. Most of the previous studies concentrate their efforts around tackling sexual harassment issues. There are few studies that encapsulate a wider range of citizen safety hazards with a more analytical approach. Previous systems are also ill-equipped to deal with the critical mass problem. They rely entirely on the users for the data, and as the system data source is empty in the early stages, users can get little to no benefit from the system. This may result in loss of users before the system is self-sustaining.

4 Methodology

To this extent, we aim to build a smartphone-based platform called "Sachetan". As we understand that fighting social problems cannot be simply achieved by self-awareness of certain individuals, what we want is social awareness or consciousness at a global level, and hence, the name Sachetan aptly symbolizes it. Sachetan is a crowdsource-based Android mobile application that will collect data related to security problems and antisocial activities. The key feature of the app is allowing the user to create reports of a wider variety of incidents and not only sexual harassment. Users can report incidents related to *mugging*, *road accident*, *sexual harassment*, and *theft*. The users can rate the incident based on the degree of its severity, from 1 to 5, where 5 indicates the most severe degree. They can optionally write a description of the incident that they want to share. After reporting the incident along with the necessary information, the data is stored in a central server. The application helps users avoid the places predicted not-so-safe by our algorithms. If the user clicks on a location, it shows the route to that location from his/her current position which is provided by Google Map, marking the risky areas in red color. Users can also see the statistical reports of a particular area by clicking on a button, which takes the users to a screen that shows incidents' frequencies with temporal information. This equips the users with the knowledge to be able to avoid certain locations at certain hours of the day when incidents are more likely to happen. To prevent the case where initial users encounter a blank app, we use newspaper archives as a supplementary data source, which is collected, processed, and stored before the system starts up, and made readily available to the users.

4.1 Newspaper Data Collection

The major newspaper websites, Prothom Alo [16], Bhorer Kagoj [17], Shomokal [18], SylhetNews24 [19], and Sylhetsangbad [20], were selected as news article sources (Table 1).

Table 1 Number of news articles collected by web crawlers

Newspaper name	Mugging	Road accident	Robbery	Sexual harassment	Theft
Prothom Alo	360	364	327	119	485
Bhorer Kagoj	722	561	670	275	1385
Shomokal	531	400	391	160	1513
SylhetNews24	67	43	43	25	98
Sylhetsangbad	38	11	26	12	46

Table 2 Keywords for each incident category

Category Name	Keywords
Mugging	ছিনিয়ে, ছিনতাই
Robbery	ডাকাত
Road Accident	সড়ক দুর্ঘটনা, মুখোমুখি সংঘর্ষ
Sexual Harassment	যৌন, হয়রানি, নিপীড়ন, শ্লীলতাহানি, ইভটিজিং, ইভ টিজিং, ইভ-টিজিং, উত্যক্ত করে
Theft	চুরি

4.2 *Newspaper Data Processing*

In order to collect newspaper data, web crawlers were programmed to systematically browse news sites and to extract the news articles based on keywords appearing in the text. Each incident type had its own set of keywords (Table 2).

4.3 *Filtering News Articles*

To clean up the noisy data not relevant to this study, each article was checked against a certain set of conditions. If the conditions were met, it was added to the refined list of articles. The conditions are as follows:

1. The article must describe one of the incidents listed above.
2. The incidents must have a specific date or must have been occurring frequently around the time the article was published.
3. The incident location must be mentioned in the article, and it must be in Sylhet Sadar Upazila, Bangladesh.

4.4 Extraction of Temporal Information from News Articles

The published date (year/month/day) of a news article is extracted by the crawler. We posit that this date most closely matches with the incident occurrence. Time of the incident is determined by manually reading the news article.

4.5 Extraction of Spatial Information from News Articles

The next challenging task was to extract spatial information from the articles. We observed that an article may have a location stated in the title/headline. In such cases, the location mentioned is highly likely to be associated with the incident. In the article text, location words relevant to the incident tend to be in close proximity with the keywords. We also noticed that in Bangla language, a location word may have suffixes attached to it but never prefixes. After taking these observations into account, we listed 112 places in Sylhet as we limit our work to Sylhet Sadar Upazila and devised the following algorithm to extract the location information.

Algorithm 1. Extraction of location information from news articles

1	Match all known locations to the prefixes of each word in the headline.
2	*IF* a match is found:
3	Tag the article with the matched location.
4	*GOTO* line 14.
5	*ELSE* scan the body of the article for keywords
6	*FOR* each keyword match, *DO*:
7	Match all known locations with the prefixes of each word in sentence
8	*IF* a match is found, add matched location to the tagged location list
9	*ELSE* scan up to N neighboring sentences forward and backward.
10	*FOR* each neighboring sentence s, *DO*:
11	*IF* s has been scanned before, *CONTINUE*.
12	*ELSE IF* location match within s is found:
13	Add the matched location to the tagged area list for the article.
14	*RETURN*

The value of N is the limit for scanning neighboring sentences back and forth and is related to optimization. If a large value of N is chosen, more neighboring sentences are scanned, and there is a greater chance of finding an area match. However, a large N may also result in inclusion of irrelevant and erroneous area data in the location information. We used 2 for value of N. A total of 73 articles were found for Sylhet

Sadar Upazila, out of which 32 articles contained location names that were extracted by the algorithm.

5 System Development and Functionality of Sachetan Mobile App

As the motto of this application is crowdsourcing, maximum user contribution has to be ensured in order for the system to be self-sustaining. We have focused on creating a user-friendly system that encourages the user to provide data. We added our collection of news articles to the application as the initial data source. The users can benefit from these and add their own reports, thus aiding the growth and reliability of the system. We have included the following features:

- User registration and authentication: Old user can sign in using e-mail and password. A new user can create an account from sign up page and proceed to sign in.
- Heat-map display of incidents: After signing in, a heat map including the recent locations of incident activity nearby a user will be displayed as seen in Fig. 1a. Areas where incidents are frequent are colored.
- Creating a report: After clicking the New Report ("+" icon) button, user can create a report by providing information like incident location, incident type, time of incident, severity, and an optional short description regarding the incident as displayed in Fig. 2a.
- Filter options: At start-up, the heat map is generated with all types of incidents informed by both Sachetan app report and newspaper article. User can toggle between them. Type of incidents and the data source can be filtered by the user.
- Indicating a route with danger points from current place to destination: User can click on the map and see the default route of the destination from his current position with unsafe areas marked in red as seen in Fig. 1b.
- Statistical reports: User can see the reports of a specific area along with statistical data and visualizations that provide key insight as displayed in Fig. 2b.
- User profile: User can view, edit, and delete all reports submitted by self. The top-right button of the page can be used for signing out.

6 Statistical Analysis and Assumption

Apart from newspaper article data, a total of 379 user reports were collected, most of which were submitted by public and private university students in Sylhet Sadar Upazila. Following the collection of data, we attempted to find patterns through data analysis (Table 3).

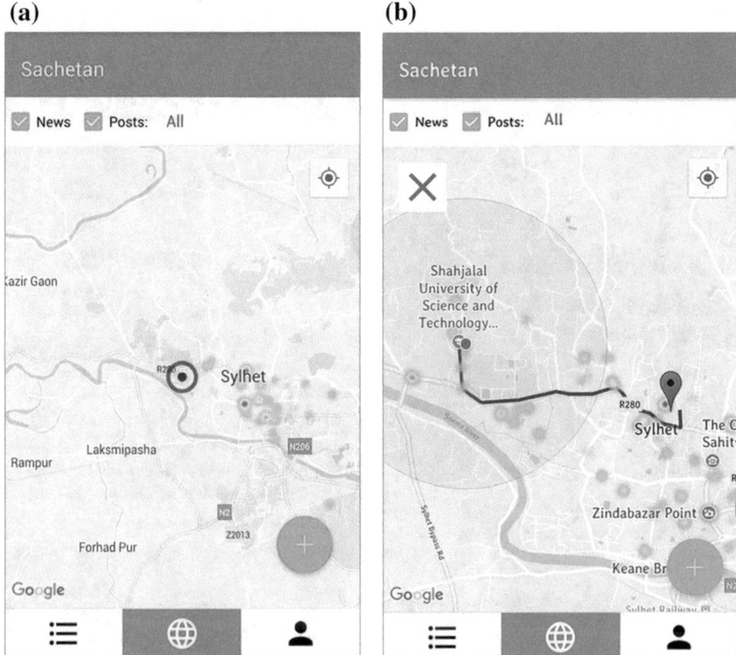

Fig. 1 **a** Heat-map screen displaying all the incidents in user surrounding area. **b** Showing a route with danger point from current place to destination

Table 3 Number of user reports for each category

Incident category	Filtered newspaper reports	User reports
Mugging	15	135
Road accident	5	113
Robbery	4	29
Sexual harassment	1	18
Theft	7	84

We only considered the years from 2014 to 2017 inclusive for our analysis, each of which contained at least 25 or more reports.

6.1 *Time-Based Analysis*

We divided the hours of the day into 12 slots with 2 h in a slot and grouped the incident reports based on their time of occurrence. We are interested in knowing if most of the occurrences of an incident type appear in or around a particular time slot. Certain incident types exhibited notable patterns.

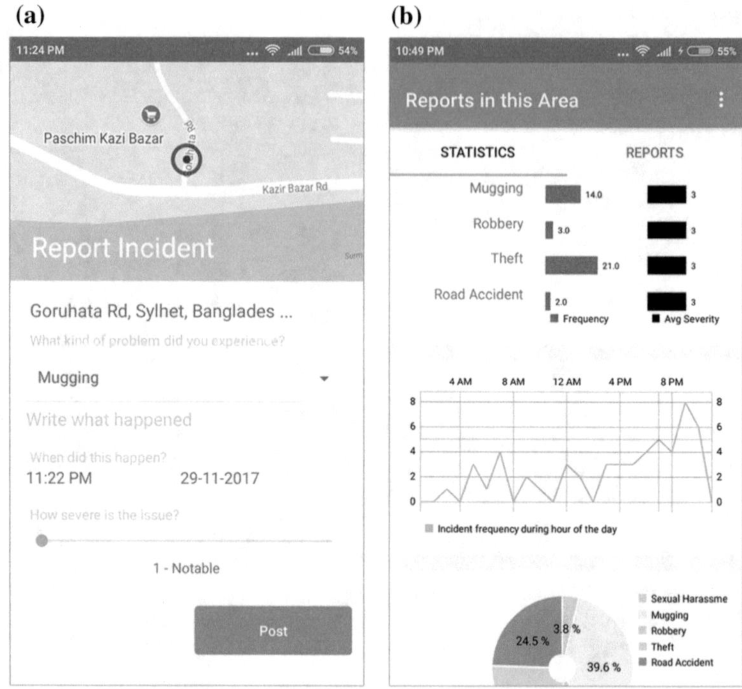

Fig. 2 **a** Creating a new report with location, description, time, and severity. **b** Showing a route with danger point from current place to destination

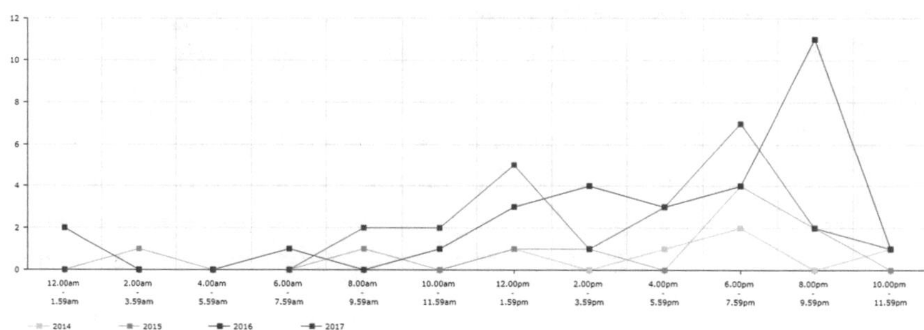

Fig. 3 Comparison of time for mugging from 2014 to 2017

The projection of mugging and theft incident frequencies for 4 years suggests that they occur more frequently within 6:00 pm and 10:00 pm (Figs. 3 and 4). It can also be noted that mugging incidents observed a spike during the last 2 years while theft rates have only increased slightly.

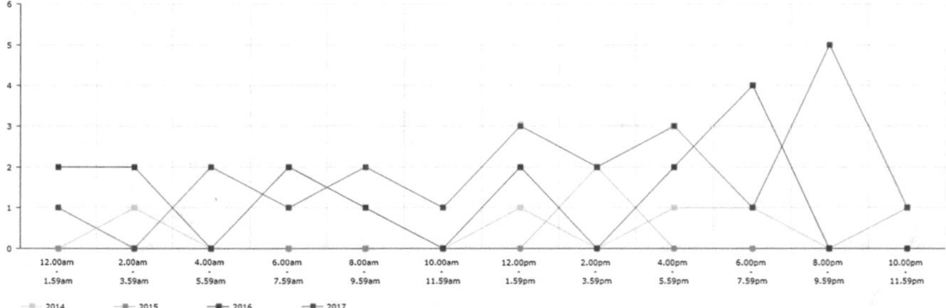

Fig. 4 Comparison of time for theft from 2014 to 2017

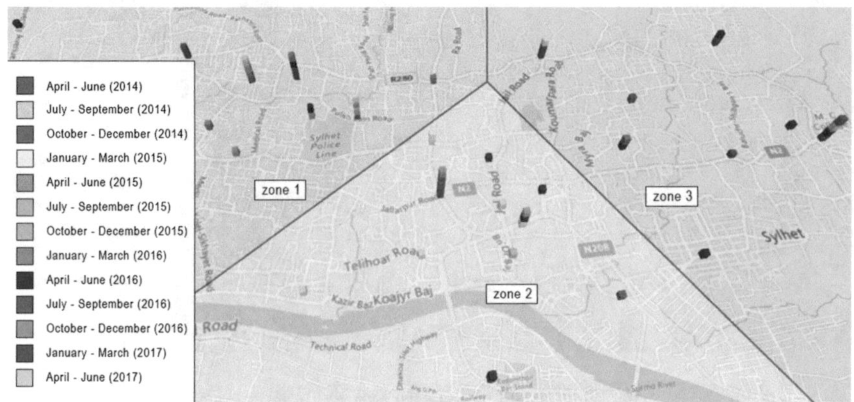

Fig. 5 Full overview of mugging locations according to user reports along with zones

6.2 *Location-Based Analysis*

Frequencies of different incident types were projected on a 3D map using latitude and longitude of the locations. We grouped multiple neighboring locations to zones and analyzed in different zones to find patterns or movements of these incidents.

Data ranging from April 2014 to June 2017 were placed in three month-sized groups to get a better understanding of the threat level for each location for each incident type (Fig. 5).

It follows from the data that Mirer Moydan is the most risk-prone area, with Subidbazar being second in the list (Fig. 6).

The data gathered on road accidents suggest that in Ambarkhana, road accidents are more recurring than the other locations (Fig. 7).

In case of theft, we noticed patterns in second zone. Here, Bandarbazar is the most frequent location for theft (Fig. 8). Its surroundings areas like Rikabibazar, Dariapara, Zindabazar, and Noyasorok are also affected individually.

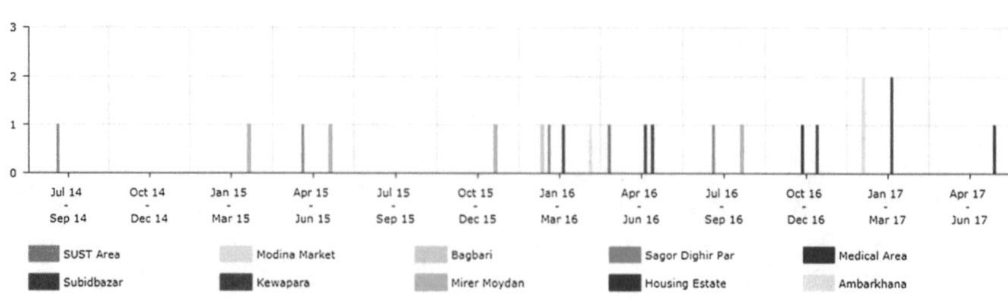

Fig. 6 Movement of mugging in first (SUST–Ambarkhana) zone

Fig. 7 Full overview of locations where road accidents took place

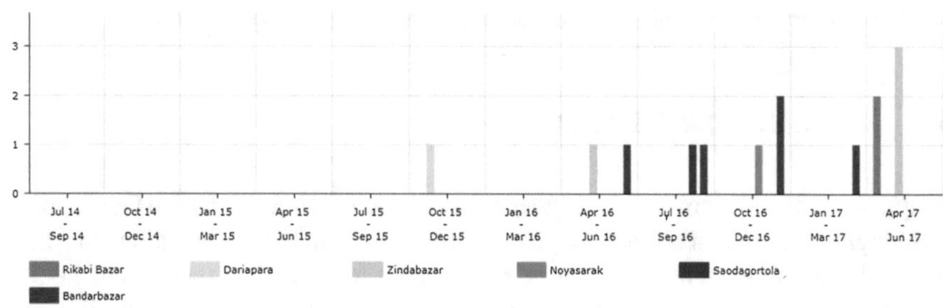

Fig. 8 Movement of theft in second (Zindabazar) zone

7 Conclusion and Future Work

If the users find the information provided through the platform helpful, they may be able to better avoid locations where incidents are more likely to occur or take necessary precautions when traveling through one of these locations. With our prediction, data collected over a longer period of time will help the authorities to pay attention to certain factors across different geographical regions and prioritize development strategies.

For future work, news article collection and filtering can be done in a fully automated manner with the help of more accurate filtering algorithms and language processing tools. User report contents can also be inspected and filtered. It can be investigated whether a correlation exists between user reports and news reports, which may shed light upon the validity of both data sources.

Acknowledgements This work was done at the Innovation Lab, Department of CSE, SUST. We would like to thank all staff, faculty, and anyone who helped us take our research forward.
Informed consent was obtained from all individual participants included in the study.

References

1. Bowman, C.G.: Street harassment and the informal ghettoization of women. Harvard Law Rev. **106**(3), 517–580 (1993)
2. SOS Response. https://globalnews.ca/news/540110/app-aims-to-increase-safety-for-b-c-women-at-risk-from-domestic-violence
3. Not Your Baby App. http://www.gender-focus.com/2013/05/18/new-apps-tackle-dating-violence-street-harassment
4. Cicleof6 website: http://www.circleof6app.com
5. Safetipin App. http://safetipin.com
6. Yarrabothu, R.S., Thota, B.: Abhaya: an Android App for the safety of women. In: 2015 Annual IEEE India Conference
7. OnWatch. http://www.onwatchoncampus.com
8. Vairaperumal, N., Saravannan, P., Kalpana, S.: SheSecure Safety App–The Hexa Umbili-cal Cord
9. Brantingham, P.L., Brantingham, P.J.: A theoretical model of crime hot spot generation. Studies Crime Crime Prev. 7–26 (1999)
10. Bendler, J., Brandt, T, Wagner, S., Neumann, D.: Investigating crime-to-twitter relationships in urban environments—faciliting a virtual neighborhood watch (2014)
11. Daily Crowdsource. https://dailycrowdsource.com/training/crowdsourcing/what-is-crowdsourcing
12. Hollaback! Together We Have the Power to End Harassment. https://www.ihollaback.org
13. Ahmed, S.I., Jackson, S.J., Ahmed, N., Ferdous, H.S., Rifat, Md.R., Rizvi, A.S.M., Ahmed, S., Mansur, R. S.: Protibadi: a platform for fighting sexual harassment in urban Bangladesh (2014)
14. HarassMap—Stop sexual harassment. http://harassmap.org/en/what-we-do/the-map/
15. Ali, M.E., Rishta, S.B., Ansari, L., Hashem, T., Khan, A.I.: SafeStreet: empowering women against street harassment using a privacy-aware location based application (2015)
16. Prothom Alo. http://www.prothom-alo.com
17. Bhorer Kagoj. http://www.bhorerkagoj.net
18. Samakal. http://samakal.com
19. SylhetNews24. http://www.sylhetnews24.com
20. Daily Sylhet er Songbad. http://www.sylhetersongbad.com

Cyberbullying and Self-harm in Adolescence, an Assessment of Detection, Prevention, and Recovery from Forum

A. Sandrine Pricilla, I. Anandhi and J. I. Sheeba

Abstract Cyberbullying is a form of bullying or harassing which is done by sending aggressive messages to a person. It can be a humiliating photo or video, an unpleasant text message, or an intolerable comment on social media. Cyberbullying has been divided into two types: direct and indirect cyberbullying. Direct cyberbullying is sending harmful material in the form of text, image, etc., to torture the victim directly. Indirect cyberbullying is when someone spreads the rumor about another person's life. In the existing techniques of direct and indirect cyberbullying, identification of user character (IUC) and identification of trustworthiness of the user (ITU) are detected using classifier techniques from the Twitter dataset. This proposed technique is going to detect cyberbullying from direct and indirect cyberbullying techniques and to prevent the cyberbullying by reflective messages and also to recover the victims using depression technique. Along with the existing technique, this proposed technique is going to detect depression from the cyberbully affected persons in order to recover from their depressive mood. The depression can be detected by using the following techniques: cognitive distortions, depressive rumination, content valence, and attitude toward others. This proposed technique also reduces the cyber harassment by showing the reflective messages to the harassing persons. The reflective message helps the users to reconsider, and it prevents them from posting harmful messages. The reflective message is going to detect from the following techniques: the potential audience, disapproval by bystanders, and harm for the receiver.

Keywords Direct cyberbullying · Indirect cyberbullying · Credibility
Depression · Reflective message

A. Sandrine Pricilla (✉) · I. Anandhi · J. I. Sheeba
Department of Computer Science and Engineering, Pondicherry Engineering College, Pondicherry 605014, India
e-mail: sandrinepricilla@gmail.com

I. Anandhi
e-mail: anandhi23011997@gmail.com

J. I. Sheeba
e-mail: sheeba@pec.edu

A. Abraham et al. (eds.), *Emerging Technologies in Data Mining and Information Security*, Advances in Intelligent Systems and Computing 813,
https://doi.org/10.1007/978-981-13-1498-8_44

1 Introduction

In the social networking sites, people interact, share, discuss, and disseminate knowledge for the benefit of others with the help of text messages, images, audios, and videos. Social networks are used by people for their benefits through sharing information but it has disadvantages too. Social networks are misused by some users for the purpose of their own enjoyment, and they harass, harm, abuse, and insult the other users. This harmful activity is known as bullying. Usually, bullying takes place between two persons or group. It is done by hurting another person physically or by verbally hurting them [1].

Cyberbullying is a harmful act which engages in sending or posting of the harmful material with the help of the Internet or any other digital technology. It has become increasingly common, especially among teenagers. Harmful bullying behavior includes posting rumors about another person, threats, or disclose of victims' personal information. Bullying or harassment can be identified by repeated behavior and intent to harm.

The victims affected by cyberbullying experience lower self-esteem, frustration, anger, and depression. Cyberbullying messages are categorized into two types. One is direct cyberbullying and another one is indirect cyberbullying. In direct cyberbullying, individuals are directly threatened by sending them abusing messages or sending inappropriate videos. Indirect cyberbullying is posting an embarrassing photo of an individual on social networking sites, which can be seen by anybody, posting a rude or abusive email about others to your contacts, spreading rumors and lies about individuals, and bullying other persons who try to interact with the affected person [1].

Reduction of cyber harassment is necessary because many health issues have been diagnosed among victims like loneliness, suicidal behavior, anxiety, and depressive symptoms. Cyberbullying is reduced by using the reflective messages, which help the users to rethink, reconsider, and prevent them from posting harmful messages [2]. By doing this, the cyberbullying can be reduced to a certain extent, but it cannot be reduced completely. So, it is necessary to recover the victims who are affected by cyberbullying by sending them consoling messages using the depression technique [3].

The objective of this paper is to detect cyberbully words (such as insulting, threatening, bad, vulgar, terrorism words) in the crime investigation forum, to prevent the users from posting cyberbullying posts using the reflective messages and to recover the victims from the depressed state using the depression technique. This paper is organized as follows. Related works are discussed in Sect. 2. Then, the designed machine learning model for cyberbully detection is presented in Sect. 3. In Sect. 3.1, the input training dataset real-time forum corpus and metrics of the proposed techniques are deliberated. In Sect. 4, concluding interpretations are given.

2 Related Works

J. I. Sheeba and B. Sri Nandhini proposed a model that focuses on detecting the cyberbullying activity in the social networking website using the Levenshtein algorithm [4]. J. I. Sheeba and B. Sri Nandhini proposed a model that focuses on detecting the cyberbullying activity in the social networking website using the fuzzy logic [5]. J. I. Sheeba and K. Vivekanandan proposed a model that focuses on the extraction of cyberbully polarity from the forum with the techniques of preprocessing, frequency calculation, noun extraction, and fuzzy C-means algorithm. The results obtained are reduced using fuzzy decision tree methods and fuzzy rules and are evaluated by the existing techniques such as naïve Bayes classifier and Mamdani fuzzy inference system [6]. A. Saravanaraj et al. proposed a model to detect rumors in the Twitter dataset. They used Twitter speech act technique, naïve Bayes classifier to detect the rumors [7]. Nafsika Antoniadou et al. proposed a model to examine the feasible common conventional and cyberbullying participants' distinct individualities that arise as the most predominant in the study of both singularities (i.e., behavior, social relationships, and social skills), as well as those that may differentiate contribution in cyberbullying (i.e., online disinhibition and internet use) [8]. Michele Di capua et al. proposed a model which facilitates the automatic detection of the traces of bullying in the social network with the help of machine learning and techniques of natural language processing [9]. Rui Zhao et al. proposed a representation, learning framework specifically for cyberbullying detection. Unique weights are assigned to words to attain the bullying attributes, and then these are combined with the bag-of-words and latent semantic features to make the final representation. Here, based on word embedding, a list of predefined insulting words is given which made the system precise [10]. Hamed Abdelhaq et al. described the keyword extraction for localized events on Twitter. The algorithmic steps involve building keyword geo-reference map, locality check, focus estimation, and topical to local keyword recovery. The work was especially to get local keywords, i.e., busty words, which contain restricted spatial range [11]. Juan Antonio Lossio-Ventura1 et al. proposed a methodology that provides many methods based on graphic, web aspects, linguistic, and statistic. For the automatic term extraction, these methods take out and rank the candidate terms with accurate outcomes and operate with different languages. The system lacks in frequency information with linguistic pattern [12]. F. Dornika et al. proposed a model for the purpose of classification and recognition tasks. They used an embedding method which is based on graph. This method was used to extract the nonlinear feature with the help of kernel tricks [13]. Sheeba J.I et al. proposed a model which is helpful in detecting the cyberbully content in social network, which is done by using Gen Lenen algorithm, and the cyberbully content which is detected is then classified with the help of fuzzy rule base [14]. Geetika Sarna and M. P. S. Bhatia proposed a model to identify the direct and indirect cyberbullying and also to control them. The credibility of the user is also found [1].

Yaakov Ophir et al. proposed a model to detect whether the depressive and non-depressive Facebook status updates differ in linguistic or stylistic aspects. In the bottom-up approach, the specific hypotheses as to which particular aspects would distinguish between the depressive and non-depressive postings are not given [3]. Kathleen Van Royen et al. proposed a model to examine how the reflective messages can be used to reduce the harassment on the social networking site. They tested with three types of reflective messages: parents as the potential audience, disapproval by bystanders, and harm to the receiver [2]. Niall McCrae et al. proposed a model to examine the relationship between the social media use and depressive symptoms in the child and adolescent population [15]. Matthew C. Morris et al. discussed the childhood sexual abuse and the prevention measures for that. But they have neither used any technique nor implemented any modules for the prevention [16]. Felix E. Eboibi discussed the cyber crimes in Nigeria and the Acts introduced by the government on cybercrime in the Nigerian polity [17]. Tata Prathyusha et al. proposed a model to identify the cyberbully words in the crime investigation forum using hybrid techniques [18]. Most of the existing systems deal only with the detection of cyberbullying words. The proper implementation techniques are not available for prevention and recovery from the cyberbullying words. To overcome the above problems, in the proposed system, the cyberbullying words are detected using the classifiers and prevented by using the reflective messages, and the depression techniques are used to recover the victims in a single system itself.

3 Proposed Framework

In Fig. 1, the proposed framework, which is given above, clearly explains the proposed system. The input dataset is the posts from www.4forums.com. The input dataset is then preprocessed to improve the quality of the data. The output of data preprocessing is sent to feature extraction which extracts important features from the preprocessed data. The output of feature extraction is then classified using classifiers, and user's character and trustworthiness of the user are found with the help of fuzzy logic. If a user is found to be harmful, a warning message will be sent to him using the reflective message module. After showing the reflective message if the user is found to depressed, it can be detected and consoled by using depression module.

3.1 Module Description

In order to detect, prevent, and recover the victims of cyberbullying, the following steps are carried out in the proposed framework:

3.1.1 Data Preprocessing
3.1.2 Feature Extraction
3.1.3 Categorization of Messages
3.1.4 Identification of User Character (IUC)

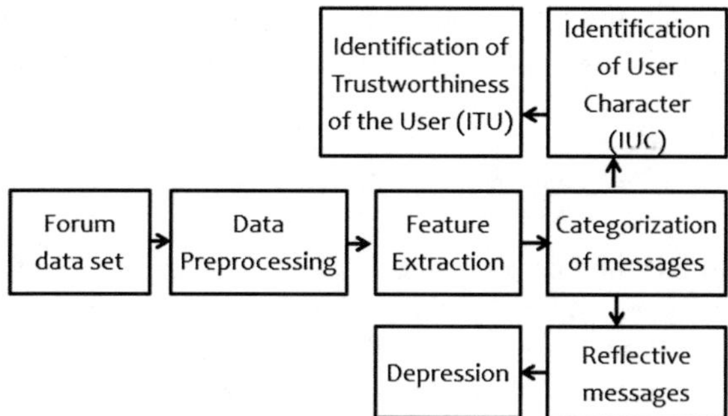

Fig. 1 Proposed framework

3.1.5 Identification of Trustworthiness of the User (ITU)
3.1.6 Reflective Messages
3.1.7 Depression

3.1.1 Data Preprocessing

Data preprocessing methods are required to reduce the huge set of irrelevant and redundant data from the forum data. A forum (newsgroup) is a place where people gather to discuss various topics and subjects. People can also discuss a common topic and share information with one another. The forum described here is www.4forums.com. This forum contains around 30 topics and five formal debates (such as techniques, styles, tournaments, open debate, and comments). Particularly, crime debate forum dataset contains 198 threads and 7621 posts.

Their application in this domain is aimed at determining, which data in the complete dataset is the most important for describing the faulty use of language in the forum. The data preprocessing is executed using the process of stop word removal.

After or before processing of natural language data, stop words are removed. Some of the examples of stop words are as follows: is, the, at, which, and on. Stop word removal is important to reduce the data redundancy and irrelevancy. The output from the data preprocessing is given as the input to the feature extraction [1].

3.1.2 Feature Extraction

Feature extraction is the process of reducing the input data into a set of features. The feature extraction process is handled using negative emotional words, positive emotional words, and proper nouns.

Negative emotions indicate bullying behavior in the messages and are used to find direct cyberbullying.

Positive emotions indicate that even though the message contains bad words, it need not be a bullying message and is used to find non-bullying messages. A proper noun is used to find indirect cyberbullying. The output from the feature extraction is given as the input to the categorization of messages [1].

3.1.3 Categorization of Messages

In this module, the inputs will be classified as direct cyberbullying, indirect cyberbullying, and no bullying using the machine learning methods such as naïve Bayes classifier, K-nearest neighbor (KNN) classifier, decision trees, and support vector machines (SVM). In direct cyberbullying, the abusing material is in the form of text, image, etc., to torture the victim directly. Indirect cyberbullying is when someone spreads the rumor about another person's life. If the input does not contain any bullying words, then it is termed as no bullying. The messages will be classified using the Weka tool. The output from the categorization of messages is given as the input to the identification of user character in order to find the behavior of the user [1].

3.1.4 Identification of User Character (IUC)

The behavior of the user is categorized into normal and abnormal behavior. The number of bullying and non-bullying words is used to find the behavior of the user. These categories of behavior depend upon the following factors:

1. The total number of non-bullying words sent by user y for words NB (y, w) indicates the normal behavior.
2. The total number of direct bullying DB (y, w) and indirect bullying words IDB (y, w) sent by user y for words w indicates the abnormal behavior.

The probability of the normal and abnormal behavior will be found from the number of words. Here, the abnormal behavior is categorized into direct and indirect bullying. From the following equations, the probability of the normal and abnormal behavior of the users will be found [1].

Probability of normal behavior of user y for the number of words w can be found by using Equation (1)

$$P\ (Nor(y,\ w)) = NB(y,\ w) \div (\ NB(y,w) + DB(y,w) + IDB(y,w)\ ). \qquad (1)$$

Probability of direct bullying of user y for the number of words w can be found by using Eq. (2)

$$P\ (DBull(y,\ w)) = DB(y,\ w) \div (\ NB(y,w) + DB(y,w) + IDB(y,w)\ ). \qquad (2)$$

Probability of indirect bullying of user y for the number of words w can be found by using Eq. (3)

$$P\ (InBull(y,w)) = IDB(y,\ w) \div (\ NB(y,w) + DB(y,w) + IDB(y,w)\ ). \qquad (3)$$

3.1.5 Identification of Trustworthiness of the User (ITU)

ITU is used to analyze the credibility of the user. It depends on the output of IUC module. The credibility of the user is found from the probability of the normal and abnormal behavior using fuzzy logic. For example, the credibility of the user is found by the given rule format.

If P (Nor (y, w)) $\geq$0.5 and P (DBull(y, w)) $\leq$ 0.4, and P (InBull(y, w)) $\leq$ 0.2), then credibility = Normal user.

Since the input is dynamic, the threshold value depends on the probability value. The threshold value of indirect bullying is set to 0.2, i.e., if P (InBull(y, w)) exceeds 0.2, risk starts, and if the threshold exceeds 0.5, then the user becomes more harmful. Similarly, the threshold value of direct bullying is set to 0.4, i.e., if P (DBull(y, w)) exceeds 0.4, risk starts. If the threshold value exceeds 0.8, then the user becomes more harmful. The threshold for normal behavior is set to 0.5, i.e., if P (Nor(y, w)) exceeds 0.5, no risk or else risk starts. From this, the credibility of the user is found [1].

3.1.6 Reflective Message

The output from the categorization of messages is given as the input to the reflective message.

The reflective messages are used to reduce the intention of the users to harass in the social networking site. The reflective message is the message which encourages the users to rethink, and it prevents them from posting harmful messages. There are three types of reflective messages.

- **Parents as potential audience**
  For youngsters, a message notifying them that the message which they are about to post could be seen by their parents can be a cause for withholding harmful message.
- **Disapproval by bystanders**
  For cyberbullying, observing a negative social burden from significant others will decrease the intention to bully.
- **Harm for the receiver**
  The objective to involve in harassment on social network sites will be decreased after revealing a message denoting the effect for the receiver [2].

Based on the above techniques, the cyberbully is reduced.

3.1.7 Depression

Depression is a state of low mood and it affects one's thoughts, behavior, feelings, and sense of well-being. The output from the reflective message is given as the input to this module. The depression can be detected by using the following techniques:

cognitive distortions, depressive rumination, content valence, and attitude toward others.

- **Cognitive Distortions**
  Cognitive distortions reflect a general cognitive bias. Cognitive distortions common in depressive disorder are overgeneralization, selective thinking, arbitrary inferences, and personalizing.
  Overgeneralization: Drawing sweeping conclusions based on a single incident.
  Selective thinking: Focusing on negative aspects while ignoring positive ones.
  Arbitrary inferences: Drawing negative conclusions without sufficient evidence.
  Personalizing: Blaming oneself for unpleasant things.
- **Depressive Rumination**
  A depressive rumination is a form of self-focus in which people engage in repetitive and negative thoughts about their depressive symptoms.
- **Content Valence**
  Depressive status updates were mostly negative and contain curses whereas non-depressive status updates were mostly positive in content.
- **Attitude Toward Others**
  Depressive status includes negative references to other people [3].

Based on the above techniques, the victims affected by cyberbullying are recovered from the depressed state.

4 Conclusion

The proposed system deals with the detection of cyberbullying words in the crime investigation forums. Cyberbullying words can be easily detected by this framework. The cyberbullying words are identified using the features such as positive emotional words, negative emotional words, and proper noun. The messages are categorized by using the classifiers such as naïve Bayes classifier, K-nearest neighbor (KNN) classifier, decision trees, and support vector machines, and this framework prevents the user from posting bullying messages by showing the reflective messages, and the victims who are in the depressed state are also recovered in this proposed framework.

References

1. Sarna, G., Bhatia, M.P.S.: Content based approach to find the credibility of user in social networks: an application of cyberbullying. Inter. J. Mach. Learn. Cybern., Springer **8**, 677–689, 682–685 (2017)
2. Van Royen, K., Poels, K., Vandebosch, H., Adam, P.: Thinking before posting? Reducing cyber harassment on social networking sites through a reflective message. Comput. Human Behav. **66**, 345–352, Elsevier, 345–350 (2017)

3. Ophir, Y., Asterhan, C.S.C., Schwarz, B.B.: Unfolding the notes from the walls: adolescents' depression manifestations on Facebook. Comput. Human Behav. **72**, 96–107, Elsevier, 101–103 (2017)
4. Sri Nandhini, B., Sheeba, J.I.: Cyberbullying detection and classification using information retrieval algorithm. In: ICARCSET '15, ACM, pp. 1–5 (2015)
5. Sri Nandhini, B., Sheeba, J.I.: Online social network bullying detection using intelligence techniques. Procedia Comput. Sci. **45**, 485–492 (2015), Elsevier, 1–8
6. Sheeba, J.I., Vivekanandan, K.: Detection of online social cruelty attack from forums. Int. J. Data Min. Emerg. Technol. 1–11 (2015). https://doi.org/10.5958/2249-3220.2014.00003.2, www.IndianJournals.com
7. Saravanaraj, A., Sheeba, J.I., Pradeep Devaneyan, S.: Automatic detection of cyberbullying from Twitter. IRACST - Int. J. Comput. Sci. Inf. Technol. Secur. (IJCSITS) **6**(6), 1–6 (2016). ISSN: 2249-9555
8. Antoniadou, N., Kokkinos, C.M., Markos, A.: Possible common correlates between bullying and cyber-bullying among adolescents. Psicologia Educativa, Elsevier **22**(27–38), 1–12 (2016)
9. Di Capua, M., Di Nardo, E., Petrosino, A.: Unsupervised cyber bullying detection in social networks. In: 2016 23rd International Conference on Pattern Recognition (ICPR), IEEE, pp. 1–6 (2016)
10. Zhao, R., Zhou, A., Mao, K.: Automatic detection of cyberbullying on social networks based on bullying features. In: 17th International Conference on Distributed Computing and Networking, ACM, p. 43 (2016)
11. Abdelhaq, H., Gertz, M., Armiti, A.: Efficient online extraction of keywords for localized events in twitter. In: GeoInformatica, pp. 1–24 (2016)
12. Lossio-Ventura, J.A., Jonquet, C., Roche, M., Teisseire, M.: Biomedical term extraction: overview and a new methodology. Inf. Retr. J. **19**(1–2), 59–99 (2016)
13. Dornaika, F., El Traboulsi, Y., Assoum, A.: Inductive and flexible feature extraction for semi-supervised pattern categorization. Pattern Recognit. **60**, 275–285 (2016)
14. Sheeba, J.I., Devaneyan, S.P.: Cyberbully detection using intelligent techniques. Int. J. Data Min. Emerg. Technol. **6**(2), 86–94 (2016)
15. McCrae, N., Gettings, S., Purssell, E.: Social media and depressive symptoms in childhood and adolescence: a systematic review. Adolesc. Res. Rev., Springer 1–16 (2017)
16. Morris, M.C., Kouros, C.D., Janecek, K., Freeman, R., Mielock, A., Garber, J.: Community-level moderators of a school-based childhood sexual assault prevention program. Child Abuse Neglect, Elsevier **63**, 295–306, 1–12 (2017)
17. Eboibi, F.E.: A review of the legal and regulatory frameworks of Nigerian Cybercrimes Act 2015. Comput. Law Secur. Rev., Elsevier 1–18 (2017)
18. Prathyusha, T., Hemavathy, R., Sheeba, J.I.: Cyberbully detection using hybrid techniques. In: International Conference on Telecommunication, Power Analysis and Computing Techniques (ICTPACT 2017). IEEE, pp. 1–6 (2017)

Visual Verification and Analysis of Outliers Using Optimal Outlier Detection Result by Choosing Proper Algorithm and Parameter

Bilkis Jamal Ferdosi and Muhammad Masud Tarek

Abstract Outlier detection is a nontrivial but important task for many of the application areas. There exist several methods in literature to find outliers. However, there is no single method that outperforms in all cases. Thus, finding proper algorithm and the value of its relevant parameter is crucial. In addition, none of the methods are perfect and verification by the domain experts can confirm if the outliers detected are meaningful or not. Proper visual representation of the detected outliers may help experts to resolve anomalies. In this paper, we proposed a visual analytic system that finds proper algorithm and value of its relevant parameter for a specific dataset using training set. Later, the chosen method is applied to the test data to obtain the outlier ranking. After that, data points are visualized with parallel coordinate plot (PCP) where colors of the lines are obtained by using the outlier factor of the data points. PCP is one of the popular high-dimensional data visualization techniques where coordinates are parallel to each other and each data point is represented by a line. Using the visual, experts can provide feedback and update the result. Experiments with different datasets ensure the strength of our system.

Keywords Outlier detection · Parallel coordinates · Human-centered computing
Visual analytics

1 Introduction

Outliers are data points that deviate substantially from the inliers. Inliers are those data points which are formed using some common processes whereas the outliers are the result of a violation of those common processes. Hence, the detection of outliers

B. J. Ferdosi (✉)
Department of CSE, University of Asia Pacific, Dhaka 1215, Bangladesh
e-mail: b.j.ferdosi@uap-bd.edu

M. M. Tarek
Department of CSE, State University of Bangladesh, Dhaka 1205, Bangladesh
e-mail: tarek@sub.edu.bd

A. Abraham et al. (eds.), *Emerging Technologies in Data Mining and Information Security*, Advances in Intelligent Systems and Computing 813,
https://doi.org/10.1007/978-981-13-1498-8_45

becomes crucial in many application areas such as fraud detection in financial data, identifying distant quasars (which are extremely rare in the universe) in astronomical data, etc.

There are several methods exist in the literature for finding outliers in full-dimensional space [1, 2] and also in subspaces [3–8]. They use different techniques such as distance-based [1, 9–12], density-based [2, 13–15], or distribution-based [16] measurements. They also vary in terms of producing outputs; few produce results by marking data points as outliers or inliers, whereas others give results in a degree of outlier-ness of each data point.

Campos et al. [17] analyzed the performance of several unsupervised distance-based outlier detection algorithms and proposed a guideline to evaluate such methods. They used precision at n (P@n) [18], average precision (AP) [19], and receiver operating characteristic (ROC) AUC for measuring the performances of the methods, where ROC AUC score is used as the primary method of performance measurement. ROC AUC curve is obtained by plotting true positive rate versus false positive rate and the ROC AUC score is measured as the area under the curve. In their paper, authors evaluated kNN [10], kNN-weight (kNNW) [12, 20], local outlier factor (LOF) [2], simplified LOF [21], local outlier probabilities (LoOP) [22], local distance-based outlier factor (LDOF) [23], outlier detection using indegree number (ODIN) [24], KDEOS [25], connectivity-based outlier factor (COF) [13], fast angle-based outlier detection (FastABOD) [26], local density factor (LDF) [27], and influenced outlier-ness (INFLO) [14] on several datasets. However, according to the performance reported in the paper on annthyroid and heart disease datasets, LDOF and FastABOD methods obtained the best result for annthyroid and heart disease datasets, respectively. However, none of the methods have ROC AUC score larger than 0.8. Thus, the methods reported several outliers with low outlier factor and inliers with high outlier factor. In Fig. 1, cardiotocography dataset is visualized with PCP [28], where along with the attributes of the dataset, outlier factor (LOF-outlier factor) of each data point obtained with LOF method and the labeling of the data point (outlier) are also plotted as axes. In the figure, we can see that the top-ranked data point (yellow line) is indeed an outlier, whereas the data point that got the next highest outlier factor (red line) is not labeled as an outlier. It is also evident from the plot that there are several outliers received low outlier factors. In their paper, Campos et al. also indicated that there is no single method that outperforms in all the cases, and finding proper parameter (value of k) of the algorithm for each dataset is also crucial.

Micenkov´a et al. in [29] argue that only identifying or ranking outliers is not sufficient; proper explanation of the outlier-ness of the data is necessary for proper understanding. In their work, they explained the outlier-ness by finding the relevant attributes, where a data point can be well separable from the rest of the data points. They used several well-established outliers finding algorithms combined with several feature selection methods to find out the relevant subspaces. However, finding the relevant subspaces for the false positive detection would not be very meaningful. In absence of ground truth, detecting false positive is only possible with human expert intervention. In that case, visualizing the data with sufficient visual aid may

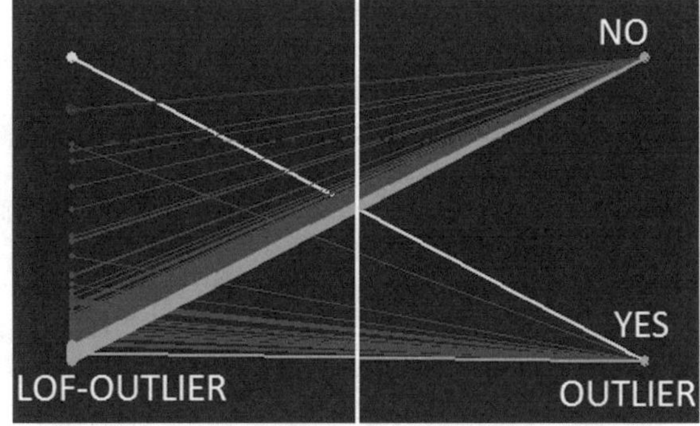

Fig. 1 Performance of LOF method on cardiotocography dataset

be helpful. In our work, we visualized the data points with colors in such a way that inliers will get less emphasize and the top-ranked outliers will have a better visibility.

In this paper, we propose a visual analytic system that includes:

- An algorithm to find proper outlier ranking algorithm among several candidates and associated parameter for a dataset.
- A visualization technique using PCP that visualizes data points with color coding such that the top-ranked outliers will have better visibility and inliers will work as a perspective with less emphasis.

We have experimented with different datasets with different levels of outlier-ness, and the results obtained ensure the effectiveness of our method.

Rest of the paper is organized as follows: In Sect. 2, we describe literature related to our work. Our proposed algorithm is stated in Sect. 3, and necessary outlier visualization technique is explained in Sect. 4. In Sect. 5, we have summarized our visual analytic system. We have a brief discussion of the datasets used in our experiments in Sect. 6. Section 7 explains the experiments and the results of our work. Conclusion and future work are summarized in Sect. 8.

2 Related Works

2.1 Outlier Detection Algorithms

In our work, we experimented with nine different algorithms: LOF method, SimplifiedLOF method, FastABOD method, COF method, intrinsic dimensionality outlier score (IDOS) [30] method, ODIN method, LDF method, kNNW outlier detection method, and subspace outlier detection (SOD) method [4]. The choices of the methods are to have variety in techniques. We have included global, local, full-dimensional, subspace, angle-based, density-based, and kernel-based outlier detection methods. Though these methods detected outliers in different ways, all of them have one thing in common; all of them utilized the notion of k-nearest neighbor distance. However, in our best algorithm finding technique, any algorithm based on k-nearest neighbor distance can be utilized.

Local outlier factor method computes a degree of outlier-ness of a point considering insolation of the point in its local neighborhood. A high value of LOF indicates outlier-ness, and the high LOF is attained when the local reachability distance (lrd) of the test point is small compared to its nearest neighbors. **Simplified LOF** method is basically the same algorithm as LOF but it uses simpler density estimates where reachability distance of LOF is replaced by kNN distance, which is faster and more efficient. **Connectivity-based outlier factor (COF) method** defines the outlier-ness of a point in a notion of "isolativity", which is measured as a degree of connectedness of the point to other objects. A low density of a point does not always imply the isolation or outlier-ness of a point. The COF method derives outlier factor in terms of average chaining distance obtained with the set-based nearest path. **Fast angle-based outlier detection (FastABOD) method** is the faster variant of the angle-based outlier detection method, which derives outlier degree of a point from the weighted variance of angles between point p and only the points in kNN, in quadratic time. The method is robust in high-dimensional data because it is independent of distances that fail in very high dimensionality. **Intrinsic dimensionality outlier score (IDOS) method** is basically an adaptation of classical LOF method where the local intrinsic dimensionality is used instead of local density estimation in LOF to obtain the outlier ranking. Intrinsic dimensionality is the number of latent variables that describe the complexity of a dataset. IDOS method utilizes the variability of intrinsic dimensionality in inliers and outliers to obtain the outlier score. **Outlier detection using indegree number (ODIN)** detects outlier using kNN graph where a vector represents a vertex in the graph, and the pointers to the neighboring vector represent the edges. If a vertex has indegree less than or equal to a certain threshold, the vertex is declared as an outlier. **Local density factor (LDF) method** is based both on statistical and variable kernel density estimation approach. Local density factor of a data point is computed as a ratio of average local density estimation (LDE) of k neighboring points and the LDE of the point itself. In **kNN-weight (kNNW) outlier detection method** unlike the kNN outlier detection method, the sum of all kNNs called as weight is used to identify the outliers that reduce the sensitivity of the outlier result on k. To compute the weight, they used a fast and efficient approach using Hilbert space-filling curve. **Subspace outlier detection (SOD) method** finds the outliers in axis parallel subspaces comparing a query point with a reference set and obtained a score of outlier-ness depending on the deviation of the point with respect to the reference set. In contrast to other methods, SOD computes outlier-ness in subspaces. Subspace outlier degree of an object is the normalized distance of the object from the subspace hyperplane defined by a set of reference point for that object. In our experiment, we used the above methods to test our algorithm; however, it is possible to incorporate other methods that have similar dependency with k.

2.2 *Outlier Visualization*

One of the oldest outlier visualization techniques is box-and-whisker plot proposed by Tuckey [31] where outliers are visualized as points further away from the median than the extremes. Though these plots summarize the statistical information efficiently, detailed behavior of outliers cannot be obtained with such visualization. The seminal work of Novotny and Hauser [32] detected outliers using binned data representation and visualized the outliers using focus+context visualization in parallel coordinates (PC). Their work is based on output-oriented approaches where 2D bin map is employed to identify the visual outliers. Whereas in our work, we proposed a data item-oriented approach keeping the idea of focus–context visualization in PC. Achtert et al. presented a visualization technique for interpreting the outlier score of each data point using histogram and bubble plot [33]. In bubble plot, visualization outlier score is scaled to the radius of the bubble centering the data point. Though it is quite intuitive to understand the outlier-ness of a point visually, such visualization does not provide any insight into how this outlier point behaves in full-dimensional space or in subspaces. Just-in-time descriptive analytics proposed by Kandogan deploys the point-based visualization technique to identify outliers [34]. They used grid-based density estimation to identify outliers only on single dimension values even though the majority of the outliers may not be obtained in one dimension.

3 Finding the Best Algorithm and Parameter

Performance of algorithms based on k-nearest neighbor distances varies with the choice of k. For some datasets, best performance can be obtained with a low value of k and for some other datasets, same algorithms may require different k values. In our experiments, we found that if we increase the value of k gradually, then after a certain value of k performance of the algorithm degrades. In addition, the performance of an algorithm varies with different datasets and different variants of a dataset. Hence, we proposed an algorithm (summarized in the pseudocode of Algorithm 1) to find the optimal outlier finding algorithm that exhibits the best performance for a specific dataset.

To measure the performance of an algorithm, a well-known approach is receiver operating characteristic (ROC) curve that provides the ratio of true positive and false positive for the top n number of points. In our work, we used ROC AUC (area under the ROC curve) that summarizes the ROC curve with a single value.

ALGORITHM 1: Finding Best Algorithm with Best K

```
DATA ←define d-dimensional dataset;
define BestKs[NrOfAlgorithms];
// value of k for which algorithm[i] exhibits best
// performance will be stored in BestKs[i]
define BestROCAUCs[NrOfAlgorithms];
// the best ROCAUC value for algorithm[i]
// will be stored in BestROCAUCs[i]
for i := 1 to NrOfAlgorithms do
      startK ←1;
      endK ←p;
      ROCnew ←0.001;  // initial values of ROCAUC
      ROCold ←0.0;
      while (ROCnew>ROCold) do
      midK← (startK+endK)/2;
      ROC1 ← Algorithm[i] (DATA, startK);
      ROC2 ←Algorithm[i] (DATA, midK);
            if (ROC1>ROC2)
                  ROCold ←ROCnew;
                  ROCnew ←ROC1;
                  kOld←kNew;
                  kNew ←startK;
                  endK←midK-1;
            else
                  ROCold ←ROCnew;
                  ROCnew ←ROC2;
                  kOld←kNew;
                  kNew ←midK;
                  startK←midK+1;
            endif
      end while
BestK[i] ←kOld;
BestROCAUC[i] ←ROCold;
end for
BestAlgorithm ←FindBestAlgo(BestROCAUC, NrOfAlgorithms);
//Finds the best algorithm with highest ROCAUC value
return BestAlgorithm;
// returns the name of the best algorithm
```

We applied the algorithms one by one on to the dataset in hand and iterate with a range of k to find the proper value of k for which the algorithm provides the best performance in terms of ROC AUC value. After we obtained the ROC AUC values for all of the outlier detecting methods (we have used the open-source implementation of the algorithms in ELKI [35] platform), we choose the algorithm with the highest ROC AUC value. In our algorithm, to find the proper k, we are influenced by the concept of binary search where after each iteration, search space is reduced to half. For a range of k ($k = 1$ to p), we compute the ROC AUC value of an algorithm for starting value of k and $mid = (startK + endK)/2$, the value of k. If the ROC AUC value of the algorithm with mid-value of k is lower than the start value, then we do not iterate for the value of k greater than the mid-value, rather we run the algorithm with the value of k in the first half.

In the seminal work of Campos et al., they tried to find the best result by iterating the value of k from 1 to 100, which is obviously very time-consuming, especially for the very large datasets. They also tried with different step sizes to reduce the number of iterations but in this way, they had the chance to miss the best k. Using our algorithm, we are reducing the number of iterations remarkably without compromising the best k. The proposed algorithm has a dependency on ROC AUC value that requires labeling of the data points. Therefore, we propose a system in Sect. 5 where the algorithm is applied to the training dataset to find the best outlier detecting algorithm, and the later chosen algorithm is applied to the test dataset.

4 Visualizing the Top-Ranked Outliers with Color

Outlier detection results obtained with the best outlier detecting algorithm chosen by our algorithm must be validated by a human expert since there is no labeling present for the test dataset. We visualize data points with outlier score in the top n% (n is a user given number) with a noticeable color whereas rest of the data points are visualized with a color which provides a base to the top-ranked outliers. In Fig. 2, we visualized WDBC dataset (each data point is labeled as "yes" if it is an outlier or "no" if it is not) where data points with outlier score in the top 1% are visualized with orange and rest of the data with gray. We can see that one of the high ranked data points (according to the outlier factor calculated by the SOD algorithm) is labeled as not outlier (circled with yellow) in the dataset. A human expert can observe the behavior of that data point among the attributes and thus may label it as an outlier or discard the result. We store the data and their color information according to the outlier factor in an XML file and invoke the GGobi [36] from our system for further analysis. The expert can change the label of a data by changing the color of the line that represents the data.

5 System for Optimal Outlier Detection

We apply our algorithm (Algorithm 1) to the labeled training dataset to find out the best algorithm. The chosen algorithm is then applied to the test dataset for finding/ranking the outliers. Obtained outlier ranking is then validated using visualization by human expert to obtain the labeling of the dataset. This test dataset, after adopting feedback from the expert, is used as a training dataset to refine the result of the proposed algorithm which in turn obtains the better outlier ranking. The system is depicted in Fig. 3.

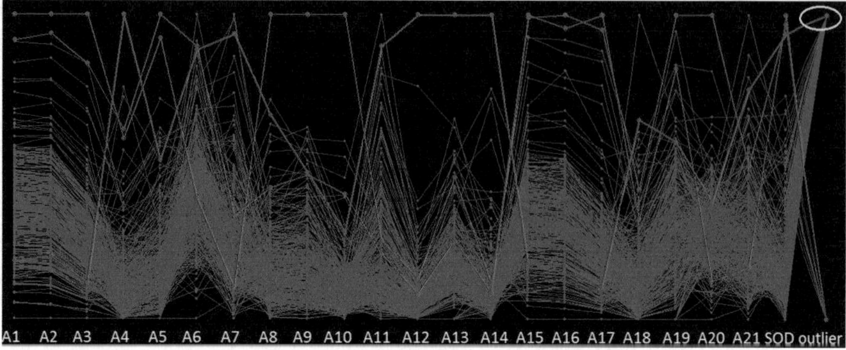

Fig. 2 Visualization of data points with outlier score in the top 1% for WDBC dataset

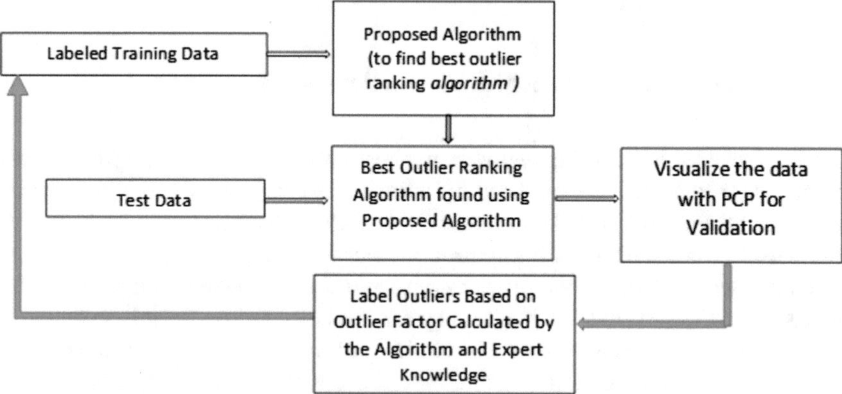

Fig. 3 Proposed system for optimal outlier detection

6 Datasets

We have used outlier datasets available in [37]. There are 12 semantically meaningful (data points with natural semantic interpretation as outliers) datasets and 11 datasets that have been used in the previous research literature. There are 1000 datasets in the repository including several variants of each of the dataset. For example, heart disease dataset has variants with 2%, 5%, 10%, 20%, and 44% of outliers; each of the variants has 10 different versions. In addition, each of the datasets has labeling of each data point as an outlier (yes) or not outlier (no). For our experiments, we separate different variants of each dataset into two groups: training and test set. The labels of the training sets are used to obtain ROC AUC value in our proposed algorithm, and the labeling of the test sets is used as experts' knowledge to validate and update the result of the chosen algorithm.

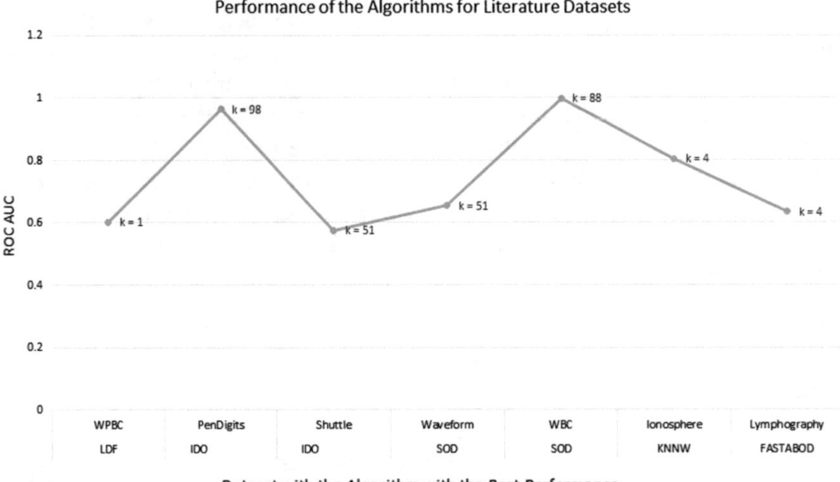

Fig. 4 Performance of the outlier detection algorithms on literature datasets

7 Experiments and Results

In Fig. 4, the performance of the algorithms on literature datasets is summarized. Here, SOD method has best performance for two datasets and for WBC dataset, it obtained the highest ROC AUC value of 0.997. These graphs prove the necessity of the proposed algorithm. Our algorithm not only finds the best algorithm and k, it also reduces the search space significantly. We also measured the performance of our system using semantic dataset and obtained similar results.

We also experimented with different versions of the same dataset, where the choice of the method and value of k also varies. For example, COF method with k = 85 outperforms (ROC AUC score = 0.89556) for the heart disease dataset with 2% of outliers, normalized, and without duplicates. For the same dataset with not-normalized version, LDF with $k = 2$ outperforms (ROC AUC score = 0.93333). Therefore, iterating the process as depicted in Fig. 3 using expert knowledge with the help of visualization and calculated outlier factor by the chosen method improves the outlier detection performance.

8 Conclusion and Future Work

Outlier detection is getting huge interest in several application areas recently. However, it is very difficult to identify outliers from real-life data. It is necessary to have automated algorithm, on the one hand, and on the other hand, it is an utmost necessity to verify the result by human experts. In this paper, we proposed a visual analytic system that includes an algorithm to find proper outlier detection algorithm from several candidates and an associated parameter for each dataset and a visualization technique to verify and improve the obtained results by a human expert. With our experiments and results, the necessity of our system is elaborated. Our proposed system could find the best algorithm for ranking outliers in a dataset in hand, using the best parameter. However, in future, we will conduct a user evaluation of real-life datasets without outlier labeling to identify how far visualization can contribute in such cases. We will also investigate if the dimension reordering of PCP can facilitate visual analysis of outliers.

References

1. Knorr, E.M., Ng, R.T.: Algorithms for mining distance-based outliers in large datasets. In: Proceedings of the 24th International Conference on Very Large Data Bases, New York, NY, pp. 392–403 (1998)
2. Breunig, M.M., Kriegel, H.-P., Ng, R., Sander, J.: LOF: identifying density-based local outliers. In: Proceedings of the SIGMOD, pp. 93–104 (2000)
3. Aggarwal, C.C., Yu, P.S.: Outlier detection for high dimensional data. In: Proceedings of the SIGMOD, pp. 37–46 (2001)
4. Kriegel, H.-P., Kröger, P., Schubert, E., Zimek, A.: Outlier detection in axis-parallel subspaces of high dimensional data. In: Proceedings of the PAKDD, pp. 831–838 (2009)
5. Müller, E., Assent, I., Steinhausen, U., Seidl, T.: OutRank: ranking outliers in high dimensional data. In: Proceedings of the ICDE Workshop DBRank, pp. 600–603 (2008)
6. Müller, E., Schiffer, M., Seidl, T.: Adaptive outlierness for subspace outlier ranking. In: Proceedings of the CIKM (2010)
7. Nguyen, H.V., Gopalkrishnan, V., Assent, I.: An unbiased distance-based outlier detection approach for high dimensional data. In: Proceedings of the DASFAA, pp. 138–152 (2011)
8. Kriegel, H.-P., Kröger, P., Schubert, E., Zimek, A.: Outlier detection in arbitrarily oriented subspaces. In: Proceedings of the 12th IEEE International Conference on Data Mining (ICDM), Brussels, Belgium (2012)
9. Knorr, E.M., Ng, R.T.: Finding intensional knowledge of distance-based outliers. In: Proceedings of the 25th International Conference on Very Large Data Bases, Edinburgh, Scotland, pp. 211–222 (1999)
10. Ramaswamy, S., Rastogi, R., Shim, K.: Efficient algorithms for mining outliers from large data sets. In: Proceedings of the SIGMOD, pp. 427–438 (2000)
11. Bay, S.D., Schwabacher, M.: Mining distance-based outliers in nearly linear time with randomization and a simple pruning rule. In: Proceedings of the KDD, pp. 29–38 (2003)
12. Angiulli, F., Pizzuti, C.: Fast outlier detection in high dimensional spaces. In: Proceedings of the PKDD, pp. 15–26 (2002)
13. Tang, J., Chen, Z., Fu, A.W.-C., Cheung, D.W.: Enhancing effectiveness of outlier detections for low density patterns. In: Proceedings of the PAKDD, pp. 535–548 (2002)
14. Jin, W., Tung, A.K. H., Han, J., Wang, W.: Ranking outliers using symmetric neighborhood relationship. In: Proceedings of the PAKDD, pp. 577–593 (2006)
15. Kriegel, H.-P., Kröger, P., Schubert, E., Zimek, A.: LoOP: local outlier probabilities. In: Proceedings of the CIKM (2009)
16. Barnett, V., Lewis, T.: Outliers in Statistical Data, 3rd edn. Wiley (1994)
17. Campos, G., Zimek, A., Sander, J., Campello, R., Micenkova, B., Schubert, E., Assent, I., Houle, M.: On the evaluation of unsupervised outlier detection: measures, datasets, and an empirical study. Data Min. Knowl. Discov. **30**(4), 891–927 (2016)

18. Craswell, N.: Precision at n. In: Liu, L., Özsu, M.T. (eds.) Encyclopedia of database systems, pp. 2127–212. Springer, Berlin (2009). https://doi.org/10.1007/978-0-387-39940-9_484
19. Zhang, E., Zhang, Y., Average precision. In: Liu, L., Özsu, M.T. (eds.) Encyclopedia of Database Systems, pp. 192–193. Springer, Berlin (2009). https://doi.org/10.1007/978-0-387-39940-9_482
20. Angiulli, F., Pizzuti, C.: Outlier mining in large high-dimensional data sets. IEEE Trans. Knowl. Data Eng. **17**(2), 203– 215 (2005). https://doi.org/10.1109/tkde.2005.31
21. Schubert, E., Zimek, A., Kriegel, H.-P.: Local outlier detection reconsidered: a generalized view on locality with applications to spatial, video, and network outlier detection. Data Min. Knowl. Discov. **28**(1), 190–237 (2014). https://doi.org/10.1007/s10618-012-0300-z
22. Kriegel, H.-P., Kröger, P., Schubert, E., Zimek, A.: LoOP: local outlier probabilities. In: Proceedings of the 18th ACM Conference on Information and Knowledge Management (CIKM), Hong Kong, pp. 1649–1652 (2009). https://doi.org/10.1145/1645953.1646195
23. Zhang, K., Hutter, M., Jin, H.: A new local distance-based outlier detection approach for scattered real world data. In: Proceedings of the 13th Pacific-Asia conference on Knowledge Discovery and Data Mining (PAKDD), Bangkok, pp. 813–822 (2009). https://doi.org/10.1007/978-3-642-01307-2_84
24. Hautamäki, V., Kärkkäinen, I., Fränti, P.: Outlier detection using k-nearest neighbor graph. In: Proceedings of the 17th International Conference on Pattern Recognition (ICPR), Cambridge, pp. 430–433 (2004). https://doi.org/10.1109/icpr.2004.1334558
25. Schubert, E., Zimek, A., Kriegel, H.-P.: Generalized outlier detection with flexible kernel density estimates. In: Proceedings of the 14th SIAM International Conference on Data Mining (SDM), Philadelphia, pp. 542–550 (2014). https://doi.org/10.1137/1.9781611973440.63
26. Kriegel, H.-P., Schubert, M., Zimek, A.: Angle-based outlier detection in high-dimensional data. In: Proceedings of the 14th ACM International Conference on Knowledge Discovery and Data Mining (SIGKDD), Las Vegas, pp. 444–452 (2008). https://doi.org/10.1145/1401890.1401946
27. Latecki, L.J., Lazarevic, A., Pokrajac, D.: Outlier detection with kernel density functions. In: Proceedings of the 5th International Conference on Machine Learning and Data Mining in Pattern Recognition (MLDM), Leipzig, pp. 61–75 (2007). https://doi.org/10.1007/978-3-540-73499-4_6
28. Inselberg, A.: Parallel Coordinates: Visual Multidimensional Geometry and its Applications. Springer-Verlag New York Inc, Secaucus, NJ, USA (2009)
29. Micenkova, B., Ng, R.T., Dang, X.-H., Assent, I.: Explaining outliers by subspace separability. In: 2013 IEEE 13th International Conference on Data Mining, vol. 00, pp. 518–527 (2013). https://doi.org/10.1109/icdm.2013.132
30. von Brünken, J., Houle, M.E., Zimek, A.: Intrinsic Dimensional Outlier Detection in High-Dimensional Data Technical Report, No. NII-2015-003E, National Institute of Informatics (2015)
31. Tukey, J.W.: Exploratory Data Analysis. Addison–Wesley (1977)
32. Novotny, M., Hauser, H.: Outlier-preserving focus+context visualization in parallel coordinates. IEEE Trans. Visual. Comput. Graphics **5**, 893-900 (2006). http://dx.doi.org/10.1109/TVCG.2006.170
33. Achtert, E., Kriegel, H.P., Reichert, L., Schubert, E., Wojdanowski, R., Zimek, A.: Visual evaluation of outlier detection models. In: Proceedings of the 15th International Conference on Database Systems for Advanced Applications (DASFAA), Tsukuba, Japan, pp. 396–399 (2010)
34. Kandogan, E.: Just-in-time annotation of clusters, outliers, and trends in point based data visualizations. IEEE TVCG 73-82 (2012)
35. Schubert, E., Koos, A., Emrich, T., Züfle, A., Schmid, K.A., Zimek, A.: A framework for clustering uncertain data. PVLDB **8**(12), 1976–1979 (2015)
36. Swayne, D.F., Lang, D.T., Buja, A., Cook, D.: GGobi: evolving from XGobi into an extensible framework for interactive data visualization. Comput. Stat. Data Anal. **43**, 423–444 (2003)
37. Datasets: http://www.dbs.ifi.lmu.de/research/outlierevaluation/DAMI/

Identification of the Recurrence of Breast Cancer by Discriminant Analysis

Avijit Kumar Chaudhuri, D. Sinha and K. S. Thyagaraj

Abstract Breast cancers enact one of the deadliest diseases that make a high number of deaths every year. It is a special type of all cancers and the primary reason for women's deaths globally (Bangal et al. Breast, carcinoma in women—a rising threat [1]). In the medical field, data mining methods are widely used in diagnosis and analysis to make decisions exclusively. Given the relationship between the degree of malignancy and recurrence of breast cancer, and given the typical model of breast cancer spread, it should be the principal goal of early detection by which cancers can be identified when they are small and node-negative. Most of the researches in the field of breast cancer have focused on predicting and analyzing the disease. There is not much evidence in study of recurrence of the disease. This paper aims at developing an approach to predict and identify the probability of recurrence of breast cancer for the patients with greater accuracy.

Keywords Breast cancers · Recurrence · Decision trees · Discriminant analysis Node-caps · Deg-malig

1 Introduction

Cancer is an unexpected expansion of cells which tends to propagate in an uncontrolled way and, in some cases, to metastasize (spread). In a country such as India, breast cancer, especially among women, is widely detected. As per a global study,

A. K. Chaudhuri (✉)
Techno India - Banipur, Banipur College Road, 743233 Habra, India
e-mail: c.avijit@gmail.com

D. Sinha
Indian Institute of Foreign Trade (IIFT), Kolkata, 1583 Madurdaha, Chowbagha Road, Kolkata 700107, India
e-mail: dsinha2000@gmail.com

K. S. Thyagaraj
Indian Institute of Technology, Dhanbad, 826004 Dhanbad, Jharkhand, India
e-mail: ksthyagaraj@gmail.com

A. Abraham et al. (eds.), *Emerging Technologies in Data Mining and Information Security*, Advances in Intelligent Systems and Computing 813,
https://doi.org/10.1007/978-981-13-1498-8_46

the number of cases of breast cancer in India is likely to increase phenomenally to two hundred thousand by 2030. The present level is around 115,000. According to World Health Organization (WHO), there are around, on an average taken over last 5 years, Over seventy thousand deaths (http://globocan.iarc.fr/Pages/fact_sheets_cancer.aspx). Further studies reveal that this disease is the second most common after cervical cancer. It is also that breast cancer affects women in India a decade before it does to those in the Western world. The detection services and medication of early years may explain poor endurance in great extent.

Over the years, there has been regular advancement pertinent to cancer research worldwide [2]. A number of methods are applied by different scientists, for example, shielding in premature stage, in pursuance of finding types of cancer before the beginning of any symptoms. Furthermore, the scientists develop new strategies regularly to predict cancer beforehand to the treatment. At the same time with the advanced new technologies in medical field, an enormous amount of data in cancer field have been collated and readily available for medical research. On the other hand, predicting the disease outcome precisely is the most demanding and fascinating job for physicians.

Data mining has been found to be a useful approach to classify and predict outcomes in medical research. In turn, this has led to the development of guidelines for further medical research, in both clinical and biological domains. The main defect of the process is that the persons who use this process are not connected with the medical science. In statistics, they explain the process of data classification from known datasets. They use many techniques to detect and classify the different types of breast cancer. Data mining techniques can enable discovery and identification of patterns and relationships between the dependent and independent variable taken from complex datasets, leading to effectively predict outcomes of a disease such as cancer. Most of the researches in the field of breast cancer have focused on predicting and analyzing the diseases. There are not many evidences in study of recurrence of the disease.

This paper presents a new approach to predict and identify the probability of recurrence of breast cancer for the patients with greater accuracy. In this paper, discriminant analysis has been used to predict the recurrence of breast cancer under different degrees of malignancy supported by the probability of such recurrence so obtained from the application of decision tree method.

2 Relevant Literature

A diverse set of algorithms and research approaches has been applied so far. These included various classifications, e.g., clustering, and prediction methods and tools. Some of them include regression, artificial intelligence, neural networks, association rules, decision trees, genetic algorithm, nearest neighbor method, and similar techniques. In medical care industry, experts take support from these techniques to strengthen the accuracy in the fields of research of breast cancer.

In a study, Delen et al. [3] pre-categorized breast cancer under two classes namely "survived" and "not survived". The dataset included records of around two hundred thousand sufferers. Their results of forecasting the survivability showed accuracy to the tune of 93%. Tan and Gilbert [4] used C4.5 to compare the performance of two methods, namely, decision tree and bagged decision tree in diagnosis of cancer. Liu et al. [5] predicted breast cancer survivability using C5 algorithm with bagging approach. Li et al. [6] used C4.5 with and without bagging to diagnose data on ovarian tumor to study cancer. A new approach was also suggested to discover the relationships in structure activity related to chemometrics in pharmaceutical industry. It used a new method combining decision tree based ensemble method with feature selection method backward elimination strategy along with bagging [7]. Tu et al. [8] suggested integration of bagging and C4.5 algorithm and naïve Bayes algorithm. He applied these combinations to diagnose cardiovascular disease. The same author [9] also applied bagging method to determine the early presence of cardiovascular disease. He gave a comparison of results so obtained from the application of decision tree with and without bagging. Tsirogiannis et al. [10] showed that, while using the methods namely classifiers, SVM, neural networks, and decision trees, bagging results were more accurate than without bagging. Pan [11] applied decision tree algorithm C4.5 with bagging to identify abnormal high-frequency electrocardiograph on data comprising ECG reports of patients. Kaewchinporn et al. [12] proposed Tree Bagging and Weighted Clustering (TBWC), a new classification algorithm for diagnosing dataset related to medical sphere. TBWC is an amalgamation of two methods, that is, decision tree with bagging and clustering. Chaurasia and Pal [12, 13] also showed the superiority of bagging techniques compared to Bayesian classification and J48.

In the year 2010, Dr. K. Usha Rani practiced neural technique by using a method based on parallel computing for detection of breast cancer. In this paper, the author considered a number of various parallelization procedures for better classification, which are based on single and multilayer neural network patterns [14].

In 2012, Ali Raad, Ali Kalakech, and MohammadAyache used two types of networks which are known as MLP (Multilayer Perceptron) and RBF (Radial Basis Function) and classified two types of breast cancer. They compared these two types of networks and came to the conclusion that the second type (RBF) is preferable to the first type (MLP) in medical science for the ascertainment of the categories of breast cancer [15].

In 2012, Bekaddour Fatima discovered a neuro-fuzzy method and used ANFIS (Adaptive Neuro-Fuzzy Inference System) for the detection of the breast cancer [16].

In 2012, Sonia Narang, Harsh K. Verma, and UdaySachdev used a network known as Adaptive Resonance Neural Networks (ARNN) and discovered an accurate system of diagnosis with the help of ART model [17]. Yasmeen M. George, Bassant Mohamed Elbagoury, Hala H. Zayed, and Mohamed I. Roushdy explained an intelligent process how to classify breast cancer. For classification of types of breast cancer, the authors stated four models with the help of the process based on computer-aided diagnosis system [18].

3 Breast Cancer Dataset

The data used in this study are collected from the UCI Machine Learning Breast Cancer Dataset. The data file contains 10 specific dependent and independent attributes and a total of 286 records. The following variables are used for testing purposes:

Age: Age of the patient when diagnosed.

Menopause: Labels the diagnostics status of patient's menopause.

Tumor-size: Refers to the size of the tumor (in mm) at the time of diagnosis.

Inv-nodes: Axillary lymph nodes, whose value ranges from 0 to 39, show the presence of the disease.

Node_caps: Finds the penetration of the tumor in the lymph node capsule or not.

Degree_of_malignancy: Labels the status (also referred as grade) of tumor in the histological investigation; the value ranges from 1 to 3. These are referred as grade 1, grade 2, and grade 3. Grade 1 refers to the tumor that comprised of cancer cells; grade 2 indicates neoplastic tumor that contains conventional characteristics of cancer cells, while grade 3 tumors are those that consist of cells that are highly affected.

Breast: Presence of cancer in either left or right breast.

Breast quadrant: The breast may be split into four quadrants with nipple considered as a central point.

Irradiation: Radiation (X-rays) history of patient suffering from breast cancer.

Class: The patients having classified under no recurrence or recurrence based on the repetition of the symptoms of breast cancer (Table 1).

Table 1 Breast cancer dataset

Sl. no.	Attributes	Values
1	Age	10–19, 20–29, 30–39, 40–49, 50–59, 60–69, 70–79, 80–89, 90–99
2	Menopause	lt40, ge40, premeno
3	Tumor-size	0–4, 5–9, 10–14, 15–19, 20–24, 25–29, 30–34, 35–39, 40–44, 45–49, 50–54, 55–59
4	Inv-nodes	0–2, 3–5, 6–8, 9–11, 12–14, 15–17, 18–20, 21–23, 24–26, 27–29, 30–32, 33–35, 36–39
5	Node-caps	Yes, no
6	Deg-malig	1, 2, 3
7	Breast	Left, right
8	Breast-quad	Left-up, left-low, right-up, right-low, central
9	Irradiation	Yes, no
10	Class	No-recurrence events, recurrence events

4 Methods

In this paper, discriminant analysis has been used to predict the recurrence of breast cancer under different degrees of malignancy supported by the probability of such recurrence so obtained from application of decision tree method.

4.1 Knowledge Models

4.1.1 Discriminant Analysis

Discriminant analysis uses continuous variable measurements on diverse associations of items to focus aspects that differentiate the groups and to use these appraisals to classify new items. Discriminant function analysis is a parametric technique to determine the weightings of quantitative variables or seer variables best discriminate between two groups of seer variables or more than two groups of seer variables [19]. The analysis generates a discriminant function which is nothing but a linear amalgamation of the weights and significant variables.

The primary reason to use multiple linear regressions is to define a function where one dependent variable is linear combination of independent variables. In discriminant analysis, a single qualitative dependent variable is a linear weighted sum of independent variable(s). The dependent variable comprises two classes or combinations in most of the cases, like, existence or nonexistence of cardiovascular disease, high blood cholesterol versus normal cholesterol, recurrence and non-recurrence of malignancy in breast cancer patients, etc. This combination of independent variables is such that it will discriminate best among the groups. This linear combination defines a discriminant function. In this discriminant function, the association of the variables is generated through correction of weights, referred as discriminant coefficients, assigned to the independent variables (Fig. 1).

The discriminant equation is

$$F = 0 + 1A1 + 2A2 + \cdots + pAp + \varepsilon \text{ or } F = \sum Ai + \varepsilon \tag{1}$$

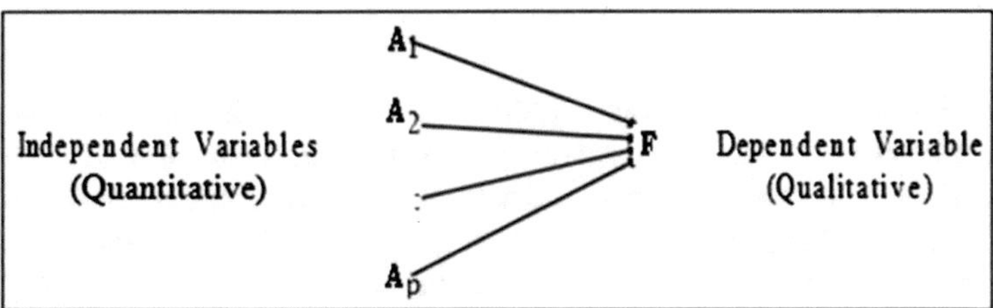

Fig. 1 The model

where

F is a dependent latent variable;

Ai represents independent variables, A1, A2,…Ap;

ε is the error term; and

0, 1, 2,…, p are the discriminant coefficients.

Assumptions

The variables A1, A2, …, Ap are not correlated.

Sizes of the mutually discriminating groups are not exclusively different.

The maximum number of independent variables will be two less than the total sample size.

The independent variables within each group of the dependent variable have similar variance–covariance structures.

By nature, the errors (residuals) are distributed and organized arbitrarily.

The distinct independent variables follow multivariate normal distribution.

There are several purposes for MDA (Multiple Discriminant Analysis):

Scrutinizing the differences between groups and identify the different groups, discarding variables which are less related to the group. Hence, MDA helps achieve parsimony without affecting the accuracy of predictability.

Creating groups by classifying different cases. The accuracy of the discriminant function is tested by observing whether test cases are classified as predicted.

It works on already defined datasets for which the groups are defined earlier.

Determine secondary relationships between a group of discriminating variables and one classification variable, contemplating to derive a correlation between dependent and independent variables.

Take out prevalent, canonical functions within the group of adjunct variables from a number of illustrations, such that discrepancy among the group is increased and discrepancy within the group is decreased toward the gradient.

An adequate number of primary variables can be condensed together to create a small set canonical function by which dimensionality of a multivariate data will be reduced.

Depending upon a collection of discriminating characteristics, the absolute inequalities between the already defined groups of sampling entities have been described.

4.1.2 Decision Trees

The decision tree is a function that constructs tree-like classification structure [20]. It segregates essential facts into classes. It also anticipates the values of a target or dependent variable constituted with the help of predictor or independent variables. These functions also produce authorization tools for preliminary and affirmatory study for classification. In our present paper, we are considering the results of prognosis for breast cancer to predict the chances of recurrences in data mining by using decision tree approach [21].

We are using two most important growing methods in decision tree algorithm that are Chi-squared Automatic Interaction Detection (CHAID) and Classification and Regression Trees (CART) analyses. These two algorithms build two different types of tree structures [22]. Collections of "if-then-else" rules are the main constructor of decision tree algorithm that shows the instructions in exhaustive form for several cases. In this method, conclusive tree will be the result that is produced from categorical input data. For categorical huge datasets, CHAID is treated as the best method among all other decision tree algorithms [23]. Though these two tree structures have a number of differences between them, they can be used for same purposes also. Especially for analyzing very large datasets, CHAID provides a much better tree structure than compared to CART [7].

CHAID [8] was designed to analyze the definitive and discontinuous target datasets, and it can break the data by using multi-way split technique; it is actually splitting of present node into more than two nodes automatically. CHAID is actually a pattern of decisive test by which relationships between dependent and independent variables can be described. The routine statistical tools like regressions cannot be used with breast cancer datasets [24]; CHAID is considered to be the perfect technique to determine the relationships between dependent and independent variables because it is more convenient to investigate the categorical data [20, 25].

5 Results and Discussions

5.1 *Classification Tree*

The tree diagram is a graphic representation of the tree model. This tree diagram (Fig. 2) shows that:

Using the CHAID method, deg-malig factor is the best predictor of breast cancer recurrence.

For value 3 for deg-malig, the next best predictor is node-caps, and for the deg-malig value 2 and 1 since there are no child nodes below it, this is considered as a terminal node (Table 2).

Table 3 shows the number of cases classified correctly and incorrectly for each category of the dependent variable that is categorical (nominal, ordinal).

Table 2 Risk estimate table

Estimate	Std. error
0.241	0.025
Growing method: CHAID	
Dependent variable: class (in numeric)	

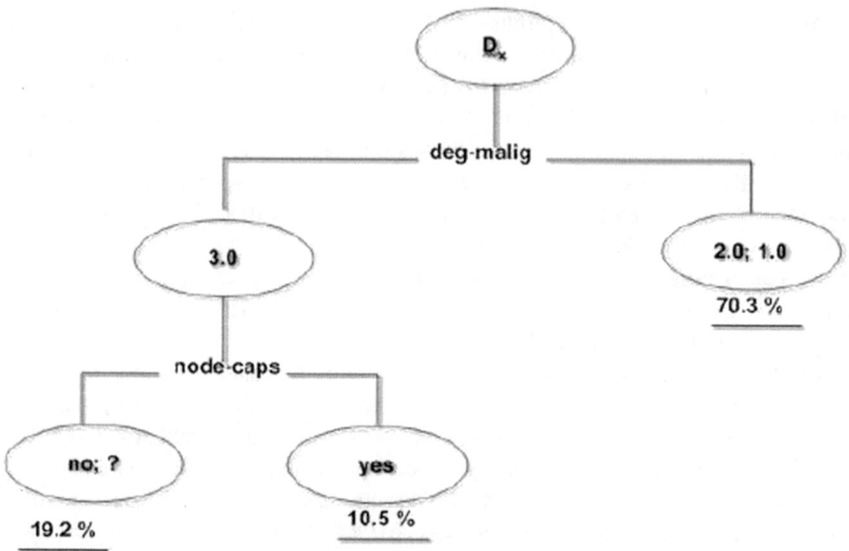

Fig. 2 Decision tree by CHAID (Chi-square automatic interaction detector) algorithm

Table 3 Classification table

Predicted			
Observed	0	1	Percent correct
0	194	7	96.5%
1	62	23	27.1%
Overall percentage			
89.5%	10.5%	75.9%	
Growing method: CHAID			
Dependent variable: class (in numeric)			

5.2 *Discriminant Analysis*

Objective:
The tangible determinants can be identified for calculation of the incidence of recurrence of breast cancer among age, menopause, tumor-size, inv-nodes, node-caps, deg-malig, breast, breast-quad, and irradiation.

The discriminant function is given by Eq. (2):

$$F = \beta 0 + \beta 1(\text{node} - \text{caps\_int}) + \beta 2(\text{deg} - \text{malig}) + \varepsilon \tag{2}$$

Table 4 shows the mean value of the node-caps and deg-malig among the no-recurrence events are lesser than the recurrence events. The p-values for tests of

Table 4 Group statistics table

				Valid (listwise)	N
Class (in numeric)		Mean	Std. deviation	Unweighted	Weighted
0	node-caps_int	0.12	0.331	201	201.000
	deg-malig	1.91	0.697	201	201.000
1	node-caps_int	0.36	0.484	85	85.000
	deg-malig	2.39	0.725	85	85.000
Total	node-caps_int	0.20	0.398	286	286.000
	deg-malig	2.05	0.738	286	286.000

Table 5 Tests of equality of group means table

	Wilks' lambda	F	df1	df2	Sig.
node-caps_int	0.923	23.564	1	284	0.000
deg-malig	0.910	27.965	1	284	0.000

Table 6 Log determinants table

Log determinants		
Class (in numeric)	Rank	Log determinant
0	2	−2.976
1	2	−2.242
Pooled within-groups	2	−2.692

equality of means are both less than 0.05 (See Table 5). Thus, perhaps both node-caps and deg-malig can be important discriminant of defaulting groups.

Log Determinants and Box's M Tables

Discriminant analysis differs from ANOVA in the way that in the latter case, it is assumed the groups have equal variances, while in case of former approach, it is assumed that the classes have equal variance–covariance matrices (between the input variables). Box's M test results table is used to test null hypothesis such that the covariance matrices will be same in between the groups formed by the dependent variable. This presumption holds true only if the equality of log determinants exists. In Box's M application, a nonsignificant M indicates similarity and lack of significant differences between groups. In our analysis, the log determinants appear similar with Box's M value as 19.057 and $F=6.291$. The value of F is significant at $p<0.000$ (Tables 6 and 7). In case of large samples, three or more groups may exist and M is significant. In such cases, a significant result is not regarded as too important. In such cases, groups with very small log determinants need to be deleted (Table 10, 11, 12, 13, 14).

The ranks and natural logarithms of determinants printed are those of the group covariance matrices.

Table 7 Box's M test results table

Test results		
Box's M		19.057
F	Approx.	6.291
	df1	3
	df2	528955.537
	Sig.	0.000

Table 8 Eigenvalues table

Eigenvalues				
Function	Eigenvalue	% of variance	Cumulative %	Canonical correlation
1	0.144[a]	100.0	100.0	0.354

[a]First 1 canonical discriminant functions were used in the analysis

Table 9 Wilks' lambda table

Wilks' lambda				
Test of function(s)	Wilks' lambda	Chi-square	df	Sig.
1	0.874	37.975	2	0.000

Table 10 Standardized canonical discriminant function coefficients table

Standardized canonical discriminant function coefficients	
	Function
	1
node-caps_int	0.581
deg-malig	0.674

Table 11 Structure matrix table

Structure matrix	
	Function
	1
deg-malig	0.828
node-caps_int	0.760

Table 12 Canonical discriminant function coefficients table

Canonical discriminant function coefficients	
	Function
	1
node-caps_int	1.520
deg-malig	0.955
(Constant)	−2.254
Unstandardized coefficients	

Table 13 Functions at group cancroids table

	Function
Class (in numeric)	1
0	−0.246
1	0.581

Table 14 Classification results table

Classification results					
			Predicted membership		Group
		Class (in numeric)	0	1	Total
Original	Count %	0	143	58	201
		1	32	53	85
		0	71.1	28.9	100.0
		1	37.6	62.4	100.0
Cross-validated[b]	Count %	0	143	58	201
		1	32	53	85
		0	71.1	28.9	100.0
		1	37.6	62.4	100.0

[a]68.5% of original grouped cases correctly classified
[b]Cross-validation is done only for those cases in the analysis. In cross-validation, each case is classified by the functions derived from all cases other than that case

Tests null hypothesis of equal population covariance matrices.

Table of eigenvalues.

There are two groups. Therefore, number of function = 1.

The eigenvalue is 0.144 (<1). Canonical correlation, rc = 0.354(>0.35) (Table 8).

WilksLamda = 0.874, p-value $\leq$ 01 (Table 9). Thus, the Function 1 explains the variation well.

Pooled within-groups correlations between discriminating variables and standardized canonical discriminant functions.

Variables ordered by absolute size of correlation within function.

The discriminant function:

$$F = -2.254 + 1.520\,(\text{node} - \text{caps\_int}) + 0.955\,(\text{deg} - \text{malig}) \tag{3}$$

Unstandardized canonical discriminant functions evaluated at group means

Between −0.246 and 0.581, the midpoint is 0.1675.

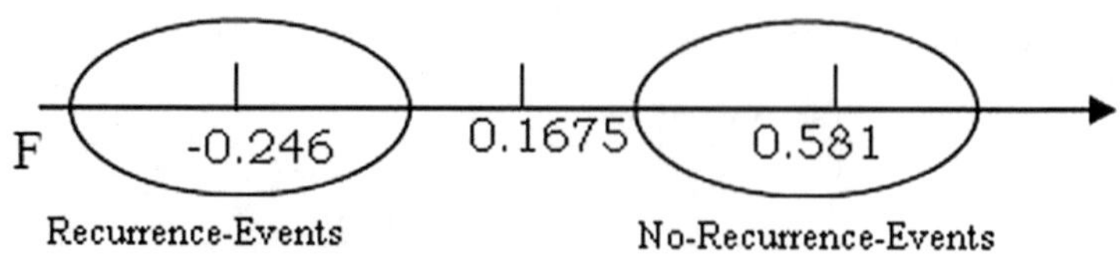

Classification Statistics

68.5% of cross-validated grouped cases correctly classified.

Analysis

The set of association rule derived from the decision tree analysis is compiled below:

If (deg-malig = 3 and node-caps = 0), then a probability of recurrency is 40%.
If (deg-malig = 3 and node-caps = 1), then a probability of recurrency is 76.7%.
If (deg-malig = 2 or deg-malig = 1), then a probability of recurrency is 19.9%.

However, decision tree fails to provide information on recurrence of breast cancer under the following scenario:

node-caps_int = 0, deg-malig = 1.
node-caps_int = 1, deg-malig = 1.
node-caps_int = 1, deg-malig = 2.

The results obtained from discriminant analysis show that it confirms to the findings of decision trees for two conditions. They are as follows:

node-caps_int = 0, deg-malig = 3 $\rightarrow$ F = $-2.254 + 1.520(0) + 0.955(3) = 0.611 > 0.1675$ $\rightarrow$ no-recurrence events: {in 60% cases, i.e., recurrency in 40% cases}.
node-caps_int = 1, deg-malig = 3 $\rightarrow$ F = $-2.254 + 1.520(1) + 0.955(3) = 2.131 > 0.1675$ $\rightarrow$ no-recurrence events: {in 23.3% cases, i.e., recurrency in 76.7% cases}.

In view of above, it can be concluded that discriminant analysis may provide the decision rule for the other three conditions not explained by decision tree. These are as follows:

node-caps_int = 0, deg-malig = 1 $\rightarrow$ F = $-2.254 + 1.520(0) + 0.955(1) = -1.299 < 0.1675$ $\rightarrow$ recurrence events.
node-caps_int = 1, deg-malig = 2 $\rightarrow$ F = $-2.254 + 1.520(1) + 0.955(2) = 1.176 > 0.1675$ $\rightarrow$ no-recurrence events.
node-caps_int = 0, deg-malig = 2 $\rightarrow$ F = $-2.254 + 1.520(0) + 0.955(2) = -0.344 < 0.1675$ $\rightarrow$ recurrence events.

Discriminant analysis has been able to correctly classify to the extent of 68.5%. Hence, the last three association rules stated above can be considered to be significant, especially being associated with malignancy.

6 Conclusion

In this paper, an analysis of dataset of breast cancer recurrence, using two different and widely accepted data mining techniques, namely, decision tree and discriminant analysis, has been carried out. These two distinct classification techniques were used for predicting the recurrence of breast cancer under all plausible scenarios. The results showed that two significant variables, deg-malig (i.e., degree of malignancy) and node-caps (i.e., tumor's penetration in the lymph node capsule or not), lead to recurrence of breast cancer. The weights of these significant variables have been obtained. The probability of such recurrence depends primarily on the degree of malignancy. However with deg-malig equal to 1 and 2, the probability varies with the presence of node-caps and could not be segregated using decision tree.

The results of discriminant analysis confirm the findings of decision tree and in addition provide the unexplained association rules under degree of malignancy equal to 1 and 2. The complete set of association rule so obtained from this study can be summarized as follows:

i. If node-caps_int = 0, deg-malig = 3, then no-recurrence events in 60% cases, i.e., recurrence in 40% cases.
ii. If node-caps_int = 1, deg-malig = 3, then no-recurrence events in 23.3% cases, i.e., recurrence in 76.7% cases.
iii. If deg-malig = 2 or deg-malig = 1, then a probability of recurrence is 19.9%.
iv. If node-caps_int = 0, deg-malig = 1, then there are recurrence events.
v. If node-caps_int = 1, deg-malig = 2, then there are no-recurrence events.
vi. If node-caps_int = 0, deg-malig = 2, then there are recurrence events.

The complete set will enable minimization of Type 1 error and enable medical practitioners to recommend tests and follow-ups according to the six different scenarios. For example, patients with node-caps_int = 1, deg-malig = 3 may be advised to undergo checkups more frequently compared to node-caps_int = 0, deg-malig = 3. Similarly, for patients with node-caps_int = 1, deg-malig = 2, the frequency of checkups may be the least. Moreover, in line with findings from literature, in cases with high probability of recurrence, patients with risk factors (such as overweight, hormone replacement therapy, alcoholic, smoking and other factors such as high BMI after menopause, radiation therapy, late pregnancy at old age, race, high bone density, etc.) may be monitored at regular intervals. A study associating attributes of recurrence with such characteristics of patient may lead to more precise conclusions.

References

1. Bangal, V.B., Shinde, K.K., Gavhane, S.P., Singh, R.K.: Breast: carcinoma in women—a rising threat. Int. J. Biomed. Adv. Res. IJBAR **04**(02) (2013)
2. Hanahan, D., Weinberg, R.A.: Hallmarks of cancer: the next generation. Cell **2011**(144), 646–674 (2011)

3. Delen, D., Walker, G., Kadam, A.: Predicting breast cancer survivability: a comparison of three data mining methods. Artif. Intell. Med. **34**(2), 113–127 (2005)
4. Tan, A.C., Gilbert, D.: Ensemble machine learning on gene expression data for cancer classification. Appl Bioinform. (3 Suppl), S75–83 (2003)
5. Liu Y.-Q., Cheng, W., Lu, Z.: Decision tree based predictive models for breast cancersurvivability on imbalanced data. In: 3rd International Conference on Bioinformatics and BiomedicalEngineering, (2009)
6. Li, J., Liu, H., Ng, S.-K., Wong, L.: Discovery of Significant Rules for Classifying Cancer Diagnosis Data, Bioinformatics, vol. 19(Suppl. 2). Oxford University Press (2003)
7. Cao, D.-S., Xu, Q.-S., Liang, Y.-Z., Chen, X.: Automatic feature subset selection for decision tree-based ensemble methods in the prediction of bioactivity. Chemometr. Intell. Lab. Syst.
8. Tu, M.C., Shin, D., Shin, D.: Effective diagnosis of heart disease through baggingapproach.In: 2nd International Conference on Biomedical Engineering and Informatics (2009)
9. Tu, M.C., Dongil, S., Dongkyoo, S.: A comparative study of medical data classificationmethods based on decision tree and bagging algorithms. In: Eighth IEEE International Conference on Dependable, Autonomic and Secure Computing (2009)
10. Tsirogiannis, G.L, Frossyniotis, D, Stoitsis, J, Golemati, S, Stafylopatis, Nikita K.S.: Classification of medical data with a robust multi-level combination scheme. In: IEEE International Joint Conference on Neural Networks
11. Pan, W.: Application of decision tree to identify a abnormal high frequency electrocardiograph. China Nat. Knowl. Infrastruct. J. (2000)
12. Kaewchinporn, C., Vongsuchoto, N., Srisawat, A.: A combination of decision tree learning andclustering for data classification. In: Eighth International Joint Conference on Computer Science and Software Engineering (JCSSE) (2011)
13. Chauraisa, V., Pal, S.: Data mining approach to detect heart diseases. Int. J. Adv. Comput. Sci. Inf. Technol. (IJACSIT) **2**(4), 56–66 (2013)
14. Usha, R.K.: Parallel approach for diagnosis of breast cancer using neural network technique. Int. J. Comput. Appl. **10**(3), 0975–8887 (2010)
15. Yasmeen, M.G., Bassant, M.E., Hala, H.Z., Mohamed, I.R.: Breast fine needle tumor classification using neural networks. Int. J. Comput. Sci. **9**(5), 2 (2012)
16. Bekaddour, F., Chikh M.A.: A neuro fuzzy inference model for breast cancer recognition. Int. J. Comput. Sci. Inf. Technol. (IJCSIT) **4**(5) (2012)
17. Narang, S., Verma, H., Sachdev, U.: A review of breast cancer detection using ART model of neural networks. Int. J. Adv. Res. Comput. Sci. Softw. Eng. **2**(10) (2012). ISSN: 2277 128X
18. Chang P.W., Liou M.D.: Comparison of three data mining techniques with genetic algorithm. In: Analysis Of Breast Cancer Data. http://www.ym.edu.tw/~dmliou/paper/compar_threedata.pdf
19. Cramer, D.: Advanced Quantitative Data Analysis. Open University Press, Buckingham (2003)
20. Mendonça, L.F., Vieira, S.M., Sousa J.M.C.: Decision tree search methods in fuzzy modelling and classification. Int. J. Approx. Reason **44**(2), 106–123 (2007)
21. IBM SPSS Decision Trees 21: IBM Corporation 1989 (2012)
22. Lakshmanan, B.C., Valarmathi, S., Ponnuraja, C.: Data mining with decision tree to evaluate the pattern on effectiveness of treatment for pulmonary tuberculosis: a clustering and classification techniques. Sci. Res. J. **3**(6), 43–48 (2015)
23. Hastie, T., Tibshirani, R., Friedman, J.: The Element of Statistical Learning Data Mining, Inference, and Prediction, 2nd edn. Springer (2013)
24. Surveillance, Epidemiology, and End Results (SEER) Program (www.seer.cancer.gov) Research Data (1973–2013): National Cancer Institute, DCCPS, Surveillance Research Program, Surveillance Systems Branch, released April 2016, based on the November 2015 submission
25. Liu Y.-Q., Cheng, W., Lu, Z.: Decision Tree Based Predictive Models for Breast Cancer Survivability on Imbalanced Data. IEEE (2009)

Automatic Multi-class Classification of Beetle Pest Using Statistical Feature Extraction and Support Vector Machine

Abhishek Dey, Debasmita Bhoumik and Kashi Nath Dey

Abstract Plant diseases have turned into a problem as it can cause substantial decrease in both quality and quantity of agricultural harvests. Image processing can help in the following issues: early detection which leads to better growth of plant, and suggestion of the type and amount of pesticides knowing the pest. Leaves and stems are the most affected part of the plants. So, they are the study of interest. The beetle can affect the leaves, which leads to severe harm in the plant. In this paper, we propose an automated method for classification of various types of beetles, which consists of (i) image preprocessing techniques, such as contrast enhancement, are used to improve the quality of image which makes advance processing satisfactory; (ii) K-means clustering method is applied for segmenting pest from infected leaves; (iii) 24 features are extracted from those segmented images by using feature extraction, mainly GLCM; and (iv) support vector machine is used for multi-classification of the beetles. The proposed algorithm can successfully detect and classify the 12 classes of beetles with an accuracy of 89.17%, which outperforms the other multi-class pest-classification algorithm by a decent margin.

Keywords Multi-class support vector machine · Pest classification · K-means clustering · Statistical feature extraction

---

A. Dey
Department of Computer Science, Bethune College, Kolkata 700006, India
e-mail: dey.abhishek7@gmail.com

D. Bhoumik (✉) · K. N. Dey
Department of Computer Science and Engineering, University of Calcutta, Kolkata 700098, India
e-mail: debasmita.ria21@gmail.com

K. N. Dey
e-mail: kndey55@gmail.com

A. Abraham et al. (eds.), *Emerging Technologies in Data Mining and Information Security*, Advances in Intelligent Systems and Computing 813,
https://doi.org/10.1007/978-981-13-1498-8_47

1 Introduction

Agriculture is the basis of Indian economy as over 75% of its population is directly or indirectly involved in this occupation. Beyond the traditional agriculture, new developments in harvesting pattern have been recognized for uplifting the status of rural community. If the plant is already affected much, then there are little things left to do. So it is necessary to monitor the plants from the beginning. The main area of interest is the leaf as about 80–90% of disease of the plant is there. To save time, efforts, labors, and use of substantial amount of pesticides, there is a need of fast and accurate pest detection [7]. Among many pests, beetle is a common one which affects mainly cotton buds and flowers, potato plants, tomato, eggplant, coconut, and capsicum. Beetles come in many shapes, sizes, and colors. Beetles can be of various classes (Fig. 1), such as asparagus beetle, bean leaf beetle, blister beetles, Colorado potato beetles, cowpea flea, Japanese, Mexican bean, spotted asparagus, spotted cucumber, striped cucumber, and white-fringed beetle [1]. Several other studies have been performed, not specifically for beetles, but for other pests and leaf diseases. Few of them have used K-means clustering algorithm for segmentation [2]. Spatial gray-level dependence matrix (SGDM) is used for texture features [13]. Artificial neural network is used to automate the classification of plant diseases [20] used. In our paper, we are using image processing to first detect then classify the beetles. The goal of the system is: given some input images of leaves, classify 12 classes of beetles. For image preprocessing, we enhance the contrast [14] to produce better results in subsequent tasks. Then, we use K-means segmentation algorithm [10] which can specify in which part the beetle exists. Methods of feature extraction, such as Gray-Level Co-occurrence Matrix (GLCM) [19], help to extract 24 features from segmented images. Classification is done using support vector machine (SVM). Parameters like accuracy, specificity, sensitivity, negative predictive accuracy, and positive predictive accuracy provide some measure of the system's performance.

The novelty of this contribution mainly lies with (i) proposing the first framework for 12 class multi-classification of beetles and (ii) extracting features only from the segmented part by K-means algorithm, which leads to smaller region of interest. Our algorithm outperforms other similar classification algorithms [12, 15, 18] by a decent margin.

The organization of the remaining part of the paper is as follows: Sect. 2 contains literature review, and Sect. 3 contains proposed methodology, which includes image preprocessing techniques, feature extraction techniques, and 12 class classifications using SVM. Experimental results are provided in Sects. 3, and 4 concludes the paper.

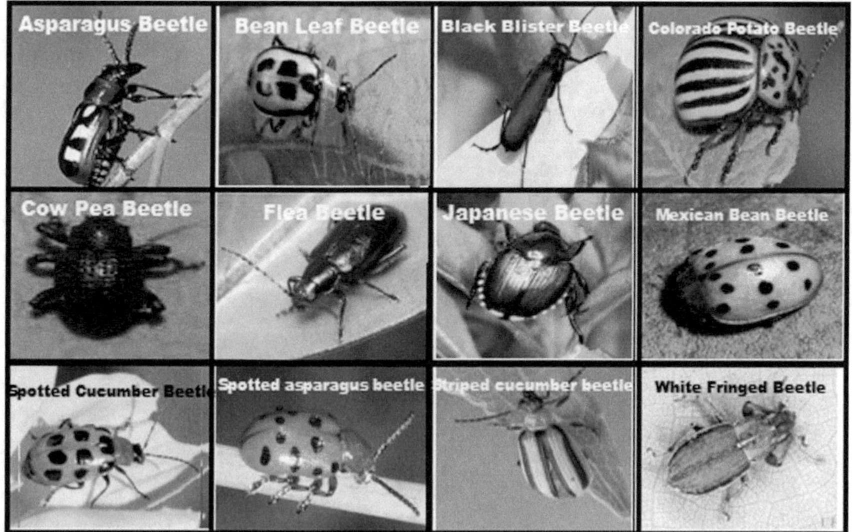

Fig. 1 Twelve classes of beetles

2 Proposed Methodology

We discuss the proposed methodology to classify the beetles into 12 classes based on some statistical features. Each step is separately described in the following subsections. However, before proceeding to the individual steps, we sum up the entire flow in Fig. 2.

2.1 Image Preprocessing

Initially, all images are resized to a fixed size, and then, noises are removed using Gaussian filter [6]. The imadjust() function is used for improving image contrast in MATLAB. In a grayscale image, this function maps the intensity values to new values so that the pixels are saturated. This is used to increase the contrast of the image. Figure 3 shows an example of image before and after preprocessing step.

2.2 Image Segmentation

Segmentation [9] partitions an image into multiple segments, where each segment consists of a set of pixels, called super pixel. Segmentation aims to extract the region of interest for entire analysis. Here, the exact location of the insect is segmented

Fig. 2 Flowchart of our proposed methodology

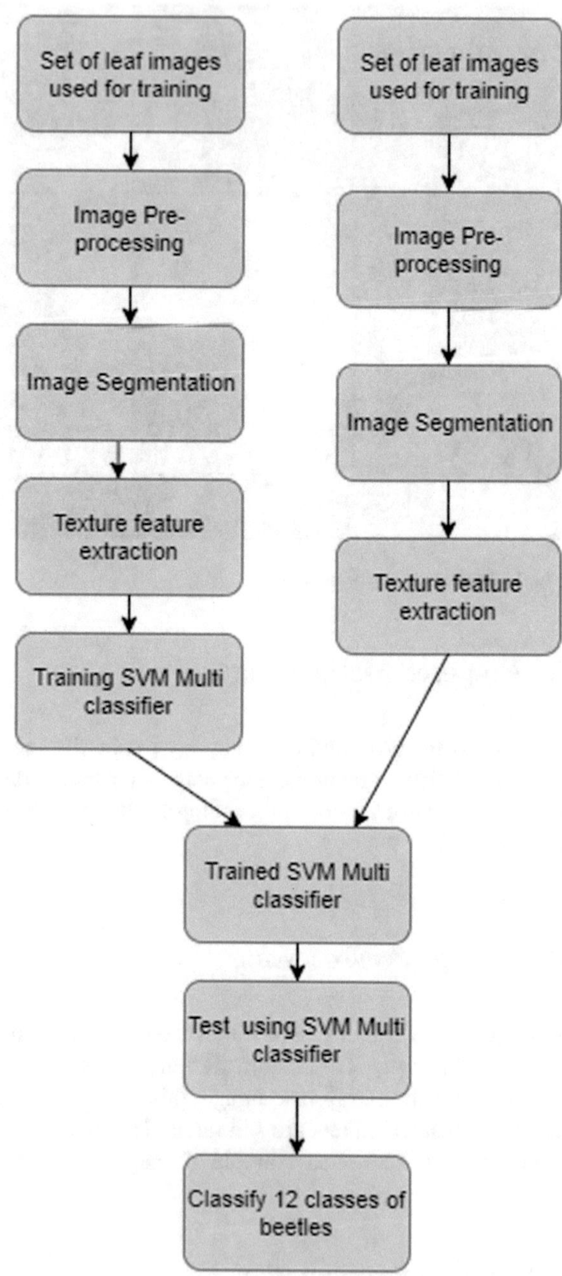

Fig. 3 Original image before preprocessing, after noise removal, and contrast enhancement

from the whole image. We have modified the K-means clustering algorithm for image segmentation and detection of the infected cluster so that the cluster in which the pest exists can be automatically detected out of all the clusters, depending on some threshold value of mean. K-means clustering [10] aims to partition a set of n observations $\{x_1, x_2, x_3, \ldots, x_n\}$ into k sets $(k \leq n)$, S = $\{S_1, S_2, S_3, \ldots, S_n\}$. The goal is to minimize the summation of distance functions of each point in the cluster to its own cluster center. The steps which we have followed here are as follows:

1. The input is read in RGB format.
2. Transformation of the image to $L * a * b*$ space.
3. K-means is used for classifying the colors.
4. All the pixels are labeled.
5. K number of images (segmented by color) are generated.
6. Eventually, the segment which contains the disease is selected.

We show the working of this algorithm in Figs. 4, 5, 6 and 7.

Fig. 4 Image after preprocessing

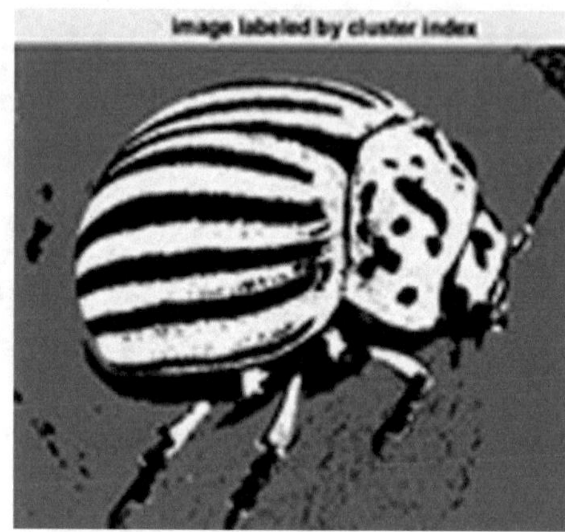

Fig. 5 Image labeled by cluster index

Fig. 6 The three clusters

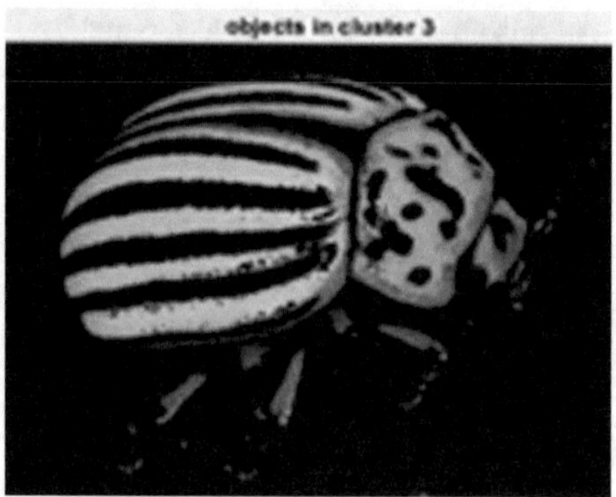

Fig. 7 Infected segment distinguished

2.3 Texture Feature Extraction Using Statistical Methods

Two criteria motivate feature selection [19]: (i) The input is too large for processing and (ii) the input is suspected to contain redundant data. In such scenarios, a reduced set of features, called feature vectors, is selected. In our proposed work, 24 statistical features are extracted from the input images, which is explained below.

In statistical texture analysis, we compute the texture features from the statistical distribution of how the pixels are residing at their position relative to the other pixels. In this paper, we have used (GLCM) [9] which is a method to extract probabilistic texture features. GLCM considers the change between two different gray levels (i and j), located distance d apart and situated at an angle θ with each other. Initially, the GLCM is designed for the input image. Henceforth, considering $d = 1$ and $\theta = 0°$, the following features are extracted from the matrix: contrast, correlation, energy, homogeneity, mean, standard deviation, entropy, root-mean-square, variance, smoothness, kurtosis, skewness, inverse different movement, cluster_prominence, cluster_shade, dissimilarity, max_prob, sum of squares, sum average, sum entropy, difference variance, difference entropy, information measure of correlation, and inverse difference normalized [21] as stated in Haralick's texture features [9].

2.4 Classification

Classification is a method to recognize and differentiate objects [11]. In supervised classification, the machine learning algorithm is first trained with a set of training data and then tested on new observations, called test set. The training set contains data with class label. After training, the algorithm is expected to identify the correct class label of test data. There are various approaches for image classification. Most of the classifiers, such as neural network [5], binary tree [16], and support vector machine (SVM) [3], require a training sample. On the other hand, there are clustering-based algorithms, e.g., K-means [10] and K-NN [4], which are unsupervised classifiers.

Here, we have used SVM classifier to classify 12 classes. It is a supervised learning algorithm in which the input is a set of the labeled training data and the output is an optimal hyperplane. SVM can be used to predict the target values of the test data, where only the test data attributes are provided, by producing a model (based on the training data). Here, the set of training images is divided into two different classes. Here, the input data set is $(x_1, y_1), (x_2, y_2), \ldots (x_m, y_m)$, where $x_i \in \mathbb{R}_d$ in the d-dimensional feature space, and the class label is $y_i \in \{+1, -1\}$, where $i = 1 \ldots m$. It can create an optimal separating hyperplane based on a mathematical function called kernel function. Each data point whose feature vectors belong to one side of the hyperplane is in the negative class, and the other points are in the positive class. The working mechanism of linear support vector machine is shown in Fig. 8.

In few cases, a straight line cannot separate all the data points. This problem is solved using the nonlinear SVM classifier. Nonlinear SVM uses the kernel function

Fig. 8 Linear SVM

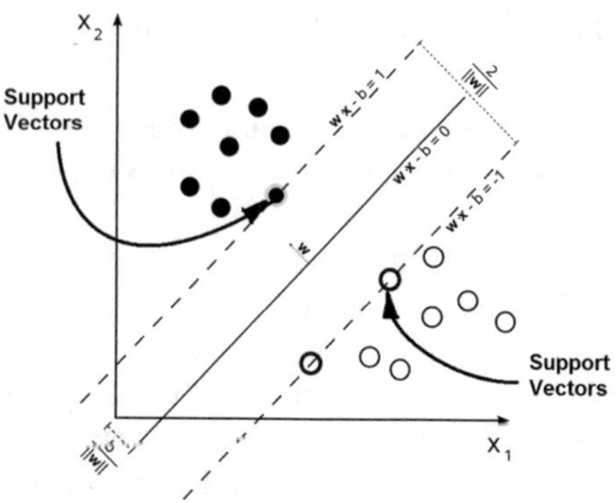

[17]. A kernel function is defined as $k(x_i, x_j) = \phi(x_i)^T\phi(x_j)$, where the feature vector is ϕ . The popular kernels are linear kernel, polynomial kernel, RBF kernel, and sigmoid kernel. RBF kernel normally performs better than other when the number of features is much lesser than the number of observations. Polynomial kernels are well suited for problems where all the training data is normalized. It is mandatory to select kernel parameters properly for achieving very high classification accuracy. Here, we have used both RBF and polynomial kernel. After comparing the results, it is concluded that the use of polynomial kernel is better in this particular application.

SVM is generally applied in binary classification. Our work requires multi-class classification since we have considered 12 classes of beetles. There are two types of method for multi-class SVM [22]:

(i) One-against-one: It combines several binary classifiers.
(ii) One-against-all: It considers all data at once.

Here, we have calculated performance for training the SVM classifier using one-against-all method. In the following section, we provide the result of classifying the images of the beetles into 12 classes using multi-class SVM.

3 Experimental Results and Discussion

The tasks of processing images for this paper are simulated in MATLAB 8.5 (R2015a). The configuration of the machine is as follows: Intel(R) Core(TM) with 4-GB RAM. We have obtained beetle images from few image databases [1]. We have trained our classifier using 180 images and tested it with 240 different images. Initially, they are preprocessed by removing noise and contrast enhancement, followed by segmentation and 24 features extraction. We have listed the average feature values of all the images in Table 1.

The feature matrices are the input of SVM classifier for training. We evaluated the performance of the classifier by the following parameters [8]:

Table 1 Feature values

Feature	Contrast	Cor	Energy	Homo	Mean	S.D.	Entropy	RMS
Values	0.716	0.860	0.679	0.950	23.683	53.195	2.080	5.536
Feature	Var	Smooth	Kurtosis	Skew	IDM	CP	CS	Diss
Values	2710.177	0.999	15.206	3.204	255	700.211	61.323	0.185
Feature	Max_prob	SOSVH	SAVGH	SUMEN	DVARH	DENTH	INF1H	INDNC
Values	0.808	5.681	3.381	0.857	0.716	0.368	0.590	0.983

1. Accuracy rate = Correctly classified samples/ All classified samples.
2. Sensitivity = Correctly classified Positive samples/TP.
3. Specificity = Correctly classified negative samples/TN.
4. Positive predictive accuracy = Correctly classified positive samples/Positive classified samples.
5. Negative predictive accuracy = Correctly classified negative samples/Negative classified samples.

where TP and TN are true positive and true negative, respectively.

These metrics for all the classes of beetles, as obtained from our classifier, are listed in Table 2. The overall average values of each of these parameters are as follows: Accuracy = 89.17% Sensitivity = 90% Specificity = 100% PPA = 100% NPA = 99%. These values are shown in the graph in Fig. 9. Hence, the analysis of the obtained results demonstrates that our proposed method can classify 12 classes of beetles with an accuracy of 89.17%, which outperforms the other similar approaches [12, 15, 18]. The comparison is listed in Table 3 and shown graphically in Fig. 10.

When an image is fed into the trained model of our proposed approach, it will show the output: that is, the beetle name. Besides, we have provided the scientific name and some control measures for the help of whoever uses this software to classify beetles.

Table 2 Class-wise metrics

Class Names	Accuracy%	Sensitivity%	Specificity%	PPA%	NPA%
Asparagus beetle	90.00	90.00	99.49	94.73	98.98
Bean leaf beetle	80.00	80.00	98.51	84.21	98.01
Black blister beetle	85.00	85.00	98.99	89.47	98.5
Colorado potato beetle	85.00	85.00	98.50	85.00	98.5
Cowpea beetle	90.00	90.00	98.99	90.00	98.98
Flea beetle	95.00	95.00	98.48	86.36	99.48
Japanese beetle	90.00	90.00	98.99	90.00	98.98
Mexican bean beetle	100.00	100.00	97.49	80.00	100.00
Spotted asparagus beetle	90.00	90.00	99.49	94.73	98.98
Spotted cucumber beetles	90.00	90.00	98.99	90.00	98.98
Striped cucumber beetles	95.00	95.00	98.00	86.36	99.49
White-fringed beetle	80.00	80.00	100.00	100.00	98.01

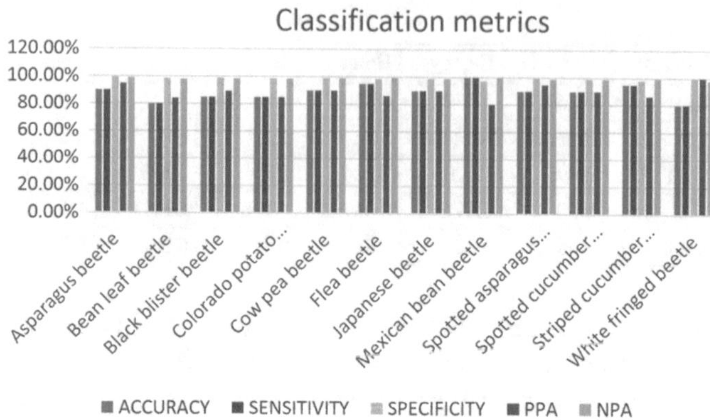

Fig. 9 Classification metrics

Table 3 Comparison with other approaches

Approach	Accuracy%
Morphology, color, and color coherence vector features (More and Nighot)	84
Texture, color, and shape features (Siricharoen and Punnarai)	88
Color, shape, and texture features (Prajapati, Jitesh, Vipul)	73.33
Our proposed method (24 statistical features)	89.17

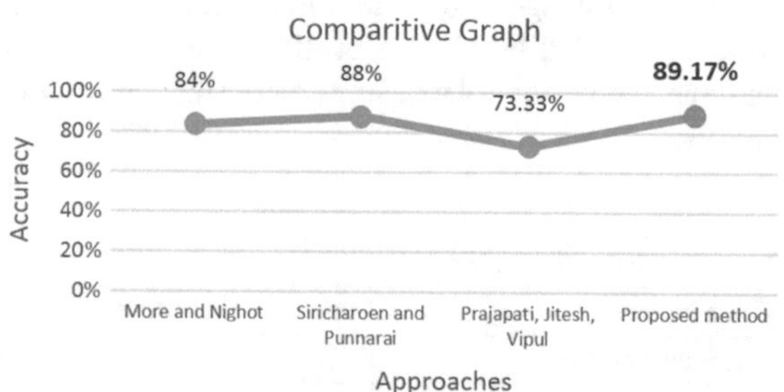

Fig. 10 Graph plotting of comparison between similar approaches

4 Conclusion

Perception of the human eye is not strong enough to differentiate slight variation in beetles. However, such differences can imply different diseases on leaves. Therefore, automatic classifier is required. In this paper, we have proposed a model which can automate multi-class classification of beetles and suggest some preventive remedies from a previously stored database. The 24 features extracted helped to form a detail training data set. A unique feature of SVM is that only the support vectors are of importance hence the remaining part of data set does not affect the classifier much. The achieved accuracy of 89.17% outperforms the accuracy of similar classifiers. Earlier detection of beetles can result in better yield in agriculture. The future scope of this work includes the following: (i) to increase the efficiency by adding more features and modifying kernel functions, and (ii) to generalize the system for other pests.

Acknowledgements Debasmita Bhoumik would like to acknowledge fruitful discussions with Ritajit Majumdar, Junior Research Fellow, Indian Statistical Institute, Kolkata.

References

1. Utilizing partnerships & information technology to advance invasive species, forestry & agriculture education. http://www.bugwood.org/ImageArchives.html
2. Al-Hiary, H., Bani-Ahmad, S., Reyalat, M., Braik, M., ALRahamneh, Z.: Fast and accurate detection and classification of plant diseases. Mach. Learn. **14**(5) (2011)
3. Cortes, C., Vapnik, V.: Support-vector networks. Mach. Learn. **20**(3), 273–297 (1995)
4. Cover, T., Hart, P.: Nearest neighbor pattern classification. IEEE Trans. Inf. Theory **13**(1), 21–27 (1967)
5. Demuth, H.B., Beale, M.H., De Jess, O., Hagan, M.T.: Neural Network Design. Martin Hagan (2014)
6. Deng, G., Cahill, L.: An adaptive gaussian filter for noise reduction and edge detection. In: Nuclear Science Symposium and Medical Imaging Conference, 1993., 1993 IEEE Conference Record, pp. 1615–1619. IEEE (1993)
7. Dey, A., Bhoumik, D., Dey, K.N.: Automatic Detection of Whitefly Pest using Statistical Feature Extraction and Image Classification Methods (2016)
8. Fawcett, T.: An introduction to roc analysis. Pattern Recogn. Lett. **27**(8), 861–874 (2006)
9. Haralick, R.M., Shapiro, L.G.: Image segmentation techniques. Comput. Vis. graph. Image Process. **29**(1), 100–132 (1985)
10. Hartigan, J.A., Wong, M.A.: Algorithm as 136: A k-means clustering algorithm. J. R. Stat. Soc. Ser. C (Applied Statistics) **28**(1), 100–108 (1979)
11. Michie, D., Spiegelhalter, D.J., Taylor, C.C.: Machine Learning, Neural and Statistical Classification (1994)
12. More, S., Nighot, M.: Agrosearch: A web based search tool for pomegranate diseases and pests detection using image processing. In: Proceedings of the Second International Conference on Information and Communication Technology for Competitive Strategies, p. 44. ACM (2016)
13. Patil, R., Udgave, S., More, S., Kasture, D.N.: Grape Leaf Disease Detection using k-means Clustering Algorithm (2016)
14. Polesel, A., Ramponi, G., Mathews, V.J.: Image enhancement via adaptive unsharp masking. IEEE trans. Image Process. **9**(3), 505–510 (2000)
15. Prajapati, H.B., Shah, J.P., Dabhi, V.K.: Detection and classification of rice plant diseases. Intell. Decis. Technol. **11**(3), 357–373 (2017)
16. Safavian, S.R., Landgrebe, D.: A survey of decision tree classifier methodology. IEEE Trans. Syst. Man Cybern. **21**(3), 660–674 (1991)
17. Schölkopf, B., Smola, A.J.: Learning with Kernels: Support Vector Machines, Regularization, Optimization, and Beyond. MIT Press (2002)
18. Siricharoen, P., Scotney, B., Morrow, P., Parr, G.: A lightweight mobile system for crop disease diagnosis. In: International Conference Image Analysis and Recognition, pp. 783–791. Springer (2016)

19. Soh, L.K., Tsatsoulis, C.: Texture analysis of sar sea ice imagery using gray level co-occurrence matrices. IEEE Trans. Geosci. Remote Sens. **37**(2), 780–795 (1999)
20. Tian, Y., Wang, L., Zhou, Q.: Grading method of crop disease based on image processing. In: International Conference on Computer and Computing Technologies in Agriculture, pp. 427–433. Springer (2011)
21. Uppuluri, A.: Glcm texture features. Matlab Central, the Mathworks (2008). Accessed 22 Mar 11
22. Weston, J., Watkins, C.: Multi-class support vector machines. Technical report, Technical Report CSD-TR-98-04, Department of Computer Science, Royal Holloway, University of London, May 1998

Probabilistic Dimension Reduction Method for Privacy Preserving Data Clustering

Hanumantha Rao Jalla and P. N. Girija

Abstract The frequent use of data mining techniques in business organizations is useful to sustain competition in real world but it leads to violation of privacy of individual customers while publishing original customer's data into real world. This paper proposes a distance preserving perturbation method for Privacy Preserving Data Mining (PPDM) using t-Stochastic Neighbor Embedding (t-SNE). The t-SNE is mainly used for dimension reduction technique; it reduces higher dimensional data sets into required lower dimensional data sets and maintains same distance in lower dimensional data. We choose K-means algorithm as knowledge-based technique; it works based on distance between data records. Setting perplexity parameter in the proposed method creates complexity to unauthorized persons to convert from low-dimensional data to original data. The proposed method is evaluated using Variation Information (VI) between original and modified data clusters. In this work, the proposed method is applied on various data sets and compared original and modified data clusters through VI.

Keywords Perturbation · t-SNE · K-means · Privacy

1 Introduction

The recent advancements in information technology invent numerous data collection tools to collect huge amount of data from customers in daily life. The data mining is essential for data analysis to sustain in competitive world to run business in commercial mode. Sometimes, the process of data analysis misuse customer's data when organizations publish original customer's data into real world. To protect the pri-

H. R. Jalla (✉)
CBIT, Hyderabad 500075, Telangana, India
e-mail: Hanu_it2007@yahoo.co.in

P. N. Girija (✉)
University of Hyderabad, Hyderabad 500046, Telangana, India
e-mail: Pn_girija@yahoo.com

A. Abraham et al. (eds.), *Emerging Technologies in Data Mining and Information Security*, Advances in Intelligent Systems and Computing 813,
https://doi.org/10.1007/978-981-13-1498-8_48

vacy of customers raises field of Privacy Preserving Data Mining (PPDM). The main objective of the PPDM is extracting valid business knowledge without revealing customer's data. The PPDM methods are preserving customer's privacy when business organizations publish data into real world for data analysis. Many researchers proposed different methods to protect the privacy of customers in the field of PPDM such as perturbation, K-anonymity, and cryptographic techniques. In perturbation techniques, the original data is converted into modified data by adding or multiplying some noise that is randomly selected from distribution to protect the privacy of customers [1]. The perturbed data gives valid data mining results because it preserves statistical properties of original data by adding or multiplying noise to the data. The perturbation techniques are simple but produce less privacy. Some perturbation techniques used transformation technique as modified technique. The transformation technique transforms original data from one domain to another domain. The distance-based data mining algorithms such as K-NN and K-means algorithms work based on the distance between records. If a transformation technique preserves the distance between records in modified data, it produces valid data mining results from modified data. The K-anonymity technique [2, 3] is used to replace a group of K data values with a data value including a special character * which has some common properties. The cryptography techniques are used in distributed environment. Secure Multiparty Computation (SMC) is mainly used in PPDM as cryptography technique [4–6]. It requires a lot of computations to provide privacy to customers.

Our proposed perturbation method using t-SNE is inspired from other dimension reduction and distance preserving methods such as Principle Component Analysis (PCA) and Nonmetric Multidimensional Scaling (NMDS) [7]. t-SNE is a probabilistic method to visualize the structure of high-dimensional data sets into lower dimensional space with distance preserving between neighbors. In this work, t-SNE is used as privacy preserving technique to maintain privacy when publishing data into real world.

2 Related Work

In the field of PPDM, numerous methods are proposed by researchers to protect the privacy of customers while performing data analysis. Ricardo Mendes and Joao P. Vollela [8] conducted an extensive survey in PPDM methods. They have discussed randomization methods such as additive noise and multiplicative noise, various K-anonymity techniques, and cryptographic techniques with merits, metrics, and applications of each method. Mamata Narwaria and Suchita Arya [9] discuss comparative analysis in state of the art in PPDM. Apart from these methods, some researchers addressed PPDM methods to preserve distances in the data using transformations for distance-based data mining techniques such as K-nearest neighbors and K-means algorithms. Shibanath Mukherjee et al. [10] proposed a method using DCT to preserve distance in modified data. The Discrete Cosine Transformation (DCT) also preserves same energy in lower dimensional as original higher dimensional data set.

The Discrete Wavelet Transformation (DWT) is applied on original data through a series of filters such as low-pass and high-pass filters to get approximation and detailed coefficients, to remove half of the coefficients to preserve distance in modified data discussed in [11]. Jen-Wei Hung et al. [12] proposed a FISIP transformation technique to preserve distance in modified data by multiplying a matrix with original data. FISIP technique also preserves first and second orders inner product in modified data. Wei Lu [13] proposed distance preserving method for PPDM by using diffusion map. Generally, diffusion maps are used for data reduction but he used as encryption technique to preserve statistical information such as distance between the data records. Somya Upadhyay et al. [14] proposed a perturbation method using 3D rotation to preserve the privacy of individual customers. Jie Wang and Jun Zhang [15] addressed a framework based on matrix factorization in the context of PPDM; they have used Singular Value Decomposition (SVD) and Nonnegative Matrix Factorization (NMF) methods.

Our proposed method using t-SNE is motivated dimensional reduction techniques using NMDS and PCA for PPDM proposed in [7, 16].

3 Stochastic Neighbor Embedding

A probabilistic method to visualize the structure of high-dimensional data sets, preserving neighbor similarities is stochastic neighbor embedding (SNE), proposed by Hinton and Roweis [17]. The SNE is mainly used to convert high-dimensional data $X = \{x_1, x_2, \ldots, x_i, x_{i+1}, \ldots, x_n\}$, where x_i is an object with m dimensions that is $x_i = \{x_{i1}, x_{i2}, \ldots, x_{im}\}$ to low-dimensional data $Y = \{y_1, y_2, \ldots, y_i, y_{i+1}, \ldots, y_n\}$, where y_i is an object with p dimensions, $yi = \{y_{i1}, y_{i2}, \ldots, y_{ip}\}$. In this process, the SNE computes conditional probability $p_{j|i}$, that is, an ith object would pick jth object as neighbor. If neighbors were selected in proportion to their probability density under a Gaussian centered at x_i. Proximate data points in high-dimensional data, $p_{j|i}$ is comparatively high, whereas for widely separated data points, $p_{j|i}$ will be almost insignificant [18]. The conditional probability is given by

$$p_{j|i} = \frac{exp\left(-\|x_i - x_j\|^2 / 2\sigma_i^2\right)}{\sum_{k \neq i} exp\left(-\|x_i - x_j\|^2 / 2\sigma_i^2\right)} \tag{1}$$

where σ_i is can either set by hand or determined by binary search on the entropy of the distribution over neighbors equal to *log k*, where k is the effective local neighbors or perplexity and here σ_i is chosen by hand. The data points represented in high-dimensional data as x_i and x_j and in low-dimensional data as y_i and y_j. For calculating conditional probabilities $q_{j|i}$, that is y_i would pick y_j as its neighbor is same as in high-dimensional data but with a fixed σ_i.

$$q_{j|i} = \frac{exp\left(-\|y_i - y_j\|^2/2\sigma_i^2\right)}{\sum_{k\neq i} exp\left(-\|y_i - y_j\|^2/2\sigma_i^2\right)} \quad (2)$$

In SNE, the cost function is used to minimize mismatches between $p_{j|i}$ and $q_{j|i}$. It is sum of Kullback–Leibler divergence of $p_{j|i}$ and $q_{j|i}$ over neighbors of each object. The cost function SNE is given by

$$C = \sum_i \sum_j p_{j|i} log \frac{p_{j|i}}{q_{j|i}} = \sum_i KL(P_i \| Q_i) \quad (3)$$

where P_i represents the conditional probability distribution over high-dimensional data X_i and Q_i represents the conditional probability distribution over embedding data points Y_i. The Kullback–Leibler divergence is used in cost function. The cost function is not symmetric, different types of errors to preserve pairwise distance in embedding data points [18]. The gradient descent applied on cost function with respect to Y_i to maintain local distances in embedding data because Y_i affects Q_i.

The crowding problem may be observed in SNE, a two-dimensional map is not a good model to maintain the distance between data points of higher dimensional data. Van der maaten and Hinton [19] proposed t-SNE algorithm which is an enhancement of SNE considering another probabilistic approach to visualize the structure of high-dimensional data sets. Instead of using conditional probabilities using joint probabilities and symmetric version of SNE with cost function which minimizes single Kullback–Leibler divergence between join probability distribution P of higher dimensional data and joint probability distribution Q of lower dimensional data points.

$$C = \sum_i \sum_j p_{ij} log \frac{p_{ij}}{q_{ij}} = KL(P_i \| Q_i) \quad (4)$$

The joint probability is defined as

$$P_{ij} = \frac{P_{j|i} + P_{i|j}}{2n} \quad (5)$$

In t-SNE, we employ a Student t-distribution with one degree of freedom as the heavy-tailed distribution in the low-dimensional map. Using this distribution, the joint probabilities q_{ij} are defined as

$$q_{ij} = \frac{\left(1 + \|y_i - y_j\|^2\right)^{-1}}{\sum_{k\neq l} \left(1 + \|y_k - y_l\|^2\right)^{-1}} \quad (6)$$

The gradient of the Kullback–Leibler divergence between P and the Student t-distribution based joint probability distribution Q (computed using Eq. 6) is given by

$$\frac{\delta C}{\delta y_i} = 4\sum_j \left((p_{ij} - q_{ij})(y_i - y_j)\left(1 + \|y_i - y_j\|^2\right)\right)^{-1} \tag{7}$$

4 Proposed Method

In this work, we proposed a method for PPDM using t-SNE, which is discussed in Sect. 3. The t-SNE is also used to reduce the dimensions for given higher dimensional data and maintains same local structure in lower dimensional space. Given data contains a set of objects $X = \{x_1, x_2, \ldots\ldots, x_n\}$ and each object has m dimensions with dissimilarities d_{ij}, $1 \leq i \leq j \leq n$. The t-SNE maps set of objects X into set of objects $Y = \{y_1, y_2, \ldots\ldots, y_n\}$ with p dimensions, $p < m$ and dissimilarities δ_{ij}, $1 \leq i \leq j \leq n$. The t-SNE maintains same dissimilarities in both higher and lower dimensional data as much as possible, that is $d_{ij} = \delta_{ij}$ and data is modified. The t-SNE is used as a transformation technique in PPDM to transform original higher dimensional set of data points X into modified lower dimensional data points and preserves the distances between data points as original data, so that it is harder to reconstruct original data from modified data [7].

The t-SNE algorithm takes two input parameters such as dissimilarity matrix of higher dimensional data or original higher dimensional data and perplexity. The performance of t-SNE depends on different settings of perplexity values. Perplexity value is density of data set. Typical values for the perplexity range between 5 and 50 [19]. If perplexity value changes from one value to other, it all these changes also reflected in modified data set. The t-SNE algorithm is presented in [20].

4.1 Data Utility

In PPDM, researchers address the issue of data utility vs. privacy of customers before publishing data into the real world. The data utility depends on data mining task to be considered for data analysis. K-means is one of simple unsupervised learning algorithm in data mining. Given data set is formed into certain clusters, i.e., K clusters. Initially, K data points are selected randomly as cluster centroids, and then each data point is assigned to a cluster based on the distance between cluster centroid and data point. After processing all data points, the K-means algorithm calculates cluster centroid for each cluster. If new cluster centroid is same as previous iteration cluster centroids then it returns K clusters. The K-means algorithm minimizes a squared error. The squared error function is given by

$$E = \sum_{j=1}^{k} \sum_{i=1}^{n} \left\| x_i^{(j)} - c_j \right\|^2 \tag{8}$$

The t-SNE maps original higher dimensional data into modified lower dimensional data and preserves the local structure, by maintaining same dissimilarity matrix as original data dissimilarity matrix. The K-means algorithm assigns a data point to a cluster based on the similarity between data point and cluster centroid. In this paper, we used variation information to compare two different clusters on same data set to evaluate data utility of modified data set. It is proposed by Marina in [21].

4.2 *Variation Information (VI)*

To evaluate data utility of modified data, we compared the clusters formed by both original data X and modified data Y. Assume that, the K-means algorithm forms clusters $C=\{C_1,C_2,\ldots\ldots,C_k\}$, from X and $C'=\{C_1,C'_2,\ldots\ldots,C'_k{}'\}$, from Y, where no data point is belonging to two clusters in both data sets and K is number of clusters. Based on information theory to compare one cluster to other VI is computed. The low value of VI indicates that two data partitions in, C and C' are quite similar and high value indicates that both C and C' are not similar [20].

To compare two data set clusters, it is necessary to construct an association matrix of the pair C, C′. The association matrix size is $k \times k'$, whose kk′th element is number of data points belonging to both cluster C_k from C and $C_k{}'$ from C′. $n_{kk'} = |C_k \cap C_{k'}|$, where $n_{kk'}$ is number of data points belonging to both cluster Ck from C and $C_k{}'$ from $C^{'}$.

VI is calculated from association matrix.

$$\mathrm{VI} = \mathrm{H(C)} + \mathrm{H}(\mathrm{C}') - 2 * \mathrm{MI}(\mathrm{C}, \mathrm{C}') \tag{9}$$

H(C) is entropy of original data clusters is given by

$$H(C) = -\sum_{k=1}^{k} p(k) \log p(k) \tag{10}$$

where $p(k) = \frac{n_k}{n}$ is probability that a data point belonging to cluster k, n_k is number of data points in cluster k, and n is the total number of data points in original data set. H(C′) is entropy of modified data clusters calculated using Eq. (11). Let p(k,k′) represent the probability that a point belonging to cluster k in C and cluster k′ in C′, that is joint probability of the random variables associated with clusterings C and C′. MI(C,C′) is mutual information between the two clusterings C and C′, is defined as

$$MI(C, C') = \sum_{k=1}^{k} \sum_{k'=1}^{k'} p(k, k') log \frac{p(k, k')}{p(k)p(k')} \tag{11}$$

Table 1 Data sets description

S.I No.	Data set name	Number of objects	Number of attributes	Number of clusters
1	BCW	699	9	2
2	Ecoli	336	7	8
3	Iris	150	4	3
4	Pima	768	8	2
5	Wine	178	13	3

5 Experimental Results

Experiments are conducted on 5 real data sets collected from UCI machine learning repository. Data set is represented as matrix format. Here, each row indicates an object and column indicates an attribute. Only numerical attributes are considered in experiments. In a data set if any missing values are present the data is preprocessed by replacing the mean value of attribute. The data sets description is shown in Table 1.

The proposed method discussed in Sect. 3 is implemented in MATLAB. WEKA tool is used for simple K-means algorithm. In this section, we present the quality of clusters formed from original and modified data sets evaluated using VI. The proposed method maps high-dimensional data set into low-dimensional data set; it takes three parameters as input such as data set or dissimilarity matrix, perplexity, and number of reduced dimensions. In our experiments, for first input parameter, we considered original data set, for second input parameter, chosen three different perplexity values based on square root of total number of records and final input parameter, the reduced number of dimensions randomly selected from 2 to m-1(m is number of attributes in original data set) attributes. The attribute values are varying from one perplexity value to another in modified data. The experimental set up for running K-means algorithm, the K value is set for each data set as number of clusters shown in Table 1.

We compared K-means accuracy on original and modified data using VI and repeat the same K value in K-means algorithm for different lower dimensions (p), that is, 2 to m-1, where $p \leq m-1$ to find out which lower dimension(p) gives maximum data utility. The VI values on BCW data set are shown in Fig. 1. Initially, the VI values on BCW data set is nearly 0.9 when $p < m/2$ then while increasing the dimensions from 5 to 8 the VI value comes down for all perplexity values. Our proposed method also produce same clusters because it satisfies $0 \leq VI \leq (2 \log k)/2$ when $p > m/2$ is mentioned in [7]. The data utility is high when modifying data using the proposed method.

The quality of original and modified data clusters on Ecoli data set is shown in Fig. 2. For this data set also proposed method maintains same clusters as well as original data. Comparing other perturbation methods, such as NMDS, PCA, SVD, and DCT, that are presented in [7] with the proposed method, VI values are very less in lower dimensions for all perplexity values. The maximum VI value on Ecoli

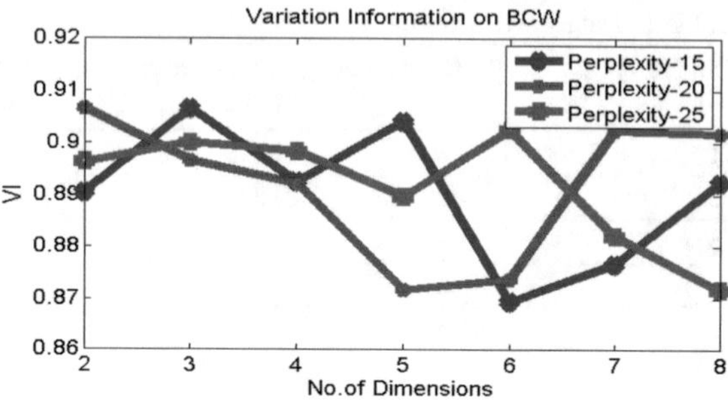

Fig. 1 VI values on BCW data set using K-means algorithm

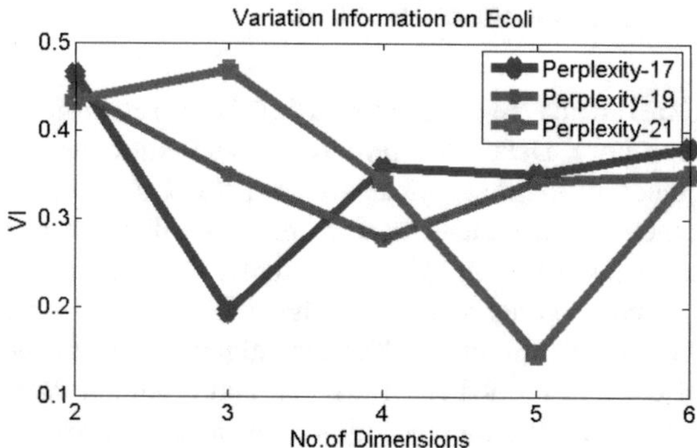

Fig. 2 VI values on Ecoli data set using K-means algorithm

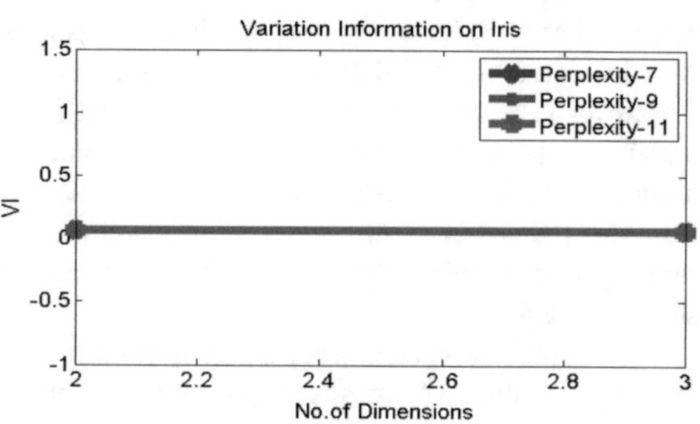

Fig. 3 VI values on Iris data set using K-means algorithm

data set is 0.46 and other perturbation methods minimum value for most of the lower dimensions is 3. Data utilization is high for the proposed method than all other perturbation methods for this data set.

The VI values for different dimensions on Iris data set are shown in Fig. 3. For this data set, our proposed method gives same VI values as other perturbation methods such as NMDS and PCA, that is, nearly zero for all lower dimensions and perplexity values. From observation of VI values on Iris data set, the proposed method provides maximum data utility. The VI values on Pima and Wine data sets are shown in Fig. 4 and Fig. 5, respectively.

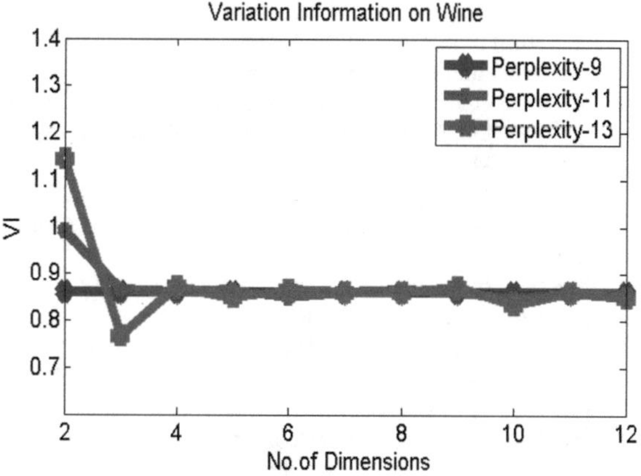

Fig. 4 VI values on Wine data set using K-means algorithm

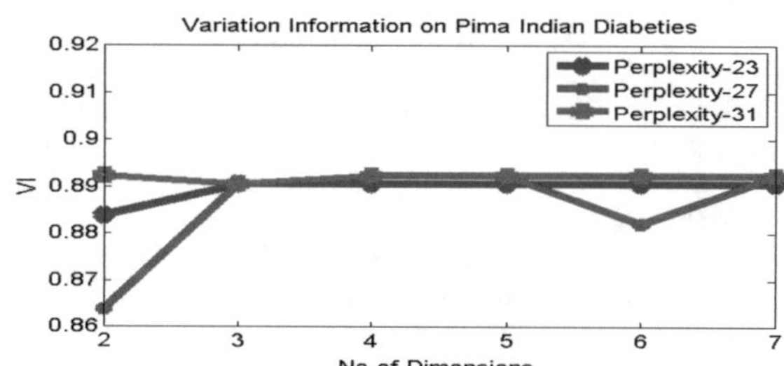

Fig. 5 VI values on Pima data set using K-means algorithm

6 Conclusion

In this work, we proposed a perturbation method for PPDM using t-SNE. The proposed method reduces the number of dimensions in original data and preserves the distance between records in modified data. It forms same clusters as original data when applying K-means algorithm. Our proposed method evaluated using variation information between original and modified clusters and compared with existing dimension reduction perturbation methods.

References

1. Traub, J.F., Yemini, Y., Woznia kowski, H.: The statistical security of a statistical database. ACM Trans. Database Syst. **9**(4), 672–679 (1984)
2. Samarati, P., Sweeney, L.: Protecting privacy when disclosing information: k-anonymity and its enforcement through generalization and suppression. In: Proceedings of the IEEE Symposium on Research in Security and Privacy, pp. 384–393 (1998)
3. Samarati, P., Sweeney, L.: Generalizing data to provide anonymity when disclosing information. In: Proceedings of the PODS, p. 188 (1998)
4. Yao, A.C.: How to generate and exchange secrets. In: Proceedings 27th IEEE Symposium on Foundations of Computer Science, pp. 162–167 (1986)

5. Vaidya, J., Clifton, C.: Privacy preserving association rule mining in vertically partitioned data. In: Proceedings of the Eighth ACM SIGKDD International Conference on Knowledge Discovery and Data Mining, pp. 639–644 (2002)
6. Vaidya, J., Clifton, C.: Privacy-preserving k-means clustering over vertically partitioned data. In: Proceedings of the Ninth ACM SIGKDD International Conference on Knowledge Discovery and Data Mining, pp. 206–215 (2003)
7. Alotaibi, K., Rayward-Smith, V.J., Wang, W., de la Iglesia, B.: Non-linear dimensionality reduction for privacy-preserving data classification. In: PASSAT-2012, IEEE, pp. 694–701 (2012)
8. Mendes, R., Vollela, J.P.: Privacy-preserving data mining: methods, metrics, and applications. IEEE Access **5**, 10562–10582 (2017)
9. Narwaria, M., Arya, S.: Privacy preserving data mining—a state of the art. In: International Conference on Computing for Sustainable Global Development (INDIACom), IEEE, pp. 2108–2112 (2016)
10. Mukharjee, S., Chen, Z., Gangopadhyay, A.: A privacy-preserving technique for euclidean distance-based mining algorithms using fourier-related transforms. VLDB J. **15**(4), 293–315 (2006)
11. Patel, V., Jain, Y.K.: Wavelet transform based data perturbation method for privacy protection. In: ICETET-2009, IEEE computer society, pp. 294–297 (2009)
12. Huang, J.-W., Su, J.-W., Chen, M.-S.: FISIP: A Distance and Correlation Preserving Transformation for Privacy Preserving Data Mining, TAAI, IEEE, pp. 101–106 (2011)
13. Lu, W.: Privacy Preserving Classification Algorithm Based Random Diffusion Map, SKG-2009, IEEE, pp. 318–321 (2009)
14. Upadhyay, S., Sharma, C., Sharma, P., Bharadwaj, P., Seeja, K.R.: Privacy preserving data mining with 3-D rotation transformation. J. King Saud Univ. Comput. Inf. Sci. 1–7 (2016). Elsevier
15. Wang, J., Zhang, J.: Addressing accuracy issues in privacy preserving data mining through matrix factorization. In: Proceedings of IEEE International Conference on Intelligence and Security Informatics, pp. 217–220 (2007)
16. Gokulnath, C., Priyan, M.K., Vishnu Balan, E., Rama Prabha, K.P., Jeyanthi, R.: Preservation of privacy in data mining by using PCA based perturbation technique. In: International Conference on Smart Technologies and Management for Computing, Communication, Controls, Energy and Materials (ICSTM), IEEE, 6–8 May 2015, pp. 202–206 (2015)
17. Hinton, G., Roweis, S.: Stochastic neighbor embedding. In: Advances in Neural Information Processing Systems, vol. 15, pp. 833–840. MIT Press (2003)
18. Bunte, K., Haase, S., Biehl, M., Villmann, T.: Stochastic neighbor embedding (SNE) for dimension reduction and visualization using arbitrary divergences. In: Neurocomputing 90, pp. 23–45 (2012). Elsevier
19. https://lvdmaaten.github.io/tsne/
20. van der Maaten, L., Hinton, G.: Visualizing data using t-SNE. J. Mach. Learn. Res. **9**, 2579–2605 (2008)
21. Meila, M.: Comparing clusterings—an information based distance. J. Multivar. Anal. **98,** 873–895 (2007). Elsevier

Spam Detection in SMS Based on Feature Selection Techniques

Aakanksha Sharaff

Abstract Short Message Service (SMS) has become the most effective and efficient means of communication. The popularity for SMS is due to their ease of use. But with the advent of this media, the problem of SMS spamming has increased. These are undesired and illegal messages which cause a lot of inconvenience to the users. So in order to get rid of these spam messages, classification algorithm along with feature selection techniques is incorporated. This technique helps us to choose better features and provides a better accuracy. The irrelevant and redundant attributes that do not contribute to the accuracy of the model are removed using the feature selection techniques. The complexity of the model is reduced as only fewer attributes are desired and the simpler model is easy to understand. A comparative study of different algorithms has also been studied in this work.

Keywords SMS · Classification · Feature selection · Spam detection

1 Introduction

Communication is an important aspect of information sharing between individuals and organizations. Today, almost all the users are using Short Message Service (SMS) as their professional and personal communication media. SMS is a text-based communication service for mobile systems. As the number of SMS sent/received day by day increases, it gave rise to the problem of spam messages. Spams are irrelevant and unsolicited messages that are sent to large number of users for the purpose of marketing or spreading malware, etc. Users receive spam messages in bulk which fill the inboxes of the users and make it difficult for them to manage their messages properly. Spam also many a times contains malware which can cause harm to the user systems. Unsolicited messages also known as spam can be separated from the

A. Sharaff (✉)
Department of Computer Science and Engineering,
National Institute of Technology Raipur, Raipur, India
e-mail: asharaff.cs@nitrr.ac.in

A. Abraham et al. (eds.), *Emerging Technologies in Data Mining and Information Security*, Advances in Intelligent Systems and Computing 813,
https://doi.org/10.1007/978-981-13-1498-8_49

actual desired messages by using SMS classification technique. The classification of messages in spam and ham makes it easy for the user to manage their messages well. The problem of classification of SMS is a popular one and there are many classification techniques but the technique of classification using feature selection is little explored. In this paper, feature selection techniques along with the classification algorithms are explored as it simplifies attributes for further classification. Feature selection technique basically is an attribute selection method. It automatically selects the attributes in the data that are relevant to the modeling problem. Feature selection method reduces the number of attributes in the data set by including or excluding attributes without changing them. The feature selection technique has three classes of algorithms: Filter, wrapper, and embedded. In this paper, filter-based feature selection methods are used. These methods are statistical. They assign a score to each attribute and then rank the attributes based on the score. This method considers the feature independently.

SMS classification is done on preprocessed data. The preprocessing of data includes tokenization, stopping, stemming, and for SMS an additional preprocessing technique of homoglyphing is also used. In tokenization, the strings in the dataset are broken down into tokens. The delimiter used in this work generally is the space. In stopping, the stop words are removed as these are the words that do not contribute significantly. Stemming is used for the conversion of any word to its root form. Homoglyphing is a technique for SMS which converts the similar looking word called homoglyphs into the actual word. After the preprocessing feature selection techniques, chi-square and information gain are used. Once the ranked attributes are obtained, classification algorithms such as Naïve Bayes, J48, and SVM are used. In this work, the problem of spam identification of SMS has been addressed and a solution using classification technique along with feature selection methods is provided. The author also performs a comparative study based on different algorithms for SMS.

2 Literature Review

There exist various machine learning algorithms which were compared by their performance to classify SMS text messages. Almeida et al. offered a real, public, non-encoded, and largest collection of SMS and also compared various machine learning methods [1]. They observed while comparing different algorithms that Support Vector Machine outperforms other algorithms. Uysal et al. comprehensively studied several feature extraction and selection mechanism and its impact on filtering spam messages from SMS collection [2]. Jain et al. proposed a Semantic Convolutional Neural Network (SCNN), a deep learning approach to detect spam messages in social network [3]. A nested bi-level classification approach has been presented by Nagwani and Sharaff to detect spam SMS and after detecting non-spam messages, a nested clustering approach has been proposed to identify threads in non-spam messages [4]. Various approaches were introduced for identifying actual features

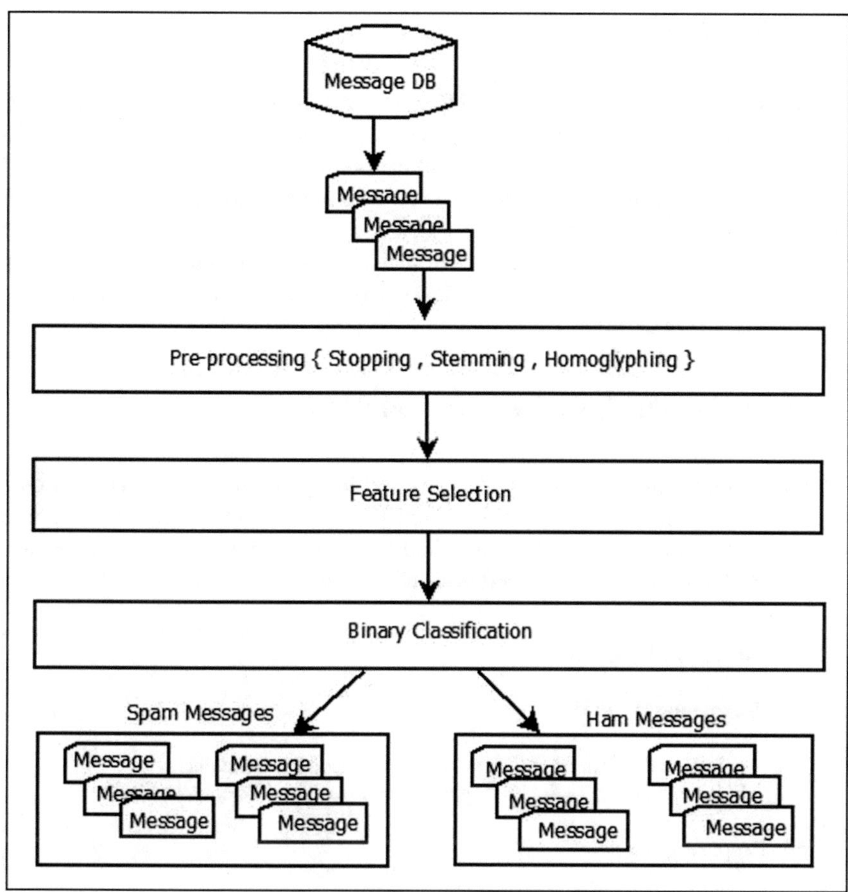

Fig. 1 Methodology for spam identification in SMS classification

which are responsible for detecting spam messages and a lot of techniques including soft computing, deep learning, and evolutionary techniques have been investigated [5–7]. Abdulla et al. developed a system for detecting worms by using KNN and Naïve Bayes algorithm [8]. Fonseca et al. [9] proposed an approach to reduce spam messages by identifying neighborhoods and classify these messages into good or bad neighborhoods and identify the source of spam. Sulaiman et al. considered the issues of SMS length, battery, and memory performances for identifying relevant features for detecting spam messages [10]. Jeong et al. developed a real synthetic social network dataset by proposing Social network Building Scheme for SMS spam detection (SBSS) algorithm without disclosing any private information [11]. Giannella et al. developed a content-based Bayesian classification approach by addressing the problem of unsupervised and semi-supervised SMS spam identification [12].

3 Methodology

The methodology for classification of SMS using feature selection is shown in Fig. 1. The initial work is to retrieve the SMS messages from the data source and gather them for analysis. To prepare the data for classification, the basic understanding of SMS structure is a prerequisite. SMS are unstructured messages as they are in particular do not follow a specific format.

Once the basic structure is understood and the messages are gathered from its data source, the data is prepared for classification. The first step in making the data ready for classification is preprocessing.

4 Algorithm for Spam Detection in SMS

In the algorithm, initially, the data is retrieved by gathering it from various data sources. Then for tokenization, word tokenizer is used to convert the document into attributes forming a document-term matrix. For preprocessing string to word vector, unsupervised attribute selection method is used in weka tool.

Algorithm: SMS Classification

Step 1: Retrieving the data

1a. Gather the data from data sources to retrieve the input data.
1b. Perform tokenization on input data to form Term Document Matrix.

Step 2: Preparing the data

2a. Apply pre-processing techniques- Stopping, Stemming and Homoglyphing in SMS.
2b. Rank the attributes according to its frequency.

Step 3: Feature selection

Apply one of the feature selection techniques (Chi-square and Information gain) to select the attributes.

Step 4: Classification

Apply any one classification techniques (Naïve Bayes, J48, and SVM) to classify messages into spam and non-spam (ham).

Step 5: Evaluation of result

Evaluate the results using performance parameters like Accuracy, Precision, Recall and F-measure.

The stopwords are removed using the standard stoplist. Lower case tokens have been considered in this work and the words are set to keep as 1000. After preprocessing, feature selection techniques have been applied namely chi-square attribute selection technique; Search method used along with chi-square method is Rankers search. Similarly, information gain method for attribute selection in supervised attribute filter has been applied. Search method used along with information gain is also Ranker search. After using either of the above feature selection techniques, classification algorithms namely Naïve Bayes Classifier in Bayes Classifier, J48 in decision tree classifier, and libSVM have been used for categorization. The classifiers categorize messages into spam and ham. Evaluation of results is then done using performance parameters like accuracy, precision, and recall.

The performance was evaluated using 10-fold cross validation technique where the input data is divided into 10 parts. Out of those 10 parts, 9 parts are used for training and remaining is used for testing. The above method is repeated for 10 times and the mean of results is calculated.

5 Experiment and Result Analysis

The experiments are performed using Java programming language.

5.1 Datasets

The experiment has been performed by applying classification on SMS. The dataset is available at http://www.dt.fee.unicamp.br/~tiago/smsspamcollection/ for SMS. The total number of unique attributes formed after tokenization was 7566 and that after preprocessing was 1833 when the words to keep count are set to 1000.

5.2 Result Analysis

The comparative study of various classification algorithms like Naïve Bayes, J48, and SVM has been studied and aims to find out the most efficient classifier. Along with these, the experiment also compares the performance of classifier with and without feature selection techniques. The most efficient classifier (Naïve Bayes, J48, and SVM) for message classification without feature selection technique and the most efficient classifier (Naïve Bayes, J48, and SVM) for message classification with feature selection technique (Chi-square, Information Gain) has been identified.

The performance analysis of result is done using four performance parameters—accuracy, precision, recall, and F-Measure. The values of these performance parameters for abovementioned cases are shown in Tables 1, 2 and 3.

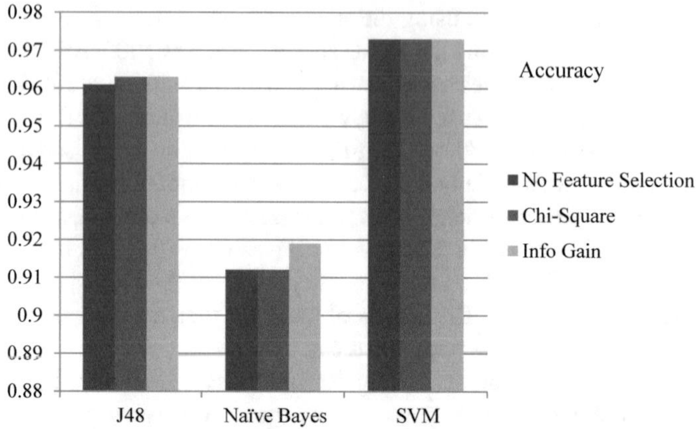

Fig. 2 SMS spam detection accuracy

From the above figures (Figs. 2. 3 and 4), it has been observed that without using any feature selection technique, SVM gives the highest accuracy of 97.3% followed by J48 which gives an accuracy of 96.1%. For the dataset used in this experiment, Naïve Bayes classifier gives the least accuracy of 91.1% among all the three classifiers. When using feature selection technique, SVM does not depict any variation in the accuracy while both J48 and Naïve Bayes show a slight improvement in the accuracy. From the two feature selection technique used, info gain performs better than chi-square in case of Naïve Bayes while for J48 and SVM, accuracy remains same for both chi-square and info gain. So overall performance was better using info gain in case of Naïve Bayes and J48 and applying no feature selection technique in case of SVM.

Table 1 Performance parameters for classification of various classifiers without feature selection

Algorithm	Accuracy	Precision	Recall	F-measure
Naïve Bayes	0.911	0.934	0.912	0.918
J48	0.961	0.960	0.961	0.960
SVM	0.973	0.974	0.973	0.972

Table 2 Performance parameters for classification of various classifiers with feature selection technique—chi-square

Algorithm	Accuracy	Precision	Recall	F-measure
Naïve Bayes	0.911	0.934	0.912	0.918
J48	0.963	0.963	0.963	0.962
SVM	0.973	0.974	0.973	0.972

Table 3 Performance parameters for classification of various classifiers with feature selection technique—information gain

Algorithm	Accuracy	Precision	Recall	F-measure
Naïve Bayes	0.918	0.935	0.919	0.924
J48	0.963	0.963	0.963	0.962
SVM	0.973	0.974	0.973	0.972

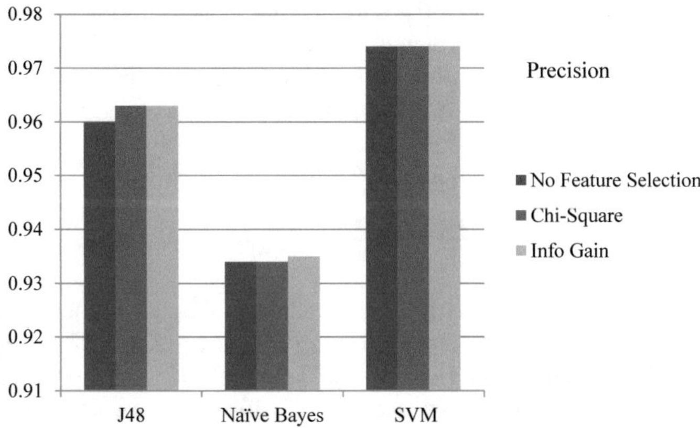

Fig. 3 SMS spam detection precision

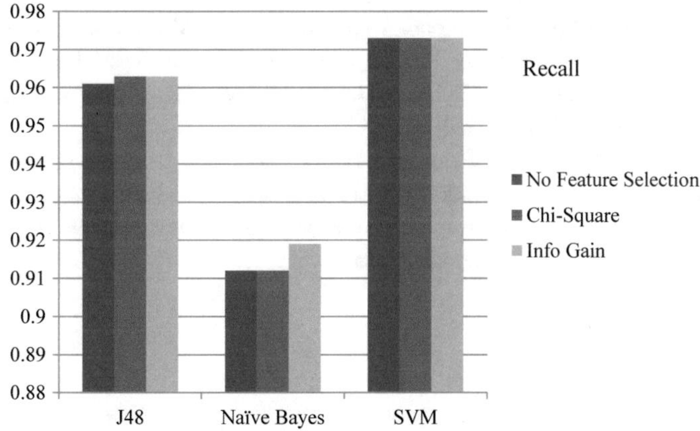

Fig. 4 SMS spam detection recall

6 Discussion on Results

The performance of classifiers most importantly depends on the type of dataset provided as input. From the analysis done above, it has been observed that SVM gives the highest accuracy with or without feature selection technique and its performance does not increase when using a feature selection technique before classification. One of the possible explanations for this can be that SVM does not depend upon the dimensionality of the feature space as for training, it is based on the dual optimization problem which does not explicitly mention the number of features. So SVM being independent of the dimensionalities of feature does not show any change even when the dimensionality of the feature is reduced by using feature selection techniques. SVM uses regularization to avoid over fitting. So until the regularization parameter C is provided appropriately, SVM will work fine even without feature selection. In the case of SVM, feature selection technique might not improve the accuracy but it makes the model more reliable. For feature selection techniques, filter-based technique is used instead of wrapper class as it is computationally more effective.

7 Conclusion and Future Work

In this work, a comparative study of various classifiers with and without feature selection techniques is presented. The results illustrated the effect of using feature selection techniques before classification. Using feature selection technique improved the performance of two classifiers while one of the classifiers remains unaffected. So the overall result shows that using feature selection technique is highly recommended in the domains of SMS classification as they are high dimensional in nature. From the results, it has been concluded that among information gain and chi-square, information gain has the most positive impact. In classification, SVM classifier outperforms all the other classifiers. Another finding from result is that SVM has no effect of feature selection technique while Naïve Bayes and J48 show slight improvement on the use of feature selection in SMS.

In future, more number of classifiers can be compared and also some work can be done on the theoretical foundations to make SMS classification more efficient. Next future direction can be on feature ordering and feature re-ranking of SMS to improve the efficiency of text classification.

References

1. Almeida,T.A, Hidalgo, J.M.G., Yamakami, A.: Contributions to the study of SMS spam filtering: new collection and results. In: Proceedings of ACM symposium on document engineering (DocEng'11), Mountain View, CA, pp. 1–4 (2011)
2. Uysal, A.K., Gunal, S., Ergin, S., et al.: The impact of feature extraction and selection on SMS spam filtering. Electron. Electr. Eng. **19**(5), 67–72 (2013)
3. Jain, G., Sharma, M., Agarwal, B.: Spam detection on social media using semantic convolutional neural network. Int. J. Knowl. Discov. Bioinform. (IJKDB) **8**(1), 12–26 (2018)
4. Nagwani, N.K., Sharaff, A.: SMS spam filtering and thread identification using bi-level text classification and clustering techniques. J. Inf. Sci. **43**(1), 75–87 (2017)
5. Suleiman, D., Al-Naymat, G.: SMS Spam Detection using H_2O Framework. Proc. Comput. Sci. **113**, 154–161 (2017)

6. Arifin, D.D., Bijaksana, M.A.: Enhancing spam detection on mobile phone Short Message Service (SMS) performance using FP-growth and Naive Bayes Classifier. In: Wireless and Mobile (APWiMob), 2016 IEEE Asia Pacific Conference on 2016, pp. 80–84
7. Choudhary, N., Jain, A.K. Towards filtering of SMS spam messages using machine learning based technique. In: Advanced Informatics for Computing Rcscarch Springer, Singapore 2017, pp. 18–30
8. Abdulla, S., Ramadass, S., Altyeb, A.A., et al.: Employing machine learning algorithms to detect unknown scanning and email worms. Int. Arab J. Inf. Technol. **11**(2), 140–148 (2014)
9. Fonseca, O., Fazzion, E., Las-Casas, P.H.B., et al.: Neighborhoods and bands: an analysis of the origins of spam. J. Internet Serv. Appl. **6**(1), 9 (2015)
10. Sulaiman, N.F., Jali, M.Z.: A new SMS spam detection method using both content-based and non content-based features. In: Advanced Computer and Communication Engineering Technology. Springer, Cham, pp. 505–514 (2016)
11. Jeong, S., Noh, G., Oh, H., Kim, C.K.: A re-configuration scheme for social network based large-scale SMS spam. J.KIISE **42**(6), 801–806 (2015)
12. Giannella, C.R., Winder, R., Wilson, B.: (Un/Semi-) supervised SMS text message SPAM detection. Nat. Lang. Eng. **21**(4), 553–567 (2015)

Recent Trends of Data Mining in Cloud Computing

Abhijit Sarkar, Abhishek Bhattacharya, Soumi Dutta and Komal K. Parikh

Abstract In recent times, data mining plays one of the crucial aspects in business intelligence tasks by extracting useful pattern and future prediction. Cloud computing, on the other hand, is a topical trend in the field of providing computing resources as a service over the network. Combining data mining in cloud computing is a recent trend in knowledge discovery field as because no large number of resolutions are effusively accomplished and accessible to the cloud clients. This paper presents the basic concepts of data mining in cloud framework along with relevant significant works done in this field. Different frameworks along with approaches for diverse data mining tasks have been surveyed and presented in detail.

Keywords Cloud data mining · Classification · Clustering

1 Introduction

The progression of extorting convenient and valuable information and knowledge from a set of unprocessed data is recognized as Data mining. Generally, data mining includes the following foremost tasks:

(1) Association Rule Learning: It is the process of finding one event as a result of occurrence of another event.

A. Sarkar (✉) · A. Bhattacharya · S. Dutta · K. K. Parikh
Department of Computer Application, Institute of Engineering & Management, Kolkata 700091, India
e-mail: abhi41001@gmail.com; abhijit.sarkar@iemcal.com

A. Bhattacharya
e-mail: abhishek.bhattacharya@iemcal.com

S. Dutta
e-mail: soumi.it@gmail.com

K. K. Parikh
e-mail: komalparikh09@gmail.com

A. Abraham et al. (eds.), *Emerging Technologies in Data Mining and Information Security*, Advances in Intelligent Systems and Computing 813,
https://doi.org/10.1007/978-981-13-1498-8_50

(2) Classification: It refers to the process of categorization where different objects are grouped into particular categories on the basis of their characteristics.
(3) Clustering: It is considered as a process to make several groups what we call as cluster maintaining a constraint that objects part of same group will hold one or more same attribute in comparison to objects in other groups.

On the note of introducing Cloud computing, it refers to everything that involves bringing computing resources as a services rather than a product over the Internet [1, 2]. In Cloud Computing facility; software, information added with resources are provided to computers and other devices as a utility over a network, typically Internet. Typical form of Cloud Computing areas:

(1) Client–Server model: It tags along a Client–server sculpt of computation.
(2) Grid computing: It provides a structure of distributed computing facility, where a cluster of networked, loosely coupled machines makes up a super along with virtual machine which is powerful enough for bulk processing.
(3) Mainframe computer: Outsized organizations use controlling machines for decisive applications, typically bulk data processing.
(4) Utility computing: Utility metered service is provided to the end-users as a package of computation and resources.

There are typically four types of cloud services as shown in Fig. 1.

(1) IaaS (Infrastructure as a Service): Computer infrastructure is provided as a utility service in this layer, typically a virtualized environment.
(2) PaaS (Platform as a Service): In this layer, a platform is provided using a cloud infrastructure or facility.
(3) DaaS (Data as a Service): Data storage as a service is provided in this layer to the user, which is a subspecies of IaaS and PaaS and sits on top of these layers.
(4) SaaS (Software as a Service): Application software is provided over the Internet or Intranet via a cloud infrastructure in this layer [1].

Mentioned below are some of the advantages and disadvantages of this technology. Cloud computing which is available at much cheaper rates is almost undoubtedly the foremost cost competent means to use, maintain. Massive storage facility is achieved by using cloud computing structure. Now that the data stays in the cloud facility, backup and restoration happens much easily. Also integrating the software can be done in cloud facility without any volunteer or assistance. On the note of Mentioning some of the disadvantages of this technology, cloud computing does face Technical Issues. But system may face grave problem at times. The other key issue is that all the company's susceptible information will be surrendered to a third-party cloud facility provider. Hence, the most reliable service provider is being chosen to store information strongly. Further in this paper, Sect. 2 describes about cloud data mining. In Sect. 3, related work in accordance with approach and framework have been presented. Section 4 concludes the article with open research.

Fig. 1 Cloud service layers

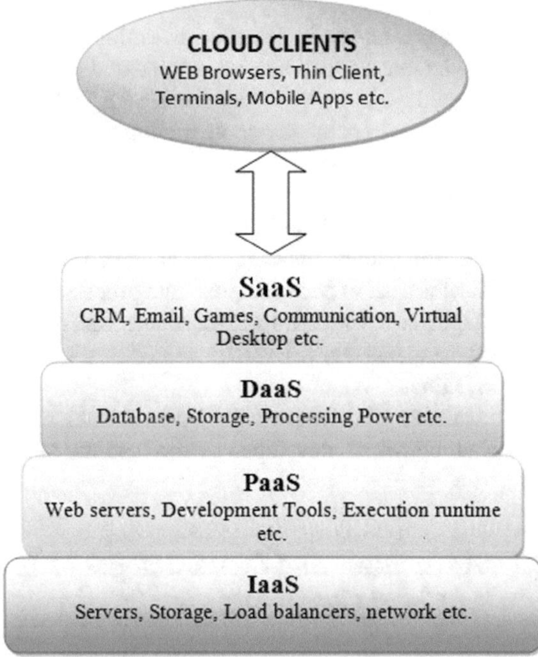

2 Cloud Data Mining

Recently data mining and its hands-on application have converted indispensable in cloud computing competence. As cloud computing is piercing significantly in all areas of scientific calculation and computation, it is eventually becoming an enormous area for practicing data mining. Clouds of servers is needed for cloud computing facility in the handling of jobs. From the web sources of data different structured information is retrieved from a pool of unstructured data sources or raw data which used different cloud servers in cloud data mining. The result assists different establishments to accomplish software, data and services in a central way maintaining reliability [1]. As a consequence, customer pays solitary for tailored data mining facility as an alternative of paying a lot of money for the uncut offline data mining suite which is not required at all. Leading cloud computing service providers are Amazon Web Services,[1] Microsoft Azure,[2] OpenStack[3] etc. On these foundations, different data mining open service frameworks based on cloud computing have been developed and made available commercially for small scale data mining.

[1] http://aws.amazon.com/.

[2] http://azure.microsoft.com/en-in/.

[3] http://www.openstack.org/.

(1) Weka4WS
Weka is extensively practiced data mining toolkit that runs on a solitary device. It is available as open-source. An updated version of it, Weka4WS,[4] supports distributed skeleton. Weka and the WSRF Technology[5] is integrated by Weka4WS. This helps to manage remote mining algorithm along with distributed workflow [2].

(2) BC-PDM
In 2009, China Mobile Institute announced the platform for development and test "Big Cloud" including the analogous data mining outfit BC-PDM [3]. BC-PDM is a set of accumulating data processing investigation and mining system. MapReduce on cloud infrastructure based BC-PDM is a SaaS tools which is used expansively [2].

(3) PDMiner
PDMiner is a equivalent dispersed data mining computing stage based on Hadoop, which developed by the Institute of Computing Technology.[6]

2.1 Advantages and Disadvantages of Using Data Mining in Cloud Computing

Cloud computing in data mining can present dominant capacity of storage, computing, and outstanding resource supervision. Cloud computing comes up with providing great facilities in the process of data mining and analysis. The crucial concern with the data mining progression is with the space constraint and which is considered to be colossal. Whereas, if data mining is combined with cloud computing, it can save a substantial quantity of space as because the client does not have to take headache about the space or computation for data mining. All the vital analysis operations are actually being performed in the service provider side. Also, data mining in cloud computing is cost beneficial. The question of the hour is "How safe is your data?" when you are performing data mining in cloud. The decisive issue in cloud data mining is that the data is not protected as cloud flair has the aggregate control of essential substructure. Also, it is totally depending on Internet connectivity. So, a possible downtime in cloud computing makes small business facility totally dependent on the service providers.

[4]http://weka4ws.wordpress.com/.

[5]https://www.oasis-open.org/.

[6]http://english.ict.cas.cn/.

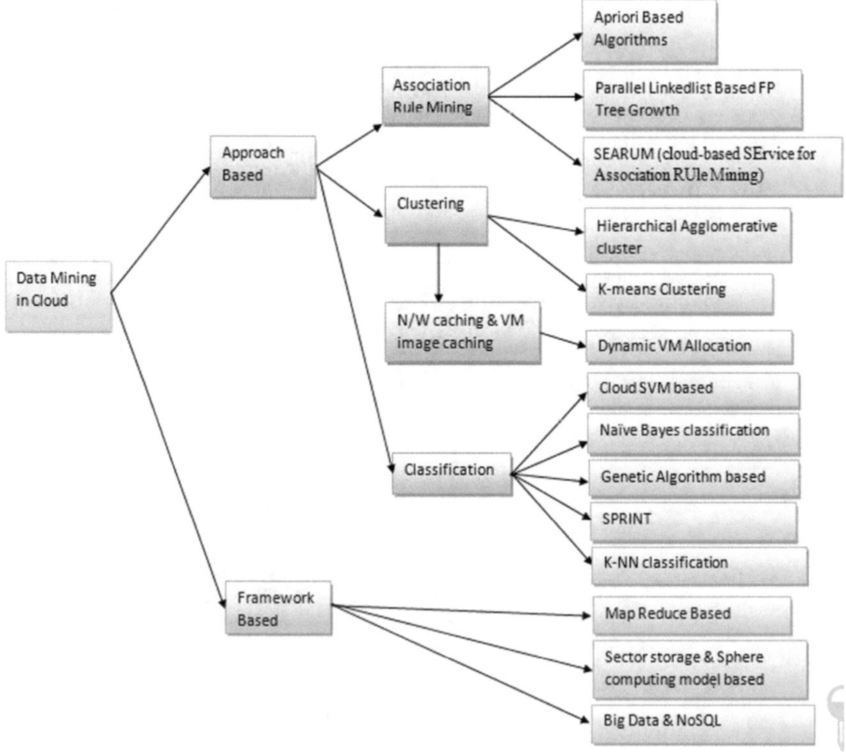

Fig. 2 Classification of different aspect of data mining in cloud

3 Related Work

In this part of the paper, different significant research work in the field of data mining in cloud computing have been surveyed and presented briefly. We have divided them into two major categories, frameworks used for cloud data mining and approaches proposed for different data mining task in cloud paradigm. A detail overview of the classification is presented in Fig. 2. In the next two subsections, we have studied the above two categories in great detail.

3.1 Cloud Framework Based

3.1.1 MapReduce-Based Framework

The very first approach that has been surveyed is the widely used MapReduce model: a parallel programming skeleton proposed by Google in 2004. This model allows

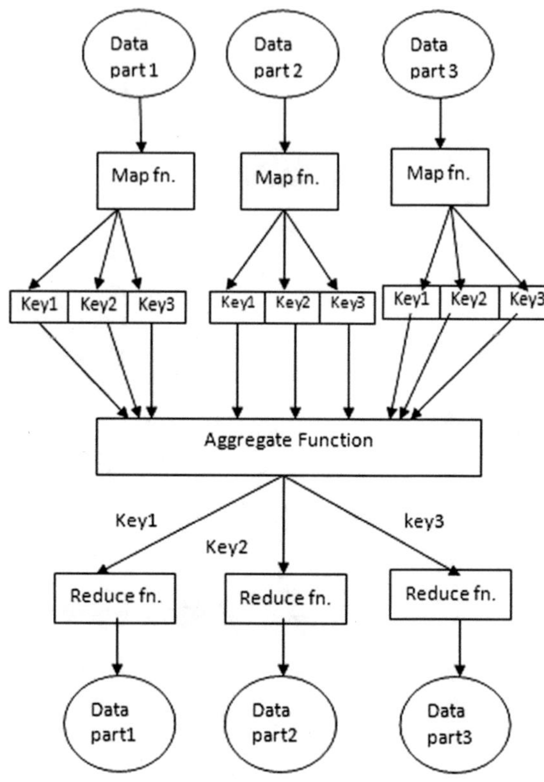

Fig. 3 MapReduce architecture

users to handle large scale data effectively and with ease. Map and Reduce forms the substantial operations:

- Map This process reads "raw data" from a scattered file system, filters and generates key-value pairs which are used in the next process.
- Reduce This process reads all possible key-value pairs generated by Map, combines all the values with respect to particular key value and present them.

A conceptual model of MapReduce framework for cloud data mining is shown in Fig. 3.

In the implementation procedure of Google MapReduce architecture, Apache Hadoop have been widely adopted. Hadoop Distributed File System (HDFS) can be deployed on reliable clusters achieving reliability through the replication of files and data. In order to accomplish effectiveness in the field of parallel data mining algorithm, Chu et al. [4] have presented MapReduce for machine learning on multi-core machine which fits Statistical Query Model. Patil et al. have also shown MapReduce Framework and its processing in detail [5]. Execution overview were shown effectively with Map step containing input reader, map function, partition function and reduce step containing comparison function, reduce function, output writer.

3.1.2 Big Data and NoSQL Framework

Robert Vrbic have presented Big Data and its related NoSQL database that proved to be a supreme technique for storing very large amounts of data in scattered systems [6]. NoSQL concept has the following properties:

- Scalability: Facility to respond involuntarily in accordance with the increase in the application.
- Replication: Data is duplicated in many nodes in order to support distributed database.
- Partitioning Data: Data distribution in a way that the different parts of the database are in different nodes.

There are several key factors that have influenced the appearance and development of NoSQL databases, including continued growth of data production, growing demands for processing semi-structured and non-structured data, avoiding composite and costly object-relational mapping, cloud computing requests etc. as mentioned in the proposed approach [6]. NoSQL databases is Key-Value paired, Document, Graph, and Column oriented. So, this can be considered to be another wing of improvement in Big Data management domain.

3.1.3 Sector and Sphere Framework

Grossman et al. [7] have presented a new concept of mining in cloud computing environment with the disentanglement of storage and computing services. The indispensable computation is performed using a message passing mechanism. Sector can work as the storage to massive datasets. On the other hand, Sphere executes different functions in parallel stream wise over the Sector managed data via user defined functions. A conceptual Sector and Sphere model for cloud data mining has been shown in Fig. 4.

3.2 Data Mining Approaches

3.2.1 Association Rule Mining Approaches

Apriori-Based Algorithms

Various schemes were projected to improve Apriori algorithm's competence of generating recurrent item set and applying it in a distributed approach which can be applied in cloud computing infrastructure. In earlier days, focus was mainly on reducing candidate itemsets, frequent itemset based on pruning and minimizing database scan. On the track of improving Apriori algorithm, Park et al. [8] described a new technique

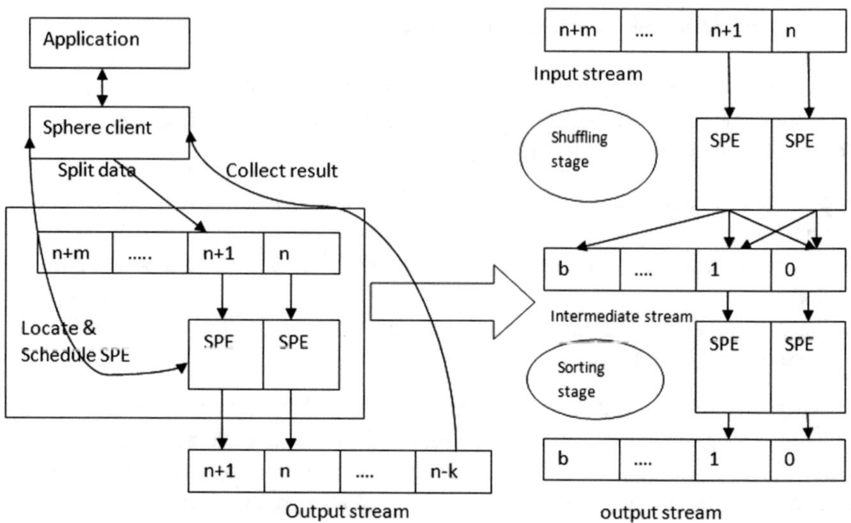

Fig. 4 Sector and sphere model

of hashing to trim the candidate set. The former the recurrent itemsets are separated, the sooner would be the mining.

Another algorithm was proposed by Li et al. in [9] have presented implementation of the popular Apriori algorithm in a parallel way named as PApriori. It pursued MapReduce programming model framework which can be used in cloud environment's distributed cluster nodes. The method of tracking the occurrence of crucial candidate of size k in parallel in controlled by map function whereas summing the occurrence count is done by reduce function. The input record dataset is stocked up on HDFS as a series file of [key, value] pairs. In accordance with the experimental results shown, the extent of parallelism increases with the augment of reducers.

Li et al. [10] have presented another version of Apriori based on supplementary experimental implementation of Apriori over MapReduce. In the first phase of the approach, dataset will be secluded. In the second phase, Apriori algorithm is applied in accordance with the MapReduce. Combined output is stored in temporary files which are used by the reducer to sum up all associated values with respect to key.

Cheung et al. [11] introduced the radical initiative of local and global recurrent itemset separation. Pruning infrequent resident support count in the foundation of resident support count have been described in the methodology along with how remaining candidate set is broadcasted to get other support count. Then on the basis of reaction, it is resulted whereas large itemsets are globally frequent or not.

FP Tree Based Algorithms

In [12], authors have proposed an approach in which candidate set creation is not requisite for getting the repeated itemset. This algorithm requires merely two complete I/O scans of the dataset which utilizes frequent-pattern tree or FP tree.

Conformist FP-Growth algorithm suffers in mining colossal datasets resulting follow up processing requires massive computer storage and computing facility which is a major drawback. Zhou et al. [13] have presented an approach based on FP tree which can be considered a very novel approach in this relevant field. A new approach of linked list based FPG algorithm on MapReduce programming model have been presented by the author which uses linked list for the purpose of recording the connections between the FP Tree nodes. As a result, it saves time as well as space complexities by a huge factor. It is found that the running efficiency of the new algorithm [13] in dealing with massive datasets is much higher.

SEARUM

Apilietti et al. [14] have presented a cloud-based essential service, named SEARUM (a cloud-based SErvice for Association RUle Mining), for mining association rules in a much effective way. The task concomitants with the SEARUM construction are as follows:

- Procuring network measurement
- Records preprocessing
- Computing Item regularity
- Itemset excavating or mining
- Extraction of potential rules
- Rule accumulation and categorization.

No matter how bulky data is there for processing, the SEARUM architecture takes a very fewer period for rule extraction, accretion, and sorting course. Both the horizontal scalability of the advancement as well as the effectiveness of the understanding generated has been addressed by the author properly.

3.2.2 Clustering-Based Approaches

Mahendiran et al. in [15] proposed the implementation approach of K-means clustering algorithm specifically in cloud computing environment. It was mainly implemented in Google Cloud using Google App Engine with cloud SQL showing significant result in cloud framework over the popular IRIS dataset.

In [16], Shindler et al. proposed fast K-means clustering algorithm applicable for huge dataset accessed consecutively. Approximate nearest-neighbor algorithms were applied to compute the facility assignment of each point in this approach. The experimental result shows the efficiency of K-means clustering in comparison to the conventional Divide and Conquer algorithm.

In [17], Dutta et al. have proposed a micro-blogging data clustering approach combining classical clustering approach K-Means and optimization technique such as genetic algorithm over micro-blogging data which is applicable on Cloud Data.

Srivastava et al. in [18] have presented a way out of implementing Hierarchical Agglomerative Clustering algorithm making it suitable for huge dataset focusing mainly on increasing the efficiency by executing tasks in parallel. The subsets of database are read and clustering algorithm is applied on these subsets. After that, results are combined with the previous samples and procedure is continued until all the data is available in main cluster. Experimental results show the efficiency of this algorithm which increases by parallelism on cloud-based framework.

More recently, Panchal et al. [19] have focused on scalability issues of data mining in cloud computing. I/O parameters were chosen for increasing the performance. The radical concept prevents excessive hits to the database. The proposed method reduces data traffic and also minimizes the database hits. In another paper [20], Bhupendra Panchal et al. proposed the concept of dynamic virtual machine allocation algorithm using Clustering in cloud computing. Proposed VM allocation uses K-means clustering for generating different significant clusters. This dynamic VM allocation with clustering algorithm shows significant result in each data center taken for experiment.

Comparisons on different clustering algorithms in cloud platform have been presented in Table 1.

3.2.3 Classification-Based Approaches

Classification is one of the major data mining technique used for extracting models, describing important classes. It is used to predict group membership in database. In this regard, some of the major approaches following classification which is suited for cloud computing were surveyed.

Ozgur CATAK has proposed in [21] the concept of CloudSVM. The approach of training a distributed SVM (Support Vector Machine) was proposed for cloud computing environment drastically.

Zhou et al. [22] have proposed classification algorithm on Cloud Computing environment on the basis of parallel implementation supported by modified Naive Bayes Classification algorithm. The Naive Bayes classification algorithm comprises of training stage making the use of Map and Reduce policy and prediction stage to predicate the data record with the output of the training model.

Ding et al. [23] have proposed a different approach for classification based on Genetic Algorithm (GA) embedded classification rule mining in cloud computing framework. The phases of GA as mentioned in the paper are encoding, initial population and fitness function, selection, crossover, and mutation operators' functionality. The experimental result over cloud resources shows remarkable improvement of classification accuracy of the proposed GA over traditional genetic classification.

Table 1 Comparison of different clustering approaches

Approach proposed by	Algorithm used	Features	Applicable platform	Micro/Macro clustering assisted
Mahendiran [15]	K-means clustering	Clustering implementation on cloud	Google cloud using Google app engine	No
Shindler [16]	K-means clustering	Fast K-Means based on approximate nearest-neighbor	For huge data	No
Dutta [17]	Optimized K-means clustering	Fast K-Means based on feature selection	Micro blogging data	No
Srivastava [18]	Hierarchical agglomerative clustering	Parallelism in cloud	Huge data set	Yes
Panchal [20]	Network caching VM image caching	Scalability, load sharing, reduction in data traffic	Cloud data	No
Panchal [19]	Dynamic VM Allocation using K-means clustering	Upgradation to network VM Image caching, load sharing by virtual machine	Cloud sim	No

On the note of further improvement, Zhang et al. [24] have proposed the idea of decision tree based improved classification mining SPRINT algorithm for suiting cloud environment. The proposed improved SPRINT algorithm firstly divides the training data set and submits to different nodes which generates attribute list in parallel and scans the attribute list. Subset data is recursively partitioned till each partition contains members from a same class or the partition is very small.

The very latest approach has been surveyed which is proposed by Wang [25] which have addressed the privacy issue. The author has proposed K-NN classification algorithm for privacy preserving in cloud computing. The focus is on using privacy matching technology to intersect the record of database between nodes. The author has proposed private matching protocol ensuring Binary Weighted Cosine (BWC) metric similarity measures.

A comparison on different classification algorithms in cloud platform have been presented in Table 2.

Table 2 Analysed classification approached comparison

Approach proposed by	Algorithm used	Addressed issues	Applicable platform
Catak [21]	Cloud support vector machine	Scalability, parallelism of split cloud data	Cloud data
Zhou [22]	Naïve Bayes classification	Parallel implementation	Hadoop
Ding [23]	Canonical genetic algorithm based classification	Improvement over traditional classification	Cloud data
Zhang [24]	SPRINT	Decision tree based improved classification mining, MapReduce based	Hadoop, Ubuntu
Wang [25]	K-NN classification	Privacy preservation following BWC metric, accuracy efficiency privacy of data has been addressed	Cloud data

4 Conclusion and Open Research

Cloud computing has been emerged as a technique to manage massive client datasets efficiently. Now with the growth of datasets, large scale of data processing is actually becoming a crucial point of information technology these days. Data mining through the cloud computing is an essential feature in today's businesses. It helps in making practical, knowledge driven decisions from the predicted future trends and behaviours from big data. This paper has provided a synopsis of data mining in cloud computing, all the momentous research work advancement relevant to this field and their overview. Here we have reviewed different approach like from earlier and overused MapReduce Model, Apache Hadoop implementation of MapReduce to the novel approach involving Linked list based FP Tree, SEARUM, Sector/Sphere, Parallel data mining, different clustering and classification algorithm to support cloud framework. Most importantly, as a survey observation, it is noticed that most research work is going on to apply mining in parallel, decomposing the main task into different independent sub-task and performing it parallel at the same time to process the huge set of data as fast as possible. A vital issue in this field of research work is security as all the user data becomes easily accessible in the service providerend. As a future advancement, along with making the process faster, supporting distributed computing effectively, securing the client data from service provider end is also crucial.

References

1. Ambulkar, B., Borkar, V.: Data mining in cloud computing. Int. J. Comput. Appl. 23–26 (2012)
2. Geng, X., Yang, Z.: Data mining in cloud computing. In: Proceedings of International Conference on Information Science and Computer Applications ISCA, 2013, pp. 1–7
3. Yu, L., Zheng, J., Shen, W.C., Wu, B., Wang, B., Qian, L., Zhang, B.R.: BC-PDM data mining, social network analysis and text mining system based on cloud computing. In Proceedings of the 18th ACM SIGKDD International Conference on Knowledge Discovery and Data Mining, pp. 1496–1499 (2012)
4. Chu, C.T., Kim, S.K., Lin, Y.A., Yu, Y.Y., Bradski, G., Ng, A.Y., Olukotun, K.: Map-reduce for machine learning on multicore. Adv. Neural Inf. Process. Syst. J. **19**, 281–287 (2007)
5. Patil, V., Nikam, V.B.: Study of Data mining algorithm in cloud computing using MapReduce Framework. J. Eng. Comput. Appl. Sci. **2**(7), 65–70 (2013)
6. Vrbic, R.: Data mining and cloud computing. J. Inf. Technol. Appl. 75–87 (2012)
7. Grossman, R., Gu, Y.: Data mining using high performance data clouds: experimental studies using sector and sphere. In: Proceedings of 14th ACM SIGKDD International Conference on Knowledge Discovery and Data Mining, Las Vegas, USA, 2008, pp. 920–927
8. Park, J.S., Chen, M., Yu, P.S.: An effective hash-based algorithm for mining association rules. In: Carey, M., Schneider, D. (eds.) Proceedings of ACM SIGMOD International Conference on Management of Data, pp. 175–186. ACM, New York, USA (1995)
9. Li, N., Zeng, L., He, Q., Shi, Z.: Parallel implementation of Apriori algorithm based on MapReduce. Int. J. Netw. Distrib. Comput. **1**(2), 89–96 (2013)
10. Li, J., Roy, P., Khan, S.U., Wang, L., Bai, Y.: Data mining using clouds an experimental implementation of Apriori over map-reduce. In: Proceedings of 12th IEEE International Conference ScalCom, Dec 2012
11. Cheung, D.W., et al.: A fast distributed algorithm for mining association rules. In: Proceedings of Parallel and Distributed Information Systems, pp. 31–42. IEEE CS Press (1996)
12. Han, J., Pei, J., Yin, Y.: Mining frequent patterns without candidate generation. Data Mining Knowl. Discov. J. USA **8**(1), 53–87 (2004)
13. Zhou, L., Wang, X.: Research of the FP-growth algorithm based on cloud environment. J. Softw. **9**(3) (2014)
14. Apiletti, D., Baralis, E., Cerquitelli, T., Chiusano, S., Grimaudo, L.: SEARUM: a cloud-based service for association rule mining. In: Proceedings of 12th IEEE International Conference ISPA, Melbourne, July 2013, pp. 1283–1290
15. Mahendiran, A., Saravanan, N., Sairam, N., Subramanian, V.: Implementation of K-Means clustering in cloud computing environment. Res. J. Appl. Sci. Eng. Technol. **4**, 1391–1394 (2012)
16. Shindler, M., Wong, A., Meyerson, A.: Fast and accurate k-means for large datasets. In: Proceedings of Advances in Neural Information Processing Systems NIPS, pp. 2375–2383 (2011)
17. Dutta, S., Ghatak, S., Ghosh, S., Das, A.K.: A genetic algorithm based tweetclustering technique. In: 2017 International Conference on Computer Communication and Informatics (ICCCI), pp. 1–6 (2017)
18. Srivastava, K., Shah, R., Swaminarayan, H., Valia, D.: Data mining using hierarchical agglomerative clustering algorithm in distributed cloud computing environment. Int. J. Comput. Theory Eng. **5**(3) (2013)
19. Panchal, B., Kapoor, R.K.: Performance enhancement of cloud computing. Int. J. Eng. Adv. Technol. **2**(5) (2013)
20. Panchal, B., Kapoor, R.: Dynamic VM allocation algorithm using clustering in cloud computing. Int. J. Adv. Res. Comput. Sci. Softw. Eng. **3**(9) (2013)
21. Catak, F.O., Balaban, M.E.: CloudSVM: training an SVM classifier in cloud computing systems. Comput. Res. Repos. J. (2013). arXiv:1301.0082
22. Zhou, L., Wang, H., Wang, W.: Parallel implementation of classification algorithms based on cloud computing environment. Indones. J. Electr. Eng. **10**(5), 1087–1092 (2012)

23. Ding, J., Yang, S.: Classification rules mining model with genetic algorithm in cloud computing. Int. J. Comput. Appl. **48**(18), 888–975 (2012)
24. Zhang, L., Zhao, S.: The strategy of classification mining based on cloud computing. In: Proceedings of International Workshop on Cloud Computing and Information Security, pp. 57–60 (2013)
25. Wang, J.: A novel K-NN classification algorithm for privacy preserving in cloud computing. Res. J. Appl. Sci. Eng. Technol. **4**(22), 4865–4870 (2012)

Impact of Data Mining Techniques in Predictive Modeling: A Case Study

Suparna DasGupta, Soumyabrata Saha and Suman Kumar Das

Abstract Data mining is the most imperative step for the unearthing of knowledge discovery in the database and has been considered as one of the noteworthy parts in the prediction. Research in data mining continues to grow in business and in the organization of learning in the coming decades. This paper represents a case study to envisage whether a person would play any outdoor games or not and by studying diverse prognostic methods on the basis of diverse weather conditions the required decision has taken care of. This article shows the significance of data mining techniques in predictive modeling. Several data mining algorithms based on classification have been used and the necessary statistics are generated based on all cataloging algorithms. This revise makes a mention of various techniques that are likely to be chosen for the prediction of time and highlights the analysis of the performance of the algorithms. It is used to predict and forecast the climate conditions of a specific region based on prehistoric data, which helps to save resources and prepare for future changes.

Keywords Data mining · Knowledge discovery · Transformation · Classification technique · Predictive modeling

1 Introduction

Data mining is the prevailing new technology that has engrossed a lot of attention in academia, business, and industrial research. Due to the colossal accessibility of hefty

S. DasGupta (✉) · S. Saha · S. K. Das
Department of Information Technology, JIS College of Engineering, Kalyani 741235, West Bengal, India
e-mail: suparna.dasgupta@jiscollege.ac.in

S. Saha
e-mail: soumyabrata.saha@jiscollege.ac.in

S. K. Das
e-mail: itsmesuman999@gmail.com

A. Abraham et al. (eds.), *Emerging Technologies in Data Mining and Information Security*, Advances in Intelligent Systems and Computing 813,
https://doi.org/10.1007/978-981-13-1498-8_51

amount of data and the requirement for whirling such data into functional information and knowledge, data mining techniques have been extensively used. Data mining has been used for extraction of concealed information from large databases and known as Knowledge Discovery in databases. Researchers have investigated that data mining can be used to envisage future trends and behavior, allowing companies to make proactive decisions based on knowledge. Data mining tools and techniques can counter business questions that conventionally were too intense in time to resolve. The results obtained from data mining are essentially used for making analysis and predictions. There are different techniques have been used in data mining, such as Association, Clustering, Classification, Prediction, Outlier Detection, and Regression.

Prediction is the most significant data mining technique that employs a set of pre-classified examples to develop a model that can classify the data and discover the relationship between independent and dependent data. Predictive modeling is a process that uses data mining and probability techniques to estimate the outcomes. Each model is composed of a series of predictors that are the variables that can influence future results. Once the data have been collected for pertinent predictors, statistical methods have been formulated for the next steps of executions. Predictive modeling is generally associated with meteorology and weather forecasting, health care, insurance, customer relationship management, stock markets, etc.

Weather forecasting is one of the extreme evils that weather departments around the world tackle. Researchers have classified these prognostic strategies into a twofold based on numerical modeling and scientific processing. Many researchers have shown that most of the changes in the climate environment are mainly due to global climate change. Therefore, if these changes are identified over time, effective forecasting techniques can be planned. Data mining techniques play an imperative role in the prophecy of global parameters with these assumptions; the discovery of knowledge is done by integrating the time series analyzed together with data mining techniques.

For achieving the objective of the proposal, the authors have compiled the Weka repository data in CSV format and the collected data have been transformed into the ARFF format. Data mining algorithm based on different classifications such as Multilayer Perception, Naive Bayes, SMO, J48, and REPTree have been applied to achieve the requirement. A knowledge flow layout has been made by using all the classification algorithms mentioned in the literature. Based on different performance metrics, the comparative analysis has been carried out to achieve the objective.

This article presents a brief review of the literature carried out in the area of data mining for the improvement of research work in this area. The rest of the article is organized as follows. Section 2 analyzes complete surveys of related data mining works. In Sect. 3, the proposed methodology and performance analysis have been presented. The final conclusion is offered in Sect. 4.

2 Related Work

There are many researches and their experimental work carried out using various data mining techniques that have demonstrated the importance of data mining to make predictions. The section provides a survey of the available literature of some algorithms employed by different researchers to utilize various techniques.

There are many researches and their experimental work is carried out using various data mining techniques that have demonstrated the importance of data mining to make predictions. The section provides a survey of the available literature of some algorithms employed by different researchers to utilize various techniques.

The authors [1] proposed a predictive model for the prediction of wind speed based on the support vector machine. The authors have used neural network support vector machine for wind speed prediction and the performance comparison has been made based on their perception of multiple layers. In this technique, the performance has been measured by the mean squared error and it has been observed that the SVM performs MLP throughout the system with orders ranging from 1 to 11. The authors [2] proposed a predictive model that provides a brief description of the SQL Server system, which deals with data mining clients for Excel 2007 and explains how actuators can use EXCEL to construct predictive models with little or no knowledge of the underlying SQL Server system. The authors [3] presented a generic methodology for the meteorological weather forecast series using incremental K-means clustering that is applied to the initial data and the incremental K-means are applied to the new data that is being used by the Manhattan metric measure. The authors [4] presented an approach to analyze and predict temperature data, moisture values for the future using the clustering technique. The closest K-neighbor algorithm is used to predict the values of the temperature and humidity parameters of the climate. To find the distances between the datasets of the absolute distance algorithm KNN and the measurement of Euclidean distance are used. The authors [5] presented a model using the K-means clustering algorithm, where the authors have used the probability density function algorithm to generate numerical results in the k-means cluster for weather-related predictions.

The authors [6] presented a methodology for mining meteorological data using artificial neural networks. The main objective of this technique is based on the extraction of predictive data through which the authors have extracted interesting patterns of a large amount of meteorological data. Here, the authors have used the propagation of the BPN neural network for initial modeling. The authors [7] presented the clustering technique to provide an analysis report of weather forecasters and climate analyzers using data mining techniques. The technique applied to climate data helps produce similar weather patterns with the deliberation of the spatial nature. The authors [8] developed a machine learning algorithm based on the hybrid SVM model to analyze the metrological data. The K-means clustering technique has been applied to the grouped data set collected and the decision tree has been used to predict the observations. The authors [9] used data mining techniques to obtain meteorological data and find veiled patterns within the large data set using k-means partitioning

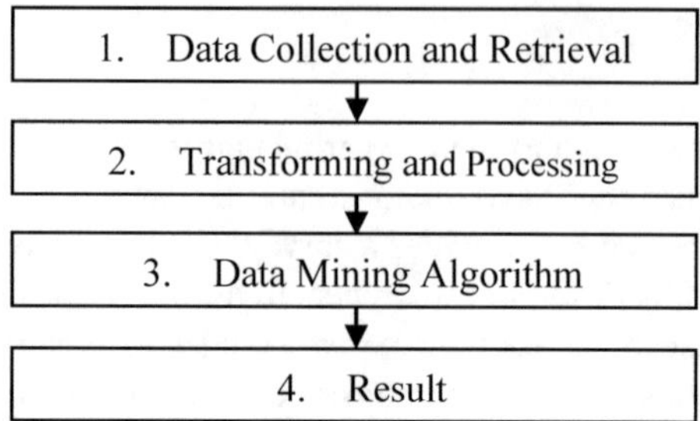

Fig. 1 Steps used for data mining applications

```
Time taken to build model: 0.27 seconds

=== Stratified cross-validation ===
=== Summary ===

Correctly Classified Instances          10               71.4286 %
Incorrectly Classified Instances         4               28.5714 %
Kappa statistic                          0.3778
Mean absolute error                      0.287
Root mean squared error                  0.5268
Relative absolute error                 60.2616 %
Root relative squared error            106.7798 %
Total Number of Instances               14

=== Detailed Accuracy By Class ===

               TP Rate   FP Rate   Precision   Recall   F-Measure   ROC Area   Class
                 0.778     0.4       0.778      0.778     0.778       0.778    yes
                 0.6       0.222     0.6        0.6       0.6         0.778    no
Weighted Avg.    0.714     0.337     0.714      0.714     0.714       0.778

=== Confusion Matrix ===

 a b   <-- classified as
 7 2 | a = yes
 2 3 | b = no
```

Fig. 2 Output generated by MLP in Weka workbench

techniques. The data preprocessing technique has been applied to the meteorological data, which consists of several necessary parameters and data transformation has been used to reduce the noise and complete the missing values. Dynamic methods of data mining are required to compile, which can be learned dynamically to match the nature of the nature of time and sudden events that change rapidly.

In this work, the authors have chosen the classification technique, since other data mining techniques, such as grouping, association, are not capable of problems related to prediction. Clustering is able to partition the related data elements into the same set and, on the other hand, the association is usually used to establish relationships between the attributes that exist in a data set.

3 Proposed Methodology and Performance Analysis

The methodology used in this paper consists of convinced steps that are usually used in data mining applications. The steps have been presented using the flow diagram which has presented in Fig. 1.

1. Data Collection and Retrieval: Data used in the literature are inserted in CSV Format.
2. Data Transformation and Data Preprocessing: The CSV file has converted to ARFF file using Java code named as weather.arff. After that the weather.arff file

Table 1 Weather-related parameters

Variable/attribute name	Domain
Temperature	Hot, mild, cool
Outlook	Sunny, rainy, overcast
Humidity	High, low
Wind	True, false

has used as a source file and then re-sampling technique was applied on it. Re-sampling technique deals with choosing instances from the data set on a random basis.

3. Data Mining Algorithm: Here the various classification based data mining technique are used.
4. Result: The output obtained after applying data mining algorithm (Table 1).

3.1 Tools and Techniques

Different Data Mining techniques are available for making prediction and in this paper, the authors have used classification technique for prediction. The output data set is tested and analyzed with five classification algorithms using cross validation, which are as follows: Multilayer Perception, Naïve Bayes, SMO, J48, and REPTree. For implementations of all these tasks, the authors have used Weka tool, which has been identified as an open-source tool used for Machine Learning and Data Mining Tasks.

3.2 Classification Rules

A generalized classification rule would be used for making prediction on the basis of our case study is mentioned below:

RULE 1:	IF OUTLOOK = SUNNY AND HUMIDITY = HIGH, THEN DECISION = NO
RULE 2:	IF OUTLOOK = SUNNY AND HUMIDITY = LOW, THEN DECISION = YES
RULE 3:	IF OUTLOOK = OVERCAST, THEN DECISION = YES
RULE 4:	IF OUTLOOK = RAINY AND WIND = TRUE, THEN DECISION = NO
RULE 5:	IF OUTLOOK = RAINY AND WIND = FALSE, THEN DECISION = YES

As the authors have used five different classifiers in the proposed work, for comparative purposes several performance metrics such as True positive rate, False positive rate, Precision, Recall, F-Measure, and ROC Area have been used to achieve the

```
Time taken to build model: 0.05 seconds

=== Stratified cross-validation ===
=== Summary ===

Correctly Classified Instances           9               64.2857 %
Incorrectly Classified Instances         5               35.7143 %
Kappa statistic                          0.186
Mean absolute error                      0.3571
Root mean squared error                  0.5976
Relative absolute error                 75      %
Root relative squared error            121.1306 %
Total Number of Instances               14

=== Detailed Accuracy By Class ===

               TP Rate   FP Rate   Precision   Recall  F-Measure   ROC Area  Class
                 0.778     0.6       0.7         0.778     0.737     0.589    yes
                 0.4       0.222     0.5         0.4       0.444     0.589    no
Weighted Avg.    0.643     0.465     0.629       0.643     0.632     0.589

=== Confusion Matrix ===

 a b   <-- classified as
 7 2 | a = yes
 3 2 | b = no
```

Fig. 3 Output generated by SMO in Weka workbench

```
Time taken to build model: 0.11 seconds

=== Stratified cross-validation ===
=== Summary ===

Correctly Classified Instances           7               50      %
Incorrectly Classified Instances         7               50      %
Kappa statistic                         -0.0426
Mean absolute error                      0.4167
Root mean squared error                  0.5984
Relative absolute error                 87.5    %
Root relative squared error            121.2987 %
Total Number of Instances               14

=== Detailed Accuracy By Class ===

               TP Rate   FP Rate   Precision   Recall  F-Measure   ROC Area  Class
                 0.556     0.6       0.625       0.556     0.588     0.633    yes
                 0.4       0.444     0.333       0.4       0.364     0.633    no
Weighted Avg.    0.5       0.544     0.521       0.5       0.508     0.633

=== Confusion Matrix ===

 a b   <-- classified as
 5 4 | a = yes
 3 2 | b = no
```

Fig. 4 Output generated by Naive Bayes in Weka workbench

```
Time taken to build model: 0.11 seconds

=== Stratified cross-validation ===
=== Summary ===

Correctly Classified Instances           7               50      %
Incorrectly Classified Instances         7               50      %
Kappa statistic                         -0.0426
Mean absolute error                      0.4167
Root mean squared error                  0.5984
Relative absolute error                 87.5    %
Root relative squared error            121.2987 %
Total Number of Instances               14

=== Detailed Accuracy By Class ===

               TP Rate   FP Rate   Precision   Recall  F-Measure   ROC Area  Class
                 0.556     0.6       0.625       0.556     0.588     0.633    yes
                 0.4       0.444     0.333       0.4       0.364     0.633    no
Weighted Avg.    0.5       0.544     0.521       0.5       0.508     0.633

=== Confusion Matrix ===

 a b   <-- classified as
 5 4 | a = yes
 3 2 | b = no
```

Fig. 5 Output Generated by J48 in Weka workbench

stated goal. Figures 3, 4, 5, 6, and 7 show the output generated by the Multilayer Prevention, SMO, Naive Bayes, J48, and REPTree, respectively, in Weka workbench. Figure 8 represents a graphic representation of all statistics obtained based on performance metrics.

Y represents Value and X represents Obtained Performance Metrics for Used Algorithms based on "YES" or "NO" class. Accuracy for each algorithm has calculated based on the obtained confusion matrix generated individually on Weka workbench as shown in Figs. 2, 3, 4, 5 and 6 for Multilayer Perception, SMO, Naive Bayes, J48, REPTree, respectively. Typical Representation of Confusion Matrix is shown below in Fig. 8.

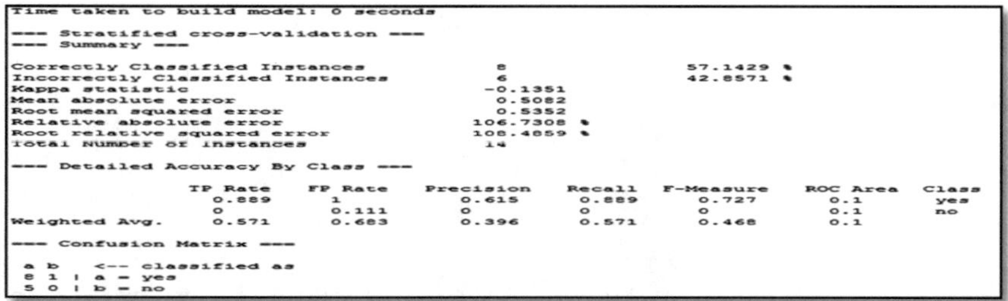

```
Time taken to build model: 0 seconds

=== Stratified cross-validation ===
=== Summary ===

Correctly Classified Instances           8               57.1429 %
Incorrectly Classified Instances         6               42.8571 %
Kappa statistic                         -0.1351
Mean absolute error                      0.5082
Root mean squared error                  0.5352
Relative absolute error                106.7308 %
Root relative squared error            108.4859 %
Total Number of Instances               14

=== Detailed Accuracy By Class ===

               TP Rate   FP Rate   Precision   Recall   F-Measure   ROC Area   Class
                 0.889     1          0.615      0.889     0.727      0.1        yes
                 0         0.111      0          0         0          0.1        no
Weighted Avg.    0.571     0.683      0.396      0.571     0.468      0.1

=== Confusion Matrix ===

 a b   <-- classified as
 8 1 | a = yes
 5 0 | b = no
```

Fig. 6 Output generated by REPTree in Weka workbench

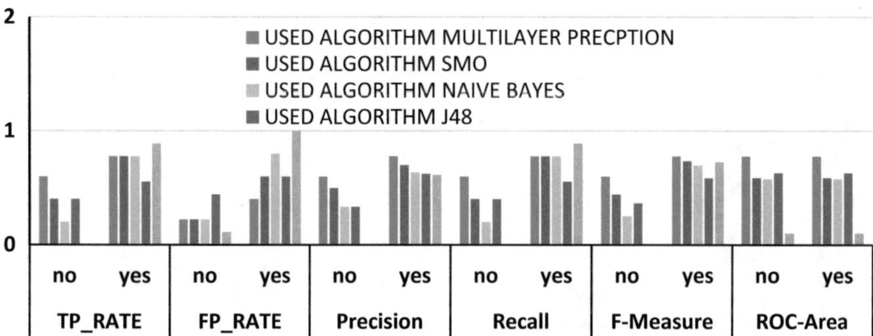

Fig. 7 Graphical representation of obtained statistics

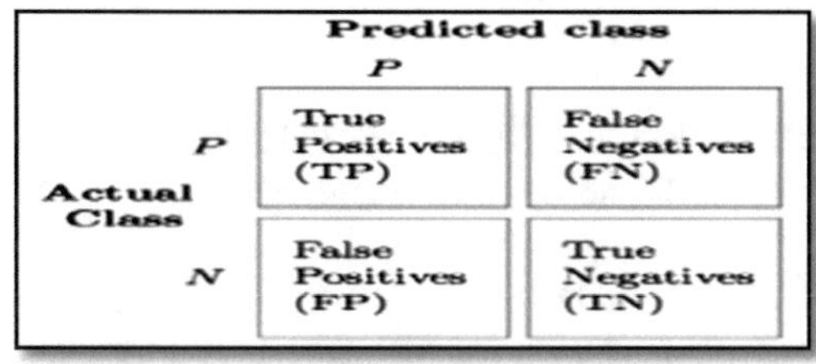

Where, P is Positive or YES outcome and N is negative or NO outcome.

Fig. 8 Representation of confusion Matrix

- From Fig. 2, we have identified True Positive (TP) as 7 and False Positives (FP) as 3 and no. of instances in our work is 14, as calculated Accuracy for Multilayer Preceptron is (7 + 3)/(14) = 0.7142 or 71.42%. (i)
- In Fig. 3, we have True Positive (TP) as 7 and False Positives (FP) as 2 and no. of instances in our work is 14, so we have Accuracy for SMO as (7 + 2)/(14) = 0.6428 or 64.28%. (ii)
- Figure 4, we have True Positive (TP) as 7 and False Positives (FP) as 1 and no. of instances in our work is 14, as calculated Accuracy for Naive Bayes as (7 + 1)/(14) = 0.5714 or 57.14%. (iii)

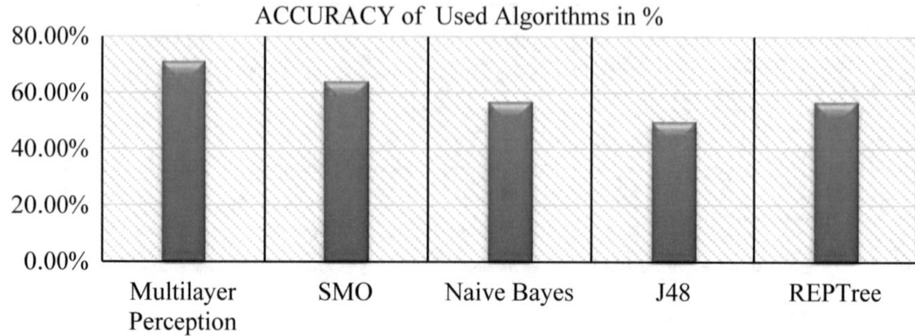

Fig. 9 Graphical representation of obtained accuracy of all used classifiers

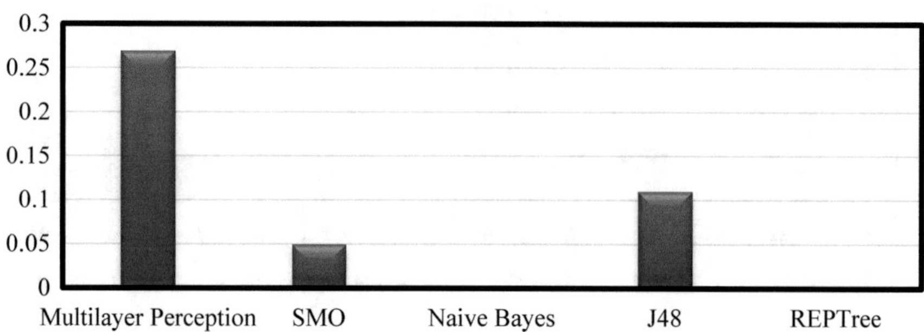

Fig. 10 Comparison of classifiers on the basis of the time requirement

- From Fig. 5, we have identified as True Positive (TP) as 5 and False Positives (FP) as 2 and no. of instances in our work is 14, so we have Accuracy for J48 as $(7+0)/(14)=0.5000$ or 50%. (iv)
- In Fig. 6, we have True Positive (TP) as 8 and False Positives (FP) as 0 and no. of instances in our work is 14, so we have Accuracy for REPTree as $(8+0)/(14)=0.5714$ or 57.14%. (v)

After receiving all the statistical results (using Eqs. (i) to (v)), comparison accuracy of all classifiers have been presented in Fig. 9 that has shown a graphical representation of the accuracy of all classifiers and it has identified that Multilayer Perception Technique performs best with an accuracy of 71.40%.

After having comparison of classifiers on the basis of statistical analysis and accuracy, the authors have compared with respect to time (in seconds) requirement to build the corresponding classifier model with the help of output shown in Figs. 2, 3, 4, 5 and 6 for Multilayer Preceptron, SMO, Naive Bayes, J48, REPTree, respectively, and Fig. 10 shows graphical representation of the same. It has identified that although Multilayer Perception technique is best, it requires more time to build the classifier model compared to other classification techniques used in this literature.

Comparison of all classifiers with the help of Weka experimenter has been conducted and in this case also Multilayer Perceptron performs best among all classifiers

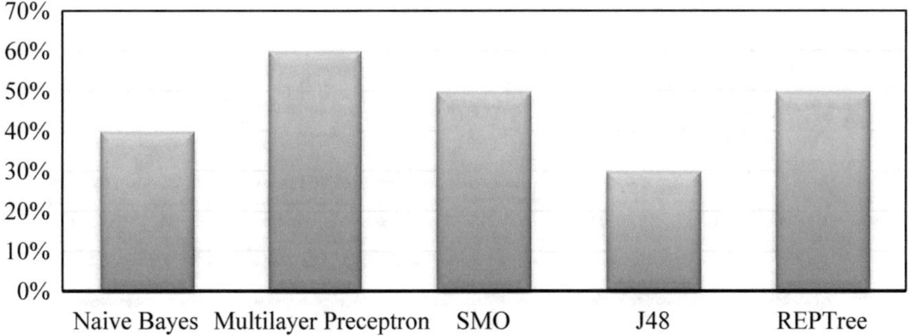

Fig. 11 Histogram representation of all classifiers on the basis of Kappa statistics

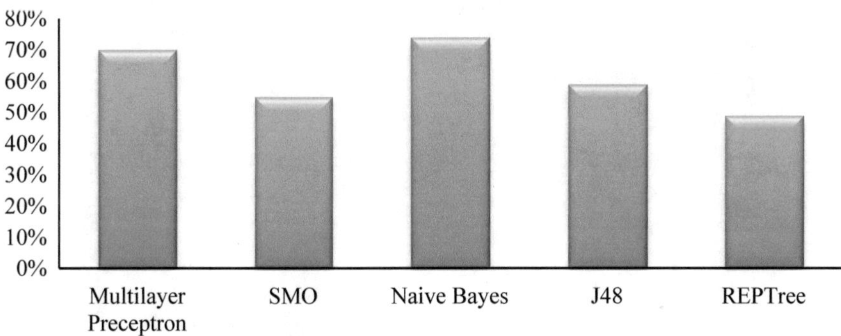

Fig. 12 Histogram representation of all classifiers on the basis of area under region of curve in percentage

with F_Measure 27%. Another comparison of all classifiers on the basis of Kappa Statistics has executed using Weka Experimenter and it has identified that Multilayer Perception performs best among all the classifiers with having Kappa Statistic 60%. Histogram representation of the same is shown in Fig. 11.

The obtained Model Performance Chart is shown in Fig. 12, which shows the Region of Convergence (ROC) curve for all the classifiers that are used in our work. Finally, the last comparison of all classifiers has executed on the basis of Elapsed Time Training, i.e., the time taken by all classifiers since its execution to train the used dataset using Weka Experimenter which has been shown in Fig. 13 and Histogram representation of the same is shown in Fig. 14.

After performing all the comparison of classifiers on the basis of various performance metrics, a knowledge flow model has been presented in Fig. 15, which shows membership tree structure. After developing the knowledge flow model, the authors have loaded the dataset (weather.arff file) using ArffLoader functionality and after running the entire model, a model performance chart has obtained. The obtained Model Performance Chart is shown in Fig. 16, which shows the Region of Convergence (ROC) curve for all the classifiers that are used in our work.

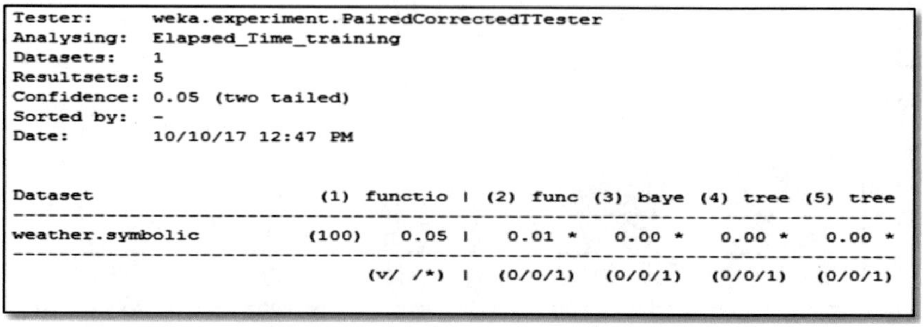

```
Tester:       weka.experiment.PairedCorrectedTTester
Analysing:    Elapsed_Time_training
Datasets:     1
Resultsets:   5
Confidence:   0.05 (two tailed)
Sorted by:    -
Date:         10/10/17 12:47 PM

Dataset                    (1) functio | (2) func (3) baye (4) tree (5) tree
-----------------------------------------------------------------------------
weather.symbolic           (100)   0.05 |   0.01 *   0.00 *   0.00 *   0.00 *
-----------------------------------------------------------------------------
                                (v/ /*) |  (0/0/1)  (0/0/1)  (0/0/1)  (0/0/1)
```

Fig. 13 Comparison of all classifiers using weka experimenter on the basis of elapsed time

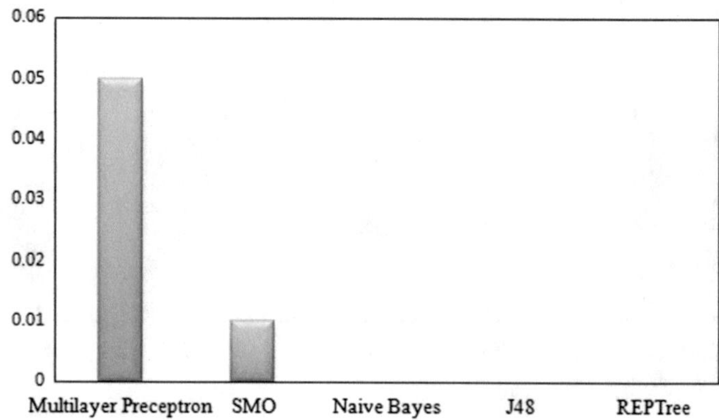

Fig. 14 Histogram representation of all classifiers on the basis of elapsed time training

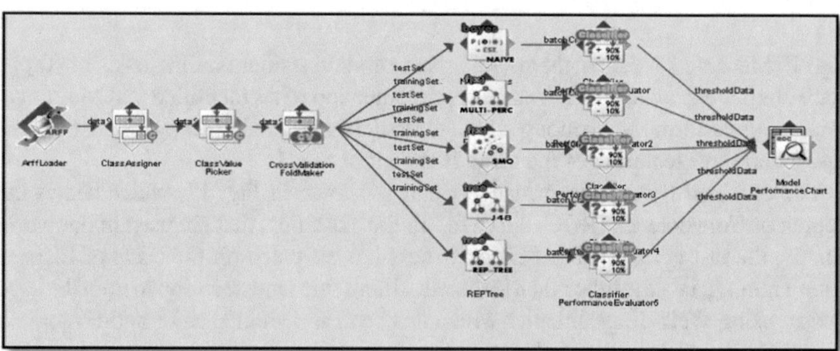

Fig. 15 Knowledge flow layout

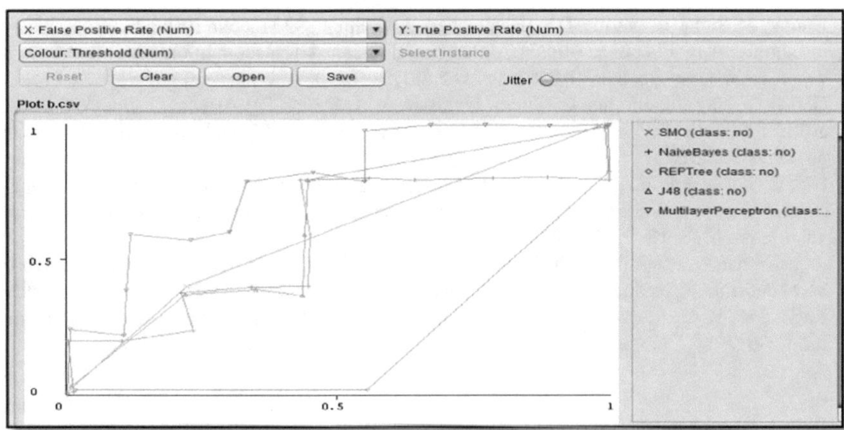

Fig. 16 Model performance chart

4 Conclusion

In this paper, classification techniques are used for the prediction on the 14 days dataset of meteorological records and the corresponding decision has been commissioned whether or not to play outdoor games. In this study, a model based on 14 days' meteorological records has been developed. Among all data mining classifiers multilayer perceptron acheived best performance with 71.40% accuracy and thus, MLP is the potentially efficient test classifier algorithm. On the other hand, the time required to construct the classifier model by MLP is 0.08 s, which is more in comparison with other classifier algorithms. The comparison between five classifiers with the help of experimenter Weka has been implemented; in this case also, MLP turns out to be the best with F-Measures of 27% and Kappa Statistics of 60%. Although the MLP is lagging behind in the time it takes to construct the classifier model, but it is better and more efficient for all the performance metrics taken in this literature, therefore MLP performance is higher than other classifiers.

References

1. Mohandes, M.A., Halawani, T.O., Rehman, S., Hussain, A.A.: Support Vector Machines for Wind Speed Prediction. Int. J. Renew. Energy **29**(6), 939–947 (2004)
2. Ganas, S.: Data mining and predictive modeling with excel 2007. Internet Electron. J. (2009)
3. Chakraborty, S., Nagwani, N.K., Dey, L.: Weather forecasting using incremental k-means clustering. In: International Conference in High Performance Architecture and Grid Computing, Vol. 169, Part-2, pp. 338–341 (2011)
4. Badhiye, S.S., Chatur, P.N., Wakode, B.V.: Temperature and humidity data analysis for future value prediction using clustering technique: an approach. Int. J. Emerg. Technol. Adv. Eng. **2**(1), 88–91 (2012). ISSN: 2250-2459

5. Kumar, A., Sinha, R., Verma, D.S., Bhattacherjee, V., Singh, S.: Modeling using k-means clustering algorithm. In: First International Conference on Recent Advances in Information Technology, vol. 4, issue-1, issue no: 4577-0697, pp. 1–5 (2012)
6. Ghosh, S., Nag, A., Biswas, D., Singh, J.P., Biswas, S., Sarkar, P.P.: Weather data mining using artificial neural network. In: 2011 IEEE Recent Advances in Intelligent Computational Systems, pp. 192–195 (2011)
7. Gokila, S., Anandkumar, K., Bharathi, A.: Clustering and classification in support of climatology to mine weather—a review. In: International Conference Computing and Intelligence Systems, vol. 04, pp. 1336–1340, Mar 2015. ISSN: 2278-2397
8. Rajashekhar, N., Rajinikanth, T.V.: Weather analysis of guntur district of andhra region hybrid SVM data mining techniques. Int. J. Eng. Adv. Technol. **3**(4), 133–136 (2014). ISSN: 2449-8958
9. Kalyankar, M.A., Alaspurkar, S.J.: Data mining technique to analyze the meteorological data. Int. J. Adv. Res. Comput. Sci. Softw. Eng. **3**(2), 114–118 (2013). ISSN:2277-128X

Distortion-Based Privacy-Preserved Association Rules Mining Without Side Effects Using Closed Itemsets

H. Surendra and H. S. Mohan

Abstract Privacy-Preserving Data Mining (PPDM) is taking more focus in the recent years due to growing concern about privacy. Association rule mining is a widely used data mining technique. The goal of privacy-preserved association rule mining is to protect sensitive rules from disclosing in the published data. It is achieved by reducing the confidence below a minimum threshold or adding noise to the original confidence of the sensitive rules. Existing data sanitization techniques suffer from side effects which reduce the utility of the data. We propose a model-based approach for preserving the privacy of sensitive rules without side effects. The proposed value distortion-based sanitization algorithm sanitizes the closed itemsets instead of transactions of the database. Also, the proposed solution is scalable for distorting the whole database without changing the relationship between attribute values or itemsets in the original database. Experimental results show that the proposed method is better than other well-known techniques based on transaction modification concerning the sanitization time and side effects.

Keywords Value distortion · Association rule hiding · Closed itemsets
Privacy-preserving data mining · Privacy-preserving data publishing

1 Introduction

Association rule mining is a widely used technique for understanding the correlation between the attributes or itemsets in the database. These association rules may disclose sensitive information which otherwise hidden in the data. Various data sanitization methods such as perturbation, generalization, sampling [1] are used to protect

H. Surendra (✉) · H. S. Mohan
SJB Institute of Technology, BGS Health and Education City,
Dr. Vishnuvardhan Road, Kengeri, Bengaluru 560060, India
e-mail: surendra.h@gmail.com

H. S. Mohan
e-mail: mohan_kit@yahoo.com

A. Abraham et al. (eds.), *Emerging Technologies in Data Mining and Information Security*, Advances in Intelligent Systems and Computing 813,
https://doi.org/10.1007/978-981-13-1498-8_52

the sensitive information from disclosing. The sensitive rules are protected either by reducing the confidence below a threshold or adding noise to the original confidence of sensitive rules. Existing well-known techniques alter the transactions in the original database or specific characteristics like frequency or attribute relationships. These techniques disturb the nonsensitive data resulting in side effects.

In this paper, we propose a model-based approach for sanitizing the original data to preserve the original confidence of sensitive rules. The database is modeled as closed itemsets. Closed itemsets provide lossless, compact representation of the original data. The proposed sanitization algorithm distorts the support of the sensitive rule itemsets in the model within a safe range and avoids any side effects.

The paper is organized as follows. In Sect. 2, different existing methods for modeling the data for privacy-preserved association rule mining is discussed. In Sect. 3 the proposed method is described in detail. The experimental result of the proposed method is presented in Sect. 4. Finally, the paper is concluded in Sect. 5 with a discussion on the future scope of the proposed work.

2 Related Work

Caiola and Reiter [2] modeled data as Random Forests and sensitive attributes are modified to protect them from disclosing. The technique is used to generate partially synthetic data for categorical data. Zhang et al. [3] proposed Bayesian network-based method to model the data. The conditional probability distribution of attributes is modeled as differentially private Bayesian Network. The sanitization is performed by adding noise to the sensitive distributions. Jumaah et al. [4] proposed a simple method of hiding sensitive rules. The support of rule itemsets is altered by adding or removing itemsets from the transactions. Cheng et al. [5] proposed blocking-based association rule masking method. An unknown symbol like ? is used to replace the sensitive itemsets so that the support and confidence of any sensitive rule lie in a range instead of single value. Tsai et al. [6] proposed k-anonymization for masking sensitive association rules. Their method yields higher privacy gain, but suffer from huge information loss and side effects.

Hong et al. [7] proposed Sensitive Items Frequency–Inverse Database Frequency (SIF-IDF), a greedy-based method derived from the concept of Term Frequency–Inverse Document Frequency (TF-IDF) used for Text Mining. The transactions are ranked based on their SIF-IDF value, and sensitive itemsets are removed from transactions with best SIF-IDF value. Modi et al. [8] proposed a hybrid technique for hiding sensitive association rules. The technique first finds association rules using apriori algorithm, and then, the list of transactions to be modified is found. The sanitization is performed by removing itemsets of the sensitive rule from these transactions. Le et al. [9] used lattice theory to hide sensitive rules. The intersection lattice of frequent itemsets is used to reduce the side effects. The sensitive itemsets are removed from selected transactions to reduce their support below a threshold. Shahsavari and Hosseinzadeh [10] proposed different factors to measure the sensitivity of the given rules. Then, these rules are listed in the decreasing order of their sensitivity. The rules are hidden one after another from the top order by removing the sensitive rule itemsets from the transactions. Cheng et al. [11] proposed an Evolutionary Multi-objective Optimization (EMO) method to hide sensitive rules. The frequent itemsets and association rules are mined using apriori algorithm. The transactions to be altered for hiding sensitive rules are identified using EMO. The itemsets of sensitive rules are removed from these transactions to reduce their support. Bonam et al. [12] proposed distortion-based association rule masking using Particle Swarm Optimization (PSO). The best transaction to remove sensitive itemset is found using PSO. The sensitive item with the highest support is selected and removed from the transactions in multiple iterations. This process is repeated for all sensitive rules.

3 Proposed Method

Most of the existing methods select specific transactions to modify them to protect sensitive itemsets. The efficiency of these methods degrades for large and high-dimensional data. It will get worse when the data is dense, and many transactions need modification for hiding a small set of rules. Also, these techniques do not scale for distorting the whole database. To overcome these limitations, we propose to model the database as closed itemsets and perform data sanitization on the model instead of transactions of the database.

3.1 Closed Itemsets as Model

We propose to model the database as closed itemsets. Closed itemsets provide a lossless compact representation of the data. The model is sanitized instead of transactions of the database. The user can query the sanitized model to find frequent itemsets and association rules. The sanitized model is published instead of the database. Also, a sample transactional database representing the sanitized data can be generated from the sanitized model at the user end if required.

3.2 Preserving Sensitive Association Rules

Association rule helps in predicting the occurrence of an item based on the occurrence of other items in the transactions. An association rule between itemsets A and B is expressed as $A \rightarrow B$. The support of the association rule $A \rightarrow B$ is the support of the combined itemsets AB. The confidence of the rule is measured by the conditional probability of occurrence of itemset B given Itemset A in the given database D.

That is,

$$Confidence(A \rightarrow B) = P(B|A) = \frac{Support(A \cup B)}{Support(A)} \quad (1)$$

From Eq. 1, the original confidence of the sensitive rule can be preserved by altering the support of itemset comprising the rule or rule antecedent. So the problem of preserving sensitive association rule can be reduced to distorting these itemsets. The itemsets that need to be sanitized are called Sensitive Itemsets (SI).

3.3 Sanitization Algorithm for Distorting Sensitive Itemsets Safely

Consider a sample closed itemsets lattice with itemsets arranged at their support levels as shown in Fig. 1. Let us consider itemset (2, 3) as sensitive itemset which needs distortion. A random number is used to replace the original support of the sensitive itemset. The random number is found within a safe range to avoid altering the support of other nonsensitive itemsets in the model. The safe range is found by taking the difference between the highest support value of the immediate supersets (Floor Limit) and lowest support value of the immediate subsets (Ceiling Limit) of the sensitive itemset. For itemset (2, 3), its superset with the highest support is (2, 3, 5) with support 0.4 which is the Floor Limit. Its subset with least support is (2) with support 0.7, which is the Ceiling Limit. So the new distorted support for the sensitive itemset (2, 3) is randomly chosen between 0.4 and 0.7. If the given sensitive itemset

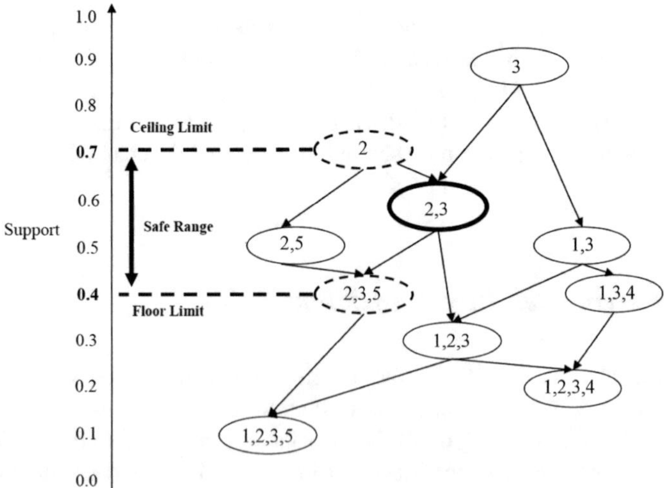

Fig. 1 Safe distortion range for sensitive itemsets in closed itemset lattice

is not a closed itemset, then it is added to the model with its support ranging from Floor and Ceiling Limits. The algorithm for safely distorting the sensitive itemset without side effects using closed itemsets is shown in Algorithm 1.

Algorithm 1 Safe Distortion of Sensitive Itemsets

Input: Closed Itemsets of the Given Database D (Model), Sensitive Itemset (SI)
Output: Sanitized Model with SI Safely Masked

```
{Clone the original model}
SanitizedClosedItemsets = ClosedItemsets.Clone
{Check if the given SI is present in the Model as Closed Itemset}
SanitizedClosedItemsets.Search(SI)
{If the SI is not present in the Model then add SI as Closed Itemset to the Model with its Support value initialized to Zero.}
if SI NotFound then
    SI.Support = 0
    SanitizedClosedItemsets.Add(SI)
end if
{Find the Floor and Ceiling Limit for the SI and Assign a Random Value between these limits as new distorted Support value.}
FloorLimit = GetFloorLimit(SI) {Find Floor Limit}
CeilingLimit = GetCeilingLimit(SI) {Find Ceiling Limit}
SafeRange = CeilingLimit − FloorLimit
SI.Support = FloorLimit + RANDOM(SafeRange)
{Update the changes of SI in sanitized model}
SanitizedClosedItemsets.Update(SI)
```

4 Results and Discussion

We tested the proposed method on four representative real datasets, namely, pumsb, mushroom, retail, and kosarak published in [13]. The implementation of the proposed method is done by extending the CHARM [14] closed itemset mining algorithm in SPMF Java Open Source Data Mining Library. The experiments are performed on a Dell laptop with 8 GB RAM and Intel Core i5 processor. All values presented in the results are the average of three repeated executions. The characteristics of the representative datasets are given in Table 1.

4.1 Model (Closed Itemsets) Generation

The first step in our proposed approach is to model the database as closed itemsets. Experimental results show that mining closed itemsets takes more time and memory compared to finding frequent itemsets in large, dense, and high-dimensional data but yields a high degree of compression of itemsets as shown in Fig. 2. For small, sparse, and low-dimensional data, mining closed itemsets takes far less time and memory

Table 1 Characteristics of representative datasets

Dataset	No. of transactions	No. of unique items	Minimum transaction length	Maximum transaction length	Average transaction length
Pumsb	49046	2113	74	74	74
Mushroom	8124	119	23	23	23
Retail	88162	16470	1	76	10
Kosarak	990002	41270	1	2498	8

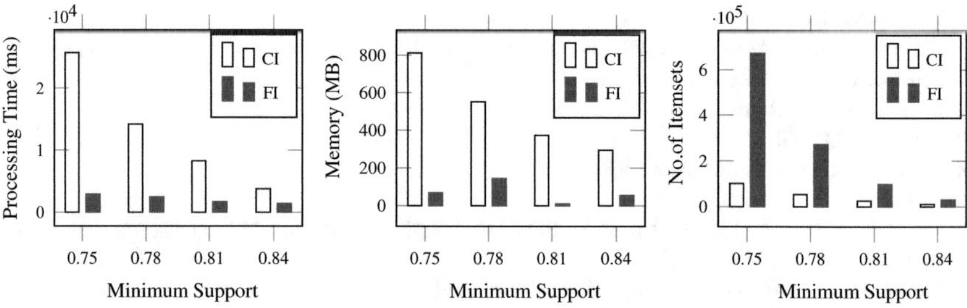

Fig. 2 Performance comparison of modeling the pumsb dataset as closed itemset and frequent itemset for different minimum support

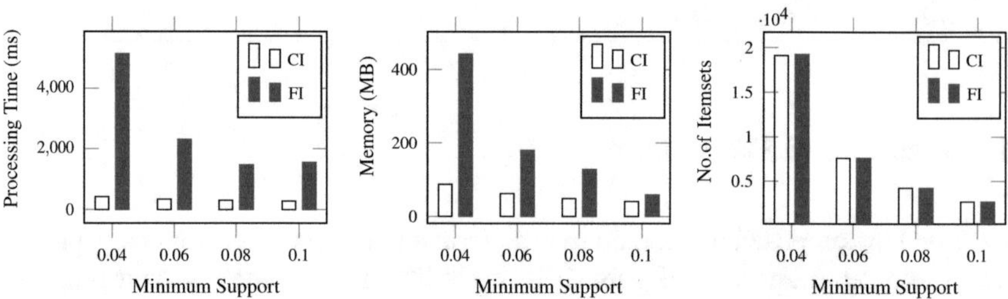

Fig. 3 Performance comparison of modeling the mushroom dataset as closed itemset and frequent itemset for different minimum support

compared to finding frequent itemset as shown in Fig. 3. Figures 4 and 5 show that for large, sparse, and high-dimensional data, mining closed itemsets takes relatively less time but more memory with little compression. Since mining closed itemsets is a one-time activity, the limitation of higher memory requirement for large data can be traded off with the performance efficiency achieved concerning processing time and data compression. Other benefits of our proposed model include faster data sanitization and personalization.

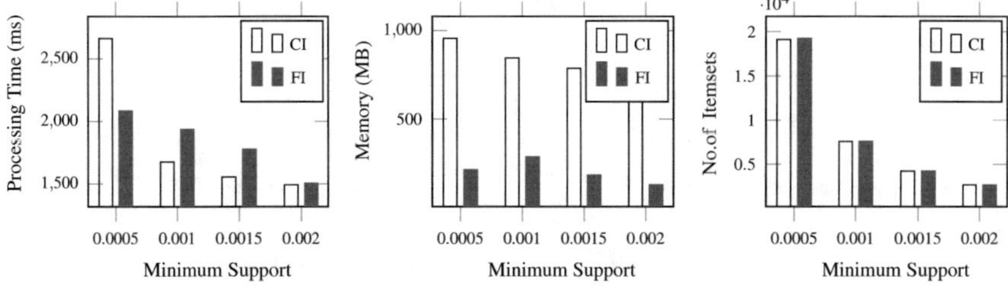

Fig. 4 Performance comparison of modeling the retail dataset as closed itemset and frequent itemset for different minimum support

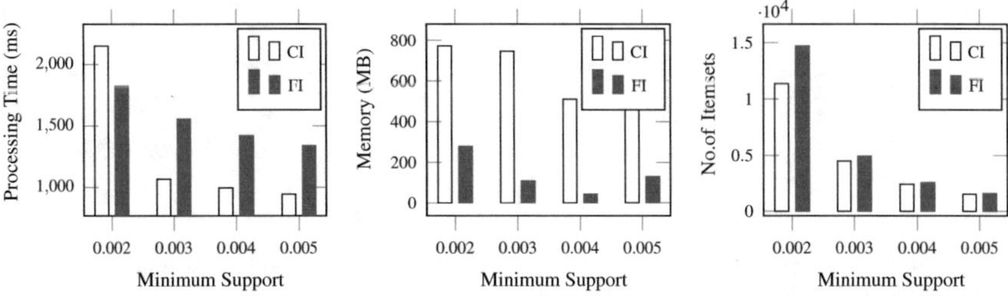

Fig. 5 Performance comparison of modeling the kosarak dataset as closed itemset and frequent itemset for different minimum support

4.2 Model Sanitization for Distorting Sensitive Itemsets

A general model is developed using frequent itemsets for transaction modification to compare with the proposed approach. The proposed sanitization algorithm is run on both proposed Closed Itemsets (CI) based model and the general, Frequent Itemsets (FI) based model to preserve the privacy of sensitive rules without side effects.

Figures 6, 7, 8 and 9 show the comparison of time taken to distort support value of sensitive itemset of different sizes on different model sizes. It is observed that, for dense data, the time taken to sanitize sensitive itemsets of smaller size (smaller value of k in k-itemsets) has no or minimal improvement but outperforms existing techniques for the large itemsets size. The improvement is due to less number of subsets, and supersets need sanitization in the proposed approach. The sanitization algorithm quickly finds the safe region and distorts the original support with a random support value within the safe range as shown in Fig. 6c, d. For sparse data, the number of supersets and subsets of the sensitive itemset is almost same in both CI and FI models. Figures 8 and 9 shows that the sanitization time is equal or nearly same in both models irrespective of the size of sensitive itemsets.

The main reason for the performance improvement in using closed itemsets is its closure property. The support of any sensitive itemsets which is closed will be different from its immediate supersets or subsets. So the support value of sensitive itemset can be safely modified without processing its supersets and subsets.

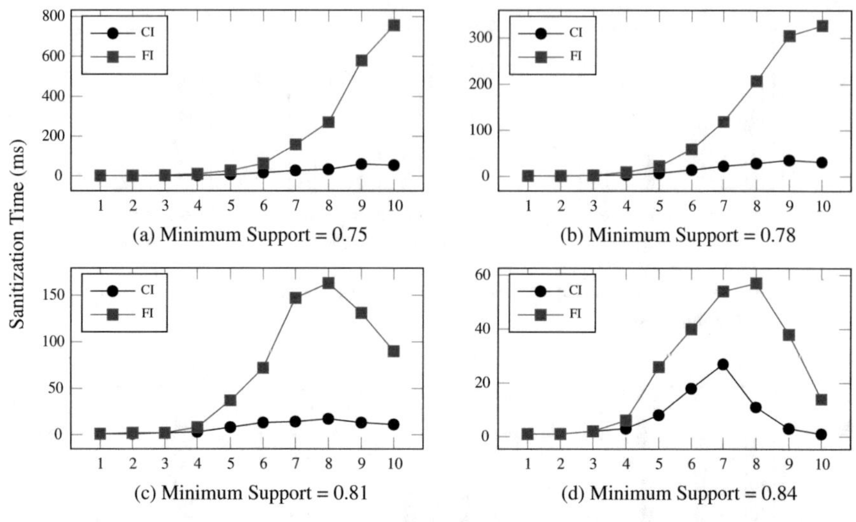

Fig. 6 Time taken to distort sensitive itemsets of different size on CI and FI models of the pumsb dataset generated with different minimum support

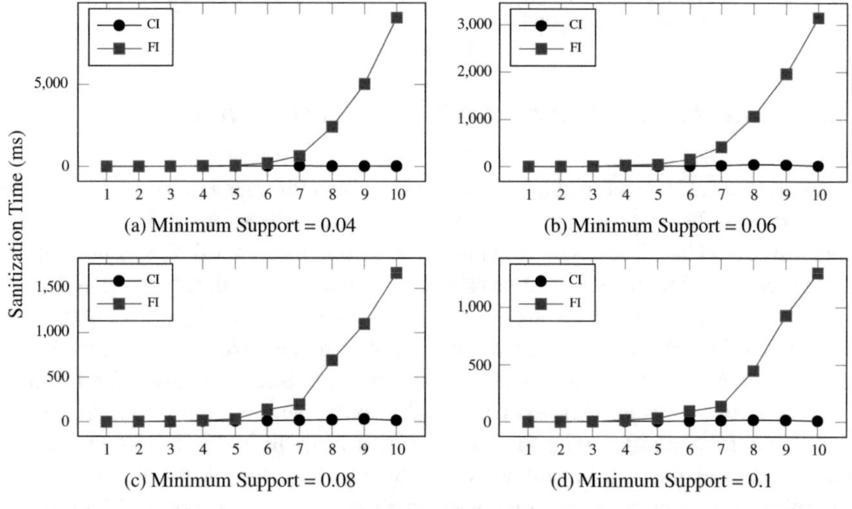

Fig. 7 Time taken to distort sensitive itemsets of different size on CI and FI models of the mushroom dataset generated with different minimum support

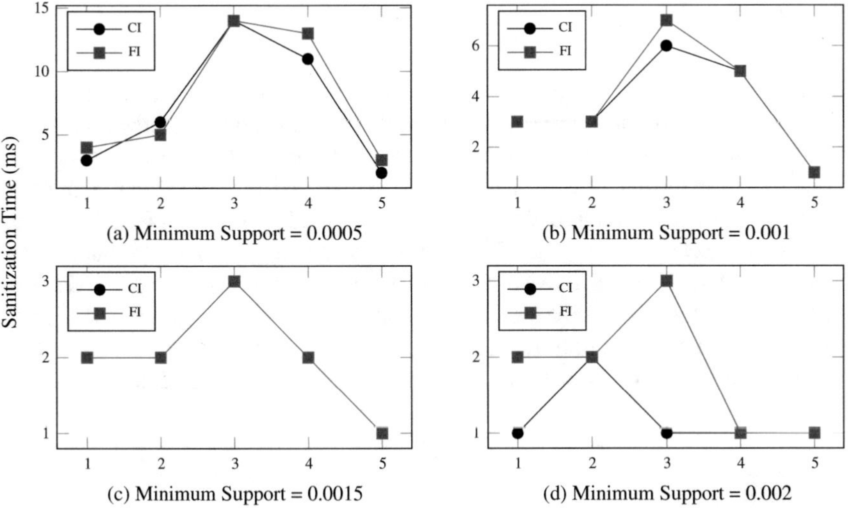

(a) Minimum Support = 0.0005 (b) Minimum Support = 0.001

(c) Minimum Support = 0.0015 (d) Minimum Support = 0.002

Sensitive Itemset Size (k)

Fig. 8 Time taken to distort sensitive itemsets of different size on CI and FI models of the retail dataset generated with different minimum support

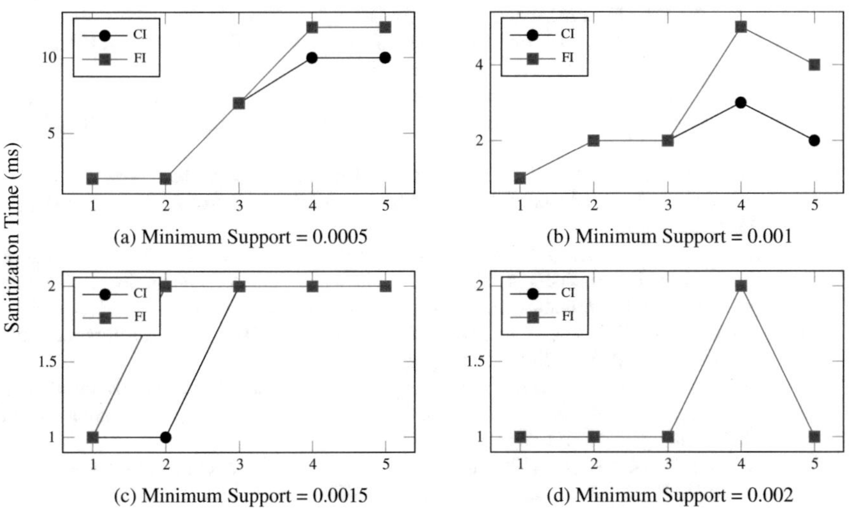

(a) Minimum Support = 0.0005 (b) Minimum Support = 0.001

(c) Minimum Support = 0.0015 (d) Minimum Support = 0.002

Sensitive Itemset Size (k)

Fig. 9 Time taken to distort sensitive itemsets of different size on CI and FI models of the kosarak dataset generated with different minimum support

5 Conclusion

In this paper, we have proposed to use closed itemsets for efficient masking of sensitive association rules. The closed itemsets in the model are sanitized instead of the transactions. Experimental results show that the proposed model-based approach is efficient in distorting the sensitive association rules and compared to other methods based on transaction modification. Also, the proposed approach safely distorts the sensitive itemsets avoiding side effects. However, creating the initial model of data with closed itemsets involves extra time and memory for the large and dense datasets; the benefits of faster and safe sanitization using the model outweigh this limitation. For very sparse data, the proposed approach incurs huge information loss due to the broad safe range for distortion which needs to be addressed in the future work.

References

1. Charu, C.A., Philip, S.Y.: Privacy-Preserving Data Mining: Models and Algorithms. In: Springer series in Advances in Database Systems, vol. 34 (2008)
2. Caiola, G., Reiter, J.P.: Random forests for generating partially synthetic, categorical data. Trans. Data Priv. 27–42 (2010)
3. Zhang, J., Cormode, G., Procopiuc, C.M., Srivastava, D., Xiao, X.: PrivBayes: private data release via Bayesian networks. In: Proceedings of ACM SIGMOD International Conference on Management of Data, June 2014, pp. 1423–1434
4. Jumaah, A.K., Al-Janabi, S., Ali, N.A.: An enhanced algorithm for hiding sensitive association rules based on ISL and DSR algorithms. Int. J. Comput. Netw. Technol. Teach. **3**(3), 83–89 (2015)
5. Cheng, P., Lee, I., Li, L., Tseng, K.-K., Pan, J.-S.: BRBA: a blocking-based association rule hiding method. In: Proceedings of the 13th AAAI Conference on Artificial Intelligence, pp. 4200–4201. ACM (2016)
6. Tsai, Y.-C., Wang, S.-L., Song, C.-Y., Ting, I.-H.: Privacy and utility of k-anonimity on association rule hiding. In: Proceedings of the 3rd Multidisciplinary International Social Networks Conference on Social Informatics, Data Science, Aug 2016, Article no. 42, pp, 42:1–42:6
7. Hong, T.-P., Lin, C.-W., Yang, K.-T., Wang, S.-L.: Using TF-IDF to hide sensitive itemsets. Appl. Intell. **38**(4), 502–510 (2013). (Springer)
8. Modi, C.N., Rao, U.P., Patel, D.R.: Maintaining privacy and data quality in privacy preserving association rule mining. In: International Conference on Computing Communication and Network Technologies. IEEE (2010)
9. Le, H.Q., Arch-Int, S., Nguyen, H.X., Arch-Int, N.: Association rule hiding in risk management for retail supply chain collaboration. Comput. Ind. **64**(7), 776–784 (2013). (Elsevier)
10. Shahsavari, A., Hosseinzadeh, S.: CISA and FISA: efficient algorithms for hiding association rules based on consequent and full item sensitivities. In: Proceedings of International Symposium on Telecommunications, pp. 977–982. IEEE (2014)
11. Cheng, P., Pan, J.-S., Harbin, C.-W.L.: Use EMO to protect sensitive knowledge in association rule mining by removing items. In: Proceedings of IEEE Congress on Evolutionary Computing, Jul 2014, pp. 1108–1115
12. Bonam, J., Reddy, A.R., Kalyani, G.: Privacy preserving in association rule mining by data distortion using PSO, ICT and critical infrastructure. In: Proceedings of the 48th Annual Convention of Computer Society of India—vol II, Advances in Intelligent Systems and Computing, vol. 249, pp. 551–558 (2014)

13. Frequent Itemset Mining Implementations Repository. http://fimi.ua.ac.be/, http://fimi.ua.ac.be/data/. Accessed 27 Mar 2017
14. Zaki, M.J., Hsiao, C.-J.: CHARM: an efficient algorithm for closed itemset mining. In: Proceedings of International Conference on Data Mining, pp. 457–473, Apr 2002

A Data Warehouse Based Schema Design on Decision-Making in Loan Disbursement for Indian Advance Sector

Ishita Das, Santanu Roy, Amlan Chatterjee and Soumya Sen

Abstract Disbursement of loan is an important decision-making process for the corporate like banks and NBFC (Non-banking Finance Corporation) those offers loans. The business involves several parameters and the data which are associated to these parameters are generated from heterogeneous data sources and also belong to different business verticals. Henceforth the decision-making on loan scenarios are critical and the outcome involve solving the issues like whether to grant the loan or not, if sanctioned what is highest amount, etc. In this paper we consider the traditional parameters of loan sanction process along with these we identify one special case of Indian credit lending scenario where the people having old loans with good repayment history get priority. This limits the business opportunities for Bank/NBFC or other loan disbursement organizations as potential good customers having no loan history are treated with less priority. In this research work we propose a data warehouse model which integrates the existing parameters of loan disbursement related decisions and also incorporates the newly identified concepts to give the priorities to the customers who don't have any old credit history.

Keywords Priority sector · Data warehouse · Loan disbursement Decision-making

1 Introduction

Banking sector [1] is a continuously growing sector and a pillar of economy of a country. The Indian banking sector is classified into scheduled banks and non-scheduled banks. Assets and liabilities are two main aspects of banking. Liabilities refer to deposits. The assets in banking sector indicates advance sector. The advance sector is further classified into priority and non-priority sector [2]. The priority sector [3] includes agriculture, micro, small and medium enterprises, export credit, education,

I. Das · S. Roy · A. Chatterjee · S. Sen (✉)
University of Calcutta, Kolkata, India
e-mail: iamsoumyasen@gmail.com

A. Abraham et al. (eds.), *Emerging Technologies in Data Mining and Information Security*, Advances in Intelligent Systems and Computing 813,
https://doi.org/10.1007/978-981-13-1498-8_53

housing, social infrastructure and others. Non-Priority Sector (NPS) is where India is self sufficient and does not depend on government to get any kind of priority.

Along with lending, there exists an inseparable factor: repayment within specified time period. In today's situation, NPA (Non Performing Asset) [5] is a huge problem in banking sector. It is considered as a potential threat that directly hampers the growth of profit earning of a bank. It indicates lending that does not get repayment in terms of interest and/or installment of principal in scheduled time and consists of phases like Sub-Standard, Doubtful, and Loss category. The Doubtful category is further classified into three subcategories: D1, D2, and D3 as per their tenure to be in the Doubtful class. The security(s) offered for a bank loan is an important aspect. It reduces the chances of a loan to become NPA. If the debt is due, then by means of security or collateral, it can be recovered. The potential customers for loan disbursement in India is often determined by CIBIL score [4].

These types of decision-making regarding loan disbursement is very complex. Analytical processing on the historical data could help decision-making process to be real time and effective. There are various research works which employs data warehouse schema [7] to design the framework for different complex applications involving huge data. In [6] a Data Warehouse based analysis has been performed on CDR to Depict Market Share of different mobile brands to propose a business model to generate useful information for customer analysis. Data cube [8] may be further formed and concept hierarchy [9] may be defined for business specific customized analysis.

In this research work, the decision-making parameters regarding loan disbursement have been considered to construct a data warehouse schema to reflect the structure of a loan disbursement process.

2 Overview of Indian Banking System

2.1 *Priority and Non-priority Sector*

The term "**Priority Sector**" indicates the sectors of national importance that needs development. Sectors like agriculture, small businesses, small industries were neglected, thus they have been marked as priority sectors for further establishment for the purpose of bank credit. Non-priority sector includes credit card, personal loan and others.

2.2 *Banking Related Terms*

A **Credit History** is a record of a borrower's responsible repayment of debts. A credit report consists of record of the borrower's credit history from a number of sources, including banks, credit card companies, collection agencies, and government.

A **Credit Report** indicates detailed report of an individual's credit history. Credit bureaus collect information and create credit reports based on that information.

A borrower's **Credit Score** is the result of a mathematical algorithm applied to a credit report and other sources of information to predict future delinquency.

Credit Rating is an assessment of the creditworthiness of a borrower in general terms. A credit rating can be assigned to any entity that seeks to borrow money. It can be an individual, corporation, state or provincial authority or sovereign government.

CIBIL (Credit Information Bureau India Limited) works as collecting and maintaining records of an individual's payments pertaining to loans and credit cards. These records are submitted to CIBIL by member banks and credit institutions, on monthly basis. This information is then utilized to create Credit Information Reports (CIR).

3 Problem Statement

This project work focuses on the decision-making capability in Loan Disbursement. In case of financial institutions, the decision-making becomes more critical due to huge amount of data, several parameters, incomplete knowledge, etc. In order to take a decision based on these different issues a framework is required which could handle these several parameters at a time. In order to analyze these several issues under a single umbrella we propose a Data Warehouse Schema to incorporate all the relevant issues of loan disbursement along with that we give a special focus to the customers who do not have any loan till now. In Indian banking context priority is given to the customers who already have existing loan and repayment history is good. We identify this assumption as a limiting factor in loan disbursement as some potential good customers get low priority as they do not have any loan/loan repayment history.

A bank can analyze its historical performance over time in Loan Disbursement to be able to plan for the future. The key performance indicator here is credit, profit in Loan Disbursement, branches, employees, etc. The nature of the customers' transactions and accounts can be monitored to include them in credit disbursement. Also the prospective loanee customers, who repay the debt on time, can be offered some new loan based on their history of credit usage. This analysis of new credit disbursement would be based on the performance of the customers over time, profit earned in Loan Disbursement, total credit of the bank in that particular sector, business per employee, etc.

CIBIL plays an important role in this process as it helps to judge the credit worthiness of a loanee. The existing loanee customers (who have taken loan earlier) come under credit scoring based on their repayment nature. There is another aspect of including fresh or small loanee customers into bigger credit disbursement for business enhancement. In future those prospective loanee customers may be included in CIBIL.

4 Proposed Methodology

The main goal of this paper is to design a data warehouse schema for the decision-making in loan disbursement. Loan disbursement process involves multiple parameters; we need to identify each dimension to correspond to this parameter. These will form dimension tables. After this fact tables are created. As we have several issues to resolve multiple fact tables are formed and they share some of the dimension tables. This results in a Fact Constellation Schema.

The schema described or proposed focuses on the following fields:

i. Decision-making in Loan Disbursement such that a loan will be granted or not.
ii. If granted, the loan period to repay the debt and preferable loan amount.
iii. It highlights the existing customers who are regular and maintaining a sound deposit along with the frequent transactions that reflect their spending nature, can be prospective loanee. Also the borrowers of small loans with a good repayment nature can opt for bigger credit. Thus enhancing the business of a bank.

This model can be applied for both priority and non-priority sector lending of a bank. Some attributes or parameter or some table may not always participate in the decision-making process depending on the sector of lending. This produces a new idea of complex business analysis. Some measures in the fact tables are participating as dimension tables, i.e., the measures are input to some other fact table. The values or range of values of each measure and dimension are assigned.

Whenever a loan application is received by a bank, the first thing comes into question is, whether the person is an existing customer of the bank or not. This comes under preprocessing of loan. Then the matters to look upon are: what type of loan, age of the person and the nature of job of the person seeking for a loan. Minimum work experience in present job is required to analyze the job profile, net salary, Income tax returns, etc. Loan period is decided considering the moratorium period, remaining job period (say if retirement of the borrower is within next 5 years then the loan cannot be granted). With all these, a factor of security arises to secure the loan. A job can be a security (say, EMI is deducted from salary account). Any collateral can be provided by the borrower. Its valuation is required which may vary with time. The entire process is very complex as several issues guide the loan disbursement process. Here a data warehouse schema is designed to take the decision regarding the loan disbursement.

4.1 Table Description

1. **Loan_Application Dim. Table**

   Age_Group (in years)
   Min_work_experience (in years)
   This dimension is related in terms of taking further decisions about sanctioning a loan.
   i. "*Age_Group*" denotes a range, i.e., people belonging to that specified age group are eligible to get a loan.
   ii. "*Min_work_experience*" (Minimum Work experience) would help to understand the regularity of the job. The minimum limit would be considered, say two years in present job is satisfactory to apply for a new credit.

2. **Loan_Period Dim. Table**

   Loan_Type (Character)
   Max_Loan_Prd (in months)
   i. "*Loan_Type*" is a parameter denoting the type of the loan like whether it is a priority or non-priority sector loan. More precisely, the category where the loan belongs to.
   ii. "*Max_Loan_Prd*" (Maximum Loan Period) is the total time period including repayment period and moratorium period. It depends on the loan amount.

3. **Security Dim. Table**

   Min_Security_value (number)
   Security_type (alphabetic value)
   Security is important to secure the loan. It would mainly have these two basic points:
   i. "*Min_Security_value*" states the lower limit on value of any security.
   ii. "*Security_type*" indicates whether the security is primary (security created out of bank loan), collateral (offered by borrower: any FD, jewelry, land or property), third party guarantee by any person.

4. **Income Dim. Table**

   Min_net_sal (number)
   Max_Expenditure (number)
   i. "*Min_net_sal*" (Minimum Net Salary) is required to depict the monthly in-hand to decide whether the borrower would be able to pay the EMI.
   ii. "*Max_Expenditure*" (Maximum expenditure) is required to analyze the spending nature of the borrower.

   The attribute "Min_net_sal" will be used in CIR fact table along with EMI amount from EMI dimension table.

5. **Loan Fact Table**

Age_group (in years)
Min_net_sal (number)
Min_security_value (number)
Max_Loan_Period (in months)
Loan_amt_UL (number) (*measure to be computed)
A loan application would be reviewed depending upon these factors:

i. "*Age_group*" denotes the age limit to decide the repayment capability of loan according to job tenure. Say, age between 21 years to 45 years is considered to be the ideal time to apply for and granting of a loan by some bank. This helps to decide the repayment period of any loan.
ii. "*Min_net_sal*" states the minimum net salary per month (after deducting PF, tax, and others) to repay the debt. It takes into considerations the monthly expenditure and savings.
iii. "*Min_security_value*" is an important parameter. In case of Priority Sector Lending; Government would secure the loan by means of subsidy. But in reality, security is mandatory for each type of loan. It can be primary, collateral, or third-party guarantee.
iv. "*Max_Loan_Period*" (Maximum Loan Period) denoting the repayment period w.r.t. loan amount, remaining job period, loan type. The loan period may differ.
v. "*Loan_amt_UL*" is the measure which states the maximum amount of loan that can be sanctioned by a bank. There is a certain upper limit of each type of loan. For, e.g., a basic home loan's upper limit is Rs. 10 Lakh. The range would be taken as a measure that can be fed to other tables for further analysis.

6. **Credit History Dim. Table**

Credit_Rating (Alphabet)
Min_Work_Experience (in years)
Tax_Income_UL (number)
Total_Household_Income_UL (number)
Credit_Score_Range (number)

i. "*Credit_Rating*" is similar to Credit Score used for any organization or institution. It can be denoted by some alphabet.
ii. "*Min_Job_Experience*" is another attribute that helps to understand the regularity of the job. It helps to analyze a loan application along with no. of years remaining in the job (if not self employed or business or any others, which do not require any age limit).
iii. "*Tax_Income_UL*" (Tax Income Upper Limit) or some range of Tax income is considered in analyzing the credit history.
iv. "*Total_Household_Income_UL*" may include rental income, interest income, and income from any business or from some other sources. This information brings clarity in credit information. The decision-making

becomes easier. All these are relevant in terms of securing a loan from being NPA.

v. "*Credit_Score_Range*" is a measure from CIR fact table denoting the calculated score assigned to some borrower to reflect the possibility of sanctioning a new loan. The range of score is considered; say score 350–800 is preferable.

Credit history maintains the above mentioned factors. Credit score and credit rating are literally same, used in different fields.

7. **EMI Dim. Table**

Max_EMI_Amount (number)
Max_No_of_EMI (number)
Rate_of_Intt (fraction)
Max_Disbursed_Amt (number)

i. "*Max_EMI_Amount*" is calculated depending on rate of interest, maximum no. of EMI, Loan period.
ii. "*Max_No_of_EMI*" denotes the total numbers of EMI to repay the loan. Generally total no. of EMI is less than the loan period. EMI amount varies with total no. of EMI.
iii. "*Rate_of_Intt*" (Rate of interest) is specified by bank. It helps in calculating EMI.
iv. "*Max_Disbursed_Amount*" (Maximum Disbursed Amount) denotes the amount disbursed at a time. The entire loan amount may not be disbursed at a time. A loan may have several phases.

8. **CIR Fact Table**

CIR is a fact table containing:
EMI_to_Income_Ratio (a ratio value in fraction)
Max_Credit_Balance (number)
Credit_Score_Range (number) (*measure to be computed)

i. "*EMI_to_Income_Ratio*" is a measure that produces the required ratio from an estimated EMI & income of the borrower. It is related to Monthly_Income attribute of Income table and EMI_Amount of the EMI dimension table.
ii. "*Max_Credit_Balance*" is the amount of credit someone owes to the lender.
iii. "*Credit_Score_Range*" is the measure denoting a limit or a range which is applicable to judge a borrower's credit worthiness, say score between 350 to 800 are more likely to get a loan. Higher the score is computed, higher the chance of being loan sanctioned.

9. **CIBIL Score Dim. Table**

Credit_Balance (number)
Credit_Usage_Period (in years)

A/c_Status (number/alphabet)
DPD (in days)

i. "*Credit_Balance*" is the amount a borrower owes to the bank. It depends on total credit limit being sanctioned and how much of it have been utilized.
ii. "*Credit_Usage_Period*" indicates the time span of servicing the loan. It is expressed in months.
iii. "*A/c_Status*" (Account status) denotes the status of the account at the time of CIBIL search for a particular customer. It can be written off, closed, NPA and others. We can assign some alphabet or number for each kind of status. For e.g. Write-off account is denoted by 1, closed account denoted as 2 and so on.
iv. "*DPD*" is the measure that can be taken from CIR fact table and shows the nature of the borrower in repaying the debt.

CIBIL is calculated by analyzing the applicant's repayment history, the credit balance, the credit usage period, new credit applied and credit mix. We would consider credit balance and credit usage period here as dimensions. The repayment history can be retrieved from the credit history table. New credit applied and credit mixes are implied here, so not shown as dimensions. Credit balance and its usage period are interlinked. A/c status is important to depict the above two dimensions. Say, for a closed account, credit balance and credit usage period does not have much importance.

10. **Repayment Dim. Table**

Max_Disbursed_Amt (number)
Max_Repayment_Period (in months)
Actual_Payment (number)

i. "*Max_Disbursed_Amt*" (Maximum Disbursed Amount) indicates the amount of credit disbursed to a borrower at a time, say in a single phase. A loan may contain several phases. Variable amount may be disbursed at each phase.
ii. "*Max_Repayment_Period*" is an attribute that can directly participate in Prospective New Loan fact table which would help to develop an idea about fresh borrowers going to be included in CIBIL. Repayment nature reflects the possibility of sanction new credit to the same borrower. Repayment period (defaulter or not) and repayment amount (whether only interest is being paid or debt including principle amount) are included in it.
iii. "*Actual Payment*" shows the exact amount which the borrower has repaid.

11. **New Loan Fact Table**

Introduction of this fact table is major contribution of this research work where we allow new loans to be sanctioned without having CIBIL score. However, if customers have old loan of small amount but no significant CIBIL score this can also be addressed here:

- It can be thought of as a fact table where new loanee customers can be added. Customers who are acquiring loan for the first time may be offered new loan based on different criteria such as DPD, Repayment Period etc. but not previous loan amount. Hence previous loan amount is an optional attribute.
- Customers having small loans can be prospective borrowers to apply for bigger loans. Customers who have taken small loans would be analyzed based on these dimensions.
  The description of the fact table is given below:
  Max_Deposit_Val (number)
  Previous_loan_amount_UL (number)
  Max_R_Period (in months)
  Max_DPD (in days)
  Max_Loan_Sanction (number) (*measure to be computed)
  i. "*Max_Deposit_Val*" (Maximum Deposit Value) is a parameter to decide the savings nature of the customer as well as repayment capability in future if a loan becomes unpaid for long.
  ii. "*Previous_loan_amount_UL*" (Previous Loan Amount Upper Limit) indicates the amount of loan being taken by a customer earlier. However, we keep this attribute optional if the customer has not borrowed any loan previously.
  iii. "*Max_R_Period*" (Maximum Repayment Period) denotes the time span taken by a customer to repay the debt.
  iv. "*Max DPD*" denotes Maximum Days Past Due. DPD means that for any given month, how many months the payment is unpaid. If DPD value is 90 for a date say 06–12, it means in June 2012, the payment is due for last 90 days, which means 3 months dues! So you can now understand the DPD in the last month (May 2012) would be 60 and for Apr 2012 would be 30.
  v. "*Max_Loan_Sanction*" is the measure which denotes the maximum loan could be sanctioned.

12. **Deposit Dim Table**

    Max_Deposit_Val (number)
    Deposit_Type (character)
    Max_Deposit_Period (in months)
    i. "*Max_Deposit_Val*" is the deposit value one is maintaining with the bank.
    ii. "*Deposit_Type*" denotes whether it is a term deposit or demand deposit.
    iii. "*Max_Deposit_Period*" denotes the maximum amount of time a deposit is kept at the bank.

4.2 Proposed Data Warehouse Schema

In Fig. 1 proposed data warehouse schema is depicted. The * mark denotes the measure for each fact table.

- In the fact table Loan the measure is chosen as Loan_Amt_UL
- In the fact table CIR the measure is chosen as Credit_Score_Range.
- In the fact table New_loan the measure is chosen as Max_Loan_Sanction.

5 New Features of Data Warehouse Schema Identified

The data warehouse schema that we have designed here reflects a complex business analysis scenario. We have identified different situations which are not depicted in standard star schema, snowflake schema or fact constellation. Here we list below the instances where new properties of data warehouse schema could be introduced to handle complex business situations like these.

1. The fact table loan has a measure "Loan_Amt_UL" (Loan amount upper limit) which is acting as a dimension value in another fact table "Fact_New Loan". That is, the fact table is acting as a dimension table to another fact table.
2. The same attribute "Min_Job_Experience" (Minimum job experience), Max_Disb_Amt (Maximum Disbursement Amount) appear in two different dimension tables. It does not follow the snowflake schema but placed so as per requirement of the decision-making process.
3. A parameter in the table CIR named as "EMI_to Income_Ratio" is computed from both Max_EMI_amount of EMI dimension table and "Min_Net_Sal" of Income dimension table. That is, the dimension "EMI_to_Income_Ratio" is an entry from two different dimension tables.

6 Conclusion and Future Work

In this research work we propose a data warehouse model for decision-making in loan disbursement as a generic approach. One of the main contributions of this research work is to include the possible loanees/borrowers who are not considered for bigger amount of loan in current context. A mechanism is proposed to compute a measurement alternative to CIBIL score (it is applicable for those who already have loan) for those who do not have any existing loan. The major feature of this research work is to include the possible loanees/borrowers who are not considered for bigger amount of loan in current context. A mechanism is proposed to compute a measurement alternative to CIBIL score (it is applicable for those who already have loan) for those who do not have any existing loan.

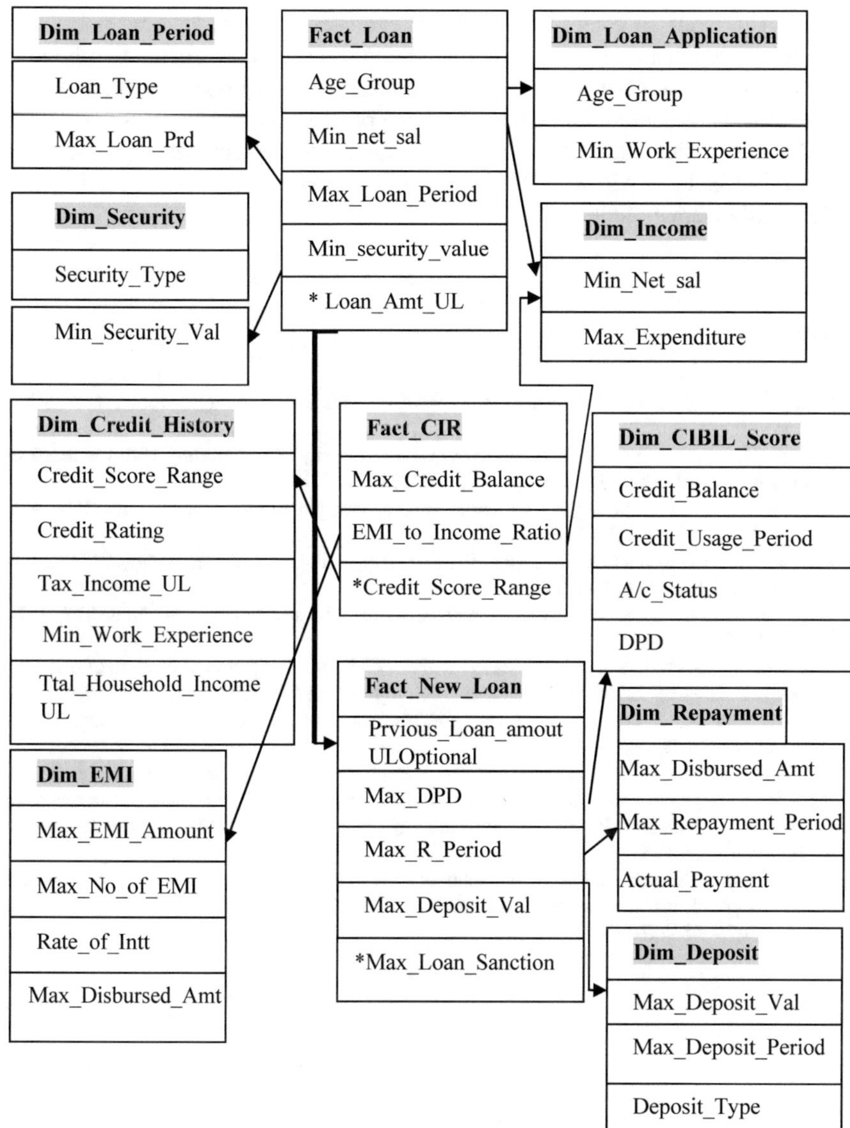

Fig. 1 A generic data warehouse schema for loan disbursement analysis

The most important contribution of this research work is discussed in previous section where we identify possible new features of data warehouse schemas which could be formalized and defined as new features of data warehouse schema.

This research work could be extended further to incorporate different concepts of data mining over the proposed data warehouse model of loan disbursement process. The model is established as the integration of several parameters related to the banking industry. Data mining methodologies such as Association rule mining over different attributes could lead to interesting knowledge generation. Other methodologies of data mining could also be applied for further improvement. This model may be redefined further for specific type of loan disbursement. These business specific analyses will contribute towards the innovation of business intelligence for loan disbursement industry.

References

1. Indian Institute of Banking & Finance: Principles and Practices of Banking, 3rd edn., pp. 267–274 (2015)
2. Throve, H.A.: Analytical study of priority and non-priority sector lending with reference to Nationalized Banks in India. Paridnya—The MIBM Res. J. **3**(1) (2015)
3. Dave, K.S.: A study of priority sector lending for selected public sector banks of India. IJRAR-Int. J. Res. Anal. Rev. **3**(3) (2016)
4. Bhiwandikar, M.: Role of CGTMSE and CIBIL in supporting the Indian Economy. ISSN-2347-7571, Volume 1, Issue 6, June 2014
5. Uppal, R.K.: Priority sector advances: trends, issues and strategies. J. Account. Tax. **1**(5), 079–089 (2009)
6. Maji, G., Sen, S.: A data warehouse based analysis on CDR to depict market share of different mobile brands. IEEE Indicon (2015). https://doi.org/10.1109/INDICON.2015.7443706
7. Peyravi, M.H.: A schema selection framework for data warehouse design. Int. J. Mach. Learn. Comput. **2**(3) (2012)
8. Sen, S., Chaki, N., Cortesi, A.: Optimal space and time complexity analysis on the lattice of cuboids using galois connections for data warehousing. In: 4th International Conference on Computer Sciences and Convergence Information Technology (ICCIT 2009). https://doi.org/10.1109/iccit.2009.185
9. Sen, S., Roy, S., Sarkar, A., Chaki, N., Debnath, N.C.: Dynamic discovery of query path on the lattice of cuboids using hierarchical data granularity and storage hierarchy. Elsevier J. Comput. Sci. **5**(4), 675–683 (2014)

Analysis and Design of an Efficient Temporal Data Mining Model for the Indian Stock Market

Cerene Mariam Abraham, M. Sudheep Elayidom and T. Santhanakrishnan

Abstract “30 years after Black Monday, can the stock market crash again?” This question is still pertinent in the minds of every stakeholder of the stock market community. Less risk are favorite words to hear in this community. Derivative market is known for low-risk investment when compared to the equity market. This is a less-explored research area in the Indian stock market. In this paper, we perform analysis on the Indian derivative market data of some major companies using data mining techniques. After discussing with experts from the equity domain, we have brought down our focus on certain factors which is impacting the Indian derivative market. We statistically verified the significance of these variables on stock price. This paper also shows how these factors help in prediction of future stock price.

Keywords Derivative market · Open interest · Deliverable quantity · Temporal data mining

1 Introduction

Temporal data mining (TDM) is the analysis on time-related data to harvest useful information out of it. The collected temporal data yields reliable conclusion when we have large volumes of time related data for analysis. Temporal data mining tasks include temporal data characterization or comparison, temporal clustering, classification, temporal pattern analysis, temporal prediction, and trend analysis [1].

C. M. Abraham (✉) · M. Sudheep Elayidom (✉)
Cochin University of Science and Technology, Kochi 682022, India
e-mail: cma@cusat.ac.in

M. Sudheep Elayidom
e-mail: sudheep@cusat.ac.in

T. Santhanakrishnan (✉)
Defence Research & Development Organization, Kochi 682021, India
e-mail: santhanpol@gmail.com

A. Abraham et al. (eds.), *Emerging Technologies in Data Mining and Information Security*, Advances in Intelligent Systems and Computing 813,
https://doi.org/10.1007/978-981-13-1498-8_54

In the financial sector, the investors and fund managers in banks and insurance companies use the information obtained from temporal data mining to channel their funds properly for better returns. There exists several prediction models, but still there is scope for improving the prediction accuracy and preserving data trend which forms the primary motivation for prediction of such financial TDM. It is very important to understand the present data to arrive at the best prediction model. On this perspective, we have collected 3 months daily stock data of four companies in the Derivative market and Cash market and tried to develop an appropriate model using statistical analysis, which would help us to predict the future unobserved values of the Indian equity market.

This paper describes the role of three variables from the stock data and models its effect on stock price in the financial derivative market. Three months data of four companies are taken into account for the analysis. The rest of the paper is organized as follows. Some prediction models existing in the literature are discussed in Sect. 2. In Sect. 3, implementation is detailed. The model is applied on National Stock Exchange data (NSE) and the results are analyzed and discussed in Sect. 4. The paper ends with conclusion in Sect. 5.

2 Related Works

One of the motivations for this work was found from the early observations of Ms. Shalini H. S. and Dr. Raveendra P. V. [2]. As per their study, they found that the Derivatives market provide an opportunity to transfer risk from one to another. They also provide details about the success of the launch of equity derivatives market in India. As per Michael Chui, Derivatives are initiated in response to some fundamental changes in the global financial system. If they are properly handled, it would lead to improvement of resilience of the system and bring economic benefits to the users [3]. Shiguang Lin quantified and analyzed the financial derivatives markets and arrived at a risks control model of the financial derivatives markets using the basic methods and principles of VAR [4]. The asymmetric relationship between the India Volatility Index and stock market returns was analyzed by M. Thenmozhi and Abhijeet Chandra [5]. Through their paper, they demonstrated that the Nifty returns are negatively related to the changes in the India Volatility Index levels. Abhi Dattasharma and Praveen Kumar Tripathi identified the interdependencies between different stocks, and they concluded in their paper that investment in one stock can be done when a related stock is performing well [6]. They found similarity based on the directions of amount of change among the related stocks. In a paper, Priti Saxena and co-authors proposes a new approach of analyzing the stock market and predicts the market using a hybrid form of linguistic-a priori concept. It provides more accurate results in stock prediction which is of great help for the investors and brokers in their decision making. This approach provides the brokers with various useful patterns and helps the clients with easy access to information status of any stock price movement immediately [7].

Aditya Gupta uses Hidden Markov Models to explain the stock market prediction [8]. The paper adopts forecasts the stock values for the next day based on historical data using Maximum a Posteriori HMM approach. The small changes in Stock value and the intra-day high and low values of the stock is used to train the continuous HMM. Maximum a Posteriori decision is taken using this HMM over all the possible stock prices for the next day. This method was compared to existing models for forecasting of stocks such as HMM-fuzzy model, ARIMA and ANN.

C. Narendra Babu and B. Eswara Reddy suggest a hybrid ARIMA-GARCH model for highly volatile financial TSD, which is suitable for multi-step ahead forecasting [9]. This model is applied on selected NSE India data sets and the results obtained are evaluated using various error performance measures such as MAPE, MaxAPE, RMSE. These values confirm improved prediction accuracy when compared to traditional models such as ARIMA, GARCH and trend-ARIMA and wavelet-ARIMA. The data trend is also preserved over the prediction horizon better than the other models.

3 Data Analysis

After discussions with some of the experienced stock exchange members and using the previous data of price variations of stocks, the factors such as Open Interest, Number of Contracts and Deliverable quantity were shortlisted to study their effect on the future price of shares.

A futures contract is a contractual agreement to buy or sell a particular commodity or financial asset at a predetermined price in the future. Open interest is the total number of options and/or futures contracts that are not closed or delivered on a particular day. It is denoted by X. Number of contracts is total number of contracts on security. It is denoted as Y. Deliverable Volume is the quantity of shares which actually move from one set of people (who had those shares in their demat account before today and are selling today) to another set of people. This is the amount of shares which actually get transacted. It is denoted by Z.

For our data analysis we use three months data of four companies from National Stock Exchange website [10]. The required data is extracted. Detailed description of these variables was given in previous paper [11]. Closing Price (P) is taken as the response variable. We have used three variables as regressors initially to check which all variables affect the Price of the stock, P. We try to fit a linear regression model. We consider the general linear model,

$$y = \beta_0 + \beta_1 x_1 + \beta_2 x_2 + \ldots\ldots \beta_{p-1} x_{p-1} + \epsilon. \quad (1)$$

This links a response variable y to predictors $x_1, x_2, \ldots, x_{p\text{-}1}$ where $\beta_1, \beta_2, \ldots, \beta_p$ are coefficients of predictor variables and ϵ is the error term.

Table 1 Coefficients of first company

Model	Unstandardized coefficients		Standardized coefficients	T	Sig.	Collinearity statistics	
	B	Std. error	Beta			Tolerance	VIF
1 (Constant)	1828.986	14.272		128.152	0.000		
X	−2.783E−005	0.000	−0.723	−5.620	0.000	0.139	7.170
Y	−0.002	0.001	−0.210	−1.496	0.141	0.117	8.517
Z	−1.233E−005	0.000	−0.050	−0.845	0.402	0.648	1.544
2 (Constant)	1818.741	7.514		242.033	0.000		
X	−2.650E−005	0.000	−0.688	−5.661	0.000	0.155	6.441
Y	−0.003	0.001	−0.268	−2.207	0.032	0.155	6.441

3.1 Significance of Variables

The hypothesis for checking the significance of the variables states that the parameters are insignificant. $H_0 : \beta_1 = \beta_2 = \cdots = \beta_{p-1} = 0$ and $H_1 : \beta_j \neq 0$ for at least one j. The result of the first company is given in Table 1.

Here we used backward elimination technique. Here the p-values corresponding to the regressors Z in model 1 is greater than 0.05. Hence we do not reject the hypothesis that the parameters are zero. The estimates for the parameters excluding Z is given in model 2. We can see that the p-values corresponding to model 2 are low for the parameters. Therefore we reject the null hypotheses. Similarly the coefficients of the other companies are calculated.

3.2 Model Diagnostics

The coefficient of determination R^2 is useful in checking the effectiveness of a given model. R^2 is defined as, general linear model,

$$R^2 = (SSR/TSS), \tag{2}$$

where TSS is the total sum of squares and SSR is the regression sum of squares. R^2 is defined as the proportion of the total response variation that is explained by the model. The R^2 value of the first company is given in the following Table 2.

The value of R^2 under model 2 suggests that 88.5% of the total response variation is explained in this linear model. Even though R^2 is an important measure for checking the effectiveness of a model, one cannot say the model is effective solely based on the value of R^2. Other factors also need to be taken into consideration.

Table 2 Model Summary

Model R	R square	Adjusted R square	Std. error of the estimate	Change statistics					Durbin–Watson
				R square change	F change	df1	df2	Sig. F change	
1 0.942[a]	0.887	0.880	42.38112	0.887	128.072	3	49	0.000	
2 0.941[b]	0.885	0.881	42.25995	−0.002	0.714	1	49	0.402	0.631

[a]Predictors: (Constant), Z, X, Y
[b]Predictors: (Constant), X, Y
[c]Dependent Variable: P

Table 3 ANOVA

Model	Sum of squares	df	Mean square	F	Sig.
1 Regression	690115.261	3	230038.420	128.072	0.000[b]
Residual	88011.817	49	1796.160		
Total	778127.078	52			
2 Regression	688831.921	2	344415.961	192.853	0.000[c]
Residual	89295.157	50	1785.903		
Total	778127.078	52			

[a] Dependent Variable: P
[b]Predictors: (Constant), Z, X, Y
[c]Predictors: (Constant), X, Y

3.3 Significance of Regression

The test for significance of regression in the case of multiple linear regression analysis is carried out using the analysis of variance (ANOVA). The test is used to check if a linear statistical relationship exists between the response variable and at least one of the predictor variables.

The hypothesis for checking the significance of the variables states that the parameters are insignificant. $H_0 : \beta_1 = \beta_2 = \cdots = \beta_{p-1} = 0$ and $H_1 : \beta_j \neq 0$ for at least one j and the result is given in Table 3.

Since p-value is too small, we reject H_0. That is, the regression is significant. Hence there is ample relevance of linearity in the relation between the regressors and the independent variable.

3.4 Checking the Normality Assumptions

We look at the residuals to check the assumption that the errors in the model are normally distributed. One approach is to prepare a normal probability plot. The advantage of this method of computing probability plots is that the intercept and slope estimates of the fitted line are in fact estimates for the location and scale parameters of the distributions. If the points are scattered around a straight line we can say that there is normality. Deviations from straight line indicate lack of normality. The normal probability plot for our model is given in Fig. 1. Here since the points are very close to the reference line, our assumption that the errors are normally distributed is valid.

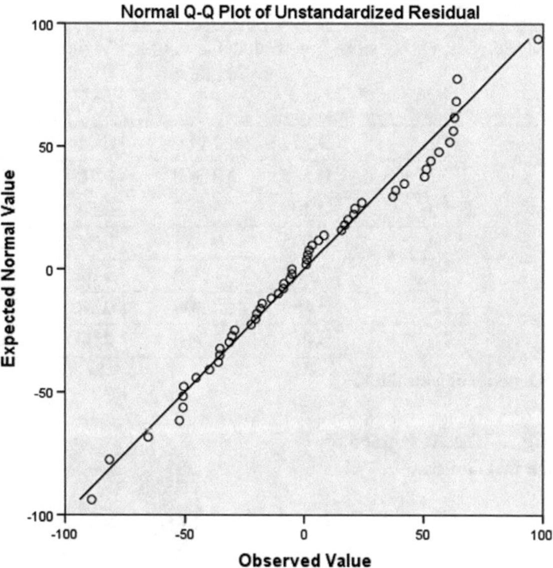

Fig. 1 The straight line in the figure shows expected normal value. Residual of this model is shown as dotted symbols

3.5 *Multicollinearity Diagnostics*

Multicollinearity is the undesirable situation where the correlations among the regressors are strong. Multicollinearity increases the standard errors of the coefficients, thereby coefficients for some regressors may be found not to be significantly different from zero.

Tolerance, Variance Inflation Factor (VIF), Eigen values and Condition index are the collinearity factors which help us to identify multicollinearity. Several eigenvalues are close to 0, indicating that the predictors are highly intercorrelated and that small changes in the data values may lead to large changes in the estimates of the coefficients. The condition indices are computed as the square roots of the ratios of the largest eigenvalues to each successive eigen values. Values greater than 15 indicate a possible problem with collinearity; greater than 30, a serious problem. Six of these indices are larger than 30, suggesting a very serious problem with collinearity.

Variance inflation factor (VIF) measure how much the variance of the estimated coefficients is increased over the case of no correlation among the regressors.

$$\mathrm{VIF} = 1/\left(1 - \mathrm{R}_{\mathrm{i}}^{2}\right), \tag{3}$$

where $\mathrm{R}_{\mathrm{i}}^{2}$ is the coefficient of determination from the regression of x_{i} on all other regressors. If x_{i} is linearly dependent on the other regressors, then $\mathrm{R}_{\mathrm{i}}^{2}$ will be large. Values of VIF larger than 10 are taken as solid evidence of multicollinearity. When the tolerances are close to 0, it indicates that there is high multicollinearity and the

Table 4 Coefficients of first company

Model	Dimension	Eigenvalue	Condition index	Variance proportions			
				(Constant)	X	Y	Z
1	1	3.272	1.000	0.01	0.01	0.01	0.01
	2	0.588	2.360	0.10	0.04	0.03	0.04
	3	0.105	5.585	0.55	0.11	0.04	0.54
	4	0.035	9.632	0.34	0.84	0.93	0.42
2	1	2.482	1.000	0.06	0.01	0.01	
	2	0.469	2.300	0.94	0.03	0.03	
	3	0.049	7.110	0.00	0.96	0.96	

[a]Dependent Variable: P

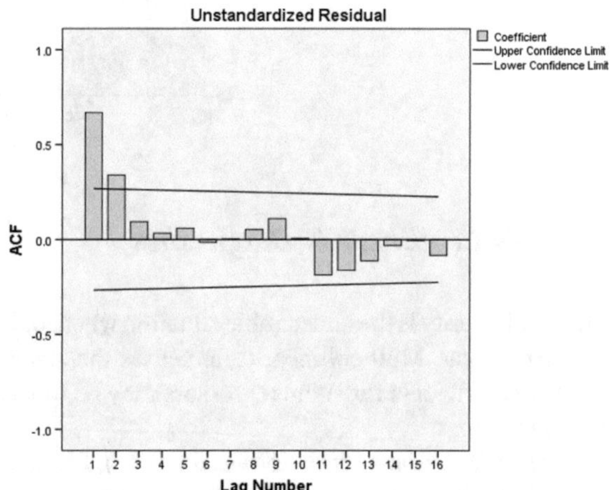

Fig. 2 The ACF graph of the first company

standard error of the regression coefficients will be inflated. Tolerance and VIF is given in Table 1. Eigen values and Conditional Index is given in Table 4.

Here we can see that the variance inflation factor is less than 10 and the tolerance is not close to 0. Also since the Eigen values are not very close to zero and conditional index are less than 15 the problem of multicollinearity does not arise in this model.

3.6 *Checking for Autocorrelation*

Autocorrelation can be checked using the Durbin–Watson test. For independent errors the Durbin–Watson test statistic is approximately 2. Durbin–Watson value of this model is given in Table 2. The ACF and PACF graph is given in Figs. 2 and 3.

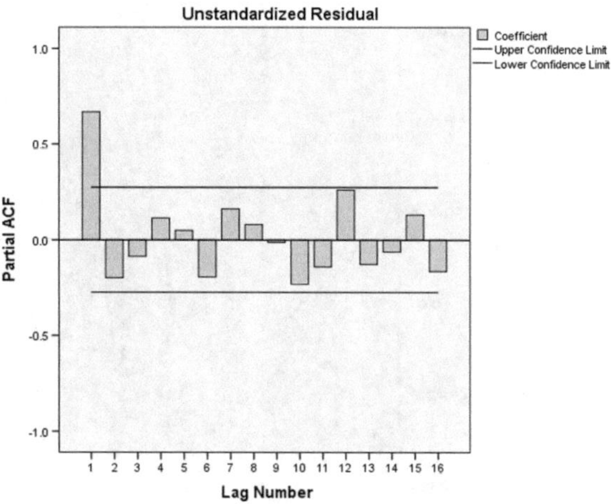

Fig. 3 The PACF graph of the first company

Table 5 One-sample Kolmogorov–Smirnov test

	Unstandardized residual	
N	53	
Normal parameters[a,b]	Mean	0.0000000
	Std. deviation	41.43928757
Most extreme differences	Absolute	0.077
	Positive	0.071
	Negative	−0.077
Kolmogorov–Smirnov Z	0.557	
Asymp. Sig. (2-tailed)	0.915	

[a]Test distribution is Normal
[b]Calculated from data

3.7 *Goodness of Fit*

A test for goodness of fit involves examining a random sample from some unknown distribution in order to test the null hypothesis that the unknown distribution function is in fact a known specific function. Normality test result is given in Table 5 and the result shows that it is normal.

4 Results and Discussions

Since the assumptions of regression are violated due to autocorrelation, we move on to time-series regression. An autoregressive integrated moving average (ARIMA) model is a generalization of an auto regressive moving average (ARMA) model.

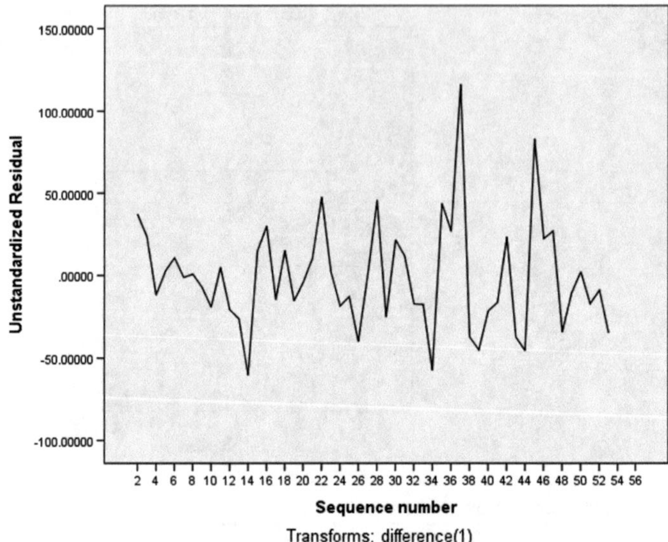

Fig. 4 Time Series plot of the first company

These models are fitted to time series data either to better understand the data or to predict the future values in the series. They are applied in some cases where the data show evidence of non-stationary, where an initial differencing step can be applied to remove the non-stationary.

4.1 Time Series Graph

Time series plot of our model is given in Fig. 4.

4.2 Prediction Model

We consider P as response variable and X, Y, Z as the regressors. It is found that the time series model of the residual from the regression is ARIMA(0,0,1). Therefore, the equation is,

$$\epsilon_t = a_t - \theta a_{t-1}. \tag{4}$$

where a_t is the white noise and θ is the moving average parameter. Moving average parameter is calculated and its value is −0.620. The residual ACF and PACF plot is given in Fig. 5.

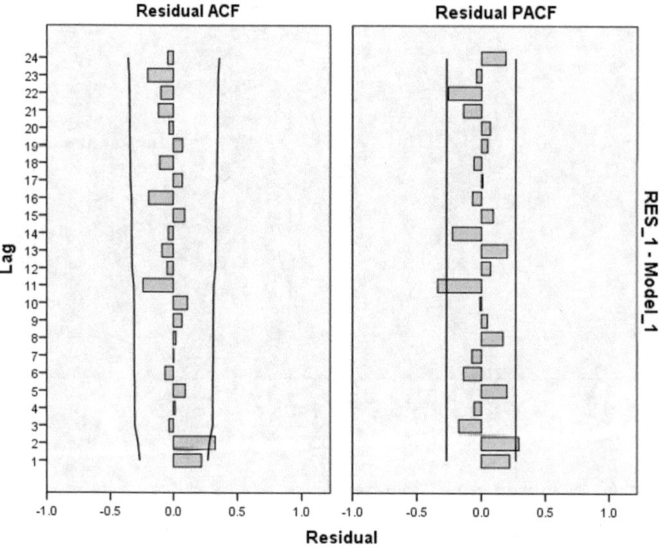

Fig. 5 Residual ACF and PACF plots of the first company

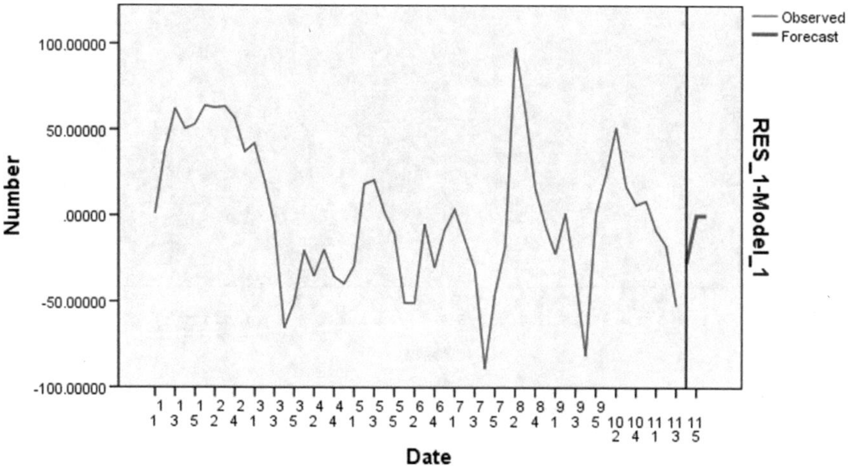

Fig. 6 Prediction plot of the first company

By combining Eqs. (1) and (4), we can get our model. Coefficients of parameters are obtained from Table 1. Therefore our new fitted regression model is in Eq. (5) and Fig. 6 gives prediction plot,

$$p = 1818.741 - 2.650 \times 10^{-005}x - 0.003y + a_t + 0.620a_{t-1}. \quad (5)$$

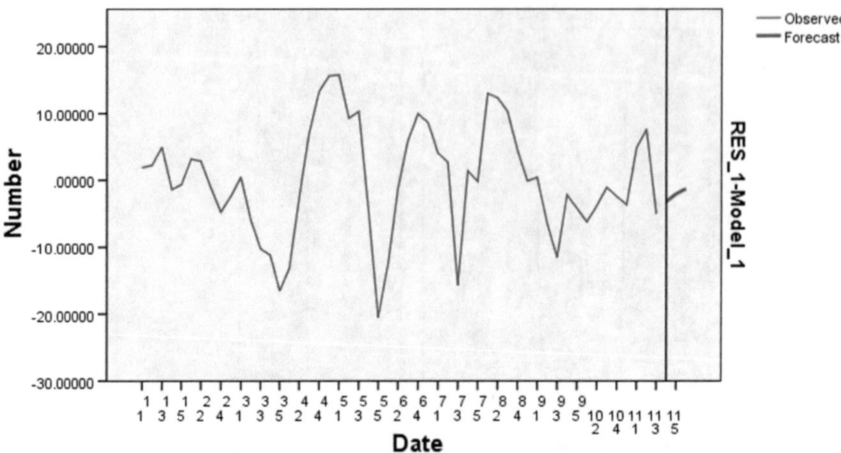

Fig. 7 Prediction plot of the second company

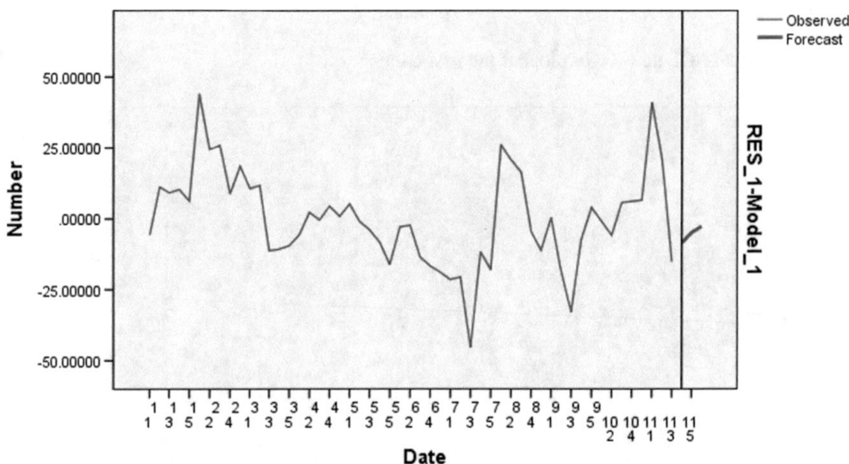

Fig. 8 Prediction plot of the third company

Similarly prediction models are generated for the other three companies. Prediction models are given in Eqs. (6) and (7). Prediction plot for the second company is in Fig. 7.

$$p = 270.956 - 4.679 \times 10^{-007}x + 6.433 \times 10^{-007}z + 0.637\epsilon_{t-1} + a_t. \quad (6)$$

Prediction model for the third company is Eq. (7) and prediction plot in Fig. 8.

$$p = 588.751 - 3.329 \times 10^{-006}x - 2.525 \times 10^{-006}z + 0.567\epsilon_{t-1} + a_t. \quad (7)$$

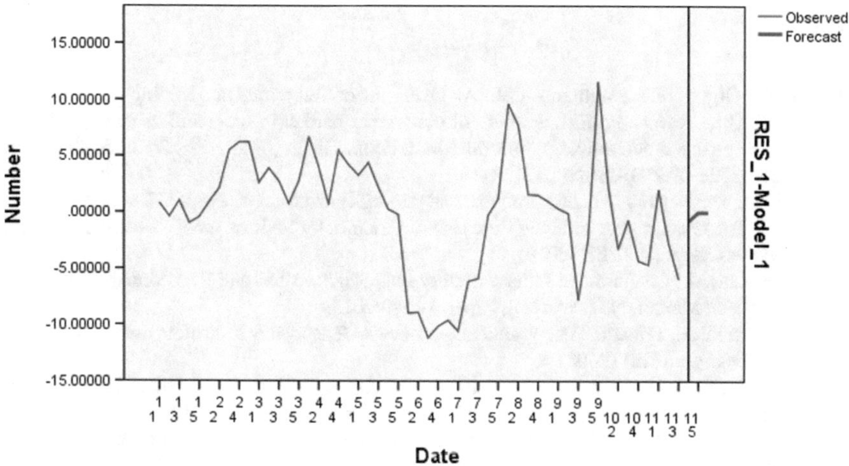

Fig. 9 Prediction plot of the fourth company

Prediction model for the fourth company is given below and plot is in Fig. 9.

$$p = 107.726 - 3.720 \times 10^{-007}x - 0.001y - 1.739 \times 10^{-006}z + a_t + 0.935a_{t-1}. \quad (8)$$

5 Conclusion

Using temporal data mining techniques, analysis of four companies for a period of 90 days is done and prediction models are obtained. From the study, it was statistically proven that the factors Open Interest, Number of Contracts, and Deliverable quantity have significant influence on the stock price. This conclusion will be of great importance to the investors and brokers in the Indian Derivatives market. We have also found a pattern among these regressors with response variable, P, which is the Price of the stock. Now this pattern recognition can be further analyzed by conducting a study on 5 years data of five large cap stocks using big data analytics, which forms the future scope of this paper.

Acknowledgements The authors wish to express their gratitude to Mr. Manoj P. Michel, member of Cochin Stock Exchange, who shared his profound knowledge of the equity market and for his useful discussions.

References

1. Lin, W., Orgun, M.A., Williams, G.J.: An Overview of Temporal Data Mining
2. Shalini, H.S., Raveendra P.V.: A study of derivatives market in India and its current position in global financial derivatives markets. IOSR J. Econ. Finan. (IOSR-JEF) **3**(3), 25–42 (2014). e-ISSN: 2321-5933, p-ISSN: 2321-5925
3. Chui, M.: Derivatives markets, products and participants: an overview. IFC Bull. **35**
4. Lin, S.: The Quantitative Analytic Research of Extenics by VAR on the Risks of the Financial Derivatives Markets. IEEE (2010)
5. Thenmozhi, M., Chandra, A.: India Volatility Index (India VIX) and Risk Management in the Indian Stock Market, NSE Working Paper, W P/9/2013
6. Dattasharma, A., Tripathi, P.K.: Practical Inter-stock Dependency Indicators using Time Series and Derivatives. IEEE (2008)
7. Saxena, P., Pant, B., Goudar, R.H., Srivastav, S., Garg, V., Pareek, S.: Future Predictions in Indian Stock Market through Linguistic-Temporal Approach. IEEE (2012)
8. Gupta, A.: Stock Market Prediction Using Hidden Markov Models. IEEE (2012)
9. Narendra Babu, C., Eswara Reddy, B.: Selected Indian stock predictions using a hybrid ARIMA-GARCH model. In: International Conference on Advances in Electronics, Computers and Communications. IEEE (2014)
10. National Stock Exchange of India Ltd. https://www.nseindia.com/
11. Abraham, C.M., Elayidom, M.S., Santhanakrishnan, T.: Implementation of correlation and regression techniques for efficient analysis of Indian stock market. Int. J. Res. Eng. Technol. 70–73. eISSN: 2319-1163 (2017)

Shilling Attack Detection in Recommender System Using PCA and SVM

Noopur Samaiya, Sandeep K. Raghuwanshi and R. K. Pateriya

Abstract Shilling Attack is where, deceptive users insert fake profiles in recommender system to bias the rating, which is termed as shilling attack. Several studies were conducted in past decade to scrutinize different shilling attacks strategies and their countermeasures mainly categorized in linear algebra, statistical, and classification approach. This paper explores two different methods for shilling attack detection namely, Principal Component Analysis (PCA) and Support Vector Machine (SVM) and compares their performance on attack detection. We had experimented with simulating various attack models like average, random, and bandwagon models. This paper further discusses the importance of detecting malicious profile from the genuine one and suggests deep insights of developing new and more efficient shilling attack detection techniques. The experiments were conducted on the Movie Lens 100 K Dataset and compared the performance of PCA technique with supervised SVM-classification method.

Keywords Shilling attack · Unsupervised learning · Supervised learning
Attack models and attack detection strategies

N. Samaiya (✉) · S. K. Raghuwanshi
Department of Computer Science and Engineering, Samrat Ashok Technological Institute, Vidisha 464001, MP, India
e-mail: noopursamaiya235@gmail.com

S. K. Raghuwanshi
e-mail: sraghuwanshi@gmail.com

R. K. Pateriya
Department of Computer Science and Engineering, Maulana Azad National Institute of Technology, Bhopal 462003, MP, India
e-mail: pateriyark@gmail.com

A. Abraham et al. (eds.), *Emerging Technologies in Data Mining and Information Security*, Advances in Intelligent Systems and Computing 813,
https://doi.org/10.1007/978-981-13-1498-8_55

1 Introduction

Peoples come from the era of information scarcity to abundant information era. Many users prefer shopping (buying/selling products) over the Internet via e-commerce sites. As the number of users is increasing, utilization of vendor and items is also increasing, because of this it becomes a challenge for user to choose the right product without wasting too much time. To help customers and surpass such challenge, a collaborative filtering algorithm is used by e-commerce sites. CF help a user select the product that they might like. In order not to frustrate users, online vendor exploit CF system which must provide recommendations efficiently, which make collaborative filtering system attacked, by injecting fake profiles [1]. Such system is vulnerable to attacks by the malicious user which termed as "shilling" or "profile injection" [2, 3]. Attack profile needs to be detected and removed accurately to ensure the credibility of recommender systems, CF system can achieve if data is of high quality. It is almost impossible to estimate accurate predictions from low-quality data. Shilling attack can be performed only if the attacker is familiar with recommender system. The information might include, but not limited to the mean rating and standard deviation for each item and user in the user-item matrix, rating distribution, and so on [3]. Typically, an attack is realized by inserting several attack profiles into a recommender system database to cause bias on selected target items. Attacks might be applied on the basis of different purposes, and they can be classified according to different dimensions [2], i.e., the intent of attack, targets, required knowledge, cost, algorithm dependence, and detectability. An attack is mainly classified according to intent and amount of knowledge required by the system. According to the intent, they are push and nuke attack. The objective of push attack is to enhance the popularity of item while the nuke is to minimize the popularity of the item. While according to the required knowledge they are high-knowledge and low-knowledge [4].

2 Related Works

The term Shilling was first used by [2]. The target of paper is first to detect and then remove fake profiles in profile injection attacks [1, 5]. These attacks contain fabricated attack profile which is difficult to identify as "Shilling" profiles designs are similar to genuine profiles. Attack model constructed by random ratings, assigned to filler items in the profile [3]. As the attack is deliberate, this may cause erosion in user trust in term of objectivity and accuracy of the system.

A bunch of research on supervised learning is being studied and implemented for shilling attack detection [1, 6]. SVM-classification are used to improve the robustness of the recommender system [7]. However, the paper undertakes classifier of SVM which are Linear and RBF (Radial Basis Function). A linear kernel is suitable for a large number of features.

Table 1 Structure of attack models

Attack model	I_S (Selected items)	I_F (Filler items)	I_T (Target items)
Random Attack	Φ	Random rating	r_{max}/r_{min}
Average Attack	Φ	Mean of each item	r_{max}/r_{min}
Bandwagon Attack	r_{max}	Random rating	r_{max}/r_{min}

2.1 Attack Model

Attacker's knowledge and purpose are to identify attack model [1]. Here three popular attack models are employed: random attack, average attack, bandwagon attack. The attacker injects attack profile to mislead the CF system. Such profiles defined by three set of items. Target item I_T for those items, the ratings are r_{max} under push attack and r_{min} under nuke attack. Set of selected items I_S Selected set is a set of widely popular items. I_F is randomly chosen, filler items in the fake profile are set of items that make the profile look normal and make a malicious profile harder to detect. Table 1 is showing the structure of attack model.

Random attacks: In this the items is rated randomly around the overall mean vote. This attack is carried out by selecting filler items I_F randomly. No rating is assigned to a selected item in this case. It requires less knowledge about the system and has less impact on the system.

Average Attack: In this the items is rated randomly around the mean vote of every item. Attacker randomly chooses filler items and provides them with a rating in the same manner as in random attack. This attack requires prior knowledge of system so difficult to implement.

Bandwagon Attack: In this the items are rated randomly among the overall mean vote and some highly popular items are selected and rated by the maximum vote. Here target items, I_T is provided with the highest rating and often rating to selected items, I_S. The selection of filler item set is same as described in random attacks.

2.2 Detection Approaches for Shilling Attacks

Here the attacks are implemented manually using various methods discussed in the last section, Hence, statistically created profiles can be differentiate from genuine users. This conflict can be viewed in many ways, which include an abnormal deviation from the average system rating or an unusual number of ratings in a profile [8]. As a result, these metrics can detect anomalies and is used in identifying attack profiles [9] presented several heuristic attributes of detecting these anomalies. These detecting methods use features for getting the number of bogus user in the rating database as they lie separately compared to legitimate users. Further, these features are used in SVM.

Degree of agreement with another user (DD): For the rating matrix $M = [rij]_{m \times n}$, let v_j be the mean rating of item j. Then, the degree to which the user i differ from the other users on item j is given by $|r_{ij} - v_j|$ and DD defined as [10] (Fig. 1):

$$DD(i) = \frac{\sum_{j \in I_i} |r_{ij} - v_j|}{|I_i|} \quad (1)$$

The above graph represents that the degree of agreement works well for all three models but for average attack model the result is much better than other two models.

Rating deviation from Mean Agreement (RDMA): The attackers are identified by assessing the profile's average change per item weighted by the inverse of a number of rating per item as in (2).

$$RDMAj = \frac{\sum_{i=0}^{Nj} \frac{|r_{i,j} - Avg_i|}{NR_i}}{N_j}, \quad (2)$$

where N_j, represent are item user j rated, r_{ij} represent the rating on item i by user–user j, NR_i is the overall rating of the system assigned to item i (Fig. 2).

The graph shows that RDMA can be used for bandwagon and random attack model, the result is not good for average attack model.

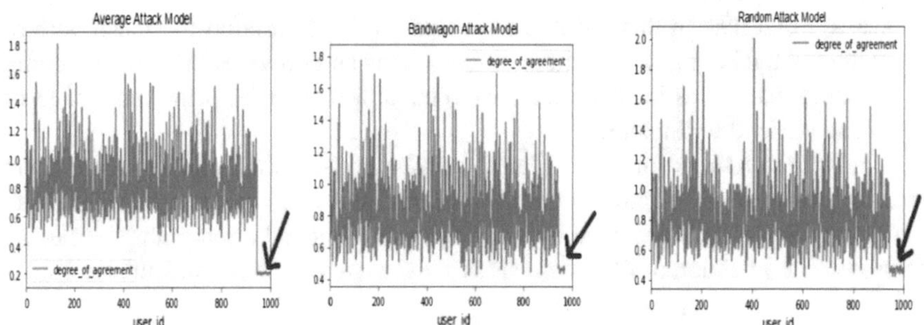

Fig. 1 For DD the marked is the bogus users while rest are normal users

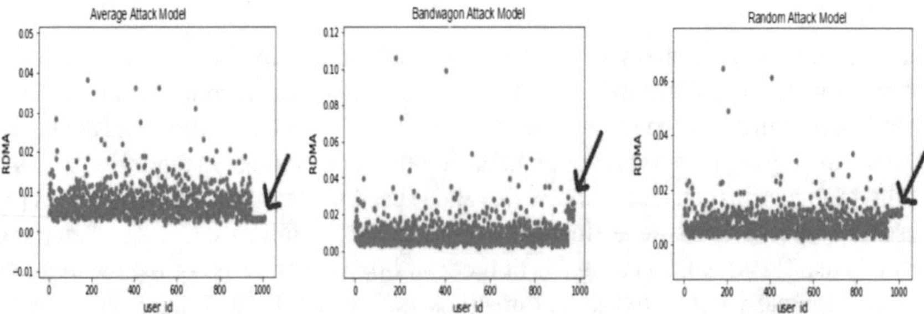

Fig. 2 For RDMA the marked is the bogus users while rest are the normal users

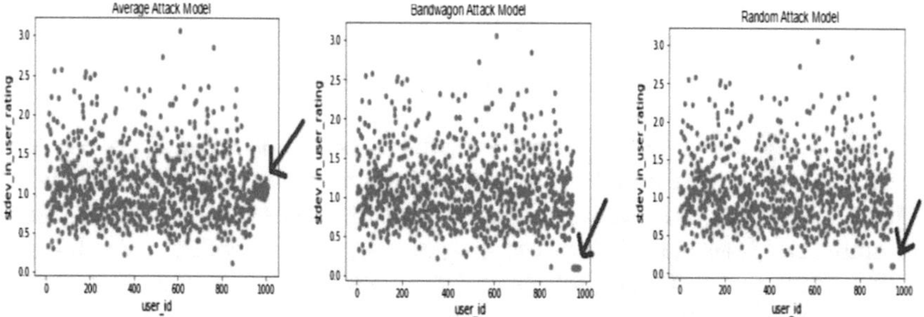

Fig. 3 For Standard deviation the marked is the bogus users while rest are normal users

Standard Deviation in User Rating: For the particular user rating can be given by standard deviation. If μ_i is the average rating of user i, and I_i is the number of items rated by that user [10], then the standard deviation σ_i is computed as follows (Fig. 3):

$$\sigma_i = \frac{\sum_{j \in Ii} \left(r_{ij} - \mu_i\right)^2}{|I_i| - 1} \tag{3}$$

The graph shows that standard deviation can be used for bandwagon and random attack model, and the result is not good for average attack model.

3 Proposed Work

Proposed model for shilling attack detection are PCA is an unsupervised method, which requires pervious knowledge rather than training samples for standardizing data reducing the feature space, whereas SVM is a supervised learning which is focused on the feature extraction of the user profile and train classifier to perform classification.

Principal Component Analysis: Shilling Attack detection using PCA had been discussed in a recommender system [5, 11]. PCA is a linear dimensionality reduction method, PCA find eigenvector of a covariance matrix and then uses those to project the data into new space of less dimensional and others are regarded as redundant information is discarded. First, we had extracted the feature and then on these feature selection processes is done before applying it to classifier. PCA detection performs well to detect shilling profiles constructed by several usual attack models, such as random attack, average attack, and bandwagon attack [11, 12].

The PCA-based algorithm is also known as PCA-Var-Select. It is the well-known unsupervised detector in shilling attack detection. PCA compute the user–user covariance matrix., let r_u and r_v, be rating vector of user u and v, the covariance denoted by $Cov_{u,v}$ is

$$\mathrm{Cov}_{u,v} = \frac{\sum_i (r_{u,i} - \bar{r}_u)(r_{v,i} - \bar{r}_v)}{\sqrt{\sum_i (r_{u,i} - \bar{r}_u)^2}\sqrt{(r_{v,i} - \bar{r}_v)^2}}, \quad (4)$$

where $r_{u,i}$ is the rating given by user u to item i, $\bar{r}_u$ is the average rating of user u.

On the covariance matrix, Eigenvalue decomposition is implemented, and principal component (PCs) is extracted from x components which have largest Eigenvalues. Generally, "x" is assigned value according to need, here the value of x is taken as 3. As a result, these user profiles are projected to x principal components, and the sum of squares on x PCs is adopted as the ranking scores. Finally, the users ranked smallest in top-k are filtered out for classification.

Support Vector Machine: Support Vector Machine (SVM) is a classification tool, it provide more efficient and accurate than other classification methods due to the implementation of Structural Risk Minimization (SRM) principle it guaranteed the lowest classification error [13]. Its purpose is to approximate the nonlinear function with good generalization performance even for high-dimensional data [14]. If the data are living in a p-dimensional space, SVM finds the separating hyperplane with maximal margin [15] and makes a point of training set distance the classification hyperplane as far as possible.

SVM method is used in two ways; firstly in the user-item matrix, all the items have been used as features. In this method, SVM is applied using linear kernel. Second, SVM is performed in two phases; the first phase involves features extraction. Here Degree of agreement, Rating deviation from the mean agreement and Standard deviation in users rating are used as features. In second phase SVM is applied to the extracted features.

4 Experimental Results and Analysis

This section, present the overall performance of the method used above for the classification and feature extraction.

Dataset: We are using a MovieLens100K dataset that is the widely used data in the for detecting shilling attack. This dataset consists of 100,000 rating on 1682 movies by 943 users.

Attacks formation: Shilling profiles or attack profiles have been constructed with the help of different attack model, including Random, Average and Bandwagon attack model with different values of attack size (1, 3, 5, 7, and 10%) and filler size is 300.

Evaluation Metrics: Here the large number of user are required to produce the attack and prior knowledge is require, if the attackers are small in number then there will be no effect on the recommender system. So, for calculating the accuracy we have adopted precision (6), recall (7) and F-measure (8) as defined below (Fig. 4):

$$\text{Precision} = \frac{\text{True positive}}{\text{True positive} + \text{False positive}} \quad (5)$$

$$\text{Recall} = \frac{\text{True positive}}{\text{True positive} + \text{Flase Negative}} \quad (6)$$

$$\text{F - measure} = \frac{2 \cdot \text{Precision} \cdot \text{Recall}}{\text{Precision} + \text{Recall}} \quad (7)$$

To measure recommendation accuracy and correctness, we used Mean Absolute Error(MAE) and Root Mean Squared Error(RMSE). Metrics are defined as:

$$\text{MAE} = \frac{1}{\text{UI}} \sum_{i=1}^{U} \sum_{j=1}^{I} \left| r_{ij} - p_{ij} \right| \quad (8)$$

$$\text{RMSE} = \sqrt{\frac{1}{\text{UI}} \sum\nolimits_{i=1}^{U} \sum\nolimits_{j=1}^{I} \left(r_{ij} - p_{ij} \right)^2}, \quad (9)$$

where r_{ij} is the real rating of the user and p_{ij} is the predicted rating (Fig. 5).

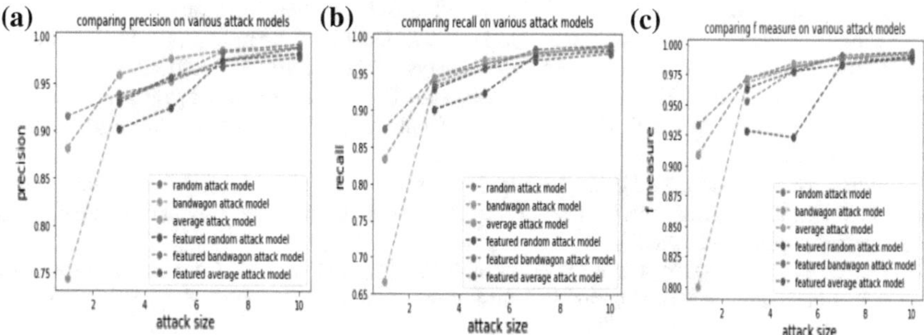

Fig. 4 **a** Precision, **b** Recall, **c** F-measure for both SVM methods used, on different attack models

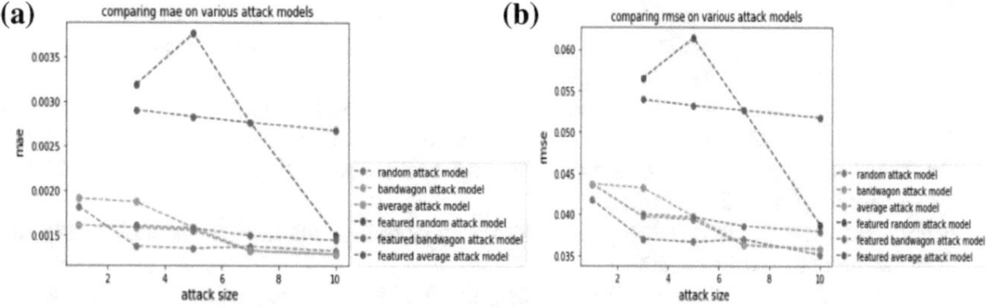

Fig. 5 **a** MAE and **b** RMSE for both SVM methods used on different attack models

5 Conclusion

Paper provides in-depth analysis of shilling attack and describes two different approaches to detect malicious profiles in collaborative filtering system. In both proposed SVM methods (Linear and RBF), the one with feature extraction is a viable option, but the performance regarding the accuracy of SVM using the item as features is slightly better compared to another one. So, for detecting the fake profiles in the recommender system SVM using the item as a feature is advisable. The supervised algorithm present here has good classification and highly accurate as profiles classified the attackers from the genuine one.

References

1. Burke, R., Mobasher, B., Williams, C., Bhaumik, R.: Classification features for attack detection in collaborative recommender systems. In: ACM SIGKDD International Conference on Knowledge Discovery and Data Mining, pp. 542–547 (2006)
2. Lam, S., Riedl, J.: Shilling recommender systems for fun and profit. In: ACM Proceedings of the 13th International Conference on World Wide Web, pp. 393–402 (2004)
3. Zhou, W., Wen, J., Koh, Y., Xiong, Q., Gao, M., Dobbie, G.: Shilling attacks detection in recommender systems based on target item analysis. PLoS ONE **10**(7), e0130968 (2015)
4. Mobasher, B., Burke, R., Bhaumikr, R., Williams, C.: Toward trustworthy recommender systems: an analysis of attack models and algorithm robustness. *ACM Trans Internet Technol* 23–60 (2007)
5. Mehta, B., Hofmann, T., Fankhauser, P.: Lies and propaganda: detecting spam users in collaborative filtering. In: Proceedings of the 12th International Conference on Intelligent User Interfaces, pp. 14–21 (2007)
6. Horng, S.-J., Su, M.-Y., Chen, Y.-H., Kao, T.W., Chen, R.-J., Lai, J.-L.: A novel intrusion detection system based on hierarchical clustering and support vector machines. Elsevier Expert Syst. Appl. **38**, 306–313 (2011)
7. Williams, C.A., Mobasher, B., Burke, R.: Defending recommender systems: detection of profile injection attacks. Serv. Oriented Comput. Appl. **1**(3), 157–170 (2007)
8. Burke, R., Mobasher, R., Williams, C., Bhaumik, R.: Detecting profile injection attacks in collaborative recommender systems. In: The 8th IEEE International Conference on E-Commerce Technology and the 3rd IEEE International Conference on Enterprise Computing, E-Commerce, and E-Services, pp. 23–30 (2006)
9. Chirita, P.-A., Nejdl, Zamfir, Z.C.: Preventing shilling attacks in online recommender systems. In: ACM International workshop on web information and data management, pp. 67–74 (2005)
10. Aggarwal, C.C.: Recommender systems-the textbook. In: ACM SIGMOD-SIGACT-SIGART symposium, pp. 1–493 (2016)
11. Mehta, B.: Unsupervised shilling detection for collaborative filtering. In: Association for the Advancement of Artificial (2007)
12. Deng, Z.-J., Zhang, Z., Wang, S.P.S.: Shilling attack detection in collaborative filtering recommender system by PCA detection and perturbation. In: International Conference on Wavelet Analysis and Pattern Recognition (2016)

13. Gayathri, K., Marimuthu, A.: Text document preprocessing with the KNN for classification using the SVM. In: International Conference on Intelligent Systems and Control (2013)
14. Al-Anazi, A.F., Gates, I.D.: Support vector regression for porosity prediction. Elsevier Comput. Geosci. 36, 1494–1503 (2010)
15. Belen, M.-B., Rosa, L., Juan, R.: Interpretable support vector machines for functional data. In: European Journal of Operational Research (2014)

A Hybrid Clustering Algorithm Based on Kmeans and Ant Lion Optimization

Santosh Kumar Majhi and Shubhra Biswal

Abstract Kmeans is one of the most-efficient hard-clustering algorithms. It has been successfully applied to a number of problems. However, the efficiency of kmeans is dependent on its initialization of cluster centres. Different swarm intelligence techniques are applied for clustering problem. In this work, we have considered Ant Lion Optimization (ALO) which is a stochastic global optimization models. In this work Kmeans has been integrated with ALO for optimal clustering. The statistical measures of different performance metrics has been calculated and compared. The proposed method performs preferably better than Kmeans and PSO-Kmeans in terms of sum of intracluster distances and F-measure.

1 Introduction

Clustering is a process of grouping a set of objects based on some similarity measure. Each group of partitioned objects is known as a cluster. The objects within a cluster have high similarity measure, but are different from the objects of other clusters. A number of clustering algorithms are there to generate clusters, which when applied on the same data set may produce different results. The partitioning is performed by clustering algorithms. Hence, clustering is advantageous because it creates the possibility of obtaining previously unknown groups within the same data. Data clustering is an effective method for discovering structure in datasets. Some clustering methods partition objects so that there is no particular boundary among the clusters, whereas some other methods partition objects into mutually exclusive clusters. Also, the distance between two objects is considered as the similarity criteria by some methods.

S. K. Majhi (✉) · S. Biswal
Veer Surendra Sai University of Technology, Burla 768018, Odisha, India
e-mail: smajhi_cse@vssut.ac.in

S. Biswal
e-mail: shubhrabiswal08@gmail.com

A. Abraham et al. (eds.), *Emerging Technologies in Data Mining and Information Security*, Advances in Intelligent Systems and Computing 813,
https://doi.org/10.1007/978-981-13-1498-8_56

There are many clustering algorithms which can be categorized into partitioning methods, hierarchical methods, grid-based methods, density-based methods [1]. Different factors that affect the results of clustering are number of clusters to be formed in a data set, clustering tendency and quality of clustering. Accessing clustering tendency determines whether a non-random structure exists in the data. The existence of a non-random structure in a data set results in meaningful cluster analysis. Determining the number of clusters to be formed in a data set is important for few clustering methods in which the number of clusters is used as parameter. To measure the quality of clustering a number of metrics are used. Some methods measure how well the clusters fit the data set, while others measure how well the clusters match the ground truth.

Kmeans algorithm developed by Stuart Lloyd in 1957, is a centroid-based hard clustering algorithm which comes under the partitioning method of clustering. In centroid-based partitioning method of clustering the centroid of a cluster is used to represent the cluster. Kmeans is computationally faster than hierarchical clustering for small k value. But the problem with kmeans is that the k value is difficult to predict [2]. Also the positions of k centres are initialized randomly, which may not result in good quality of clustering. So to ensure the quality, kmeans has been integrated with different optimization algorithms.

2 Related Works

A comparison of different clustering algorithms belonging to partitional clustering, hierarchical clustering, grid-based clustering, density-based clustering, model based clustering has been described [3]. It compares the different clustering algorithms based on the four important properties of big data, i.e., volume, velocity, variety and value. Another survey has been done on clustering algorithms in [4] to provide better data aggregation and scalability for large wireless sensors networks. The challenges in clustering algorithms are intercluster communication, cluster formation and cluster head selection. A comparative study of different clustering algorithm is presented in [5]. A method for getting better results by application of sorted and unsorted data into the Kmeans, Kmeans++ and fuzzy C-means algorithms has been proposed. Different factors such as elapsed time and total number of iterations has been considered in [5] for the analysis of behavioural patterns. The data is first sorted and given into the algorithms. Passing sorted data decreases the elapsed time and also the number of iterations to get the desired results. Kmeans has been successfully implemented in a number of problems but this method has its own drawbacks. The efficiency of kmeans is dependent on its initialization cluster centres. To enhance the quality of kmeans different optimization methods has been integrated [6]. A population based clustering [7] combines several PSO-based methods with kmeans in which to enhance exploration, PSO-based methods help some particles escape from local optima and kmeans is employed to improve the partitioning results for accelerating convergence. Another hybrid clustering algorithm known as intertwined kmeans PSO [8] applies the advantages of both kmeans and PSO to increase the quality of clustering. In kmeans the mean point of members of the cluster replaces the old cluster centre. In [8] this advantage is used in the PSO to speed up its convergence rate.

3 Materials and Methods

Kmeans Clustering Kmeans clustering is an unsupervised hard partitioning clustering method. The objective is to find k clusters from the data based on the objective function J given in Eq. (1).

$$J = \sum_{i=1}^{k}\sum_{j=1}^{N} d^2(C_i - X_j), \tag{1}$$

where $d^2(C_i - X_j)$ is the squared Euclidean distance between ith cluster centroid and jth data point. N is the total number of data points. The aim of kmeans algorithm is to minimize the objective function.

In kmeans clustering the position of the k cluster centres are being initialized randomly, which may not result in good quality of clusters. So to ensure the quality of clustering we have integrated kmeans clustering algorithm with different optimization algorithms such as Ant lion optimization and Particle swarm optimization.

Particle Swarm Optimization Particle swarm optimization is a population based heuristic algorithm. It is a nature inspired algorithm which impersonates the process of flocking of group of birds [9]. The population consists of set of particles called as swarm, each particle representing a solution. Like any other optimization technique, the goal of PSO is to find the best solution among set of solutions. The objective function determines the performance or fitness of each particle. The particle which shows optimal performance among the entire swarm is considered as the global best (gbest). If the performance of a particle in a position is better than its previous recorded position, then it is called as local best (pbest). The velocity and position of every particle is calculated using Eqs. (2) and (3).

Velocity of the ith particle is given by the equation,

$$V_i(t+1) = wV_i(t) + C_1 \cdot rand(0,1) \cdot (pbest_i - X_i(t)) + C_1 \cdot rand(0,1) \cdot (gbest - X_i(t)) \tag{2}$$

Position of the ith particle is given by the equation,

$$X_i(t+1) = X_i(t) + V_i(t+1), \tag{3}$$

$V_i(t+1)$ is the velocity of ith particle in the current location, $V_i(t)$ is the velocity of ith particle in previous location. W is the inertia factor, varies from 0 to 1. C_1 is the local acceleration constant and C_2 is the global acceleration constant. The values of C_1 and C_2 are selected from 0 to 2.

$X_i(t)$ is the position of ith particle at tth iteration.

Ant Lion Optimization Ant Lion Optimization method is also a nature inspired algorithm which follows the hunting behaviour of antlion larvae [10]. Nowadays it has been used in many engineering domains [11, 12]. An ant lion larva creates a conical-shaped hole by moving along a circular path in the sand and throwing the sand with huge jaw. After digging the trap, larvae hide at the bottom of the cone and waits for the ants to be trapped in the pit. Once the ant lion realizes that a prey has been caught in the trap, the ant lion throws sand outwards and slips its prey into the pit. When a prey is caught into thejaw, the ant lion pulls the prey toward itself and consumes. This process is mathematically designed to perform optimization.

3.1 Algorithms

Kmeans clustering algorithm

1. Select k random points as cluster centres
2. WHILE the end criterion is not satisfied
    - 2.1. FOR each point
        - 2.1.1. Find the Euclidean distance of each point from the cluster centres
        - 2.1.2. Assign the point to the cluster with minimum Euclidean distance
        - 2.1.3. END FOR
    - 2.2. Compute the mean of all points in each cluster
    - 2.3. Assign the mean values as the new cluster centres

    END WHILE
3. Return k clusters

KMeans-PSO Clustering algorithm

1. Select k random points as cluster centres
2. WHILE the end criterion is not satisfied
    - 2.1. FOR each point
        - 2.1.1. Find the Euclidean distance of each point from the cluster centres
        - 2.1.2. Assign the point to the cluster with minimum Euclidean distance

        END FOR
    - 2.2. Compute the mean of all points in each cluster
    - 2.3. Assign the mean values as the new cluster centres

    END WHILE
3. Return k clusters
4. FOR each point in the cluster
    - 4.1. Initialize the position and velocity
    - 4.2. Initialize the global best cost as ∞

4.3. Calculate the fitness value using the Eq.

4.4. Update the global best cost with the particle cost *if* particle cost $<$ global cost

END FOR

5. WHILE end criterion is not satisfied

5.1.1. FOR each particle in the cluster

5.1.2. Update the velocity by using Eq. (2)

5.1.3. Update the position by using Eq. (3)

5.1.4. Calculate the fitness value

5.1.5. Update the global best with the particle if particle cost $<$ global best cost

END FOR

5.2. Return global best position

END WHILE

6. Select the global best position as the new centre for the cluster

KMeans-ALO Clustering algorithm

1. Select k random points as cluster centres
2. WHILE the end criterion is not satisfied

2.1. FOR each point

2.1.1. Find the Euclidean distance of each point from the cluster centres

2.1.2. Assign the point to the cluster with minimum Euclidean distance

END FOR

2.2. Compute the mean of all points in each cluster

2.3. Assign the mean values as the new cluster centres

END WHILE

3. Return k clusters
4. FOR each cluster

4.1. Initialize the first population of ants from the dataset

4.2. Initialize the first population of ant lions randomly

4.3. Calculate the fitness of ants and ant lions using the objective function

4.4. Select the ant lion with minimum fitness value as elite

4.5. WHILE the end criterion is not satisfied

4.5.1. FOR every ant

4.5.1.1. Select an ant lion using Roulette wheel

4.5.1.2. Update minimum of all variables and maximum of all variables at tth iteration

4.5.1.3. Create a random walk as

$$X[t] = \left[0, cumsum(2r(t_1) - 1),\ cumsum(2r(t_2) - 1), \ldots\ldots cumsum(2r(t_n) - 1)\right],$$

where $r(t) = 1$ if $rand > 0.5$ or 0 if $rand \leq 0.5$

4.5.1.4. Normalize the random walk using Eq. (4) of min-max normalization

$$X_i^t = \frac{\left(X_i^t - a_i\right) \times \left(b_i - C_i^t\right)}{\left(d_i^t - a_i\right)} + C_i, \tag{4}$$

4.5.1.6. Update the position of ant by Eq. (5)

$$Ant_i^t = \frac{R_A^t + R_E^t}{2}, \tag{5}$$

END FOR

4.5.2. Find the fitness value of all ants

4.5.3. Replace an ant lion with its corresponding ant if f (Ant$_i$)<f (Antlion$_j$)

4.5.4. Update elite if f (Antlion$_j$)<f (elite)

END WHILE

4.6. Return elite

END FOR

5. Select elite as the new centre of the cluster d_i^t is the maximum of i-th variable at t-th iteration. ai is the minimum of random walk of i-th variable and bi is the maximum of random walk of i-th variable at t-th iteration. R_A^t is the random walk around the antlion selected by the roulette wheel at t-th iteration and R_E^t is the random walk around the elite at t-th iteration.

3.2 Data Sets

To implement the algorithms mentioned in Algorithms section, six data sets are considered. The data sets are iris, wine, breast cancer, seed, diabetes, vehicle.

3.2.1 Iris Data Set

Iris data set is collected from three different species of Iris flowers, i.e., *Iris Versicolour*, *Iris Setosa*, *Iris Virginica*. Each species has 50 samples in the data sets. Every sample has four attributes: sepal length, petal length, sepal width, petal width. The attributes are measured in cm. All the attributes are numeric and predictive attributes.

3.2.2 Wine Data Set

This data set has been produced from the chemical analysis of wine from three different cultivars, grown at the same region in Italy. This data set has 178 instances: class 1 has 59 instances, class 2 has 71 instances, class 3 has 48 instances. Each instance has 13 continuous attributes: alcohol, ash, malic acid, alkalinity of ash, total phenols, magnesium, flavonoids, non-flavonoid phenols, colour intensity, proanthocyanins, hue, proaline, OD280/OD315 of diluted wines.

3.2.3 Breast Cancer Data Set

This data set contains two categories of data:benign and malignant. There are 569 instances with 32 attributes.

3.2.4 Seed Data Set

This dataset comprises of information about kernels collected from three different varieties of wheat: Rosa, Kama and Canadian. Each class contains 70 elements. Each wheat kernel has seven real-valued continuous attributes: area, parameter, length of kernel, width of kernel, compactness, length of kernel groove, asymmetry coefficient.

3.2.5 Diabetes Data Set

This data set has two classes; tested_positive and tested_negative. This data set is a subset of a larger data set. For instance, all patients are females at least 21 years old. The attributes of this data set are Plasma glucose concentration a 2 h in an oral glucose tolerance test, Number of times pregnant, Triceps skin fold thickness (mm), Diastolic blood pressure (mm Hg), 2-h serum insulin (mu U/ml), Diabetes pedigree function, Body mass index (weight in kg/(height in m)^2), Class variable (0 or 1), Age (years).

3.2.6 Vehicle Data Set

Vehicle dataset contains 946 example of four classes: OPEL, SAAB, BUS, VAN. There are 240 instances in each of the classes of OPEL, SAAB & BUS; class VAN contains 226 instance. Hundred examples of the data set are being kept by University of Strathclyde for validation, So the file contains 846 examples with 18 attributes.

3.3 Performance Metrics

Different performance metrics has been used for the proposed methods in order to evaluate the quality of clustering. The metrics are average of sum of intracluster distances, intercluster distance and F-measure.

3.3.1 Average of Sum of Intracluster Distances

Several methods are there to calculate the intracluster distance of a cluster, such as complete diameter, average diameter, centroid diameter.

Here the sum of intracluster distance has been measured using the centroid diameter method of finding the intracluster distance. Centroid diameter method calculates the intracluster distance by calculating the average distance of the cluster centroid from all the points of that cluster. To get high quality of clustering, the sum of intracluster distance should be small.

3.3.2 Intercluster Distance

Different methods to calculate intercluster distance are single linkage, complete linkage, average linkage, centroid linkage, average to centroid methods. In this work complete linkage method has been considered. In Complete linkage method the maximum distance between two points belonging to separate clusters is considered as the intercluster distance. Objects belonging to different clusters should be as far as possible i.e. the intercluster distance should be maximum in order to get better quality of clustering.

3.3.3 F-Measure

F-measure is calculated using the concepts of precision and recall from information retrieval.

Each class i of the data set is regarded as the set of n_i items desired for a query. Each cluster j is considered as the set of n_j items retrieved for a query. n_{ij} represents the number of elements of class i within cluster j. For each class i and cluster j, Eq. (6) and Eq. (7) define the precision and recall respectively.

$$p(i, j) = \frac{n_{ij}}{n_j}, \tag{6}$$

$$r(i, j) = \frac{n_{ij}}{n_i}, \tag{7}$$

$$F(i, j) = \frac{\left(b^2 + 1\right) \cdot p(i, j) \cdot r(i, j)}{b^2 \cdot p(i, j) + r(i, j)}, \tag{8}$$

Table 1 Results obtained by kmeans, kmeans-PSO, kmeans-ALO algorithms for 10 different runs on Iris data set for 100, 500 and 1000 iterations

Methods	Iterations	Best value	Average value	Worst value	F-measure
Kmeans	100	2.3857	7.5463	12.0822	3.8666
Kmeans-PSO	100	3.1826	7.2351	11.3276	4.8600
Kmeans-ALO	100	2.3826	6.85515	11.3277	4.8767
Kmeans	500	2.3996	7.5463	11.3272	3.8722
Kmeans-PSO	500	2.8845	6.88605	11.3277	4.7863
Kmeans-ALO	500	2.3826	6.85515	12.3276	4.8767
Kmeans	1000	3.1067	7.0013	11.9000	3.7622
Kmeans-PSO	1000	2.9871	6.9956	11.3266	4.7688
Kmeans-ALO	1000	2.3829	6.8661	11.3077	4.8745

Table 2 Results obtained by kmeans, kmeans-PSO, kmeans-ALO algorithms for 10 different runs on Wine data set for 100, 500 and 1000 iterations

Methods	Iterations	Best value	Average value	Worst value	F-measure
Kmeans	100	1.9396	2.0857	2.2319	1.0803
Kmeans-PSO	100	1.1680	1.6895	2.2110	1.0810
Kmeans-ALO	100	1.1634	1.58115	1.9989	1.0832
Kmeans	500	1.9976	2.11475	2.2319	1.0807
Kmeans-PSO	500	1.9393	1.96455	1.9898	1.0809
Kmeans-ALO	500	1.1638	1.52675	1.8897	1.0880
Kmeans	1000	2.0101	2.5543	3.76781	1.0809
Kmeans-PSO	1000	1.1680	1.6990	1.9998	1.0880
Kmeans-ALO	1000	1.1633	1.4908	1.8789	1.0890

Then F-measure is calculated by Eq. (9).

$$F = \sum_{i=1}^{k} \frac{n_i}{N} \max\{F(i, j)\}, \tag{9}$$

4 Results and Discussion

All the algorithms discussed in the materials and methods section are implemented with MatlabR2016a on a Windows platform using Intel(R)Core(TM) i3-2310 M, 2.10 GHz, 4 GB RAM computer. The experimental results for the average of sum of intracluster distance are calculated on all the six data sets discussed in the materials & methods section are provided in Tables 1, 2, 3, 4, 5 and 6. The results are collected over 10 different runs, for 100 and 500 iterations.

Table 3 Results obtained by kmeans, kmeans-PSO, kmeans-ALO algorithms for 10 different runs on Breast cancer data set for 100, 500 and 1000 iterations

Methods	Iterations	Best value	Average value	Worst value	F-measure
Kmeans	100	39.3858	44.9354	50.4850	0.7578
Kmeans-PSO	100	39.3443	44.8995	50.4465	1.2109
Kmeans-ALO	100	39.3384	44.8895	50.4406	1.2341
Kmeans	500	39.3993	44.9264	50.4535	0.7589
Kmeans-PSO	500	39.3717	44.93705	50.5024	0.9976
Kmeans-ALO	500	39.3417	44.8976	50.4535	0.9989
Kmeans	1000	39.4300	44.9986	50.5654	0.7568
Kmeans-PSO	1000	39.3828	44.8999	50.5678	1.0917
Kmeans-ALO	1000	39.3623	44.8821	50.5567	1.2108

Table 4 Results obtained by kmeans, kmeans-PSO, kmeans-ALO algorithms for 10 different runs on Seed data set for 100, 500 and 1000 iterations

Methods	Iterations	Best value	Average value	Worst value	F-measure
Kmeans	100	6.0466	7.6571	10.8509	1.3411
Kmeans-PSO	100	5.8491	7.0211	10.6661	1.3902
Kmeans-ALO	100	4.8504	6.70661	9.6586	1.4392
Kmeans	500	6.0458	7.8653	10.6590	1.3421
Kmeans-PSO	500	5.8490	7.5846	10.6552	1.9834
Kmeans-ALO	500	4.3500	6.8732	8.9592	1.9980
Kmeans	1000	7.0321	8.0987	10.7644	1.3422
Kmeans-PSO	1000	4.9876	7.3211	9.98654	1.8794
Kmeans-ALO	1000	6.0466	7.6571	10.8509	1.3411

Table 5 Results obtained by kmeans, kmeans-PSO, kmeans-ALO algorithms for 10 different runs on Diabetes data set for 100, 500 and 1000 iterations

Methods	Iterations	Best value	Average value	Worst value	F-measure
Kmeans	100	467.8321	919.0833	1757.1	8.0433
Kmeans-PSO	100	463.9216	918.6566	1757.1	9.7834
Kmeans-ALO	100	459.7836	529.0613	573.0715	9.9765
Kmeans	500	466.3097	932.9001	1755.9088	8.0431
Kmeans-PSO	500	463.8342	900.0612	1757.1	9.2863
Kmeans-ALO	500	459.8307	530.2311	572.3219	9.6534
Kmeans	1000	468.9076	909.8728	1757.1011	9.9870
Kmeans-PSO	1000	463.3766	896.6735	1748.9811	9.9908
Kmeans-ALO	1000	459.7587	530.0134	578.9824	9.9982

Table 6 Results obtained by kmeans, kmeans-PSO, kmeans-ALO algorithms for 10 different runs on Vehicle data set for 100, 500 and 1000 iterations

Methods	Iterations	Best value	Average value	Worst value	F-measure
Kmeans	100	581.4685	583.3787	586.7656	1.6166
Kmeans-PSO	100	580.8282	581.0879	585.0977	1.7233
Kmeans-ALO	100	579.7258	579.9120	580.6224	1.7765
Kmeans	500	581.9878	581.9978	583.0976	1.6324
Kmeans-PSO	500	580.8282	581.9938	583.9176	1.6873
Kmeans-ALO	500	579.8263	579.9199	581.4280	1.7021
Kmeans	1000	582.0987	583.3827	585.9865	1.6650
Kmeans-PSO	1000	582.2276	583.9873	584.0979	1.6902
Kmeans-ALO	1000	579.0972	579.9753	581.0921	1.7903

From the values in Tables 1, 2, 3, 4, 5 and 6 it can be observed that the results obtained by Kmeans-ALO is better than the other two methods for the iris, wine, breast cancer, seed, diabetes and vehicle data sets. Kmeans-ALO provides the optimal values of 2.3926, 1.1633, 39.3384, 4.3598, 459.7587, 579.0972 for the iris, wine, breast cancer, seed, diabetes and vehicle data set respectively.

5 Conclusion

This work integrates kmeans clustering with Ant Lion Optimization. The quality of clustering is measured using the intracluster distance and the F-measure. The simulation results validate that Kmeans-ALO performs better than the kmeans and kmeans-PSO algorithm.

References

1. Jain, A.K.: Data clustering: 50 years beyond K-means. Pattern Recogn. Lett. **31**(8), 651–666 (2010)
2. Erisoglu, M., Calis, N., Sakallioglu, S.: A new algorithm for initial cluster centers in k-means algorithm. Pattern Recogn. Lett. **32**(14), 1701–1705 (2011)
3. Sajana, T., Rani, C.S., Narayana, K.V.: A survey on clustering techniques for big data mining. Indian J. Sci. Technol. **9**(3) (2016)
4. Atabay, H.A., Sheikhzadeh, M.J., Torshizi, M.: A clustering algorithm based on integration of K-Means and PSO. In: 2016 1st Conference on Swarm Intelligence and Evolutionary Computation (CSIEC), pp. 59–63. IEEE (2016)
5. Kapoor, A., Singhal, A.: A comparative study of K-Means, K-Means++and Fuzzy C-Means clustering algorithms. In: 2017 3rd International Conference on Computational Intelligence & Communication Technology (CICT), pp. 1–6. IEEE (2017)
6. Cui, X., Potok, T.E.: Document clustering analysis based on hybrid PSO+K-means algorithm. J. Comput. Sci. (special issue) **27**, 33 (2005)

7. Chen, S., Xu, Z., Tang, Y.: A hybrid clustering algorithm based on fuzzy c-means and improved particle swarm optimization. Arabian J. Sci. Eng. **39**(12), 8875–8887 (2014)
8. Niu, B., Duan, Q., Liu, J., Tan, L., Liu, Y.: A population-based clustering technique using particle swarm optimization and k-means. Nat. Comput. **16**(1), 45–59 (2017)
9. Poli, R., Kennedy, J., Blackwell, T.: Particle swarm optimization. Swarm Intell. **1**(1), 33–57 (2007)
10. Mirjalili, S.: The ant lion optimizer. Adv. Eng. Softw. **83**, 80–98 (2015)
11. Pradhan, R., Majhi, S.K., Pradhan, J.K., Pati, B.B.: Performance evaluation of PID controller for an automobile cruise control system using ant lion optimization. Eng. J. **21**(5), 347–361 (2017)
12. Pradhan, R., Majhi, S.K., Pradhan, J.K., Pati, B.B.: Ant lion optimizer tuned PID controller based on Bode ideal transfer function for automobile cruise control system. J. Indus. Inf. Integr. (2018)

Flight Arrival Delay Prediction Using Gradient Boosting Classifier

Navoneel Chakrabarty, Tuhin Kundu, Sudipta Dandapat, Apurba Sarkar and Dipak Kumar Kole

Abstract The basic objective of the proposed work is to analyse arrival delay of the flights using data mining and four supervised machine learning algorithms: random forest, Support Vector Machine (SVM), Gradient Boosting Classifier (GBC) and k-nearest neighbour algorithm, and compare their performances to obtain the best performing classifier. To train each predictive model, data has been collected from BTS, United States Department of Transportation. The data included all the flights operated by American Airlines, connecting the top five busiest airports of United States, located in Atlanta, Los Angeles, Chicago, Dallas/Fort Worth, and New York, in the years 2015 and 2016. Aforesaid supervised machine learning algorithms were evaluated to predict the arrival delay of individual scheduled flights. All the algorithms were used to build the predictive models and compared to each other to accurately find out whether a given flight will be delayed more than 15 min or not. The result is that the gradient boosting classifier gives the best predictive arrival delay performance of 79.7% of total scheduled American Airlines' flights in comparison to kNN, SVM and random forest. Such a predictive model based on the GBC potentially can save huge losses; the commercial airlines suffer due to arrival delays of their scheduled flights.

N. Chakrabarty (✉) · T. Kundu · S. Dandapat · D. K. Kole
Jalpaiguri Government Engineering College, Jalpaiguri 735102, West Bengal, India
e-mail: nc2012@cse.jgec.ac.in

T. Kundu
e-mail: tk1910@cse.jgec.ac.in

S. Dandapat
e-mail: sd2036@cse.jgec.ac.in

D. K. Kole
e-mail: dipak.kole@cse.jgec.ac.in

A. Sarkar
Indian Institute of Engineering Science and Technology Shibpur,
Shibpur 711103, West Bengal, India
e-mail: as.besu@gmail.com

A. Abraham et al. (eds.), *Emerging Technologies in Data Mining and Information Security*, Advances in Intelligent Systems and Computing 813,
https://doi.org/10.1007/978-981-13-1498-8_57

Keywords American airlines · Flight arrival delay · Gradient boosting classifier k-nearest neighbour · Random forests · Supervised machine learning · Support vector machine

1 Introduction

The United States Department of Transportation's Bureau of Transportation Statistics (BTS) approximated in its report that twenty percent of all commercial scheduled flights are delayed due to several reasons such as weather, security and air carrier, costing billions of dollars of losses and severe inconvenience to passengers [1].

Late arriving aircraft are mostly due to inclement weather while it is not considered as weather induced delays. It is well known that weather plays a vital role in delaying scheduled flights due to its extremities in nature, but is not the only causality. The National Aviation System includes the rerouting of commercial flights due to tempestuous weather [1].

American Airlines, Inc. (AA) is the world's largest American Airline when measured by number of destinations served, revenue, fleet size and scheduled passenger-kilometres flown. American Airlines together with its regional partners operates an extensive international and domestic network with an average of nearly 6,700 flights per day to nearly 350 destinations in more than 50 countries. Through the airline's parent company, American Airlines Group, it is publicly traded under NASDAQ: AAL with a market capitalization of about $25 billion as of 2017 [2].

Hence, intelligent systems are needed to predict and propose the delay of scheduled flights, along with its causality, for airline companies such as American Airlines, which operates passenger aircraft in such a large scale on a daily basis. Arrival delay, as earlier suggested, is a root cause to multibillion dollar losses suffered by airlines operating commercial flights, due to various reasons, and hence, developing intelligent machinery which are automated in nature, promises to save losses from the financial books of commercial aircraft operators, and potentially, American Airlines.

The model demonstrates various supervised machine learning algorithms in the effort to achieve the prediction of arrival delays of flights operated by American Airlines connecting the aforementioned, top five busiest airports by number of passengers, according to [3].

The paper has been structured as follows: Sect. 2 provides a brief review of the literature to understand the problem and to build the model. Section 3 describes the proposed methodology for the model. In Sect. 4, performance of the classifiers are compared and concluded in terms of training accuracy, testing accuracy, precision, recall, AUROC, F1 score, mean error, RMS error and confusion matrix. Finally, it is concluded with future work in Sect. 5.

2 Literature Review

2.1 *Model Evaluation*

2.1.1 Receiver Operating Characteristic (ROC) Curve

It is a plot of True Positive Rate (TPR) versus False Positive Rate (FPR) for a set of threshold, τ, where TPR is known as the sensitivity performance metric. Then, TPR and FPR are computed as follows as shown in Eqs. 1 and 2, respectively [4]:

$$TPR = \frac{TP}{(TP + FN)} \approx p(\hat{y} = 1|y = 1) \tag{1}$$

$$FPR = 1 - sensitivity = \frac{FP}{(TN + FP)} \approx p(\hat{y} = 1|y = 0) \tag{2}$$

The ideal point on the ROC curve is (0, 1) that includes all the positive examples which are classified correctly and no negative examples are misclassified as positive [5]. The Area Under ROC (AUROC) curve is also a strict measurement of prediction performance, where greater the AUROC, better the performance.

2.2 *Supervised Machine Learning Algorithms*

2.2.1 Support Vector Machine (SVM)

The classifier aims by creating a model that predicts the class labels of unknown data or validation data instances consisting only of attributes, as given in Eq. 3. Kernels of SVM are generally used for mapping non-linearly separable data into higher dimensional feature space instance, which consists of only attributes [6].

$$f(x) = sign(\sum_{i \epsilon SV} \alpha_i^* y_i \cdot k(x_i, x_{SV}) + b^*) \tag{3}$$

where $k(x_i, x_{SV})$ is a kernel function and the equation for kernel k for Radial Basis Function (RBF) as mentioned in Eq. 4.

$$k(x, y) = exp(-\frac{||x - y||^2}{2\sigma^2}) \tag{4}$$

2.2.2 Gradient Boosting Classifier (GBC)

Gradient boosting is a method used to develop classification and regression models to optimize the learning process of the model, which are mostly non-linear in nature and are more widely known as decision or regression trees.

A group of weak prediction models, for example, regression decision trees, are modelled by adding new learner in a gradual sequential manner. It consists of nodes and leaves which yield predictive results based on the decision nodes. Regression trees, individually, are weak models, but when viewed as an ensemble, their accuracy was much improved. Therefore, the ensembles are built gradually in an incremental manner such that every ensemble rectifies the error in the previous ensemble, mathematically as Eq. 5 [7].

$$f_k(x) = \sum_{m=1}^{k} \gamma_m h_m(x) \tag{5}$$

2.2.3 Random Forest

Random forests are an ensemble of many individual decision trees [8]. Mathematically, let $\hat{C}_b(x)$ be the class prediction of the bth tree, the class obtained from random forest $\hat{C}_{rf}(x)$ is

$$\hat{C}_{rf}(x) = majorityvote\{\hat{C}_b(x)\}_1^B \tag{6}$$

2.2.4 k-Nearest Neighbour Algorithm

It is a machine learning algorithm based on Euclidean distance between instances. KNN predicts class labels for different instances by measuring its shortest Euclidean distance from other instances where the Euclidean distances are calculated considering all the features or attributes as dimensions as given in the following equation:

$$d(x_i, x_j') = ||x_i - x_j||^2 = \sum_{k=1}^{d} (x_{ik} - x_{jk})^2 \tag{7}$$

3 Proposed Methodology

3.1 Data Collection

To extract the data of the flights operated by American Airlines, first the airline ID was looked upon from [3] and then, the data was collected from Bureau of Transportation

Statistics (BTS), U.S Department of Transportation [9] through the API using the airline ID of American Airlines. While collecting the data through the BTS API, the Origin Airport ID and Destination Airport ID features were filtered to include the top five busiest airports in US in 2015 and 2016 as per [3] which included Hartsfield–Jackson Atlanta International Airport, Los Angeles International Airport, O'Hare International Airport, Dallas/Fort Worth International Airport, and John F. Kennedy International Airport, located in Atlanta, Los Angeles, Chicago, Dallas/Fort Worth, and New York, respectively.

3.2 Data Processing

The dataset collected included 97,360 instances of flights connecting the aforementioned 5 major airports through flights operated by American Airlines. Of the total 97,360 instances, 1602 instances had data missing in them, and hence, they were removed. The final dataset included 95,758 instances with selected attributes (Year, Quarter, Month, Day of Month, Day of Week, Flight Num, Origin Airport ID, Origin World Area Code, Destination Airport ID, Destination World Area Code, CRS Departure Time, CRS Arrival Time, Arr Del 15), after the removal of redundant parameters, as follows which were a causality to the arrival delay of flights.

3.3 Classification Model

Here, following the rules of the BTS, flights arriving at the airport gate within 15 minutes beyond the scheduled arrival time are considered to be on-time and delayed otherwise [10].

Shuffling was performed on the dataset and the training set comprised of 80% of the 95,758 instances while the testing set included the rest of the instances from the dataset.

Feature scaling was performed for algorithms kNN and random forest classifiers for easy computation and better performance [11].

For the analysis of the supervised machine learning algorithms used to classify the predictive delay of the American Airlines' flights, eight metrics have been primarily considered which are confusion matrix, accuracy, precision, recall, F1 score, AUROC, mean absolute error and root mean square error.

The process of predicting the arrival delay of flights operated by American Airlines has been depicted in Fig. 1.

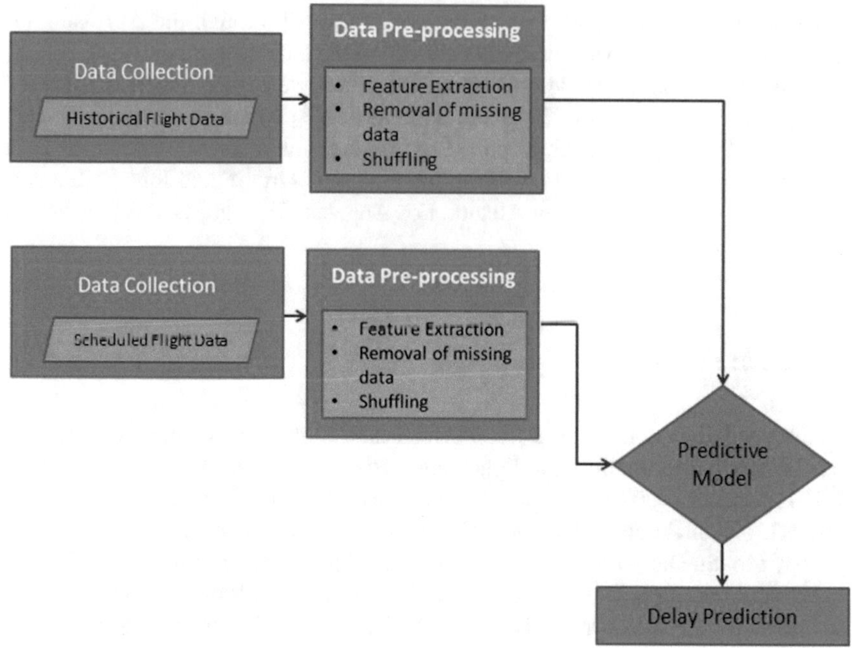

Fig. 1 Flowchart of the model developed

4 Results

Random forests, Gradient Boosting Classifier (GBC), k-Nearest Neighbour (kNN) and Support Vector Machine (SVM) were used on the training dataset to develop the predictive model to accurately predict the arrival delay on the American Airlines' flights, for which the training and testing accuracies are given in Table 1 while precision, recall, AUROC, F1 Score, mean error and RMS error are mentioned in Table 2. The Area Under ROC Curve (AUROC) gives out the predictive performance of every algorithm in Fig. 2.

Table 1 Training and testing accuracy

Classifier	Training accuracy (%)	Test accuracy (%)
Random forest	99.9882	78.7019
Support vector machine (RBF kernel)	84.7962	78.2947
KNN	83.8133	77.924
Gradient boosting	81.0641	79.7201

Table 2 Precision, recall, AUROC, F1 score, mean error and RMS error

Classifier	Precision	Recall	AUROC	F1 score	Mean error	RMS error
Random forest	0.74	0.79	0.56	0.75	0.213	0.4615
Support vector machine (RBF kernel)	0.72	0.78	0.53	0.72	0.217	0.4659
KNN	0.74	0.78	0.56	0.74	0.2208	0.4698
Gradient boosting	0.76	0.8	0.54	0.74	0.2028	0.4503

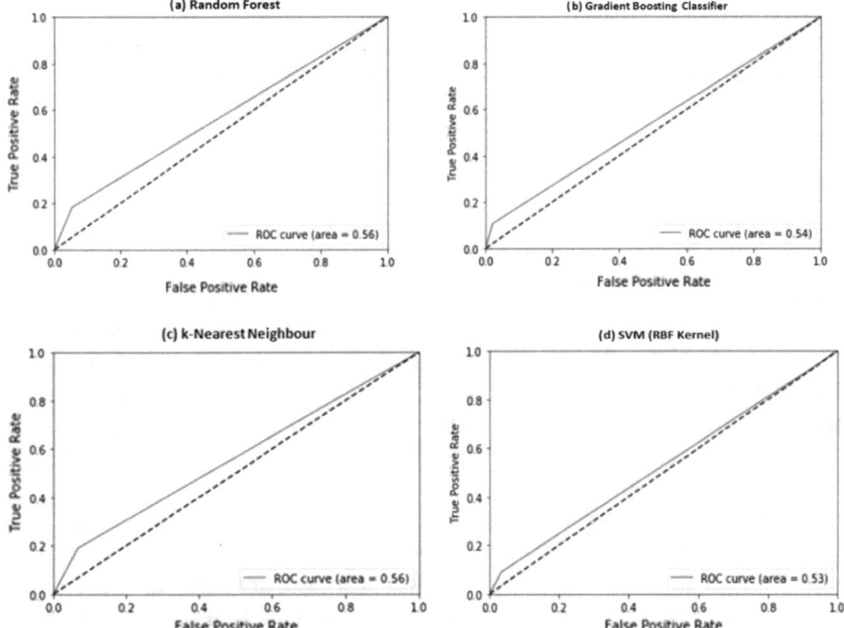

Fig. 2 ROC curves for the model: **a** random forests, **b** gradient boosting classifier, **c** k-nearest neighbour, and **d** support vector machine (RBF kernel)

The random forests classifier was applied after normalizing all the features of the dataset and it can be concluded that it is an overfitting model given the training accuracy exceeds the testing accuracy beyond all limits. The confusion matrices of the 4 models given in Table 3.

The application of k-nearest neighbour algorithm also required the normalization of the datasets while the other two, SVM and GBC, did not require the same. All three SVM, GBC and kNN, when used as the algorithm to run the predictive model

Table 3 Confusion matrix for random forest, SVM, kNN and gradient boosting classifier

Classifier		Predicted on-time	Predicted delay
Random forest	Actual on-time	14343	809
	Actual delay	3270	730
Support vector machine (RBF kernel)	Actual on-time	14633	519
	Actual delay	3638	362
k-nearest neighbour	Actual on-time	14153	999
	Actual delay	3229	771
Gradient boosting	Actual on-time	14836	316
	Actual delay	3568	432

to classify whether the American Airlines' flights were delayed or not, resulted into 'good fit' models, where the training accuracy exceeds the testing accuracy within allowed limits.

5 Conclusion

The experimental results show that the gradient boosting classifier performs the best with a testing accuracy of 79.7%, considering the four most important factors in deciding the best predictive model which are testing accuracy, nature of fit, number of false negatives present in the confusion matrix and AUROC. GBC has a good fit model with 0.54 AUROC and having the minimal number of false negatives in its confusion matrix (which indicates the number of on-time flights predicted as delayed ones).

The problem of development of real-time flight arrival delay prediction promises to play an important role in the drastic reduction of aircraft arrival delay loses for large commercial airliners such as American Airlines, and also the identification of factors that could possibly be eliminated for the reduction of such operational losses. Predictive models with greater accuracy may be developed with the usage of larger datasets and more complex hybrid predictive models on appropriate amount of computer processing capabilities. This research article showcase that supervised machine learning predictive models can be used to development intelligent systems of corporate use to multinational commercial passenger aircraft operator in their quest to offer on-time quality service to their passengers at industry competitive airfares.

References

1. U. D. of Transportation, February 2016 on-time performance up from previous year (2016)
2. Wikipedia contributors: American Airlines. Wikipedia, The Free Encyclopedia. https://en.wikipedia.org/w/index.php?title=American_Airlines&oldid=812987658. Accessed 09 Nov 2017
3. List of Top 40 Airports in US. World Airport Codes. https://www.world-airport-codes.com/us-top-40-airports.html. Accessed 09 Nov 2017
4. Murphy, K.P.: Machine Learning: A Probabilistic Perspective. MIT press (2012)
5. Chawla, N.V., Bowyer, K.W., Hall, L.O., Kegelmeyer, W.P.: SMOTE: synthetic minority over-sampling technique. J. Artif. Intell. Res. **16**, 321–357 (2002)
6. Babu, N.R., Mohan, B.J.: Fault classification in power systems using EMD and SVM. Ain Shams Eng. J. (2015)
7. Aler, R., Galvn, I.M., Ruiz-Arias, J.A., Gueymard, C.A.: Improving the separation of direct and diffuse solar radiation components using machine learning by gradient boosting. Solar Energy **150**, 558–569 (2017)
8. Breiman, Leo: Random forests. Mach. Learn. **45**(1), 5–32 (2001)
9. OST_R | BTS | Transtats. OST_R | BTS | Transtats. http://www.transtats.bts.gov/. Accessed 10 Nov 2017
10. Choi, S., Kim, Y.J., Briceno, S., Mavris, D.: Prediction of weather-induced airline delays based on machine learning algorithms. In: 2016 IEEE/AIAA 35th Digital Avionics Systems Conference (DASC), pp. 1–6. IEEE (2016)
11. About Feature Scaling and Normalization. Sebastian Raschka's Website. July 11, 2014. http://sebastianraschka.com/Articles/2014_about_feature_scaling.html. Accessed 10 Nov 2017

The Prediction of Traffic Flow with Regression Analysis

Ishteaque Alam, Dewan Md. Farid and Rosaldo J. F. Rossetti

Abstract Traffic data mining applying machine learning algorithms is necessary to analyse and understand the road traffic flow in busy cities. Also, it is very essential for making smart cities. Mining traffic data helps us to reduce travel delays and improve the city life. Currently, many cities in the developed countries use different sensors to collect the real-time traffic data and apply machine learning algorithms on the traffic data to improve the traffic condition. In this paper, we have collected the real-time traffic data from the city of Porto, Portugal, and applied five regression models: Linear Regression, Sequential Minimal Optimisation (SMO) Regression, Multilayer Perceptron, M5P model tree and Random Forest to predict/forecast the traffic flow of Porto city. Also, we have tested the performance of these regression models. The experimental results show that the M5P regression tree outperforms the other regression models.

Keywords Historical traffic data · Predictive model · Regression analysis
Traffic flow forecast

1 Introduction

Mining traffic big data becomes an inevitable part in the way of building smart cities nowadays [3]. Every day we face a tremendous amount of trouble for the road traffic inconvenience. To improve the road traffic condition many cities are

I. Alam · D. Md. Farid (✉)
Department of Computer Science and Engineering, United International University,
Dhaka, Bangladesh
e-mail: dewanfarid@cse.uiu.ac.bd
URL: http://cse.uiu.ac.bd

I. Alam
e-mail: ishteaque.ark@gmail.com

R. J. F. Rossetti
LIACC - Artificial Intelligence and Computer Science Laboratory, Faculty of Engineering,
Department of Informatics Engineering, University of Porto, Porto, Portugal
e-mail: rossetti@fe.up.pt

A. Abraham et al. (eds.), *Emerging Technologies in Data Mining and Information Security*, Advances in Intelligent Systems and Computing 813,
https://doi.org/10.1007/978-981-13-1498-8_58

collecting road traffic information using modern technologies. Then, this information is analysed to understand the traffic flow and its characteristics to improve real-time Traffic Management System (TMS). Knowledge extraction from big traffic data applying machine learning and data mining algorithms has become an essential part for building Intelligent Transportation System (ITS) [13, 22]. We need real-time traffic information to take smart decision to minimise the traffic delay that also saves our time to move from one place to another [12]. Several types of sensors and wireless communication devices such as Wi-Fi, Bluetooth and GPS are used to collect real-time traffic data recently, and then the traffic data is analysed for traffic flow forecasting and decision-making for Traffic Management Systems by applying different machine learning algorithms [25].

Controlling the road traffic is always a challenging task. Still in many cities, the traditional traffic control devices like traffic lights and human traffic police are using to maintain the traffic flow [15]. However, the traditional traffic control approaches are not sufficient enough to improve road traffic condition in the busy cities like New York, Paris, London, etc. To improve the traffic management systems, intelligent computational researchers are applying several machine learning and data mining tools and algorithms for mining traffic data nowadays. Analysing historical traffic data helps us to minimise road accidents and travel delays. A traffic accident prediction model for Western Desert Road at Aswan city was developed using fuzzy modelling to examine and build an accident prediction model based on historical traffic data collected from the year 2011 to 2015 [14]. This model was also used for transportation management and planning on the desert roads to reduce highway accidents. Luis et al. [17] conducted a survey at Quito Metropolitan District and trained an artificial neural network (ANN) to evaluate the willingness to pay among people to minimise the road traffic noises. Oruc et al. [4] collected floating car data from Ankara City and extracted critical traffic patterns for urban traffic. Sasan et al. [5] came up with a real-time traffic control architecture stand on big data analytics. Although the study could not access the real traffic data, the architecture is capable to accommodate different data storage settings and analytical engines.

This work extends our previous work [3], where we developed a traffic data visualisation tool and applied regression models to understand the relationship among the features in traffic dataset. In this paper, we have taken a real-time traffic data from City of Porto, Portugal for 3 years from 2013 to 2015. Then, we have applied five regression models: Linear Regression, Sequential Minimal Optimisation (SMO) Regression, Multilayer Perceptron, M5P model tree and Random Forest on the data to forecast the real traffic flow of Porto city. The 23 sensors that are placed at the different streets of the Porto city collected the traffic data. Initially, we have used Python high-level programming language to visualise and pre-process the traffic data. Then, we have used Weka (Waikato Environment for Knowledge Analysis) libraries to develop regression models. Finally, we compared the predicted results of regression models in comparison with the actual real-time traffic flow.

The rest of the paper is organised as follows: Sect. 2 presents the related works regarding road traffic flow prediction. Section 3 presents the regression models that used for forecasting traffic flow. Section 4 presents the traffic data that collected from city of Porto, Portugal, and the experimental results. Finally, conclusion and future work is discussed in Sect. 5.

2 Related Work

Predicting short-term and long-term traffic flow is necessary for building smart cities. Xiaobo et al. [11] presented a short-term traffic flow prediction model applying hybrid genetic algorithm-based LSSVR (Least Squared Support Vector Regression) model using the dataset collected from 24 observation sites from freeways of Portland, United States. One advantage of the method is that it can interpret the relationship among the spatiotemporal variables.

Nicholas et al. [19] proposed an innovative architecture based on deep learning to predict traffic flow. The architecture was combined with a linear model with another sequence of tenth layers. The study identifies that deep learning performs better over linear models for prediction. Yalda et al. [21] used adaptive model to develop a real-time data flow prediction model. A two-step approach was employed to capture uncertainty. The generality of their method was tested through an open-access database called PeMS. PeMS is a performance measurement system that also provides a variety of support for historical data analysis. Their proposed method was executed for imputation of missing data and their results presented that the method is more efficient in comparison with PPCA and k-NN. Xianyao et al. [16] also proposed an algorithm (APSO-MSVM) based on Adaptive Particle Swarm Optimisation and Multi-kernel Support Vector Machine (MSVM) for short-term traffic flow prediction. The study collected real data from roadside units (RSU) and the algorithm had significantly lower error rate on both freeway and urban road.

In 2014, Marcin et al. [8] presented an approach to boost the performance of k-Nearest Neighbour (kNN) for predicting road traffic condition. The paper investigated the data segmentation method to figure out useful neighbours for better prediction. Their experimental result had a better accuracy by searching the nearest neighbour among the useful neighbours. In 2015, Afshin et al. [1] proposed a methodology to predict road traffic till 30 min ahead by linking their estimated demand with a limited real-time data. The study used Monte Carlo method for the evaluation of the algorithm in San Francisco and California road network. In 2016, Jinyoung et al. [2] presented a method to predict traffic flow in their paper using Bayesian Classifier and SVR. The performance of their method was later tested on the traffic data collected from Gyeongbu Expressway. Dick et al. [7] recommended a road traffic estimation model for less occupied roads in Wyoming. The study proposed a linear and another logistic regression models for estimating the traffic level.

In 2017, Hubert and Jerome [6] explained a methodology for an Ensemble Kalman Filter depending on the static detector measurements and vehicle data to estimate future traffic velocity and density. Their methodology was applied over a single road to simplify the task. Xi et al. [10] studied traffic demand by conducting an experimental case study on a school bus and people's psychological characteristics. Finally, they made a forecast model on traffic demand for different sunny, rainy and exam days based on the survey.

3 Regression Analysis

Regression analysis is often performed in data mining to study or estimate the relationship among several predictors or variables. Generally, a regression model gets created to perform analysis over a dataset. Then, it uses least squared method to estimate the parameter that fits the best. The model gets verified using one or more hypothesis tests. In this paper, we have applied Linear Regression, SMO Regression, Multilayer Perceptron, M5P model tree and Random Forest to forecast the future traffic flow for the fourth week of a month.

3.1 Linear Regression

Linear model predicts one variable based on another using statistical method [18]. It uses the equation of a line to draw an estimation line to predict result of a dependent variable from the result over an explanatory variable. The objective of a simple linear regression is to create best fit line that minimises the sum of the squared residuals. This is used to establish the relationship between two variables or to find a possible statistical relationship between the two variables.

3.2 SMO Regression

Sequential Minimal Optimisation (SMO) regression is an algorithm that deals solving the quadratic programming (QP) problem. The algorithm is widely used to find solutions to Support Vector Machine (SVM). The algorithm continues running the iteration loop to reach the most promising combination of pairs to optimise weights. The model also optimises time-series forecasting by reducing the operation runtime than general SVM algorithm as it avoids performing quadratic programming [24].

3.3 Multilayer Perceptron

The perceptron consists of weights and summation process along with activation function. Multilayer perceptron has the same functionality as a single perceptron, which is a class of artificial neural network (ANN) with one or more hidden layers.

One hidden layer gets connected to the input layer and using the inputs and weights, it forward passes the outputs from one layer to the next. The output gets computed utilising the sigmoid function as like nonlinear transfer functions.

3.4 *M5Base Regression Tree*

M5Base also known as M5P is basically a method for creating regression tree by using m5 algorithm. It implements a base routine for generating m5 model tree by using the separate-and-conquer technique to create decision lists. The tree is generated by employing a base routine, just as applying piecewise function over several linear models [20]. Later, an improved version of the algorithm was developed to handle missing feature values effectively by adding them artificially [23].

3.5 *Random Forest*

Random Forest is an often used algorithm in data science to make decision generated from random trees [9]. The algorithm combines a large number of decision trees to make a classification using bagging technique. The technique combines a number of learning models in order to increase the classification accuracy. This powerful supervised machine learning algorithm is capable to perform both classification and regression tasks. It chooses the classification, having the most votes among the generated decision trees. And, in the case of regression, it takes the average of the outputs calculated by different trees.

4 Experimental Analysis

4.1 *Traffic Dataset*

A real-time traffic data was collected from the city of Porto, which is the second largest city in Portugal. The city is very well known for its impressive bridges and wine production. However, the road traffic data was collected via 23 sensors placed on the city roads. Most of the sensors were positioned in Via de Cintura Interna (VCI). This road has a length of 21 Km portrayed as a ring-shaped motorway. The road is also referred as tambm chamada IC23. All the data collected through each sensor gets stored in PostgreSQL. This is an open-source SQL standard object-relational database management system, which supports a variety of data types. Students are allowed to use the dataset for research purpose without any cost. Table 1 shows the detail explanation of each column attributes of the collected dataset sensed through

Table 1 Feature description of traffic dataset

Name	Description
Equipment ID	Sensor ID
LANE_NR	Lane number
LANE_DIRECTION	Lane direction
TOTAL_VOLUME	Number of vehicles
AVG_SPEED_ARITHMETIC	Arithmetic average speed of vehicles
AVG_SPEED_HARMONIC	Harmonic average speed of vehicles
AVG_LENGTH	Average length of the vehicles
AVG_SPACING	Average spacing between vehicles
OCCUPANCY	Occupancy
INVERTED_CIRCULATION	No. of vehicles pass through with opposite direction
LIGHT_VEHICLE_RATE	The % is the ratio of light vehicles
VOLUME_CLASSE_A	No. of vehicles of class A
VOLUME_CLASSE_B	No. of vehicles of class B
VOLUME_CLASSE_C	No. of vehicles of class C
VOLUME_CLASSE_D	No. of vehicles of class D
VOLUME_CLASSE_0	No. of unidentified vehicles
AXLE_CLASS_VOLUMES	No. of vehicles per vehicle axis number
AGGREGATE_BY_LANEID	Is a unique ID fusing two IDs, the lane ID and the lane direction
AGG_PERIOD_LEN_MINS	Is the time between each measurement the sensors made. It is always 5 min
AGG_PERIOD_START	Time interval
NR_LANES	No. of lanes in the street of the event
AGGREGATE_BY_LANE_BUNDLEID	Similar to AGGREGATE_BY_LANEID
AGG_ID	Street ID

each sensor. Our initial dataset contained 2,83,97,311 instances starting from the year 2013 to 2015. In this paper, we present the forecast mechanism and experimental results performed focusing on the sub-dataset of the month April 2014. The study predicts future traffic flow of the month utilising previous historical road traffic data. The sub-dataset April 2014 has more than 10 lakh (10,39,077) instances.

4.2 Data Pre-processing

Our collected dataset had 23 attributes. We applied feature selection method to exclude some redundant attributes that are not useful features for forecasting the traffic flow, such as, AGG_PERIOD_LEN_MINS, INVERTED_CIRCULATION,

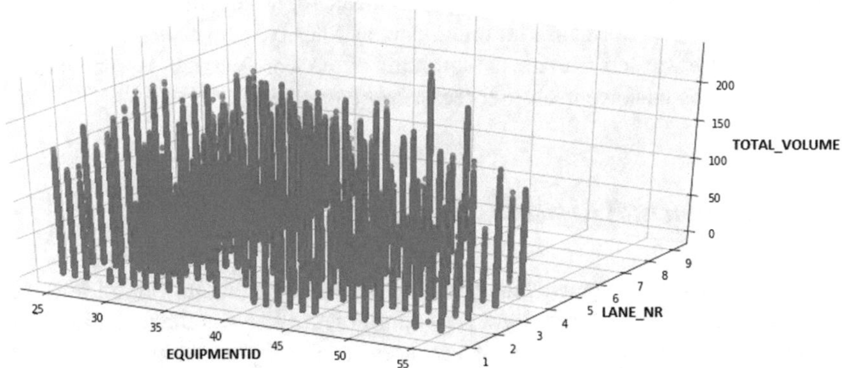

Fig. 1 Total volume of each EquipmentID and lane number for April 2014 of Porto City

etc. The main objective of this paper was to apply machine learning algorithms to forecast the Total Volume of the traffic flow. Therefore, other volume classes were not important to consider. The dataset had no missing values, but yet it was difficult to visualise for the randomness in between. To clear the randomness of the dataset, we serialised it utilising Pandas library from Python programming. In recent times, Pandas is a very dominant data analysis module provided by Python, which is also open source. Next, we plotted some figures to make sense of the dataset. Figure 1 explains that lane number varies with respect to each Equipment id. Equipment id represents the sensor's id that is placed on different roads of the city. So, we break the dataset with respect to each equipment id to forecast the traffic flow for each of the roads. Afterward, we merge together total volume of all the lanes of each road for a particular time. In this approach, we take the first 3 weeks data of the month and try to make a model that can predict the fourth week's traffic flow for the roads of Porto city. We have used five different regression models to perform the prediction.

4.3 Experimental Setup

This study has used WEKA which has a vast collection of machine learning algorithms, and it is widely popular in the field of data mining. A new package was installed in the workbench named 'timeseriesForecasting', which is specially made for time-series analysing. This new framework uses machine learning approaches to make models for time series. It transforms the data into a processable format for standard propositional learning algorithms.

Our train dataset contained the first three weeks preprocessed data for the month April 2014 of Porto city. This was use to train each of the regression models. Then, we set the number of units to forecast to 7 with a daily periodicity for predicting traffic flow for the upcoming week. We generated a linear model where m5 method was used for attribute selection with a ridge parameter set to default 1.0e−8. For SMO regression, we used normalised training data as filter type for better data integrity. Everything else was left with the default setup of the workbench. A lag variable was generated by the package itself to forecast based on the time series.

4.4 Experimental Results

The forecasting models generated by each of the machine learning algorithms were implemented for future traffic flow prediction. Figure 2 shows the patterns yielded by each of those algorithms. Figure 3 focuses specifically on the predicted pattern (From the date 22 to 28). Afterwards, the accuracy of the algorithms was measured by comparing the mean absolute errors. Table 2 describes the error rate for each of the regression models. The worst performance was found by the linear model with the highest error rate and the best performance was found by the M5P regression model with the lowest error rate of 2.88%. We have only presented the experimental result for a single road in Porto city in this paper. However, we applied the same technique over other roads of the city and find out that M5P performs the best over all the other algorithms.

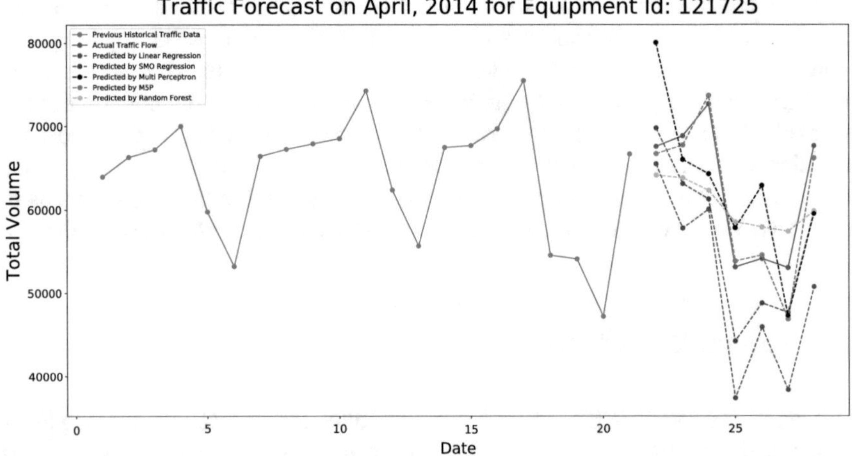

Fig. 2 Prediction of traffic flow by regression models for the fourth week of April 2014 based on previous 3 weeks of historical traffic data

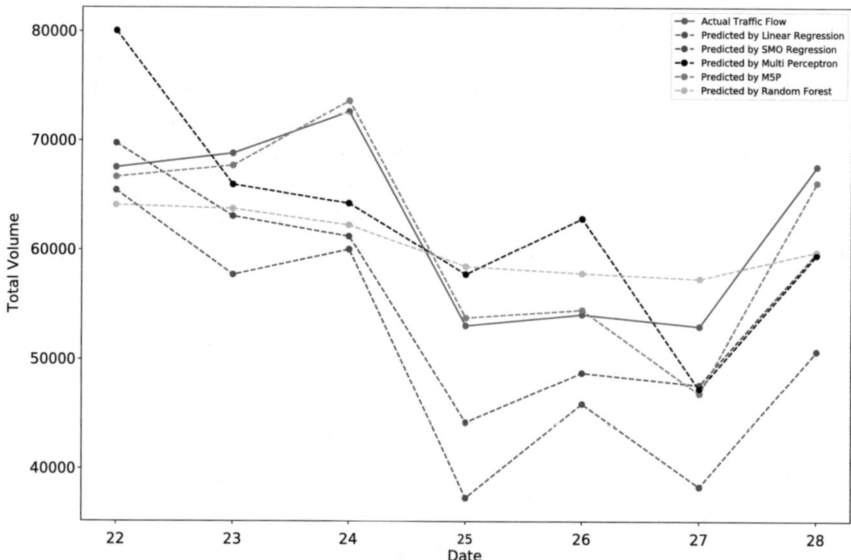

Fig. 3 Prediction comparison of the traffic flow of models with real traffic data for the fourth week of April 2014

Table 2 Result

Day	Linear regression (%)	SMO regression (%)	Multilayer perceptron (%)	M5P model tree (%)	Random forest (%)
22nd Apr	3.08	3.26	18.48	1.28	5.08
23nd Apr	16.12	8.32	4.12	1.60	7.32
24nd Apr	17.34	15.67	11.51	1.40	14.24
25nd Apr	29.69	16.68	8.84	1.35	10.19
26nd Apr	15.12	9.86	16.26	0.78	6.96
27nd Apr	27.69	10.10	10.81	11.60	8.19
28nd Apr	24.98	11.63	11.95	2.16	11.51
MAE	19.15	10.79	11.71	2.88	9.07

5 Conclusion

Predicting road traffic flow is one of the major challenges over the past decade in the field of data mining. This paper presented a new innovative approach to predict real traffic flow in a long term utilising popular machine learning algorithms. The study used Linear Regression, SMO Regression, Multilayer Perceptron, M5P model tree, and Random Forest. The traffic dataset from Porto city was collected and processed using Python programming. Later on, we applied the regression algorithms to generate patterns of traffic flow for the fourth week of each month using historical traffic data of the previous 3 weeks of that month. The results of the models were compared with the actual traffic flow. M5P model tree outperformed every other models with the lowest mean absolute error rate. For a particular road that we showed in this work, the error rate was 2.88%. This paper establishes high possibility of M5P algorithm in terms of forecasting real-world traffic flow. The same approach can be applied to forecast the traffic flow for a whole month also. In future, we would like to present a traffic flow visualising and forecasting system for the capital of Bangladesh, Dhaka City.

Acknowledgements We would like to thank "Armis" (http://www.armis.pt/) for sharing the traffic data.

References

1. Abadi, A., Rajabioun, T., Ioannou, P.A.: Traffic flow prediction for road transportation networks with limited traffic data. IEEE Trans. Intell. Transp. Syst. **16**(2), 653–662 (2015)
2. Ahn, J., Ko, E., Kim, E.Y.: Highway traffic flow prediction using support vector regression and Bayesian classifier. In: 2016 International Conference on Big Data and Smart Computing (BigComp), pp. 239–244. IEEE (2016)
3. Alam, I., Ahmed, M.F., Alam, M., Ulisses, J., Farid, D.M., Shatabda, S., Rossetti, R.J.F.: Pattern mining from historical traffic big data. In: IEEE Technologies for Smart Cities (TENSYMP 2017), pp. 1–5. IEEE (2017)
4. Altintasi, O., Yaman, H.T., Tuncay, K.: Detection of urban traffic patterns from floating car data (FCD). Transp. Res. Procedia **22**, 382–391 (2017)
5. Amini, S., Gerostathopoulos, I., Prehofer, C.: Big data analytics architecture for real-time traffic control. In: 2017 5th IEEE International Conference on Models and Technologies for Intelligent Transportation Systems (MT-ITS), pp. 710–715. IEEE (2017)
6. Andre, H., Ny, J.L.: A differentially private ensemble Kalman filter for road traffic estimation. In: 2017 IEEE International Conference on Acoustics, Speech and Signal Processing (ICASSP), pp. 6409–6413. IEEE (2017)
7. Apronti, D., Ksaibati, K., Gerow, K., Hepner, J.J.: Estimating traffic volume on Wyoming low volume roads using linear and logistic regression methods. J. Traffic Transp. Eng. (Engl. Ed.) **3**(6), 493–506 (2016)
8. Bernaś, M., Płaczek, B., Porwik, P., Pamuła, T.: Segmentation of vehicle detector data for improved k-nearest neighbours-based traffic flow prediction. IET Intell. Transp. Syst. **9**(3), 264–274 (2014)
9. Breiman, L.: Random forests. Mach. Learn. **45**(1), 5–32 (2001)
10. Chen, X., Peng, L., Zhang, M., Li, W.: A public traffic demand forecast method based on computational experiments. IEEE Trans. Intell. Transp. Syst. **18**(4), 984–995 (2017)
11. Chen, X., Wei, Z., Liu, X., Cai, Y., Li, Z., Zhao, F.: Spatiotemporal variable and parameter selection using sparse hybrid genetic algorithm for traffic flow forecasting. Int. J. Distrib. Sens. Netw. **13**(6), 1550147717713,376 (2017)
12. Csikós, A., Charalambous, T., Farhadi, H., Kulcsár, B., Wymeersch, H.: Network traffic flow optimization under performance constraints. Transp. Res. Part C: Emerg. Technol. **83**, 120–133 (2017)
13. Dell'Orco, M., Marinelli, M.: Modeling the dynamic effect of information on drivers' choice behavior in the context of an advanced traveler information system. Transp. Res. Part C: Emerg. Technol. **85**, 168–183 (2017)

14. Gaber, M., Wahaballa, A.M., Othman, A.M., Diab, A.: Traffic accidents prediction model using fuzzy logic: Aswan desert road case study. J. Eng. Sci. Assiut Univ. **45**, 2844 (2017)
15. Ghorghi, F.B., Zhou, H.: Traffic control devices for deterring wrong-way driving: historical evolution and current practice. J. Traffic Transp. Eng. **4**, 280–289 (2017)
16. Ling, X., Feng, X., Chen, Z., Xu, Y., Zheng, H.: Short-term traffic flow prediction with optimized multi-kernel support vector machine. In: 2017 IEEE Congress on Evolutionary Computation (CEC), pp. 294–300. IEEE (2017)
17. Moncayo, L.B., Naranjo, J.L., García, I.P., Mosquera, R.: Neural based contingent valuation of road traffic noise. Transp. Res. Part D: Transp. Environ. **50**, 26–39 (2017)
18. Montgomery, D.C., Peck, E.A., Vining, G.G.: Introduction to Linear Regression Analysis, vol. 821. Wiley (2012)
19. Polson, N.G., Sokolov, V.O.: Deep learning for short-term traffic flow prediction. Transp. Res. Part C: Emerg. Technol. **79**, 1–17 (2017)
20. Quinlan, J.R., et al.: Learning with continuous classes. In: 5th Australian Joint Conference on Artificial Intelligence, vol. 92, pp. 343–348. Singapore (1992)
21. Rajabzadeh, Y., Rezaie, A.H., Amindavar, H.: Short-term traffic flow prediction using time-varying Vasicek model. Transp. Res. Part C: Emerg. Technol. **74**, 168–181 (2017)
22. Talebpour, A., Mahmassani, H.S., Hamdar, S.H.: Effect of information availability on stability of traffic flow: percolation theory approach. Transp. Res. Procedia **23**, 81–100 (2017)
23. Wang, Y., Witten, I.H.: Inducing model trees for continuous classes. In: Proceedings of the Ninth European Conference on Machine Learning, pp. 128–137 (1997)
24. Yang, J.F., Zhai, Y.J., Xu, D.P., Han, P.: SMO algorithm applied in time series model building and forecast. In: 2007 International Conference on Machine Learning and Cybernetics, vol. 4, pp. 2395–2400. IEEE (2007)
25. Zhou, M., Qu, X., Li, X.: A recurrent neural network based microscopic car following model to predict traffic oscillation. Transp. Res. Part C: Emerg. Technol. **84**, 245–264 (2017)

SBi-MSREimpute: A Sequential Biclustering Technique Based on Mean Squared Residue and Euclidean Distance to Predict Missing Values in Microarray Gene Expression Data

Sourav Dutta, Mithila Hore, Faraz Ahmad, Anam Saba, Manuraj Kumar and Chandra Das

Abstract Due to technical problems in DNA microarray technology, the output matrix named gene expression data contains a huge number of missing entries. These missing entries create problems when analysis algorithms like classification, clustering, etc. are applied on microarray gene expression data as these methods require complete data matrix. To solve the above-mentioned problem, several missing value imputation techniques have been developed. Among them clustering- and biclustering-based missing value prediction methods are most popular due to their simplicity. In this regard, here a new biclustering-based sequential imputation method is proposed. In this method, for every missing position, a bicluster is formed in a novel manner using the concept of mean squared residue (MSR) score and Euclidean distance. Then the imputation is carried out sequentially by computing the weighted average of the neighbour genes and samples present in the bicluster. To evaluate the

---

S. Dutta · M. Hore · F. Ahmad · A. Saba · M. Kumar · C. Das (✉)
Netaji Subhash Engineering College, Garia, Kolkata 700152, India
e-mail: daschandra08@gmail.com

S. Dutta
e-mail: dutt.sourav@gmail.com

M. Hore
e-mail: horemithila@gmail.com

F. Ahmad
e-mail: faraza72@gmail.com

A. Saba
e-mail: anam.saba378@gmail.com

M. Kumar
e-mail: raj.manuraj124@gmail.com

A. Abraham et al. (eds.), *Emerging Technologies in Data Mining and Information Security*, Advances in Intelligent Systems and Computing 813,
https://doi.org/10.1007/978-981-13-1498-8_59

performance, the proposed method is rigorously tested and compared with the well-known existing methods. The effectiveness of the proposed method, is demonstrated on different microarray data sets including time series, non-time series, and mixed.

1 Introduction

In the last few decades, DNA Microarray technology has become a very powerful workhorse to biological researchers. With the aid of this high throughput biotechnology, it is possible to measure expression level of several thousands of genes under hundreds of experimental samples simultaneously [1, 2]. The outcome of microarray technology is a gene expression data matrix where a row carries information about a gene and a column carries information about a sample and a cell carries information about a gene for a specific sample. Analysis of microarray gene expression data is very crucial as it has widespread applications over various domains of biological and medical fields [3, 4], etc. Unfortunately, some data in this matrix may be lost due to technical problems in microarray experiments (like image corruption, dust or scratches on the slides etc.) [5]. The missing entries must be filled up before subsequent analysis as several analysis algorithms [6–8] require complete data matrices. Repetition of same microarray experiment to acquire the missing data is not feasible as it requires immense cost and time. Removing gene expression profiles with missing data is also not feasible as it may lose valuable information. Replacing missing values by row averages or zeros may distort the relationship among the variables. Thus computational techniques for missing value estimation are necessary and crucial.

A huge number of computational techniques for missing value estimation already exists in the literature. These methods are divided into four categories [9, 10]: (a) global methods (b) local methods (c) hybrid methods (d) knowledge assisted methods. Global methods [11, 12] use information from the entire data matrix. Local methods [9, 10] impute missing values by taking information from the local substructures. These methods are divided into two categories: (1) clustering based approaches (2) numerical method based approaches. Traditional K nearest neighbour imputation (KNNimpute) [11], sequential K nearest neighbour imputation (SKNNimpute) [13], iterative K nearest neighbour imputation (IKNNimpute) [14], iterative sequential K nearest neighbour imputation (ISKNNimpute) [15] are clustering-based popular and well-established methods. Least square and its several versions, EM-based methods [16–18] are examples of numerical approach based local methods. In Hybrid methods [19, 20] global and local information from the matrix are used for imputation. Knowledge assisted approaches [21–23] use external domain knowledge information to predict missing values.

Although a huge number of missing value estimation techniques exist in the literature, it has been found that no single algorithm exists which can give best results for all types of data and all types of distribution of missing entries. So, still there is room to develop new missing value imputation techniques for gene expression data.

Among all existing missing value estimation techniques, clustering-based imputation techniques are most popular due to their simplicity but their prediction accuracy are low compared to numerical methods. Like clustering based imputation techniques, biclustering (a special kind of clustering) based imputation techniques [24–28] are also simple and their prediction accuracy are better than clustering based imputation techniques but lower compared to numberical techniques. Biclustering based already developed imputation techniques are divided into two categories: (1) biclustering based imputation techniques [25, 27] (2) integrated imputation techniques combining biclustering and numerical approach [26, 28]. Among them prediction accuracy of integrated imputation techniques are far better than biclustering-based imputation techniques but former methods are popular due to their simplicity. In this regard, here, the emphasis is given toward improving the prediction accuracy of biclustering-based imputation techniques and so, a novel biclustering-based sequential imputation method named SBi-MSREimpute has been proposed to predict missing values more accurately in microarray gene expression data than existing clustering- and biclustering-based imputation techniques.

2 Proposed Method

The proposed biclustering-based missing value estimation technique named SBi-MSREimpute is applicable for all categories of microarray datasets (time series, non-time series and mixed). This method is proposed based on Cheng and Church's residue score and Euclidean distance and is a modification of bsimpute method. In the following subsections, Cheng and Church's residue score, Euclidean distance, bsimpute method and its drawbacks are described first and finally the proposed SBi-MSREimpute method is explained. Before that the following notations are described which are used throughout this paper.

2.1 Notations

$D = (X, Y)$ is a gene expression data matrix, with m number of genes and n number of samples/conditions/experiments and $m \gg n$. Here, X represents the set of genes and Y represents the set of samples or conditions of the matrix D such that $X = \{x_1, x_2, \ldots, x_m\}$ and $Y = \{y_1, y_2, \ldots, y_n\}$ respectively. Each entry in the data matrix D is represented as x_{ij} which signifies the expression level of ith gene (row) for jth sample or condition (column). Any missing position at the qth sample of the pth gene is denoted by α that means $x_{pq} = \alpha$. To remember each missing position a missing indicator matrix $F_{m\times n}$ is used here. Each entry in F is denoted by f_{ij}. If $f_{ij} = 0$ that signifies the corresponding position in D is missing and if $f_{ij} = 1$ then the value of corresponding position is present. A gene for which imputation procedure is carried out is considered as target gene and remaining genes are considered as candidate genes. A sample for which imputation procedure is carried out is considered as target sample and remaining samples are considered as candidate samples.

2.2 Cheng and Church's Biclustering Model

In gene expression data biclustering technique was introduced by Cheng and Church [24]. Biclustering [24] is one kind of clustering technique in which rows and columns are clustered simultaneously. In case of gene expression data a bicluster is a sub-matrix of the input gene expression data matrix in which a subset of genes show similar behaviour under a subset of samples and vice versa. A bicluster $P = (I, J)$ in the gene expression matrix $D = (X, Y)$ can be defined as a sub-matrix of D containing $I \in X$, a subset of genes and $J \in Y$, a subset of samples. Here, $|I|$ and $|J|$ represent the number of genes and number of samples respectively.

Cheng and Church had developed a method for finding coherent value based biclusters from the gene expression data matrix. They had introduced a measure known as mean squared residue (MSR) score to measure the compactness among genes and samples present in the bicluster. They defined $P = (I, J)$ as a δ bicluster in the input gene expression matrix $D = (X, Y)$ if MSR score $H(P) \leq \delta$, where δ is a given threshold and $\delta \geq 0$. The MSR score of an entry p_{ij} is denoted by $R(p_{ij})$ and is defined in Eq. (1).

$$R(p_{ij}) = p_{ij} - p_{iJ} - p_{Ij} + p_{IJ} \quad (1)$$

where p_{ij} represents the value present at ith row and jth position in the bicluster $P \cdot p_{iJ} = \frac{1}{|J|}\sum_{j \in J} p_{ij}$, $p_{Ij} = \frac{1}{|I|}\sum_{i \in I} p_{ij}$, $p_{IJ} = \frac{1}{|I||J|}\sum_{i \in I, j \in J} p_{ij}$ are the ith row mean, jth column mean and bicluster mean in the bicluster P respectively. The MSR score of the bicluster P is defined in Eq. (2).

$$H(P) = H(I, J) = \frac{1}{|I||J|} \sum_{i \in I, j \in J} R^2(p_{ij}) \quad (2)$$

The residue of ith row and jth column in the bicluster P is defined as in Eqs. (3) and (4) respectively.

$$H_{i \in I}(P) = \frac{1}{|J|} \sum_{j \in J} R^2(p_{ij}) \quad (3)$$

$$H_{j \in J}(P) = \frac{1}{|I|} \sum_{i \in I} R^2(p_{ij}) \quad (4)$$

2.3 Euclidean Distance

Euclidean distance [15] is most frequently used distance measure. The Euclidean distance between the objects $p = (p_1, p_2, \ldots, p_n)$ and $q = (q_1, q_2, \ldots, q_n)$ of n dimension is defined as given in the Eq. (5). This distance works well when two objects have similar values for the corresponding attribute/dimension.

$$E(p, q) = E(q, p) = \sqrt{\sum_{i=1}^{n} (q_i - p_i)^2} \quad (5)$$

2.4 bsimpute Method

In bsimpute [27] method, initially genes are sorted according to their missing rates and then imputation is carried out sequentially starting from the gene that has lowest missing rate. For each missing position in that gene, the proposed method first forms a bicluster by selecting a subset of similar genes and a subset of similar samples or conditions using a novel distance measure and then predicts the missing value by computing the weighted average of those subset of genes and samples. This value is imputed in the gene expression matrix D and then used for latter imputation in sequential manner. The proposed novel distance is a combination of MSR score and Euclidean distance and is defined in Eq. (6).

$$HE(x_t, x_c) = H(x_t, x_c) + E(x_t, x_c) \quad (6)$$

This distance has several shortcomings and in the proposed SBi-MSREimpute these shortcomings are overcome.

2.5 Detail Description of the Proposed Method

Like bsimpute method, in the proposed SBi-MSREimpute method, genes are arranged in descending order according to their missing rates and imputation procedure starts from the gene having lowest missing rate. For each missing entry in that gene, the proposed method first forms a bicluster by choosing a subgroup of similar genes and a subgroup of similar samples or conditions in a novel manner and then predicts the missing value by computing the weighted average of those subset of genes and samples. This value is imputed in the given gene expression matrix and used for latter imputation in sequential manner. The difference between SBi-MSREimpute and bsimpute is in the process of gene and sample selection for formation of bicluster. In bsimpute genes and samples are selected based on HE distance while in SBi-MSREimpute genes and samples are selected using a novel approach.

(1) A New Method to Overcome the Drawback of HE distance: In case of gene expression data, it has been found that pattern based similar genes (co-expressed) have similar biological functions. For weighted average based missing value estimation procedure, if co-expressed and magnitude wise closer neighbour genes are chosen with respect to a target gene, then missing values can be predicted more accurately.

Due to this phenomenon, Euclidean distance, Manhattan distance are not suitable measure for gene expression data. According to these distances two objects are similar if they are physically close. Using Pearson correlation coefficient [28] similarity measure, it is possible to capture the pattern based similarity between two objects. The drawback of Pearson correlation coefficient is that it cannot differentiate among scaling, shifting, and magnitude wise closer co-expressed patterns. So, using Pearson correlation coefficient both scaling or shifting pattern based co-expressed neighbour genes as well as magnitude wise co-expressed genes of a target gene can be selected.

It has been already proved that shifting pattern has no effect on MSR score but scaling pattern has a lot. Due to this reason in bsimpute [27] method, MSR score is used as a component of the proposed HE distance. In bsimpute approach, the proposed distance is the summation of Euclidean distance and MSR score. If the HE distance between a target gene and candidate gene is low then that means candidate gene is highly co-expressed and magnitude wise closer with respect to the target gene because low residue value will discard the chance of selection of scaling patterns while low Euclidean distance will reject possibility of selection of highly shifted or highly scaled candidate genes. So, using this distance pattern based similar (co-expressed) and magnitude wise closer gene selection is possible. But the drawback of this distance is that, it may happen that a candidate gene which has higher mean square residue with respect to the target gene, is still selected due to lower Euclidean distance and also vice versa that means the gene which have higher Euclidean distance is still selected due to low mean square residue which may affect the imputation process.

To remove the above-mentioned difficulties, at first, a set (U) of w genes according to the lowest mean squared residue value and another set (V) of w genes according to lowest Euclidian distance with respect to a target gene are selected. Then the intersection operation between these two sets U and V is carried out forming a set Z of u number of genes. This set consists of finally selected neighbour genes with respect to that target gene.

$$Z = U \cap V \tag{7}$$

This situation is demonstrated in Fig. 1. In Fig. 1a–d G_tg is considered as the target gene while all other genes are candidate genes. If k number of nearest neighbour genes are chosen and k is set at 4 then according to MSR score G4, G6, G1, and G8 will be selected respectively. This is shown in Fig. 1a where selected neighbour genes are marked using black colour. If Euclidean distance wise neighbour genes are chosen then G1, G7, G8 and G3 will be selected respectively which is shown in Fig. 1b. In Fig. 1c, G1, G4, G7 and G8 are selected as neighbour genes according to HE distance. In Fig. 1d, according to intersection of MSR score and Euclidean distance G1 and G7 are selected. From Fig. 1d G1 and G7 are magnitude wisc most closest and co-expressed neighbour genes of G_tg.

(2) SBi-MSREimpute Method: In the proposed SBi-MSREimpute method, to recover a missing value in the tth sample (i.e. x_{st}), of a target gene x_s in D, first the target gene x_s is considered as the initial bicluster B. Then u number of nearest neighbour genes of x_s are selected according to the above mentioned criteria and are added in the bicluster B as shown below in Eq. (8). Now the bicluster B contains $u + 1$ number of genes and n number of samples.

$$B = \begin{pmatrix} x_s \\ x_{n1} \\ x_{n2} \\ \vdots \\ x_{nu} \end{pmatrix} = \begin{pmatrix} x_{s1} & x_{s2} & \dots & \alpha & \cdots & x_{sn} \\ x_{n11} & x_{n12} & \cdots & x_{n1t} & \cdots & x_{n1n} \\ x_{n21} & x_{n22} & \cdots & x_{n2t} & \cdots & x_{n2n} \\ \vdots & \vdots & \dots & \vdots & \dots & \vdots \\ x_{nu1} & x_{nu2} & \cdots & x_{nut} & \cdots & x_{nun} \end{pmatrix} \tag{8}$$

After formation of this bicluster B, the MSR score of this bicluster $H(B)$ is calculated. Then any one sample/column from the bicluster B except the target sample (here t) is ignored and mean squared residue of the remaining bicluster is calculated. This operation is repeated for all the samples/columns present in the bicluster B (except for the target sample). After that, the ignored sample (let q), for which MSR score of the bicluster has decreased most is deleted permanently from the bicluster B. This operation is repeated until there exists no such columns for which MSR score of the bicluster B is decreased or number of samples present in the bicluster $\leq \varphi$, where φ is a threshold and here it is set to $n/2$. In this way, v number of nearest neighbour samples/columns with respect to target sample/column in the bicluster B are selected and other samples/columns from the bicluster B are deleted. This modified bicluster B is renamed as C which contains $u + 1$ number of genes and $v + 1$ number of samples and is shown in the Eq. (9). In this bicluster C first row is the target row and lth column is the target column. c_i and $\ddot{c}_j$ represents ith row and jth column of the bicluster C.

$$C = \begin{pmatrix} x_{s1} & x_{s2} & \dots & \alpha & \cdots & x_{s(v+1)} \\ x_{n11} & x_{n12} & \dots & x_{n1l} & \cdots & x_{n1(v+1)} \\ x_{n21} & x_{n22} & \dots & x_{n2l} & \cdots & x_{n2(v+1)} \\ \vdots & \vdots & \dots & \vdots & \dots & \vdots \\ x_{nu1} & x_{nu2} & \dots & x_{nul} & \cdots & x_{nu(v+1)} \end{pmatrix} \tag{9}$$

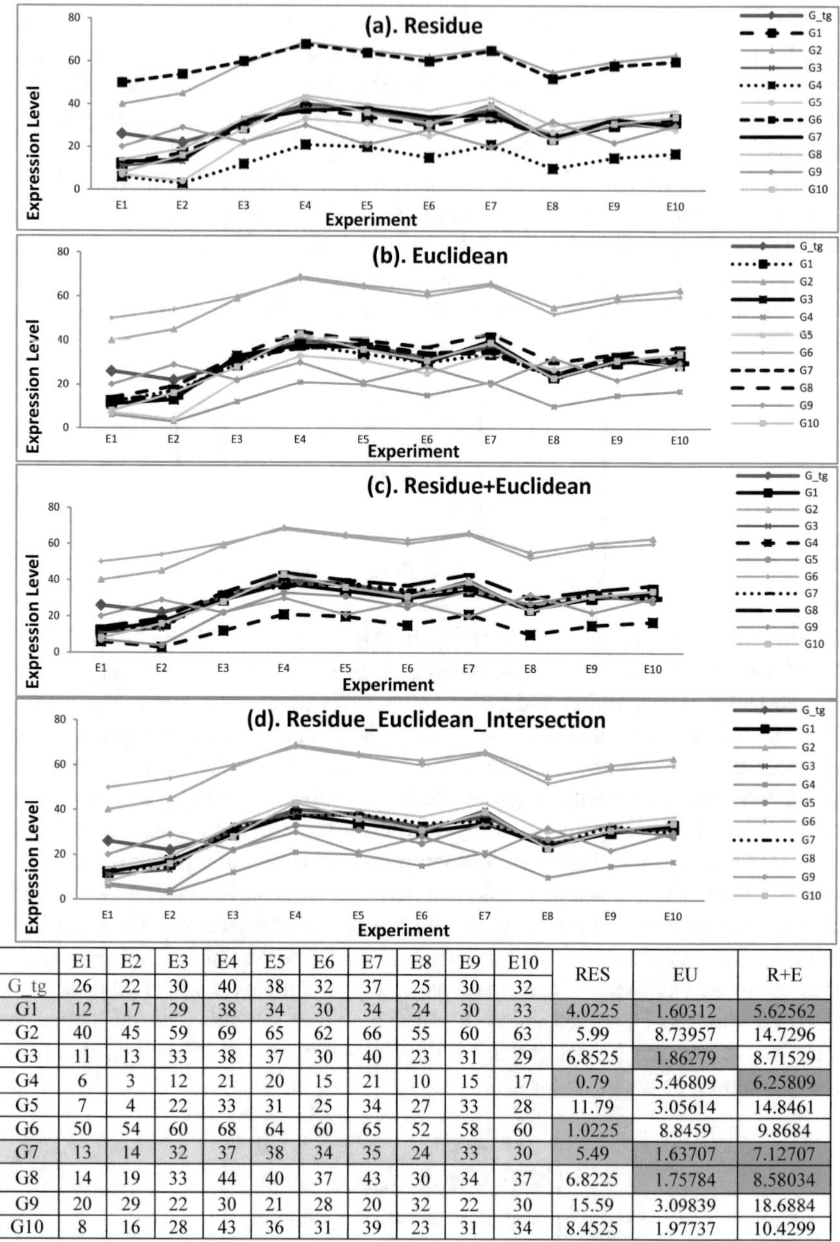

	E1	E2	E3	E4	E5	E6	E7	E8	E9	E10	RES	EU	R+E
G_tg	26	22	30	40	38	32	37	25	30	32			
G1	12	17	29	38	34	30	34	24	30	33	4.0225	1.60312	5.62562
G2	40	45	59	69	65	62	66	55	60	63	5.99	8.73957	14.7296
G3	11	13	33	38	37	30	40	23	31	29	6.8525	1.86279	8.71529
G4	6	3	12	21	20	15	21	10	15	17	0.79	5.46809	6.25809
G5	7	4	22	33	31	25	34	27	33	28	11.79	3.05614	14.8461
G6	50	54	60	68	64	60	65	52	58	60	1.0225	8.8459	9.8684
G7	13	14	32	37	38	34	35	24	33	30	5.49	1.63707	7.12707
G8	14	19	33	44	40	37	43	30	34	37	6.8225	1.75784	8.58034
G9	20	29	22	30	21	28	20	32	22	30	15.59	3.09839	18.6884
G10	8	16	28	43	36	31	39	23	31	34	8.4525	1.97737	10.4299

Fig. 1 Gene selection based on **a** Mean Square Residue (MSR); **b** Euclidean distance; **c** MSR + Euclidean; **d** intersection of MSR and Euclidean

After formation of the final bicluster C, weight (W_{i1}) of every row i (except for first row/target gene) with respect to target gene/first row and weight $\left(W_{jl}\right)$ of every column (except target column (here l)) with respect to target column (here l) in the matrix C are calculated. Finally, missing value (α) is replaced by $\left(R_{avg}+C_{avg}\right)/2$ where R_{avg} is the weighted average of target gene (here first row) and C_{avg} is the weighted average of the target column (here l). This value is placed in the x_{st} position of the gene expression matrix D and corresponding position of the indicator matrix is set to 1. This process is repeated for other missing positions present in that target gene as well as for other genes also. R_{avg} and C_{avg} are calculated as in Eq. (10) and Eq. (11) respectively.

$$R_{avg} = \sum_{j=1\, and\, j\neq l}^{v+1} W_{jl} C_{1j} \tag{10}$$

$$C_{avg} = \sum_{i=2}^{u+1} W_{i1} C_{il}, \tag{11}$$

where W_{i1} is the weight of the ith row with respect to target gene (here first row) in the bicluster C and W_{jl} is the weight of jth column with respect to target column l. W_{i1} and W_{jl} are calculated as shown in Eqs. (12) and (13) respectively.

$$W_{i1} = E(c_i, c_1) / \sum\nolimits_{k=2}^{u+1} E(c_k, c_1) \tag{12}$$

$$W_{jl} = E\left(\dddot{c}_j, \dddot{c}_l\right) / \sum\nolimits_{k=1 and\, k\neq l}^{v+1} E(\dddot{c}_k, \dddot{c}_l) \tag{13}$$

Algorithm: SBi-MSREimpute

Input: The microarssray gene expression matrix D with missing entries, where m is the number of genes and n is the number of samples/columns and $m \gg n$.

Output: The missing entries are predicted and complete gene expression matrix D is formed.

(1) Replace all the MVs in the given gene expression matrix D by the estimates given by row averages.
(2) Sort all m number of genes in ascending order according to their missing rate and consider for imputation one after another.
(3) For each missing position t in every target gene x_s do:
    (a) Consider the target gene x_s as the initial bicluster B.
    (b) Add set Z (formed according to Eq. (7)) containing u number of nearest neighbour genes of x_s in the bicluster B as shown in Eq. (8).
    (c) Repeat from step (i) to step (iv) until the no. of columns in the bicluster $B \leq \varphi$ or no decrease in the MSR score of the bicluster B due to deletion of any column.
        (i) Calculate MSR score $H(B)$ of the bicluster B and store it in $Temp$.
        (ii) $Max := Temp$

(iii) For every column (p) in the bicluster B except for the target column do:
   (A) Calculate MSR score $H(B)$ of the bicluster B but ignoring column p and store it in $Temp1$.
   (B) If $(Temp1 < Max)$ then

$$Max := Temp1$$
$$Pos := p$$

(iv) Delete Pos column from the bicluster B and reduce no. of columns of B by 1.

(d) Rename the modified bicluster B as C.
(e) Replace missing value α by $(R_{avg} + C_{avg})/2$ where R_{avg} is the weighted average of target gene (here first row) as shown in Eq. (10) and C_{avg} is the weighted average of the target column (here l) as shown in Eq. (11).
(f) Place calculated value for α at the x_{st} position in the input gene expression matrix D and set the corresponding position of the indicator matrix F to 1.

(4) End.

3 Experimental Results

In this paper, performance of the proposed SBi-MSREimpute method is compared with some of the existing well-known clustering- and biclustering-based imputation algorithms using normalized root mean squared error (NRMSE) [28] as a metric. To analyse the performance of different algorithms, the experimentation is done on different microarray gene expression datasets. The information about different datasets (SP.AFA, ROSS, Tymchuk, YOS) is given in [28]. The missing datasets are prepared as in [14]. All the above-mentioned methods are implemented in C using Linux environment in a machine with a 4 GB RAM, and 3.2 GHz core i3 processor.

3.1 Comparative Performance Analysis Based on NRMS Error

The efficiency of the newly proposed SBi-MSREimpute method is compared with *K*NNimpute, S*K*NNimpute, BICimpute and bsimpute, by applying them to different types of microarray datasets with different missing rates.

Figure 2 plots the performance of different methods as a function of various percentages (1, 5, 10, 15, 20% and unequal) of missing entries on different datasets. The performance is judged by the NRMS error value. From Fig. 2, it can be said that

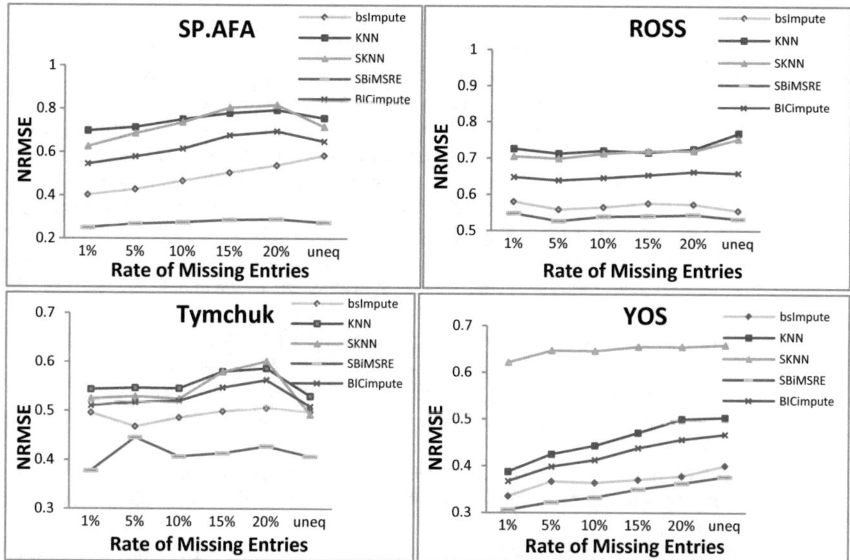

Fig. 2 Comparison of different clustering- and biclustering-based imputation methods with respect to NRMSE for different microarray gene expression datasets

SBi-MSREimpute method shows outstanding performance for all types of missing rates in every type of dataset compared to other clustering- and biclustering-based missing value prediction methods.

4 Conclusion

In this paper, a new sequential biclustering technique named SBi-MSREimpute has developed to predict missing values more accurately in microarray gene expression data. To select co-expressed and magnitude wise closer genes, information from mean squared residue and Euclidean distance has taken. From the results it has been observed that the proposed method is superior compared to traditional clustering- and biclustering-based imputation techniques.

References

1. Lockhart, D.J., Winzeler, E.A.: Genomics, gene expression and DNA arrays. Nature **405**, 827–836 (2000)
2. Schulze, A., Downward, J.: Navigating gene expression using microarrays—a technology review. Nat. Cell Biol. **3**, E190–E195 (2001)

3. Golub, T.R., Slonim, D.K., Tomayo, P., et al.: Molecular classification of cancer: class discovery and class prediction by gene expression monitoring. Science **286**, 531–537 (1999)
4. Maji, P., Das, C.: Biologically significant supervised gene clusters for microarray cancer classification. IEEE Trans. Nanobiosci. **11**(2), 161–168 (2012). Nanobioscience
5. Alizadeh, A.A., Eisen, M.B., Davis, R.E., Ma, C., Lossos, I.S., Rosenwald, A., Boldrick, J.C., Sabet, H., Tran, T., Yu, X., Powell, J.I., Yang, L., Marti, G.E., Moore, T., Hudson, J.J., Lu, L., Lewis, D.B., Tibshirani, R., Sherlock, G., Chan, W.C., Greiner, T.C., Weisenburger, D.D., Armitage, J.O., Warnke, R., Staudt, L.M.: Distinct types of diffuse large B-cell lymphoma identified by gene expression profiling. Nature **403**, 503–511 (2000)
6. Raychaudhuri, S., Stuart, J.M., Altman, R.B.: Principal component analysis to summarize microarray experiments: application to sporulation time series. In: Pacific Symposium on Biocomputing, pp. 455–466 (2000)
7. Alter, O., Brown, P.O., Bostein, D.: Singular value decomposition for genome-wide expression data processing and modeling. Proc. Natl. Acad. Sci. U.S.A. **97**, 10101–10106 (2000)
8. Sehgal, M.S.B., Gondal, I., Dooley, L.: Statistical neural networks and support vector machine for the classification of genetic mutations in ovarian cancer. In: Proceedings of IEEE, CIBCB"04, USA, 2004
9. Liew, A.W., Law, N.F., Yan, H.: Missing value imputation for gene expression data, computational techniques to recover missing data from available information. Brief Bioinform. **12**(5), 498–513 (2011)
10. Moorthy, K., Mohamad, M.S., Deris, S.: A review on missing value imputation algorithms for microarray gene expression data. Curr. Bioinform. **9**, 18–22 (2014)
11. Tryosanka, O., Cantor, M., Sherlock, G., et al.: Missing value estimation methods for DNA microarrays. Bioinformatics **17**, 520–525 (2001)
12. Oba, S., Sato, M.A., Takemasa, I., Monden, M., Matsubara, K., Ishii, S.: A Bayseian missing value estimation method for gene expression profile data. Bioinformatics **19**, 2088–2096 (2003)
13. Kim, K.Y., Kim, B.J., Yi, G.S.: Reuse of imputed data in microarray analysis increases imputation efficiency. BMC Bioinform. **5**(160) (2004)
14. Bras, L.P., Menezes, J.C.: Improving cluster-based missing value estimation of DNA microarray data. Biomol. Eng. **24**, 273–282 (2007)
15. Das, C., Bose, S., Chattopadhyay, M., Chattopadhyay, S.: A novel distance based iterative sequential KNN algorithm for estimation of missing values in microarray gene expression data. IJBRA **12**(4), 312–342 (2016)
16. Bo, T.H., Dysvik, B., Jonassen, I.: LSimpute: accurate estimation of missing values in microarray data with least squares methods. Nucleic Acids Res. **32**(3) (2004)
17. Kim, H., Golub, G.H., Park, H.: Missing value estimation for DNA microarray expression data: local least square imputation. Bioinformatics **21**, 187–198 (2005)
18. Cai, Z., Heydari, M., Lin, G.: Iterated local least squares microarray missing value imputation. J. Bioinform. Comput. Biol. **4**, 1–23 (2006)
19. Jornsten, R., Wang, H.Y., Welsh, W.J., et al.: DNA microarray data imputation and significance analysis of differential expression. Bioinformatics **21**(22), 4155–4161 (2005)
20. Pan, X., Tian, Y., Huang, Y., Shen, H.: Towards better accuracy for missing value estimation of epistatic miniarray profiling data by a novel ensemble approach. Genomics **97**(5), 257–264 (2011)
21. Tuikkala, J., Elo, L., Nevalainen, O.S., Aittokallio, T.: Improving missing value estimation in microarray data with gene ontology. Bioinformatics **22**(5), 566–572 (2006)
22. Kim, D.W., Lee, K.Y., Lee, K.H., Lee, D.: Towards clustering of incomplete microarray data without the use of imputation. Bioinformatics **23**(1), 107–113 (2007)
23. Hu, J., Li, H., Waterman, M.S., Zhou, X.J.: Integrative missing value estimation for microarray data. BMC Bioinform. (2006)
24. Cheng, Y., Church, G. M.: Biclustering of gene expression data. In: Proceedings of the International Conference on Intelligent Systems for Molecular Biology, pp. 93–103 (2000)
25. Ji, R., Liu, D., Zhou, Z.: A bicluster-based missing value imputation method for gene expression data. J. Comput. Inf. Syst. **7**(13), 4810–4818 (2011)

26. Cheng, K.O., Law, N.F., Siu, W.C.: Iterative bicluster-based least square framework for estimation of missing values in microarray gene expression data. Pattern Recogn. **45**(4), 1281–1289 (2012)
27. Bose, S., Das, C., Chattopadhyay, M., Chattopadhyay, S.: A novel biclustering based missing value prediction method for microarray gene expression data. In: Proceedings of MAMMI (2016)
28. Das, C., Bose, S., Chattopadhyay, S., Chattopadhyay, M., Hossain, A.: A bicluster-based sequential interpolation imputation method for estimation of missing values in microarray gene expression data. Curr. Bioinform. **12**(2) (2017). Bentham Science

FTD Tree Based Classification Model for Alzheimer's Disease Prediction

Bala Brahmeswara Kadaru, M. Uma Maheswara Rao and S. Narayana

Abstract Due to the increasing demand on Alzheimer's, the continuous monitoring of health and characteristics are significant for maximizing the yields. Even though many physicochemical parameters are available for monitoring the Alzheimer's, the knowledge of domain experts are expected to analyze these parameters to find the final decision about the Alzheimer's disease. In order to utilize the knowledge of the domain experts for Alzheimer's disease, we have developed a functional tangent decision tree algorithm which, predict the disease based on the physiochemical parameters. The proposed method of predicting the disease consists of three important steps such as, uncertainty handling, feature selection using reduce and core analysis, classification using the functional tangent decision tree.

1 Introduction

The amount of knowledge with in the world, in our lives appears to be increasing and there's no end in view present. Because the volume of knowledge will increase, the proportion of it is that people "understand" decreases alarmingly. Lying hidden altogether this knowledge is information, probably helpful data, that is never expressed or taken advantage of individuals which are seeking patterns within the knowledge. Because the flood of the knowledge swells and machines which will undertake the finding out data becomes common place, the opportunities for data processing will increase (Fig. 1).

B. B. Kadaru · M. Uma Maheswara Rao (✉) · S. Narayana
CSE Department, Gudlavalleru Engineering College, Gudlavalleru, Krishna district, India
e-mail: umalu537@gmail.com

B. B. Kadaru
e-mail: balukadaru2@gmail.com

S. Narayana
e-mail: satyala1976@gmail.com

A. Abraham et al. (eds.), *Emerging Technologies in Data Mining and Information Security*, Advances in Intelligent Systems and Computing 813,
https://doi.org/10.1007/978-981-13-1498-8_60

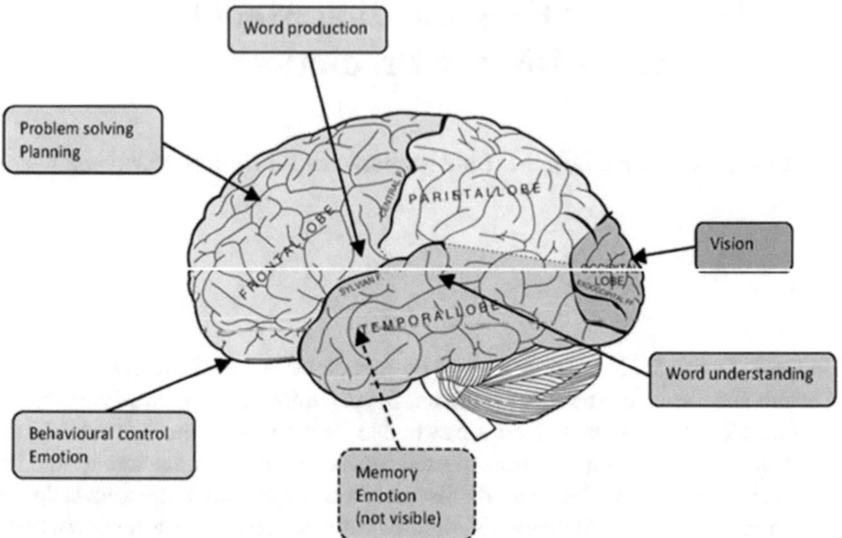

Fig. 1 Cognitive functions associated with each part in the brain

1.1 *Episodic Memory*

Alzheimer's disease, Mild cognitive impairment, amnesic type, Dementia with Lewy bodies, Encephalitis (most commonly, herpes simplex encephalitis), Frontal variant of frontotemporal dementia, Korsakoff's syndrome, Transient global amnesia, Vascular dementia, Multiple sclerosis.

1.2 *Semantic Memory*

Alzheimer's disease, Semantic dementia (temporal variant of frontotemporal dementia), Traumatic brain injury.

1.3 *Procedural Memory*

Parkinson's disease, Huntington's disease, Progressive supranuclear palsy, Olivopontocerebellar degeneration, Depression, Obsessive–compulsive disorder.

1.4 Working Memory

Normal aging, Vascular dementia, Frontal variant of frontotemporal dementia, Alzheimer's disease, Dementia with Lewy bodies, Multiple sclerosis, Traumatic brain injury, Attention deficit–hyperactivity disorder, Schizophrenia, Parkinson's disease, Huntington's disease, Progressive supranuclear palsy.

1.5 Dementia of the Alzheimer's Type

Alzheimer's disease (AD) was first delineated by the German specialist, Alois Alzheimer, in 1907. The unwellness appeared less common within the early decades of the twentieth century. Nowadays, however, dementedness could be a quite common unwellness within the older. According to the Alzheimer's association, AD is that the commonest explanation for dementia within the older, i.e., or so common fraction of all cases of dementedness. There are around 35 million patients laid low with Alzheimer's unwellness everywhere the globe, out of that U.S. of America alone has around four. Five million patients. AD is a neurodegenerative disorder touching major brain areas together with the cortex and limbic system, and is characterized by progressive decline in memory with impairment of a minimum of one alternative psychological feature operate. AD usually begins with symptoms like immediate memory loss, and continues with more widespread psychological feature and emotional pathology. Another type called late-onset AD (LOAD) happens when age 65. AD options current deterioration of patients' functioning which ends up in substantial and lasting incapacity over the approximate 7–10 years from identification to ultimate death. Even though AD sometimes shows no symptom on motor or sensory alterations, sure atypical clinical presentations area unit often found in some patients.

2 Machine Learning Methods

We will compare the most common decision tree algorithms which are implemented serially.

2.1 Classification

Classification will be delineated as a supervised learning algorithmic program within the Machine learning method. It assigns category labels to knowledge objects supported previous knowledge of sophistication that the information records belong. It is a knowledge mining technique, has made it attainable to co-design and co-develop package and hardware, and hence, such elements. However, the integration deals with information extraction from database records and prediction of sophistication label from unknown knowledge set of records. In classification a given set of records is split into small and individual check data sets.

2.2 Decision Tree Algorithm

Decision tree induction represents an easy and powerful methodology of classification which generates a tree and a collection of rules, representing the model of various categories, from a given dataset. The decision tree may be a flowchart-like tree structure, where each internal node denotes a take a look at on associate degree attribute, every branch represents associate degree outcome of the test and every leaf node represents the category. The highest most node in a very tree is that the root node. For DT induction, ID3 algorithmic rule and its successor C4.5 algorithmic rule by Quinlan (1993) ar wide used. Decision tree algorithmic rule can be a data processing technique which recursively partitions an information set of records mistreatment.

Depth-first greedy approach or breadth-first approach till all the information things belong to a selected category.

2.3 ID3 [1]

ID3 decision tree algorithmic rule was introduced in 1986 by Quinlan Ross. It supported Hunt's algorithmic rule and is serially enforced. Like other alternative decision tree algorithms the tree is built in two phases; tree growth and tree pruning. The information is sorted at each node throughout the tree building section in order to select the simplest splitting single attribute. IDE3 uses info gain live in choosing the splitting attribute. It solely accepts categorical attributes in building a tree model. IDE3 does not offer correct result once there's too much noise or details in the coaching information set, so we have an associate degree intensive preprocessing of information is taken out before building a call tree model with IDE3.

2.4 C4.5 [2]

C4.5 formula is associate degree improvement of IDE3 formula, developed by Quinlan Ross (1993). It's supported Hunt's formula and additionally, like IDE3, it's serially implemented. Pruning takes place in C4.5 by substitution the inner node with a leaf node thereby reducing the error rate. In contrast to IDE3, C4.5 accepts each continuous and categorical attributes in building the choice tree. Its associate degree increased methodology of tree pruning that reduces misclassification errors attributable to noise or too many details in the training dataset. Like IDE3 the information is sorted at each node of the tree so as to see the best splitting attribute. It employs gain magnitude relation impurity methodology to judge the splitting attribute.

2.5 CART [3]

CART (Classification and regression trees) was introduced by Breiman (1984). It builds each classification and regressions trees. The classification tree construction by CART relies on binary rending of the attributes. It conjointly supported Hunt's model of call tree construction and may be enforced serially. It uses Gini index rending live in choosing the rending attribute. Pruning is completed in CART by employing a portion of the training information set. CART uses each numeric and categorical attributes for building the choice tree and has inbuilt options that manage missing attributes. CART is exclusive from alternative Hunt's based mostly rule because it is also used for multivariate analysis with the assistance of the regression trees. The regression analysis feature is employed in the statement of a variable (result) given a collection of predictor variables over a given amount of your time. It mainly utilizes several single variable rending criteria such as Gini index, symgini, etc. and one multi-variable (linear combinations) in determining the simplest split purpose and information is sorted at each node to work out the simplest splitting purpose. The linear combination rending criteria is employed throughout regression analysis. SALFORD SYSTEMS enforced a version of CART referred to as CART® using the first code of Breiman (1984). CART® has increased the options and the capabilities that address the short-comings of CART giving rise to a contemporary Decision tree classifier with high classification and prediction accuracy.

2.6 Naïve Bayes

Bayesian classifiers are applied mathematics classifiers. They will predict category membership probabilities, like the chance that a given sample belongs to a specific category. The Bayesian classifier is predicated on Bayes' theorem. Naive Bayesian classifiers assume that the impact of AN attribute price on a given category is freelance of the values of the other attributes. This assumption is named category conditional independence. It is made to modify the computation concerned and, during this sense, is taken into account "naïve".

2.7 Neural Network

Multilayer perceptron square measure feed forward Neural networks trained with the quality Back propagation algorithmic rule. They are supervised networks in order that they need a desired response to be trained. They find out how to rework computer file into a desired response, in order that they square measure wide used for pattern classification. With one or two hidden layers, they will approximate nearly any input–output map. Examples square measure typically provided one at a time. For every example, the particular vector is computed and compared to the desired output. Then, weights and thresholds square measure adjusted, proportional to their contribution to the error created at the individual output. One among the foremost used strategies is the Back propagation methodology, within which within the unvaried manner, the errors square measure propagated (error = the distinction between desired output and therefore the output of actual ANN) into the lower layers, to be used for the difference of weights.

3 Proposed Model

3.1 FTD Tree Algorithm for Classification of Alzheimer's Data

The FTD Tree could be a recreate of ancient decision tree technique. This method is employed for regression and classification [4, 5]. Its the kind of non-parametric supervised learning methodology. The key advantage of decision tree over the opposite learning algorithmic program is easy to visualize the training datasets and straightforward to interpret and additionally the price is minimum. Within the decision tree, the rending is finished by the economical algorithmic program known as hunt's algorithmic program. Within the algorithmic program, the Gini price is calculated for every and each node like parent and kid node. Relying upon the Gini price the split is chosen, and therefore the minimum Gini price is desirable to separate. The projected system work of FTD Tree algorithmic program contains two major steps like (i) FTD Tree construction and (ii) activity classification through an FTD Tree. The definitions of construction and classification of the FTD Tree are given below.

Definition 1 The functional tangent entropy is given as follows:

$$FE(a_j) = -\sum_{j=i}^{u(a_j)} prob^j \, f(prob^j) \tag{1}$$

$$f(prob^j) = \frac{1}{2}\left[\log(prob^j) + \frac{-1}{a\ \tanh(prob^j)}\right] \tag{2}$$

where, $FE(a_j)$ is the functional tangent entropy of a_j, $u(a_j)$ is the number of unique values in the attributes, log(•) is the logarithmic function and a tanh(•) is the inverse hyperbolic tangent function.

Definition 2 The functional information gain $FIG(a_j, a_b)$ is calculated as follows.

$$FIG(a_j, a_b) = FE(a_j) - FCE(a_j, a_b) \tag{3}$$

where $FIG(a_j, a_b)$ is the functional information gain, $FE(a_j)$ is functional entropy of the attribute a_j, $FCE(a_j, a_b)$ is the functional conditional entropy of a_j and a_b.

$$FCE(a_j, a_b) = \sum_{j=1}^{u(a_j)} prob(a_j = j, a_b = j) cf\left(prob(a_j = j, a_b = j)\right) \tag{4}$$

$FCE(a_j, a_b)$ is the functional conditional entropy of a_j and a_b, *cf* is the functional tangent conditional entropy.

$$cf\left(prob(a_j = j, a_b = j)\right) = \frac{1}{2}\left[\begin{array}{c} \log\left[prob(a_j = j, a_b = j)\right] \\ -1 \\ + \dfrac{}{a \tanh\left(prob(a_j = j, a_b = j)\right)} \end{array}\right] \tag{5}$$

Cf is the functional tangent conditional entropy, log(•) is the logarithmic function and a tanh(•) is the inverse hyperbolic tangent function.

3.2 Construction of FTDT

The major step within the projected system is that the construction of FTDT. The FTDT takes the input as selected options and produces the output as a decision tree. For choosing the options, tangent entropy is employed. These forms of tangent entropy use the inverse hyperbolic tangent perform. The global disorder of a dataset is measured by the useful tangent entropy. It conjointly finds the info variable uncertainty value of the dataset. The knowledge distribution and therefore the sequent chance of the info objects area unit wont to realize the inverse hyperbolic tangent perform. To seek out the importance of attributes entropy measures provides the hidden correlation which can mix each knowledge distribution moreover as sequent chance. Finally, the decision tree construction has four necessary steps like (i) choice of attributes, (ii) rending rule, (iii) Calculation of serious live for the leaf node.

3.3 Selection of Attributes

The construction of FTD Tree structure, leaf node represents the category worth, and also the interior node represents the split worth with the attributes. The decision tree construction starts with the foundation node. The root node acts as the prime of the tree. After the creation of root node, there's a desire to form leaf nodes. The root node more splits into the leaf node. During this stage, the primary task is to search out the simplest attribute to be crammed among the inside node. And also the second task is to search out the proper split worth. A new necessary task is performed to develop the FTD Tree is finding the foremost suited parent interior node. Usually within the decision tree, the parent interior node is acknowledged supported the Gini values in the Hunt's algorithmic rule. In the decision, tree entropy performs, criteria choice utilizes the entropy perform.

3.4 Splitting Rule

In the decision tree construction, when the choice of interior nodes for the parent node, it is necessary to seek out the simplest split value for the decision tree. The rending value is sorting out by using the worth of functional information gain that, have the conditional useful tangent entropy. To seek out the split node price of explicit attributes, first, calculate the distinctive value for all the attributes and so assumes to separate the node value. After that, the functional information gain is calculated for the assumed split values. At last, the simplest split node relying upon the functional information gain is obtaining.

3.5 Block Diagram

See Figs. 2, 3 and 4.

4 Results and Discussion

Experimental results are carried out on Alzheimer's dataset [6].

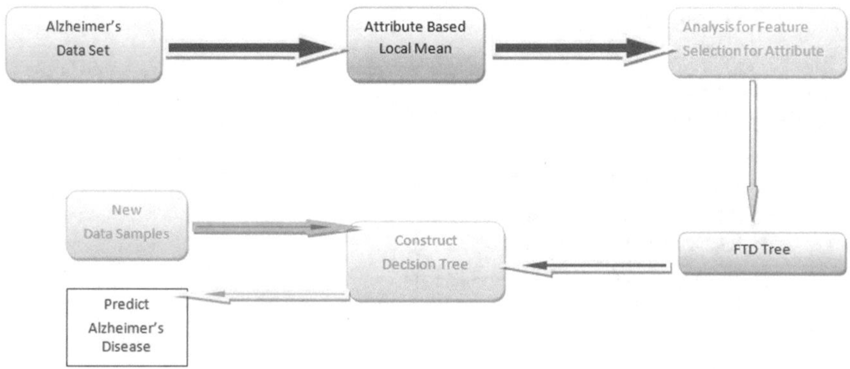

Fig. 2 Block diagram of FTD tree

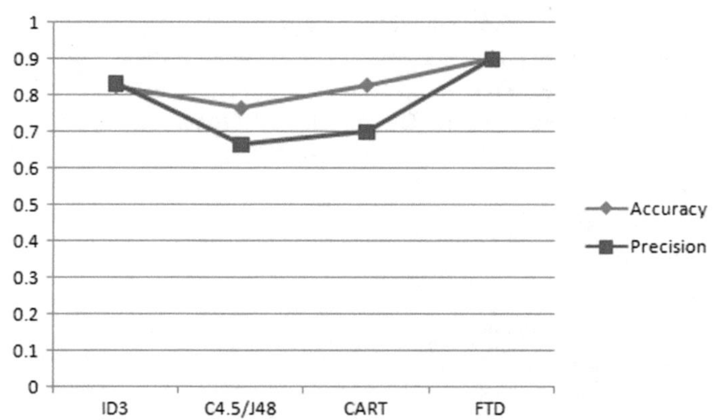

Fig. 3 Comparison of models in terms of accuracy and precision

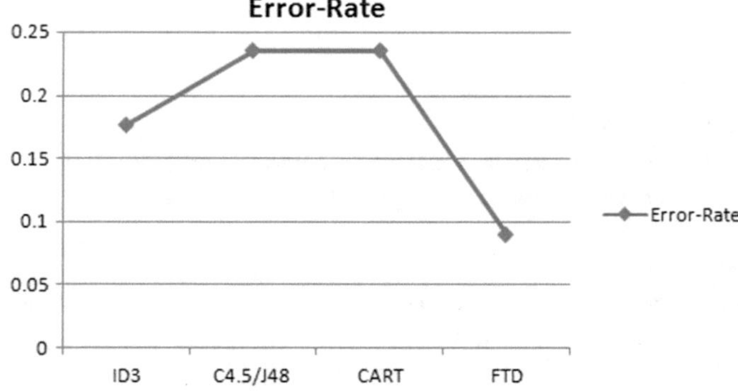

Fig. 4 Comparison in terms of error rate

4.1 FTD Prune Tree

```
Ab_42 <= 11.286488
|  AXL <= -0.4: Control
|  AXL > -0.4
|  |  Kidney_Injury_Molecule_1_KIM_1 <= -1.213217
|  |  |  MIP_1alpha <= 4.968453: Control
|  |  |  MIP_1alpha > 4.968453: Impaired
|  |  Kidney_Injury_Molecule_1_KIM_1 > -1.213217
|  |  |  EGF_R <= -1.087383: Control
|  |  |  EGF_R > 1.087383
|  |  |  |  E2 <= 0
|  |  |  |  |  IL_16 <= 2.420557: Impaired
|  |  |  |  |  IL_16 > 2.420557
|  |  |  |  |  |  Pancreatic_polypeptide <= -0.198451: Control
|  |  |  |  |  |  Pancreatic_polypeptide > -0.198451
|  |  |  |  |  |  |  ENA_78 <= -1.346117
|  |  |  |  |  |  |  |  Tamm_Horsfall_Protein_THP <= -3.130927: Control
|  |  |  |  |  |  |  |  Tamm_Horsfall_Protein_THP > -3.130927: Impaired
|  |  |  |  |  |  |  ENA_78 > -1.346117: Control
|  |  |  |  E2 > 0: Impaired
Ab_42 > 11.286488
|  PAI_1 <= 0.538879
|  |  Lipoprotein_a <= -2.733368
|  |  |  NT_proBNP <= 4.919981
|  |  |  |  Leptin <= -1.877005
|  |  |  |  |  Glutathione_S_Transferase_alpha <= 1.049512: Control
|  |  |  |  |  Glutathione_S_Transferase_alpha > 1.049512: Impaired
|  |  |  |  Leptin > -1.877005: Control
|  |  |  NT_proBNP > 4.919981
|  |  |  |  E4 <= 0
|  |  |  |  |  MIP_1beta <= 2.833213: Control
|  |  |  |  |  MIP_1beta > 2.833213
|  |  |  |  |  |  B_Lymphocyte_Chemoattractant_BL <= 2.296982: Control
|  |  |  |  |  |  B_Lymphocyte_Chemoattractant_BL > 2.296982: Impaired
|  |  |  |  E4 > 0: Impaired
|  |  Lipoprotein_a > -2.733368
|  |  |  TTR_prealbumin <= 2.833213: Impaired
|  |  |  TTR_prealbumin > 2.833213: Control
|  PAI_1 > 0.538879
|  |  E2 <= 0
|  |  |  PAPP_A <= -2.902226: Control
|  |  |  PAPP_A > -2.902226: Impaired
|  |  E2 > 0: Control
```

4.2 *Performance Measure Alzheimer's Disease*

Models	ID3	C4.5/J48	CART	FTD
Accuracy	82.35	76.47	82.74	90.1
Precision	0.833	0.667	0.7	0.9
Recall	0.7142	0.857	1	0.89
Fmeasure	0.769	0.75	0.824	0.9
Error rate	0.1765	0.2353	0.2353	0.09

5 Conclusion

The uncertain data is used to predict the Alzheimer's Disease from the Alzheimer's dataset. In reduct and core analysis are used for features selection. After the identification of important features, the classification is performed. The FTD Tree introduced two new subroutine called functional tangent entropy and splitting rules. Once the FTD Tree is constructed for predicting the disease, the physicochemical parameters are analyzed. For scientific research, the 133 parameters are taken from Disease Data. The performance of the overall proposed system is examined with the existing classifier algorithm with a help of the parameters, such as Precision, Recall and accuracy. The FTD Tree system reaches the maximum classification accuracy of 95% as compared with the existent work. To obtain optimization algorithm for predicting the Alzheimer's Disease in future the best split point can be found out. The best split point for the FTD Tree algorithm may improve the prediction quality.

References

1. Quinlan, J.R.: Induction on decision tree. Mach. Learn. **1**, 81–106 (1986)
2. Soman, K.P., Diwakar, S., Ajay, V.: Data Mining Theory and Practice. Chia Machine Press (2009)
3. Kadaru, B.B., Narni, S.C., Raja Srinivasa Reddy, B.: Classifying Parkinson's disease using mixed weighted mean. IJAER **12**(1) (2017)
4. Meenakshi, S., Venkatachalam, V.: FUDT: a fuzzy uncertain decision tree algorithm for classification of uncertain data. Arab. J. Sci. Eng. **40**, 3187–3196 (2015)
5. Tsang, S., Kao, Ben, Yip, K.Y., Ho, W.-S., Lee, S.D.: Decision trees for uncertain data. IEEE Trans. Knowl. Data Eng. **23**(1), 64–78 (2009)
6. http://course1.winona.edu/bdeppa/Stat%20425/Data/Alz%20Train%20(Final).csv
7. Raja Srinivasa Reddy, B., Kadaru, B.B.: An integrated hybrid feature selection based ensemble learning model for Parkinson and Alzheimer's disease prediction. IJAER **12**(22) (2017)

Data Mining in Frequent Pattern Matching Using Improved Apriori Algorithm

Bhukya Krishna and Geetanjali Amarawat

Abstract The progressions in the field of database innovation have made it conceivable to store a colossal measure of data. Data mining plans have been extremely utilized for deleting non-paltry data facts from such large volume of data. It is helpful in numerous applications like key basic leadership, monetary conjecture, and medicinal conclusion and so forth can be connected either as a elucidating or as a prescient device. Affiliation Data mining manage mining is one of the functionalities of data mining. This kind proposes a link of plans for sending forward affiliation mining, affiliation administer managing. The way toward creating affiliation rules includes the errand of finding the set of all the successive thing sets and creating promising standards. This proposal proposes a procedure for determining the continuous thing sets with a solitary output of the exchanges database in the circle. Amid this single database examine data about thing sets and their events are caught in a table kept in the principle memory. While determining the regular thing sets, this table is examined rather than the plate. Sometimes, the quantity of incessant thing sets to be produced is vast. All these thing sets are hard to deal with and oversee. This matter can be examination by data mining Maximal Frequent Sets as of individual. A MFS of length m suggests the nearness of 2 m^{-2} visit thing sets. Subsequently, all the continuous thing sets can be promptly surmised from the MFS. Along these lines, the era of MFS lessens the time taken for acquiring all the continuous thing sets. Another quicker procedure is proposed in this proposition for mining the MFS, which decreases the quantity of database sweeps to the most extreme of two and furthermore keeps away from the era of competitors. Despite the fact that data mining has a ton of benefits, it has a couple of bad marks. Moreover, Delicate data contained in the database might be brought out by the data mining devices. One strategy accomplishes the decrease of time many-sided quality by altering the structure of the Frequent Pattern Growth Tree. A model is proposed in this theory for finding substantial standards from data wellsprings of various sizes where the substantial

B. Krishna (✉) · G. Amarawat (✉)
CSE Department, Madhav University, Abu Road, Sirohi, Rajasthan, India
e-mail: krish685@gmail.com

G. Amarawat
e-mail: gits.princess@gmail.com

A. Abraham et al. (eds.), *Emerging Technologies in Data Mining and Information Security*, Advances in Intelligent Systems and Computing 813,
https://doi.org/10.1007/978-981-13-1498-8_61

tenets are high-weighted guidelines. These guidelines can be acquired from the high recurrence rules created from each of the neighborhood data sources. Bolster Equalization is another strategy proposed which concentrates on disposing of the low recurrence rules at the neighborhood locales themselves, in this way lessening the standards by a noteworthy sum.

Keywords Data mining · Maximal frequent sets · Frequent pattern growth tree Neighborhood locales

1 Introduction

Now the world treats to have wealth of data or information, preserved all over the planet but it is useful to be gaining that data. It has been initiated that the volume of data doubles probably every nineteen months. This is actually true from the use of PCs and electronic database intigrity sets. The volume of data easily overcomes what a human can comprehend and some tools are needed to comprehend as possible as much to gain the useful data. Out of the several available tools and techniques the domain of data mining become one of the popular and hotly utilized. The role of data mining is simple and has been described as "extracting knowledge from large amounts of data". Association rule mining is one of the dominating data mining technologies. Association rule is a process for identifying possible relations in between data items in massive datasets. It permits regular patterns and correlations, among patterns to be detected with small human risk, bringing most common information to the surface for use. Association rule has been concluded to be a resultant technique for retrieving useful information from massive datasets. Different algorithms were implemented many of which have been embeded in several application areas that permits telecommunication field. The success of applying the retrieved rules to facing real world complexities are very frequent stopped by the quality of the rules. Calculating the quality of association rules is tough and present modes appear to be unmatching , especially when multi-level rules are participated. To improvement of information innovation has begin gigantic measure of database and enormous data in different fields. The examinations fields in database and information innovation have offered ascend to a way to deal with store and control this valuable data for encourage basic leadership. The Data Mining is a process obtained for retrival and important study from large quantity of data. As this process its simple to divide not discovered patterns. At the final day, it is same as called information projection processes.

2 Apriori Algorithm

Apriori algorithm is an algorithm proposed by Agrawal et al for mining continuous itemsets and is appeared in Fig. 1. The name of the algorithm depends on the way that the algorithm utilizes earlier information of regular itemset properties. Apriorialgorithm utilizes an iterative approach known as level-wise inquiry, where

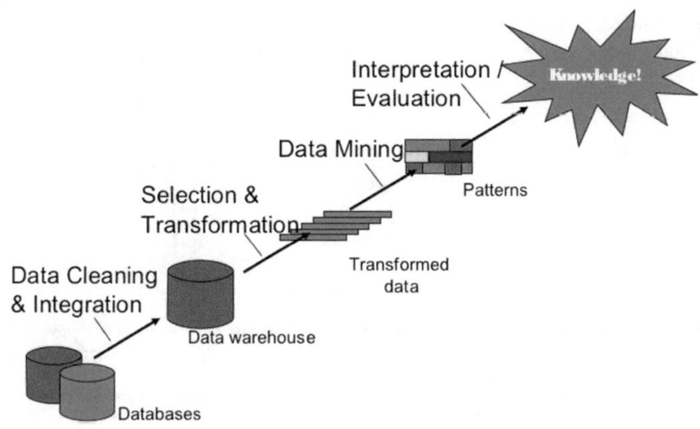

Fig. 1 Data mining process (KDD)

k-itemsets are utilized to get (k+1)—thing sets. At first an arrangement of regular 1-Itemset is found by examining the database for the events of number of everything and gathering those things that fulfills the base help check.

```
procedure APRIORI_FREQUENTITEMSETS(min_sup, S)
    L1 ← itemsets
    for k = 2; L(k−1) ≠ ∅; k + + do
        Ck = aprioriGen(L(k−1)) ▷ Create the candidates
        for each c ∈ Ck do
            c.count ← 0
        end for
        for each I ∈ S do
            Cr ← subset(Ck, I) ▷ Identify candidates that
belong to I
            for each c ∈ Cr do
                c.count + +▷ Counting the support values
            end for
        end for
        if c.count ≥ min_sup then
            Lk = Lk ∪ c
        end if
    end for
    return Lk
end procedure
```

The subsequent set L1 is utilized to discover L2, the arrangement of incessant 2-Itemsets, which is utilized to discover L3, et cetera, until not any more continuous k-itemsets can be found. Each Lk requires one full sweep of the database to discover thing sets and to enhance the productivity of the level-wise era of successive itemsets, an imperative property called the Apriori property utilized for decreasing the inquiry

Fig. 2 Transaction Database for Apriori Algorithm

TID	List of items
T1	I1, I2, I4
T2	I2, I4
T3	I3, I4
T4	I1, I2, I5
T5	I1, I4
T6	I1, I3, I5
T7	I1, I2, I3, I5

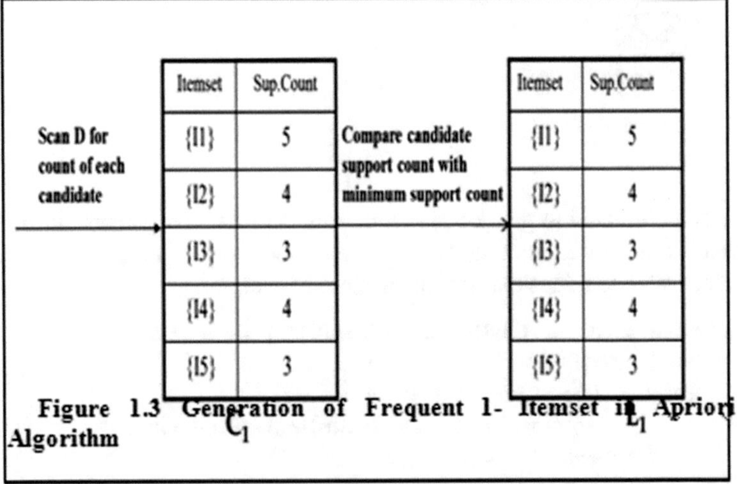

Scan D for count of each candidate
Itemset
Sup.Count
{I1} 5
{I2} 4
{I3} 3
{I4} 4
{I5} 3
Compare candidate support count with minimum support count
Itemset
Sup.Count
{I1} 5
{I2} 4
{I3} 3
{I4} 4
{I5} 3
Figure 1.3 Generation of Frequent 1- Itemset in Apriori Algorithm
C1
L1

Fig. 3 Generation of frequent 1 Itemset in Apriori Algorithm

space. Apriori property: All non-discharge subsets of an incessant itemset should likewise be visit.

There are two possibilities to recognize the most frequent itemsets:

Join step: To identify L_k, a group of candidate k-itemsets C_k, is produced by combining L_{k-1} with itsown.

Prune step: A parsing of the database to decide they include of each competitor Ck carries about the believe of Lk and all support having a tally the base support tally are approach and in this path have a location with Lk. To reduce the metrics of Ck, the apriori state is utilized as takes before that, Any (K − 1) Items kne K-Item group (Fig. 2).

The transaction T1 contains the itemsets of I1, I2, I3 and transaction T2 contains the itemsets of I2, I4. Along these lines all the seven transactions have its own particular itemsets. Apriori algorithm is utilized for the era of frequent 1-Itemset from the databases D and is appeared in the Fig. 3. In the main cycle of the algorithm, the whole accessible thing in the database is an individual from the applicant 1-itemsets, C1.

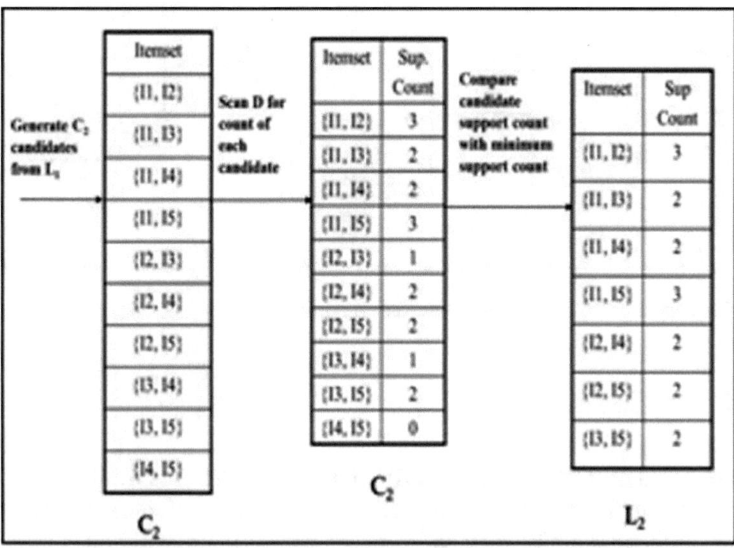

Generate C_2 candidates from L_1
Itemset
{I1, I2}
{I1, I3}
{I1, I4}
{I1, I5}
{I2, I3}
{I2, I4}
{I2, I5}
{I3, I4}
{I3, I5}
{I4, I5}
C_2
Scan D for count of each candidate
Itemset
Sup. Count
{I1, I2} 3
{I1, I3} 2
{I1, I4} 2
{I1, I5} 3
{I2, I3} 1
{I2, I4} 2
{I2, I5} 2
{I3, I4} 1
{I3, I5} 2
{I4, I5} 0
C_2
Compare candidate support count with minimum support count
Itemset
Sup Count
{I1, I2} 3
{I1, I3} 2
{I1, I4} 2
{I1, I5} 3
{I2, I4} 2
{I2, I5} 2
{I3, I5} 2
L_2

Fig. 4 Generation of frequent 2 Itemset in Apriori Algorithm

The algorithm checks the quantity of events of everything by filtering every one of the transactions. The base help check is 2, Frequent 1-Itemsets, L1, can be dictated by the greater part of the hopefuls in C1 that fulfills least help.

The applicant 2 itemsets are created by joining L1 with L1, and amid pruning, no hopefuls are expelled from C2 in light of the fact that every subset of the competitors set is likewise visit. Database is checked for producing the help include of every applicant itemset C2. The arrangement of Frequent 2-Itemsets, L2, is then decided, comprising of those hopeful 2-Itemsets in C2 having least help.

Visit 3 itemsets from database D is produced utilizing Apriori algorithm and is appeared. The applicant 3 itemsets are created by joining L2 with L2 and apply pruning operation on that competitor 3 itemsets. The transactions in D are examined to decide L3, comprising of those candidate3-itemsets in C3 having least support. L3 is having 2 sets of 3 itemsets which fulfills the base support number are {I1, I2, I5} and {I1, I3, I5}. Visit 4 Itemsets from database D is created by joining L3 with L3 to acquire hopeful 4 itemsets and brings about {I1, I2, I3, I5} and it is pruned in light of the fact that its subset is not visit. Subsequently C4 = ¢ and algorithm ends. Let I = {i1, i2, … im} be the arrangement of things. Let, the errand pertinent data D, be an arrangement of database transactions where every transaction T is an arrangement of things with the end goal that T and I. An affiliation run is a ramifications of the frame. A->B where A, I, AΩB = null. The run A->B holds in the transaction set D with support s, where s is the rate of transactions in D that contain AUB. This is taken to be the likelihood, P (AUB). The control A=>B has confidence c in the transaction set D, where c is the rate of transactions accessible in D, containing A that likewise contain B. This is taken to be the contingent likelihood, P (A|B).

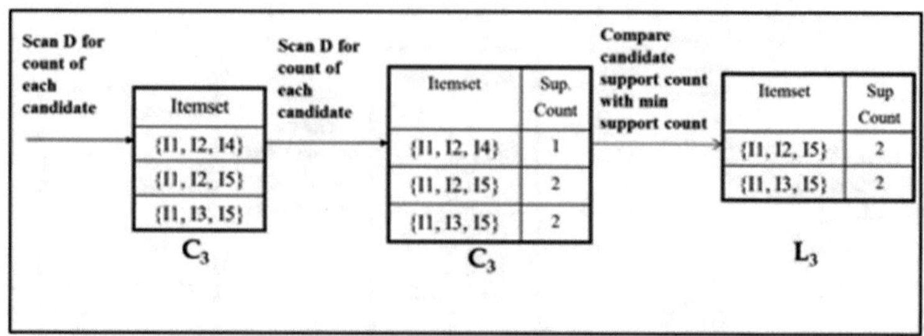

Fig. 5 Generation of frequent 3 Itemset in Apriori Algorithm

2.1 Formulas

(A=>B) = P (AUB) Confidence (A=>B) = P (B | A) Tenets that fulfill both the base support edge (min sup) and the base confidence edge (min conf) are called solid principles. The solid affiliation govern created for Table 1 transaction is

$$|2^{\wedge}|5 => |1|3 \wedge |5 => |1$$

$$|1^{\wedge}|3 => |5$$

The above affiliation decide determines that, in a sales domain, the client who is getting I2 and I5 will likewise get I1, the client who is purchasing I3 and I5 will likewise purchase I1 and the client who is buying I1 and I3 will likewise buy I5.

3 Literature Survey

The critical pattern in information innovation is to distinguish the significant information from the colossal measure of data put away in documents, databases, and different vaults and to grow effective means for analysis and translation of such

Table 1 Transaction database for apriori algorithm

TID	List of items
T1	I1, I2, I4
T2	I2, I4
T3	I3, I4
T4	I1, I2, I5
T5	I1, I4
T6	I1, I3, I5
T7	I1, I2, I3, I5

data for the extraction of fascinating knowledge that could help in basic leadership. The two principle purposes behind utilizing data mining are, having little information with an excessive amount of data and another is to extricate helpful information from gigantic measure of data. Data mining helps for finding significant new connection, designs and conceivably valuable precious stones of knowledge from huge measure of data utilizing statistical and numerical techniques. The idea of data mining has developed since numerous years, however the term data mining have been presented in 1990. Data mining have experienced a few research and formative stages for a long time, and due to three distinct families, it has achieved its frame and those families are measurements, manmade brainpower (AI) and machine learning. Data mining is based on insights which are the establishment of most advances, e.g., standard deviation, standard fluctuation, relapse analysis, standard circulation, bunches analysis and so forth. Manmade brainpower is additionally the base for data mining, which tries to reenact human point of view or human knowledge in statistical issues. Another center region for data mining is machine learning, and it is a mix of insights and AI. Data mining is fundamentally the adjustment of machine learning techniques to business applications. Data mining is the accumulation of verifiable and late.

3.1 Frequent Patterns Mining Evaluation

Frequent examples are thing sets, sub-successions, or sub-structures that show up in a data set that fulfills the client indicated least support check with recurrence at the very least a client determined edge. Agrawal et al. [1] proposed a productive algorithm that creates all the association administers between the things in the customer transaction database and this algorithm consolidates novel estimation, support administration and pruning techniques. This algorithm is connected on sales data of an extensive retailing organization and viability of the algorithm has been demonstrated. Agrawal et al. [1] proposed two algorithms Apriori and AprioriTid for tackling the issue of mining frequent examples, and afterward joined both of these algorithms. The Apriori and AprioriTid algorithms produce the hopeful thing sets to be numbered in a go by utilizing just the itemsets discovered huge in the past pass not considering the transactions in the database. This system brings about era of a substantially more modest number of competitor itemsets. The AprioriTid algorithm has the property that the database is not utilized at all to count the support of applicant itemsets after the primary pass. Srikant et al. presents the mining of summed up association rules from an expansive database of transactions [2], which indicates the association between the data things accessible in the transaction. Jiawei Han et al. [3] portray mining association rules at numerous idea levels bringing about the discovery of more particular and solid knowledge from data. By upgrade of existing association manage mining strategies; a top-down dynamic extending technique is displayed for mining different level association rules from huge transaction databases. Ming-syanchen et al. presents a similar study and classification of the accessible data mining techniques. Improvement of effective techniques for mining various sorts of knowledge at numerous deliberation levels are accessible. Han created DBMiner a social data mining framework at SimonFraser University, for mining numerous sorts of principles at various reflection levels, similar to trademark rules, discriminate rules, association rules, classification rules and so forth. Show-Jane Yen et al. presents an effective chart based algorithm to mine the frequent data things in the database [4] and this algorithm develops association diagram to demonstrate the associations amongst things and after that navigate the diagram to produce expansive thing sets and arrangements. This algorithms need to examine the database just once there by enables diminishing output to time and enhancing the execution. Srikant et al. [5] presented another algorithm GSP that finds the summed up successive examples. GSP scales straightly with the quantity of data arrangements, and has great execution with the normal data grouping size. Observational assessment utilizing engineered and genuine data demonstrates that GSP is significantly quicker than the Apriori algorithm.

3.2 Improved Technique for Retrieving Frequent Patterns

Jiawei Han et al. proposed FP-development, for mining the total arrangement of frequent examples, the productivity of mining is accomplished with three techniques: a substantial database is compacted into an exceptionally dense significantly littler data structure, which maintains a strategic distance from expensive, rehashed database checks, FP-tree-based mining [6] receives an example section development strategy to keep away from the exorbitant era of countless sets. Cheng-Yue Chang et al. and other authors [7, 8] investigate another algorithm called Segmented Progressive Filter which portion the database into sub-databases such that things in each sub-database will have either the basic beginning time or the basic closure time. For each sub-database, SPF dynamically channels hopeful 2-itemsets with combined sifting edges either forward or in reverse in time. This component permits SPF of embracing the output decrease strategy by producing all applicant k-itemsets (k>2) from competitor 2-itemsets specifically. William Cheung et al. [9] elites CATs tree broadened the possibility of FP-Tree to enhance the capacity pressure and frequent example mining without creating the hopeful itemsets and permit single ignore the database and inclusion and erasure of transactions in the tree should be possible productively. Chung-Ching Yu et al., proposed two proficient algorithms to find consecutive examples from d-dimensional arrangement data. To begin with algorithm Apriori algorithm is a refreshed Apriori algorithm to mine successive examples from a multidimensional database. It utilizes tree structures called competitor tree like hash tree. Second algorithm is prefix traverse algorithm which is the refreshed adaptation of prefix span algorithm. The established Apriori and AprioriTid algorithm, are utilized to build the frequent itemset, the primary burden is that it devours more opportunity for checking the database. Keeping in mind the end goal to evade this, Zhi-Chao Li et al. creates a profoundly proficient AprioriTid algorithm [10], for diminishing the checking time. Jaw Chuan et al., propose a single pass algorithm called incspam [11] to mine the arrangement of consecutive examples over the flood of itemset succession. It gives a transaction-delicate sliding window model in a data stream condition. It gets the transaction from the data stream and uses the idea of Bit-Vector, Customer Bit- vector exhibit with sliding window, to store the information of things for every customer. A weighted capacity is adjusted in this sliding window model. It lessens the time and memory expected to slide the window Lei Ji et al. [12] enhances the Apriori algorithm, by utilizing the division, administration system and presentation of Hash innovation, the examining time of the database is diminished.

4 Methodology

The present algorithm is dependent to identify the most occurent items on the basis partitioning of database in n possibilities. It overcomes the memory problem for large database which do nofit into main memory because small parts of database easily fit into main memory. This algorithm divides into two passes,

1. As per the first pass database is categorized into n number of states.
2. Each divided database is stored into MM (main memory) step by step and local occurrence of data items are found.
3. Combine the all locally frequent data items and make it to the whole world candidate set.
4. Find the whole world often data items from this candidate group (Tables 2 and 3).

Minimum support for a particular partition is the multiplication of minsup and the total number of transactions in that partition. Above algorithm builds a special data structure called TIDLISTs. For every itemset a tid list is generated. A tidlist for an itemset contains the TIDs of all the transactions that contain that itemset in the

Table 2 Tid lists for 1-itemsets

Item1	Count	Tidlist
1	2	100, 300
2	3	200, 300, 400
3	3	100, 200, 300
4	1	100
5	3	200, 300, 400

Table 3 Tid lists for 2-itemsets

Item1	Count	Tidlist
1	2	300
1	3	100, 300
1	5	300
2	3	200, 300
2	5	200, 300, 400

partition. The TIDs in the tidlist are maintained in the sorted order. Tidlists are used for counting the support for the candidate items sets. Cardinality of the tidlists of an itemset divided by the total number of transactions in a partition gives the support for that itemset in that partition. Initially the tidlist for the entire partition is generated. From here we find out tidlists corresponding to the 1-itemsets. Higher level tidlists are generated by the intersection of tidlists.

The algorithm is formulated below:

```
P = partition database (D) n
= Number of partitions
// Phase I

For i = 1 to nloop

    Read-in partition (pi ∈ P)
    Li = gen_large_itemsets (pi) End
loop;
// generating global candidate itemsets
    For (i = 2; L^j i≠ φ, j = 1, 2.....n: i++)loop Ci^G
    = ∪j = 1, 2,...nLi^j
End loop;

// Phase II

    For i = 1 to n loop
    Read-in partition (pi ∈ P)
        For all candidates c ∈CG generate count (c, pi) End loop;
LG = {c ∈ CG | c.count>minsup} Return LG;
```

4.1 Assure Sampling Algorithm

The sample can be a partition of the database. In that case the partition is treated just like a random sample chosen. The algorithm picks up a random sample form the database and then finds out frequent itemsets in the sample using support that is less than the user specified minimum support for the database. These frequent itemsets are denoted by S. Then the algorithm finds out the negative border of these itemsets denoted by NBd (S). The negative border is the set of itemsets those are candidate itemsets but did not satisfy minimum support. Simply NBd (Fk) = Ck − Fk. After that for each data itemset X in S U NBd (S) it examines whether X is often item group in whole database by parsing the database. If NBd (S) having no often items in a set then the most often items are detected. If NBd (S) having most often itemsets then the algorithm procedure builds a group of candidate itemsets C_G by elaborating the no positive border of S U NBd (S) until the nonpositive border is not occupied. In the best case when all the frequent itemsets are found in the sample this algorithm requires only one scan over the database. In the worst case it requires two scans over the database.

The sampling algorithm is depicted below:

s = Draw_random_sample (*D*);

//Generate frequent itemsets for the sample drawn. S = generate_frequent_itemsets (*s*, *low support*);

//Counting support for the itemsets and their negative border generated in the sample, in the database D.F={*X* ∈ *S U NBd (S)* | *X.count>=minsup*};

References

1. Agrawal, R.: Fast algorithms for mining association rules in large databases. In: Proceedings of the 20th International Conference on Very Large Data Bases, VLDB, pp. 487–499, Santiago, Chile, Sept 1994
2. Srikant, R., Agrawal, R.: Mining generalized. In: Proceedings of the 21st International Conference on Very Large Databases, Zurich, Switzerland, pp.407–419, (1995)
3. Han, J., Pei, J., Yin, Y.: Mining frequent patterns without candidate generation. ACM. SIGMOD. Rec. **29**(2), 1-12.U (2000)
4. Yen, S.J., Chen A.L.: A graph-based approach for discovering various types of association rules. IEEE. Trans. Knowl. Data Eng. **13**, 839–845 (2001)
5. Srikant, R., Agrawal, R.: Apriori-based method (GSP: Generalized Sequential Patterns). Advances in database technology EDBT '96 (1996)
6. Han, J., Pei J., Yin Y., Mao R.: Mining frequent patterns without candidate generation: A frequentpattern tree approach. Data mining and knowledge discovery, Kluwer Academic Publishers, Manufactured in The Netherlands (2001)
7. Chang, C.Y., Chen, M.S., Lee, C.H.: Mining general temporal association rules for items with different exhibition periods. In: Proceedings of the 2002 IEEE International Conference on Data Mining (ICDM'02), pp. 59–66, Maebashi City, Japan (2002)
8. Junheng-Huang WW.: Efficient algorithm for mining temporal association rule. Int. J. Comp. Sci. Netw. Secur. IJCSNS **7**(4) 268–271 (2007)
9. Cheung, W., Zaiane O.R.: Incremental miningof frequent patterns without candidate generation or support constraint.In: Proceedingsof IEEEInternationalConferenceon Database Engineering and Applications Symposium, pp. 111–116 (2003)
10. Li, Z.C., He, P.L., Lei, M.: A high efficient aprioritid algorithm for mining association rule. In: International Conference on Machine Learning and Cybernetics, vol. 3, pp. 1812–1815 (2005)
11. Ho, C.C., Li, H.F., Kuo, F.F., Lee, S.Y.: Incremental mining of sequential patterns over a stream sliding window. In: Proceedings of IEEE International Conference on Data Mining, pp. 677–681 (2006)
12. Ji, L., Zhang, B., Li, j.: A new improvement on apriori algorithm. In: Proceedings of IEEE International Conference on Computational Intelligence, vol. 1, pp. 840–844 (2006)

A Model for Predicting Outfit Sales: Using Data Mining Methods

Mohammad Aman Ullah

Abstract The objective of this study is the development of a model that will use data mining techniques to predict the sum of outfit sales. Data mining is the method of extracting appropriate information from the dataset and converts them into a valuable one. In this research, I have adopted the data mining procedures such as preprocessing (e.g., Data cleaning, missing values replacement and reduction of Data, etc.) on the data of outfit sales, which enhance the performance up to 20%. I have then train the different classifiers, for example, Naive Bayes, Multilayer Perceptron, and K-star. Finally, assessed the performance of every classifier via tenfold cross validation and compared the outcomes, where, Multilayer perception classifier showed the highest performance with 83.8% accuracy and Nave Bayes showed the lowest performance with 71.6% accuracy. This research also showed the standard level of improvement in the performance (up to 9%), as compared to the other predicted models of outfit sales. The total research was implemented in a Data mining software tool, WEKA.

Keywords Data mining · Preprocessing · Classifier · Classification · Weka

1 Introduction

Data mining (DM) is a method of using statistical, machine learning, and knowledge-based approaches to analyze the vast data sets to detect patterns, relationships, and associations among its features. Detecting and predicting pattern is the classification problem. Classification in Data mining is the generation of a model with some predicting attributes to predict the class to which each observation of the data set should be placed [1]. Classification methods may be unsupervised or supervised. Supervised classification considers one attribute with predetermined values from the dataset that represent classes. In unsupervised classification, the aim is to groups or clusters the data instances of the dataset based on various logical associations that

M. A. Ullah (✉)
International Islamic University Chittagong, Chittagong 4203, Bangladesh
e-mail: ullah047@yahoo.com

A. Abraham et al. (eds.), *Emerging Technologies in Data Mining and Information Security*, Advances in Intelligent Systems and Computing 813,
https://doi.org/10.1007/978-981-13-1498-8_62

holds between them but that must yet be detected [2]. So, predicting of data pattern such as outfit sales could be done through a classification technique.

There are many algorithms in data mining to help garment manufacturing industries in their better decision making. But, an outfit sales prediction system can tell them what types of outfit should be produced for selling, that helps in reducing the time, reduce costs, ensuring the best-selling capability and make the decisions making easy. Correctly Predicted performance also answers many other questions:

1. What are the feedbacks of the customer for the particular outfit?
2. What are the factors that influence sales?
3. What will be the amount of production of particular outfits?
4. When to produce some special outfits?
5. What aged people's outfits should be mostly produced?

In this paper, I have proposed a prediction system in order to address above mentioned issues. This system considers sales as a predictive attribute. In doing so, I have collected data from the UCI machine Learning Repository [3]. This data set contains various attributes on outfit sales. Each attribute could be used to predict different issues of outfit sales. I have then preprocessed the data set. Finally, applied the proposed approach, which includes application of binary classification and multiclass classification. Total experiment was conducted and tested in WEKA- Data mining Software.

The remainder of the paper is organized as follows: in the next section I have provided a brief overview of the related works. In Sect. 3, I have discussed about the methodology of the proposed system. Experimental results are shown in Sect. 4. Finally, I have provided the conclusion and future work of this research in Sect. 6.

2 Literature Review

Data mining and statistical tools are broadly used for sales data analytics and to uncover the significant issues regarding sales. This paper presents a genetic algorithm based methods to classify outfit sales data. In this section, I have summarized some related works relating to classification techniques that are used in predicting domain. The only research work done on the Dataset I have used were by [4], three mining techniques were used to recommend sales of the dresses and compare them on the basis of accuracy, where the authors have applied the attribute selection method to find the better classifier among Decision tree, Naive Bayes, and Multilayer Perceptron in analyzing dress sales. A study on "Product Sales Prediction based on Sentiment Analysis Using Twitter Data" was done by [5], where they have examined the sampling techniques for categorizing among two heads one was Probability Sampling and other was Non-Probability Sampling. But in their study, random sampling was the best method among others to obtain a representative sample of the population.

In order to find the performance of SVM, J48, Nave Bayes, Multilayer Perceptron, decision table and Conjunctive rule in predicting chronic kidney disease, Distributed Classification Algorithms of Data Mining were used in [6, 7]. In [8], five classification algorithms were studied in WEKA to find the best classifier and found J48 as best classifier with an accuracy of 70.59% that takes 0.29 s for training.

In [9], five data mining algorithms such as Nave Bayes, j48, SMO, REF tree, and Random tree were applied to Diabetes Data Set in order to find best classifier on the scale of accuracy, correctly and incorrectly classified instances. They have implemented these algorithms using the WEKA data mining tool and finally recommended best classifier for this sort of analysis. A research on "Comparison of Different Classification Techniques Using WEKA for Hematological Data" analyzed the outcome of Nave Bayes, J48 decision tree, and Multilayer perceptron classifiers based on their accuracy; total time taken in building the model and the lowest average error. As per them, there were huge differences between all features accuracy and lessen features accuracy, with respect to execution time in building the model [10].

3 Proposed Approach

Figure 1 represents the proposed model for outfit sale prediction. In the very first step, I have collected different data sets on outfits, then preprocessed those data sets. In the next step, I have folded those data into two folds, where one fold was used as a training data set and other data were used as a test data set. In order to recommend the outfit sales, I have then classified those data into two classes. On the basis of positive recommendation, the classes are then further classified into multiple classes, through which sales quantity could be predicted.

3.1 Data Source

In this research I have used the dataset collected from UCI [3] which was developed by Muhammad Usman and Adeel Ahmed of Air University, which consists of 500 instances and 13 attributes (i.e.; Neckline, Material, Decoration, Style, Price, Waiseline, Sleeve Length, Rating, Size, Season, Pattern, Type, Class, and Fabric Type). Some of the attributes are categorical and some of them are numerical. The same data set were used by only few researchers. But, the most promising result was found by Jasim [4]. I have analyzed this dataset with the use of Weka software that have a huge numbers of built-in algorithms and visualization tools which could predict and analysis the data.

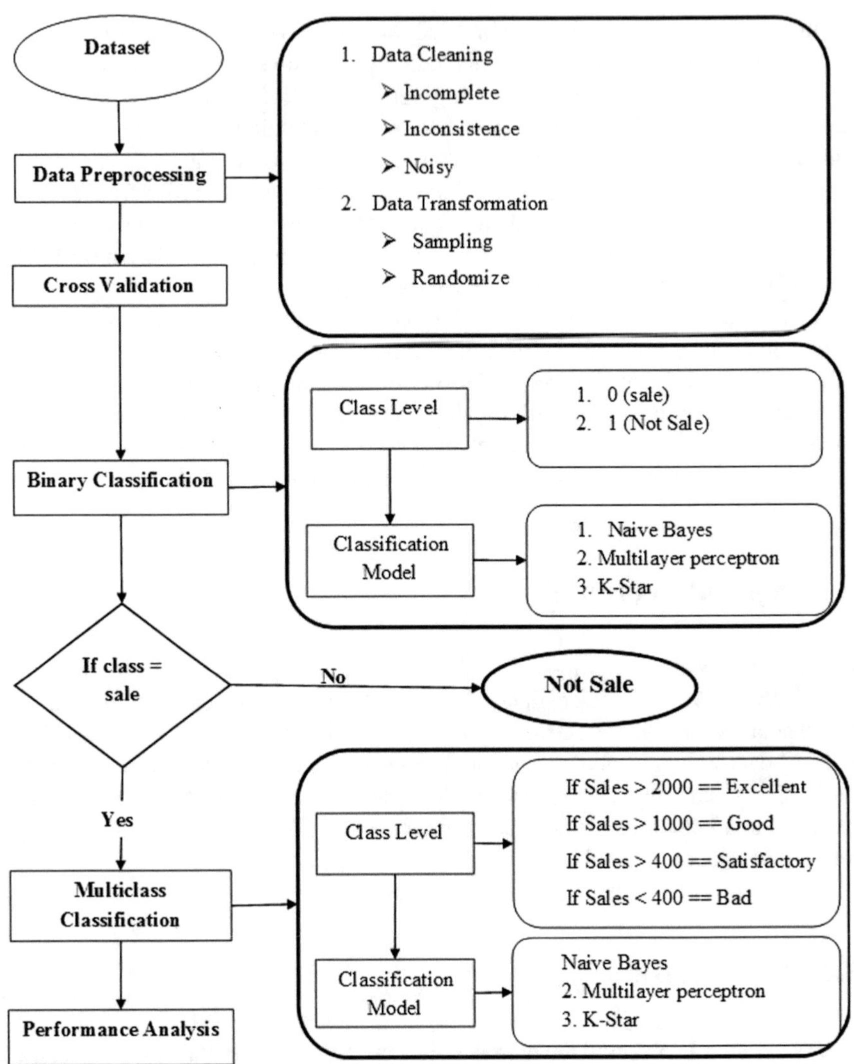

Fig. 1 Proposed model

3.2 Data Preprocessing

The data generally collected from the real world are either noisy or poor in nature. Well-organized data set has huge impact on the result. So, before applying the classification algorithm, it is necessary to preprocess those data. Though in real many types of preprocessing could be done, but I have applied few of them (such as eliminating noisy information, including missing values, eliminating outliers, adjusting unbalanced data, and finally normalizing) in this research.

3.2.1 Data Cleaning

The missing values in the data set was replaced by a constant symbols (?) and all duplicate values of the attributes were replaced by its original value.

3.2.2 Sampling

Sampling is the way of reducing the data volume size of very large data sets. It is obvious fact that, if I use whole data set for my classification purpose then the cost of building and testing those classification model will increase to a large extend. So, using only the representative sample of the data instances can lower this cost. In this work, the Resample filter was applied, that is the reproduction of the random subsample of the data set by replacement of sample.

3.2.3 Randomize

The data set is then randomized, which is the shuffling of the order of data instances. For doing this, I have reset the random number generator with the seed value and when a new set of data instances passed on, it can then generate identical series of non-repeating numbers which can further be used to synchronize external sources of data.

3.3 Classification

In this research, I have applied two kinds of classification such as binary classification and multiclass classification. And used three kinds of classification model for classification.

3.3.1 Binary Classification

The Dataset was divided into two classes, one was used to mean sale and another one for not sale. I have then applied my proposed binary classification system to predict whether the outfits will be sold or not sold. Binary classification were carried out using three classification model such as K-Star, Naive Bayes, and Multilayer perceptron.

Table 1 Description for multiclass classification at class-level

Class-level	Description
Excellent	If the sales>2000
Good	If the sales>1000
Satisfactory	If the sales>400
Bad	If the sales<400

Table 2 Performance of binary class classification algorithms

Algorithm	Accuracy (%)	Precision (%)	Recall (%)	F-score (%)
Naïve Bayes	72	74	81	77
Multilayer perceptron	84	86	88	87
K-Star	81	85	83	84

3.3.2 Multiclass Classification

If the binary classifier predicts that the outfit will be sold, only then the multiclass classifier will become active. It is being applied to forecast the amount of outfits should be produced. Multiclass classifier has four level such as Bad, Satisfactory, Good, and Excellent. This scale is determined on the amount of outfit sales. Table 1 shows different classes of multiclass classifier with description.

Classifier Description: As stated in the previous section, I have used three classifier model for binary classification such as K-Star (A lazy learning classifier for dealing with symbolic and real-valued attributes, and missing values [9]), Naive Bayes (it is one of the fastest statistical classification model which works well on the probability of the attributes included in the data sample individually for classifying them accurately [9]), and Multilayer perceptron (with input, output, and hidden layer, it is a feed forward NN model, which allows each node of the layer to be connected to nodes of the adjacent layers [11]).

4 Experiment Result

I have implemented my experiments with the use of data sources and classifier discussed in the previous section. I have measured the performance of the classifier either they are correctly classifying or not, with the use of confusion matrix. The parameter of the matrix used in evaluation of the performance of the classifier were Accuracy, F-score, Precision, and Recall. All the classifier were implemented in a data mining software WEKA. Table 2 shows the performance of the classifier on the basis of those papramemters.

4.1 Performance Analysis of the Binary Classifier

Accuracy: Accuracy of the confusion matrix measures how often the classifier is correct [12]. From Table 2, it is observed that the Multilayer perceptron classifier provides strong classification accuracy with 84% comparing to other classifiers used for this analysis. However, Naïve Bayes classifier shows poor classification accuracy with 72%. It is also observed that, K-star classifier shows moderate accuracy with 81.40%.

Precision: Precision of the confusion matrix measures how often the classifiers predicted yes is truely yes [12]. Table 2 shows that, the Precision value is also higher for Multilayer perceptron. Whereas, Naïve Bayes shows poor precision value of 74%, and K-star shows moderate precision value of 85.22%.

Recall: Also known as sensitivity which measures the ratio between true positive and actual yes [12]. It is observed from the Table 2 that, Recall value of Multilayer perceptron classifier is 88%, where as, K-star and Naïve Bayes shows 83% and 81% respectively.

F-Score: The F-score is the weighted average of the precision and recall [12]. From Table 2, it is obvious that the Multilayer perceptron classifier provides strong F-score value with 87% comparing to other classifiers used for this analysis. However, Naïve Bayes classifier shows poor F-score with 77%. It is also observed that, K-star classifier shows moderate F-score value with 84%.

Moreover, in this Experiment I have used tenfold cross validation along with each classifier. That is I have divided the dataset into ten parts, which contains nine training sets and one test set. And then applied the above mentioned algorithms for determining performance. Among all the algorithms, the performance of the Multilayer perceptron was relatively higher in all parameters than the performance of Naïve Bayes and K-Star.

4.2 Performance Analysis of the Multiclass Classifier

In this paper, I have also used multiclass classification algorithm for outfit sales prediction. On the basis of the sales, I have classified the sale data into four classes such as Excellent, Good, Satisfactory and Bad. The multiclass dataset contains 500 outfit sales data. Table 3 shows accuracy of multiclass classification with their rank (measured according to accuracy).

I have measured the Accuracy Performance of Classification Algorithms for multiclass data. From Table 3, it is clear that, K-star algorithm gives more classification accuracy, i.e., 68.6% comparing to other classifiers used. However, the only algorithm perform poorly is Naïve Bayes with classification accuracy of 73.80%.

In all the classifiers, I have experimented Accuracy, Precision, Recall and F-score for each defined class, the result have been defined in the Tables 4, 5, 6 and 7.

Table 3 Performance of multiclass classification algorithms

Algorithm	Accuracy (%)	Rank
K-Star	69	1
Multilayer perceptron	65	2
Naïve Bayes	64	3

Table 4 Result for excellent class

Algorithm	Precision (%)	Recall (%)	F-score (%)
Naive Bayes	14.30	5.60	8
Multilayer perceptron	11.10	5.60	7.40
K-Star	12.50	11.10	11.80

Table 5 Result for bad class

Algorithm	Precision (%)	Recall (%)	F-score (%)
Naive Bayes	73	82	78
Multilayer perceptron	75	82	78
K-Star	80	83	82

Table 6 Result for good class

Algorithm	Precision (%)	Recall (%)	F-score (%)
Naive Bayes	11.10	3	4.80
Multilayer perceptron	19.00	12.10	14.80
K-Star	46.70	21.20	29.20

On the basis of sales amount, K-Star algorithm shows better performance on confusion matrix parameters (Accuracy, Precision, Recall and F-score) for Multiclass (Excellent, Good, Satisfactory, and Bad). However, Naïve Bayes performs very weakly on outfit sales prediction. From both the classification experiments, it was clearly proved that, Naïve Bayes algorithm is not suitable for mining this set of data in predicting outfit sales.

Table 7 Result for satisfactory class

Algorithm	Precision (%)	Recall (%)	F-score (%)
Naive Bayes	31.10	28.00	29.50
Multilayer perceptron	36.00	31.00	33.30
K-Star	41.50	44.00	42.70

Table 8 Relative performance of the proposed model versus state-of-art model

Proposed model		State-of-art model [4]	
Algorithm	Accuracy (%)	Algorithm	Accuracy (%)
Naive Bayes	71.60	Naive Bayes	72.00
Multilayer perceptron	83.80	Multilayer perceptron	76.00
K-Star	81.40	Decision tree	69.13

5 Validation of the Model

Only a research work was done on the dataset I have used by Jasim [4]. They have also conducted the experiment using WEKA. Have conducted some preprocessing. But they have adopted separate approach and algorithms for prediction. Table 8 shows relative performance of the proposed model and state-of-art model. From the table it is observed that, state-of-art work on this dataset used three data mining algorithm for prediction such as Naïve Bayes, Multilayer perceptron, and Decision tree, where as I have used Naïve Bayes, Multilayer perceptron, and K-Star. From the two common algorithm, Naïve Bayes, Multilayer perceptron it is clear that, the proposed model in this research out perform their model by 7%. Also, considering K-star in place of Decision tree proved helpful, as K-star out perform decision tree by 12%. Finally, it could be recommended that, the approach I have adopted is most suitable for predicting sales of outfits that state-of-art work.

6 Conclusions and Future Works

The main purpose of this paper is to forecast or predict outfit sales. It could also predict the total sales amount of the outfits. I have proposed a new model, where I have included two types of classification method such as binary and multiclass. Also, implemented the experiments in three famous classifiers such as Naive Bayes, Multilayer Perceptron, and K-star. The experiments were conducted in a famous mining software tool WEKA. I have then measured the performance of these classifiers using confusion matrix. In this research, among all the algorithms, the performance of the Multilayer perceptron was relatively higher in all parameters than the performance of Naïve Bayes and K-Star. Finally, I have validated the results of the proposed model by comparing the results with the state-of-art model. From the two common algorithm, Naïve Bayes, Multilayer perceptron it is clear that, my proposed model out perform their model by 7%. Also, considering K-star in place of Decision tree proved helpful, as K-star out perform decision tree by 12%. In future, I have aimed to develop an alarming system that could help sales individual in making decisions and finding opportunities. I am also planning for including more attributes in this sort of analysis.

References

1. Padhy, N., Mishra, P.: The survey of data mining applications and feature scope. Int. J. Comput. Sci. Eng. Inf. Technol. (IJCSEIT) **2**(3) (2012)
2. Guerra, L., McGarry, M., Robles, V.: Comparison between supervised and unsupervised classifications of neuronal cell types: a case study. Dev. Neurobiol. **71**(1), 71–82 (2011)
3. Machine Learning Repository (UCI). https://archive.ics.uci.edu/ml/index.html
4. Jasim, D.S.: Data mining approach and its application to dresses sales recommendation. In: Research Gate (2015)
5. Gaikar, D., Marakarkandy, B.: Product sales prediction based on sentiment analysis using twitter data. Int. J. Comput. Sci. Inf. Technol. (IJCSIT) **6**(3), 2303–2313 (2015)
6. Kusiak, B., Shah, S.: Predicting survival time for kidney dialysis patients: a data mining approach. Comput. Biol. Med. **35**(4), 311–327 (2005)
7. Chetty, N., Vaisla, K.S., Sudarsan, S.D.: Role of attributes selection in classification of chronic kidney disease patients. In: 2015 International Conference on Computing, Communication and Security (ICCCS), pp. 1–6. IEEE (2015)
8. Wang, P.L.W., Chen, X., Gaikar, D.D., Marakarkandy, B.: Using twitter data to predict the performance of bollywood movies. Ind. Manag. Data Syst. **115**(9), 1604–1621 (2015)
9. Yasodha, P., Ananthanarayanan, N., et al.: Comparative study of diabetic patient datas using classification algorithm in WEKA tool. Int. J. Comput. Appl. Technol. Res. **3**(9), 554–558
10. Amin, M.N., Habib, M.A.: Comparison of different classification techniques using WEKA for hematological data. Am. J. Eng. Res. **4**(3), 55–61 (2015)
11. Jagdish, N., Kumar, D., Rajput, A.: An empirical comparison by data mining classification techniques for diabetes data set. Int. J. Comput. Appl. **131**(2), 6–11 (2015)
12. Confusion matrix. https://www.google.com/amp/www.dataschool.io/simple-guide-to-confusion-matrix-terminology/amp/. Accessed 24 Nov 20

A Novel Feature Selection Technique for Text Classification

D. S. Guru, Mostafa Ali and Mahamad Suhil

Abstract In this paper, a new feature selection technique called Term-Class Weight-Inverse-Class Frequency is proposed for the purpose of text classification. The technique is based on selecting the most discriminating features with respect to each class. Nevertheless, the number of selected features by our technique is equal to the multiples of the number of classes present in the collection. The vectors of the document have been built based on varying number of selected features. The effectiveness of the technique has been demonstrated by conducting a series of experiments on two benchmarking text corpora, viz., *Reuters-21578* and *TDT2 using KNN* classifier. In addition, a comparative analysis of the results of the proposed technique with that of the state-of-the-art techniques on the datasets indicates that the proposed technique outperforms several techniques.

Keywords Feature selection · TCW-ICF · Term-class weighting · Text classification · KNN · TF-IDF

1 Introduction

Text Classification (*TC*) is the process of routing each document to its labeled class. The classification results are always affected by the features, which are selected to represent each class and forming the vectors of each document. There are many feature selection techniques, which are used to select features for forming the vectors of the documents in the corpus. The techniques such as *IG*, *Gini Index*, *BNS*, and

---

D. S. Guru · M. Ali (✉) · M. Suhil
Department of Studies in Computer Science, University of Mysore, Manasagangothri, Mysuru 570006, India
e-mail: mam16@fayoum.edu.eg

D. S. Guru
e-mail: dsg@compsci.uni-mysore.ac.in

M. Suhil
e-mail: mahamad45@yahoo.co.in

A. Abraham et al. (eds.), *Emerging Technologies in Data Mining and Information Security*, Advances in Intelligent Systems and Computing 813,
https://doi.org/10.1007/978-981-13-1498-8_63

WLLR have been used by many researchers in the task of feature selection, but some researchers claimed that there are few drawbacks like redundancy, irrelevancy, and stability, which affect the results of classification.

In this direction, a new feature selection technique is proposed in this paper to overcome the drawbacks of the existing techniques mentioned above. Our proposed technique for feature selection evaluates the importance of a feature for each class. So, the selected features from our technique are the most discriminating features with respect to each class. Each class is represented by a set of features S that characterizes the class. The new unlabeled document can be tested according to every class set and based on that, it is classified to the most nearest class. The varying number of features in the set S, which are selected in the training stage will be equal to the multiples of the number of classes. Finally, the major contribution of this work is proposing a new feature selection technique that helps in selecting the most discriminating features with respect to each class. The new technique is experimentally tested and outperforms four of the state-of-the-art techniques.

This paper is organized as follows: Sect. 2 presents a survey of the literature on feature selection techniques. Section 3 briefs the four feature selection techniques considered for comparison. Section 4 gives a view of our new technique and its derivation from *TF-IDF*. Section 5 refers to the formation process of document vectors. Section 6 defines the two standard corpuses used in experiments. Section 7 shows the results with evaluation and comparison.

1.1 The Term Weighting Method TF-IDF

TF-IDF is used to estimate the importance of a term for a given document of interest with respect to all other documents in the corpus. As a weighting term, *TF-IDF* is considered as an unsupervised technique [1]. The *TF-IDF* computes the weight for a term t based on its frequency within the document d and its weight within the whole corpus D. Hence, the ability of the term t in discriminating a class from the other is not captured during the computation of its weight. The value of *TF-IDF* will be highest, if the term t has many occurrences within a small number of documents and this gives high discriminating power to those documents. It will become lower if the term t has a few occurrences in a document or if occurs in many documents. The term *TF-IDF* can be shown in Eq. (1).

$$TF(t_i, d) = \left\{\frac{f(t_i, d)}{|N_d|}\right\}, IDF(t_i, D) = \log\left\{\frac{N}{|d \in D : t_i \in d|}\right\} \quad (1)$$

where N_d is the total number of term frequencies in the document and N is the total number of documents in the corpus $N = |D|$.

2 Previous Work

Many researchers have tried to address the problem of text classification through feature selection. The authors in [2] applied *tf * idf*. *IG* and *MI* are claimed to be powerful techniques for text data in [3]. In [4], a new measure called term_class relevance to compute the relevancy of a term in classifying a document into a particular class is introduced. A method known as within class popularity (*WCP*) has been proposed in [5] to explore the relative distribution of a feature among different classes in English language to deal with feature selection based on the concept of Gini coefficient of inequality. The authors claimed that the term *WCP* outperforms the other feature selection methods. The authors in [6] calculated the normalized Gini index of each word in the lexicon in order to calculate its significance to the lexicon. Then, they treat the words with the highest value with respect to predefined value as removable items. Then, they applied seed-based technique to categorize the text document. A number of novel feature selection algorithms based on genetic algorithm are proposed in [7]. In [8], the authors introduce two new nonlinear feature selection methods, namely Joint Mutual Information Maximization (*JMIM*) and Normalized Joint Mutual Information Maximization (*NJMIM*) based on MI to address the problem of the selection of redundant and irrelevant features using Information theory. The authors in [9] recommended adaptive keyword, and in [10] recommended global ranking. Term frequency is used in [11] and there are some researchers who tried to classify text through feature reduction in [12, 13].

It is observed for some of the previous techniques that affected as the number of selected features increased; hence this instability will affect the performance of classifiers. Meanwhile, some researchers claimed that most of the feature selections techniques have problems like selecting redundant features or selecting irrelevant features [8].

3 Background Studies

In this section, initially, a brief introduction to several feature selection methods considered for our study is presented.

3.1 *Information Gain (IG)*

IG [3] is defined as in Eq. (2).

$$IG(t) = -\sum \Pr(c_i)\log \Pr(c_i) + \Pr(t)\sum \Pr(c_i|t)\log \Pr(c_i|t) + \Pr(\bar{t})\sum_i \Pr(c_i|\bar{t})\log \Pr(c_i|\bar{t}) \quad (2)$$

It is used in machine learning as a term goodness criterion. By knowing the presence or the absence of a term in the document, the number of bits required is measured for predicting the category.

3.2 Binormal Separation (BNS)

BNS is defined as the difference between the inverse cumulative probability function of true positive ratio (*tpr*) and false positive ratio (*fpr*) [14]. This measures the separation between these two ratios and is given by Eq. (3)

$$BNS(t, c) = F1(tpr) - F1(fpr) \tag{3}$$

where t is a feature and c is a category, and *F1* is the normal inverse cumulative distribution function. It is considered as a feature selection metric.

3.3 Weighted Log Likelihood Ratio (WLLR)

WLLR is proportional to the frequency measurement and the logarithm of the ratio measurement [15]. It is computed as in Eq. (4)

$$WLLR(t, c) = \frac{A_i}{N_i} \log \frac{A_i * (N_{all} - N_i)}{B_i * N_i} \tag{4}$$

where t is a feature, c is the category label, *Ni* is the number of documents belonging to class c in the training set, A_i is the number of times t, and c is the co-occurence in the number of documents belonging to class c and B_i is the number of times t occurs without c in the documents belonging to class c.

3.4 Gini-Index Algorithm

Gini index algorithm is introduced based on the Gini Index theory for text feature selection. It is designed with a new measure function of the Gini index and has been used to select features in the original feature space. The authors in [16] considered Gini-Index algorithm, which is Complete Gini-Index Text (*GIT*) feature selection algorithm for text classification. It can be shown in Eq. (5).

$$Gini(W) = P(W)(1 - \sum_{i=1}^{m} P(C_i|W)^2) + P(\bar{W})(1 - \sum_{i=1}^{m} P(C_i|\bar{W})^2 \tag{5}$$

4 A Novel Feature Selection Technique

In our new technique, it is important to extract the terms which can help in discriminating the documents of different classes. For this type of problem, we have proposed a new feature selection technique called as Term-Class Weight-Inverse-Class Frequency (*TCW-ICF*), which not only measures the ability of the term in preserving the content of a class but also computes the weight of a term as a function of its ability to discriminate the document of different classes. The proposed technique consists of two parts; the first one is Term-Class weight (*TCW*) defined as shown in Eq. (6).

$$TCW = \left\{ \frac{f(t, c)}{n_c} \right\} \tag{6}$$

where $f(t, c)$ is the frequency of the term t in class c and $n_c = \sum_{i=1}^{n} f_i(t, c)$ is the sum of the frequencies of all the n terms in class c. *TF* term is used to weight the term with respect to the document but, TCW captures the weight of the term t in preserving the content of class c.

As *IDF* is the inverse relative document frequency of the term with respect to other documents D in the corpus, the second part is *ICF* means the inverse relative class frequency of the term with respect to other classes C is defined as shown in Eq. (7).

$$ICF(t,\ C) = \log\left\{ \frac{K}{k = |c \in C : t \in c|} \right\} \tag{7}$$

where K is the total number of classes in the corpus and k is the number of classes which contain the term t. The term *ICF* is previously suggested as a supervised weighting term by [1], but here is a part of a technique for feature selection.

The value of *ICF* will have the highest value if a term is present in only one class and it will be 0 if the term is present in all the classes. Thus, if the value of *ICF* is high for a term then the term will be a discriminating term for the classes in which it is present. *TCW-ICF* is shown in Eq. (8)

$$TCW - ICF(t_i, c) = \frac{f(t_i, c)}{\sum_{j=1}^{n} f(t_j, c)} * \log \frac{K}{\mathrm{k} = |\mathrm{c} \in \mathrm{C} : \mathrm{t}_i \in \mathrm{c}|} \tag{8}$$

5 Forming Vectors for Documents

It is clear that the feature selection techniques try to capture different prosperities with class discriminating ability to represent each document in the dataset. So it will be better to select the most discriminating features that help classifiers to result in high classification performance.

In view of this, our model *TCW-ICF* selects the highest discriminating features based on two partial terms; *TCW*, which weights the feature within each class and *ICF*, which calculates the inverse of this term feature with respect to other classes. After calculating the *TCW-ICF* for each term, the whole matrix that contains all terms with respect to all classes is sorted in descending order by the value of TCW-ICF. The result of sorting is an index contains the sorted term feature values. The index will be a matrix of size ($K * N_f$), where N_f is the number of all features. Table 1 shows an example of a small training sample consists of 15 documents which are categorized into four classes with nine features i.e., K=4 and N_f=9. The table is filled with the calculated *TCW-ICF* value for each term t_i with respect to each class c_j. Table 2 shows the sorted index for the feature terms within each class based on their scores.

In index table, for each class, the value of *TCW-ICF* of feature f_i is greater than the value of f_{i+1} as following:

$$TCW - ICF(f_i, c_j) \geq TCW - ICF(f_{i+1}, c_j), \ where\ i = 1, 2, 3, \ldots NF\ and\ j = 1, 2, 3, \ldots K.$$

Hence the relevant features (which have high value) only are used to represent and discriminate each class and redundant features which have a low value of *TCW-ICF* are not selected. But with the increase in the number of selected features, the redundancy starts to increase in low rate.

By applying our technique, the Bag of Words (BoW) will contain a number of features equal to multiples of the number of classes so, the $BoW=\{f_i c_j, with\ i=1, 2, 3 \ldots m\ and\ j=1, 2, 3, \ldots K\}$, where m is the number of features that represent each class. For example, if m is selected to be 2 in the previous example, then the BoW will contain $\{t_4, t_3, t_7, t_1, t_9, t_8, t_5, t_6\}$ by taking f_1 and f_2 from all classes respectively. Finally, after selecting the term features, they will be represented by their respective *TF.IDF*. So, we can say that *TCW-ICF* is calculated for the purpose of selecting the most discriminating features.

6 Text Corpora

For an assessment of our new feature selection technique, the experiments have conducted on two commonly used text corpora viz., *Reuters-21578* and *TDT2*. The Reuters corpus is a collection of 8293 news articles taken from Reuters newswire *(available at* http://www.daviddlewis.com/resources/testcollections/reuters21578/*)*. The total number of topics is 65 where the total number of features is 18,933.

The *TDT2* corpus (Nist Topic Detection and Tracking corpus) consists of data collected during the first half of 1998 and taken from 6 sources including 2 newswires (*APW*, *NYT*), 2 radio programs (*VOA*, *PRI*) and 2 television programs (*CNN*, *ABC*). It consists of 10212 on-topic documents which are classified into 96 semantic categories. The total number of features is 36,771.

Table 1 Term-class matrix with TCW-ICF value

Class	t_1	t_2	t_3	t_4	t_5	t_6	t_7	t_8	t_9
c_1	0	0.12	0	0.43	0	0.11	0.09	0	0.25
c_2	0.1	0.01	0.38	0	0.17	0	0	0.2	0.14
c_3	0.15	0	0	0.1	0.21	0.1	0.44	0	0
c_4	0.49	0.12	0.13	0.04	0	0.21	0	0	0.01

Table 2 Index matrix for term-class matrix

Class	f_1	f_2	f_3	f_4	f_5	f_6	f_7	f_8	f_9
c_1	t_4	t_9	t_2	t_6	t_7	t_3	t_1	t_5	t_8
c_2	t_3	t_8	t_5	t_9	t_1	t_2	t_4	t_6	t_7
c_3	t_7	t_5	t_1	t_4	t_6	t_2	t_3	t_8	t_9
c_4	t_1	t_6	t_3	t_2	t_4	t_9	t_5	t_7	t_8

Table 3 Classification results of Reuters-21578 corpus with various feature selection techniques

# of selected features from each class (m)	# features	TCW-ICF	IG	BNS	WLLR	Gini index
1	65	71.74	78.69	67.02	63.49	73.92
2	130	77.19	81.65	72.47	67.61	75.93
3	195	79.75	82.77	75.06	68.97	76.39
4	260	81.71	83.46	75.98	69.81	78.21
5	325	83.11	83.95	76.96	70.96	78.54
6	390	83.58	84.32	77.29	71.57	78.45
7	455	**83.59**	**84.2**	77.3	72.13	79.08
8	520	83.15	83.78	77.22	72.88	79.52
9	585	83.12	83.35	77.35	73.21	**79.72**
10	650	83.02	83.11	77.2	74.62	79.21
15	975	82.84	81.32	77.56	**78.76**	77.56
20	1300	82.86	80.05	**77.75**	77.84	77.11
25	1625	82.78	78.43	77.45	77.55	76.25
30	1950	82.62	77.22	76.89	76.74	75.02
40	2600	82.4	74.38	75.7	75.37	72.91
45	2925	82.35	73.53	74.88	74.27	71.93
50	3250	82.25	72.74	73.63	73.19	71.51
55	3575	82.02	71.94	72.66	72	71.01
60	3900	81.85	71.24	72.01	71.19	70.84

7 Experiments

The performance of our new technique has been assessed using classification accuracy. *KNN* [17] classifier with $k=5$ is applied to evaluate the performance of feature selection based on the technique used and based on the number of selected features. *KNN* is applied because of its simplicity and accuracy [18]. In addition to *TCW-ICF*, the feature selection techniques *IG*, *Gini Index*, *BNS* and *WLLR* are also used as feature selectors with a different number of features equal to multiples of the number of classes. For *Reuters-21578*, the number of selected features is multiple of 65 based on the number of selected features from each class m, but for *TDT2*, the number of selected features is multiples of 96 based on the number of selected features from each class m. Classification step then is executed with 10 different trials by 50:50 for training stage and classification stage. Table 3 shows the value of chosen m and the produced number of selected features for all applied techniques, also, it presents the classification performances for all feature selectors which applied on *Reuters-21578*. Figure 1 presents a visual comparison of the accuracy results for all used feature selectors.

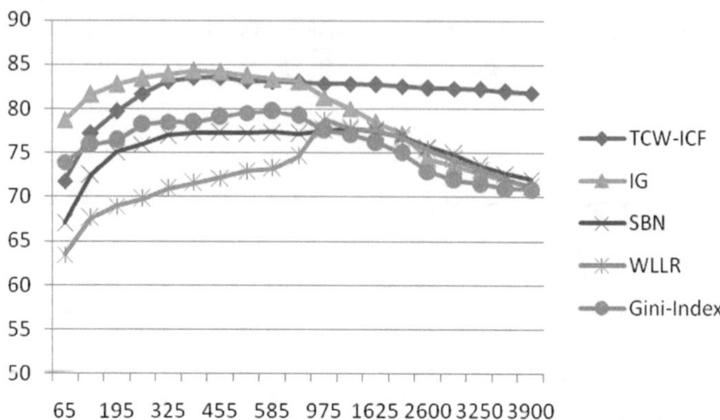

Fig. 1 Visualized results of Reuters-21578 corpus with various feature selection techniques

As we can see from the results in Table 3, in *Reuters-21578*, all techniques start with low accuracy values and the accuracies increased as the number of features increased. The highest accuracy values for *TCW-ICF, IG, BNS, WLLR* and *Gini Index* are 83.59, 84.2, 77.75, 78.76, 79.72 at the number of features equal to 544, 455, 1300, 975 and 585, respectively. *IG* has the highest classification accuracy but as the number of features increased, its accuracy starts to go down with high rate. On the other hand, our technique has the goodness of high stability in accuracy values as the number of features increased. At the number of features equal to 3900, the accuracy value of our technique is 81.85 compared to the accuracy of *IG* which is 71.24. The accuracy values of *BNS*, *WLLR* and *Gini Index* start to decrease after reaching the maximum of each of them which means instability.

Table 4 presents the classification performances for all feature selectors which are applied on *TDT2*. It is observed that our technique has the highest performance than all other technique. The highest performance for *TCW-ICF*, *IG*, *BNS*, *WLLR*, and *Gini Index* are 91.57, 89.54, 80.25, 83.43, 83.08 at the number of features equal to 960, 576, 5760, 3840, and 5760, respectively. The same behavior for *IG* of going down as the number of features increased appears. *BNS* and Gini Index have increasing accuracy values from beginning till the highest number of selected features but, the accuracy value of *WLLR* start going down after the number of features is equal to 3840. Figure 2 presents a visual comparison of the accuracy results for all used feature selectors.

Finally, from the previous experiments, we can observe that our technique outperforms other techniques in selecting the most discriminating features, which are the best to represent each class and come out with high performance in classifying the unlabeled documents. Moreover, the new technique addresses the problem of selecting redundant features or selecting irrelevant features.

Table 4 Classification results of *TDT2* corpus with various feature selection techniques

# of selected features from each class (m)	# features	TCW-ICF	IG	BNS	WLLR	Gini index
1	96	84.19	76.25	54.08	45.01	57.8
2	192	89.41	85.05	59.76	56.09	65.47
3	288	90.27	87.92	63.2	61.07	72.19
4	384	90.56	88.92	66.64	63.67	73.6
5	480	91.33	89.37	68.26	66.04	73.84
6	576	91.29	**89.54**	69.95	67.99	75.31
7	672	91.45	89.38	71.34	70.18	75.87
8	768	91.5	88.95	72.26	71.52	76.44
9	864	91.49	88.67	73	73.03	76.39
10	960	**91.57**	88.38	73.89	73.73	76.44
15	1440	90.56	87.12	75.94	78.91	77.62
20	1920	90.43	85.62	76.09	82.53	80.02
25	2400	89.74	85.51	76.98	82.37	81.83
30	2880	88.9	84.91	77.82	83.05	81.24
40	3840	87.49	81.92	78.8	**83.43**	80.34
45	4320	87.64	81.25	79.85	81.89	80.6
50	4800	87.32	80.7	79.36	81.21	79.85
55	5280	86.98	80.19	78.86	80.83	82.62
60	5760	86.76	79.69	**80.25**	79.97	**83.08**

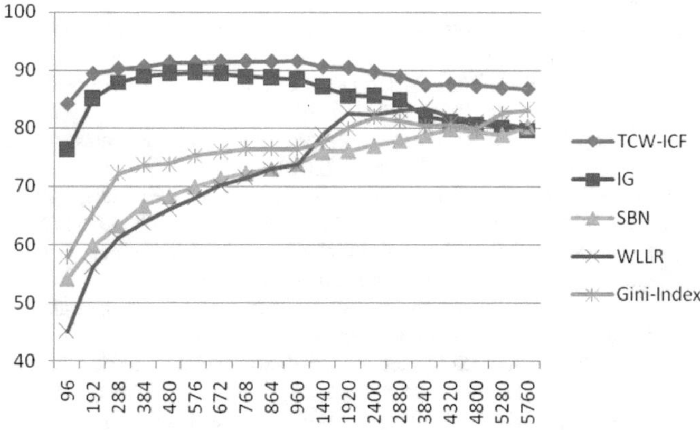

Fig. 2 Visualized results of *TDT2* corpus with various feature selection techniques

8 Conclusion

In this work, a new feature selection technique is proposed by changing the attitude of *TF-IDF* term weighting technique. The new technique *TCW-ICF* weights the term feature based on each class and that gives the technique the ability to select the most discriminating features with respect to each class. The new technique is evaluated by conducting experimentation on 2 standard corpuses *Reuters-21578* and *TDT2* using *KNN* classifier. The new technique is then compared with four feature selection techniques of the state of the art. It is observed that our new technique outperforms the other techniques in stability of accuracy as the number of selected features increased. The new technique also outperforms other techniques in handling the problems of selecting the redundant or the irrelevant features.

Acknowledgements The second author of this paper acknowledges the financial support rendered by the Indian Council for Cultural Relations (ICCR) and the Egyptian Cultural Affairs and Mission Sector.

References

1. Wang, D., Zhang, H., Wu, W., Lin, M.: Inverse-category-frequency based supervised term weighting schemes for text categorization. J. Inf. Sci. Eng. **29**, 209–225 (2013)
2. Azam, N., Yao, J.: Comparison of term frequency and document frequency based feature selection metrics in text categorization. Expert Syst. Appl. **39**, 4760–4768 (2012)
3. Yang, Y., Pedersen, J.O.: A comparative study on feature selection in text categorization. In: Proceedings of the Fourteenth International Conference on Machine Learning (ICML-97), pp. 412–420 (1997)
4. Guru, D.S., Suhil, M.: A novel Term_Class relevance measure for text categorization. Procedia Comput. Sci. **45**, 13–22 (2015)
5. Aggarwal, C.C., Gates, S.C., Yu, P.S.: On using partial supervision for text categorization. IEEE Trans. Knowl. Data Eng. **16**(2), 245–255 (2004)
6. Singh, S.R., Murthy, H.A., Gonsalves, T.A.: Feature selection for text classification based on Gini coefficient of inequality. Fsdm **10**, 76–85 (2010)
7. Bharti, K.K., Singh, P.K.: Opposition chaotic fitness mutation based adaptive inertia weight BPSO for feature selection in text clustering. Appl. Soft Comput. **43**, 20–34 (2016)
8. Bennasar, M., Hicks, Y., Setchi, R.: Feature selection using joint mutual information maximization. Expert Syst. Appl. Int. J. **42**(22), 8520–8532 (2015)
9. Tasci, S., Gungor, T.: Comparison of text feature selection policies and using an adaptive framework. Expert Syst. Appl. **40**, 4871–4886 (2013)
10. Pinheiro, R.H.W., Cavalcanti, G.D.C., Ren, T.I.: Data-driven global-ranking local feature selection methods for text categorization. Expert Syst. Appl. **42**, 1941–1949 (2015)
11. Wang, D., Zhang, H., Li, R., Lv, W., Wang, D.: t-Test feature selection approach based on term frequency for text categorization. Pattern Recogn. Lett. **45**, 1–10 (2014)
12. Corrêa, G.N., Marcacini, R.M., Hruschka, E.R., Rezende, S.O.: Interactive textual feature selection for consensus clustering. Pattern Recogn. Lett. **52**, 25–31 (2015)
13. Zong, W., Wu, F., Chu, L.K., Sculli, D.: A discriminative and semantic feature selection method for text categorization. Int. J. Prod. Econ. **165**, 215–222 (2015)
14. Forman, G.: An extensive empirical study of feature selection metrics for text classification. J. Mach. Learn. Res. **3**(1), 1289–1305 (2003)

15. Nigam, K., McCallum, A., Thrun, S., Mitchell, T.: Text classification from labeled and unlabeled documents using EM. Mach. Learn. **39**(2/3), 103–134 (2000)
16. Park, H., Kwon, S., Kwon, H.-C.: Complete gini-index text (git) feature-selection algorithm for text classification. In: 2010 2nd International Conference on Software Engineering and Data Mining (SEDM), pp. 366–371 (2010)
17. Cover, T.M., Hart, P.E.: Nearest neighbor pattern classification. IEEE Trans. Inf. Theory **13**(1), 21–27 (1967)
18. Lam, W., Han, Y.: Automatic textual document categorization based on generalized instance sets and a metamodel. Proc. IEEE Trans. Pattern Anal. Mach. Intell. **25**(5), 628–633 (2003)

Prediction of Death Rate Using Regression Analysis

Satya Ranjan Dash and Rekha Sahu

Abstract Feature prediction is necessary in different areas like Death Rate, Thunder Basin Antelope Study, Systolic Blood Pressure Data, General Psychology, child health, examination result, etc. For that, we can collect previous information and study the relationship between the different facts. Using the neurons concept, we can train our system to predict what must happen according to different situations. Multiple regression models can be used to predict the death rate. The key information of the article is the mathematical formulation for the forecast linear equation that estimates the multiple regression model calculation. The value of dependent variable forecast under the influence of independent variables is explained.

Keywords Feature prediction · Multiple regression · Linear regression analysis Independent variable

1 Introduction

The discovery of the death rate or survivability of a certain disease is possible by extracting the knowledge from the data related to the population. The characteristics of a population can be observed to establish the factors associated with a specific outcome. Observational studies, such as statistical learning and data mining, can establish the association of the variables to the outcome, but they do not always establish the cause-and-effect relationship of the association. Data-driven statistical research is becoming a common complement to many scientific areas like medicine and biotechnology. According to Yo-Lun Chow, "Regression analysis is used to find

S. R. Dash (✉)
School of Computer Applications, KIIT, Deemed to be University,
Bhubaneswar 751024, Odisha, India
e-mail: sdashfca@kiit.ac.in

R. Sahu
Academy of Management & Information Technology, Khurda 752057, Odisha, India
e-mail: sahu_r@rediffmail.com

A. Abraham et al. (eds.), *Emerging Technologies in Data Mining and Information Security*, Advances in Intelligent Systems and Computing 813,
https://doi.org/10.1007/978-981-13-1498-8_64

out the relationship between the variables and through which we find out the functional relationship between the variables and provide the mechanism for prediction or forecasting." It is found that regression analysis concept can be used in different areas for prediction. Extracting knowledge from different paperworks, it is found that regression analysis can be done for forecasting or predicting in the ground of health care. In this paper, we present data mining techniques (Regression Analysis) to predict the death rate based on some heath-and treatment-related data.

The death rate depends on various factors like hospital availability, expert doctors availability, per capita income, population densities, etc. [1]. Many techniques are there to predict the death rate. Through multiple regression also we can analyze and predict some values for feature consideration [2, 3]. Multiple regression technique is used in many places such as it is used as the principal component analysis to forecast the returns for the stock exchange of Thailand [4]. The adult obesity rate through regression analysis is analyzed in [5] to find economic and behavioral risk factor among adults. A technology on flash flood warning system using SMS with advanced warning information based on prediction algorithm and two independent variables were considered to generate the regression equation [6] through which current and forthcoming risk of the flood can be computed.

Traditional forecasting methods cannot deal with forecasting problems in which old data are present linguistically. Using various other techniques like fuzzy time series [7] and regression techniques, we can overcome the drawbacks. Again stock price prediction is a very difficult task as there is no certain variable present which can precisely predict the customer's demand. Naïve Bayes and Random Forest algorithm are used to classify the tweets to calculate sentiment and through linear regression analysis trying to predict the stock market [8]. Customer Churn prediction is very difficult in telecom industry [9], again in this emerging era customers switch over from one operator to another and from landline to mobile very frequently. The data extracted from telecom industry can help analyze the reasons of customer churn and use that information to retain the customers so through data mining and machine learning techniques with logistic regression we can predict the churn [10].

2 Literature Survey

Noise creates annoyance, hampers mental and physical peace, and even causes severe damage to health. Due to motorized traffic, noise pollution plays a vital role. Influence of noise pollution impacts five to seven percent of the patients of the Sylhet Osmani Medical College and as a result they are suffering from deafness. Alam et al. [11] generates predictive equations to measure the noise of density.

Gold price depends upon various factors like inflation, currency price movements, and others. Again economical status is generally evaluated through gold price. Selecting some factors that influence gold price, Ismail et al. [12] generated a multiple linear regression equation. It predicts the price of gold which is not seriously biased.

de Castro et al. [7] suggests a prediction model based on multiple regression analysis. It was found that Philippines is suffering from frequent flash flooding, which causes panic and fear to the people. For precaution and alarm purposes, Joel T. presents a multiple regression analysis model for an advanced prediction. Two factors such as water level and speed are taken as two independent variables and considering dependent variable as flashflood, the model is generated. This model is used to forecast the likelihood of flashflood using the reading on current data.

Sopipan [5] forecasts the returns for the Stock Exchange of Thailand Index. Some data are used for analyzing the data. Twelve numbers of independent variables are considered for one dependent variable. A set of data of Thailand is considered for forecasting using multiple regression analysis based on the principal component analysis. MSE and MAE are also considered and found the best performance.

The procedures, “multiple regression analysis” as well as “multilevel analysis”, structure hierarchical and possible to incorporate variables from all levels which leads to accurate analysis and suitable interpretation of the data. It deals with proportional results under multiple regression analysis and multilevel analysis in relation to the association of some factors with a number of children ever born in Uttar Pradesh [13].

Stock price prediction is a very difficult task. Linear regression analysis is used to solve this difficulty [10]. Stock prices of different companies of Indonesia are collected with their opinions. Twitter took a vital role.

The death rate in Canada, the United State, and worldwide is due to third leading cause that is COPD (Chronic Obstructive Pulmonary Disease). It can be preventable and treatable, which depends upon various factors. From those factors, doctor availability, hospital availability, annual per capita income, and population density can be taken into consideration [14, 15].

3 Problem Specification

Considering the health records and data available in [16], we have considered X_1 as the dependent variable and (X_2, X_3, X_4, X_5) as the independent city-wise variables are as follows:

X_1 Death rate per 1000 residents
X_2 Doctor availability per 100,000 residents
X_3 Hospital availability per 100,000 residents
X_4 Annual per capita income in thousands of dollars
X_5 Population density of people per square mile.

Using the dataset from [16] as input, we have applied multiple regression analysis to generate the death rate on the basis of Availability of Doctors, Hospital's Availability, Annual per Capita Income, and Population Density.

3.1 Our Approach

The multiple regression analysis can be used to predict the death rate. In case of the multiple regression analysis, two or more independent variables are used to estimate the values of a dependent variable. Through multiple regression analysis, we derive an equation which estimates the dependent variable from two or more independent variables' values.

According to our problem, we find out one dependent variable's value that is death rate from four independent variables' values doctor's availability, hospitals' availability, annual per capita income, and population density. So the regression equation of X_1 on X_2, X_3, X_4, and X_5 has the following form:

$$X_1 = a_1 + b_2X_2 + b_3X_3 + b_4X_4 + b_5X_5 \tag{1}$$

Subject to the following conditions:

$$\begin{aligned}\sum X_1 = Na_1 + b_2 \sum X_2 + b_3 \sum X_3 \\ + b_4 \sum X_4 + b_5 \sum X_5\end{aligned} \tag{2}$$

$$\begin{aligned}\sum X_1X_2 = a_1 \sum X_2 + b_2 \sum X_2 + b_3 \sum X_2X_3 \\ + b_4 \sum X_2X_4 + b_5 \sum X_2X_5\end{aligned} \tag{3}$$

$$\begin{aligned}\sum X_1X_3 = a_1 \sum X_3 + b_2 \sum X_2X_3 + b_3 \sum X_3^2 \\ + b_4 \sum X_3X_4 + b_5 \sum X_3X_5\end{aligned} \tag{4}$$

$$\begin{aligned}\sum X_1X_4 = a_1 \sum X_4 + b_2 \sum X_2X_4 \\ + b_3 \sum X_3X_4 + b_4 \sum X_4^2 + b_5 \sum X_4X_5\end{aligned} \tag{5}$$

$$\begin{aligned}\sum X_1X_5 = a_1 \sum X_5 + b_2 \sum X_2X_5 \\ + b_3 \sum X_3X_5 + b_4 \sum X_4X_5 + b_5 \sum X_5^2\end{aligned} \tag{6}$$

where

X_1 is the dependent variable

$X_2, X_3, X_4,$ *and* X_5 are independent variables.

N stands for the number of data studied for generating the equation. The input source contains $N = 53$ number of datasets.

The algorithm for generating the constraints described above for any number of independent variables for one dependent variable is as follows:

Algorithm 1: Constraints Generation for variables
Step-1: **Input**: $X_1, X_2, X_3, \ldots, X_n$
Enter the factors as independent variables that consider the dependent variable.
Step-2: Consider $f = a_0 + a_1 * X_1 + a_2 * X_2 + \cdots + a_n * X_n$

(f is the dependent variable)
Step-3: **Generate** equations for finding the value of
$a_0, a_1, a_2, \ldots, a_n$.
Step-3.1: $f=(f_1, f_2, \ldots\ldots\ldots, f_N)$
Enter N numbers of f values
for $i=1$ **to** N **do**
Enter the values of $X_{i1}, X_{i2}, \ldots, X_{in}$ as factors values to evaluate ith value of f.
{N number of data considered as INPUT}
end for
Step-4: **Compute**

$$\sum_{i=1}^{N} f_i = N * a_0 + a_1 * \sum_{i=1}^{N} x_{i1} + a_2 * \sum_{i=1}^{N} x_{i2} + \cdots + a_n * \sum_{i=1}^{N} x_{in}$$

for $k=1$ **to** n **do**

$$\sum_{i=1}^{N} f_i * x_{ik} = a_0 * \sum_{i=1}^{N} x_{ik} + a_1 * \sum_{i=1}^{N} (x_{i1} * x_{ik}) + a_2 * \sum_{i=1}^{N} (x_{i2} * x_{ik}) + \cdots + a_n * \sum_{i=1}^{N} (x_{in} * x_{ik})$$

end for
return

Using the dataset, we can get the five system of Eqs. (2–6) in matrix form, as follows:

$$X = \begin{bmatrix} 1.0\text{e} + 005 * 0.0049 \\ 1.0\text{e} + 005 * 0.5764 \\ 1.0\text{e} + 005 * 2.9407 \\ 1.0\text{e} + 005 * 0.0464 \\ 1.0\text{e} + 005 * 0.5344 \end{bmatrix} X \tag{7}$$

$$A = 1.0e + 007 * \begin{bmatrix} 0.0000 & 0.0006 & 0.0031 & 0.0001 & 0.0006 \\ 0.0006 & 0.0789 & 0.3823 & 0.0059 & 0.0679 \\ 0.0031 & 0.3823 & 2.4189 & 0.0295 & 0.3611 \\ 0.0001 & 0.0059 & 0.0295 & 0.0005 & 0.0056 \\ 0.0006 & 0.0679 & 0.3611 & 0.0056 & 0.0765 \end{bmatrix} \tag{8}$$

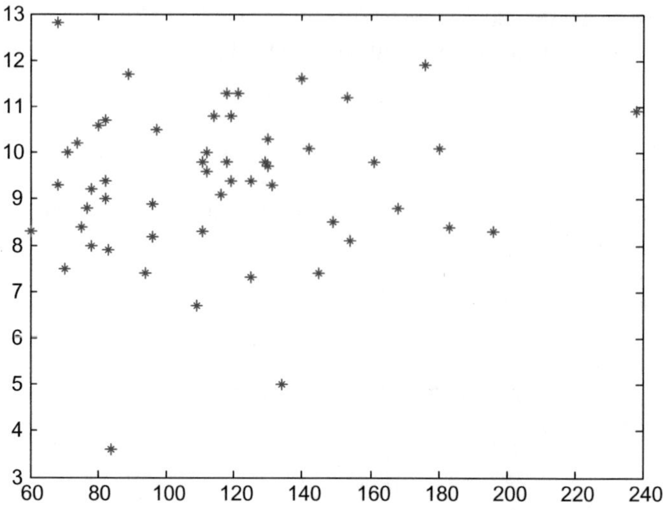

Fig. 1 Plotted points X_2 versus X_1

Equations 7 and 8 can be solved using Guass Elimination method to find the values of a_1, b_2, b_3, b_4, and b_5.

Using the Algorithm-1, we can find the values of

$a_1 = 2934.34$,
$b_2 = 7283.9$,
$b_3 = 273523.453$,
$b_4 = 5732.37$,
$b_5 = 97423.721$

Thus, we have the equation of regression, from Eq. 1 as follows:

$$X_1 = 2934.34 + 7283.9\, X_2 + 273523.453\, X_3 + 5732.37\, X_4 + 97423.721\, X_5 \tag{9}$$

We have represented the relation between X_1 (x-axis) the dependent variable with the independent variables, i.e., X_2, X_3, X_4, and X_5 in (y-axis). The plotted scatter points are shown in Figs. (1, 2, 3, and 4) that depicts the relationship between variables X_2 & X_1, X_3 & X_1, X_4 & X_1, and X_5 & X_1.

The final graph shown in Fig. (5) for the values of X_1 representing Y-axis with the different corresponding values of X_2, X_3, X_4, and X_5 is represented as follows:

Using Eq. 1, we predict the death rate according to doctors' availability, hospitals availability, annual per capita income, and population density. If the death rate depends on any additional other factors we can also generate the regression equation correspondingly using the same process only altering the number of independent variables.

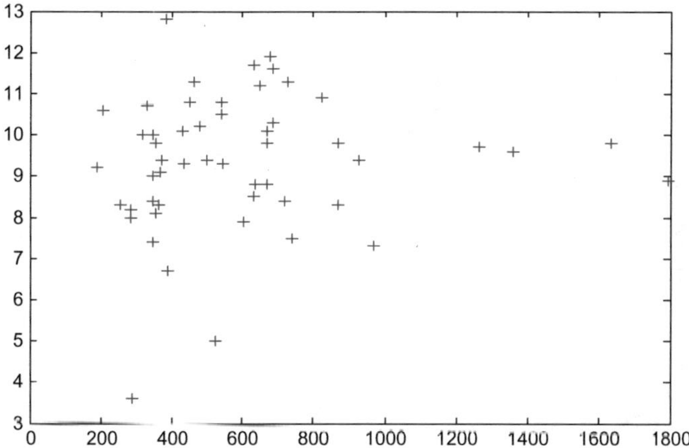

Fig. 2 Plotted points X_3 versus X_1

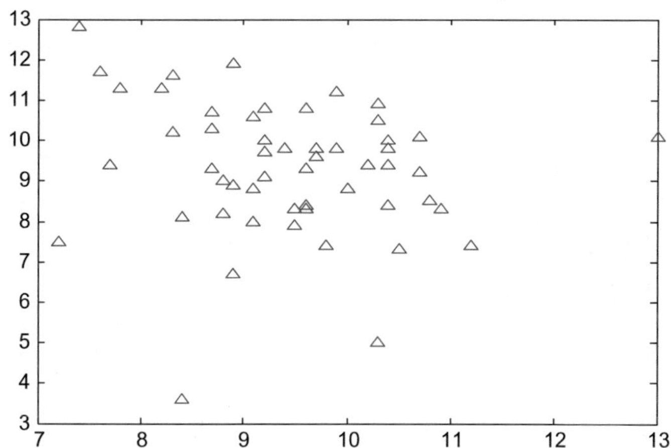

Fig. 3 Plotted points X_4 versus X_1

4 Outcome Analysis

To predict the death rate we have, the above mentioned dependent factors. But we have to consider whether this dependency is correctly considered or not. For that, we can use analysis of variance technique in the following way.

The ANOVA technique can be used whether all independent variables X_i, where $i = 2–5$ taken together significantly explains the variability observed in the depended variable X_1. Here, *F-test* is used with the following hypothesis:

Null hypothesis:

H_0: X_1 does not depend on X_i, where $i = 2–5$

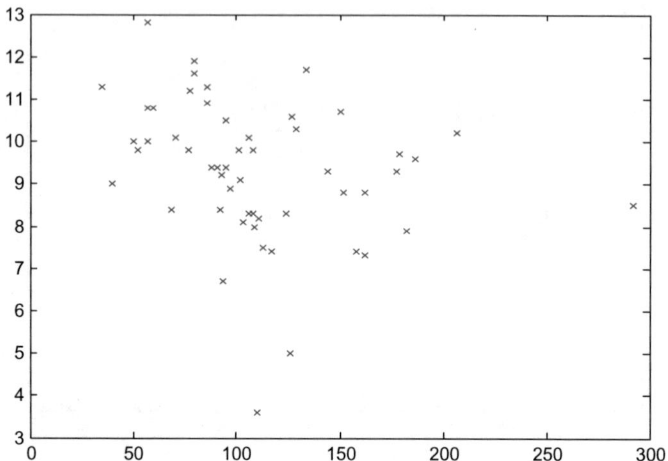

Fig. 4 Plotted points X_5 versus X_1

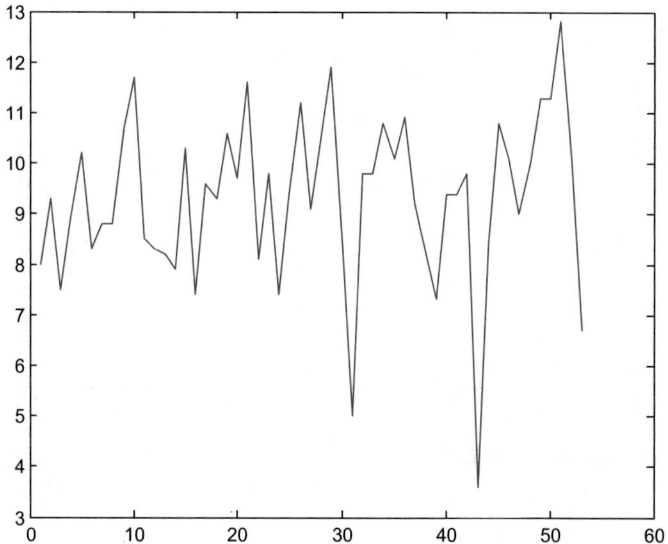

Fig. 5 y-axis $\rightarrow X_1 = f(X_2, X_3, X_4, X_5)$ death rate, x-axis $\rightarrow$ Based on the values of X_2, X_3, X_4, X_5

Alternative hypothesis:
H_1: X_1 depends on at least one X_i, where $i = 2$–5

The significance of regression effect is tested by computing the *F-test* statistics, which is given in Table 1.

Table 1 Computation of the F-test

Source of variation	Sum of squares	Degrees of freedom	Mean square	F-ratio
Regression	SSR	K	MSR = SSR/k	F = MSR/MSE
Residual (or Error)	SSE	N – (k + 1)	MSE = SSE/ (N – (k + 1))	

where

$$SSR = \sum_{i=1}^{N} ((Y_1 - \mathrm{mean(X_1)})^2 \tag{10}$$

and

$$SSE = \sum_{i=1}^{N} (X_1 - \mathrm{mean(X_1)})^2 \tag{11}$$

If the calculated value F is more than its F-table value at a given level of significance and degree of freedom k for the numerator and $(N - k - 1)$ for the denominator, then H_0 is rejected and H_1 is accepted. Otherwise, H_0 is accepted and H_1 is rejected.

It is practically impossible to give a perfect prediction by using regression equation. Hence, we further analyzed to find out the correctness in the prediction so as to identify the errors that may arise and whether this dependency is correctly considered or not. For that, we have applied the analysis of variance technique in following way.

The standard error of estimate measures the accuracy of the estimated figures. The smaller the value of the standard error of estimate, the closer the dots to the regression line and better the estimates based on the equation. If standard error of estimate is zero, there is no variation about the line and the equation is the perfect one. We can estimate the standard error (S) using the following formula given as Eq. 12.

$$S_{1.2345} = \sqrt{\frac{\sum (X_1 - mean(Y_1))^2}{N - (k + 1)}} \tag{12}$$

where

X_1 → Sample value of the dependent variable,
Y_1 → Estimated value of dependent variable from the regression equation,
N → Number of observations in the sample, and
k → Number of independent variables.

The standard error $S_{1:2345}$ represents the extent of variation of observation in the dataset with respect to a hyperplane of five variables.

5 Conclusion

This paper has outlined, discussed, and resolved the issues, algorithms, and techniques for the problem of death rate or survivability prediction. We have considered here the death rate that depends upon the doctors' availability, hospitals' availability, annual per capita income, and population density and tested its correctness using ANOVA technique. The paper also contains the estimate, how accurate the regression equation is, by calculating the standard error. Regression analysis can be used in a discussed way for feature prediction of the death rate.

This concept and technique can be used in different areas. Generally, when the given data depends upon two or more than two factors in different scenarios.

References

1. Dash, S.R., Dehuri, S., Rayaguru, S.: Discovering interesting rules from biological data using parallel genetic algorithm. In: 2013 IEEE 3rd International Advance Computing Conference (IACC). IEEE (2013)
2. Skrepnek, G.H.: Regression methods in the empiric analysis of health care data. J. Manag. Care Pharm. **11**(3), 240–251 (2005)
3. Thomas, G.S.: The Rating Guide to Life in America's Small Cities. Prometheus Books, 59 John Glenn Drive, Amherst, NY 14228-2197 (1990)
4. Witten, I.H., et al.: Data Mining: Practical Machine Learning Tools and Techniques. Morgan Kaufmann (2016)
5. Sopipan, N.: Forecasting the financial returns for using multiple regression based on principal component analysis. J. Math. Stat. **9**(1) (2013)
6. Wang, Y., Beydoun, M.A.: The obesity epidemic in the United States—gender, age, socioeconomic, racial/ethnic, and geographic characteristics: a systematic review and meta-regression analysis. Epidemiol. Rev. **29**(1), 6–28 (2007)
7. de Castro, J.T., et al.: Flash flood prediction model based on multiple regression analysis for decision support system. In: Proceedings of the World Congress on Engineering and Computer Science, vol. 2 (2013)
8. Chen, S.-M., Hwang, J.-R.: Temperature prediction using fuzzy time series. IEEE Trans. Syst. Man Cybern. Part B (Cybernetics) **30**(2), 263–275 (2000)
9. Dalvi, P.K., et al.: Analysis of customer churn prediction in telecom industry using decision trees and logistic regression. In: Symposium on Colossal Data Analysis and Networking (CDAN). IEEE (2016)
10. Cakra, Y.E., Trisedya, B.D.: Stock price prediction using linear regression based on sentiment analysis. In: 2015 International Conference on Advanced Computer Science and Information Systems (ICACSIS). IEEE (2015)
11. Alam, J.B., et al.: Study on traffic noise level of sylhet by multiple regression analysis associated with health hazards. J. Environ. Health Sci. Eng. **3**(2), 71–78 (2006)
12. Ismail, Z., Yahaya, A., Shabri, A.: Forecasting gold prices using multiple linear regression method (2009)
13. Dwivedi, S.N., Rajaram, S.: Some factors associated with number of children ever born in Uttar Pradesh: a comparative results under multiple regression analysis and multilevel analysis. Indian J. Community Med. **29**(2), 72 (2004)

14. Zhang, J., et al.: Predicting COPD failure by modeling hazard in longitudinal clinical data. In: 2016 IEEE 16th International Conference on Data Mining (ICDM). IEEE (2016)
15. WHO. Chronic Obstructive Pulmonary Disease (COPD) (2015). www.who.int/mediacentre/factsheets/fs3
16. http://college.cengage.com/mathematics/brase/understandablestatistics/7e/students/datasets/mlr/frames/mlr07.html

An Investigative Prolegomenon on Various Clustering Strategies, Their Resolution, and Future Direction Towards Wireless Sensor Systems

Sushree Bibhuprada B. Priyadarshini, Amiya Bhusan Bagjadab, Suvasini Panigrhi and Brojo Kishore Mishra

Abstract In the current era of recent technological proliferation, the sensor network is a highly alluring field of research, thus, finding large-scale applications ranging from health care to military applications. In such networks, various protocols and approaches are used over time and again for prolonging the lifetime of the network. With the pace of time, various clustering approaches are used effectively for minimizing the energy expenditure of nodes. Basically, during the occurrence of any event, the scalars ensnare the event and they collectively report the data to the base station. Basically, such scalars are the sensors those can trap the textual information and they combinedly report their data either individually or collectively to the desired sensor(s) or base station for further processing. The current article presents an introduction to clustering and several issues associated with it. Afterward, it investigates various clustering approaches and collaborates briefly the various operations on those algorithms considering several attributes of clustering in mind. Our current article basically discusses various clustering approaches used in diversified fields of technology and proposes a cluster sensing model as a future direction of research toward wireless sensor systems.

Keywords Cluster head · Cluster member · Life time · Hierarchical clustering

---

S. B. B. Priyadarshini (✉) · B. K. Mishra
C. V. Raman College of Engineering, Bhubaneswar 752054, India
e-mail: bimalabibhuprada@gmail.com

B. K. Mishra
e-mail: brojokishoremishra@gmail.com

A. B. Bagjadab · S. Panigrhi
Veer Surendra Sai University of Technology, Burla,
Samabalpur 768018, India
e-mail: amiya7bhusan7@gmail.com

S. Panigrhi
e-mail: suvasini26@gmail.com

A. Abraham et al. (eds.), *Emerging Technologies in Data Mining and Information Security*, Advances in Intelligent Systems and Computing 813,
https://doi.org/10.1007/978-981-13-1498-8_65

1 Introduction

In the modern arena of technological peregrination, the wireless sensor network (WSN) has emerged as an effective technology in many application domains. Various prevalent applications incorporate predictive maintenance, habitat monitoring, forest fire detection, micro-climate prediction, traffic monitoring, defense operation monitoring, assembly line control, healthcare monitoring, disaster management, consumer electronics, industrial process tracking, fault detection and diagnosis, tactical surveillance, intrusion detection, etc. Such networks consist of tiny sensors those are sprinkled over a huge area of interest and are densely deployed either inside the concerned phenomenon of interest or very close to it [1]. The fundamental components of node involve processor, battery, energy, transceiver, and memory having limited power. Moreover, the nodes deployed in sensor networks may be homogeneous or heterogeneous. First, homogeneous nodes represent the nodes that are equipped with the same amount of sensing, processing as well as communication capabilities. On the contrary, heterogeneous nodes possess varying sensing, processing as well as communication features.

The wireless sensor systems can be assessed on the ground of various specifications. They incorporate the following:

(i) **Lifetime of Network**: Every node pertaining to the network has to be constructed for admonishing the local energy supply for the purpose of prolonging the network lifetime.
(ii) **Scalability**: During the time, new nodes are joined in the network, there must not be any change in the corresponding network.
(iii) **Coverage**: It is attained when the nodes are sprinkled in all the area to be tracked.
(iv) **Response Time**: It is the whole time taken from sending the inquiry, and getting the response.
(v) **Cost**: The aim is to diminish the comprehensive cost associated with the network without hampering the effective performance of the network.
(vi) **Security of network**: This is the most important feature of a network once the system is deployed. In the sensor system, a node at the individual level can carry out encryption as well as authentication on the basis of requirements.
(vii) **Flexibility**: The architecture of concerned nodes must be adaptive as well as flexible to be used in multiple applications.
(viii) **Fault tolerance**: The deployed nodes must be fault tolerant so that the sensor system tolerates individual node failure.
(ix) **Communication range of node**: The communication range of the node should be high enough for increasing the performance effectively.

Figure 1 portrays a scenario of data collection and its transmission to the destined client. In the sensor system, data are collected from various sensing zones and are forwarded to the base station and from the base station data are transferred to the client over the Internet. In case of wireless sensor systems, an aggregated value is

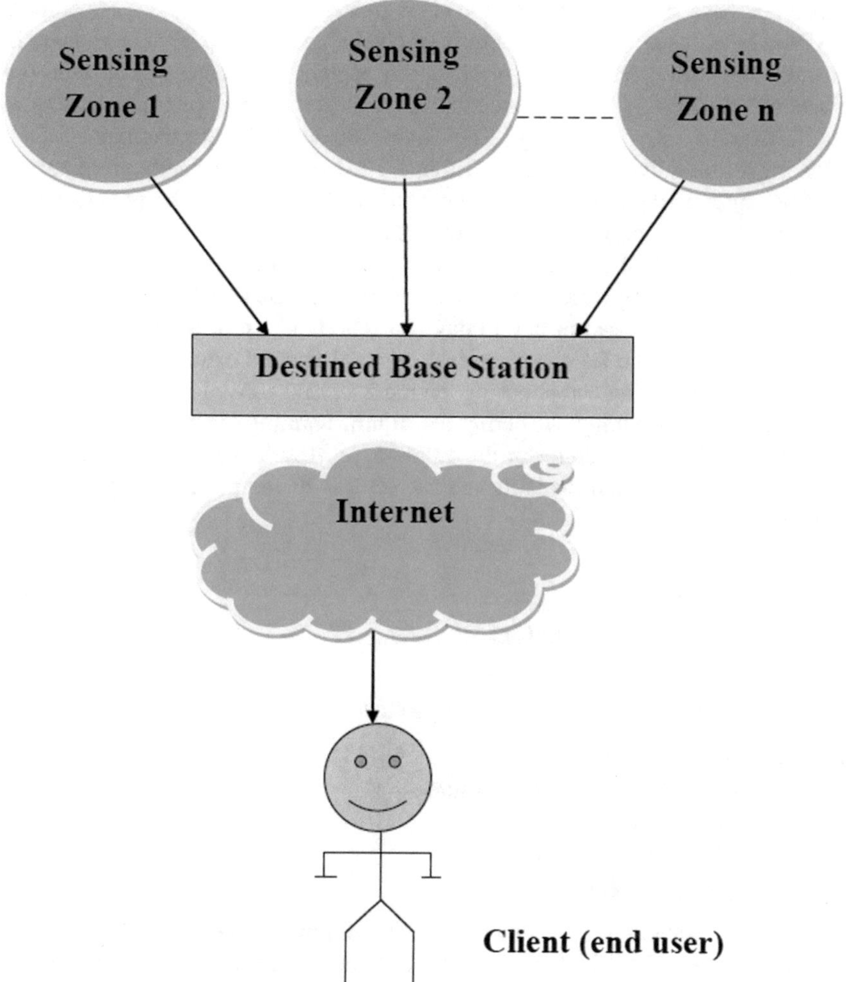

Fig. 1 Data gathering from various sensing zones in wireless sensor system

always created for end users and the conglomeration of data assists in diminishing the transmission overhead as well as energy expenditure. For supporting such feature, the nodes are arranged in groups known as clusters. It is used for prolonging the network lifetime as well as energy efficiency. The administrator of the whole cluster that is liable for data aggregation, clustering as well as transmission of data to the desired base station is regarded as Cluster Head (CH) and the rest of the nodes pertaining to the cluster are the Cluster Members (CMs).

Basically, the cluster members sense the data and forward the data to the corresponding cluster head. Afterward, the CH aggregates the received data and performs desired processing and reports it to either other nodes for further processing or to the concerned base station. The CH expends a greater amount of energy as compared to CMs. In a clustering system, both inter-cluster and intra-cluster communication prevail. The various compasses of clustering include the number of clusters (i.e., cluster count), uniformity of cluster size, intra-cluster routing, inter-cluster routing, etc. In the context of clustering [2], distance and similarity are the major foundations for forming the clustering algorithms [3]. Distance is normally preferable for establishing the relationship among the concerned data and similarity is taken into account while considering the qualitative aspects of data.

With the advent of technology, clustering is applied not only in case of sensor networks but also in various domains. The current article collaborates with various clustering strategies used in day-to-day life. The whole process of clustering can be segregated into the following several steps (a) Feature extraction and selection that involves extracting and choosing the most representative features from the initial dataset, (b) Designing Clustering algorithm: Forming the clustering algorithm according to the main features of the problem, (c) Result assessment: assessing the results of clustering and deciding the correctness of the preferred algorithm, (d) Explaining the results.

2 Constituents of Clustering Task

A typical clustering activity involves the following tasks:

(a) Pattern representation
(b) Defining pattern proximity measure
(c) Grouping or Clustering
(d) Data Abstraction (if desired)
(e) Assessing the output

Basically, "pattern representation" refers to the total number of available patterns, the total number of available classes, the scale of features associated with the clustering algorithm. Further, "feature selection" represents the process of determining the most effective subset of the initial original feature employed in the algorithm [4]. Further, the pattern proximity [5] is assessed by employing a distance function defined on pairs of the corresponding pattern. Various types of distance measures are employed in distinct communities [6, 7].

Similarly, grouping can be carried out in various ways. Likewise, hierarchical clustering forms a nested series of partitions based on the criteria for splitting the cluster or merging it based on the prevailing condition [4]. Moreover, data abstraction is the process of getting the compact and simpler representation of data. However, the cluster validity analysis involves assessment of the output obtained from the clustering and this is conducted to make sure that the obtained output is quite meaningful [6].

3 The Role of Expertise in Resolving Use Dilemma

Cluster analysis is a very crucial technique used in exploratory data analysis and sensor networks. Normally, a huge collection of algorithms in the literature confuses the use while choosing a specific algorithm. A set of admissible criteria are, therefore, defined by [8] in [9]. The admissibility criteria involve the following:

(a) The way normalization is carried out on data
(b) Similarity measure to be used for specific scenario
(c) The way large data set can be clustered effectively
(d) The way of utilization of domain knowledge

Actually, no clustering algorithm exists that is applicable universally in all circumstances. Further, a large number of clustering techniques is available in the literature. Therefore, it becomes essential for the concerned user to know the whole technicality of the data gathering process while attaining the associated domain expertise. Greater is the amount of information available with the user, more effectively, he or she can be successful in judging its true class structure [10].

Such domain information can enrich the quality of similarity computation as well as feature extraction and cluster representation [11]. While considering clustering several constraints are also taken into account. In this context, mixture resolving [12] stands as a very significant process, where it is considered that data are drawn from a huge number of unknown densities. The objective is to find out the number of mixture components. The idea of density clustering and a technique for decomposing the feature space is being used with traditional clustering methodology [13].

4 Techniques of Clustering

The various methods for clustering of data are diagrammatically represented in Fig. 2. Clustering is broadly classified into two types namely: (a) hierarchical and (b) partitional clustering.

4.1 Hierarchical Clustering

Hierarchical clustering is fundamentally the type of clustering, where a set of nested clusters are arranged in the form of a tree [4]. Further, hierarchical clustering is of two types as follows:

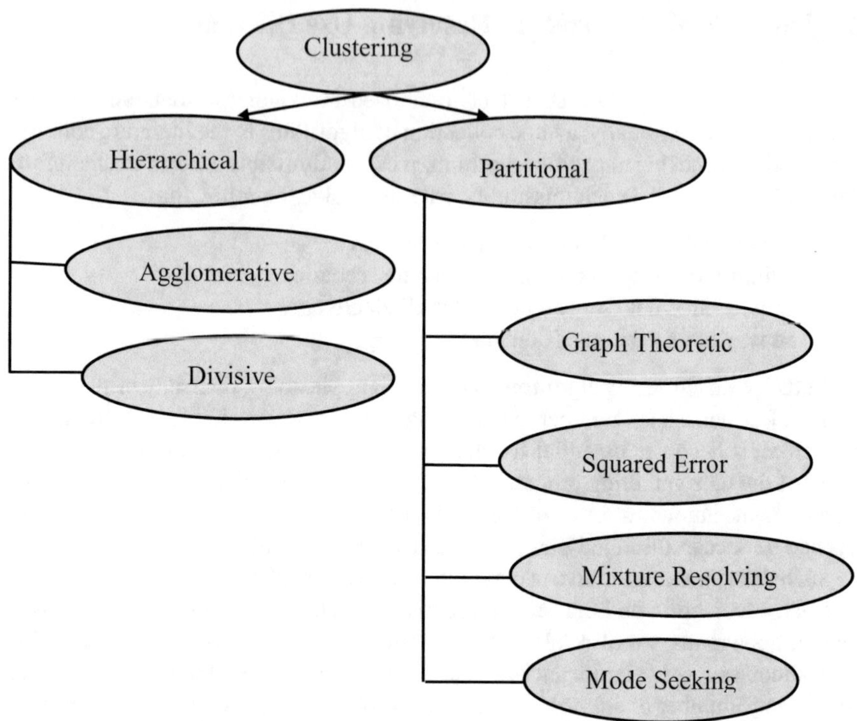

Fig. 2 Classification of basic clustering techniques

4.1.1 Agglomerative Clustering

In case of agglomerative clustering, we start with the points as individual clusters and in every subsequent stage, we merge the closest pair of clusters. In this strategy, the two closest clusters are merged. The proximity matrix gets updated each time for reflecting the relationship between new and older clusters. Agglomerative clustering employs bottom-up approach. In terms of storage and computational requirements, agglomerative hierarchical clustering is preferable [14, 15].

a. **Agglomerative Single-Link Clustering**. In this strategy, each pattern is put in its own cluster. A list of inter-pattern distance gets computed. For all the distinct unordered pair of patterns. Now, the list is sorted in ascending order. A graph is constructed considering each distinct dissimilarity value, where the pair of patterns closer than the dissimilarity value are connected by an edge [4]. Finally, a nested hierarchy of graph is obtained as the corresponding output.
b. **Agglomerative Complete-Link Clustering**. In this strategy, each pattern is put in their own cluster. For each of the distinct unordered pairs of patterns, a list of inter-pattern distance gets computed. Thereafter, the list is sorted in ascending sequence. For each of the distinct dissimilarity value, a graph is constructed. In

this graph, pairs of patterns closer than dissimilarity value are connected by an edge. During the time, when all the patterns become the member of the completely connected graph, we need to stop [4].

4.1.2 Divisive Clustering

A divisive approach initiates with all patterns, within one cluster and applies splitting till a stopping criterion is met. It employs top-down strategy. They produce more accurate hierarchies as compared to bottom-up strategies [15].

4.2 Partitional Clustering

In case of partitional clustering, a set of data objects are segregated into nonoverlapping clusters so that every data object belongs to exactly one subset. Such clustering is classified into four types as follows:

4.2.1 Graph-Theoretic Clustering

The very well-known graph-theoretic clustering strategy is based on the construction of Minimal Spanning Tree (MST) and then deleting the MST edges with the corresponding maximal length for producing a greater number of clusters. These are somehow related to hierarchical approaches. Single-link clusters represent the sub-graphs of the MST of data [16], whereas complete-link clusters represent the maximum complete sub-graphs [17] and are closely concerned with node colorability of the concerned graph [18]. The maximal complete sub-graph represents the strictest representation of a cluster in [19, 20]. Further, a Delaunay graph is obtained by associating all the pairs of points, which are Voronoi neighbors [21]. Similarly, another Distributed Collaborative Camera Actuation based on Scalar Count (DCA-SC) is presented in [22], where the camera sensors collaboratively decide which among them have to be actuated.

4.2.2 Squared Error Clustering

An initial partition of the pattern is carried out with a fixed count of clusters and cluster centers. Each of the patterns is assigned to its closest cluster center and new cluster centers are computed as the centroids of the clusters. This process gets repeated until convergence is attained. This means until the cluster member attains stability. Depending on the heuristic information, clusters are merged or split over time [4].

4.2.3 Mixture Resolving

In case of mixture resolving, the main aim is to identify the parameters and their associated numbers.

4.2.4 Mode Seeking

In case of clustering, the scores of the patterns estimate their likelihood of being drawn from a specific component of the concerned mixture that can be viewed as hints at the class of the concerned pattern.

Some other frequently used clustering strategies are listed as follows:

- Clustering algorithm for large-scale data
- Clustering algorithm based on density and distance
- Clustering algorithm based on affinity propagation
- Clustering algorithm based on kernel

5 Proposed Sensor Model

After analyzing various clustering approaches, we have developed a cluster-based sensing system that consists of scalar sensors those can entrap the textual information. In the proposed sensor model, we propose a system where some scalar cluster heads are effectively selected such as those operate as scalar representatives for event information reporting; which leads to less amount of message exchange to take place among the sensors by preventing multi-event reporting by multiple scalars. Further, since scalar cluster heads are selected, a lesser amount of energy consumption will take place for reporting of scalar cluster head data to the base station. Furthermore, myevent message is exchanged in between the individual cluster members and event report message will be sent from the cluster head to the concerned base station for conveying regarding the prevailing event.

In the beginning, the entire sensing region is segregated into a number of virtual compartments and in each virtual compartment, a scalar cluster head is selected. The scalar that is surrounded by a maximum number of sensing neighbors is regarded as the scalar cluster head of a particular virtual compartment. Further, the sensing regions are determined such that the length of each sensing region is the same as that of the sensing range of the scalar sensors deployed. During the occurrence of any event, the scalar sensors such as those residing in the event region detect the event and the event information is communicated to their respective cluster heads. The selected cluster heads report this information to their respective camera sensors provided they reside within the DOF of the corresponding camera sensor. The camera sensors being informed from the cluster heads decide collaboratively who among them have to be actuated.

Table 1 Initial approach versus proposed approach

NCD	NCA (initial approach)	NCA (proposed approach)	EECA (joule) (Initial approach)	EECA (joule) (Proposed approach)
80	13	10	16.64	12.80
100	17	14	21.76	17.92
120	23	21	29.44	26.88
140	26	23	33.28	29.44
160	30	25	38.40	32

6 Experimental Results

We have carried out experimentation by varying the number of cameras deployed (NCD) and observing its impact on the number of camera sensors actuated (NCA) as well as energy expenditure for camera activation in both initial DCA-SC and the proposed approach as shown in Table 1. It is evident from the table that the number of cameras activated in both the approaches rises with the rise in a number of cameras deployed. It is found to be lesser in case of prefered approach since only the scalar cluster heads report the prevailing event information. Since the number of cameras activated is lesser in the proposed scheme, hence, obviously the energy expenditure for camera actuation (EECA) is also lesser as compared to initial approach.

7 Conclusions and Future Direction

This article discusses elaborately various clustering approaches used in various fields of technical applications. The approaches used involve the cluster member and cluster head determination approach. We have also proposed a novel approach that involves the selection of scalar cluster heads, followed by collaborative camera actuation. All the elected cluster heads act as the chief representatives of the respective sensors of virtual compartments and they report the data entrapped in lieu of all the cluster members, thus, diminishing multi-event reporting. We have implemented the paper and from experimental outcomes, it is crystal clear that the number of cameras actuated as well as the energy consumption for camera activation is minimized in our proposed approach as compared to DCA-SC approach. Further, we want to extend our idea towards finding the coverage and redundancy incurred in our proposed approach as our further investigation in this research.

References

1. Akyildiz, I.F., Su, W., Sankarasubramaniam, Y., Cayirci, E.: Wireless sensor networks: a survey. Comput. Netw. **38**(4), 393–422 (2002)
2. Xu, D., Tian, Y.: A Comprehensive Study of Clustering Algorithms **2**(2), 165–193 (2015). https://link.springer.com/content/pdf/10.1007%2Fs40745-015-0040-1.pdf
3. Jain, A.K., Dubes, R.C.: Algorithms for Clustering Data. https://homepages.inf.ed.ac.uk/rbf/BOOKS/JAIN/Clustering_Jain_Dubes.pdf. Accessed 14 Nov 2017
4. https://www.cs.utah.edu/~piyush/teaching/clustering-survey.pdf. Accessed 14 Nov 2017
5. Anderberg, M.R.: Cluster Analysis for Applications. Academic Press Inc., New York (2000)
6. Diday, E., Simon, J.C.: Clustering analysis. In: Fu, K.S. (ed.) Digital Pattern Recognition. Springer, New York
7. Jain, A.K., Dubes, R.C.: Algorithms for Clustering Data. Prentice Hall. https://homepages.inf.ed.ac.uk/rbf/BOOKS/JAIN/Clustering_Jain_Dubes.pdf. Accessed 18 Nov 2017
8. Fisher, L., Van Ness, J.W.: Admissible clustering procedures. Biometrika **58**, 91–104 (1971)
9. Dubes, R.C., Jain, A.K.: Clustering techniques: the user's dilemma. Pattern Recogn. **8**, 247–260 (1976)
10. Jain, A.K., Dubes, R.C.: Algorithms for Clustering Data. Prentice Hall (1988)
11. Murty, M.N., Jain, K.: Knowledge based clustering scheme for collection management and retrieval of library books. Pattern Recogn. **28**, 949–964 (1995)
12. Titterington, D.M., Smith, A.F.M., Makov, U.E.: Statistical Analysis of Finite Mixture Distributions. Wiley, NY (1985)
13. Bajcsy, P.: Ph.D. Dissertation, University of Illinois
14. https://www3.nd.edu/~rjohns15/cse40647.sp14/www/content/lectures/13%20-%20Hierarchical%20Clustering.pdf. Accessed 14 Nov 2017
15. https://nlp.stanford.edu/IR-book/html/htmledition/divisive-clustering-1.html. Accessed 14 Nov 2017
16. Gower, J.C., Ross, G.J.S.: Minimum spanning trees and single linkage cluster analysis. Appl. Stat. **18**, 54–64 (1969)
17. Gotlieb, G.C., Kumar, S.: Semantic clustering of index terms. J. ACM **15**, 493–513 (1968)
18. Backer, F.B., Hubert, L.J.: A graph theoretic approach to goodness of fit in complete link hierarchical clustering. J. Am. Stat. Assoc. **71**, 870–878 (1976)
19. Augustson, J.G., Minker, J.: An analysis of some graph theoretical clustering techniques. J. ACM **17**, 571–588 (1970)
20. Raghavan, V.V., Yu, C.T.: A comparison of the stability characteristics of some graph theoretic clustering methods. IEEE Trans. Pattern Anal. Mach. Intell. **3**, 393–402 (1981)
21. Toussaint, G.T.: The relative neighborhood graph of a finite planar set. Pattern Recogn. **12**, 261–268 (1980)
22. Newell, A., Akkaya, K.: Distributed collaborative camera actuation for redundant data elimination in wireless multimedia sensor networks. Ad Hoc Netw. **9**(4), 514–527 (2011)

A Comparison Study of Temporal Signature Mining Over Traditional Data Mining Techniques to Detect Network Intrusion

Sharmishtha Dutta, Tanjila Mawla and Md. Forhad Rabbi

Abstract Network intrusion means any attempt to compromise the confidentiality or availability of a computer network. The growing speed of data transmission poses a challenge in the detection of such intrusions. Most of the existing systems for detecting network intrusion ignore a significant feature associated with every sort of data, time. On the other hand, some systems have taken the temporal aspect of data into account (Ahmed in Online network intrusion detection system using temporal logic and stream data processing [1] and Hogo in 2014 International Carnahan Conference on Security Technology (ICCST), pp. 1–6, 2014 [2]). These systems show better accuracy, low false alarm rates, higher bandwidth with full coverage rate, and system availability in case of data flow rate close to 1 GB/s. The concept of temporal data mining in network intrusion detection is being more popular as it is providing more promising results than traditional mining techniques.

1 Introduction

To ensure effective and efficient use of rapidly growing volumes of data that we have nowadays, data mining was introduced. The process by which significant information from a large set of data is extracted is called data mining [3]. The useful information gained by data mining could be an interesting pattern, trend, signature, exceptions, anomalies, or dependencies. Data mining methods consist of database systems, artificial intelligence, machine learning, and statistics.

S. Dutta (✉)
Metropolitan University, Sylhet 3100, Bangladesh
e-mail: shoron.dutta321@gmail.com

T. Mawla
North East University Bangladesh, Sylhet 3100, Bangladesh
e-mail: tani109.bd@gmail.com

Md. F. Rabbi
Shahjalal University of Science and Technology, Sylhet 3114, Bangladesh
e-mail: post2rabbi@gmail.com

A. Abraham et al. (eds.), *Emerging Technologies in Data Mining and Information Security*, Advances in Intelligent Systems and Computing 813,
https://doi.org/10.1007/978-981-13-1498-8_66

To extract significant and useful information of temporal data from huge collections of temporal data is usually referred to by temporal signature mining [4]. Temporal signature mining techniques have some modifications to handle temporal relationships, such as before, after, during, and so on. By attaching a time duration to the data, it becomes possible to store different states of the database.

There are many advantages of data mining over regular data analysis. It performs operations on numeric and categorical properties of data to find out hidden relationships. But when it comes to handling data associated with time interval or data that changes its dimension slowly over time, traditional data mining does not help. Because it ignores a significant feature that is associated with every sort of data *time*. Temporal data is data associated with time. Using temporal signature mining, the extracted information becomes more meaningful.

Temporal signature mining can extract patterns during a time interval that repeats itself after a certain period. Temporal signature mining gives an advantage over traditional mining techniques when dealing with multiple event ordering, duration, and heterogeneity. To extract information from event sequences, temporal mining is necessary.

For example, a system can raise an alarm if it detects an unusual behavior at a particular time. A temporal database deals with both valid time and transaction time. This way it preserves not only the updated data, but also provides a snapshot of data during a certain time interval in the past. This feature of a temporal database can be proved very useful if utilized in a proper manner in case of network intrusion detection. Temporal signature mining can be suitable to address intrusions because often they are sequences of events.

All the existing systems in network intrusion detection focus on finding the optimum solution, but so far there are not any. This is why we think much work can be done to improve the state of the art of this field. And using a temporal approach, we can use a different perspective on this problem.

In this paper, we are going to discuss why temporal signature mining is becoming more and more popular and its effectiveness in network intrusion detection. The rest of this paper is organized as follows: Sect. 2 represents the works done on network intrusion detection using traditional and temporal signature mining techniques. In Sect. 3, we discuss why temporal signature mining is a better solution than traditional data mining techniques and we draw a conclusion in Sect. 4.

2 Works on Network Intrusion Detection

2.1 Using Traditional Data Mining Techniques

A lot of research works have been done on intrusion detection in order to identify various attacks in network traffic. There are all based on traditional data mining.

Intrusion detection system can be classified into three categories [5]. These are as follows:

1. Signature-based intrusion detection
2. Classification-based intrusion detection
3. Anomaly-based intrusion detection.

Steps of network intrusion detection are as follows:

1. Monitoring and analyzing the network traffic
2. Detecting abnormality
3. Raising an alarm if assessed to be severe.

The properties of previous known attacks' signatures are used in signature-based approach, whereas in anomaly-based approach [6–8] previously undocumented events are also detected. An effective Intrusion Detection System must have high accuracy and detection rate and low false alarm rate. Two very common data mining techniques are classification and clustering. These two algorithms can detect only known attacks and in this case of detection clustering is more efficient than classification [9].

One decision tree model-based IDS (Intrusion Detection System) is proposed by Caulkins et al. [10]. The system reduces false negative and false positive alarm rates and accurately detects probable attack data sessions. The major limitation of this approach is that it can detect only TCP packets and is not suitable for use in real time.

Reference [11] uses the alerts generated by Snort, one of the most widely used open-source IDS. Then, the alerts are filtered using clustering and association rules to predict and visualize multistage attacks. The system performs well in case of finding known attacks in small scale, while fails to detect unknown attacks and to manage a real-time environment.

Transductive Confidence Machines (TCM) [12] uses k nearest neighbors algorithm to detect anomalies in the network traffic and calculates confidence using algorithmic randomness theory. This leads to good performance even with noise in the training dataset. But the major drawback of the system is that it can identify only purely clean attack data, which is impractical to expect in real-time situations.

Reference [13] provides a data preprocessing method before they enter the intrusion detection system. They propose an approach combining rule-based classifier and density-based clustering and show better results in case of memory efficiency, high accuracy, and low false alarm rates than simple k-means and expectation maximization. But it fails to provide results as fast as the others compared with it.

A database of known attacks is used in [5], which is previously loaded by attacks considered by intrusion analysts and experts. The method proposes to use honeypot [14] to obtain low false alarm rates and data warehouse for efficient memory management. But the proposed system is not in working condition, thus we cannot gather knowledge on its performance in real-life scenario.

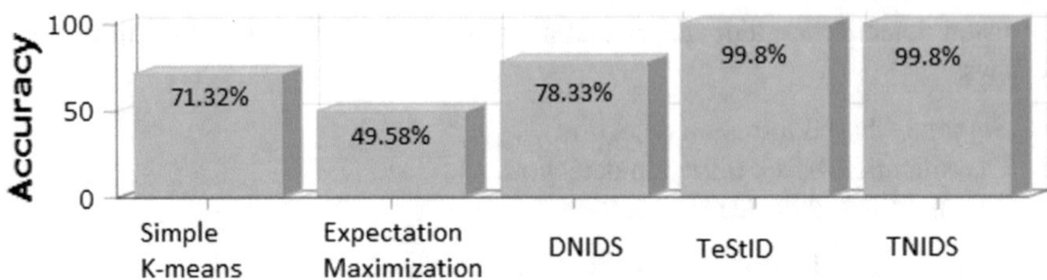

Fig. 1 Comparative analysis of network intrusion detection approach

2.2 *Using Temporal Signature Mining*

To the best of our knowledge, many works have been done using temporal signature mining in different fields especially on healthcare data [15–17], but the intrusion detection field is still in need of using the temporal aspect of data.

Reference [2] presented a network intrusion detection system that considers the time dimension of data and built a classifier using the concepts of naïve Bayes networks. He used the KDD99 data to train and test the system. For the classifier, the training data is segmented into 12 slices. The work shows surprisingly accurate results ($\geq$99.8% in all the cases). The segmentation process in his method leads to high accuracy and low false alarm rates.

Reference [1] proposed an online network-based intrusion detection system—Temporal Stream Intrusion Detection (*TeStID*), for high-speed networks using temporal logic. The author used stream data processing technology for fast processing of huge amounts of high-speed data. The system was trained and tested on the well-known DARPA 1999 data files. The use of temporal logic gives the system an intuitive way of expressing the attacks. The use of stream data processing leads to a highly scalable system.

A comparative analysis of detection accuracy among simple k-means algorithm, expectation maximization, DNIDS [13], *TeStID* [1], and TNIDS [2] is shown in Fig. 1.

3 Discussion

As reviewed in the previous section, many works have been done to detect network intrusion, using traditional data mining. But still, there are some unsolved problems.

First, the existing systems using traditional mining techniques talk about their high intrusion detection rates but do not mention the false alarm rate that increases proportionally. This is a usual limitation of most of the existing intrusion detection systems. The works of [2] using temporal analysis of data runs the data through all the 12 classifiers in the system, where each classifier classifies the data separately with a specific accuracy rate. The model with highest accuracy rate is chosen to indicate probable network intrusion. This way there remains an option for a trade-off between accuracy and false alarm rate.

Second, the time efficiency is a major factor in comparing intrusion detection systems. The use of temporal logic in [1] provides a simple illustration of attacks leading to a time-efficient approach.

Third, to address the memory efficiency, [1] stores the sessions in memory only when the events hold a temporal relationship among them. This way, it has been proved a very efficient technique in terms of storage management. The complexity of the system is reduced by using this approach.

Again, in case of detecting unknown attacks, most traditional mining techniques fail, whereas [1] can detect unknown attacks and then add it to the database and recompile for future purpose.

Moreover, considering the scalability of a system, the system proposed in [1] is highly scalable compared to other systems that use traditional mining techniques. The system outperformed two standard network-based open-source NIDs: Snort [18] and Bro [19] considering performance in real-time environment.

In case of detecting intrusions, false alarms create an unnecessary denial of essential data packets that are no threat at all. Scalability is also a significant issue in this matter. Because a system may perform well in a limited set of data, where in reality the continuous network traffic is enormous. Detecting unknown threats is a key requirement in any system. Without it, the system remains vulnerable toward newer attacks that are not yet known of. When an intrusion detection system is run on real-time data, it faces many challenges. The high-speed flow of data occupies the available storage of the system. If the storage is not handled properly, the system may crash. If these data are not handled time efficiently, it creates a lag in detection which is the last thing one wants in an intrusion detection system. A 100% accurate intrusion detection system is of no use if it detects a threat after the network has been attacked by it.

Clearly, it is seen that several works on intrusion detection using temporal signature mining provides solutions to the problems of existing methods. The system is not prone to the attacks as the temporal approach handles it all. The effectiveness and usability in a real-time environment, reducing false alarm rates, dealing with the high-speed data flow, scalability, and storage management—every purpose is served by temporal approaches. It proves the significance of using temporal aspect of data in mining techniques and encourages more work to obtain a complete intrusion detection system using temporal signature mining.

4 Conclusion

The information security issue regarding network intrusion requires fast detection of unusual behavior from large-scale continuous web log data. Comparison between traditional and temporal data mining on intrusion detection helps us to prepare traffic data for temporal signature mining, handle them using a temporal database, perform the mining techniques, address the issues that the implementation faces, and finally to assess the efficiency of the method. In this paper, we discussed the advantages and disadvantages of traditional data mining approaches to detect intrusion. Though very little work has been done on network intrusion detection using temporal signature mining, we represented their approaches. Comparing various techniques, we reached a conclusion that we can apply different frameworks of temporal signature mining on network intrusion detection to gain better solutions of this yet unsolved issue.

References

1. Ahmed, A.: Online network intrusion detection system using temporal logic and stream data processing. Ph.D. dissertation, University of Liverpool (2013)
2. Hogo, M.A.: Temporal analysis of intrusion detection. In: 2014 International Carnahan Conference on Security Technology (ICCST). IEEE, pp. 1–6 (2014)
3. Tan, P.-N.: Introduction to Data Mining. Pearson Education India (2006)
4. Lin, W., Orgun, M.A., Williams G.J.: An overview of temporal data mining. In: Proceedings of the 1st Australian Data Mining Workshop (2002)
5. Katkar, V.: Intrusion detection system. In: Fourth International Conference on Neural, Parallel & Scientific Computations (2010)
6. Wang, K.: Network payload-based anomaly detection and content-based alert correlation. Ph.D. dissertation, Columbia University (2006)
7. Mahoney, M.V., Chan, P.K.: Learning nonstationary models of normal network traffic for detecting novel attacks. In: Proceedings of the Eighth ACM SIGKDD International Conference on Knowledge Discovery and Data Mining. ACM, pp. 376–385 (2002)
8. Mahoney, M.V.: Network traffic anomaly detection based on packet bytes. In: Proceedings of the 2003 ACM Symposium on Applied Computing. ACM, pp. 346–350 (2003)
9. Wankhade, K., Patka, S., Thool, R.: An overview of intrusion detection based on data mining techniques. In: 2013 International Conference on Communication Systems and Network Technologies (CSNT). IEEE, pp. 626–629 (2013)
10. Caulkins, B.D., Lee, J., Wang, M.: A dynamic data mining technique for intrusion detection systems. In: Proceedings of the 43rd Annual Southeast Regional Conference, vol. 2. ACM, pp. 148–153 (2005)
11. Brauckhoff, D., Dimitropoulos, X., Wagner, A., Salamatian, K.: Anomaly extraction in backbone networks using association rules. IEEE/ACM Trans. Netw. (TON) **20**(6), 1788–1799 (2012)
12. Li, Y., Fang, B., Guo, L., Chen, Y.: Network anomaly detection based on TCM-KNN algorithm. In: Proceedings of the 2nd ACM Symposium on Information, Computer and Communications Security, pp. 13–19, Mar 2007
13. Sivaranjani, S., Pathak, M.R.: Network intrusion detection using data mining technique
14. Khosravifar, B., Bentahar, J.: An experience improving intrusion detection systems false alarm ratio by using honeypot. In: 2008 IEEE, 22nd International Conference on Advanced Information Networking and Applications
15. Wang, F., Lee, N., Hu, J., Sun, J., Ebadollahi, S., Laine, A.F.: A framework for mining signatures from event sequences and its applications in healthcare data. IEEE Trans. Pattern Anal. Mach. Intell. **35**(2), 272–285 (2013)
16. Henriques, R., Pina, S., Antunes, C.: Temporal mining of integrated healthcare data: methods, revealings and implications. In: Proceedings of SDM IW on Data Mining for Medicine and Healthcare, pp. 52–60 (2013)

17. Gotz, D., Wang, F., Perer, A.: A methodology for interactive mining and visual analysis of clinical event patterns using electronic health record data. J. Biomed. Inform. **48**, 148–159 (2014)
18. Sourcefire. SNORT (2010). http://www.snort.org/. Accessed 02 Jan 2013
19. Lawrence Berkeley National Laboratory. Bro Intrusion Detection System (2011). http://www.bro-ids.org/. Accessed 02 Jan 2013

An Optimized Random Forest Classifier for Diabetes Mellitus

N. Komal Kumar, D. Vigneswari, M. Vamsi Krishna and G. V. Phanindra Reddy

Abstract Machine learning-based classification algorithms help in diagnosing the symptoms at early stages by prior diagnosing of symptoms and taking medications according to it. Combining of genetic algorithm with the random forest classifier can optimize the results obtained only by the random forest classifier. In this proposed system, genetically optimized random forest classifier is used for the classification of diabetes mellitus. *Aims*. To develop an optimized random forest classifier by genetic algorithm for diabetes mellitus. *Methods.* A genetic algorithm is used in the first stage for optimizing random forest, and the optimized outputs are fed into the fine-grained random forest to diagnose the symptoms of diabetes mellitus. *Results*. In this analysis, the proposal of hybrid optimized random forest classifier (GA-ORF) with a genetic algorithm is made. In this evaluation, the various performance metrics of classifiers, GA-ORF has achieved accuracy higher than of the previously proposed classifiers for diabetes mellitus.

Keywords Multi-class decomposition · Genetic algorithm · Optimization GA-ORF · Machine learning

N. Komal Kumar (✉) · M. Vamsi Krishna · G. V. Phanindra Reddy
QIS College of Engineering and Technology, Ongole, AP, India
e-mail: komalkumarnapa@gmail.com

M. Vamsi Krishna
e-mail: vamsi.join@gmail.com

G. V. Phanindra Reddy
e-mail: nanimoonrockers30@gmail.com

D. Vigneswari
QIS Institute of Technology, Ongole, AP, India
e-mail: vigneswari121192@gmail.com

A. Abraham et al. (eds.), *Emerging Technologies in Data Mining and Information Security*, Advances in Intelligent Systems and Computing 813,
https://doi.org/10.1007/978-981-13-1498-8_67

1 Introduction

Class decomposition is defined as the process of breaking down the huge dataset into a number of subsets by applying clustering to the attributes present in the dataset that belongs from time to time. Let D = {D_1, D_2, D_3, … D_n) be a dataset which contains records, that contain various attribute values. The dataset is decomposed based upon the condition "C" by clustering; the dataset D is decomposed as D_c, containing multiple classes of attribute values. Class decomposition [1] can be seen back in 2003 where the decomposition takes by clustering technique employed in classes of attributes. In order to apply the clustering to a medical database, first, the data has to be preprocessed for supervised learning, and it takes two stages for data preprocessing. The first stage [2] applies to only the positive classes of the datasets and the second stage [3] comprises of generalizing the decomposition to negative and positive classes, respectively. Even the diversification of the dataset came from the processes, can increase the performance further, however, class decomposition must have proper parameters set. The random forest comes with its own parameter setting such as a number of trees and number of features. Realizing that the parameter setting of the random forest has a greater influence on optimizing the random forest is an optimization problem. Genetic algorithm is superior to the random forest classifiers containing parameter like local optima. The motive of the genetic algorithm is to detect the original subclasses of the random forest classifier. Ensemble on the subclasses of the random forest classifier provides an optimal and separatability of classes.

2 Related Work

Random forest is considered as a superior of all classifiers, considered as a state-of-the-art ensemble methods [4, 5], various comparisons have identified that random forest classifier is superlative [6] including gradient boosting trees. Random forest adopts models such as data replicas and bootstrap sampling called bagging [7]. The random forest has two main parameters such as a number of trees and number of features as discussed earlier. By default, the number of trees is set between 100 and 500 and number of features as $\log_2(n)$. Random forest extension has been proposed in [8]. Various problems of random forest haven been identified in [9] such as oversampling, under sampling, and sensitivity. Most recently, random forest has vision on machine learning for classification tasks [10–12]. Genetic algorithms are envisioned in recent times for hard optimization problems [13–15]. It starts with providing the solution for each individual by a chromosome, and then each chromosome is evaluated on the fitness. The fitness is used to make the chromosome to survive in the entire population. Two basic adoptions are applied, and crossover and mutation are used to generate the randomness in the solution area. Many varieties of these are proposed in [16]. Class decomposition was first introduced in [1] with high bias and less variance. The goal of the clustering process is not only clustering but also the cluster separation, the class decomposition is applied to a medical diagnosis in [2], and the clustering is done by the separation of positive and negative classes, respectively. Random forest's high-bias classifiers perform well when the clusters are remerged [1], and the class decomposition method using the very fast neural network is presented in [17]. A genetic algorithm has been widely used in optimizing random forest classifier in [18] where each chromosome is considered as a variety of trees, and also the variable length chromosome in the solution space. Recently, [19] have a thoroughly experimented a number of support vector machines and concluded that LPSVM is superior in diagnoses.

The proposed research is organized as the following sections, Sect. 3 describes the description of the dataset considered for evaluation, Sect. 4 describes the research methodology, Sect. 5 contains the experimental study, and is followed by a conclusion and further outlook in Sect. 6.

3 Dataset Description

The diabetes dataset details have been described in this section. The dataset is obtained for the repository of the University of California at Irvine. The detailed attribute value and its type are described in Table 1. The dataset contained 50 attributes with more than a lakh of observation of various patients associated with encounter id.

The dataset contained more than a lakh observations of several patients after data preprocessing and for simplicity, the observations were reduced to 2000.

4 Methodology

The main objective of this system is to optimize the inputs of random forest classifier to produce an efficient result by genetic algorithm. The first step involves in handling missing values and normalizing the values in accordance with [20]. The detailed scenario is depicted in Fig. 1. The workflow starts from the preprocessed data, where the data are subjected to be normalized, the preprocessed dataset is divided into a training set and a test/validation set, and the splitting is based upon the percentage split. The training set is given to the genetic algorithm where it will decompose into two classes such as positive and negative classes, the important parameters of random forest is kvalues, ntrees, and mtry, the random forest classifier produces number of trees according to the experimenter, these ntrees and mtry values are decomposed by class decomposition, by the way, the test/validation set is given to the fit random forest classifier and finally, the optimized random forest is obtained, and the obtained classifier is used in the prediction of diabetes mellitus.

Table 1 Attribute, class, and value of diabetes mellitus dataset

S. no	Attribute	Values
1	encounter_id	Continuous
2	patient_nbr	Continuous
3	race	Discrete
4	gender	Male or Female
5	age	Continuous
6	weight	Continuous
7	admission_type_id	Continuous
8	discharge_disposition_id	Continuous
9	admission_source_id	Continuous
10	time_in_hospital	Continuous
11	payer_code	Continuous
12	medical_specialty	Continuous
13	num_lab_procedures	Continuous
14	num_procedures	Continuous
15	num_medications	Continuous
16	number_outpatient	Continuous
17	number_emergency	Continuous
18	number_inpatient	Continuous
19	diag_1	Continuous
20	diag_2	Continuous
21	diag_3	Continuous
22	number_diagnoses	Continuous
23	max_glu_serum	Yes/No/Steady
24	A1Cresult	Yes/No/Steady
25	metformin	Yes/No/Steady
26	repaglinide	Yes/No/Steady
27	nateglinide	Yes/No/Steady
28	chlorpropamide	Yes/No/Steady
29	glimepiride	Yes/No/Steady
30	acetohexamide	Yes/No/Steady
31	glipizide	Yes/No/Steady
32	glyburide	Yes/No/Steady
33	tolbutamide	Yes/No/Steady
34	pioglitazone	Yes/No/Steady
35	rosiglitazone	Yes/No/Steady
36	acarbose	Yes/No/Steady
37	miglitol	Yes/No/Steady
38	troglitazone	Yes/No/Steady

(continued)

Table 1 (continued)

S. no	Attribute	Values
39	tolazamide	Yes/No/Steady
40	examide	Yes/No/Steady
41	citoglipton	Yes/No/Steady
42	insulin	Yes/No/Steady
43	glyburide–metformin	Yes/No/Steady
44	glipizide–metformin	Yes/No/Steady
45	glimepiride–pioglitazone	Yes/No/Steady
46	metformin–rosiglitazone	Yes/No/Steady
47	metformin–pioglitazone	Yes/No/Steady
48	changc	Yes/No/Steady
49	diabetesMed	Yes/No/Steady
50	readmitted	Yes/No

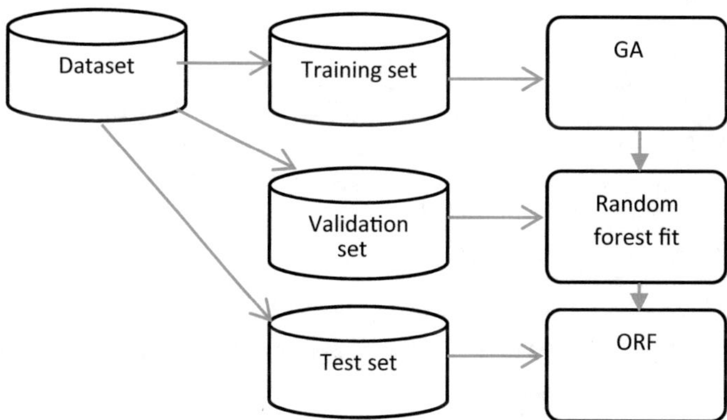

Fig. 1 Optimizing RF framework

Algorithm 1: Multi class decomposition

```
Input: minK, maxK, minNTree, maxNTree, treeIncrement
RF best ← 0
RFC best ← 0
k ← minK
i ← 0
while k < maxK do
Dc ← kmeans(D, k)
for (n = minNTree, n < maxNTree, n = n + treeIncrement) do
        RF ← randomForest(D, n)/2
        RFC ← randomForest(DClust, n)/2
if(RF > RF best) then
          RF best ← RF
end if
if (RFC > RFC best) then
          RFC best← RFC
        end if
end for
k ← (k + 1)
end while
```

Algorithm 2: Genetic Algorithm

```
Input: (it, n, GA Parameters )
begin
Initialize c=0 and i=0,
Generation: generate random n solutions ;
      Compute Fitness(s) and Generation c ;
While fitness not reached compute for i iterations do
Generation c+1   evolve(Generation c) ;
fitness   computeFitness(s) and Generation c;
            i =i + 1;
end
return(solution fitness)
end
```

Algorithm 3: Fitness computation

```
Input: Dataset(D), Chromosome
Output: Accuracy of the Random Forests
begin
D – Dataset;
Compute kvalues, ntrees, mtry  by decoding (Chromosome);
Dc – decompose the set as (D, kvalues);
Fitnessmodel – RF fit(Dc, ntrees, mtry);
Accuracy – evaluate(model)
return(accuracy);
end
```

Algorithm 4: Optimized Random Forest

```
Input: minK, maxK, minNTree, maxNTree, treeIncrement,RF best,RF fit
Output: Optimized RF
begin
Compute computeFitness(s) and Generation c
Evaluate fitness and return fitness,
Accuracy(Fit RF best)
Fit RF best= Optimized RF
Optimized RF(D)= solution
end
```

The main steps involved in the output of optimized random forest are as follows:

Step 1: Multi-class decomposition

In this step, the classes of the diabetes patient are decomposed into two classes such as positive and negative classes.

Step 2: To train multi-class decomposed approach and evaluate the fitness with GA

In this step, the obtained classes are optimized based on the genetic algorithm, and genetic algorithm gets the input as random forest, computes the fitness of each and every tree in the forest, and iterates till the optimized trees are found.

Step 3: Fitness computation

In this step, fitness of the trees is evaluated based on the values of kvalues, ntrees, and mtry by decoding the chromosomes, and the decomposed class from step 2 are evaluated for fitness.

Step 4: Optimized RF classifier

In this step, the optimized RF classifier is obtained from the combination of multi-class decomposition and genetic algorithm.

Step 5: Termination

Terminate the operation, if the optimal RF is obtained.

Step 6: Result

This step contains the classified data for predicting diabetes mellitus.

The accuracy, specificity, and sensitivity of the classifiers are compared with the optimized RF, and the results are obtained.

5 Experimental Study

The aim of this experimental study is to establish a usefulness of genetic algorithm with the combination of class decomposition to get an optimized random forest classifier applied on a medical dataset. In order to achieve this aim, the class is divided by multi-class decomposition, and the inputs are given to the genetic algorithm with inputs as random forest parameters such as kvalues, ntrees, and mtry, thus producing an optimized random forest classifier. The details of implementation environment and results are discussed in this section.

The confusion matrix of the PCA-RFC, Fisher-RFC, GA-RFC, Relief-RFC, and GA-ORFC as shown in the Table 2 is obtained in the analysis of diabetes mellitus. Accuracy, specificity, sensitivity, MCC, ROC, and Kappa Statistics (KS) are computed based on the confusion matrix. As shown in Table 3, PCA-RFC achieved an accuracy of 0.839, which is slightly more than the relief-RFC, which achieved 0.812, Fisher-RFC achieved 0.755 while GA-RFC managed to achieve 0.921 and finally, GA-ORFC achieved an accuracy of 0.945 being the topmost accurate classifier. Being very specific to the inputs, GA-ORF achieved 0.924 while PCA-RFC as 0.831, Fisher-RFC as 0.748, relief-RFC as 0.821, and GA-RFC as 0.895. On the sensitivity rates, GA-ORF achieved 0.901, PCA-RFC as 0.845, Fisher-RFC as 0.785, GA-RFC as 0.899, and relief-RFC as 0.811. ROC rates of PCA-RF as 0.899, Fisher-RFC as 0.745, GA-RFC as 0.645, relief-RFC as 0.865, and GA-ORF as 0.874, respectively, as shown in Fig. 2.

The Kappa Statistics were also calculated for the classifiers under analysis for an inter-rater agreement for categorical items.

Table 2 Confusion matrix

Algorithm	TP	TN	FP	FN
PCA-RFC	47	4	18	39
Fisher-RFC	48	3	14	43
GA-RFC	45	6	10	47
Relief-RFC	45	6	12	45
GA-ORFC	42	9	25	32

Confusion matrix is obtained with a percentage split of 65%

Table 3 Performance metrics

Algorithm	Accuracy	Specificity	Sensitivity	MCC	ROC	KS
PCA-RFC	0.839	0.831	0.845	0.678	0.899	0.677
Fisher-RFC	0.755	0.748	0.785	0.762	0.745	0.764
GA-RFC	0.921	0.895	0.899	0.877	0.645	0.674
Relief-RFC	0.812	0.821	0.811	0.832	0.865	0.812
GA-ORFC	0.945	0.924	0.901	0.886	0.874	0.879

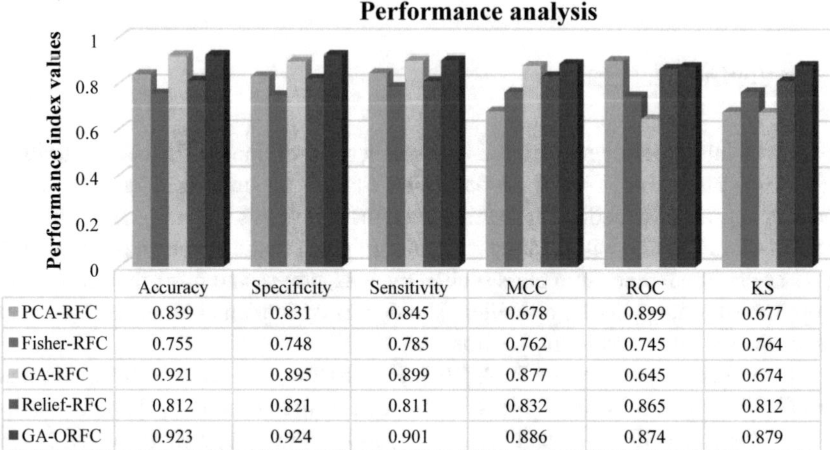

Fig. 2 Performance metrics comparisons for optimized random forest classifier with genetic algorithm

6 Conclusion and Future Outlook

An optimized random forest classifier with the genetic algorithm is proposed in this paper. The main idea of the medical diagnosis is to extract the valuable information from the symptoms, and providing an appropriate medication in a lesser amount of time. The proposed system is compared with the existing hybrid classifiers, and the results are obtained. The proposed GA-ORF classifier outperformed with an accuracy of 0.923, specificity of 0.924, the sensitivity of 0.901, and Kappa Statistics of 0.879, which are higher than the existing classifier approaches for diabetes mellitus. The future investigation can take done in combining other classifier algorithms with hybrid genetic algorithms for greater accuracy.

References

1. Vilalta, R., Achari, M.K., Eick, C.F.: Class decomposition via clustering: a new framework for low-variance classifiers. In: Third IEEE International Conference on Data Mining, 2003. ICD 2003, pp. 673–676. IEEE (2003)
2. Polaka, I.: Clustering algorithm specifics in class decomposition. In: Proceedings of the 6th International Scientific Conference on Applied Information and Communication Technology, 2013, pp. 29–36 (2013)
3. Elyan, E., Gaber, M.M.: A fine-grained random forests using class decomposition: an application to medical diagnosis. Neural Comput. Appl. 1–10 (2015)
4. Breiman, L.: Random forests. Mach. Learn. **45**(1), 5–32 (2001)
5. Cutler, D.R., Edwards Jr., T.C., Beard, K.H., Cutler, A., Hess, K.T., Gibson, J., Lawler, J.J.: Random forests for classification in ecology. Ecology **88**(11), 2783–2792 (2007)
6. Fernańdez-Delgado, M., Cernadas, E., Barro, S., Amorim, D.: Do we need hundreds of classifiers to solve real world classification problems? J. Mach. Learn. Res. **15**, 3133–3181 (2014)
7. Breiman, L.: Bagging predictors. Mach. Learn. **24**(2), 123–140 (1996)
8. Fawagreh, K., Gaber, M.M., Elyan, E.: Random forests: from early developments to recent advancements. Syst. Sci. Control Eng. Open Access J. **2**(1), 602–609 (2014)
9. del Ro, S., Lpez, V., Bentez, J.M., Herrera, F.: On the use of mapreduce for imbalanced big data using random forest. Inf. Sci. **285**, 112–137 (2014). Processing and Mining Complex Data Streams
10. Liu, X., Song, M., Tao, D., Liu, Z., Zhang, L., Chen, C., Bu, J.: Random forest construction with robust semi supervised node splitting. IEEE Trans. Image Process. **24**(1), 471–483 (2015)
11. Ristin, M., Guillaumin, M., Gall, J., Gool, L.V.: Incremental learning of random forests for large-scale image classification. IEEE Trans. Pattern Anal. Mach. Intell. **38**(3), 490–503 (2016)
12. Li, T., Ni, B., Wu, X., Gao, Q., Li, Q., Sun, D.: On random hyper-class random forest for visual classification, Neurocomputing 281–289 (2016)
13. BoussaïD, I., Lepagnot, J., Siarry, P.: A survey on optimization metaheuristics. Inf. Sci. **237**, 82–117 (2013)
14. Eiben, A.E., Smith, J.E.: Introduction to Evolutionary Computing. Springer Science & Business Media (2003)
15. Whitley, D.: A genetic algorithm tutorial. Stat. Comput. **4**(2), 65–85 (1994)
16. Davis, L.D., De Jong, K., Vose, M.D., Whitley, L.D.: Evolutionary Algorithms, vol. 111. Springer Science & Business Media (2012)
17. Jaiyen, S., Lursinsap, C., Phimoltares, S.: A very fast neural learning for classification using only new incoming datum. IEEE Trans. Neural Netw. **21**(3), 381–392 (2010)
18. Azar, T., Elshazly, H.I., Hassanien, A.E., Elkorany, A.M.: A random forest classifier for lymph diseases. Comput. Methods Programs Biomed. **113**(2), 465–473 (2014)
19. Azar, A.T., El-Said, S.A.: Performance analysis of support vector machines classifiers in breast cancer mammography recognition. Neural Comput. Appl. **24**(5), 1163–1177 (2014)
20. Stekhoven, D.J.: missForest: Nonparametric Missing Value Imputation using Random Forest, r package version 1.4 (2013)

Development of Cardiac Disease Classifier Using Rough Set Decision System

B. Halder, S. Mitra and M. Mitra

Abstract This paper describes the rough set classifier for cardiac disease classification over the medical dataset obtained from characteristics feature of ECG signals. The sets of characterizes feature are used as an information system to find minimal decision rules that may be used to identify one or more diagnostic classes. After **gathering** knowledge from various medical books as well as feedback from well-known cardiologists, a knowledge base has been developed. The rough set-based **degree of attributes dependency** technique and their significance predicted the universal least decision rules. Such rule has the least number of attributes so that their combination defines the largest subset of a universal decision class. Hence, the minimal rule of an information system is adequate for predicting probable complications. Lastly, the performance parameters such as accuracy and sensitivity have been expressed in the form of confusion matrix by ROSETTA software which yields information about actual and predicted classification achieved by the proposed system.

Keywords Information system · Rough set · ECG · Differential histogram · MI

B. Halder (✉)
Department of Computer Science & Engineering, Neotia Institute of Technology, Management & Science Jhinga, D.H. Road, 24-parganas(s), Kolkata 743368, West Bengal, India
e-mail: basudev_h@yahoo.com

S. Mitra
Department of Electronics, Netaji Nagar Day College (Affiliated to University of Calcutta), 170/436, NSC Bose Road, Regent Estate, Kolkata 700092, India
e-mail: s_mitrasarkar@rediffmail.com

M. Mitra
Faculty of Technology, Department of Applied Physics, University of Calcutta, 92, APC Road, Kolkata 700009, India
e-mail: mmaphy@caluniv.ac.in

A. Abraham et al. (eds.), *Emerging Technologies in Data Mining and Information Security*, Advances in Intelligent Systems and Computing 813,
https://doi.org/10.1007/978-981-13-1498-8_68

1 Introduction

The concept of rough sets theory introduced by Z. Pawlak [1, 2] in 1982 is a relatively new intelligent numerical tool to manage vague, imprecise, or uncertain information that has been wielded to the medical field. It is used for the innovation of data dependencies, reduces all redundant objects and features, and recognizes and classifies objects in medical field [3]. In addition, one of the significant application of rough set "is the generation of decision rules from a given information system for classification of known objects" [4].

Many researchers have endeavored to develop "rough set approach for feature reduction and creation of classification rules from a set of health datasets." Mitra et al. [5, 6] presented an idea of "rule-based rough-set decision system" for the development of a disease "inference engine for ECG classification" from the different standard time-plane feature. In [7], implemented a comparatively innovative application of rough sets to reduce the detection rules for the features in ECG, and the reduced rules are used as constraint circumstances of eigenvalue determination arithmetic to recognize characteristic points in ECG. Tsumoto [8] implemented the rough sets-based knowledge discovery system and feature-oriented generalization and its use in medicine. In [9], implemented a rough set technique where from a medical dataset, characteristic points are reduced and the invention of classification rules. In Bayes' theorem, Pawlak [10] applied the rough set theory that can be used for generating rule base to recognize the presence or absence of disease.

Development of a rough set-based minimal decision rules for the diagnosis of the cardiac diseases from the features of ECG signal is the main involvement of this paper. For this purpose, knowledge base has been developed from the essential features of ECG signal [11] after consultation of various medical books as well as feedback from well-known cardiologists. The rough set-based degree of attributes dependency and their significance predicted the universal minimal decision rules. Set of decision rules has been computed from the reduct of information table and classifier is developed using rule-based rough set method. Finally, classification accuracy, coverage, and true positive rate have been expressed in the form of confusion matrix which yields information about predicted and actual classification achieved by a classification system.

The rest of this paper is organized as follows. Section 2 describes rough set theory. Section 3 describes the ECG interpretation. Development of disease classification system is described in Sect. 4. Section 5 describes the experimental results. Finally, concludes the paper in Sect. 6.

2 Rough Set Preliminaries

The fundamental concept behind rough set theory is the upper and the lower approximation of the vague concept. The *lower approximation* of a set of attributes is defined by objects that will certainly belong to the concept and the *upper approximation* of a set of attributes is defined by objects that will possibly belong to the concept. Observably, the difference between the upper and the lower approximation of a set of attributes form the boundary region of the vague concept. If the upper and lower approximation is equal, i.e., boundary region is empty, then the set is **crisp**, otherwise, it is **rough or inexact**.

In rough sets theory, an "*information system* is a pair I = (*U, A*)", where *U* is a non-empty, finite set of objects called the universe and *A* is a non-empty finite set of attributes; any attribute $\boldsymbol{a}$ is a map $a : U \rightarrow V_a \; for \; a \in A$. The set V_a is called the *value set* (domain) of *a*. A *decision table* is an information system of the form I = $(U, A \cup \{d\})$, where $d \notin A$ is a distinguished attribute called *decision*. The elements of *A* are called *conditions*. Let $P \subseteq A$ be a subset of the attributes, then the set of P—indiscernible objects is the set of objects having the same set of attribute values are defined as follows:

$$\boldsymbol{IND_I(P)} = \{(\boldsymbol{x}, \boldsymbol{y}), \boldsymbol{x}, \boldsymbol{y} \in \boldsymbol{U} | \forall \boldsymbol{a} \in \boldsymbol{A}, \boldsymbol{x}(\boldsymbol{a}) = \boldsymbol{y}(\boldsymbol{a})\} \quad (1)$$

In simple words, two objects are indiscernible if they cannot be discerned between them, because they do not differ enough. So, an indiscernibility relation partitions the set of objects into a number of equivalent classes. For any concept, $X \subseteq U$ and $B \subseteq A$, X could be approximated by the lower and upper approximations, denoted by $\underline{B}(X)$ and $\bar{B}(X)$, are defined as follows:

$$\bar{\boldsymbol{B}}(\boldsymbol{X}) = \{\boldsymbol{x} \in \boldsymbol{U} : [\boldsymbol{x}]_{\boldsymbol{B}} \subseteq \boldsymbol{X}\} \quad (2)$$

$$\underline{\boldsymbol{B}}(\boldsymbol{X}) = \{\boldsymbol{x} : [\boldsymbol{x}]_{\boldsymbol{B}} \cap \boldsymbol{X} \neq \emptyset\} \quad (3)$$

The boundary region of a set *X*, is defined as:

$$\boldsymbol{BN_B}(\boldsymbol{X}) = \bar{\boldsymbol{B}}(\boldsymbol{X}) - \underline{\boldsymbol{B}}(\boldsymbol{X}) \quad (4)$$

If *X* is the empty set, i.e., $BN_B\,(X) = \emptyset$, then *X* is crisp (exact) set with respect to *B*. On the other hand, if $BN_B\,(X) \neq \emptyset$, *X* is ***rough (inexact) set*** with respect to *B*.

The positive region, denoted by $POS_C(D)$ is defined as follows:

$$\boldsymbol{POS_C}(\boldsymbol{D}) = \bigcup\nolimits_{\boldsymbol{X} \in \boldsymbol{U}/\boldsymbol{D}} \underline{\boldsymbol{C}}(\boldsymbol{X}) \quad (5)$$

of the partition U/D with respect to C, i.e., the set of all objects in U that can be uniquely classified by elementary sets in the partition U/D by means C. A subset $R \subseteq C$ is said to be a D-reduct of *C* if $\boldsymbol{POS_R}(\boldsymbol{D}) = \boldsymbol{POS_C}(\boldsymbol{D})$. A reduct with minimal cardinality (the number of elements in a set) is called a ***minimal reduct***. It is a fundamental concept which can be used for knowledge reduction.

One of the significant features of the information system is the dependency of attributes in the indiscernibility relation. Attribute D depends totally on attribute C, denoted $C \Rightarrow D$, if all values of attributes D are uniquely determined by values of attributes C. In other words, D depends totally on C, if there exists a functional dependency between values of D and C. *Degree of dependency* of attribute D on attribute C, denoted $C \Rightarrow_K D$ is defined by:

$$K = \gamma(C, D) = \frac{|POS_C D|}{|U|} = \frac{\sum_{X \in U/D} |\underline{C}(X)|}{|U|} \tag{6}$$

where D depends totally on C, if K = 1, D depends partially on C, if 0 < K < 1 and D does not depend on C, if K = 0.

3 ECG Interpretation

An electrocardiogram (ECG), shown in Fig. 1, is an electrical view of the heart which produces waveforms with specific shapes and duration that occurs at a certain rate and regularity. Every ECG consists of three distinct waves labeled by P, QRS, and T followed by a conditional U-wave. Accurate and complete extraction of ECG signal and its characteristic features is the premise to assist doctors with their task of diagnosing of many cardiac diseases [12]. The most prominent feature ECG signal is QRS complex and identification of QRS complex is the basis of almost all automated ECG analysis algorithms. It can afford useful information about the heart condition. Additional ECG components like P, QRS onset–offset, PR interval, T-wave, ST segment, and QT interval, etc., are used subsequently for automatic ECG analysis.

Information table has been generated for this classification system from the extracted features of ECG signal [11]. The whole module has been applied to various ECG data of all the 12 leads taken from PTB diagnostic database (PTB-DB) of PhysioNet (www.physionet.org).

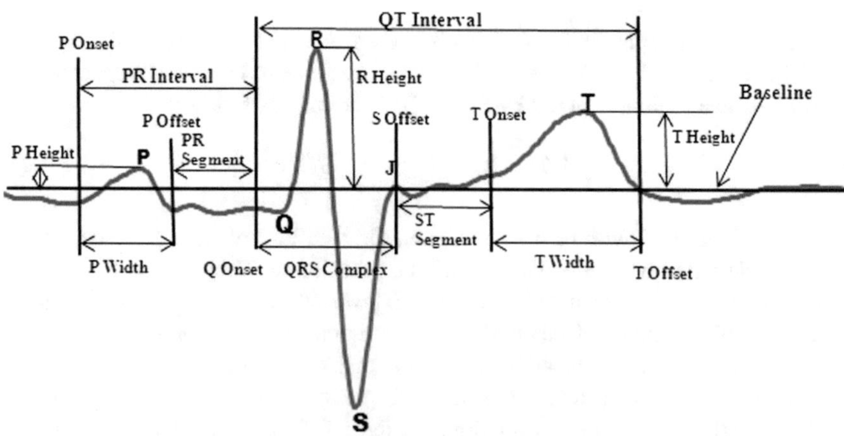

Fig. 1 Typical ECG signal with its distinctive points

Table 1 A knowledge base for cardiac diseases

Localization wall	ST elevated	ST depressed	T inverted	T upright	Q pathologic
Anterior MI	V2, V3, V4	II, III, VF	V2, V3, V4	II, III, VF	V2, V3, V4
Inferior MI	II, III, VF	I, VL	II, III, VF	I, VL	II, III, VF

4 Development of Knowledgebase and Disease Classification System

In this research, we concentrated on analysis and classification of cardiac disease. For this, a knowledge base (shown in Table 1) about the characteristic features of ECG signal has been expanded from the different medical books [13, 14] and the comment of specialist cardiologists. With the help of differential histogram technique [11], all the features have been extracted from each of the 12-lead ECG signals. Table 1 is the **Knowledgebase for Cardiac diseases** generated mainly the lead positions where the specific irregularity occurs. For example, when infarction or necrosis occurs in the *Inferior wall* of the heart, then it is classified as *Inferior wall infarction* and this infarction pattern is replicated in leads II, III, and aVF. Similarly, damage in *anterior* wall of the heart is referred to as *anterior wall infarction* and reflected leads are V2, V3, and V4. We have selected some time-plane features that depend on the knowledge base for identification of the damaged area. In this paper, we have concentrated on the location of the damage in the Inferior Wall and Anterior Wall MI.

In this paper, the classification system is tested with three types of ECG data, namely—*normal*, *Inferior MI*, and A*nterior MI*. We have computed the degree of dependency (k) using the rough set technique. For this purpose, the most significant conditional attributes with respect to the decision attributes have been found. According to their significance, all the attributes are arranged and finally, reducts serve the purpose of inducing minimal decision rules. One of popular and widely used rough set software toolbox is ROSETTA [15]. This free tool supports different selection for generating decision tables, discretization techniques, reducts, decision algorithms, and classifications. In this research, we have used this free tool.

5 Experimental Results

5.1 Rules Generation Using Rough Set Technique

Now, we describe a method for generating the least number of rules (i.e., rules with a minimum number of descriptors on left-hand side) in information systems and decision tables. The technique is based on the idea of rough set which is based on indiscernibility relation, applied to compute the degree of attributes dependency.

Table 2 Information table

Patient ID	Patient Name	Lead II				Lead III				Lead aVF				Lead V2				Lead V3				Lead V4									
		ST segment	T Wave	Q Pathologic	R progression	ST segment	T Wave	Q Pathologic	R progression	ST segment	T Wave	Q Pathologic	R progression	ST segment	T Wave	Q Pathologic	R progression	ST segment	T Wave	Q Pathologic	R progression	ST segment	T Wave	Q Pathologic	Q Pathologic	HR (b pm)	QRS width (sec)	R height (mv)	QTc (sec)	P height (mv)	Disease (D)
		C1	C2	C3	C4	C5	C6	C7	C8	C9	C10	C11	C12	C13	C14	C15	C16	C17	C18	C19	C20	C21	C22	C23	C24	C25	C26	C27	C28	C29	C30
1	s0015lre	D	U	N	N	D	U	N	N	D	U	Y	N	E	I	Y	Y	E	I	Y	Y	E	I	Y	Y	NO	NO	AN	NO	AN	anterior
2	s0034_re	D	U	N	N	D	U	N	N	D	U	N	N	E	I	N	Y	E	I	N	Y	E	I	N	Y	NO	NO	NO	NO	AN	anterior
3	s0046lre	D	U	Y	N	D	U	Y	N	D	U	Y	N	E	U	N	N	E	I	N	N	E	I	N	N	NO	NO	AN	NO	AN	anterior
4	s0182_re	D	U	N	N	D	U	N	N	D	U	N	N	E	I	N	Y	E	I	N	Y	E	I	N	Y	NO	NO	NO	NO	AN	anterior
5	s0029lre	D	U	N	N	D	U	N	N	D	U	N	N	E	I	N	Y	E	I	Y	Y	E	I	Y	Y	NO	NO	AN	NO	AN	anterior
6	S0511_RE	D	U	N	N	D	U	N	N	D	U	N	N	E	I	N	Y	E	I	N	Y	E	I	N	Y	NO	NO	NO	NO	AN	anterior
7	s0028lre	E	I	Y	N	E	I	Y	N	E	I	Y	N	D	U	N	Y	D	U	N	Y	D	U	N	N	NO	NO	AN	NO	AN	inferior
8	s0037lre	E	I	Y	N	E	I	Y	N	E	I	Y	N	D	U	N	Y	D	U	N	Y	D	U	N	N	NO	NO	AN	NO	AN	inferior
9	s0039lre	E	U	N	N	E	I	Y	N	E	I	Y	N	D	U	N	Y	D	U	N	Y	D	U	N	Y	NO	NO	NO	NO	AN	inferior
10	s0044lre	E	U	N	N	E	I	Y	N	E	I	Y	N	D	U	N	Y	D	U	N	Y	D	U	N	Y	NO	NO	AN	NO	AN	inferior
11	s0049lre	E	I	Y	N	E	I	Y	N	E	I	Y	N	D	U	N	Y	D	U	N	Y	D	U	N	Y	NO	NO	AN	NO	AN	inferior
12	s0067lre	D	I	Y	N	E	I	Y	N	E	I	Y	N	D	U	N	Y	D	U	N	Y	D	U	N	Y	NO	NO	AN	NO	AN	inferior
13	s0474_re	D	U	N	N	D	I	Y	N	D	U	N	N	E	U	N	Y	E	U	N	Y	E	U	N	Y	NO	NO	AN	NO	AN	normal
14	s0499_re	D	U	Y	N	D	I	Y	N	D	U	N	N	E	U	N	Y	E	U	N	Y	E	U	N	Y	NO	NO	AN	NO	AN	normal
15	s0480_re	D	U	Y	N	D	I	Y	N	D	U	N	N	E	U	N	Y	E	U	N	Y	E	U	N	Y	NO	NO	AN	NO	AN	normal
16	s0311lre	D	U	N	N	D	I	N	N	D	U	N	N	iso	I	N	Y	iso	U	N	Y	iso	U	N	Y	NO	NO	NO	NO	AN	normal
17	s0301lre	D	U	N	N	D	U	N	N	D	U	N	N	iso	U	Y	Y	iso	U	Y	Y	iso	U	N	Y	NO	NO	AN	NO	AN	normal

Where D: Depressed, E: Elevated, U: Upward, I: Inverted, N: No, Y:Yes, NO: Normal, AN:Abnormal, iso: isoelectric,

Example 1 illustrates the process of rule generation from the features of ECG signal as an information system.

Example 1 Let us consider a medical information system I = (U,A) given in Table 2, where U = {X_k : k = 1…17} is a non-empty finite set of objects called the ***universe*** of I and A is a non-empty finite set of attributes; any attribute ***a*** is a map $a : U \rightarrow V_a for\ a \in A$. The set V_a is called the *value set* (domain) of *a*. We might partition the attribute set A into C = {C_j: j = 1 … 29} and D = {C30}, called condition and decision attributes, respectively. The first step of this technique is to obtain the sets of attribute values of each attribute induced by discernible attribute value. For example, in Table 2, the sets of attribute values are: V_{c1} = {E,D}, V_{c2} = {I,U}, V_{c3} = {Y,N}, V_{c5} = {E,D}, V_{c6} = {I,U}, V_{c7} = {Y,N}, V_{c9} = {E,D}, V_{c10} = {U,I}, V_{c11} = {Y,N}, V_{c13} = {E,D,ISO}, V_{c14} = {U,I}, V_{c15} = {Y,N}, V_{c16} = {Y,N}, V_{c17} = {E,ISO,D}, V_{c18} = {U,I}, V_{c19} = {Y,N}, V_{c20} = {Y,N}, V_{c21} = {E,ISO,D}, V_{c22} = {U,I}, V_{c23} = {Y,N}, V_{c24} = {Y,N}, V_{c27} = {AN,NO}.

Here, we eliminated the conditional attribute C4, C8, C12, C25, C26, C28, and C29, since all the samples are indiscernible with respect to these conditions. Hence, out of 29 attributes, 22 attributes are now considered.

The second step of this technique is to compute the indiscernible objects and the partitions as shown in below from the information Table 2.

P1: X(C1 = E) = {7,8,9,10,11}, X(C1 = D) = {1,2,3,4,5,6,12,13,14,15,16,17}
IND_I (C1) = {{7,8,9,10,11}, {1,2,3,4,5,6,12,13,14,15,16,17}}
P2: X(C2 = I) = {7,8,11,12}, X (C2 = U) = {1,2,3,4,5,6,9,10,13,14,15,16,17}
IND_I (C2) = {{7,8,11,12}, {1,2,3,4,5,6,9,10,13,14,15,16,17}}

P3: X(C3 = Y) = {7,8,9,10,11,12}, X (C3 = N) = {1,2,3,4,5,6,13,14,15,16,17}
IND_I (C3) = {{7,8,9,10,11,12}, {1,2,3,4,5,6,7,13,14,15,16,17}}
P4: X(C5 = D) = {1,2,3,4,5,6,13,14,15,16,17}, X(C5 = E) = {7,8,9,10,11,12}
IND_I (C5) = {{1,2,3,4,5,6,13,14,15,16,17}, {7,8,9,10,11,12}}
P5: X(C6 = I) = {7,8,9,10,11,12,13,14,15,16,17}, X(C6 = U) = {1,2,3,4,5,6,17}
IND_I (C6) = {{7,8,9,10,11,12,13,14,15,16,17}, {1,2,3,4,5,6,17}}
P6: X(C7 = Y) = {3,4,7,8,9,10,11,12,13,14,15}, X (C7 = N) = {1,2,4,5,6,16,17}
IND_I (C7) = {{3,4,7,8,9,10,11,12,13,14,15}, {1,2,4,5,6,16,17}}
P7: X (C9 = D) = {1,2,3,4,5,6,13,14,15,16,17}, X (C9 = E) = {7,8,9,10,11,12}
IND_I (C9) = {{1,2,3,4,5,6,13,14,15,16,17}, {7,8,9,10,11,12}}
P8: X(C10 = I) = {7,8,9,10,11,12}, X (C10 = U) = {1,2,3,4,5,6,13,14,15,16,17}
IND_I (C10) = {{7,8,9,10,11,12}, {1, 2,3,4,5,6,13,14,15,16,17}}
P9: X(C11 = Y) = {1,3,7,8,9,10,11,12}, X (C11 = N) = {2,4,5,6,13,14,15,16,17}
IND_I (C11) = {{1,3,7,8,9,10,11,12}, {2, 4,5,6,13,14,15,16,17}}
P10: X(C13 = D) = {7,8,9,10,11,12}, X (C13 = E) = {1,2,3,4,5,6,13,14,15}, X (C13 = ISO) = {16,17}.
IND_I (C13) = {{7,8,9,10,11,12}, {1,2,3,4,5,6,13,14,15}, {16,17}}
P11: X(C14 = I) = {1,2,4,5,6,16}, X (C14 = U) = {3,7,8,9,10,11,12,13,14,15,17},
IND_I (C14) = {{1,2,4,5,6,16}, {3,7,8,9,10,11,12,13,14,15,17}}
P12: X(C15 = Y) = {1,17}, X (C15 = N) = {2, 3,4,5,6,7,8,9,10,11,12,13,14,15,16}
IND_I (C15) = {{1,17}, {2, 3,4,5,6,7,8,9,10,11,12,13,14,15,16}}
P13: X(C16 = N) = {3}, X (C16 = N) = {1,2,4,5,6,7,8,9,10,11,12,13,14,15,16,17}
IND_I (C16) = {{3}, {1,2,4,5,6,7,8,9,10,11,12,13,14,15,16,17}}
P14: X(C17 = D) = {7,8,9,10,11,12}, X (C17 = E) = {1,2,3,4,5,6,13,14,15}, X (C17 = ISO) = {16,17},
IND_I (C17) = {{7,8,9,10,11,12}, {1,2,3,4,5,6,13,14,15}, {16,17}}
P15: X(C18 = I) = {1,2,3,4,5,6}, X (C18 = U) = {7,8,9,10,11,12,13,14,15,16,17}
IND_I (C18) = {{1,2,3,4,5,6}, {7,8,9,10,11,12,13,14,15,16,17}}
P16: X(C19 = Y) = {1,5,17}, X (C19 = N) = {2,3,4,6,7,8,9,10,11,12,13,14,15,16}
IND_I (C19) = {{1,5,17}, {2,3,4,6,7,8,9,10,11,12,13,14,15,16}}
P17: X(C20 = N) = {3}, X (C20 = Y) = {1,2,4,5,6,7,8,9,10,11,12,13,14,15,16,17}
IND_I (C20) = {{3}, {1,2,4,5,6,7,8,9,10,11,12,13,14,15,16,17}}
P18: X(C21 = D) = {7,8,9,10,11,12}, X (C21 = E) = {1,2,3,4,5,6,13,14,15}, X (C21 = ISO) = {16,17},
IND_I (C21) = {{7,8,9,10,11,12}, {1,2,3,4,5,6,13,14,15}, {16,17}}
P19: X(C22 = I) = {1,2,3,4,5,6}, X (C17 = U) = {7,8,9,10,11,12,13,14,15,16,17}
IND_I (C22) = {{1,2,3,4,5,6}, {7,8,9,10,11,12,13,14,15,16,17}}
P20: X(C23 = Y) = {1,5}, X (C23 = N) = {2,3,4,6,7,8,9,10,11,12,13,14,15,16,17}
IND_I (C23) = {{1,5}, {2,3,4,6,7,8,9,10,11,12,13,14,15,16,17}}
P21: X(C24 = N) = {3,7,8}, X (C24 = Y) = {1,2,4,5,6,9,10,11,12,13,14,15,16,17}
IND_I (C24) = {{3,7,8}, {1,2,4,5,6,9,10,11,12,13,14,15,16,17}}
P22: X(C27 = AN) = {1,3,5,7,8,10,11,12,13,14,15,17}, X (C27 = NO) = {2,4,6,9,16}
IND_I (C27) = {{1,3,5,7,8,10,11,12,13,14,15,17}, {2,4,6,9,16}}

P23: X(C30 = anterior) = {1,2,3,4,5,6}, X(C30 = inferior) = {7,8,9,10,11,12},X (C30 = normal) = {13,14,15,16,17}. IND_I (C30) = {{1,2,3,4,5,6},{7,8,9,10,11,12}{13,14,15,16,17}}

The next step of this technique is to compute the degree of dependency of attribute D on attribute C. Let us consider C = {C1} and disease = {D}. We already have IND_I (disease) = {{1,2,3,4,5,6},{7,8,9,10,11,12}{13,14,15,16,17}} and IND_I (C1) = {{7,8,9,10,11}, {1,2,3,4,5,6,12,13,14,15,16,17}}. Since{7,8,9,10,11} ⊆ {7,8,9,10,11,12}{7,8,9,10,11,12}, but{1,2,3,4,5,6,12,13,14,15,16 ⊈ {1,2,3,4, 5,6}, {1,2,3,4,5,6,12,13,14,15,16,17} ⊈{7,8,9,10,11,12}, and {1,2,3,4,5,6,12,13, 14, 15, 16,17} ⊈{13,14,15, 16, 17}, then attribute dependency of Disease (D) on C1 in a degree K(0 K 1) is computed based on Eq. (6), and is defined as follows:

$$K_{P1} = \frac{|\{7, 8, 9, 10, 11\}|}{|U|} = \frac{5}{17} = 0.29, \quad \text{Hence } C_1 \Rightarrow_{0.29} (D)$$

This implies that the attribute D is partially dependent on the attribute C_1. In a similar manner, the other attribute dependencies for D are calculated and given below:

$$K_{P2} = \frac{|\{7, 8, 11, 12\}|}{|U|} = \frac{4}{17} = 0.23,\ K_{P3} = \frac{|\{7, 8, 9, 10, 11, 12\}|}{|U|} = 0.35,\ K_{P4} = \frac{|\{7, 8, 9, 10, 11, 12\}|}{|U|} = 0.35,$$

$$K_{P5} = \frac{|0|}{|U|} = 0,\quad K_{P6} = \frac{|0|}{|U|} = 0,\ K_{P7} = \frac{|\{7, 8, 9, 10, 11, 12\}|}{|U|} = 0.35,\ K_{P8} = \frac{|\{7, 8, 9, 10, 11, 12\}|}{|U|} = 0.35,$$

$$K_{P9} = \frac{|0|}{|U|} = 0,\quad K_{P10} = \frac{|\{7, 8, 9, 10, 11, 12\}| + |\{16, 17\}|}{|U|} = 0.47,\quad K_{P11} = \frac{|0|}{|U|} = 0\quad K_{P12} = \frac{|0|}{|U|} = 0,$$

$$K_{P13} = \frac{|\{3\}|}{|U|} = 0.0,\quad K_{P14} = \frac{|\{7, 8, 9, 10, 11, 12\}| + |\{16, 17\}|}{|U|} = 0.47,\quad K_{P15} = \frac{|\{1, 2, 3, 4, 5, 6\}|}{|U|} = 0.35,$$

$$K_{P16} = \frac{|0|}{|U|} = 0,\quad K_{P17} = \frac{|\{3\}|}{|U|} = 0.05,\quad K_{P18} = \frac{|\{7, 8, 9, 10, 11, 12\}| + |\{16, 17\}|}{|U|} = 0.47,$$

$$K_{P19} = \frac{|\{1, 2, 3, 4, 5, 6\}|}{|U|} = 0.35,\quad K_{P20} = \frac{|\{1, 5\}|}{|U|} = 0.11,\quad K_{P21} = \frac{|0|}{|U|} = 0,\quad K_{P22} = \frac{|0|}{|U|} = 0.$$

In the process of inducing decision rules, **reducts** play an important role as they generate minimal length of decision rules. It is found from these calculations that there are partial degree of dependency between attributes disease (D) and conditional attributes (C). Here, attribute (D) depends partially on the attributes (C). Hence, degree of dependency of those attributes (C) are greater than 0, and these sets of attributes is called a reduct. Here, reduct is {C1,C2,C3,C5,C9,C10, C13,C16, C17,C18, C20, C21,C22,C23}.

For the calculation of minimum reduct, all the reduct attributes have been arranged shown in Table 3, according to their highest degree of dependency. The maximum dependency degree of each attributes implies the more accuracy for selecting minimum reduct.

Now, we consider all the conditional attributes according to highest degree of dependency. If we consider C = {C13} = IND_I (C13) = {{7,8,9,10,11,12},

Table 3 Reduct with degree of dependency value

Index i	1	2	3	4	5	6	7	8	9	10	11	12
Conditional attributes	C13	C17	C21	C5	C9	C10	C18	C22	C2	C23	C16	C20
Degree of Dependency	Kp10	Kp14	Kp18	Kp4	Kp7	Kp8	Kp15	Kp19	Kp2	Kp20	Kp13	Kp17
Dependency value K	0.47	**0.47**	47	0.35	0.35	0.35	0.35	0.35	0.23	0.11	0.05	0.05

{1,2,3,4,5,6,13,14,15}, {16,17}} and D = {C30} = IND_I(C30) = {{1,2,3,4,5,6}, {7,8,9, 10,11,12}{13,14,15,16,17}}, then K = 0.47.

If C = {C13,C17} = IND_I (C13) ⊗ IND_I (C17) = {{7,8,9,10,11,12}, {1,2,3,4,5,6,13, 14,15}, {16,17}} ⊗ {{7,8,9,10,11,12}, {1,2,3,4,5,6,13,14,15}, {16,17}} = {{7,8,9,10,11,12}, {1,2, 3,4,5,6,13,14,15}, {16,17}}, then attribute dependency of Disease (D) = {{1,2,3,4,5,6}, {7,8,9, 10,11, 12}{13,14,15,16,17}} on C = {C13,C17} in a degree K($0 \le K \le 1$) is again 0.47.

In an analogous manner, if the numbers of conditional attributes are added step by step based on the highest degree of dependency (shown in Table 3), the value of k also increases and becomes 1. The calculations of different values of k for different conditions are given below:

If C = {C13, C17, C21} k = 0.47
If C = {C13, C17, C21, C3} k = 0.47
If C = {C13, C17, C21, C3, C5} k = 0.47
If C = {C13, C17, C21, C3, C5, C9} k = 0.47
If C = {C13, C17, C21, C3, C5, C9, C10} k = 0.47
If C = {C13, C17, C21, C3, C5, C9, C10, C18} k = 0.47
If C = {C13, C17, C21, C3, C5, C9, C10, C18, C22} k = 1.00

The set is crisp with respect to C, if degree of dependency (K) is equal to 1. So, the set {C13,C17,C21,C3,C5,C9,C10,C18,C22} is crisp set and inclusion of C22, widely changes the value of K. It should be the most significant condition of this set. Hence, **minimum reduct** {C13,C17,C21,C3,C5,C9,C10,C18,C22} **from the set of reducts is obtained**.

5.2 *Rules Generation Using Rosetta Software*

In this paper, we have used Rosetta software for reduct computation, rules generation, and classification. A genetic algorithm is used for reduct computation. The rules are sorted by RSH coverage, i.e., RHS or THEN part of the rule and LHS or IF part of the rule. After experimenting with genetic algorithm, we have seen that the experiment generates 206 rules with 42 reduct as shown in Fig. 2.

	Rule	LHS Support	RHS Support	RHS Accuracy	LHS Coverage	RHS Coverage	RHS Stability	LHS Length	RHS Length
1	T Wave L V2(U) AND R progression L V2(Y) AND ST segment L V3 (D) => Disease (D)(MI)	18	18	1.0	0.352941	1.0	1.0	3	1
2	Q Pathologic L aVf(Y) AND T Wave L V3 (U) => Disease (D)(MI)	18	18	1.0	0.352941	1.0	1.0	2	1
3	ST segment LIII (D) AND T Wave L V3 (I) => Disease (D)(ISC)	18	18	1.0	0.352941	1.0	1.0	2	1
4	ST segment LIII (E) AND T Wave L V3 (U) => Disease (D)(MI)	18	18	1.0	0.352941	1.0	1.0	2	1
5	T Wave LIII (U) AND ST segment L V3 (E) => Disease (D)(ISC)	18	18	1.0	0.352941	1.0	1.0	2	1
6	T Wave LIII (I) AND ST segment L V3 (D) => Disease (D)(MI)	18	18	1.0	0.352941	1.0	1.0	2	1
7	T Wave LIII (I) AND Q Pathologic L aVf(Y) AND T Wave L V2(U) => Disease (D)(MI)	18	18	1.0	0.352941	1.0	1.0	3	1
8	T Wave LIII (I) AND Q Pathologic L aVf(Y) AND Q Pathologic L V2(N) => Disease (D)(MI)	18	18	1.0	0.352941	1.0	1.0	3	1
9	Q Pathologic LIII (Y) AND Q Pathologic L aVf(Y) AND ST segment L V4(D) => Disease (D)(MI)	18	18	1.0	0.352941	1.0	1.0	3	1
10	T Wave L aVf(U) AND T Wave L V3 (I) => Disease (D)(ISC)	18	18	1.0	0.352941	1.0	1.0	2	1
11	T Wave L aVf(I) AND T Wave L V3 (U) => Disease (D)(MI)	18	18	1.0	0.352941	1.0	1.0	2	1
12	ST segment L aVf(D) AND T Wave L V3 (I) => Disease (D)(ISC)	18	18	1.0	0.352941	1.0	1.0	2	1
13	ST segment L aVf(E) AND T Wave L V3 (U) => Disease (D)(MI)	18	18	1.0	0.352941	1.0	1.0	2	1
14	T Wave L V3 (I) AND ST segment L V4(E) => Disease (D)(ISC)	18	18	1.0	0.352941	1.0	1.0	2	1
15	T Wave L V3 (U) AND ST segment L V4(D) => Disease (D)(MI)	18	18	1.0	0.352941	1.0	1.0	2	1
16	Q Pathologic L aVf(Y) AND T Wave L V2(U) AND ST segment L V3 (D) => Disease (D)(MI)	18	18	1.0	0.352941	1.0	1.0	3	1
17	T Wave LIII (I) AND ST segment L V2 (D) AND Q Pathologic L V2(N) => Disease (D)(MI)	18	18	1.0	0.352941	1.0	1.0	3	1
18	R progression LII(N) AND Q Pathologic LIII (Y) AND ST segment L V2 (D) => Disease (D)(MI)	18	18	1.0	0.352941	1.0	1.0	3	1
19	T Wave L V2(U) AND ST segment L V3 (D) AND R progression L V3(Y) => Disease (D)(MI)	18	18	1.0	0.352941	1.0	1.0	3	1

Fig. 2 A portion of generated rule set

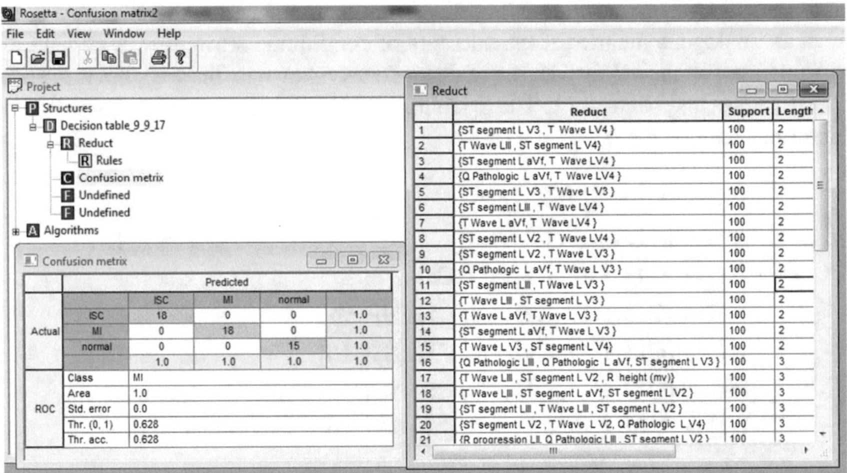

		Predicted			
Actual		ISC	MI	normal	
	ISC	18	0	0	1.0
	MI	0	18	0	1.0
	normal	0	0	15	1.0
		1.0	1.0	1.0	1.0
ROC	Class	MI			
	Area	1.0			
	Std. error	0.0			
	Thr. (0, 1)	0.628			
	Thr. acc.	0.628			

	Reduct	Support	Length
1	{ST segment L V3 , T Wave LV4 }	100	2
2	{T Wave LIII , ST segment L V4}	100	2
3	{ST segment L aVf, T Wave LV4 }	100	2
4	{Q Pathologic L aVf, T Wave LV4 }	100	2
5	{ST segment L V3 , T Wave L V3 }	100	2
6	{ST segment LIII , T Wave LV4 }	100	2
7	{T Wave L aVf, T Wave LV4 }	100	2
8	{ST segment L V2 , T Wave LV4 }	100	2
9	{ST segment L V2 , T Wave L V3 }	100	2
10	{Q Pathologic L aVf, T Wave L V3 }	100	2
11	{ST segment LIII , T Wave L V3 }	100	2
12	{T Wave LIII , ST segment L V3 }	100	2
13	{T Wave L aVf, T Wave L V3 }	100	2
14	{ST segment L aVf, T Wave L V3 }	100	2
15	{T Wave L V3 , ST segment L V4}	100	2
16	{Q Pathologic LIII , Q Pathologic L aVf, ST segment L V3 }	100	3
17	{T Wave LIII , ST segment L V2 , R height (mv)}	100	3
18	{T Wave LIII , ST segment L aVf, ST segment L V2 }	100	3
19	{ST segment LIII , T Wave LIII , ST segment L V2 }	100	3
20	{ST segment L V2 , T Wave L V2, Q Pathologic L V4}	100	3
21	{R progression LII, Q Pathologic LIII , ST segment L V2 }	100	3

Fig. 3 Confusion matrix output and reduct

The confusion matrix contains information about actual and predicted classification done by the classification system. Each column represents the instances in the predicted class, while each row represents the instances in the actual class. It also shows the overall accuracy as well as sensitivity and accuracy for each class. In this experiment, 51 objects are classified according to the genetic algorithm.

The confusion matrix (Fig. 3) shows cent percent accuracy for all the three set of trained data, which is the most excellent result that has been received from ROSETTA software for decision table. Hence, much better results can be obtained using rough set approach in the cost of more number of rules.

6 Conclusion

In this paper, we have proposed rough set-based ECG classification over the medical dataset obtained from the extracted feature of ECG signal. The proposed technique finds the major factor affecting the decision. A rough set quantitative measure, a degree of dependency (k), has been done to arrange the set of attributes according to their significance in order to make the decision. In addition to this, the identical objects in the training dataset were reduced in order to avoid unnecessary analysis. Currently, three types of ECG data, namely normal, Inferior MI, and Anterior MI have been tested. Hence, experimental results showed that of proposed method extracts specialist knowledge accurately.

Rule generation of rough set is specifically designed to extract human comprehensible decision rules from nominal data. Heart disease is one of the most common causes of death all over the world. Different statistical surveys indicate that heart disease, especially myocardial infarction (MI), has become a major health issue also in India, where these surveys show a steady increase of MI throughout the country. Therefore, advanced research is needed as a preventive measure against this silent killer.

References

1. Pawlak, Z.: Rough sets. Int. J. Comp. Inform. Sci. **11**, 341–356 (1982)
2. Pawlak, Z., Skowron, A.: Rudiments of rough sets. Inf. Sci. **177**, 3–27 (2007)
3. Hassanien, A.E., Abraham, A, Peters. J.F., Kacprzyk, J.: Rough sets in Medical Imaging: Foundations and Trends. CRC Press LLC (book chapter 1) (2001)
4. Polkowski, L., Tsumoto, S., Lin, T.Y.: Rough Set Methods and Applications. Physica-Verlag (a Springer-Verlag company-book) (2000)
5. Mitra, S., Mitra, M., Chaudhuri, B.B.: A rough-set-based inference engine for ECG classification. IEEE Trans. Instrum. Meas. **55**(6), 2198–2206 (2006)
6. Mitra, S., Mitra, M., Chaudhuri, B.B.: A rough-set-based approach for ECG classification. In: Peters, J.F., et al. (eds.) Transactions on Rough Sets IX, LNCS 5390, pp. 157–186 (2008)
7. Huang, X., Zhang, Y.: A new application of rough set to ECG recognition. In: International Conference on Machine Learning and Cybernetics, vol. 3, pp. 1729–1734 (2003)
8. Tsumoto, S.: Mining diagnostic rules from clinical databases using rough sets and medical diagnostic model. Inf. Sci. Int. J. **162**(2), 65–80 (2004)
9. Hassanien, A.E., Ali, J.M., Hajime, N.: Detection of speculated masses in mammograms based on fuzzy image processing. In: 7th International Conference on Artificial Intelligence and Soft Computing. Springer Lecture Notes in AI, vol. 3070, pp. 1002–1007 (2004)
10. Coast, D.A., Stem, R.M., Cano. G.G., Briller, S.A.: An approach to cardiac arrhythmia analysis using hidden Markov models. IEEE Trans. Biomed. Eng. **37**, 826–836 (1990)
11. Halder, B., Mitra, S., Mitra, M.: Detection and Identification of ECG waves by Histogram approach. In: Proceeding of CIEC16, IEEE, Held at University of Calcutta, pp. 168–72 (2016)
12. Schamroth, L.: An introduction to Electro Cardiography, 7th edn. Blackwell Publisher, India (2009)
13. Goldberger, A.L.: Clinical Electrocardiography, 6th edn., mosby 14
14. Lippincott Williams & Wilkins: ECG Interpretation Made Incredible Easy, 5th edn., Kluwer
15. Polkowski, L., Skowron, A.: Rough Sets in Knowledge Discovery. Physica – Verlag, Wurzburg (1998)

Decision Support System for Prediction of Occupational Accident: A Case Study from a Steel Plant

Sobhan Sarkar, Mainak Chain, Sohit Nayak and J. Maiti

Abstract Decision support system (DSS) is a powerful tool which helps decision-makers take unbiased and insightful decisions from the historical data. In the domain of occupational accident analysis, decision-making should be effective, insightful, unbiased, and more importantly prompt. In order to obtain such decision, development of DSS is necessary. In the present study, an attempt has been made to build such DSS for accident analysis in an integrated steel plant. Two classifiers, i.e., support vector machine (SVM) and random forest (RF) have been used. RF produces better level of accuracy, i.e., 99.34%. The developed DSS has full potential in making insightful decisions and can be used in other domains like manufacturing, construction, etc.

Keywords Occupational accident analysis · Decision support system SVM · Random forest

S. Sarkar (✉) · J. Maiti
Department of Industrial & Systems Engineering, IIT Kharagpur, Kharagpur, India
e-mail: sobhan.sarkar@gmail.com

J. Maiti
e-mail: jhareswar.maiti@gmail.com

M. Chain · S. Nayak
Department of Metallurgical & Materials Engineering, IIT Kharagpur, Kharagpur, India
e-mail: mainakchain21@gmail.com

S. Nayak
e-mail: sohit.nayak@gmail.com

A. Abraham et al. (eds.), *Emerging Technologies in Data Mining and Information Security*, Advances in Intelligent Systems and Computing 813,
https://doi.org/10.1007/978-981-13-1498-8_69

1 Introduction

Decision support system (DSS) is an interactive software-based system. It intends to help decision-makers in taking decision by accessing a large volume of data gathered from different information systems involved in organizational business process, such as automation system, safety management systems, etc. It is developed basically to solve the semi-structural or nonstructural problems (which appears to be very difficult to be modeled) by improving decision-making effectiveness and efficiency through integrating information resources and analytical tools. The use of DSS started since mid-1960s. Ferguson and Jones [1] reported the first experimental study using a computer-aided decision system in 1969. The ability of DSS to provide decision-based solutions makes DSS unique in quick decision-making scenarios, as opposed to its traditional counterpart.

1.1 Importance of DSS in Occupational Safety

Occupational safety of workers in the industry is one of the important aspects since accidents are occurring at workplace globally in each year. There are numerous factors attributable to the accidents. Some of them include carelessness of human behavior, inadequate training in safety, unsafe conditions, etc. Consequently, the occupational accidents such as injuries, fatalities, material and/or environmental damages lead to economical losses. Hence, safety managers try to find other possible solutions to reduce potential hazards through the analysis of workplace accidents. Conventional data analysis approaches using machine learning techniques provide satisfactory results but at the cost of time and reliability of results [2–11]. DSS deals with numerous application domains. Therefore, it is not also surprising that the uses of DSS are also diverse. In a nutshell, the aim of DSS is to provide a better decision-making process. It automates the tedious tasks, thus allowing an analyst to explore the problem in a rigorous way. This additional exploration capacity is made possible by the virtue of DSS, which consequently improves insights of the problem by the decision-maker and thus it is able to come up with a more viable solution. This system can enhance the process of decision-making, can increase the organizational control, and can improve the personal efficiency and interpersonal communication. Nevertheless, it quantitatively and qualitatively aids to the satisfaction of the decision-maker as well. In addition, the decisions taken have the effects of reduction in costs and risks, increasing revenue, and improving assets usage efficiency.

1.2 Applications of DSS

There are numerous applications of DSS in and around the industrial, business, and research sectors. A self-predicting system was made aiming to assess accidental risks in building project sites. The results from the same was astonishingly significant in the field of risk assessment. In order to mitigate the risks in maintenance activities and make safety certain, another support system is developed. Yet another proposition of a web-based DSS (WDSS) to predict potential risks at shipyards [12] has been made that aims to provide a set of preventive policies to avoid risks at the site. The clinical field is not left untouched by the revolutionary impact of support systems. A clinical decision system is used to suggest appropriate rehabilitation for injured workers [13] and a WDSS suggesting automatic diet prescription for the patient [14] has been built. It has also large impact on environment. A knowledge-based decision support system for risk assessment of groundwater pollution has been developed by [15]. Therefore, it is observed that the DSS has importance in occupational accident analysis domain.

1.3 Research Issues

Based on the review of literature on DSS mentioned above, some of the research issues have been identified, which are cited below.

(i) In most of the cases of decision-making, information and overall understanding of the system play an important role. For human, it is a very difficult task to derive decisions from the information lying within a huge amount of data with a stipulated time.
(ii) Decision-making often demands a substantial amount of human labor which consequently consumes time and manpower.
(iii) Decisions taken by humans might be biased and inaccurate in some specific conditions, and in most of the time, they depend largely on human judgments or experience.

1.4 Contributions

Realizing the need for research on some of the issues as identified above, the present work attempts to contribute as follows:

(i) A DSS has been developed in accident analysis at work.
(ii) Decisions from the accident data for the prediction of incident outcomes, i.e., injury, near miss, and property damage have been made automated.
(iii) Development of DSS for occupational accident analysis can be regarded as an initiative in the steel industry.

The remainder of this study is structured in three sections and are as follows. Development of DSS has been discussed along with the methods or models embedded in the system in Sect. 2. Section 3 presents a case study of an integrated steel plant.

In Sect. 4, the results obtained from the DSS are presented and discussed. Finally, conclusions are drawn with recommendations for further research in Sect. 5.

2 Methodology

In this section, a short description of the development of DSS and the models, i.e., algorithms embedded in the system have been provided in the following section. The overall proposed DSS structure has been depicted in Fig. 1.

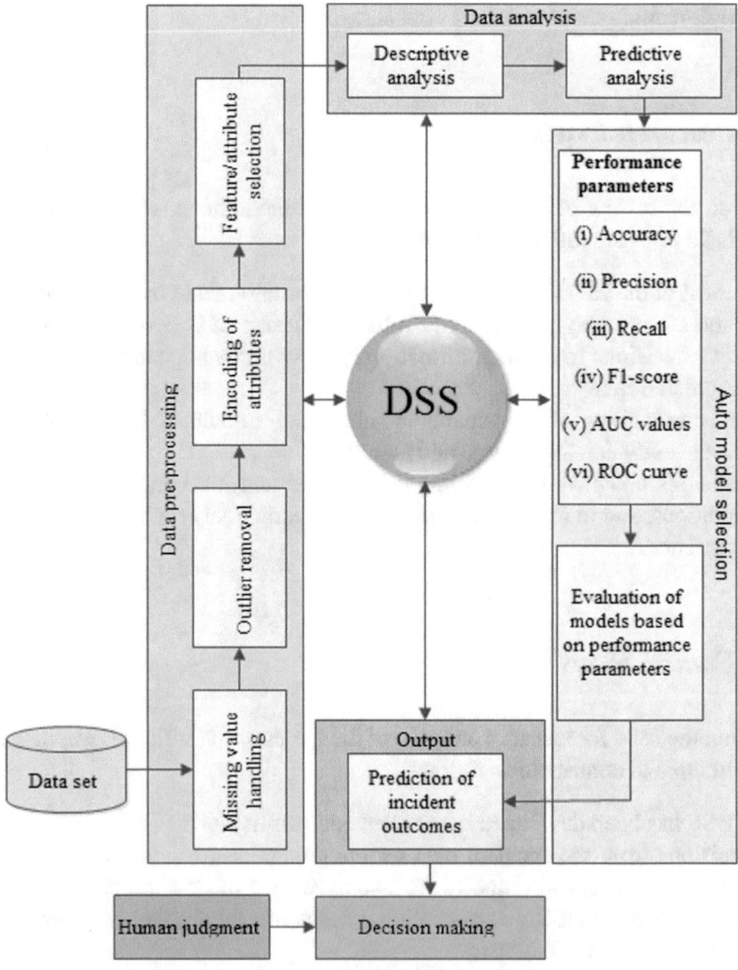

Fig. 1 Overall proposed decision support system (DSS) structure

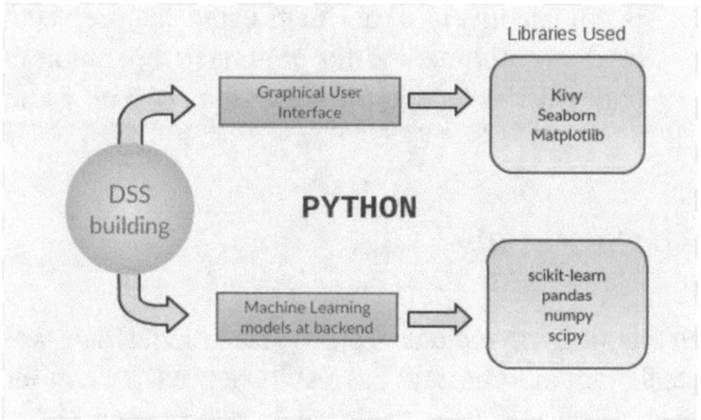

Fig. 2 Development of DSS using Python libraries

2.1 DSS Building

The DSS framework is built entirely using Python, an open-source, interactive, object-oriented, and high-level programming language. It is based on a graphical user interface (GUI), which includes visual aids such as graphs. It consists of a series of back-end programming codes of machine learning algorithms intended for the analysis of accident data (refer to Fig. 2).

2.2 Methods Used

(i) **Random Forest**: Random forest (RF) is a popular classification algorithm, which can handle nonlinear data efficiently along with linear ones. It belongs to the broader class of decision tree family called *ensemble learning* algorithm. Ensemble learning involves the combination of several classifiers or models to solve a single prediction problem. The performance of prediction of RF becomes better than that of single decision tree. Finally, the class is selected based on the maximum votes provided by all trees in forest. However, in regression approach, the average of outputs from the trees is usually considered. One of the major benefits of this algorithm is the ability to handle large amount of data with higher dimensionality. The detailed description of the algorithm is kept beyond the scope of the paper. Interested readers may refer to [16–18].

(ii) **Support Vector Machine**: Support vector machine (SVM) is a supervised learning algorithm, which can be used for both classification and regression types of problems. This algorithm plots each data point in n-dimensional space, where n being equal to the number of features. Each of the features is denoted by a coordinate in the feature space. Thereafter, classification task is performed by determining optimal separating hyperplane which can differentiate the two classes clearly. The SVM algorithm uses kernels which are basically the functions taking low-dimensional input space and transforming it to a high-dimensional space, i.e., it converts non-separable data to separable data. Basically, using kernel is comparatively easier than using feature vectors. That is why it is found to be the useful method for nonlinear separation problem. For understanding the mathematics behind this algorithm, one may refer to [19–21].

3 Case Study

In this study, to validate the proposed model, data were retrieved from the integrated steel plant in India, and the results of prediction of incident outcomes were obtained from the classifiers, namely random forest and SVM embedded in the DSS. A short description on data set and its preprocessing technique have been provided below.

Table 1 A short description of data set used

Attributes	Description	Data type
Date of incident	Date on which the incident took place	Text
Month	Month of the occurrence of incident	Catagorical
Division	Division in which the incident occurred	Catagorical
Department	Department involved in the incident	Catagorical
Incident outcomes (injury, nearmiss and property damage)	Outcomes of the incident	Catagorical
Injury types	Types of injury occurred to victim(s)	Catagorical
Primary causes	Cause of the occurrence of the incident	Catagorical
Brief description of incident	Description of how the incident happened	Text
Status	Current status of the incident	Catagorical
Event	Events which led to the incident	Text
Working condition (single or in group)	Condition in which the victim was working Catagorical	
Machine condition (working or not working state)	Condition of the machine when the incident occurred Catagorical	
Observation type	Type of obseravtion	Catagorical
Employee type	Contractor/Employee	Catagorical
Serious process incident score	Score corresponding to the seriousness of the incident	Numerical
Injury potential score injury to the victim	Score corresponding to the seriousness of numerical	
Equipment damage score	Score corresponding to the damage to equipment	Numerical
Safety standards time of occurrence	Standard of safety at the incident scenario at the catagorical	
Incident type	Type of incident	Catagorical
Combined_SOP	Combination of SOP adequacy, SOP compliance, SOP availability, SOP not availability	Catagorical

3.1 Data Set

To validate the results from the developed DSS, data collected from a steel plant were used. The data set consists of 9478 incident reports describing accident consequences, i.e., near miss, property damage, and injury. The attributes of the data set used in this study are listed in Table 1 with description and types.

3.2 Data Preprocessing

Once the data were obtained, they were preprocessed. Missing data, and other inconsistencies were removed from the raw data set. Thereafter, the data were coded for easy handling during and after the analysis.

4 Results and Discussions

In this section, the results of the two classification algorithms, i.e., SVM and RF are discussed. Based on their performance in terms of classification accuracy, the classifiers are compared and automatically a better one is selected in DSS platform, which is later used for the prediction of incident outcomes.

The analytical platform for this study is DSS, which is built by Python. There were seven basic libraries used for the building of DSS structure. They are 'Kivy', 'Seaborn', "Matplotlib", "Scikit-learn", "Pandas", "numpy", and "Scipy". Using

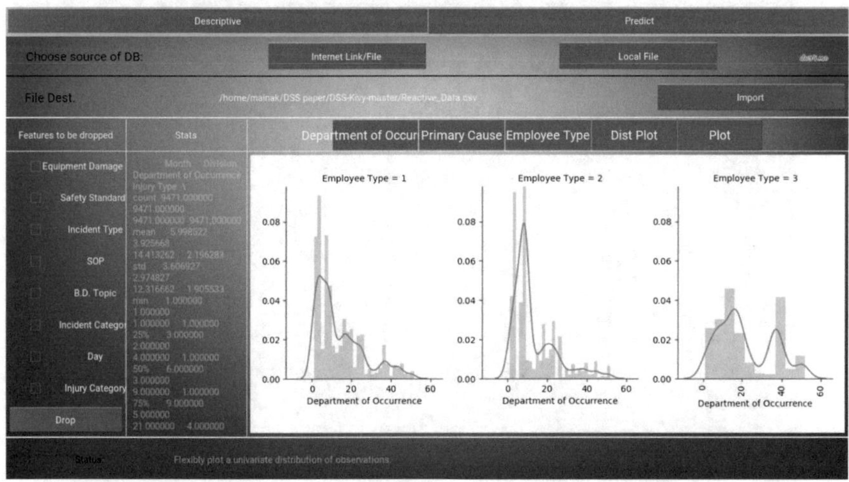

Fig. 3 Display of results window of DSS showing the frequency of "Primary cause" to accident across various "Department of Occurrence" considering different employee types

these libraries, the "Homepage" of the DSS is developed so that it is able to import the data set from the drive. Once the data is loaded, features or attributes of the data set can be selected manually. In addition, for feature selection, algorithms including chi-square, and RF are also embedded into the system. Indeed, there are two basic operations are included as data preprocessing tasks in DSS. First, is the descriptive analysis and second, is the predictive analysis. Once the features are selected for analysis, descriptive analyses were performed for each of them. For example, in Fig. 3, graphs from descriptive analysis are displayed which explains the frequency of primary cause happened in each of the department with employee types.

In the window of predictive analysis in DSS, two basic classifiers, i.e., SVM and RF are included. Using 10-fold cross-validation, the classification accuracies of both SVM and RF algorithm are calculated, which are 70.56% and 99.34% respectively. A

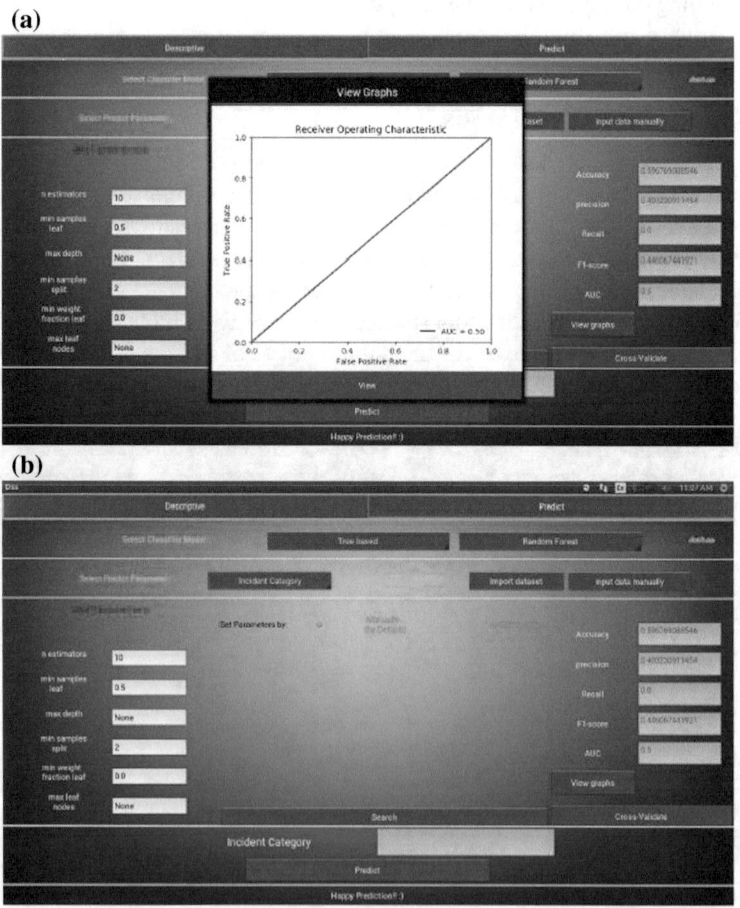

Fig. 4 **a** Display of results window of DSS showing ROC curve for incident outcome prediction; **b** Display of results window of DSS showing accuracy, F1-score, and AUC values

provision has been made automatically to select the best one from a set of classifiers. As a result, in this case, RF has been selected. Apart from accuracy, F1-score, area under curve (AUC) values, and receiver operating characteristics curve have been shown in Fig. 4a, b. It is noteworthy to mention that provisions are so made in DSS that comparison among classifiers can also be performed using different performance parameters like F1-score, AUC values, etc.

5 Conclusions

The present study attempts to develop a DSS for occupational accident analysis and prevention. The developed DSS can reduce a substantial amount of human labor during the task of data preprocessing including missing value handling, outlier or inconsistency removal, and data analysis through automatic selection of model (i.e., classifier) based on a certain performance parameter usually set by the user. The developed DSS can make an unbiased estimation of results in terms of classification accuracy of incident outcomes from the historical accident data. Out of the two classifiers used in this study, i.e., SVM, and RF, the later one performs better with accuracy 99.34% obtained from 10-fold cross-validation. Therefore, the initial attempt to build such a DSS for accident analysis in steel industry has full potential to help decision-makers take more prudent and insightful decisions.

As the scope for future study, one might opt for automating data analysis process entirely from importing of data to the exporting of results. Another important challenging task, which can be made DSS more strengthened, is to use optimization algorithms for determining optimal value of parameters of the classifiers. Rule-based analysis, which is very important for occupational accident analysis, can also be embedded in the DSS structure.

References

1. Ferguson, R.L., Jones, C.H.: A computer aided decision system. Manag. Sci. **15**(10), B–550 (1969)
2. Sarkar, S., Vinay, S., Maiti, J.: Text mining based safety risk assessment and prediction of occupational accidents in a steel plant. In: 2016 International Conference on Computational Techniques in Information and Communication Technologies (ICCTICT), pp. 439–444. IEEE (2016)
3. Krishna, O.B., Maiti, J., Ray, P.K., Samanta, B., Mandal, S., Sarkar, S.: Measurement and modeling of job stress of electric overhead traveling crane operators. Safety Health Work **6**(4), 279–288 (2015)
4. Gautam, S., Maiti, J., Syamsundar, A., Sarkar, S.: Segmented point process models for work system safety analysis. Safety Sci. **95**, 15–27 (2017)
5. Sarkar, S., Patel, A., Madaan, S., Maiti, J.: Prediction of occupational accidents using decision tree approach. In: 2016 IEEE Annual India Conference (INDICON), pp. 1–6. IEEE (2016)

6. Sarkar, S., Vinay, S., Pateshwari, V., Maiti, J.: Study of optimized SVM for incident prediction of a steel plant in india. In: 2016 IEEE Annual India Conference (INDICON), pp. 1–6. IEEE (2016)
7. Sarkar, S., Lohani, A., Maiti, J.: Genetic algorithm-based association rule mining approach towards rule generation of occupational accidents. In: International Conference on Computational Intelligence, Communications, and Business Analytics, pp. 517–530. Springer (2017)
8. Sarkar, S., Verma, A., Maiti, J.: Prediction of occupational incidents using proactive and reactive data: a data mining approach. In: Industrial Safety Management, pp. 65–79. Springer (2018)
9. Verma, A., Chatterjee, S., Sarkar, S., Maiti, J.: Data-driven mapping between proactive and reactive measures of occupational safety performance. In: Industrial Safety Management, pp. 53–63. Springer (2018)
10. Sarkar, S., Pateswari, V., Maiti, J.: Predictive model for incident occurrences in steel plant in india. In: 8-th ICCCNT, 2017, pp. 1–5. IEEE (2017)
11. Sarkar, S., Vinay, S., Raj, R., Maiti, J., Mitra, P.: Application of optimized machine learning techniques for prediction of occupational accidents. Comput. Oper. Res. (2018)
12. Cebi, S., Akyuz, E., Sahin, Y.: Developing web based decision support system for evaluation occupational risks at shipyards. Brodogradnja **68**(1), 17–30 (2017)
13. Gross, D.P., Zhang, J., Steenstra, I., Barnsley, S., Haws, C., Amell, T., McIntosh, G., Cooper, J., Zaiane, O.: Development of a computer-based clinical decision support tool for selecting appropriate rehabilitation interventions for injured workers. J. Occup. Rehabil. **23**(4), 597–609 (2013)
14. Caballero-Ruiz, E., García-Sáez, G., Rigla, M., Villaplana, M., Pons, B., Hernando, M.E.: A web-based clinical decision support system for gestational diabetes: automatic diet prescription and detection of insulin needs. Int. J. Med. Inform. **102**, 35–49 (2017)
15. Uricchio, V.F., Giordano, R., Lopez, N.: A fuzzy knowledge-based decision support system for groundwater pollution risk evaluation. J. Environ. Manag. **73**(3), 189–197 (2004)
16. Breiman, L.: Random forests. Mach. Learn. **45**(1), 5–32 (2001)
17. Wang, B., Gao, L., Juan, Z.: Travel mode detection using gps data and socioeconomic attributes based on a random forest classifier. IEEE Trans. Intell. Transp. Syst. (2017)
18. Ahmed, M., Rasool, A.G., Afzal, H., Siddiqi, I.: Improving handwriting based gender classification using ensemble classifiers. Expert Syst. Appl. **85**, 158–168 (2017)
19. Poria, S., Peng, H., Hussain, A., Howard, N., Cambria, E.: Ensemble application of convolutional neural networks and multiple kernel learning for multimodal sentiment analysis. Neurocomputing (2017)
20. Dumais, S., et al.: Using SVMs for text categorization. IEEE Intell. Syst. **13**(4), 21–23 (1998)
21. Chen, T., Lu, S.: Subcategory-aware feature selection and SVM optimization for automatic aerial image-based oil spill inspection. IEEE Trans. Geosci. Remote Sens. (2017)

Community Detection Based Tweet Summarization

Soumi Dutta, Asit Kumar Das, Abhishek Bhattacharya, Gourav Dutta, Komal K. Parikh, Atin Das and Dipsa Ganguly

Abstract In today's world, online social media is one of the popular platforms of information exchange. Twitter is one of the most popular social networking sites wherein people post and interact with the help of short messages known as tweets. During huge number of messages are exchanged between a large set of users, huge information are exchanged daily in microblogs during important events such as cricket match, election, budget, movie release, natural disaster and so on. So, this huge source of information open doors to the researchers to handle different challenges. Twitter employs the searching felicity based on trending/non-trending topic. So, topic-wise grouping of information is required for these tweets so that the stories go under specific groups or 'clusters' of relevant title or heading. Along with this, summarization is another challenging task due to information overload and irrelevant information in many of the posts. In this work, a graph-based approach has been proposed for summarizing tweets, where a tweet similarity graph is first constructed, and then community detection methodology is applied on the similarity graph to

S. Dutta (✉) · A. K. Das
Indian Institute of Engineering Science and Technology Shibpur, Shibpur, India
e-mail: soumi.it@gmail.com

A. K. Das
e-mail: akdas@cs.becs.ac.in

S. Dutta · A. Bhattacharya · K. K. Parikh · A. Das · D. Ganguly
Institute of Engineering and Management, Kolkata 700091, India
e-mail: komalparikh09@gmail.com

A. Das
e-mail: atindas.scm@gmail.com

D. Ganguly
e-mail: gangulydipsa@gmail.com

A. Bhattacharya
Birla Institute of Technology, Kolkata 700050, India
e-mail: abhishek.bhattacharya@iemcal.com

G. Dutta
Cognizant Technology Solutions, Pune, India
e-mail: cse.gourav@gmail.com

A. Abraham et al. (eds.), *Emerging Technologies in Data Mining and Information Security*, Advances in Intelligent Systems and Computing 813,
https://doi.org/10.1007/978-981-13-1498-8_70

cluster similar tweets. Finally, to represent the summarization of tweets, from each cluster, important tweets are identified based on different graph centrality measures such as degree centrality, closeness centrality, betweenness centrality, eigenvector centrality and PageRank. The proposed methodology achieves better performance than Sumbasic, which is an existing summarization method.

Keywords Graph clustering · Feature selection · Microblogs · Centrality measure · Louvain community detection

1 Introduction

Twitter is one of the most indelible microblogging sites wherein people can tweet about facts or stories or anything in general. It is a platform where people can gather information about various events or news stories. At present, there are over 360 million users on Twitter. As such, a huge amount of data is generated. Thus, there is a lot of research work going on in the world to meet several challenges in the analysis of these social networking data. Some fields on research includes trending topic analysis, tweet summarization, user sentiment analysis, spam detection, event detection, impact of stopwords on clusters, theme-based clustering, etc.

Microblogging services like Twitter has become very popular means of dissemination of news, ideas and opinions as well as a platform for marketing, sales and public relations. These microblogging sites are the source of vast and crucial information and news updates for crisis situations such as Mumbai terror attacks or Iran protests. Around thousands of tweets are being posted on Twitter per second, daily. Twitter facilitates the searching based on hashtags which are also tagged as trending or non-trending topic. The search outcome always returns all relevant tweets matched with the searched keyword in reverse chronological order. Tweets are shorter in nature (at most 140 characters in length) and often informally written, e.g. using abbreviations, colloquial language and so on. Moreover, many of the tweets contain very similar information, so this redundant information may cause information overload for the user. To comprehend knowledge of any topic, it is not possible all the time to go through all the tweets. In this scenario, summarization of tweets are essential to generate a brief overview of the topic. Due to irrelevant and short nature of the tweets, summarization is one of the challenging tasks for the researchers. Summarization of tweets is one of the motivations of the clustering task, where similar tweets are grouped into multiple clusters and selecting important tweets from each cluster, the summary can be generated.

Several types of summaries can be inferred from tweets such as extractive (some selected tweets which can represent the entire dataset) and abstractive (re-generating the extracted content of the tweets). In this work, a graph-based methodology has been developed for summarization of tweets. In this proposed methodology, a community detection-based extractive summarization algorithm has been proposed choosing important tweets which are having the highest amount of relevant information.

The rest of the paper is represented as follows. Section 2 contains a literature survey on clustering and summarization of microblogging dataset. Section 3 includes the background studies. Section 4 describes the proposed summarization approach. Section 5 discusses the results of the comparison among the various summarization algorithms. In Sect. 6, the paper is concluded with some potential future research directions.

2 Related Work

Automatic text summarization [1, 7, 11, 16] is one of the challenging area of research nowadays. Several researches are ongoing on automatic summarization problem of single or multiple large text documents as well as for extractive and abstractive summarization problem. Luhn [10] proposed a method for automatic extraction of the technical articles using surface features like word frequency in 1950s.

Considering a hierarchical text graph as an input document the TOPIC system [15] proposed a topics-based summary of the text. Soumi et al. [4] proposed a graph-based Infomap-based community detection algorithm for tweet clustering, as well as for summarization. Another graph-based approach is proposed in the TextRank algorithm [2] which generates an adjacency matrix to find the most highly ranked sentences based on keywords in a document, evaluate using the PageRank algorithm. In SCISOR system [14], a concept graph is constructed from the dataset Down Jones newswire stories to generate conceptual summary. In all above methods, graph-based approaches have been adopted where another motivation task is clustering of similar tweets before summary selection. Apart from graph-based method, Soumi et al. [6] proposed a genetic algorithm based clustering technique. In another work feature selection based clustering methodology [5] is also proposed which can increase the effectiveness of clustering as well as summarization.

Another important algorithm, proposed by Radev et al. [3] for text summarization include the centroid algorithm and SumBasic [17], which is publicly available. Among publicly available programmes, MEAD [13] is one of the flexible proposal for multilingual and multidocument text summarization. MEAD has been proposed for summarization of large text documents. In SumFocus, Vanderwende et al. [17] is proposed to a dynamically topic changing summary generation algorithm implementing a set of rules on topic changes during summary generation. Kupiec [9] proposed supervised learning and statistical methods for multidocument summarization considering human created summaries. The centroid algorithm proposed single document summarization algorithm evaluating centrality measure of a sentence in relation to the overall topic of the document cluster.

The goal of the present work is to develop an extractive graph-based summarization approach for tweets considering several centrality measures to identify the most important tweets in a dataset.

3 Background Studies

This paper presents an extractive summarization approach followed by community detection. First, a tweet similarity graph has been generated considering tweets as nodes and tweet similarity as the weight of the graph. In the next step, communities are identified from the graph using Louvain modularity approach [12]. This section briefly discusses Louvain modularity approach that is used as community detection algorithm in the graph.

3.1 Community Detection in Graphs

Community in a graph are the set of similar nodes. Detection of communities in graphs is done based on the network structure to identify the groups or clusters of data that are connected due to similarities. Intra-community edges are the connections that exist between nodes of the same community, whereas intercommunity edges are connections that exist between nodes of different communities. In the proposed work, tweets are considered as nodes and the edges are the similarities among the tweets. There are several community detection methodologies such as Infomap [8], Louvain [12] which are used to detect community for finding the modules contains similar nodes from a graph. For instance, InfoMap sees community detection as a compression problem whereas label propagation sees it as a diffusion process. In this work, Louvain modularity approach is considered which relies on optimization of an objective function called modularity. Modularity is a metric that quantifies the quality of nodes assignment to communities by evaluating how much more densely connected the nodes within a community are compared to how connected they would be, on average, in a suitably defined random environment. The Louvain method detects communities in networks for maximizing the modularity. This method consists of a repetition of application of two steps, first being the greedy assignment of nodes to communities that abets local optimizations of modularity and second, the definition of a new coarse-grained network in terms of communities already found. These two steps are repeated in a loop until no further modularity increasing reassignments of communities are possible.

4 Proposed Methodology

A set of tweets are given as input for the proposed methodology, as an outcome which identifies a subset of the tweets as a summary of the entire set of tweets, as the proposed algorithm has been designed for extractive summary generation. This section describes the experimental dataset and the proposed methodology.

Table 1 A sample dataset of the floods in the Uttaranchal region of India in 2013

Tweet ID	Tweets
1	@iCanSaveLife: #Uttarakhand flood helpline numbers: 0135-2710334, 0135-2710335, 0135-2710233. Please RT
2	58 dead, over 58,000 trapped as rain batters Uttarakhand, UP http://t.co/6nJibt87Pm
3	Rahul Gandhi promises to reach out to the victims in Kedarnath once Congress confirms that it would be secular to do so
4	Uttarakhand rains: Harbhajan Singh turns counsellor for stranded pilgrims, tourists http://t.co/YJ2rxtMO6V
5	@DDNewsLive: #ITBP : News reports that 3 of its personnel died in Kedarnath are WRONG; Out of 6 jawans posted in Kedarnath 5 have been lo

4.1 Data Set

For the experiment purpose, a dataset of 2921 tweets is collected from Twitter API using crawler. Twitter provides the facility to acquire 1% random sample of tweets by keyword matching for the public which is mainly used by the researchers. The challenge of using Twitter API is its limitation with respect to the number of application requests one can make per hour. For unauthenticated calls, the restriction was 150 requests per hour and generally for authenticated calls, the count is 350. Here, using the keyword 'Uttarakhand' AND 'flood' a set of real-time events(the floods in the Uttaranchal region of India in 2013) data are collected. A few sample tweets of the dataset is shown in Table 1.

4.2 Data Preprocessing

Due to noisy nature of microblog dataset, preprocessing is required. This process includes removal of stopwords, non-textual characters like smileys, @usernames, exclamation/question marks. Special characters are also removed, except '#' symbol, as hashtag plays a significant role in microblogging.

4.3 Evaluation of Tweet Similarity

In this approach, between every pair of tweets similarity are evaluated considering term-level similarity and semantic similarity. The term-level similarities are measured upon the word-count similarity between two tweets, similarities based upon similiar URLs, common hashtags, common usernames and also, the cosine similarity.

4.4 Word-Count Similarity

If there are more than three common words between a pair of tweets, then that is considered as a frequency of similarity between them is considered as 1, otherwise 0. We can consider even a higher number of word count for formulating the similarity quotient, but for this work, we have fixed it as three.

4.5 URL Count Similarity

If two tweets contain the same URL, then their similarity frequency is 1, else it is 0. This means that if a pair of tweets contains a common URL, which is easy to find, then they are considered as similiar and their similarity quotient based on the URL count increases. For instance, if tweets has a URL 'https://www.facebook.com' and another tweet has the exactly same URL, then they are considered to have similarity frequency as 1.

4.6 Hashtag Count Similarity

This is almost similar to the word-count similarity in the sense that the approach is the same, but with an exception. Here, these are not simple words, rather these are keywords, basically, words following the hashtag, e.g. #subway. If another tweet also contains the same keyword or hashtag, then their similarity count is 1. Obviously, more the number of similar hashtags, more the hashtag similarity count between a pair of tweets.

4.7 Username Count Similarity

If a pair of tweets has a common username in it, say @sonia, then it is considered as a similarity between the two tweets.

4.8 Cosine Similarity

The cosine similarity between a pair of tweets is defined as the cosine of the angle between a pair of non-zero vectors. In collections library, there is a dict subclass. Counter which is used to count hashtable objects. This dict subclass processes words to convert them from text to vector which is finally sent as a parameter to this cosine similarity calculating function. The formula for finding the cosine similarity between two vectors is shown in Eq. 1,

$$cos(\boldsymbol{x}, \boldsymbol{y}) = \frac{\boldsymbol{x} \cdot \boldsymbol{y}}{||\boldsymbol{x}|| \cdot ||\boldsymbol{y}||} \tag{1}$$

where A_i and B_i are components of vector A and B, respectively.

4.9 Semantic Similarity

Many tweets have similar meaning in the dataset, but they are having different set of words which are semantically similar. WordNet web database is considered to identify the words, which are linked together by their semantic relationship. If two words are having common WordNet synset, then two words are considered as semantically similar.

4.10 Graph Generation

In the next step, an undirected, weighted tweet similarity graph is generated considering tweets as the nodes and tweet similarity scores as edge weight for each pair of tweets. A graph is a mathematical representation of sets of objects using nodes/vertices and edges/arcs between pairs of nodes. So, it is safe to say that a graph is represented as $\mathrm{G} = (\mathrm{V}, \mathrm{E})$, where V is the set of vertices and E is the set of edges. As for each pair of nodes, edge has been added in the graph, so for n number of nodes (tweets) there will be $\mathrm{n} * (\mathrm{n} - 1)/2$ number of edges. Considering the sample dataset shown in Table 1, a graph is generated as shown in Fig. 1a, where each node is the tweet and the edges carry the similarity weight between a pair of nodes. The minimum spanning tree of the graph has been computed, to retain only the important edges. Figure 1b shows the minimum spanning graph for the tree in Fig. 1a.

4.11 Summary Selection

To find the correlation between the nodes, communities are identified from the weighted graph. Communities are also named as clusters in this work. Communities within a graph/network are regions of densely connected nodes with higher weights of similarities between each pair of nodes within them. To detect communities, Louvain modularity algorithm is applied on the tweet similarity graph. After detection of the communities, important nodes are identified for each community using centrality measures of the graph, such as degree centrality, closeness centrality,

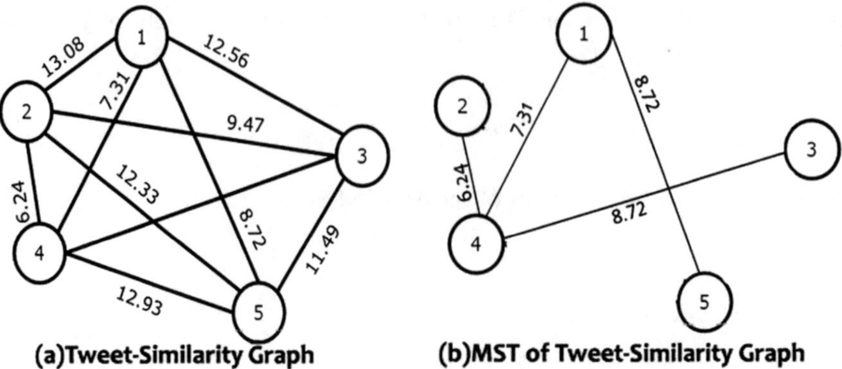

Fig. 1 **a** Example of tweet similarity Graph for the sample dataset shown in Table 1, **b** Minimum spanning graph of tweet similarity graph for the sample dataset shown in Table 1

betweenness centrality, eigenvector centrality, PageRank. Also for each community, longest length tweet is identified for summary generation as these tweets can have maximum information. There are different types of centrality measures, discussed which are considered to identify important nodes for the summary generation.

4.11.1 Degree Centrality Summary

The in-degree measure is the number of edges that enter into a node, i.e. how many nodes point to a particular node. For each community, highest degree centrality nodes are selected to generate Degree Centrality Summary.

4.11.2 Closeness Centrality Summary

Closeness centrality simply means that nodes that are central are the shortest distance away from all other nodes. The more central a node is, the shorter its total distance to all other nodes. Closeness can be regarded as a measure of how long it will take to spread information from the central node to all other nodes sequentially. For each community, highest closeness centrality nodes are selected to generate Closeness Centrality Summary.

4.11.3 Betweenness Centrality Summary

Betweenness means that the central nodes are those that show up in the shortest paths between two nodes during traversal. Betweenness is quite closer to closeness in the sense of meaning, but betweenness is able to differentiate nodes structure much better than closeness centrality. For each community, highest betweenness centrality nodes are selected to generate betweenness centrality summary.

4.11.4 Eigenvector Centrality Summary

The centrality is proportional to the sum of neighbours centralities. Hence, it is a measure of the influence of a node in a graph and thus, helps distinguish the most influential nodes. It assigns relative scores to all nodes in the network based on the concept that connections to high-scoring nodes contribute more to the score of the node in question than equal connections to low-scoring nodes. For each community, highest eigenvector centrality nodes are selected to generate Eigenvector Centrality Summary.

4.11.5 PageRank Summary

PageRank means that nodes which are central are the ones that would occur the most, on traversal between any two nodes through any path. For each community, the highest PageRank centrality nodes are selected to generate PageRank Centrality Summary.

4.11.6 Longest Length Summary

For each community, the longest length nodes are selected to generate longest length summary as tweets that are longer and contain more information.

4.11.7 Ensemble Summary

Considering all types of centrality measures, an ensemble summary has been generated. Then, those summaries are chosen finally which are evaluated by any two of the centrality measures.

5 Experimental Results

From the proposed approach, seven different types of summaries on tweet dataset (degree centrality summary, closeness centrality summary, betweenness centrality summary, eigenvector centrality summary, PageRank Summary, longest length summary and ensemble summary) have been generated. To compare the performance of the proposed summaries, a set of human-generated summaries using human experts are collected. As an alternative approach, a popular method is also considered, such

Table 2 Performance comparison of proposed summarization algorithms and sumbasic

Method	Precision	Recall	F-Measure
Sumbasic	0.838	0.827	0.821
Length summary	0.897	0.878	0.865
Degree summary	0.884	0.885	0.872
Closeness summary	0.868	0.861	0.853
Betweenness summary	0.853	0.849	0.840
Eigenvector summary	0.846	0.836	0.839
PageRank summary	0.825	0.815	0.830
Ensemble summary	**0.957**	**0.937**	**0.931**

as SumBasic algorithm. So, here the evaluation of the performance of the proposed approach is compared with the above-mentioned approaches. The design of SumBasic is based upon the observation that words occurring frequently in the document occur with higher probability in the human summaries than words occurring less frequently.

The performance of the proposed methodology is measured using mostly used standard metrics such as precision (P), recall (R) and F-measure (F). The metrics are calculated as follows:

Let True be the number of tweets extracted by both the algorithms and the experts, False be the number of tweets extracted by the algorithm but not by the experts and missed be the number of tweets extracted by the human experts but not by the algorithm. The precision (P) of the proposed algorithm for the True class is: Precision (P) $=$ True$/$(True $+$ False).

In other words, P captures what fraction of the tweets chosen by the algorithm was also chosen by the human experts. The recall (R) of the proposed algorithm for the positive class is Recall (R) = True/(True + missed). Thus, R captures what fraction of the tweets chosen by the human experts could be identified by the algorithm. Finally, the F-measure is the harmonic mean of precision (P) and recall (R). F-measure (F) $= 2 * \mathrm{P} * \mathrm{R}/(\mathrm{P} + \mathrm{R})$.

The experimental result shown in Table 2 demonstrates that proposed approaches give better result than Sum basic, whereas Sumbasic performs better than PageRank Summary. It can be seen that ensemble summary comes with better result than other approaches.

6 Conclusion

This research work proposed a simple and effective methodology for Tweet Summarization based on syntactic and centrality measures of text-based microblog dataset. Adopting methods to find degree centrality, closeness centrality, betweenness centrality, eigenvector centrality, PageRank and length of the tweets, text similarity finding task has become more effective. As future work, other algorithms can be developed which can help predict certain patterns and trends on Twitter after the vast amount of tweets have been clustered with better complexity.

References

1. Barzilay, R., Elhadad, M.: Using lexical chains for text summarization. In: orkshop on Intelligent calable Text Summarization, pp. 10–17 (1997)
2. Brin, S., Page, L.: The anatomy of a large-scale hypertextual web search engine. Comput. Netw. ISDN Syst. **30**(1–7), 107–117 (1998). https://doi.org/10.1016/S0169-7552(98)00110-X. http://dx.doi.org/10.1016/S0169-7552(98)00110-X
3. Cormen, T.H., Leiserson, C.E., Rivest, R.L., Stein, C.: Introduction to Algorithms, 3rd edn. The MIT Press (2009)
4. Dutta, S., Ghatak, S., Roy, M., Ghosh, S., Das, A.K.: A graph based clustering technique for tweet summarization. In: 2015 4th International Conference on Reliability, Infocom Technologies and Optimization (ICRITO) (Trends and Future Directions), pp. 1–6 (2015). https://doi.org/10.1109/ICRITO.2015.7359276
5. Dutta, S., Ghatak, S., Das, A.K., Dasgupta, S.: Feature selection based clustering on microblogging data. In: International Conference on Computational Intelligence in Data Mining (ICCIDM-2017) (2017)
6. Dutta, S., Ghatak, S., Ghosh, S., Das, A.K.: A genetic algorithm based tweet clustering technique. In: 2017 International Conference on Computer Communication and Informatics (ICCCI), pp. 1–6 (2017). https://doi.org/10.1109/ICCCI.2017.8117721
7. Hassel, M.: Exploitation of named entities in automatic text summarization for swedish. In: Proceedings of NODALIDA03 14th Nordic Conferenceon Computational Linguistics, Reykjavik, Iceland, May 3031 2003gs of, p. 9 (2003). QC 20100712
8. Infomap—community detection. http://www.mapequation.org/code.html
9. Kupiec, J., Pedersen, J., Chen, F.: A trainable document summarizer. In: Proceedings of the 18th Annual International ACM SIGIR Conference on Research and Development in Information Retrieval, SIGIR '95, pp. 68–73. ACM, New York, NY, USA (1995). https://doi.org/10.1145/215206.215333. http://doi.acm.org/10.1145/215206.215333
10. Luhn, H.P.: The automatic creation of literature abstracts. IBM J. Res. Dev. **2**(2), 159–165 (1958). https://doi.org/10.1147/rd.22.0159. http://dx.doi.org/10.1147/rd.22.0159
11. Mani, I.: Advances in Automatic Text Summarization. MIT Press, Cambridge, MA, USA (1999)
12. Meo, P.D., Ferrara, E., Fiumara, G., Provetti, A.: Generalized louvain method for community detection in large networks. In: CoRR (2011). http://arxiv.org/abs/1108.1502
13. Radev, D.R., Allison, T., Blair-Goldensohn, S., Blitzer, J., Çelebi, A., Dimitrov, S., Drábek, E., Hakim, A., Lam, W., Liu, D., Otterbacher, J., Qi, H., Saggion, H., Teufel, S., Topper, M., Winkel, A., Zhang, Z.: MEAD—A platform for multidocument multilingual text summarization. In: Proceedings of the Fourth International Conference on Language Resources and Evaluation, LREC 2004, 26–28 May 2004, Lisbon, Portugal (2004). http://www.lrec-conf.org/proceedings/lrec2004/pdf/757.pdf
14. Rau, L.F., Jacobs, P.S., Zernik, U.: Information extraction and text summarization using linguistic knowledge acquisition. Inf. Process. Manag. **25**(4), 419–428 (1989). https://doi.org/10.1016/0306-4573(89)90069-1. http://www.sciencedirect.com/science/article/pii/0306457389900691
15. Reimer, U., Hahn, U.: Text condensation as knowledge base abstraction. In: Proceedings of the Fourth Conference on Artificial Intelligence Applications, vol. 1988, pp. 338–344 (1988). https://doi.org/10.1109/CAIA.1988.196128

16. Saggion, H., Poibeau, T.: Automatic Text Summarization: Past, Present and Future, pp. 3–21. Springer, Berlin, Heidelberg (2013). https://doi.org/10.1007/978-3-642-28569-1_1. https://doi.org/10.1007/978-3-642-28569-1_1
17. Vanderwende, L., Suzuki, H., Brockett, C., Nenkova, A.: Beyond sumbasic: task-focused summarization with sentence simplification and lexical expansion. Inf. Process. Manag. **43**(6), 1606–1618 (2007). https://doi.org/10.1016/j.ipm.2007.01.023. http://dx.doi.org/10.1016/j.ipm.2007.01.023

Heart Diseases Prediction System Using CHC-TSS Evolutionary, KNN, and Decision Tree Classification Algorithm

Rishabh Saxena, Aakriti Johri, Vikas Deep and Purushottam Sharma

Abstract Heart diseases are now one of the most common diseases in the world. Heart disease dataset (HDD) of Cleveland incorporate challenges to research societies in the form of complexity and efficient analysis of patients. Cardiovascular disease (CVD) is the scientific name given to heart diseases. The dataset extracted from the clinical test results of 303 patients which were under angiography at the Cleveland Clinic, Ohio, was implemented to a group of 425 patients in Hungarian Institute of Cardiology in Budapest, Hungary (with a frequency of 38%). Feature selection is done to reduce the irrelevant and redundant number of attributes. Further, the efficiency or the accuracy of the classifier used is calculated using few of the data mining algorithms.

Keywords KEEL · K-nearest neighbor (KNN) classification algorithm
CHC-TSS evolutionary algorithm · Friedman aligned test · Decision tree algorithm

1 Introduction

Data is considered as an important repository for every person and vast amount of data is available around the globe. Nowadays, various repositories are available which can be used to store the data into data warehouses, databases, information repository etc. [1]. To get useful and meaningful knowledge, we need to process this huge amount of available data. The technique of processing raw data into a

R. Saxena (✉) · A. Johri (✉) · V. Deep (✉) · P. Sharma (✉)
Department of Information Technology, Amity University Uttar Pradesh, Noida, India
e-mail: rishabhrhea95@gmail.com

A. Johri
e-mail: aakriti.johri9@gmail.com

V. Deep
e-mail: vikasdeep8@gmail.com

P. Sharma
e-mail: psharma5@amity.edu

A. Abraham et al. (eds.), *Emerging Technologies in Data Mining and Information Security*, Advances in Intelligent Systems and Computing 813,
https://doi.org/10.1007/978-981-13-1498-8_71

useful knowledge is known as data mining. There are various tools which are used for data mining and analysis example such as Wreka, R programming, RapidMiner, Orange, etc. One such tool used for data mining is KEEL [2]. KEEL helps the user to predict the behavior of evolutionary learning and soft computing-based techniques such as regression, classification, clustering, pattern mining, etc., that are considered as different finds of DM problem [3]. It is knowledge extraction and evolutionary learning. There are four modules in this software. First, data management which allows users to create new datasets and helps to import and export it [4].

The major role of data management is to create partitions from an existing one. Second, experiments which aims to create attractive interfaces by implementing the desired algorithm. Third, educational which is considered as the teaching part whose main objective is to design the desired experiment using a graphical user interface and an online execution of those experiments, being achievable to resume and stop them as you need. Last, modules which basically contains all the extending functionalities that are needed in the KEEL tool [5, 6]. The tool is used to build and create different data models. All data mining algorithms can be implemented on datasets. One of the features of KEEL is that it contains preprocessing algorithms like transformation, discretization, instance and feature selection, and many more. It also contains knowledge extraction algorithms like library, supervised, and unsupervised, remarking the incorporation of multiple evolutionary learning algorithms [7]. It provides with a user-friendly interface that can be used to obtain analysis of the algorithm. KEEL has various versions including KEEL version (2015–03–23), KEEL version (2014–01–29), etc.

The tool used for the analysis is KEEL 2.0. Various machine learning algorithms are being used nowadays to improve the human intervention. One such algorithm is classification algorithm. Large amount of data can be processed using classification algorithm. Predicting group members for data instances is one of the major objectives of classification algorithm [8].

2 Materials and Methods

The technique of processing raw data into a useful knowledge is known as data mining. There are various tools which are used for data mining and analysis example such as Wreka, R programming, RapidMiner, Orange, etc. One such tool used for data mining is KEEL [9]. KEEL helps the user to predict the behavior of evolutionary learning and soft computing-based techniques such as regression, classification, clustering, pattern mining, etc., that are considered as different finds of DM problem. It is knowledge extraction and evolutionary learning [10]. There are four modules in this software. First, data management which allows users to create new datasets and helps to import and export it. The major role of data management is to create partitions from an existing one. Second, experiments which aims to create attractive interfaces by implementing the desired algorithm. Third, educational which is considered as the teaching part whose main objective is to design the desired experiment

Fig. 1 Attribute used

Age: Age in years

Thalach: maximum heart rate achieved

Sex: sex (1 = male; 0 = female)

Exang: exercise induced angina (1 = yes; 0 = no)

Cp: chest pain type
-- Value 1: typical angina
-- Value 2: atypical angina
-- Value 3: non-anginal pain
-- Value 4: asymptomatic

Oldpeak: ST depression induced by exercise relative to rest

Trestbps: resting blood pressure

Slope: the slope of the peak exercise ST segment
-- Value 1: upsloping
-- Value 2: flat
-- Value 3: downsloping

Chol: serum cholestoral in mg/dl

Ca: number of major vessels

Fbs: (fasting blood sugar > 120 mg/dl) (1 = true; 0 = false)

Thal: 3 = normal; 6 = fixed defect; 7 = reversable defect

Restecg: resting electrocardiographic results

using a graphical user interface and an online execution of those experiments, being achievable to resume and stop them as you need. Last, modules which basically contains all the extending functionalities that are needed in the KEEL tool. The tool is used to build and create different data models [11]. All data mining algorithms can be implemented on datasets. One of the features of KEEL is that it contains a number of preprocessing algorithms like transformation, discretization, instance, and feature selection, etc. It also contains knowledge extraction algorithms like library, supervised, and unsupervised, remarking the incorporation of multiple evolutionary learning algorithms [12]. It provides with a user-friendly interface that can be used to obtain analysis of the algorithm. KEEL has various versions including KEEL version (2015–03–23), KEEL version (2014–01–29), etc. The tool used for the analysis is KEEL 2.0. There are four modules in this software:

- Data management.
- Experiments.
- Educational.
- Modules.

The heart diseases data has been used for the analysis. The attributes covered in the dataset are as follows (Fig. 1).

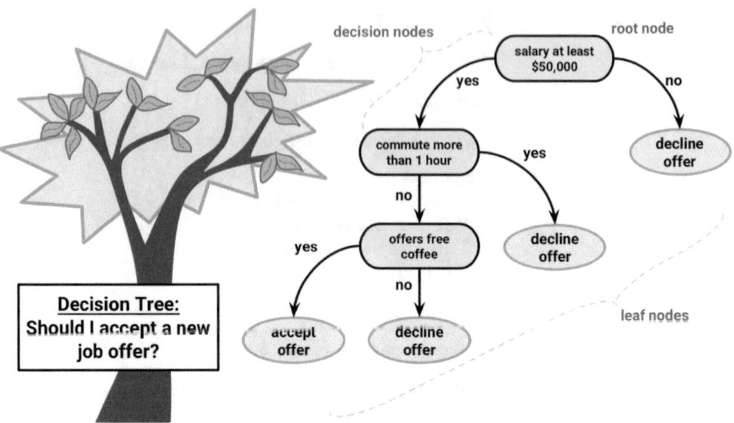

Fig. 2 Example to show how a decision tree algorithm actually works

Various machine learning algorithms are being used nowadays to improve the human intervention. One such algorithm is classification algorithm. Large amount of data can be processed using classification algorithm. Predicting group members for data instances is one of the major objectives of the classification algorithm. In the experiment module, k-nearest neighbor (KNN) classification algorithm has been used as a data mining method [2]. For the preprocessing algorithm, CHC-TSS evolutionary algorithm is applied to instant selection problem. For the statistical result, Friedman Aligned test has been used which helped to align the data before applying the Friedman test. For the education module, decision tree algorithm has been used to get the required result. After doing all the above, we were able to get the accuracy of the algorithms used as well as the confusion matrix in the result [13]. This matrix is in the form of a table that describes the efficiency or the accuracy of the classifier used in the test data of the heart diseases.

The algorithm used for analysis of the dataset is classification algorithm. In classification algorithm, the main objective is to predict the target class by analyzing the training dataset. This can be easily done by searching proper partition for each and every target class present in it [14]. In a general way, one should use training dataset to get better partition conditions which could be used so as to analyze each target class. Once the conditions over the boundary values are properly analyzed, the forthcoming step is to efficiently predict the target class. This process is known as classification. Few of the classification algorithms used are: support vector machines, k-nearest neighbor, decision trees, neural network, etc. [4].

Decision tree [15] belongs to the division of supervised learning algorithms. Decision tree algorithm solves the difficulty using its objective-type tree representation. The internal nodes of the tree belong to the attributes, and leaf nodes belong to a class labels (Fig. 2).

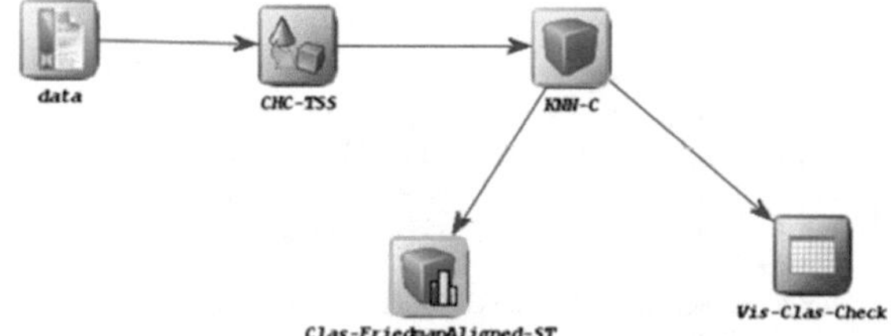

Fig. 3 The figure depicts the experiment module in KEEL which includes the KNN and CHC-TSS algorithm and a Friedman aligned test for analysis

For getting the analysis done, we have used the experiment and the educational modules of the KEEL tool. In the experiment module, k-nearest neighbor (KNN) classification algorithm has been used as a data mining method. For the preprocessing algorithm, CHC-TSS evolutionary algorithm is applied to instant selection problem. For the statistical result, Friedman Aligned test has been used which helped to align the data before applying the Friedman test. For the education module, decision tree algorithm has been used to get the required result [11]. After doing all the above, we were able to get the accuracy of the algorithms used as well as the confusion matrix in the result. This matrix is in the form of a table that describes the efficiency or the accuracy of the classifier used in the test data of the heart diseases.

3 Result and Discussion

3.1 Experiment

The experiment section is one of the modules in the KEEL tool. In the experiment module, k-nearest neighbor (KNN) classification algorithm has been used as a data mining method. For the preprocessing algorithm, CHC-TSS evolutionary algorithm is applied to instant selection problem. For the statistical result, Friedman Aligned test has been used which helped to align the data before applying the Friedman test. The desired result based on the experiment is shown here. On the given dataset, the following experiment has been done (Fig. 3).

1. ...\keel\dist\data\ExpDesign_1\results\Vis-Clas-Check\TSTKNN-C

```
TEST RESULTS
============
Classifier= cleveland
Fold 0 : CORRECT=0.7333333333333334 N/C=0.0
Fold 1 : CORRECT=0.5666666666666667 N/C=0.0
Fold 2 : CORRECT=0.6666666666666667 N/C=0.0
Fold 3 : CORRECT=0.6333333333333333 N/C=0.0
Fold 4 : CORRECT=0.5 N/C=0.0
Fold 5 : CORRECT=0.5666666666666667 N/C=0.0
Fold 6 : CORRECT=0.6774193548387097 N/C=0.0
Fold 7 : CORRECT=0.5806451612903225 N/C=0.0
Fold 8 : CORRECT=0.6666666666666667 N/C=0.0
Fold 9 : CORRECT=0.4516129032258065 N/C=0.0
Global Classification Error + N/C:
0.3956989247311828
stddev Global Classification Error + N/C:
0.08258456531914024
Correctly classified:
0.6043010752688172
Global N/C:
0.0
```

```
TRAIN RESULTS
============
Classifier= cleveland
Summary of data, Classifiers: cleveland
Fold 0 : CORRECT=0.6043956043956045 N/C=0.0
Fold 1 : CORRECT=0.641025641025641 N/C=0.0
Fold 2 : CORRECT=0.63003663003663 N/C=0.0
Fold 3 : CORRECT=0.6153846153846154 N/C=0.0
Fold 4 : CORRECT=0.6336996336996337 N/C=0.0
Fold 5 : CORRECT=0.6336996336996337 N/C=0.0
Fold 6 : CORRECT=0.6286764705882353 N/C=0.0
Fold 7 : CORRECT=0.6433823529411764 N/C=0.0
Fold 8 : CORRECT=0.6263736263736264 N/C=0.0
Fold 9 : CORRECT=0.6544117647058824 N/C=0.0
Global Classification Error + N/C:
0.3688914027149321
stddev Global Classification Error + N/C:
0.013400631646329524
Correctly classified:
0.6311085972850679
Global N/C:
0.0
```

2. ...\keel\dist\data\ExpDesign_1\results\Clas-FriedmanAligned-ST\TSTKNN-C

```
Friedman Test, Classification
Classification error in each fold:
Algorithm = KNN-C
Fold 0 : 0.26666666666666666
Fold 1 : 0.43333333333333335
Fold 2 : 0.3333333333333333
Fold 3 : 0.36666666666666664
Fold 4 : 0.5
Fold 5 : 0.43333333333333335
Fold 6 : 0.3225806451612903
Fold 7 : 0.41935483870967744
Fold 8 : 0.3333333333333333
Fold 9 : 0.5483870967741935
Mean Value: 0.3956989247311828
```

3. ...\keel\dist\data\ExpDesign_1\results\CHC-TSS.KNN-C.Cleveland

```
Result0e0

Accuracy: 0.7333333333333333
Accuracy (Training): 55.0
Kappa: 0.4936708860759492
Kappa (Training): 0.9845537757437071
Unclassified instances: 0
Unclassified instances (Training): 0
Reduction (instances): 0.989010989010989
Reduction (features): 0.0
Reduction (both): 0.989010989010989
Model time: 0.0 s
Training time: 0.007 s
Test time: 0.0 s
Confussion Matrix:
17   4    1    0    1
0    0    0    0    0
0    0    0    0    0
0    1    1    5    0
0    0    0    0    0
Training Confussion Matrix:
142  34   13   7    3
0    0    0    0    0
0    0    0    0    0
5    16   21   23   9
0    0    0    0    0
```

Result2e0				
Accuracy: 0.6666666666666666				
Accuracy (Training): 34.4				
Kappa: 0.411764705882353				
Kappa (Training): 0.9720679735063892				
Unclassified instances: 0				
Unclassified instances (Training): 0				
Reduction (instances): 0.9816849816849816				
Reduction (features): 0.0				
Reduction (both): 0.9816849816849816				
Model time: 0.0 s				
Training time: 0.003 s				
Test time: 0.0 s				
Confussion Matrix:				
17	2	0	1	0
0	3	2	4	0
0	0	0	0	0
0	0	0	0	1
0	0	0	0	0
Training Confussion Matrix:				
139	26	8	2	2
6	16	14	11	4
0	0	0	0	0
2	8	12	17	6
0	0	0	0	0

Result4e0				
Accuracy: 0.5				
Accuracy (Training): 57.666666666666664				
Kappa: 0.19210053859964094				
Kappa (Training): 0.9844507454495565				
Unclassified instances: 0				
Unclassified instances (Training): 0				
Reduction (instances): 0.989010989010989				
Reduction (features): 0.0				
Reduction (both): 0.989010989010989				
Model time: 0.0 s				
Training time: 0.002 s				
Test time: 0.0 s				
Confussion Matrix:				
14	5	1	2	1
0	0	0	0	0
0	0	0	0	0
0	3	3	1	0
0	0	0	0	0
Training Confussion Matrix:				
147	32	16	6	3
0	0	0	0	0
0	0	0	0	0
3	15	16	26	9
0	0	0	0	0

After opening the results folder in the CHC-TSS algorithm, ten results for each testing data and training data were shown. But here, only two results have been shown. Each fold in the dataset has an independent result.

Fig. 4 The figure depicts the educational module in which decision tree algorithm has been used

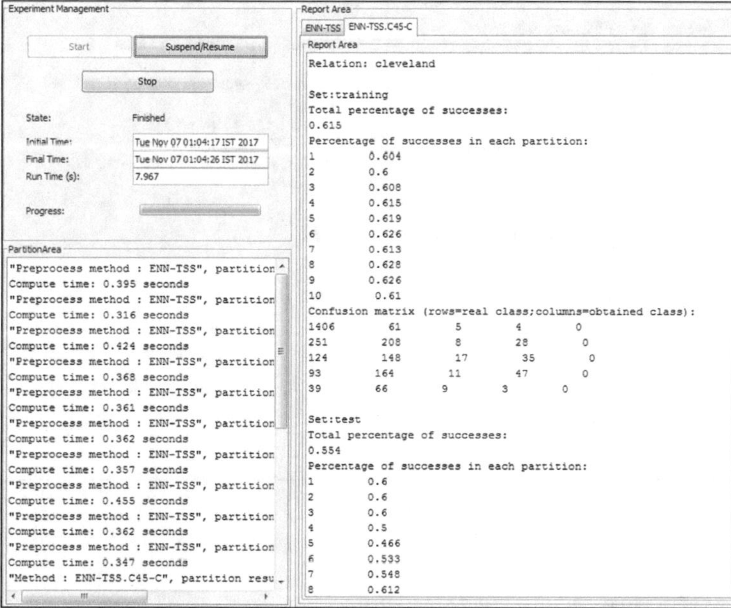

Fig. 5 After running the education section, the above result is shown. It includes the percentage of success in each partition and also the confusion matrix

3.2 *Educational*

For the education module, decision tree algorithm has been used to get the required result. After doing all the above, we were able to get the accuracy of the algorithms used as well as the confusion matrix in the result. This matrix is in the form of a table that describes the efficiency or the accuracy of the classifier used in the test data of the heart diseases (Figs. 4 and 5).

3.2.1 Decision Tree Algorithm

Decision tree algorithm is a type of classification algorithm. It has a flowchart-like structure where each internal node represents a test on the attribute, each branch represents an outcome or result of the applied test and each leave node (also called the terminal node) holds a class label, and the topmost node is the root node.

```
if ( exang <= 0.000000 ) then
{
    if ( ca <= 1.000000 ) then
    {
        num = "0"
    }
    elseif ( ca > 1.000000 ) then
    {
        if ( restecg <= 1.000000 ) then
        {
            num = "0"
        }
        elseif ( restecg > 1.000000 ) then
        {
            if ( thalach <= 172.000000 ) then
            {
                num = "1"
            }
            elseif ( thalach > 172.000000 ) then
            {
                num = "0"
            }
        }
    }
}
elseif ( exang > 0.000000 ) then
{
    if ( thal <= 3.000000 ) then
    {
        if ( ca <= 1.000000 ) then
        {
            num = "0"
        }
        elseif ( ca > 1.000000 ) then
        {
            num = "1"
        }
    }
    elseif ( thal > 3.000000 ) then
    {
        if ( age <= 50.000000 ) then
        {
            num = "3"
        }
        elseif ( age > 50.000000 ) th
        {
            num = "1"
        }
    }
}
```

4 Conclusion

In this paper, all the necessary concept of data mining has been discussed. There are various tools which are used for data mining and analysis example such as Wreka, R programming, RapidMiner, Orange, etc.. One such tool used for data mining is KEEL. Here, KEEL has been used as the tool for analysis as it helps the user to predict the behavior of evolutionary learning and soft computing-based techniques such as regression, classification, clustering, pattern mining, etc., that are considered as different finds of DM problem. First, we imported the necessary file that will be needed during the analysis. Second, the analysis is done using the appropriate algorithm. For getting the analysis done, we have used the experiment and the educational modules of the KEEL tool. In the experiment module, k-nearest neighbor (KNN) classification algorithm has been used as a data mining method. For the preprocessing algorithm, CHC-TSS evolutionary algorithm is applied to instant selection problem. For the statistical result, Friedman Aligned test has been used which helped to align the data before applying the Friedman test. For the education module, decision tree algorithm has been used to get the required result. After doing all the above, we were able to get the accuracy of the algorithms used as well as the confusion matrix in the result. This matrix is in the form of a table that describes the efficiency or the accuracy of the classifier used in the test data of the heart diseases. The exit value as 0 shows the success, which means that the classification has been correctly done. The report concludes that the effectiveness, as well as the efficiency of the various algorithms, used the two modules, i.e., education and experiment.

Acknowledgements All procedures performed in studies involving human participants were in accordance with the ethical standards of the institutional and/or national research committee and with the 1964 Helsinki declaration and its later amendments or comparable ethical standards.

References

1. Cano, J.R., Herrera, F., Lozano, M.: A study on the combination of evolutionary algorithms and stratified strategies for training set selection in data mining
2. Bouckaert, R.R., Frank, E.: Evaluating the replicability of significance tests for comparing learning algorithms. In: Proceedings of PAKDD (2004)
3. Alcalá-Fdez, J., Sánchez, L., García, S., del Jesus, M.J., Ventura, S., Garrell, J.M., Otero, J., Romero, C., Bacardit, J., Rivas, V.M., Fernández, J.C., Herrera, F.: KEEL: a software tool to assess evolutionary algorithms to data mining problems. Soft. Comput. **13**(3), 307–318 (2009). https://doi.org/10.1007/s00500-008-0323-y
4. http://www.keel.es/
5. Eggermont, J., Kok, J.N., Kosters, W.A.: Genetic programming for data classification: partitioning the search space. In: Proceedings of SAC (2004)
6. Alcalá-Fdez, J., S´anchez, L., Garc´ıa, S., del Jesus, M.J., Ventura, S., Garrell, J.M., Otero, J., Romero, C., Bacardit, J., Rivas, V.M., Fern´andez, J.C., Herrera, F.: KEEL: a software tool to assess evolutionary algorithms for data mining problems
7. https://www.mindtools.com/dectree.html
8. http://www.saedsayad.com/k_nearest_neighbors.htm
9. Alcalá-Fdez, J., Fernandez, A., Luengo, J., Derrac, J., García, S., Sánchez, L., Herrera, F.: KEEL data-mining software tool: data set repository, integration of algorithms and experimental analysis framework. J. Mult. Valued Logic Soft Comput. **17**(2–3), 255–287 (2011)
10. Esposito, F., Malerba, D., Semeraro, G.: A comparative analysis of methods for pruning decision trees. IEEE Trans. Pattern Anal. Mach. Intell. **19** (1997)
11. http://dataaspirant.com/2017/01/30/how-decision-tree-algorithm-works/
12. http://archive.ics.uci.edu/ml/index.php
13. https://i0.wp.com/dataaspirant.com/wp-content/uploads/2017/01/B03905_05_01-compressor.png?resize=768%2C424
14. Narwal, M., Keel, P.M.: A data mining tool KEEL a data mining tool: analysis with genetic. IJCSMS Int. J. Comput. Sci. Manag. Stud. **12,** 2231–5268 (2012)
15. https://media.licdn.com/mpr/mpr/AAEAAQAAAAAAAAueAAAAJGRmZDAwOTUyLTkzMTctNGE5Yi1iMWQzLTA4MzUyNjA2YTI0MQ.png

Electrofacies Analysis of Well Log Data Using Data Mining Algorithm: A Case Study

Sanjukta De, Gandhar Pramod Nigudkar and Debashish Sengupta

Abstract The unsupervised data mining methods help to extract the hidden patterns within a large multidimensional dataset of subsurface well log measurements. The well logs are continuous measurements of some parameters related to the physical properties of the rocks in subsurface with respect to depth. Log data are used by oil and gas industries to understand the lithology of the formation, and to detect potential hydrocarbon-bearing zones if any. The present study has utilized principal component analysis (PCA) to consider the contribution of each of the well logs, and to reduce the dimensionality of the multivariate logs. The first four principal components from PCA, which define around 90% variance of the dataset, were utilized for clustering using k-means clustering technique. The elbow method was employed and the optimum number of clusters was obtained to be four for the present dataset. The obtained clusters were interpreted to determine the electrofacies associated with each of the clusters. The electrofacies reflect the lithology and the potential hydrocarbon-bearing zones, in the study area.

Keywords Electrofacies · Principal component analysis · K-Means algorithm Elbow method

1 Introduction

Well logging is a technique which records the physical properties of the rocks in the subsurface against depth. Many logging tools are now used for data acquisition in the oil industries. Different logging tools have different applications. Logging tools that

S. De (✉) · G. P. Nigudkar · D. Sengupta
Department of Geology and Geophysics, Indian Institute of Technology Kharagpur, Kharagpur 721302, India
e-mail: sde@iitkgp.ac.in

G. P. Nigudkar
e-mail: gandharnigudkar@iitkgp.ac.in

D. Sengupta
e-mail: dsgg@gg.iitkgp.ernet.in

A. Abraham et al. (eds.), *Emerging Technologies in Data Mining and Information Security*, Advances in Intelligent Systems and Computing 813,
https://doi.org/10.1007/978-981-13-1498-8_72

are developed over years, record the electrical, acoustic, radioactive, electromagnetic, and many related properties of rocks and fluids contained in the rock.

An electrofacies, in a sedimentary basin, is defined as the set of log responses that describes a sedimentary rock and permits the sedimentary rock to be distinguished from other [6]. Electrofacies can be determined manually from well log and core data. But the present study provides a computerized unsupervised technique by selecting, weighing, and combining well log data for determination of the electrofacies. The obtained electrofacies from the process provides significant geological information of the region where the well is located.

Well log data used in this study has been taken from a siliciclastic sedimentary basin of India. This is a proven petroliferous basin located in the western part of India. This is an intra-cratonic rift basin of Late Cretaceous to Palaeogene age.

2 Data Description and Methodology

2.1 Data

The well log data presented in this analysis was obtained from a well in a petroliferous basin of India. More than 6000 sampling points were recorded from a thick formation encountered in this well, which was considered for the present study. The parameters considered for electrofacies analysis were the gamma ray log (GR), deep resistivity log (RT), sonic log (DT), neutron porosity log (NPHI), density log (RHOB), and clay content from the processed elemental capture spectroscopy log (ECS). In addition to this, lithology, pore fluid type, and saturation were available from core data in certain depths of this well.

2.2 Principal Component Analysis

Principal component analysis (PCA) has been used to reduce the dimensions of the multivariate logs without losing too much of the information. For "n" number of data variables, "n" numbers of principal components (PCs) are required to explain 100% of the variance. However, a lot fewer PCs can explain most of the variance in the data. The orientations of the PCs are calculated from covariance or correlation matrix of the data. The correlation matrix is generally preferred as the well logs are recorded in vastly different units. To avoid artificial and undue weightage for any particular log, the original data needed to be standardized [1, 3].

2.3 K-Means Method

Clustering is done such that similarities within the clusters are high, but similarities between the clusters are low. According to MacQueen [5], in k-means algorithm, "k" number of points from the data are randomly selected as initial centroids of the classes. Each point in the data is then assigned to its closest centroid. Once all the points are classified in this manner, new centroids are calculated. The process is iteratively repeated either for a set number of iterations or until there is no separation between the current centroids and the preceding ones.

2.4 Elbow Method

To determine the optimum number of clusters in the dataset, elbow method was used. It is a visual technique based upon observing a plot of number of clusters against a cost function which determines the distance between cluster centers and points belonging to that cluster [2, 4]. In this study, sum of squared distances was used as a cost function. In the plot, where the slope of the line abruptly changes, indicating an insignificant reduction of the cost function with the increasing number of clusters is the "elbow" [2, 4].

2.5 Flow Chart of the Methodology

Methodology followed is depicted in the flow chart as shown in Fig. 1.

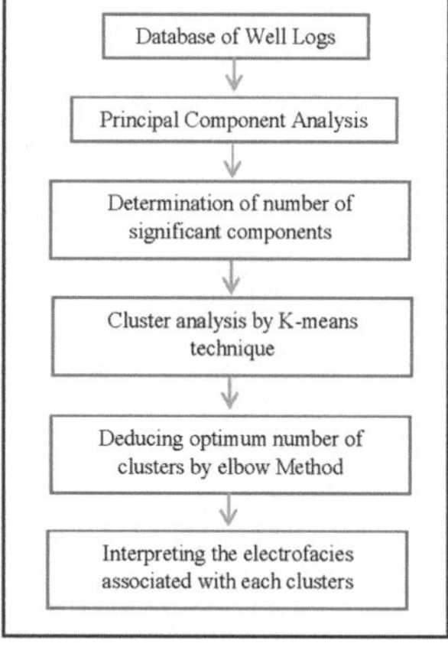

Fig. 1 Flow chart of electrofacies determination

3 Results and Discussion

The current study utilizes unsupervised methods for determining the optimum number of electrofacies from multivariate log data. Initially, PCs were computed from the correlation matrix of log measurements. The first four principal components collectively accounted for 89.52% of the total variability. The scree plot (bar plot) of the variance contribution of the PCs has been shown in the Fig. 2.

The first four PCs were utilized to perform the cluster analysis for electrofacies identification. To determine the optimum number of clusters for our data, elbow method was used. The number of clusters, k was iteratively increased in steps of one from 2 to 10. On plotting as shown in Fig. 3, the elbow was observed at k = 4. Therefore, four clusters were considered as the representations of the electrofacies as shown in Fig. 4.

From the core information of this well, the lithologies in the study area, namely shale, shaly sand (/silty sand), and claystone, were known a priori. Upon analyzing the well log responses of each cluster, the electrofacies associated grossly reflected the lithologies as shale, shaly sand (/silty sand), heavy mineral bearing sediments, and the prospective hydrocarbon (HC) zones. Table 1 summarizes the well log responses of each cluster. The methodology adopted for this study did not classify claystone as a separate lithology and bed boundaries as separate units. Moreover, some HC-bearing zones may be missed out due to low contrast in resistivity response.

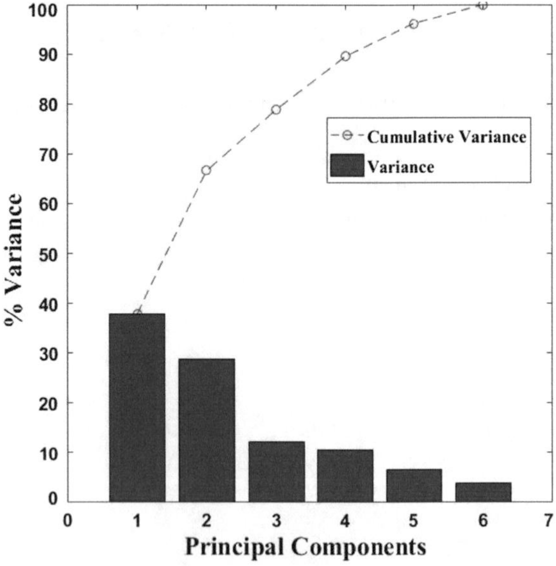

Fig. 2 Relationship between principal components and cumulative variance

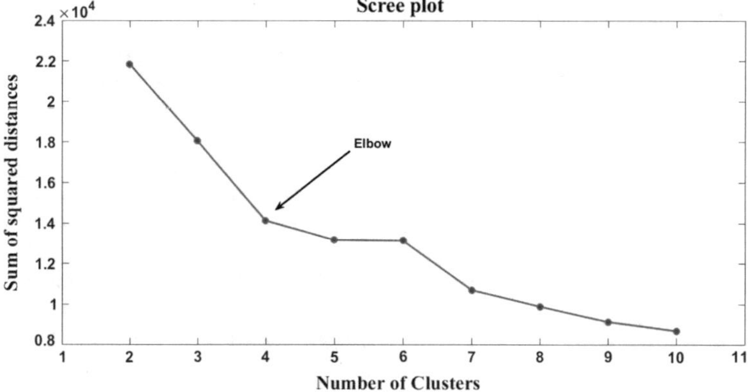

Fig. 3 Scree plot to find the optimum number of clusters

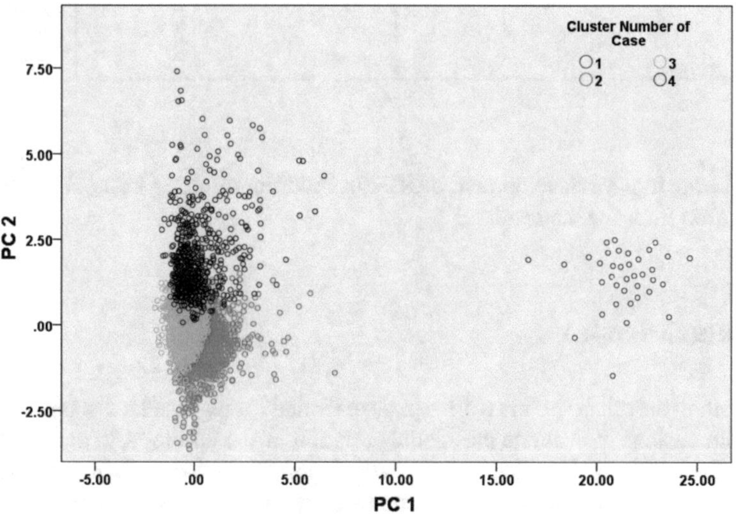

Fig. 4 Four clusters observed after cluster analysis

4 Conclusions

Unsupervised data mining techniques have been used to determine the electrofacies from the well log data. PCA was performed for dimensionality reduction of multivariate well log data, prior to clustering. For clustering, k-means algorithm was employed, which is simple, faster, and can handle large datasets efficiently. The results of the study demonstrate that the knowledge of lithology and prospective HC zones based on the electrofacies classification works well with high coherency with the core data. However, claystone got merged with shale and shaly sand clusters. Also, some of the HC-bearing zones may not be distinguished in the cluster

Table 1 Summary of the clusters

Cluster	Range of RT (Ω-m)	Range of RHOB (g/cm^3)	Range of GR (API)	Range of NPHI (fraction)	Range of clay content from ECS (fraction)	Range of DT (μs/ft)	Interpretation
1	13.00–26.61	2.48–2.70	66.09–92.42	0.29–0.39	0.33–0.57	88.22–99.11	Potential HC-bearing zones
2	0.44–9.40	2.18–2.65	45.33–148.63	0.29–0.57	0.28–0.70	88.07–168.70	Shaly sand (/silty sand)
3	0.49–3.45	2.36–2.68	45.95–191.39	0.24–0.60	0.27–0.74	89.19–161.29	Shale
4	0.67–9.10	2.49–3.10	46.81–136.28	0.29–0.51	0.30–0.68	75.00–144.07	Heavy mineral bearing sediments (Sand and shale)

analysis due to poor development of RT log. Addition of more logs (as parameters) will further improve the result.

5 Future Scopes

The method described in this work can be extended for regionalized classification in three dimensions to evaluate the spatial variation of the lithology, in the study area. Also, other unsupervised data mining techniques like self-organizing maps can be applied to the dataset and the results can be compared with the present method.

Acknowledgements The authors would like to thank Oil and Natural Gas Corporation Limited for providing the requisite dataset for the present study.

References

1. El Sharawy, M.S., Nabawy, B.S.: Determination of electrofacies using wireline logs based on multivariate statistical analysis for the Kareem formation, Gulf of Suez, Egypt. Environ. Earth Sci. **75**(21), 1394 (2016)
2. Kodinariya, T., Dan Makwana, P.: Review on determining of cluster in k-means clustering. Int. J. Adv. Res. Comput. Sci. Manag. Stud. **1**, 90–95 (2013)
3. Lim, J.S., Kang, J.M., Kim, J.: Multivariate statistical analysis for automatic electrofacies determination from well log measurements. In: SPE Asia Pacific Oil and Gas Conference and Exhibition, pp. 109–113. Society of Petroleum Engineers (1997)

4. Lopez, A.S., Martnez, F.C.: Determination of the number of cluster a priori using a k-means algorithm. Int. J. Eng. Tech. Res. **6**, 93–95 (2016)
5. MacQueen, J.: Some methods for classification and analysis of multivariate observations. In: Proceedings of the Fifth Berkeley Symposium on Mathematical Statistics and Probability, vol. 1: Statistics, pp. 281–297. University of California Press, Berkeley, California (1967)
6. Serra, O.: Fundamentals of well-log Interpretation 1. The Acquisition of Logging Data. In: Developments in Petroleum Science, vol. 15, Part A. Elsevier (1984)

A Dynamic Spatial Fuzzy C-Means Clustering-Based Medical Image Segmentation

Amiya Halder, Avranil Maity, Apurba Sarkar and Ananya Das

Abstract This chapter presents a novel method for segmentation of normal and noisy MRI medical images. In the proposed algorithm, Genetic Algorithm-based FCM-S1 (GAFCM-S1) is used which takes into consideration the effect of the neighborhood pixels around a central pixel and exploits this property for noise reduction. While spatial FCM (SFCM) also considers this feature, this method is still preferable over the others as it is a relatively faster method. Moreover, GAFCM-S1 is self-starting and the centroids are completely independent of the user inputs. Turi's Validity Index has been used as a measure of proper segmentation.

Keywords Image segmentation · MRI images · Genetic algorithm · FCM-S1

1 Introduction

Image segmentation is the process of partitioning an image into disjoint regions based on characteristics such as intensity, texture, and tone. Brain image segmentation [1–3] is an important field in medical image analysis, as proper detection of brain anomalies can lead to timely medical intervention. An MR image of brain consists

A. Halder (✉) · A. Maity · A. Das
STCET, Kidderpore, Kolkata 700023, West Bengal, India
e-mail: amiya.halder77@gmail.com

A. Maity
e-mail: avranilmaity97@gmail.com

A. Das
e-mail: ananya15.das@gmail.com

A. Sarkar
IIEST Shibpur, Howrah 711103, India
e-mail: as.besu@gmail.com

A. Abraham et al. (eds.), *Emerging Technologies in Data Mining and Information Security*, Advances in Intelligent Systems and Computing 813,
https://doi.org/10.1007/978-981-13-1498-8_73

of gray matter (GM), white matter (WM), and cerebrospinal fluid (CSF). Manual segmentation of these images is a daunting task as the images are often noisy and can lead to inhomogeneity in the images. Thus, there is a need for efficient and automatic segmentation algorithms, which can aid the medical practitioners with proper and timely diagnosis. K-means, a hard clustering technique [4, 5] is one of the earliest known segmentation algorithms. Subsequently, fuzzy clustering methods were introduced, the most important of which is the Fuzzy C-means (FCM) [6]. Over time, a lot of improved variants of FCM have been introduced [7–10]. Halder and Pathak introduced the concepts of the genetic algorithm in clustering techniques [14], while Maji and Pal incorporated the rough set [11] into it. Halder and Guha [12] recently proposed an algorithm for the segmentation of medical images using Rough Kernelized FCM.

The rest of the paper is organized as follows. Genetic algorithm and FCM and its variants are introduced in Sects. 2.1 and 2.2 as they are required to explain the proposed work. Section 3 presents the proposed method with an algorithm in pseudocode. Experimental results are compared and analyzed in Sect. 4. Finally, the paper is concluded in Sect. 5.

2 Ground Work

2.1 Genetic Algorithm

Genetic algorithm (GA) [13, 15] is a heuristic search and optimization method based on the biological principles of natural selection and evolution. This process consists of the following steps.

2.1.1 Initialization

GA is initialized with a set of chromosomes over which different genetic operators are applied for further processing. For image segmentation, cluster centers are randomly selected from the image to form the chromosomes.

2.1.2 Fitness Calculation

The second step of GA involves the determination of which chromosome is fitter than the others by calculating the fitness function, which quantifies the optimality of the chromosomes.

- Assigning pixels to appropriate clusters: If $||\rho_i - \vartheta_j|| < ||\rho_i - \vartheta_p||$, $p = 1, 2, \ldots, K$, then pixel ρ_i is assigned to cluster ϑ_j.
- Recalculation of centroid: Cluster centers are updated as $\vartheta_j = \frac{1}{N_j} \sum_{\rho_i \in C_j} \rho_i$

- Calculation of clustering metric and fitness value: Clustering metric is calculated as $\phi = \sum_{i=1}^{N} \phi_i, \phi_i = \sum_{\rho_j \in C_i} ||\rho_j - \vartheta_i||$ Fitness function is given as $f = \frac{1}{\phi}$, which needs to be maximized.

2.1.3 Selection

In this paper, we have used the Roulette Wheel selection for selecting the fittest chromosomes for further propagation. This method prefers those chromosomes with a higher fitness value, using the formula $\psi_i = \frac{f_i}{\sum_{j=1}^{N} f_j}$.

2.1.4 Crossover

Crossover is the next genetic operator, where genetic materials are exchanged between a pair of chromosomes to produce genetically stronger offspring. In this chapter, we have used single-point crossover.

2.1.5 Mutation

This is the final genetic operator where the genetic constituents are randomly altered as $r \pm p * r; r \neq 0$ and $r \pm p; r = 0$, where r is the genetic constituent and p is a random number in the range [0, 1].

2.2 *FCM and Its Variants*

Fuzzy C-means (FCM) is an unsupervised segmentation technique introduced by Dunn and modified by Bezdek [6]. The main aim of this algorithm is to minimize the cost of the objective function iteratively as follows:

$$J_{fcm} = \sum_{j=1}^{N} \sum_{i=1}^{c} \nu_{ij}^{m} ||\rho_j - \vartheta_i||^2, \tag{1}$$

where ν_{ij} = membership value of a pixel, ρ_j= pixel, or data point of an image, ϑ_i = cluster center of the ith cluster, m = fuzziness index (>1). But the major disadvantage of this method is that it does not take into account the effect of the neighborhood pixels on a central pixel, thus making this method noise sensitive. Ahmed et al. introduced a variant of FCM, called the spatial FCM, SFCM [8], taking into account the aforementioned factor, whose objective function is given by

$$J_{sfcm} = \sum_{i=1}^{c}\sum_{j=1}^{N} \nu_{ij}^{m}||\rho_j - \vartheta_i||^2 + \frac{\alpha}{N_R}\sum_{i=1}^{c}\sum_{j=1}^{N} \nu_{ij}^{m} \sum_{r\in N_j} ||\rho_r - \vartheta_i||^2, \tag{2}$$

where α = constant parameter N_R= cardinality (i.e., the number of neighboring pixels is taken into consideration). Even though SFCM succeeded in reducing the effect of noise significantly, it was still slow and time consuming, as the neighborhood term is computed in each step. Overcoming these factors, Chen and Zhang [10] introduced yet another method, FCM-S1, which computes the mean of the neighborhood pixels, thus saving the cost of computing it at every iteration. Its objective function is given as follows:

$$J_{fcm-s1} = \sum_{i=1}^{c}\sum_{k\in N_k} \nu_{ik}^{m}||\rho_k - \vartheta_i||^2 + \alpha\sum_{i=1}^{c}\sum_{k\in N_k} \nu_{ik}^{m}||\overline{\rho_k} - \vartheta_i||^2 \tag{3}$$

3 Proposed Method

In this paper, we propose an improved version of the FCM-S1 algorithm (described in the previous section), by incorporating the genetic algorithm into it. The set of cluster centers obtained by the FCM-S1 algorithm forms the basis for the genetic algorithm, which helps in the further calibration of the centroids, using the genetic operators. The proposed algorithm is described in the Algorithm 1.

4 Experimental Results

The proposed algorithm has been applied to a large number of normal and different type of noisy brain images [16]. In this chapter, we have used Turi's Validity Index [17], which needs to be minimized to obtain the optimal number of clusters. Figure 2 compares the effect of proposed and other methods existing methods such as K-means, FCM, FCM-S1, Rough FCM (RFCM), Kernelized FCM (KFCM), Rough Kernelized FCM (RKFCM), and Genetic Algorithm-based FCM(GAFCM) [15] and in the plot of Fig. 2, the validity index of the proposed method with those of the existing method as mentioned above are compared. Through these results, it can be concluded that hard clustering is not a feasible option for the segmentation of MR images of brain. Soft clustering gives comparatively better results as the validity index decreases. Among these, the proposed algorithm gives superior results than the remaining methods, as can be justified by the low value of its validity index. Advantages of the proposed algorithm are, the algorithm effectively removes the mixed noise from the images, it is relatively faster than SFCM, as the neighborhood term need not be calculated in every iteration here and the proposed method is self-starting and is completely independent of user inputs. In the proposed algorithm, we

Algorithm 1: GA BASED FCM- S1

Input: A Medical Image Im of size $N = P \times Q$
Output: Segmented image

for $i = 1......N$ **do**
  Calculate mean for each x_i
for $n = 2......K_{max}$ **do**
  // n is the number of cluster
  Initializing the set of chromosomes
  for $\tau = 1.....\omega$ **do**
    for $j = 1.....n$ **do**
      Initialize the cluster centers randomly
  // Applying FCM-S1
  for $l = 1.....\omega$ **do**
    // l is the chromosome number
    for $j = 1.....N$ **do**
      for $\tau = 1.....n$ **do**
        $\nu_{\tau j} = \frac{[||\rho_j - \vartheta_\tau||^2 + \alpha||\overline{\rho}_j - \vartheta_\tau||^2]^{\frac{-1}{m-1}}}{\sum_{k=1}^{C}[||\rho_j - \vartheta_\tau||^2 + \alpha||\overline{\rho}_j - \vartheta_\tau||^2]^{\frac{-1}{m-1}}}$
    for $\tau = 1.....n$ **do**
      $\vartheta_\tau = \frac{\sum_{j=1}^{N} \nu_{\tau j}^{m}(\rho_j + \alpha\overline{\rho}_j)}{(1+\alpha)\sum_{j=1}^{N} \nu_{\tau j}^{m}}$
  // Applying GA
  repeat
    Calculation of fitness function
    for $l = 1.....\omega$ **do**
      // l is the chromosome number
      for $\tau = 1.....N$ **do**
        for $j = 1.....n$ **do**
          if $||\rho_\tau - \vartheta_j|| < ||\rho_\tau - \vartheta_p||, p = 1, 2, \ldots, K$ **then**
            Assign ρ_τ to cluster ϑ_j
      for $j = 1.....n$ **do**
        $\vartheta_j = \frac{1}{N_j}\sum_{\rho_\tau \in C_j} \rho_\tau$
      for $\tau = 1.....N$ **do**
        for $j = 1.....n$ **do**
          $\Phi = \sum_{\rho_\tau \in C_j} ||\rho_\tau - \vartheta_j||$
      $\chi = \frac{1}{\Phi}$
    for $l = 1.....\omega$ **do**
      Apply Roulette Selection
      Crossover
      Mutation
  until $|\vartheta_{new} - \vartheta_{old}| < \delta, \delta \in [0, 1]$;

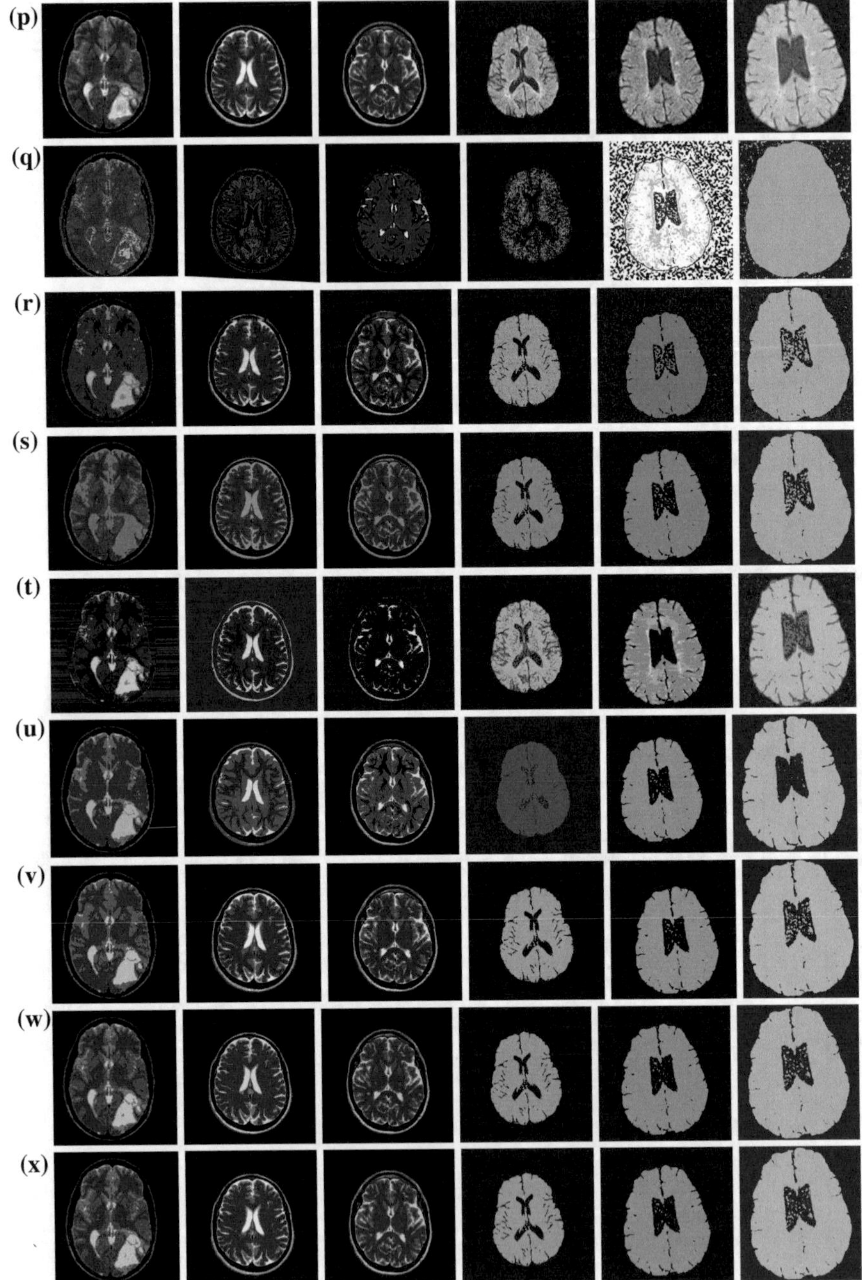

Fig. 1 **p** Original image and output segmented image using the methods **q** K-means **r** FCM **s** FCM-S1 **t** RFCM **u** KFCM, **v** RKFCM, **w** GAFCM, and **x** Proposed method (GAFCM-S1)

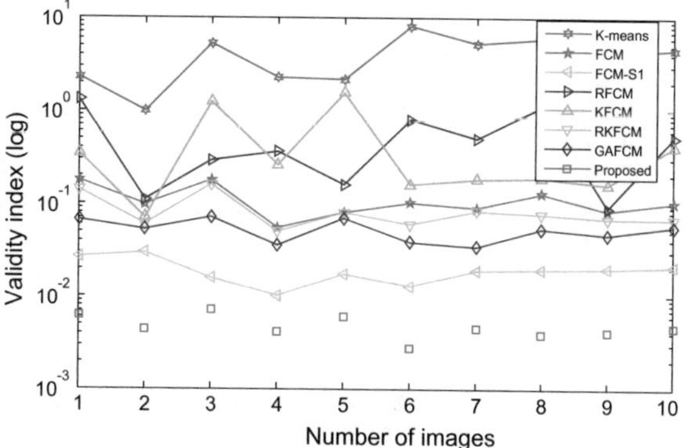

Fig. 2 Compare validity index of the normal and noisy brain images using K-means, FCM, FCM-S1, RFCM, KFCM, RKFCM, GAFCM, and Proposed method

have taken the value of m as 2, ω as 10 and $K_{max} = 8$. For most of the brain images, the optimal clusters obtained by the proposed method is 2. Figure 1 presents the segmented brain images using K-means, FCM, FCM-S1, RFCM, KFCM, RKFCM, GAFCM, and the proposed method for comparative analysis.

5 Conclusion

In this chapter, we presented a unique and novel method of segmentation and noise removal in brain images. This method provides superior results than previously proposed methods and can be proven to be an efficient aid in detecting brain anomalies.

References

1. Selvaraj, D., Dhanasekaran, R.: MRI brain image segmentation techniques—a review. Indian J. Comput. Sci. Eng. (2013)
2. Despotovi, I., Goossens, B., Philips, W.: MRI segmentation of the human brain: challenges, methods, and applications. Comput. Math. Methods Med. (2015)
3. Balafar, M.A., Ramli, A.R., Saripan, M.I., Mashohor, S.: Review of brain MRI image segmentation methods. Artif. Intell. Rev. **33**(3), 261–274 (2010)
4. Na, S., Xumin, L., Yong, G.: Research on K-means clustering algorithm: an improved K-means clustering algorithm. In: Third International Symposium on Intelligent Information Technology and Security Informatics, pp. 63–67 (2010)
5. Wagstaff, K., Cardie, C., Rogers, S., Schrdl, S.: Constrained K-means clustering with background knowledge. In: ICML vol. 1, pp. 577–584 (2001)
6. Bezdek, J.C., Ehrlich, R., Full, W.: FCM: The fuzzy c-means clustering algorithm. Comput. Geosci. **10**(2–3), 191–203 (1984)
7. Cai, W., Chen, S., Zhang, D.: Fast and robust fuzzy c-means clustering algorithms incorporating local information for image segmentation. Pattern Recogn. **40**(3), 825–838 (2007)

8. Ahmed, M.N., Yamany, S.M., Mohamed, N., Farag, A.A., Moriarty, T.: A modified fuzzy c-means algorithm for bias field estimation and segmentation of MRI data. IEEE Trans. Med. Imag. **21**(3), 193–199 (2002)
9. Chuang, K., Tzeng, H., Chen, S., Wu, J., Chen, T.: Fuzzy C-means clustering with spatial information for image segmentation. Comput. Med. Imag. Graph. **30**(1), 9–15 (2006)
10. Chen, S., Zhang, D.: Robust image segmentation using FCM with spatial constraints based on new kernel-induced distance measure. IEEE Trans. Syst. Man Cybern. **34**(4), 1907–1916 (2004)
11. Maji, P., Pal, S.K.: Rough set based generalized fuzzy *C*-means algorithm and quantitative indices. IEEE Trans. Syst. Man Cybern. Part B (Cybernetics), vol. 37, no. 6, pp. 1529–1540 (2007)
12. Halder, A., Guha, S.: Rough kernelized fuzzy C-means based medical image segmentation. In: International Conference on Computational Intelligence, Communications, and Business Analytics. pp. 466–474 (2017)
13. Bhattacharjya, R.K.: Introduction to Genetic Algorithms. IIT Guwahati (2012)
14. Halder, A., Pathak, N.: An evolutionary dynamic clustering based colour image segmentation. Int. J. Image Process. **4**(6), 549–556 (2011)
15. Halder, A., Pramanik, S., Kar, A.: Dynamic image segmentation using fuzzy c-means based genetic algorithm. Int. J. Comput. Appl. **28**(6), 15–20 (2011)
16. http://brain-development.org/ixi-dataset
17. Turi, R.H.: Clustering-based colour image segmentation. Ph.D. thesis, Monash University (2001)

Text-Based Emotion Analysis: Feature Selection Techniques and Approaches

Prathamesh Apte and Saritha Sri Khetwat

Abstract Emotion analysis is a field of research, which is sharply related to the sentiment analysis. Sentiment analysis mainly focuses on positive, negative sentiments while emotion analysis helps us recognize all sorts of emotions like happiness, sadness, anger, surprise, disgust, fear, and many more. Emotion analysis is really an important field of research as it can find many applications in many different industries. This paper will cover many different aspects of emotion analysis including properties of emotions, feature extraction and reduction techniques, different methods used for emotion analysis, challenges we faced in emotion analysis.

Keywords Emotion · Analysis · Context · Neural · Social · Network
Emoticons · Rule-based · Lexicons

1 Introduction

Emotion analysis is considered as one of the challenging tasks. Over the years, many different computer scientists have devised many different methods and ways to achieve this goal. Emotion analysis can be done using facial recognition systems. Some authors have used speech patterns to observe the tone of the voice to do this task. While in neuroscience, MRI brain scans are used to detect the emotion of a person. In this paper, we shall only discuss text-based emotion analysis.

In today's life, emotion analysis can help us to build the better world for the people. Emotion analysis finds many different applications in various domains. Emotion analysis can help us in building the review summarization, dialogue system, and

P. Apte (✉) · S. Sri Khetwat
Computer Science and Engineering Department, Maulana Azad National Institute of Technology Bhopal, Bhopal 462003, MP, India
e-mail: prathameshapte105@gmail.com

S. Sri Khetwat
e-mail: sarithakishan@gmail.com

A. Abraham et al. (eds.), *Emerging Technologies in Data Mining and Information Security*, Advances in Intelligent Systems and Computing 813,
https://doi.org/10.1007/978-981-13-1498-8_74

review ranking systems. Emotion analysis and sentiment analysis both share the common application domains. 464511

In Sect. 2, we shall see the fundamentals of emotion includes emotion properties, emotion dimensions. In Sect. 3, we shall see the different features and linguistic models such as BOW model, Skip-grams, and n-grams. In Sect. 4, we shall see the various approaches used for emotion analysis. In any recognition task, there are mainly four approaches, machine learning based, lexicon based, rule based, and hybrid. Over the years, emotion analysis has found various applications in various domains. It can be used to predict the outcome of elections, getting people's reviews on movies and products. To make better future decisions, companies use this key information.

2 Emotion Analysis Fundamentals

Before we discuss the approaches, it is only right if we discuss fundamentals of emotion first. In 2015, Schwarz-Friesel stated that text could capture only part of an emotion. To be able to completely identify one's emotion we need to take speech and facial pattern into the consideration. In this paper, we shall only discuss about text-based emotion analysis.

2.1 Modes of Emotions

Nowadays, so many people are expressing their viewpoints online. Therefore, we have plenty of information to deal with. Now, let us begin with modes of emotion [1, 2].

1. Emotion words: Here, emotions are expressed with the help of words like hate, love, anger, etc. All the words in this category carry emotional weightage.
2. Interjections and slangs: It also contains interjections (Hmm, damn, wow, oh), slangs (awesome, nope, brill), sometimes along with the punctuation such as brackets, or asterisks. They are used as action indicators.
3. Emoticons: This thing is really getting popular these days. Most of the people express their views using emoticons only. Even Facebook has updated their like–dislike button to something called "react" button. Now, we can express other emotions like anger, disgust, fear, and love. It is suffice to say that Facebook has switched to emotion analysis from sentiment analysis.
4. Hashtags: Hashtags also depicts emotions. In some experiments, the authors have classified the sentences based on hashtag labels.

2.2 Properties of Emotions

There are mainly four properties of emotions [1]. We can only focus on antecedent and response or maybe coherence if we are comparing or calculating emotional equivalence of two different people. One cannot capture signal without facial recognition system.

1. Antecedent: It is about the situation or an event, which triggers the given emotion.
2. Signal: It is related to human physiology. It is about how we express our emotions, e.g. in case of sadness through tears.
3. Response: It shows how we react to a particular situation. If we are doing tweet emotion analysis then say particular the tweet is an antecedent, where all the retweets are responses.
4. Coherence: It is about calculating similarity. It states that certain things in the world have similar antecedent, signal, and responses across different human beings.

2.3 Emotion Dimensions

Emotions can be arranged in two dimensions; valence and arousal. Valence depicts how pleasant or unpleasant emotion is. Emotions such as joy, excitement are pleasant emotions whereas emotions such as boredom, sadness, weariness are all unpleasant emotions. When it comes to arousal, different authors have come up with different definitions. Some said arousal indicates whether the individual experiencing an emotion is in control of the emotions or not [1]. According to them, it is not related to the magnitude of the emotion but it is more about level of control. Some have stated that it is about magnitude and we can score them on a scale of 1–10 [3].

3 Feature Representation

To process the given document first, we have to convert our text into numerical vectors. This process is called as word embedding. It also goes by the name of distributional semantic model, distributed representation, semantic vector space model, and vector space model. There are mainly two ways with which we can build our vector space model. In the traditional method [4], documents are encoded into the numerical vectors. First, we tokenize the entire document. Then stemming and lemmatization has been applied to find the root form of the word. We remove all the stop words before we apply lemmatization and stemming. We build the dictionary. If a word is present in a given document then the entry is one otherwise, it is zero. In second method [5], documents are encoded into the string vectors. All the w words of the documents are arranged in decreasing order based on their frequencies.

Bag-of-words [6] model can be considered while building the feature vectors. If a particular word appears in a sentence then it is one; otherwise, it is zero. BOW model does not consider the word order, e.g., "John is smarter than Bran" is the same as saying, "Bran is smarter than John."

The words remain after the tokenization, stemming and removing stop words are called as the feature candidates. The number of features can be very large. It all depends upon the size of the vocabulary. To reduce the size of the feature set, feature selection techniques are used. So many feature selection techniques are there to choose. Now, we will see some feature selection techniques.

3.1 Document Frequency Method

This process works in 3 steps [7].

1. We categorize all the documents. Documents belonging to the same category are put into the same group, e.g., all the business documents will go under "business" category, sports-related documents would go under "sports" category. Here, we can have as many categories as we want.
2. Ranking of the terms is done based on the document frequency of each term within a group. So here, we are saying that important terms tend to appear more in the document group.
3. Now parameter t is defined. Then, top t important terms of each group are selected. With this parameter t, we will be able to control the dimensionality of the feature vector.

3.2 Category Frequency-Document Frequency Method

This process works in two phases [7]. Here, the term called category frequency is introduced over DF. In the first method, we can have multiple document frequencies for the same term as we were calculating document frequency for every term and for every group separately.

1. Here, we will get only one CF value for every term. Term d appears in that group if any one of the documents in that group contains the term d. In the same way, we shall calculate CF for every term. Now, we decide the parameter t such that term is selected only if the term's CF value is below the parameter t. Once again, parameter t is used to control the dimensionality of the feature vector. We calculate CF because the most important, i.e., frequently occurring terms have very low discrimination value. They will not help us in differentiating the documents of the different categories, e.g., the term "important" itself has very low discriminating value as it appears everywhere.
2. DF method is further applied to get the reduced feature set.

3.3 Term Frequency-Inverse Document Frequency

This is another good measurement system that can be used to rank the terms. $TF_{t,d}$ means thc number of times term t appears in document d. To calculate the IDF value, we can use the following formula:

$$IDF_t = \log\left(\frac{N}{DF_t}\right) \tag{1}$$

where N is the size of our corpus, DF_t stands for the number of documents in which the term t appears. TF-IDF is the combined measure of TF-IDF that can be given by the following formula:

$$TF - IDF_{t,d} = TF_{t,d} * \log\left(\frac{N}{DF_t}\right) \tag{2}$$

TF-IDF value is used to rank the terms. Just like first two methods, here also we will select top k terms.

3.4 N-Gram Model

N-grams are contiguous sequence of n items from given sentences. Items can be anything words, letters, syllables. N-grams of size 1 are known as uni-grams, size 2 bi-grams, and size 3 tri-grams. Bi-grams and tri-grams are most commonly used. It was shown in the research before that using the combination of both two (bi-gram, tri-gram) is better [8].

E.g., Time lost is never found.

Here, if uni-gram is used we will not get any idea about the context. The author has shown us that if we consider bi-gram say "time lost", here we are getting the idea of what is actually lost which we cannot predict from only word "lost". Moreover, we can also work with negations like in the above example we are actually looking for "never found" which is the antonym of "found". The author has performed some experiments and shown us how using bi-grams and tri-grams can actually improve the accuracy of the classifier [8].

3.5 Skip-Gram Model

It is mainly used in the speech processing part. The core idea is based upon the n-gram model, but here we allow adjacent words to be skipped [9].

E.g., I hit the tennis ball.

Here, if we use the tri-gram model then we can only get the phrases like "hit the tennis", "the tennis ball". Nevertheless, the phrase "hit the ball" holds significant importance, which we will never get from our traditional n-gram model. K-skip-n-grams means our result includes zero to maximum k-skips.

E.g., Assumption is the mother of all failures.

2-skip-bi-grams = {Assumption is, Assumption the, Assumption mother, is the, is mother, is of, the mother, the of, the all, mother of, mother all, mother failures, of all, of failures, all failures}.

4 Approaches

In any recognition task, there are mainly 4 approaches, machine learning based, lexicon based, rule based, and hybrid [10]. First, we shall give you a brief introduction of the famous lexicons that are available in the market right now. Then, we shall discuss rule-based method and then machine learning based. We shall not discuss hybrid method in this paper.

4.1 Lexicons

Lexicons can be considered as a knowledge vault containing textual data labeled with emotions. It contains a list of words, e.g., for emotion joy word list may contain words like joy, pleased, happy, beautiful, fortunate, delight, enjoy, etc.

4.1.1 LIWC

It stands for Linguistic Inquiry and Word Count [1]. It contains totally 4500 words. All the words are organized into four categories. It contains conjunctions, pronouns, preposition, adjectives, cognitive words (possibility, certainty), affective words (sad, anger), interjections (Hmm, oh), etc. In generation of this lexicon, first categories are established. Then, three judges do the job of labeling. Then majority vote is taken and emotion category for the word is decided.

4.1.2 ANEW

It stands for A New English Word List [1]. It contains 1000 words. Each word is to be annotated with three labels valence, arousal, and activation. Valence states that whether the given word is positive or not. Arousal is to measure the intensity of the given emotion on a scale of 1 to 10. Activation depicts whether the given emotion is in control or not, e.g., for afraid 3-labeled tuple would be negative, 3, not. About 25 annotators are involved in the generation process. Unlike LIWC, ANEW is created in the bottom-up manner.

4.1.3 ANEW for Spanish

All the words in the ANEW are converted into Spanish [1]. Here, the assumption is made that words are likely to depict same emotion across all languages. Although magnitude and activation may differ. In order to check the quality of the newly generated lexicon, they have performed some operation in which they have observed correlation of 0.72 on activation, 0.92 on valence, 0.75 on arousal.

4.1.4 Emo-Lexicon

This lexicon is created through crowdsourcing portals [1]. Amazon's Mturk played major role in the creation of this lexicon. It contains about 10000 words. All the words are annotated and word–sense pairs are generated. For that purpose general inquirer and WordNet is used. Sense part of word–sense pair is called a "textitword-sense" portion. It gives a little idea about the context, e.g., "killer dude" means awesome whereas "killer question" means difficult. Although phrase "killer dude" can be mentioned in a negative context.

4.1.5 WordNet-Affect

WordNet-Affect is derived from WordNet [11]. Here, something called synset is used. Synset contains set of synonym words that represents the given concept. It contains 2874 synsets and 4787 words with affective-labels (a-label). It is created in semi-automated manner. In WordNet, each synset is associated with at least one domain label, e.g., sports, business, etc. Here, we added one extra hierarchy of a-label irrespective of their domain. To build the WordNet-Affect first lexical database AFFECT was realized. It contains 1903 items indirectly or directly referring to mental states, cognitive state, feelings, attitude, behaviors, and personality traits. It contains 517 adjectives, 238 verbs, 539 nouns, and 15 adverbs. For each term, a frame is created to add the affective and lexical information. Lexical information includes POS tags, antonyms, synonyms, and semantic definitions. It has several parameters such as POSR, ORTONY. POSR parameter describes the term having the different

POS tags but pointing to the same psychological category. ORTONY parameter is used to state the affective category. To see the procedure in detail, refer the references below [11].

4.1.6 Chinese Emotion Lexicon

It contains about 50000 words [12]. It was created in a semi-automated manner. First, core list of words was selected. Then, these words were labeled. Then based on the similarity matrix words are appended. Three kinds of similarities are used.

1. Syntagmatic: If the two words appear frequently in larger text then they tend to be similar.
2. Paradigmatic: It deals with the semantic similarity. Two words being synonyms, antonyms of each other are belonging to the similar domain.
3. Textbtlinguistic Peculiarity: This works in a similar way the wildcard patterns work. Just look for the syllable overlap, e.g., happy, happily, happiness they all depict the same emotion of joy. All words mentioned before, start with the same prefix "happ".

4.2 Rule-Based Approach

Most of the authors focus on detecting explicit emotions but emotions exist in our language in multiple forms. The authors of this paper [3] mainly focused on detecting implicit emotions. Explicit emotions are formed at the conscious level. They are deliberately formed. Implicit emotions are formed at the unconscious level. They are generally involuntarily formed and typically unknown to us. Emotions can be expressed in a sentence without using any emotion bearing word, e.g., "The outcome of my test brings me so much joy". The given sentence depicts happiness. Some will say the same thing in a different way e.g., "I cracked the exam". This sentence depicts the same emotion as well. Here, the authors of this paper have used the model developed by Ortony, Clore and Collins (OCC) given in [13]. OCC model describes the hierarchy of 22 emotions. OCC model depends upon the variables. OCC model is more complex and has many ambiguities removed by the authors [14]. Based on this model, rules have been devised by authors [3]. First, perform all the preprocessing tasks tokenization, stemming, POS tagging then values to the variables can be assigned. Totally five variables are used. The value for the direction variable must be "Other" or "Self". This value has been calculated using dependency parser. Dependency parser is used to calculate the value of the pronoun. Tense can have three values "Present", "Future", "Past". Remaining variables are either "Positive" or "Negative". Value of the tense can be found out from POS tags. Overall sentence polarity is calculated by polarity detection through majority vote. Dependency parsing and lexicon matching methods are used together to calculate the event polarity and action polarity. In case of event polarity, verb–object relationship is used. In case of action polarity, subject–verb relationship is used.

Experiment was carried out on three datasets (SemEval, ISEAR, Alm's). NRC lexicon for sentence-level emotion recognition is used in lexicon-based method and NB classifier with fivefold cross validation from Weka is used to compare the results with rule-based method. Quality of the text affects the results. SemEval and ISEAR datasets are made of informal text. They contain slangs and ill-structured grammar. On the other hand, Alm's strictly follows all the grammar rules. Conclusion is rule-based approach, which is good if the text is formal.

4.3 Machine Learning Based

Here, we shall get the glimpse of the work various authors have done in this area. Here, [10] the authors have used deep learning. The experiment has been carried out using deep neural networks. A multi-layered neural network is used in which they have used three hidden layers consisting of 125, 25, 5 neurons, respectively. For feature representation, bi-grams are used. To see the setting of this experiment, please see in the references section [10]. Average accuracy of this experiment is 64.47%.

Here, the authors have used supervised machine learning approach [6]. They have performed their experiments on various datasets (SemEval, Alm's, and Aman's). Weka tool has been used for the experiment [15]. They have gone with the J48 decision tree algorithm, Naive Bayes Bayesian classifier, and SMO implementation of Support Vector Machine. Bag of words, N-grams, and WordNet-Affect was used to build the feature sets of the following experiments. To reduce the size of the feature set Lovins Stemmer is used. Another way for reducing the feature set while using BOW model is not to use the contracted form of the grammar. Use "I am" instead of "I'm". ZeroR classifier from Weka is used as a baseline algorithm. For Aman's dataset, they managed to have achieved the accuracy of 81.16%. Previous best was 73.89% reported in [16]. Apart from this, different authors have used various datasets and techniques for emotion analysis. The authors have used Roget's Thesaurus as lexical resource to do the emotion analysis task [17]. The major problem with this method is lexical affinity is partial toward specific genre. They are not reusable, as each word will have a different sense in different domain. The authors of this paper [16] have used paradigmatic features. They have also considered punctuations and interjections, while accumulating a feature set. Some authors have used latent semantic analysis for classification [18]. Although statistical methods have very little use in modern social networks where language lacks delicacy and precision. Here, the authors have used conditional random fields for the very first time to do the task of emotion analysis [19]. The authors [20] have done another important work. They have used hierarchical classification. For two levels, classification accuracy is bumped up from 61.67 to 65.5. Datasets given by [6] are used for the experiments.

5 Conclusion and Future Scope

In this paper, we have described some of the major works, which have been done in emotion analysis area. When it comes to building the feature set, there are so many techniques available for us to choose from, e.g., N-grams, BOW, etc. We have mentioned only three approaches for emotional analysis in this paper. In recent years, many people have devised new techniques, approaches, and datasets. We have tried to cover the some of the work. This field also has some challenges. Context dependence is one of the major challenges in determining emotion from text, e.g., "Shut up!" can depict "anger" or can depict "surprise" as well. The given sentence is ambiguous. To be able to assign the sentence with correct emotion label, one must have to consider the context. It can be very difficult to capture the whole context when given sentence is mentioned on Twitter. Current emotion analysis systems are not advanced enough to detect the hyperbolic expressions. Sometimes humans deliberately exaggerate the sentence, e.g., "My manager asks me to work 40 h a day". Hyperbolic expressions can only be captured when we resolve first issue. Emotion analysis systems are not advanced enough to deal with the sarcasm. Sarcasm is when people means opposite of what they actually say, e.g., "I love being ignored all the time". To capture the sarcasm successfully, we have to consider emoticons, hashtags, slangs, context, and so many other things. We can start with sarcasm detection in sentiment analysis. In [2], the authors have shown how detecting the sarcasm can enhance the accuracy of our classifier.

References

1. Tripathi, V., Joshi, A., Bhattacharyya, P.: Emotion Analysis from Text: A Survey. Center for Indian Language Technology Surveys (2016)
2. Bouazizi, M., Ohtsuki, T.: Opinion mining in Twitter: how to make use of sarcasm to enhance sentiment analysis. In: 2015 IEEE/ACM International Conference on Advances in Social Networks Analysis and Mining (ASONAM). IEEE (2015)
3. Udochukwu, O., He, Y.: A rule-based approach to implicit emotion detection in text. In: International Conference on Applications of Natural Language to Information Systems. Springer, Cham, (2015)
4. Duncan, B., Zhang, Y.: Neural networks for sentiment analysis on twitter. In: 2015 IEEE 14th International Conference on Cognitive Informatics & Cognitive Computing (ICCI* CC). IEEE (2015)
5. Jo, T.: NTC (Neural Text Categorizer): neural network for text categorization. Int. J. Inf. Stud. **2**(2), 83–96 (2010)
6. Chaffar, S., Inkpen, D.: Using a heterogeneous dataset for emotion analysis in text. Adv. Artif. Intell. 62–67 (2011)
7. Lam, S.L.Y., Lee, D.L.: Feature reduction for neural network based text categorization. In: Proceedings, 6th International Conference on Database Systems for Advanced Applications, 1999. IEEE (1999)
8. Feature representation for text analyses. https://www.microsoft.com/developerblog/2015/11/29/feature-representation-for-text-analyses/
9. Guthrie, D., et al.: A closer look at skip-gram modelling. In: Proceedings of the 5th international Conference on Language Resources and Evaluation (LREC-2006). sn (2006)

10. Emotion detection and recognition from text using deep learning. https://www.microsoft.com/developerblog/2015/11/29/emotion-detection-and-recognition-from-text-using-deep-learning/
11. Strapparava, C., Valitutti, A.: WordNet affect: an affective extension of wordNet. In: LREC, vol. 4 (2004)
12. Xu, J., et al.: Chinese emotion lexicon developing via multi-lingual lexical resources integration. In: International Conference on Intelligent Text Processing and Computational Linguistics. Springer, Berlin (2013)
13. Ortony, A., Clore, G.L., Collins, A.: The Cognitive Structure of Emotions. Cambridge University Press (1990)
14. Steunebrink, Bas, R., Dastani, M., Meyer, J.-J.C.: The OCC model revisited. In: Proceedings of the 4th Workshop on Emotion and Computing (2009)
15. Witten, I.H., Frank, E.: Practical Machine Learning Tools and Techniques (2005)
16. Aman, S., Szpakowicz, S.: Identifying expressions of emotion in text. Text, Speech and Dialogue. Springer, Berlin, Heidelberg (2007)
17. Aman, S., Szpakowicz, S.: Using Roget's Thesaurus for Fine-grained Emotion Recognition. In: IJCNLP (2008)
18. Strapparava, C., Mihalcea, R.: Learning to identify emotions in text. In: Proceedings of the 2008 ACM Symposium on Applied Computing. ACM (2008)
19. Yang, C., Lin, K.H.-Y., Chen, H.-H.: Emotion classification using web blog corpora. In: IEEE/WIC/ACM International Conference on Web Intelligence. IEEE (2007)
20. Ghazi, D., Inkpen, D., Szpakowicz, S.: Hierarchical versus flat classification of emotions in text. In: Proceedings of the NAACL HLT 2010 Workshop on Computational Approaches to Analysis and Generation of Emotion in Text. Association for Computational Linguistics (2010)

Community Detection Methods in Social Network Analysis

Iliho and Sri Khetwat Saritha

Abstract Extracting insights from Social Network has been on trend for its contribution to various research domains to solve real-world applications such as in business, bioscience, marketing, etc. The ability to structure social network as a graph model with nodes as vertices and edges as links has made easier to understand the flow of information in a network and also figure out the different types of relationship existing between the nodes. Community detection is one of the ways to study and uncover the nodes exhibiting similar properties into a separate cluster. With certainty, one can deduce that nodes with similar interest and properties are likely to have frequent interactions and are also in close proximity with each other forming a community, such community can be represented as functional units of the huge social network system making it easier to study the graph as a whole. Understanding the complex social network as a set of communities can also help us to identify meaningful substructures hidden within it, which are often predominated by the superior communities to excavate people's views, track information propagation, etc. This paper will present the different ways in which one can discover the communities existing in social network graphs based on several community detection methods.

Keywords Social network analysis · Community detection · Dynamic community · Overlapping community

Iliho (✉) · S. K. Saritha
Computer Science and Engineering Department,
Maulana Azad National Institute of Technology Bhopal, Bhopal 462003, MP, India
e-mail: ilihokiho@gmail.com

S. K. Saritha
e-mail: sarithakishan@gmail.com

A. Abraham et al. (eds.), *Emerging Technologies in Data Mining and Information Security*, Advances in Intelligent Systems and Computing 813,
https://doi.org/10.1007/978-981-13-1498-8_75

1 Introduction

Analysis based on the structures of a graph often termed as social network analysis derived a class of analytics, consider combining the topological and structural attributes and retrieving the useful data from various ties formed among actors of a network. The graph of a social network is represented with a set of nodes representing partakers and edges representing relationships. An edge between nodes in a social graph indicates the presence of a link or connection between the corresponding entities. Such graphs can be further dig in to identify communities or ascertain hubs. Network analysis phenomena can be applied in any web field, where a graph may be constructed. As social networks will continue to change, uncovering communities and modelling social graphs from large-scale social networks will go on and be a challenging subject in the field of research. Great interest is being developed for research in this field for the extraction of information from social network to identify the distinct communities existing in the social network, testing and examining the network evolution to predict future behaviour and also the identification of knowledge experts in respective domains.

With the quick expansion of the web, there is a tremendous growth in online interaction of users creating a network. These real-world social networks have interesting patterns and properties, which may be analyzed for numerous useful purposes. The process of discovering the cohesive groups or clusters in the network is known as community detection. Detection of communities proves beneficial for various applications like finding a common research area in collaboration networks, finding a set of users having common taste for marketing and recommendations, epidemic spreading and finding protein interaction networks in biological networks [1]. Detecting the correct communities is of prime importance and so the algorithm that detects communities in complex networks should aim to accurately group the communities as expected.

2 Community Detection Methods

The conceptual differences between different perspectives on community detection need to be first understood to have a better understanding of what the network is and most importantly to select the most suitable community detection algorithms for the given purposes. To clarify the concepts about different community detection algorithms, four broad dimensions can be a guide for choosing an appropriate method for a specific purpose [2].

- Cut-based community detection: Cut-based or graph partitioning is the process of partitioning a graph into a predefined number of smaller groups with specific properties. The edges connecting the groups are denoted as the cut. This method minimizes the cut while maintaining a balanced group. It is mandatory to explicitly state the number of components one wishes to get in case of graph partitioning.

It is certainly not possible to specify the number of communities in real-world networks, and hence become its limitations.

- Clustering-based community detection. It partitions the points into several groups such that points from the same group are close to each other and points in separate groups move away from each other. The cut- and clustering-based community detections are almost the same but the difference arises in the later where there is no need of prior knowledge on the number of groups to be clustered and, moreover, the groups formed need not be a balanced group. An example is the hierarchical clustering mainly used to examine the similarity of connections between nodes in the network such as in analyzing social, biology and marketing networks. It can be classified into two types: (i) Agglomerative clustering (ii) Divisive algorithms.
- Stochastic block model-based community detection: Identifying nodes within a network that exhibit similar structural role in terms of their connections. In [3] also, it was stated that nodes are equivalent if they connect to equivalent nodes with equal probability. Block model aims to identify nodes in a community that are similar to nodes in another community. SBM is defined by affinity matrix Ω. The equation below gives the probability between two nodes i, j of class c_i and c_j.

$$p_{i,j} := P(A_{i,j}) = \Omega_{c_i,c_j} \tag{1}$$

- Dynamical community detection: To clearly track the underlying changes occurring in a dynamic network, considering the topology of graph alone is not self-sufficient to draw conclusions so by examining the graph at each time steps better explains the behaviour of the network system. Considering the dynamic features like the flow of information which changes and behaves none uniformly to study the network helps deliver the real-time information at different time step. This property helps in proper understanding of the complex real-time network.

3 Literature Review

3.1 Dynamic Community Detection

People with similar interests are more likely to communicate with each other more often and thereby establishing interactive patterns between the communicating parties. The issue arises when one further wishes to identity such community structure in large-scale network and uncover the natural community in real-world networks.

Several methods have been proposed to discover communities based on different standards. Community detection based on dynamic interaction is among one to spot the groups or communities in a network. Social network exhibits more of a dynamic nature rather than a static one. Real-world networks are naturally dynamic, made up of communities that are regularly changing its members. Static community detection method cannot find the new structure whenever the network evolves. Therefore, to understand the activities and interactions of users over time one has to carefully examine the changes occurring in a network. Dynamic community detection has been of great importance in examining the large amount of temporal data to reveal the network structure and its evolution. In [4], a local dynamic interaction approach for community detection called Attractor has been proposed to automatically identify the communities formed in a network. This algorithm outperforms the traditional dynamical method [5, 6] by considering the topology-driven interactions and examining the change of distance among linked nodes rather than considering only the node dynamics alone.

The attractor algorithm work on the notion that all nodes get attracted to their connected nodes which results in moving together nodes in the same community into one, while the nodes in a different community are kept far away from each other. Jaccard similarity is used to assign each edge of complex network an initial distance. They simulated the distance dynamics based on three different interaction patterns namely: Influence from the direct linked node, influence from common and influence from exclusive neighbours. Based on the different influence from each neighbour the node gets attracted to nodes with maximum influence and form a community. Using this approach great accuracy level has been achieved by identifying the small and hidden communities unlike the modularity-based community detection, which fails to detect the small communities [7] and also there is no need to specify the desired number of clusters beforehand unlike the graph partitioning-based community detection, which requires the number of clusters to be predefined.

With an influence in the way the local interaction model behaves, the author in [8] proposed a new algorithm replacing the modularity-based approach with the local interaction model. In [4], the author used the static algorithms, whereas in [8] the author has used the dynamic algorithm. Unlike the static community detection algorithms that neglect the changes in a network at different time steps, it misses to capture the evolutionary patterns as the time of interaction is not considered. Dynamic algorithms could better capture the transitions that the communities undergo between any consecutive time steps. The transitions the community undergoes is because of the resultant of the four different interaction ways that take place between two consecutive time steps, which can be addition of nodes or edges and deletion of nodes or edges. In addition to answering the first issue in state-of-the-art methods namely resolution limit problem, where uncovering the hidden communities is not easy and second issue, i.e. the dependence of the network structure on the sequence of network increments arrival, a new factor called disturbance factor was introduced to lower the range of disturbed edges in order to lower the unlimited spread of disturbed edges to yield an efficient computing time as more disturbed edges lead to more edge distance calculation while strictly maintaining the quality. Experimental evaluation was performed on real-world networks and synthetic network for comparing the performance and effectiveness of this algorithm with other algorithms and conclusions were drawn that the proposed algorithm draws better result.

Despite several improvements and advantages over other algorithms, the algorithm in [8] had loopholes by not considering some additional factors that have been researched and specified in [9]. The author designed two improved interaction patterns replacing the traditional three interaction patterns. The additional factors considered in this paper could improve the accuracy and time efficiency of distance dynamics method to a great extent. The work in [8] the traditional distances dynamic method (Attractor algorithm) do not consider the different cohesiveness of each neighbour node, i.e. it does not consider the degree of a neighbour node. The influence of each neighbour to the distance is only determined by the distance difference so when two neighbours have same distance to the same edge, then the influence from two neighbours to the distance is considered equal and influence tends to zero and lowers the speed of attraction. This limitation is overcome by considering the degree, i.e. for one distance, the cohesiveness of one neighbour is determined by his degree, in case of less degree the probability of a node to interact to any one neighbour is higher, so the interaction number is larger and the speed of attracting one neighbour to move towards itself is faster, thus the influence of this node is bigger on the contrary if one node has more degree, the interaction number from this node to any one neighbour is smaller, the speed of attracting any neighbour to move towards itself is slower, thus the influence of this node is less. Neighbour cohesion was introduced to measure the attraction strength from each neighbour to one distance. The second limitation is the situation of slow convergence and non-convergence. After exploring and experimenting it, the cause of slow and non-convergence was highlighted and a convergence coefficient to increase the convergence speed was defined and hence the situation of non-convergence is avoided. However, the distance dynamics alone to determine the relatedness of a community seems insufficient and so considering more additional information like node attributes one could increase the accuracy level of community detection.

3.2 *Community Detection Using Distance and Node Attributes*

The ever growing interactions of users on a social platform results in huge generation of data demanding the need for more sophisticated techniques and approach for studying the complex network structure created as a result of the interactions and identifying the communities present in it. Although several community detection algorithms have been proposed and almost all algorithms group the nodes based either solely on the relation or distance between the nodes. However, in real world the relatedness or commonness of users not only depend on the way they are related or connected but also considers some additional characteristic features, i.e. more precise relatedness between nodes can be obtained by considering interesting additional feature of users, i.e. the different attributes of a user that may help in defining the interest of a user. This way more accurate communities will be identified. Considering only the interest to measure the closeness would be vague as it would mean that all nodes that are nearby are strictly related to each other.

In [10], an algorithm which divides the nodes in social networks into communities considering both the geodesic location and the similarity between their interests has been put forth. To give more precise and refined estimate of the relatedness between nodes the relatedness of two nodes is determined by counting not only the number of hops between them but also the interest of each nodes. The new algorithm known by the name GEOSIM clusters the nodes based on how similar the two nodes are in terms of their connections and their interests and thought of it as a single community. The algorithm concludes with the assumption that though two nodes have very similar interests but if they are widely separated from each other they would still have a very low chance to get grouped in one community. Similarly, two nodes close to each other in terms of distance and having very different interests have a very low probability of them being in the same community.This method tries to detect those communities, which have both distance and node interest similarity. The algorithm is divided into three stages, namely centroids selection, distances and similarities calculation and finally clustering. The metrics used in this algorithm is GEOSIM score to calculate the distance similarity and the similarity between two interests as well. Based on the results obtained from the experiments detailed in [10], this algorithm produces considerably good results. In addition to distance and interest similarity other additional features of a user or nodes as per requirements can be considered to cluster the network. More related works can be found in [11, 12]. The following section explains the overlapping detection algorithms, which comparatively discover nodes having multiple properties.

3.3 Overlapping Community Detection

An overwhelming challenge faced when detecting communities in a social network is the presence of nodes that show up multiple interests taking part in different social groups and overlapping between different communities in the complex network graph. Such user represented by a node in a graph may not be part of just one community or a group; it may be a part of many closely associated or different groups existing in the network. Nodes that are shared by several communities play a crucial role as intermediate between communities and these nodes predict dynamic behaviours of individual nodes in the network. The overlapped nodes are interesting to investigate. Applying disjoint algorithms may miss the important information about an individual's concurrent interaction with the numerous social parties; this reason has driven the interest in designing overlapping community detection algorithms. The work in [13] was the extension of Girvan–Newman algorithm to detect the overlapping communities called the clique percolation algorithm. It assumed that the internal edges of a community were likely to form cliques due to their high density. Here, k-clique indicated a complete graph with k vertices and two k -cliques were considered adjacent if they shared k-1 vertices.

A community was formed with the maximal union of k cliques that could reach out from each other through a series of adjacent k cliques. To understand the direction of flow of information, directed CPM was introduced and later the link weights between two nodes were also considered for a better understanding of the overlapped structure. However, the CPM proposed failed to discover the hierarchical structure along with the overlapping attribute. To overcome this limitation, the work in [14] introduced EAGLE (agglomerative hierarchical clustering based on maximal clique) to detect hierarchical and overlapping community structure of networks as well. A non-clique-based approach called Order Statistics and Local Optimization method (OSLAM) was the first algorithm to detect clusters considering for edge directions, edge weights, overlapping communities, hierarchy and community dynamics. To evaluate the clusters OSLAM uses significance as a fitness function. It consists of three phases. First, it searches for significant clusters until convergence next it analyzes the resulting set of clusters by carefully studying the internal structure, and finally, the hierarchical structure of the clusters was captured. The OSLAM algorithm performed better on directed graphs and also in the detection of strongly overlapped communities. It could also detect the presence of randomness in graphs or absence of community structure. But the limitation of this algorithm was that it performed a lot of iterations to give more accurate results, which increased the complexity of the algorithm. Also, label-based propagation algorithm [15] was further extended to detect overlapping communities, in this approach nodes in a graph were assigned labels and the nodes with same labels form communities but it showed some non-deterministic nature and gave no unique solution. Lately, focus has been moved towards local interactions which perform comparatively lesser iterations as it considers the neighbouring information of a node rather than the global information thereby the efficiency of computation is maintained as social network being of large and dynamic nature. Unlike other local expansion methods [16] which follows random selection of seeds, this algorithm [17] follows an effective approach while selecting the seeds for expansion. It considers local density and relative density between nodes as well while selecting the structural centres in the community structure. In [16], the local discovery of communities is used by first identifying the central structures in a graph to uncover the natural overlapping communities. These central structures have comparatively higher density and are more distant to other high-density nodes and, therefore, can help to intuitively determine the number of clusters with the help of the central structures and hence outperforms other methods as the number of clusters need to be mentioned beforehand. Then, it iteratively adds the neighbours of the central structures to form a community. Hence, showing better results than the label propagation algorithm and the other local expansion methods.

3.4 Metrics Used in Community Detection

The accuracy of the community detection algorithms is tested by using either the real-world networks or the synthetic networks. Since synthetic networks are artificially generated its ground truth is known beforehand. The actual communities to be expected from the network are known from the ground truth values. Due to the availability of millions of nodes in real networks and heterogeneous distribution of node degrees, the researchers started to develop benchmarks to tackle the large-scale networks an, e.g. is LFR benchmark.

To verify that the communities detected are the correct ones evaluation metrics for measuring the quality of a community are used. The communities detected using the community detection algorithms are further evaluated to check its accuracy. For comparing the performance of different community detection algorithms a stable quality function score is essential to measure the quality of communities obtained by the different community detection algorithms. The quality function depends on the specific type of community to be detected whether it is disjoint community, overlapping community, evolving communities, etc.

- Normalized Mutual Information (NMI): the most commonly used metrics for disjoint communities. The identified communities and the ground truth are compared, and their similarity and relatedness are measured using this metric. NMI value is equal to 1 if the identified community and the community with ground truth are identical and equal to zero if they are totally distinct. It has been defined both for disjoint and overlapping communities as well. Generally, it is represented as follows:

$$NMI(X,Y) = \frac{2I(X,Y)}{H(X)+H(Y)} \tag{2}$$

$$I(X,Y) = \sum_{x \in X} \sum_{y \in Y} p(x,y) log(\frac{p(x,y)}{p(x)p(y)}) \tag{3}$$

$$H(X) = -\sum_{x \in X} p(x) log(p(x)) \tag{4}$$

$H(X)$ is the Shannon entropy of partition X, and $I(X,Y)$is the mutual information that captures the similarity between two partitions X and Y. For a set of communities that partition a network, $p(x) = |x|$ and $p(x,y) = |x \cap y|$.
- Modularity: The best known and most used quality function is Modularity Q. It was proposed by Newman and Girvan. It measures the quality of a partition of a network into communities by comparing the number of links inside a given community with the expected value for a randomized network of the same size and degrees and expressed as

$$Q = \sum_{k=1}^{k} Q_{k} = \sum_{k=1}^{k} [\frac{e_{k,k}}{e} - (\frac{d_k}{2e})^2]. \tag{5}$$

where k denotes the number of communities in the partitioning, e represent the number of edges in the graph, d_k is the total degree of the nodes in community C_k and e_{kk} is the number of edges within community C_k. When Q is high, then it denotes good communities. Modularity is always less than one and gives a negative value when each node is itself a community except for some graphs like random graphs large modularity value may not necessarily mean that the graph possesses a community structure. Some limitations of modularity metric are requiring global knowledge of graph topology and little knowledge of the number of communities, inability to identify very minor communities. Modularity metric was further extended by Nicosia in [18] to detect overlapping communities.

4 Future Work

Very few studies and research work have been done on identification and detection of a particular type of community called the hidden community also called the weak communities in a network of dominant or strong communities. Identifying such hidden communities can help in areas, where one's objective is to find the smallest groups that exist, find all different communities formed irrespective of the size of the community. First, this area of research can be beneficial for scientific learning, for example in protein–protein interaction to study the protein structures. Identifying all the communities can help gain better knowledge about the overlapping communities as well. Second, the phenomenon of overlapping communities can be better understood by capturing the interaction dynamics occurring in a number of relatively short time intervals to identify the nodes that are frequently changing its membership between communities and also the nodes that are concurrently participating in multiple communities.

5 Conclusion

There is no algorithm that can serve a general purpose for all datasets, because there are different perspectives on how to form a community, the methods to be used depends on what type of network is to be analyzed and also considering its applications as different motivations can discover different communities even for the same network. Social Network Analysis has been one of the core research domains, many research has been going on to analyze and understand the large-scale data and the complex network generated as a result of continuous human interactions. Through the detection of communities, there is an ease to study the relation among the nodes in the network, extracting more meaningful information from the large complex network by forming them into groups or communities. Communities with dynamic nature and communities with overlapping members still need an in-depth investigation as not much has been done concerning this.

References

1. Maxwell, C.: A Treatise on Electricity and Magnetism, 3rd ed., vol. 2, pp. 68–73. Oxford, Clarendon (1892). https://wires.wiley.com/WileyCDA/WiresJournal/wisId-WIDM.htmlJ
2. Schaub, M.T., et al.: The many facets of community detection in complex networks. Appl. Netw. Sci. **2.1**, 4 (2017)
3. Holland, P.W., Laskey, K.B., Leinhardt, S.: Stochastic block-models: first steps. Soc. Netw. **5**(2), 109–137 (1983)
4. Shao, J., Han, Z., Yang, Q.: Community Detection via Local Dynamic Interaction (2014). arXiv:1409.7978
5. Kernighan, B.W., Lin, S.: An efficient heuristic procedure for partitioning graphs. Bell Syst. Tech. J. **49**(2), 291–307 (1970)
6. Newman, M.E.J.: Spectral methods for community detection and graph partitioning. Phys. Rev. E **88**(4), 042822 (2013)
7. Newman, M.E.J.: Modularity and community structure in networks. In: Proceedings of the National Academy of Sciences, vol. 103.23, pp. 8577–8582 (2006)
8. Shao, J., et al.: Community detection based on distance dynamics. In: Proceedings of the 21th ACM SIGKDD International Conference on Knowledge Discovery and Data Mining. ACM (2015)
9. Meng, T., et al.: An improved community detection algorithm based on the distance dynamics. In: 2016 International Conference on Intelligent Networking and Collaborative Systems (INCoS). IEEE (2016)
10. Hutair, M.B., Aghbari, Z.A., Kamel, I.: Social community detection based on node distance and interest. In: Proceedings of the 3rd IEEE/ACM International Conference on Big Data Computing, Applications and Technologies. ACM (2016)
11. Yang, J., McAuley, J., Leskovec, J.: Community detection in networks with node attributes. In: 2013 IEEE 13th International Conference on Data Mining (ICDM). IEEE (2013)
12. Fengli, Z., et al.: Community detection based on links and node features in social net-works. In: International Conference on Multimedia Modeling. Springer, Cham (2015)
13. Palla, G., et al.: Uncovering the overlapping community structure of complex networks in nature and society. Nature **435.7043**, pp. 814–818 (2005)
14. Shen, H., et al.: Detect overlapping and hierarchical community structure in networks. Phys. A: Stat. Mech. Appl. **388.8**, 1706–1712 (2009)
15. Gregory, S.: Finding overlapping communities in networks by label propagation. New J. Phys. **12**(10), 103018 (2010)
16. Lancichinetti, A., Fortunato, S., Kertsz, J.: Detecting the overlapping and hierarchical community structure in complex networks. New J. Phys. **11**(3), 033015 (2009)
17. Wang, X., Liu, G., Li, J.: Overlapping Community Detection Based on Structural Centrality in Complex Networks. IEEE Access (2017)
18. Nicosia, V., et al.: Extending the definition of modularity to directed graphs with overlapping communities. J. Stat. Mech. Theory Exp. **2009.03**, P. 03024 (2009)

Summarizing Microblogs During Emergency Events: A Comparison of Extractive Summarization Algorithms

Soumi Dutta, Vibhash Chandra, Kanav Mehra, Sujata Ghatak, Asit Kumar Das and Saptarshi Ghosh

Abstract Microblogging sites, notably Twitter, have become important sources of real-time situational information during emergency events. Since hundreds to thousands of microblogs (tweets) are generally posted on Twitter during an emergency event, manually going through every tweet is not feasible. Hence, summarization of microblogs posted during emergency events has become an important problem in recent years. Several summarization algorithms have been proposed in the literature, both for general document summarization, as well as specifically for summarization of microblogs. However, to our knowledge, there has not been any systematic analysis on which algorithms are more suitable for summarization of microblogs posted during disasters. In this work, we evaluate and compare the performance of 8 extractive summarization algorithms in the application of summarizing microblogs posted during emergency events. Apart from comparing the performances of the algorithms, we also find significant differences among the summaries produced by different algorithms over the same input data.

Keywords Summarization · Twitter · Microblogs · Extractive · Comparison Evaluation · Rouge

---

S. Dutta (✉) · V. Chandra · K. Mehra · A. K. Das · S. Ghosh
Indian Institute of Engineering Science and Technology Shibpur, Shibpur, India
e-mail: soumi.it@gmail.com

V. Chandra
e-mail: vibhashchandra2010@gmail.com

K. Mehra
e-mail: kanav.mehra6@gmail.com

A. K. Das
e-mail: akdas@cs.becs.ac.in

S. Dutta · S. Ghatak
Institute of Engineering and Management, Kolkata 700091, India
e-mail: ghatak.sujata07@gmail.com

S. Ghosh
Indian Institute of Technology Kharagpur, Kharagpur 721302, India
e-mail: saptarshi.ghosh@gmail.com

A. Abraham et al. (eds.), *Emerging Technologies in Data Mining and Information Security*, Advances in Intelligent Systems and Computing 813,
https://doi.org/10.1007/978-981-13-1498-8_76

1 Introduction

Microblogging sites like Twitter have become extremely important sources of real-time information on ongoing events, including socio-political events, natural and man-made emergencies, and so on. Especially during emergency events, such as disasters, microblogging sites are very important sources of situational information [1]. During such events, microblogs are usually posted so rapidly, and in such large volumes, that it is not feasible for human users to go through all the posts. In such scenario, it is critical to summarize the microblogs (tweets) and present informative summaries to the people who are attempting to respond to the disaster.

Automatic document summarization is a well-established problem in Information Retrieval, and many algorithms have been proposed for the problem. The reader is referred to [2, 3] for surveys on summarization algorithms. Summarization methods are broadly of two types—abstractive and extractive. While extractive algorithms generate summaries by extracting certain portions of the input data (e.g. certain sentences that are deemed important), abstractive algorithms attempt to generate summaries by paraphrasing parts of the input data. Out of these, a majority of the algorithms proposed in literature are extractive in nature [3].

With the recent popularity of microblogs as a source of information, a number of summarization algorithms have been recently proposed specifically for microblogs as well (see Sect. 2 for details). The problem of microblog summarization is inherently a multi-document summarization problem. However, algorithms for single- document summarization are also applicable, by considering the input set of microblogs to make up a single document. Microblog summarization has some distinct challenges, primarily due to the small size of individual microblogs, and the noisy, informal nature of microblogs, which make it difficult to interpret the semantic similarity of microblogs.

Several summarization algorithms exist in the literature, both for general documents, as well as specifically for microblogs. However, to our knowledge, there has not been any systematic analysis on how effective these algorithms are in the application of summarizing microblogs posted during disaster events. In this work, we evaluate and compare *eight* off-the-shelf extractive summarization algorithms for the said application. We perform experiments over microblogs related to five recent disaster events. We observe that different off-the-shelf algorithms generate vastly different summaries from the same input set of microblogs, with very few tweets in common between the summaries generated by different algorithms. Additionally, we evaluate the performance of the different algorithms using the standard ROUGE measure. We observe that the LUHN [4] and COWTS [5] algorithms achieve relatively high ROUGE scores, as compared to the other algorithms considered here.

The rest of the paper is organized as follows. A short literature survey on summarization of microblogs is presented in Sect. 2. Section 3 describes the summarization algorithms chosen for the comparative analysis. The microblog datasets used for the comparison are described in Sect. 4, while Sect. 5 discusses the results of the comparison among the various algorithms. The paper is concluded in Sect. 6, which also states some potential future research directions.

2 Related Work

A large number of document summarization algorithms have been proposed in the literature. The reader can refer to [2, 3] for surveys on summarization algorithms. Since the present work is specifically on summarization of microblogs/tweets, we focus on summarization algorithms for microblogs in this section.

2.1 Summarization of Microblogs

Several algorithms for microblog summarization have been proposed in recent years [6–9]. For instance, Shou et al. [7] propose a system based on initially clustering similar tweets and then selecting few representative tweets from each cluster, finally ranking these according to importance via a graph-based approach (LexRank) [10]. Extracting bigrams from the tweets are considered as the graph-nodes, Olariu [6] proposed a graph-based abstractive summarization System. Some other authors have also proposed graph-based methods for summarization of tweets [11, 12].

Some other works have proposed methodologies to summarize microblogs posted during specific events, such as sports events [13–17]. Considering greedy summarization Osborne et al. [18] proposed a real event tracking system.

Along with general microblog summarization approaches, a few recent studies have also focused specifically on summarization of news articles and tweets posted during emergency events [5, 19, 20]. In particular, our prior work [5] proposed a classification–summarization technique to extract and summarize situational information from tweet streams. An approach based on Integer Linear Programming was applied to optimize for the presence of certain important terms (called "content words") in the summary.

2.2 Prior Works on Comparing Algorithms for Microblog Summarization

There have been two notable works on comparing different algorithms for microblog summarization [21, 22].

In a study, Mackie et al. [21] evaluated the performance comparison of 11 summarization approaches for 4 microblog datasets. The 11 summarization methods include a random baseline, temporal approaches (e.g. those that rank tweets by time), approaches based on term statistics such as *tf-idf*, and approaches based on term statistics, novelty, and cohesiveness [23].

Inouye et al. [22] compared algorithms for extractive summarization of Twitter posts. Two types of algorithms were considered which select tweets to produce summaries from a given set—(i) hybrid *tf-idf*-based algorithm, and (ii) a clustering-based algorithm. The performances of these algorithms are compared with manually produced summaries (gold standard) and summaries produced by eight different summarizers such as random, most recent, MEAD, TextRank, LexRank, cluster, Hybrid TF-IDF and SumBasic. The comparison showed that frequency- based summarizers (hybrid TF-IDF and SumBasic) achieve the best results in terms of ROUGE F-score.

The present work is different from these prior works as follows. First, unlike the other works, we focus on summarization of microblogs posted during disasters, which is a problem of practical interest. Second, neither of the prior works compare as many off-the-shelf summarization algorithms (eight) as we do in this work. Additionally, we also check the overlap between the summaries generated by various algorithms—which none of the prior works have done.

3 Summarization Algorithms

In this section, we outline the extractive summarization algorithms that we considered for comparison in the present work. Note that some of these algorithms were originally proposed for summarization of a single document, where the sentences of the given document are ranked according to some importance measure, and then few important sentences are selected for the summary. These algorithms can be easily applied to summarization of a set of microblogs, where each microblog is analogous to a sentence.

(1) ClusterRank: ClusterRank [24] is an unsupervised, graph-based approach which was originally designed for extractive summarization of meeting transcripts. ClusterRank algorithm is the extension of another algorithm named TextRank, which is also a graph-based method for extracting sentences from news articles. ClusterRank first segments the transcript into clusters which are represented as nodes in a graph. The similarity between all pairs of adjacent clusters is then measured, and the pair with the highest similarity is merged into a single cluster. Following this, a centroid-based approach is used to score each sentence within an important cluster. Relevancy of the sentences is also measured in addition to handling ill-formed sentences with high redundancy. Finally, the algorithm selects the highest scoring sentence and includes sentences in the summary until the length constraint is satisfied.
(2) COWTS: COWTS [5] is specifically designed for summarizing microblogs that are posted during disaster situations. The proposed approach is a classification–summarization framework for extracting meaningful situational information from microblog streams posted during disaster scenarios. Using low-level lexical and syntactic features, the classifier first distinguishes between situational and

non-situational information. Due to use of vocabulary-independent features, the classifier functions accurately in cross-domain scenarios. In COWTS, initially the classifier is trained over tweets posted during earlier disaster events and then deployed on tweets posted during a later disaster event. Then, the situational tweet stream is summarized by optimizing for the coverage of important *content words* (nouns, verbs, numerals) in the summary, using an Integer Linear Programming (ILP) framework.
(3) Frequency Summarizer: This is a simple summarization algorithm, which attempts to extract a subset of sentences which cover the main topics of a given document. The algorithm works on the simple idea that sentences which contain the most recurrent words in the text, are likely to cover most of the topics of the text.
(4) LexRank: LexRank [25] is stochastic graph-based method for computing relative importance of textual units in a document. In this method, a graph is generated which is composed of all sentences in the input corpus. Each sentence is represented as a node, and the edges denote similarity relationships between sentences in the corpus. An intra-sentence cosine similarity measure is used as edge weight in the graph representation of sentences by considering every sentence as bag-of-words model. A connectivity matrix or similarity matrix is generated using the similarity measure, which can be used as a similarity graph between sentences. A thresholding mechanism is applied (i.e. edges having weights below the threshold are removed) to extract the most important sentences from the resulting similarity matrix. A scheme based on Eigenvector centrality is also used to rank the sentences (nodes). The sentences are then included in the summary based on their importance values.
(5) LSA: LSA [26] is a generic extractive text summarization method to identify semantically important sentences for generating the summary. It is an unsupervised method of deriving vector space semantic representations from large documents, and does not need any training or external knowledge. Considering context of the input document, LSA extracts information such as which words are used together and which common words are seen in different sentences. High number of common words among sentences means that the sentences are more semantically related. LSA is based on mathematical technique which is named Singular Value Decomposition (SVD) [27] that is used to find out the interrelations between sentences and words. The input text document is first converted into a matrix, where each row represents a word and each column represents a sentence. Each cell value represents the importance of the word. SVD is then applied on this matrix to select the sentences to generate the summary.
(6) LUHN: Luhn's algorithm [4] works on the perception that some words in a document are descriptive of its content, and the sentences that express the most significant information in the document are the ones that contain many such descriptive words close to each other. The words that occur often in a document are likely to be associated with the main topic of the document. However, an exception to this observation is stopwords. Hence, Luhn proposed the idea of stopwords such as determiners, prepositions and pronouns, which do not have much value in informing about the topic of the document. So he suggested removing these words from consideration. Luhn identified descriptive words using empirically determined high- and low-frequency thresholds. The high-frequency thresholds filter out the words that occur very

frequently throughout the article. Similarly, the low-frequency thresholds filter out the words that occur too infrequently. The remaining words in the document are the descriptive words, which indicate that content which is important.

On a sentence level, a ‘significance factor’ is computed for each sentence, which could be calculated for a sentence by bracketing the significant words in the sentence, squaring the number of significant words and then dividing by the total number of words. Sentences are identified as important and included in the summary based on the significance factor values.

(7) Mead: Mead [28, 29] is a centroid-based multi-document summarizer. First, topics are detected by agglomerative clustering that operates over the tf-idf vector representations of the documents. Second, a centroid-based methodology is used to identify sentences in each cluster that are central to the topic of the entire cluster. For each sentence, three different features are computed, which are its centroid value, positional value and first-sentence overlap. A composite score of each sentence is generated as a combination of the three scores. The score is further refined after considering possible cross-sentence dependencies, e.g. repeated sentences, chronological ordering, source preferences.) Sentences are finally selected based on this score.

(8) SumBasic: SumBasic [30] is a frequency-based multi-document summarizer. SumBasic uses a multinomial distribution function to compute the probability distribution over the words in a sentence. Based on average probability of occurrence of the words in the sentence, scores are assigned to each sentence. Then the sentences with the best scores are selected. Successively, the word probabilities and sentence scores are updated until the desired summary length is reached. The updation of word probabilities gives a natural way to deal with the redundancy in the multi-document input.

It can be noted that we selected the algorithms described above because either their implementations are readily available off the shelf, or they are easy to implement.[1]

4 Dataset for Comparison of Summarization Algorithms

This section describes the dataset we use to compare the various summarization algorithms.

[1] Availability of implementations: Frequency Summarizer (http://glowingpython.blogspot.in/2014/09/text-summarization-with-nltk.html), Mead (http://www.summarization.com/mead/), SumBasic (https://github.com/EthanMacdonald/SumBasic), LexRank, LSA and LUHN are available as part of the Python Sumy package (https://pypi.python.org/pypi/sumy). COWTS (proposed in our prior work [5]) and ClusterRank were implemented by us.

4.1 Emergency Events Considered

We considered tweets posted during the following emergency events.

1. **HDBlast**—two bomb blasts in the city of Hyderabad, India [31],
2. **SHShoot**—an assailant killed 20 children and 6 adults at the Sandy Hook elementary school in Connecticut, USA [32],
3. **UFlood**—devastating floods and landslides in the Uttaranchal state of India [33],
4. **THagupit**—a strong cyclone code named Typhoon Hagupit hit Philippines [34],
5. **NEquake**—a devastating earthquake in Nepal [35].

The dataset used for experimental purpose, are the selected events occurred during natural and man-made disasters in various regions of the world. Hence, the vocabulary/linguistic style in the tweets can be predictable to be dissimilar as well.

4.2 Developing the Dataset

For experiment, we have collected relevant tweets posted during each event through the Twitter API [36] using keyword-based matching. For instance, to identify tweets related to the HDBlast event the keywords such as 'Hyderabad', 'bomb' and 'blast' are used and to collect tweets related to the SHShoot event the keywords 'Sandyhook' and 'shooting' are considered.

We initially considered the chronologically earliest 1,000 tweets for each event. Due to frequent retweeted/re-posted by multiple users [37] Twitter often contains duplicates and near-duplicates tweets. Sincc such near-duplicates are not useful for the purpose of summarization, we removed them using a simplified version of the techniques suggested in [37], as follows.

Each tweet was considered as a bag of words (excluding standard English stopwords and URLs), and the similarity between two tweets was measured as the Jaccard similarity between the two corresponding bags (sets) of words. The two tweets were considered near-duplicates if the Jaccard similarity between two tweets was found

Table 1 Datasets used for the experiments. 1,000 chronologically earliest tweets were initially considered for each event, and near-duplicates were removed using methods in [37]. The last column shows the number of distinct tweets, after removing near-duplicates

Bomb blasts in the city of Hyderabad, India	95
Earthquake in Nepal in April 2015	146
Floods and landslides in the Uttaranchal state of India	173
Sandy Hook elementary school shooting in USA	252
Typhoon Hagupit in Philippines	484

to be higher than a threshold value (0.7) and only the longer tweet (potentially more informative) was retained. Table 1 shows the number of distinct tweets in each dataset after removal of duplicates and near-duplicates.

5 Experimental Results

We describe our experimental results in this section. Apart from comparing the performances of different algorithms, we also check whether different summarization algorithms produce very different summaries from the same input data.

5.1 *Do Different Algorithms Produce Very Different Summaries?*

Extractive summarisation algorithms for microblogs will select a subset of the tweets for inclusion in the summary. We first investigate whether different algorithms select a common set of tweets for the summaries, or whether the sets of tweets selected by different algorithms (for inclusion in summary) vary significantly.

Interestingly, we observed that *different summarization algorithms usually select very different sets of tweets in the summaries*. To demonstrate this observation, Table 2 shows the overlap between the summaries generated by the different algorithms, for the Nepal earthquake dataset. The entry (i, j), $1 \leq i, j \leq 8$ in Table 2 shows the number of common tweets included in the summaries generated by the two algorithms A_i and A_j. Similarly, Table 3 shows the overlaps for the Sandy Hook dataset.

Table 2 Overlap of tweets in the summaries (of length 25 tweets each) generated by different base summarization algorithms, for the Nepal earthquake dataset. Other datasets also show very low overlap

Algorithm	CR	CW	FS	LR	LS	LH	MD	SB
ClusterRank (CR)	–	7	4	0	3	2	4	4
COWTS (CW)	7	–	4	0	2	2	1	5
FreqSum (FS)	4	4	–	3	2	4	2	5
LexRank (LR)	0	0	3	–	1	2	0	2
LSA (LS)	3	2	2	1	–	7	3	4
LUHN (LH)	2	2	4	2	7	–	1	1
MEAD (MD)	4	1	2	0	3	1	–	0
Sumbasic (SB)	4	5	5	2	4	1	0	–

Table 3 Overlap of tweets in the summary (of length 25 tweets) generated by diffrent base summarization algorithms, for the Sandy Hook school shooting dataset

Algorithm	CR	CW	FS	LR	LS	LH	MD	SB
ClusterRank (CR)	–	1	0	0	2	1	2	4
COWTS (CW)	1	–	2	0	3	0	1	7
FreqSum (FS)	0	2	–	1	5	7	0	2
LexRank (LR)	0	0	1	–	1	2	0	0
LSA (LS)	2	3	5	1	–	13	3	3
LUHN (LH)	1	0	7	2	13	–	4	1
MEAD (MD)	2	1	0	0	3	4	–	1
Sumbasic (SB)	4	7	2	0	3	1	1	–

Table 4 Examples of tweets that were selected by at least four algorithms (out of eight) for inclusion in the summaries

Event	Tweet text
Hagupit	@EarthUncutTV: Latest 06z/2pm Philippines time JMA forecast track for #typhoon #Hagupit #RubyPH [url]
Hdblast	FLASH: 9 killed, 32 injured in serial powerful #blast in Dilshuknagar area in #Hyderabad: Police
SHshoot	Powerful picture RT @HeidiVoight Kids crying, evacuating Sandy Hook Elementary in NEWTOWN [url] via @BKnox88 via NewtownBee
UFlood	Really sad to hear news Uttarakhand floods my prayers with you Lord Shiva plz help them & plz take care n come back home Mumbai
UFlood	INDIAN ARMY IN FLOOD RELIEF OPERATIONS Uttarakhand Flood Helpline numbers 0135-2710334, 0135-2710335. [url]

It is evident from these tables that there is very low overlap between summaries generated by various base algorithms (similar trends are observed for all the datasets). The overlap is slightly higher in few specific cases, e.g. for the LUHN and LSA algorithms, possibly because these algorithms work on similar principles. However, most algorithms produce summaries that have very few tweets in common with summaries produced by other algorithms.

In spite of the low overlap between summaries produced by different algorithms, we find a few specific tweets which are selected by multiple algorithms for inclusion in the summaries. Table 4 shows examples of tweets that were selected at least four algorithms (out of eight) for inclusion in the summary. We observed that these tweets contain several terms which have high document frequency, i.e. terms which occur in many of the tweets in the particular dataset. As a result, these tweets are deemed (by multiple algorithms) to be good representatives of the whole dataset.

5.2 Evaluation of Summarization Algorithms

Next we focus on the evaluation of the performance of the different algorithms. To evaluate the quality of a summary (produced by an algorithm), we follow the standard procedure of generating gold standard summaries by human annotators, and then comparing the algorithm-generated summary with the gold standard ones. We employed three human annotators, each of whom is proficient in English and is a habitual user of Twitter, and has prior knowledge of working with social media content posted during disasters. Each annotator was asked to independently summarize each of the five datasets, and prepare extractive summaries of length 25 tweets each.

We executed all the summarization algorithms (described in Sect. 3) on each dataset, and obtain summaries of length 25 tweets each. We used the standard ROUGE measure [38] for evaluating the quality of the summaries generated by different algorithms, based upon their match with the gold standard summaries. Due to the informal nature of tweets, we considered the Recall and F-score of the ROUGE-1, ROUGE-2, and ROUGE-L variants.

Table 5 reports the performance of the different summarization algorithms, averaged over all the five datasets. We find that the LUHN algorithm performs the best for all the measures, followed closely by the COWTS algorithm.

To qualitatively demonstrate the differences between the summaries which obtain high ROUGE scores and those that obtain low ROUGE scores, Table 6 and Table 7, respectively, show the summaries generated by the LUHN algorithm (which obtained highest ROUGE score) and the LexRank algorithm (which obtained the lowest ROUGE score) for the same dataset—the Hyderabad blast dataset. It is evident that

Table 5 Performance of the summarization algorithms, averaged over all five datasets. The best performance is by the LUHN algorithm (highlighted in boldface) followed by the COWTS algorithm

Algorithm	Rouge-1		Rouge-2		Rouge-L	
	Recall	F-score	Recall	F score	Recall	F-score
ClusterRank	0.459	0.467	0.230	0.233	0.448	0.456
COWTS	0.546	0.518	0.326	0.308	0.533	0.506
FreqSum	0.405	0.411	0.191	0.192	0.393	0.398
LexRank	0.278	0.371	0.124	0.164	0.273	0.365
LSA	0.515	0.486	0.284	0.267	0.503	0.475
LUHN	**0.563**	**0.531**	**0.331**	**0.313**	**0.549**	**0.518**
Mead	0.489	0.499	0.270	0.276	0.477	0.488
SumBasic	0.423	0.453	0.207	0.219	0.408	0.437

Table 6 Summary generated by the LUHN algorithm (having highest ROUGE score), for the Hyderabad blast dataset

Summary
RT @krajesh4u: Reports of explosion from busy commercial area in Hyderabad [url]
RT @SRIRAMChannel: Bomb blast in dilsukhnagar (hyderabad) near venkatadri theatre.many feared dead
RT @abpnewstv: BREAKING: 7 feared dead in Hyderabad blast - Reports
RT @ndtv: Bomb blast in Hyderabad: 50 injured, say officials
RT @abpnewstv: BREAKING: Twin blast in Hyderabad's Dilsukh Nagar suburb - reports of 15 deaths, over 50 injured
RT @BreakingNews: 2 blasts reported near bus stand in southern Indian city of Hyderabad; 10 people feared dead, at least 40 others injur
RT @abpnewstv: BREAKING: 9 killed, 32 injured in serial blasts in Hyderabad: PTI quoting official sources
RT @ibnlive: #Hyderabadblasts: High alert declared across Andhra Pradesh #IBN-news
RT @IndianExpress: FLASH: 9 killed, 32 injured in serial powerful #blast in Dil-shuknagar area in #Hyderabad: Police
[url] wrote: Hyderabad blast: High alert declared across Andhra Pradesh
RT @SkyNewsBreak: UPDATE: AFP - police say seven people have died and 47 people hurt in bomb blasts in Indian city of #Hyderabad
'9 killed in Hyderabad blast; 5 in police firing' [url] #BengalRiots #HyderabadBlast #HinduGenocide
RT @ndtv: Hyderabad serial blasts: Mumbai, Karnataka put on high alert
RT @IndiaToday: 7 feared dead, 20 others injured in 5 blasts in Hyderabad: Report: The blasts took place in a cro
RT @ndtv: Hyderabad serial blasts: at least 15 dead, 50 injured [url]
RT @SkyNews: Hyderabad Blast: 'Multiple Deaths' [url]
#india #business : Seven killed in Hyderabad blast, several injured: Times Now: Seven killed in Hyderabad blast
"@SkyNews: #Hyderabad Blast: 7 Feared Dead [url]" what's happening now
RT @bijugovind: Screen map of #Hyderabad blast are" [url]
Seven killed in Hyderabad blast, 7 feared dead: Times Now: Seven killed in Hyder-abad blast, 7 feared dead: Tim
Blasts rocked Hyderabad city many Killed & Injured [url] #Hyderabad #Blast #Dilsukh Nagar #Police. RT @SaradhiTweets: list of hospitals in hyderabad [url] #hyderabadblasts
Bomb blast in Hyderabad in busy commercial area [url]
RT @dunyanetwork: (Breaking News) Twin blasts in #Hyderabad, #India 7 people killed, 20 injured Casualties expected to rise
Explosions Rock Hyderabad; At Least 20 Killed #Hyderabadblast #hyderabad blast [url]

the summary generated by LexRank is dominated by only one type of information—regarding casualties—in which the summary generated by LUHN has much more diverse information.

It should also be noted that the best performing algorithm achieves a ROUGE-1 Recall score of 0.563, and ROUGE-2 Recall score of 0.331, which roughly implies

Table 7 Summary generated by the LexRank algorithm (having lowest ROUGE score), for the Hyderabad blast dataset

RT @krajesh4u: Reports of explosion from busy commercial area in Hyderabad [url]
RT @SRIRAMChannel: Bomb blast in dilsukhnagar (hyderabad) near venkatadri theatre.many feared dead.
RT @Iamtssudhir: Explosion took place near venkatdri theatre in dilsukhnagar
RT @abpnewstv: BREAKING: 7 feared dead in Hyderabad blast - Reports
RT @ndtv: Bomb blast in Hyderabad: 50 injured, say officials.
#Hyderabad blast took place around 7 p.m. local time; not believed to be gas cylinder explosion, @timesnow reporting
RT @khaleejtimes: Breaking News: Seven killed in Hyderabad blast [url]
Hyderabad Blast.
RT @EconomicTimes: #Hyderabad blast: Seven killed, several injured [url]
15 killed 50 injured in Hyderabad blast More Photos: [url] [url]
'9 killed in Hyderabad blast; 5 in police firing' [url] #BengalRiots #HyderabadBlast #HinduGenocide
RT @IndiaToday: 7 feared dead, 20 others injured in 5 blasts in Hyderabad: Report: The blasts took place in a cro...
RT @ndtv: Hyderabad serial blasts: at least 15 dead, 50 injured [url]
Blasts at Hyderabad [url] #News
Screen map of Dilsukhnagar #Hyderabad blast [url]
RT @ndtv: Alert in all major cities across India after serial blasts in Hyderabad.
Three blasts in Hyderabad.Fuck.
Reports of Blasts from Hyderabad @ndtv
Blasts rocked Hyderabad city many Killed & Injured [url] #Hyderabad #Blast #Dilsukh Nagar #Police
RT @timesofindia: Hyderabad Police: Two bomb blasts
11 people were killed and 50
RT @anupamthapa: Seven feared killed, 20 injured in Hyderabad blast
TimesNow : 7 feared Killed 20 injured #Hyderabadblasts thats very horryfying news
Bomb blasts near Dilsukhnagar bus stand in Hyderabad; at least 7 people injured
Explosions Rock Hyderabad; At Least 20 Killed #Hyderabadblast #hyderabad blast [url]

that the algorithmic summaries can capture only about half of the unigrams and 33% of the bigrams in the gold standard summaries. These moderate ROUGE values reiterate that summarization of microblogs posted during emergency events is a challenging problem, for which improved algorithms need to be developed in future.

6 Conclusion

Summarization of microblogs posted during emergency situations is an important and practical problem. While a large number of summarization algorithms have been proposed in the literature, to our knowledge, there has not been any systematic comparison of how effective different algorithms are in summarizing microblogs related to disaster events. In this work, we perform such a comparison of eight extractive summarization algorithms, over microblogs posted during five disaster events. We find that different algorithms generate vastly different summaries, and while some algorithms (e.g. LUHN, COWTS) achieve relatively high ROUGE scores, some other algorithms such as LexRank do not appear so effective.

We believe that the present work indicates several research directions for the future. First, given that even the best performing methods achieve ROUGE recall scores of less than 0.6, it is evident that better algorithms are needed for effectively summarizing microblogs during disaster events. Second, since different summarization algorithms produce very different summaries from the same input data, a promising direction can be to investigate whether outputs from multiple summarization algorithms can be combined to produce summaries that are better than those produced by the individual algorithms. We plan to pursue these directions in future.

References

1. Imran, M., Castillo, C., Diaz, F., Vieweg, S.: Processing social media messages in mass emergency: a survey. ACM Comput. Surv. **47**(4), 67:1–67:38 (2015)
2. Das, D., Martins, A.F.: A survey on automatic text summarization. Lit. Surv. Lang. Stat. II Course CMU **4**, 192–195 (2007)
3. Gupta, V., Lehal, G.S.: A survey of text summarization extractive techniques. IEEE J. Emerg. Technol. Web Intell. **2**(3), 258–268 (2010)
4. Luhn, H.P.: The automatic creation of literature abstracts. IBM J. Res. Dev. **2**(2), 159–165 (1958)
5. Rudra, K., Ghosh, S., Goyal, P., Ganguly, N., Ghosh, S.: Extracting situational information from microblogs during disaster events: a classification-summarization approach. In: Proceedings of ACM CIKM (2015)
6. Olariu, A.: Efficient online summarization of microblogging streams. In: Proceedings of EACL(short paper), pp. 236–240 (2014)
7. Shou, L., Wang, Z., Chen, K., Chen, G.: Sumblr: continuous summarization of evolving tweet streams. In: Proceedings of ACM SIGIR (2013)
8. Wang, Z., Shou, L., Chen, K., Chen, G., Mehrotra, S.: On summarization and timeline generation for evolutionary tweet streams. IEEE Trans. Knowl. Data Eng. **27**, 1301–1314 (2015)
9. Zubiaga, A., Spina, D., Amigo, E., Gonzalo, J.: Towards real-time summarization of scheduled events from twitter streams. In: Hypertext(Poster) (2012)
10. Erkan, G., Radev, D.R.: LexRank: Graph-Based Lexical Centrality as Salience in Text Summarization, pp. 457–479 (2004)
11. Dutta, S., Ghatak, S., Roy, M., Ghosh, S., Das, A.K.: A graph based clustering technique for tweet summarization. In: 2015 4th International Conference on Reliability, Infocom Technologies and Optimization (ICRITO)(Trends and Future Directions), pp. 1–6. IEEE (2015)
12. Xu, W., Grishman, R., Meyers, A., Ritter, A.: A preliminary study of tweet summarization using information extraction. In: Proceedings of NAACL **2013**, 20 (2013)
13. Chakrabarti, D., Punera, K.: Event summarization using tweets. In: Proceedings of AAAI ICWSM, pp. 340–348 (2011)
14. Gillani, M., Ilyas, M.U., Saleh, S., Alowibdi, J.S., Aljohani, N., Alotaibi, F.S.: Post summarization of microblogs of sporting events. In: Proceedings of International Conference on World Wide Web (WWW) Companion, pp. 59–68 (2017)
15. Khan, M.A.H., Bollegala, D., Liu, G., Sezaki, K.: Multi-tweet summarization of real-time events. In: Socialcom (2013)

16. Nichols, J., Mahmud, J., Drews, C.: Summarizing sporting events using twitter. In: Proceedings of ACM International Conference on Intelligent User Interfaces (IUI), pp. 189–198 (2012)
17. Takamura, H., Yokono, H., Okumura, M.: Summarizing a document stream. In: Proceedings of ECIR (2011)
18. Osborne, M., Moran, S., McCreadie, R., Lunen, A.V., Sykora, M., Cano, E., Ireson, N., Macdonald, C., Ounis, I., He, Y., Jackson, T., Ciravegna, F., OBrien, A.: Real-time detection, tracking, and monitoring of automatically discovered events in social media. In: Proceedings of ACL (2014)
19. Kedzie, C., McKeown, K., Diaz, F.: Predicting salient updates for disaster summarization. In: Proceedings of ACL (2015)
20. Nguyen, M.T., Kitamoto, A., Nguyen, T.T.: Tsum4act: a framework for retrieving and summarizing actionable tweets during a disaster for reaction. In: Proceedings of PAKDD (2015)
21. Inouye, D.I., Kalita, J.K.: Comparing twitter summarization algorithms for multiple post summaries. In: Proceedings of IEEE SocialCom/PASSAT, pp. 298–306 (2011)
22. Mackie, S., McCreadie, R., Macdonald, C., Ounis, I.: Comparing algorithms for microblog summarisation. In: Proceedings of CLEF (2014)
23. Rosa, K.D., Shah, R., Lin, B., Gershman, A., Frederking, R.: Topical Clustering of Tweets
24. Garg, N., Favre, B., Riedhammer, K., Hakkani-Tr, D.: Clusterrank: a graph based method for meeting summarization. In: INTERSPEECH, pp. 1499–1502. ISCA (2009)
25. Erkan, G., Radev, D.R.: Lexrank: Graph-based lexical centrality as salience in text summarization. J. Artif. Int. Res. **22**(1), 457–479 (2004)
26. Gong, Y., Liu, X.: Generic text summarization using relevance measure and latent semantic analysis. In: SIGIR, pp. 19–25 (2001)
27. Ozsoy, M.G., Alpaslan, F.N., Cicekli, I.: Text summarization using latent semantic analysis. J. Inf. Sci. **37**(4), 405–417 (2011). http://dx.doi.org/10.1177/0165551511408848
28. Radev, D.R., Allison, T., Blair-Goldensohn, S., Blitzer, J., elebi, A., Dimitrov, S., Drbek, E., Hakim, A., Lam, W., Liu, D., Otterbacher, J., Qi, H., Saggion, H., Teufel, S., Topper, M., Winkel, A., Zhang, Z.: MEAD—a platform for multidocument multilingual text summarization. In: LREC. European Language Resources Association (2004)
29. Radev, D.R., Hovy, E., McKeown, K.: Introduction to the special issue on summarization. Comput. Linguist. **28**(4), 399–408 (2002)
30. Nenkova, A., Vanderwende, L.: The impact of frequency on summarization. Technical report, Microsoft Research (2005)
31. Hyderabad blasts—Wikipedia (2013). http://en.wikipedia.org/wiki/2013_Hyderabad_blasts
32. Sandy Hook Elementary School shooting–Wikipedia (2012). http://en.wikipedia.org/wiki/Sandy_Hook_Elementary_School_shooting
33. North India floods—Wikipedia (2013). http://en.wikipedia.org/wiki/2013_North_India_floods
34. Typhoon Hagupit—Wikipedia (2014). http://en.wikipedia.org/wiki/Typhoon_Hagupit
35. 2015 Nepal earthquake—Wikipedia (2015). http://en.wikipedia.org/wiki/2015_Nepal_earthquake
36. REST API Resources, Twitter Developers. https://dev.twitter.com/docs/api
37. Tao, K., Abel, F., Hauff, C., Houben, G.J., Gadiraju, U.: Groundhog day: near-duplicate detection on twitter. In: Proceedings of Conference on World Wide Web (WWW) (2013)
38. Lin, C.Y.: ROUGE: A package for automatic evaluation of summaries. In: Proceedings of Workshop on Text Summarization Branches Out, ACL, pp. 74–81 (2004)

A Comparative Study on Cluster Analysis of Microblogging Data

Soumi Dutta, Asit Kumar Das, Gourav Dutta and Mannan Gupta

Abstract In today's world, Online Social Media is one of the popular platforms for information exchange. During any important occurrence in society such as World Cup, election, budget, natural calamity microblogs take a significant and flexible platform for public as communication media. Millions of messages are exchanged every day by a large set of users in several microblogs, which cause information overload. These phenomena open up numerous challenges for the researchers. Due to the noisy and precise nature of messages, it is a challenging task to cluster data and mine meaningful information to summarize any trending or non-trending topic. In association with this, growing of data in huge volume is another challenge of clustering. Several researchers proposed different methods for clustering. This work focuses on a comparative study of the different proposed clustering approaches using community detection and genetic algorithm-based techniques on microblogging data. The analysis also shows the comparative performance study of different clustering methods for the similar dataset.

Keywords Clustering · Feature selection · Microblogs · Genetic algorithm
Community detection

S. Dutta (✉) · A. K. Das
Indian Institute of Engineering Science and Technology Shibpur, Shibpur, India
e-mail: soumi.it@gmail.com

A. K. Das
e-mail: akdas@cs.becs.ac.in

G. Dutta
Cognizant Technology Solutions, Kolkata, India
e-mail: cse.gourav@gmail.com

M. Gupta · S. Dutta
Institute of Engineering and Management Kolkata, Kolkata 700091, India
e-mail: mownon89@gmail.com

A. Abraham et al. (eds.), *Emerging Technologies in Data Mining and Information Security*, Advances in Intelligent Systems and Computing 813,
https://doi.org/10.1007/978-981-13-1498-8_77

1 Introduction

Nowadays, Twitter (https://twitter.com/) is one of the real time, massive popular microblogging phenomena. Due to microblogging nature, Twitter post contains at most 140 characters which are known as "Tweets". Every day more than 360 million active Twitter users post 500 million tweets on trending or non-trending topics. This huge data consist of many noisy, redundant and irrelevant information, which challenges the researchers to find significant facts from the data.

Based on the real-time popularity of discussion on topics, Twitter declares a list of trending topics, which are having a huge rate of information exchange. It is difficult for any person to go through all the posts/tweets to understand any real-time trending topic. So, to handle information overload, one of the effective solutions is to group similar tweets into sub-topics. If any person will go through few tweets from each sub-topic or group, the purpose will be fulfilled. Also, clustering is the efficient approach to address the summarization problem. This clustering approach can be static and dynamic because of the incremental nature of the dataset. To focus the issue of large volume of dataset, few researchers also combine the feature reduction problem with clustering to boost the clustering performance and computational time effectively.

In this work, several clustering approaches [13] are discussed and performances are compared considering similar dataset. The performance of these approaches is also compared with existing classical clustering classical methods such as K-means and hierarchical clustering. Using the standard clustering index measure [16], the performances of the clustering techniques are compared.

The rest of the paper is organized as follows. A short literature survey on clustering approaches of microblogging dataset is presented in Sect. 2. Section 3 describes few clustering approaches. The microblog datasets used for the algorithm are described in Sect. 4, while Sect. 5 discusses the comparative analysis of the results of the various clustering algorithm. Section 6 concludes the paper with some potential future research directions.

2 Related Work

In recent times, many research [14] works are currently going on on microblogging data clustering. Using bursty keywords and their co-occurrences, Mathioudakis et al. [17] proposed an algorithm to identify the trending topics. Michelson et al. [18] proposed a novel approach to identify user's topical interest in leveraging the Wikipedia as a knowledge base. To identify the real-world incident on Twitter, Hila [1] proposed an algorithm based on clustering concept. To categorize emerging topics in the Twitter, Cataldi et al. [4] used topic-graph based on term frequency and users authority. Using a machine learning-based clustering tool, Cheong [5] proposed an unsupervised algorithm called self-organizing feature map (SOM) to perceive the hidden pattern in Twitter intra-topic user and message clustering. In another approach to categorize the tweets [19], Ramage et al. used labeled latent Dirichlet allocation (LDA) methods. Apart from clustering, Genc et al. [9] used Wikipedia-based classification technique to categorize tweets considering semantic distance as a classification metric to evaluate the distance between their closest Wikipedia pages. Using optimization technique such as genetic algorithm, Soumi et al. [7] proposed a method to categorize microblogging data.

In another work, Soumi et al. [6] proposed a feature selection- based clustering algorithm, which can increase the effectiveness of clustering. Using dynamic modeling, a hierarchical clustering algorithm has been proposed by Karypis et al. [15] to showcase dynamic modes of clusters and the adaptive merging decision. This work can discover natural clusters of various shapes and sizes. Using temporal patterns of propagation Yang and Leskovec [26] have proposed a method for clustering.

Several researchers [27] used graph-based approaches for clustering of microblogging data. Based on graph-based community detection approach, Soumi et al. [8] proposed a clustering algorithm for Twitter data which performs summarization as well. To identify the most highly ranked sentences based on keywords in a document, a graph-based approach, known as TextRank algorithm [3] is used where ranks are determined by PageRank Algorithm. To generate conceptual summarization method, in SCISOR system [20], an idea of concept graph has been proposed for the dataset—Down Jones Newswire stories.

3 Microblog Clustering Algorithms

Several researches are currently going on currently on microblogging data. This work is a review of the literature which identified three studies that implemented different clustering methods such as community detection-based graph clustering using Infomap, optimization technique-based clustering using genetic algorithm, and feature selection-based microblog data clustering. The comparison includes detailed methodology of the clustering methods, experimental dataset, and performance analysis of those methods.

3.1 Graph-Based Clustering Algorithm

This methodology [8] uses graph-based Infomap [12] clustering algorithm to identify communities or cluster from microblogging data. In this approach, a weighted similarity graph has been constructed considering all the distinct tweets as nodes/vertices and weights between the edges are the similarity between different pairs of tweets. To evaluated the similarity between tweets, term-level similarity, and semantic similarity are considered.The term-level similarity includes Levenshtein distance, cosine similarity, frequency of common hashtags, username, and URL, whereas semantic similarity includes WordNet [25] Synset Similarity and WordNet Path Similarity. The community detection technique, Infomap is applied on the tweet similarity graph to categorize similar tweets. In this proposed approach, hidden patterns and similar groups of information in tweets are identified using models of unsupervised learning. Apart from clustering, the methodology also generates a summary of the tweets, where a representative tweet is selected from each community, which is identified by Infomap to be included in the summary. From the tweet-similarity graph, three different types of summaries are generated such as—Total Summary, Total Degree Summary, Total Length Summary.

Other three types of summaries are proposed in this method, where threshold mechanism has been applied to the tweet similarity graph. From the threshold tweet-similarity graph, another three different types of summaries are generated such as—Thresholded Summary, Thresholded Degree Summary, Thresholded Length Summary.

3.2 Genetic Algorithm-Based Tweet Clustering

In the proposed methodology [7], a novel clustering algorithm has been approached for microblogging data such as tweets. This approach combines a traditional clustering algorithm (K-means) with Genetic Algorithms (GA). After dataset preprocessing, a set of distinct terms are identified, and a document-term matrix has been generated where the rows represent individual tweets, and the columns represent distinct terms. The entries in the matrix is represented by 1 and 0, where 1 signifies the presence of a term in a tweet and 0 signifies the absence. To apply Genetic Algorithm, initially, a population has been formed consisting of 100 chromosomes with randomly selected binary strings. To evaluate fitness values of each chromosome in the population, individual Projection-Matrix designed obtaining projection from the document-term matrix then K-means clustering algorithm is applied on the projected dataset, using then the fitness value is estimated for each chromosome. Then, traditional GA steps are followed such as selection, crossover, mutation. The GA is iterated until it converges (new better fitness- valued chromosome is generated) and best-fit solutions are combined to select final chromosomes for the final population. Output of the clustering corresponding to the fittest chromosome of the final population is choosen as the final clustering of the tweets.

3.3 Feature Selection-Based Clustering Algorithm

This is an efficient approach [6] to maximize the clustering efficiency due to inclusion of the feature selection mechanism. Here, latent Dirichlet allocation (LDA) [2], a topic modeler is used to identify intra-document topic on the experimental dataset. Initially, dataset is preprocessed and then the tweets are tokenized to identify possible distinct terms/words. From the preprocessed dataset, a document-term matrix is generated where the rows represent individual tweets, and the columns represent

Table 1 Dataset description

Algorithm	Data source
Graph-based clustering algorithm [8]	The floods and landslides in the Uttaranchal region of India, in 2013
Genetic algorithm (GA)-based tweet clustering technique [7]	The devastating landslides and floods in the Uttarakhand state of, India in June 2013
Feature selection-based clustering algorithm [6]	1. Bomb blasts in the city of Hyderabad [11]
	2. Shoot out at the sandy hook elementary school in connecticut [21]
	3. Devastating floods and landslides in the Uttaranchal state of India [10]
	4. Typhoon Hagupit at Philippines [23]

distinct terms. In the next step, for each individual term, the p-values are computed based on Bayes theorem. Then, applying Shannon's theorem, terms that yield less information, then average are discarded and the document-term matrix (DM) are reformed based on the new reduced corpus.

Next, a set of important features or terms are identified from intra-document topic wise using the topic wise, LDA which can represent each topic and the topic-feature matrix is generated. Then, a topic-feature matrix is generated evaluating the average Hamming distance for each topic considering the mean distance between all tweet vectors in document-feature matrix. Using greedy-based approach, each tweet is assigned to a particular topic or category. Each topic is assigned to the corresponding category for which the tweet-topic Hamming distance is less than or equal to the average distance.

4 Dataset for Clustering Algorithms

All the algorithms described in Sect. 3, experimented with microblogging dataset which are collected from Twitter using API [22] with matching keywords considering the 1% random sample of extracted tweets. Most of the data are collected during the emergency events, natural disaster such as Uttarakhand Flood in 2013, the HDBlast event, bomb blasts in the city of Hyderabad, etc. The dataset descriptions are given as follows (shown in Table 1).

5 Results and Discusion

All the above-mentioned approaches are compared here with several classical clustering methods [14] such as K-means, density based, and hierarchical clustering algorithms. Using a set of standard metrics, the efficiency of these methods are

Table 2 Comparative results for "Uttarakhand flood" dataset

Methods	Metrics for evaluating clustering						No. of cluster
	S-Index	CH	DB	I-Index	XB	D	
Infomap	0.515	0.847	0.912	0.968	101.8	0.498	16
Graph-based clustering	0.764	4.247	0.32	3.71	9.654	10.102	6
GA-based clustering	0.989	35.692	**0.05**	3.245	**0.045**	10.102	6
Feature-based clustering	**0.9918**	**37.711**	0.459	4.018	0.0994	**12.172**	5
K-means (#Cluster = 5)	0.498	3.988	0.688	3.675	52.415	0.981	5
K-means (#Cluster = 6)	0.512	3.81	0.788	3.403	5.36	1.102	6
Density based (#Cluster = 5)	0.799	5.875	0.787	3.989	1.212	1.933	5
Density based (#Cluster = 6)	0.401	3.405	0.89	3.271	9.136	0.998	6
Hierarchical (#Cluster = 5)	0.812	5.017	0.517	**4.59**	0.345	1.312	5
Hierarchical (#Cluster = 6)	0.874	4.998	0.478	3.789	0.305	1.301	6

demonstrated. The used set of standard metrics [16] to evaluate the superiority of clustering are Calinski–Harabasz index (CH), Davies–Bouldin index (DB), Silhouette index (S), Dunn index (D), I-Index (I) and Xie–Beni index (XB). Higher values for Silhouette index (S), Calinski–Harabasz index (CH), Dunn Index (D), and I-index (I) indicate better clustering performance, where as Smaller values for Xie–Beni index (XB) and Davies–Bouldin Index (DB) indicate better clustering performance.

The graph-based clustering algorithm considers Infomap Community detection approaches, which do not require any prior information of cluster number. In this approach also, the summarization method is compared with existing summarization method, Sumbasic [24]. The result shows that Total Length Summary performs better than all other approaches.

The Genetic Algorithm (GA) based tweet clustering technique uses K-means algorithm, so to fix the cluster number K, experts opinion are considered. The experimental result shows that considering 5 or 6 clusters, the proposed approach can perform the best clustering if it runs till 100 iterations of the genetic algorithm. For the same experimental dataset K-means, density based and hierarchical clustering algorithms are evaluated with 5 or 6 clusters. The standard metrics outcome shows that except I-index, the Genetic Algorithm (GA)-based tweet clustering technique performs better than other methods.

Feature Selection-based Clustering Algorithm has been experimented on four different datasets, and also evaluated the clustering performance for K-means and hierarchical clustering algorithms. In this work, the experimental result shows that the hierarchical clustering achieves the better performance than Feature Selection-based Clustering Algorithm, whereas rest all the metrics performs better than all the base-line approaches.

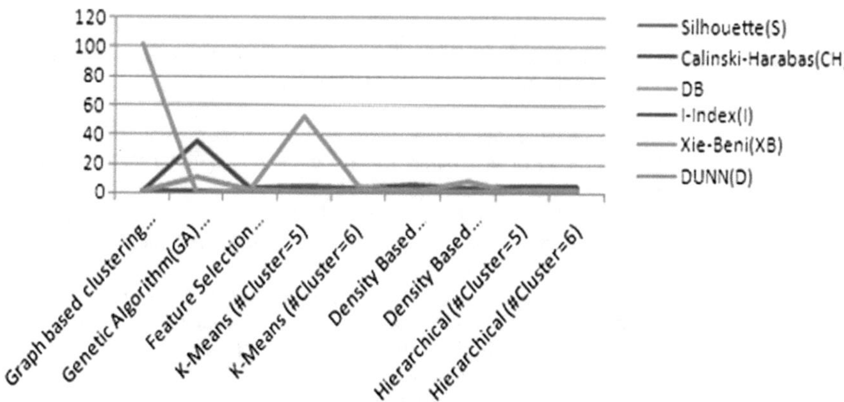

Fig. 1 Example of comparative measure of clustering index for Table 2

In Table 2, the metrics for evaluating clustering and number of clusters identified by different approaches for "Uttarakhand Flood" Dataset are listed. This comparative study, shown in Table 2 can concludes that Genetic Algorithm (GA) based clustering technique is efficient on, but for larger dataset this approach is not well fitted, whereas Feature Selection-based Clustering Algorithm can perform better in this case as feature reduction approaches are considered here. Figure 1 shows the comparative measure of clustering index for different methods listed in Table 2.

6 Conclusion

The conclusion of this study can clearly notify that clustering social media data is one of the challenging tasks, which can result in pattern recognition as well as identification of user potentials and interests. Another challenge to categorize social media data is incremental nature of the dataset. However, future research must demonstrate the effective dynamic clustering methodology for the incremental dataset to discovering knowledge.

References

1. Becker, H., Naaman, M., Gravano, L.: Beyond trending topics: real-world event identification on twitter. In: Fifth International AAAI Conference on Weblogs and Social Media (2011)
2. Blei, D.M., Ng, A.Y., Jordan, M.I.: Latent Dirichlet allocation. J. Mach. Learn. Res. **3**, 993–1022 (2003). http://dl.acm.org/citation.cfm?id=944919.944937
3. Brin, S., Page, L.: The anatomy of a large-scale hypertextual web search engine. Comput. Netw. ISDN Syst. **30**(1–7), 107–117 (1998). http://dx.doi.org/10.1016/S0169-7552(98)00110-X

4. Cataldi, M., Di Caro, L., Schifanella, C.: Emerging topic detection on twitter based on temporal and social terms evaluation. In: Proceedings of the Tenth International Workshop on Multimedia Data Mining, MDMKDD '10, pp. 4:1–4:10. ACM (2010). https://doi.org/10.1145/1814245.1814249
5. Cheong, M., Lee, V.: A study on detecting patterns in twitter intra-topic user and message clustering. In: 2010 20th International Conference on Pattern Recognition (ICPR), pp. 3125–3128 (2010). https://doi.org/10.1109/ICPR.2010.765
6. Dutta, S., Ghatak, S., Das, A., Gupta, M., Dasgupta, S.: Feature selection based clustering on micro-blogging data. In: International Conference on Computational Intelligence in Data Mining (ICCIDM-2017) (2017)
7. Dutta, S., Ghatak, S., Ghosh, S., Das, A.K.: A genetic algorithm based tweet clustering technique. In: 2017 International Conference on Computer Communication and Informatics (ICCCI), pp. 1–6 (2017). https://doi.org/10.1109/ICCCI.2017.8117721
8. Dutta, S., Ghatak, S., Roy, M., Ghosh, S., Das, A.K.: A graph based clustering technique for tweet summarization. In: 2015 4th International Conference on Reliability, Infocom Technologies and Optimization (ICRITO) (Trends and Future Directions), pp. 1–6 (2015). https://doi.org/10.1109/ICRITO.2015.7359276
9. Genc, Y., Sakamoto, Y., Nickerson, J.V.: Discovering context: classifying tweets through a semantic transform based on wikipedia. In: Proceedings of the 6th International Conference on Foundations of Augmented Cognition: Directing the Future of Adaptive Systems, FAC'11, pp. 484–492. Springer (2011)
10. Typhoon Hagupit—Wikipedia (2014). http://en.wikipedia.org/wiki/Typhoon_Hagupit
11. Hyderabad blasts—Wikipedia (2013). http://en.wikipedia.org/wiki/2013_Hyderabad_blasts
12. Infomap—community detection. http://www.mapequation.org/code.html
13. Jain, A.K.: Data clustering: 50 years beyond k-means. Pattern Recogn. Lett. **31**(8), 651–666 (2010). https://doi.org/10.1016/j.patrec.2009.09.011
14. Jain, A.K., Murty, M.N., Flynn, P.J.: Data clustering: a review. ACM Comput. Surv. **31**(3), 264–323 (1999). https://doi.org/10.1145/331499.331504
15. Karypis, G., Han, E.H.S., Kumar, V.: Chameleon: hierarchical clustering using dynamic modeling. Computer **32**(8), 68–75 (1999). https://doi.org/10.1109/2.781637
16. Liu, Y., Li, Z., Xiong, H., Gao, X., Wu, J.: Understanding of internal clustering validation measures. In: Proceedings of the 2010 IEEE International Conference on Data Mining, ICDM '10, pp. 911–916. IEEE Computer Society, Washington, DC, USA (2010). https://doi.org/10.1109/ICDM.2010.35
17. Mathioudakis, M., Koudas, N.: Twittermonitor: trend detection over the twitter stream. In: Proceedings of the 2010 ACM SIGMOD International Conference on Management of Data, SIGMOD '10, pp. 1155–1158. ACM (2010). https://doi.org/10.1145/1807167.1807306
18. Michelson, M., Macskassy, S.A.: Discovering users' topics of interest on twitter: a first look. In: Proceedings of the Fourth Workshop on Analytics for Noisy Unstructured Text Data, AND '10, pp. 73–80. ACM (2010). https://doi.org/10.1145/1871840.1871852
19. Ramage, D., Dumais, S.T., Liebling, D.J.: Characterizing microblogs with topic models. In: Proceedings of ICWSM. The AAAI Press (2010). http://dblp.uni-trier.de/db/conf/icwsm/icwsm2010.html#RamageDL10
20. Rau, L.F., Jacobs, P.S., Zernik, U.: Information extraction and text summarization using linguistic knowledge acquisition. Inf. Process. Manag. **25**(4), 419–428 (1989). https://doi.org/10.1016/0306-4573(89)90069-1. http://www.sciencedirect.com/science/article/pii/0306457389900691
21. Sandy Hook Elementary School shooting—Wikipedia (2012). http://en.wikipedia.org/wiki/Sandy_Hook_Elementary_School_shooting
22. REST API Resources, Twitter Developers. https://dev.twitter.com/docs/api
23. North India floods—Wikipedia (2013). http://en.wikipedia.org/wiki/2013_North_India_floods
24. Vanderwende, L., Suzuki, H., Brockett, C., Nenkova, A.: Beyond sumbasic: task-focused summarization with sentence simplification and lexical expansion. Inf. Process. Manag. **43**(6), 1606–1618 (2007). http://dx.doi.org/10.1016/j.ipm.2007.01.023

25. Wordnet—a lexical database for English. http://wordnet.princeton.edu/
26. Yang, J., Leskovec, J.: Patterns of temporal variation in online media. In: Proceedings of the Fourth ACM International Conference on Web Search and Data Mining, pp. 177–186. ACM (2011)
27. Zhang, C., Lei, D., Yuan, Q., Zhuang, H., Kaplan, L.M., Wang, S., Han, J.: Geoburst+: effective and real-time local event detection in geo-tagged tweet streams. ACM TIST **9**(3), 34:1–34:24 (2018)

Author Index

A. Abraham et al. (eds.), *Emerging Technologies in Data Mining and Information Security*, Advances in Intelligent Systems and Computing 813,
https://doi.org/10.1007/978-981-13-1498-8

Printed in the United States
By Bookmasters